Achieving sustainable cultivation of cocoa

It is widely recognised that agriculture is a significant contributor to global warming and climate change. Agriculture needs to reduce its environmental impact and adapt to current climate change whilst still feeding a growing population, i.e. become more 'climate-smart'. Burleigh Dodds Science Publishing is playing its part in achieving this by bringing together key research on making the production of the world's most important crops and livestock products more sustainable.

Based on extensive research, our publications specifically target the challenge of climate-smart agriculture. In this way we are using 'smart publishing' to help achieve climate-smart agriculture.

Burleigh Dodds Science Publishing is an independent and innovative publisher delivering high quality customer-focused agricultural science content in both print and online formats for the academic and research communities. Our aim is to build a foundation of knowledge on which researchers can build to meet the challenge of climate-smart agriculture.

For more information about Burleigh Dodds Science Publishing simply call us on +44 (0) 1223 839365, email info@bdspublishing.com or alternatively please visit our website at www.bdspublishing.com.

Related titles:

Achieving sustainable cultivation of coffee
Print (ISBN 978-1-78676-152-1); Online (ISBN 978-1-78676-154-5, 978-1-78676-155-2)

Global tea science
Print (ISBN 978-1-78676-160-6); Online (ISBN 978-1-78676-162-0, 978-1-78676-163-7)

Achieving sustainable cultivation of sugarcane Volume 1:
Cultivation techniques, quality and sustainability
Print (ISBN 978-1-78676-144-6); Online (ISBN 978-1-78676-146-0, 978-1-78676-147-7)

Achieving sustainable cultivation of sugarcane Volume 2: Breeding, pests and diseases
Print (ISBN 978-1-78676-148-4); Online (ISBN 978-1-78676-150-7, 978-1-78676-151-4)

Chapters are available individually from our online bookshop: https://shop.bdspublishing.com

climate•SMART•publishing
IN AGRICULTURAL SCIENCE

BURLEIGH DODDS SERIES IN AGRICULTURAL SCIENCE

NUMBER 43

Achieving sustainable cultivation of cocoa

Edited by: Prof. Pathmanathan Umaharan
Cocoa Research Centre – The University of the West Indies, Trinidad and Tobago

burleigh dodds
SCIENCE PUBLISHING

Published by Burleigh Dodds Science Publishing Limited
82 High Street, Sawston, Cambridge CB22 3HJ, UK
www.bdspublishing.com

Burleigh Dodds Science Publishing, 1518 Walnut Street, Suite 900, Philadelphia, PA 19102-3406, USA

First published 2018 by Burleigh Dodds Science Publishing Limited

Library of Congress Control Number: 2017960359

British Library Cataloguing in Publication Data
A catalogue record for this book is available from the British Library

ISBN 978-1-78676-168-2 (Print)
ISBN 978-1-78676-171-2 (PDF)
ISBN 978-1-78676-170-5 (ePub)
ISSN 2059-6936 (print)
ISSN 2059-6944 (online)

DOI 10.19103/AS.2017.0021

Typeset by Deanta Global Publishing Services, Chennai, India
Printed by Lightning Source

Contents

Part 2 Cultivation techniques

Part 4 Safety and sensory quality

Part 5 Sustainability

Series list

Title	Series number
Achieving sustainable cultivation of maize - Vol 1 From improved varieties to local applications Edited by: Dr Dave Watson, CGIAR Maize Research Program Manager, CIMMYT, Mexico	001
Achieving sustainable cultivation of maize - Vol 2 Cultivation techniques, pest and disease control Edited by: Dr Dave Watson, CGIAR Maize Research Program Manager, CIMMYT, Mexico	002
Achieving sustainable cultivation of rice - Vol 1 Breeding for higher yield and quality Edited by: Prof. Takuji Sasaki, Tokyo University of Agriculture, Japan	003
Achieving sustainable cultivation of rice - Vol 2 Cultivation, pest and disease management Edited by: Prof. Takuji Sasaki, Tokyo University of Agriculture, Japan	004
Achieving sustainable cultivation of wheat - Vol 1 Breeding, quality traits, pests and diseases Edited by: Prof. Peter Langridge, The University of Adelaide, Australia	005
Achieving sustainable cultivation of wheat - Vol 2 Cultivation techniques Edited by: Prof. Peter Langridge, The University of Adelaide, Australia	006
Achieving sustainable cultivation of tomatoes Edited by: Dr Autar Mattoo, USDA-ARS, USA & Prof. Avtar Handa, Purdue University, USA	007
Achieving sustainable production of milk - Vol 1 Milk composition, genetics and breeding Edited by: Dr Nico van Belzen, International Dairy Federation (IDF), Belgium	008
Achieving sustainable production of milk - Vol 2 Safety, quality and sustainability Edited by: Dr Nico van Belzen, International Dairy Federation (IDF), Belgium	009
Achieving sustainable production of milk - Vol 3 Dairy herd management and welfare Edited by: Prof. John Webster, University of Bristol, UK	010
Ensuring safety and quality in the production of beef - Vol 1 Safety Edited by: Prof. Gary Acuff, Texas A&M University, USA & Prof. James Dickson, Iowa State University, USA	011
Ensuring safety and quality in the production of beef - Vol 2 Quality Edited by: Prof. Michael Dikeman, Kansas State University, USA	012
Achieving sustainable production of poultry meat - Vol 1 Safety, quality and sustainability Edited by: Prof. Steven C. Ricke, University of Arkansas, USA	013
Achieving sustainable production of poultry meat - Vol 2 Breeding and nutrition Edited by: Prof. Todd Applegate, University of Georgia, USA	014
Achieving sustainable production of poultry meat - Vol 3 Health and welfare Edited by: Prof. Todd Applegate, University of Georgia, USA	015

Achieving sustainable production of eggs - Vol 1 016
Safety and quality
Edited by: Prof. Julie Roberts, University of New England, Australia

Achieving sustainable production of eggs - Vol 2 017
Animal welfare and sustainability
Edited by: Prof. Julie Roberts, University of New England, Australia

Achieving sustainable cultivation of apples 018
Edited by: Dr Kate Evans, Washington State University, USA

Integrated disease management of wheat and barley 019
Edited by: Prof. Richard Oliver, Curtin University, Australia

Achieving sustainable cultivation of cassava - Vol 1 020
Cultivation techniques
Edited by: Dr Clair Hershey, formerly International Center for Tropical Agriculture (CIAT), Colombia

Achieving sustainable cultivation of cassava - Vol 2 021
Genetics, breeding, pests and diseases
Edited by: Dr Clair Hershey, formerly International Center for Tropical Agriculture (CIAT), Colombia

Achieving sustainable production of sheep 022
Edited by: Prof. Johan Greyling, University of the Free State, South Africa

Achieving sustainable production of pig meat - Vol 1 023
Safety, quality and sustainability
Edited by: Prof. Alan Mathew, Purdue University, USA

Achieving sustainable production of pig meat - Vol 2 024
Animal breeding and nutrition
Edited by: Prof. Julian Wiseman, University of Nottingham, UK

Achieving sustainable production of pig meat - Vol 3 025
Animal health and welfare
Edited by: Prof. Julian Wiseman, University of Nottingham, UK

Achieving sustainable cultivation of potatoes - Vol 1 026
Breeding improved varieties
Edited by: Prof. Gefu Wang-Pruski, Dalhousie University, Canada

Achieving sustainable cultivation of oil palm - Vol 1 027
Introduction, breeding and cultivation techniques
Edited by: Prof. Alain Rival, Center for International Cooperation in Agricultural Research for Development (CIRAD), France

Achieving sustainable cultivation of oil palm - Vol 2 028
Diseases, pests, quality and sustainability
Edited by: Prof. Alain Rival, Center for International Cooperation in Agricultural Research for Development (CIRAD), France

Achieving sustainable cultivation of soybeans - Vol 1 029
Breeding and cultivation techniques
Edited by: Prof. Henry T. Nguyen, University of Missouri, USA

Achieving sustainable cultivation of soybeans - Vol 2 030
Diseases, pests, food and non-food uses
Edited by: Prof. Henry T. Nguyen, University of Missouri, USA

Achieving sustainable cultivation of sorghum - Vol 1 031
Genetics, breeding and production techniques
Edited by: Prof. William Rooney, Texas A&M University, USA

Acknowledgements

We wish to acknowledge the following for their help in reviewing particular chapters:

- Chapter 1: Dr Dapeng Zhang, USDA-ARS, USA; and Dr Rey Loor, INIAP, Ecuador
- Chapter 3: Dr Dapeng Zhang, USDA-ARS, USA; and Dr Gustavo Adolfo Enriquez Calderón, CATIE, Costa Rica
- Chapter 4: Dr Michelle End, INGENIC (The International Group for Genetic Improvement of Cocoa), UK
- Chapter 5: Dr Smilja Lambert, Mars Inc., Australia
- Chapter 7: Mr Agung Susilo, Indonesian Coffee and Cocoa Research Institute, Indonesia
- Chapter 9: Dr Paul E. Aikpokpodion, Cocoa Research Institute of Nigeria
- Chapter 14: Dr Aurelio Ciancio, CNR, Italy
- Chapter 15: Dr Désiré Poku, CNRA-LVB, Ivory Coast
- Chapter 17: Prof. Marina Venturini Copetti, Universidade Federal de Santa Maria, Brazil
- Chapter 19: Dr Aidan Farrell, University of the West Indies, Trinidad and Tobago
- Chapter 20: Dr Stephane Saj, CIRAD, France; and Dr Denis Sonwa, CIFOR, Cameroon
- Chapter 21: Dr Johanna Jacobi, Centre for Development and Environment (CDE), Bolivia
- Chapter 22: Dr Soetanto Jember, Indonesian Coffee and Cocoa Research Institute, Indonesia; and Mr Henk Gilhuis, UTZ, The Netherlands

Introduction

There is a growing global demand for cocoa. However, cultivation is dependent on ageing trees with low yields and increasing vulnerability to disease. There is also growing concern about the environmental impact of cultivation in such areas as soil health and biodiversity. There is therefore an urgent need to make cocoa cultivation more efficient and sustainable to ensure a successful future. These challenges are addressed in *Achieving sustainable cultivation of cocoa*.

Part 1 reviews genetic resources and developments in breeding. Part 2 discusses optimising cultivation techniques to make the most of new varieties. Part 3 summaries the latest research on understanding and combatting the major fungal and viral diseases affecting cocoa. Part 4 covers safety and quality issues whilst the final part of the book looks at ways of improving sustainability, including the role of agroforestry, organic cultivation and ways of supporting smallholders.

Part 1 Genetic resources and breeding

The first part of the volume reviews genetic resources and developments in breeding. Chapter 1 discusses the origins of cacao as well as the taxonomy and classification of the varieties of *Theobroma*. The chapter then reviews the distinctive characteristics of the three main varieties of *Theobroma*: Criollo, Forastero and Trinitario.

Moving on from classification and taxonomy to genetics, the subject of Chapter 2 is the key challenges of conserving and exploiting cocoa genetic resources. The future of the world cocoa economy depends on the availability of genetic diversity and the sustainable use of this broad genetic base to breed improved varieties. Decreasing cacao genetic diversity (whether conserved *in-situ*, on farms or in *ex-situ* collections) is a serious problem which needs to be urgently addressed. A Global Strategy was published in 2012 to optimize the conservation and maximize the use of cacao genetic resources as the foundation of a sustainable cocoa economy. The chapter describes the key challenges in delivering this strategy, how they are being addressed and the priorities for further research and actions. These include securing existing *ex-situ* cacao genetic resources, developing a global strategic cacao collection, and collecting and gap filling in *ex-situ* collections to reflect genetic diversity. The chapter also looks at the importance of ensuring the *in-situ* and on-farm conservation of important diversity, strengthening the distribution and safe movement of germplasm, improving the use of cacao genetic resources, and improving documentation and sharing of information. Some of these issues are discussed in more detail in the following three chapters.

Building on the previous chapter, Chapter 3 focuses specifically on the role of gene banks in preserving the genetic diversity of cacao. The centre of diversity of cacao in South America is characterized by genetic erosion from deforestation, but fortunately a wealth of genetic diversity exists in global cacao collections. The chapter discusses the role and types of gene banks to capture genetic diversity. The chapter deals with the distinction between cacao gene banks and other gene banks and examines how this affects the management and estimation of genetic diversity. The chapter advocates a SNP panel for fingerprinting, and addresses molecular marker-assisted management with the objective of comparing global collections and formulating a core collection.

Complementing the themes of the preceding two chapters, Chapter 4 concentrates on the safe handling and movement of cocoa germplasm for breeding. Movement of cocoa germplasm is often required in breeding programmes to increase the genetic diversity pool or for the testing of clones/progeny in the field. However, such movement risks the spread of pests and diseases, many of which are confined to particular geographical locations. It is therefore critical that movement of germplasm is conducted within a quarantine framework. The chapter reviews the risks associated with the movement of cocoa germplasm. It considers international governance of plant movement before discussing the work of the International Cocoa Quarantine Centre at the University of Reading as a hub for international movement of cocoa germplasm.

The final chapter of the section, Chapter 5, covers developments in cacao breeding programmes in Africa and the Americas. The chapter explores the main developments obtained as a result of cacao breeding programmes in Trinidad, Brazil, Ecuador and Costa Rica in the Americas and in Ghana, Côte d'Ivoire, Nigeria and Cameroon in Africa. The chapter describes the different types of commercial cacao cultivars and the breeding objectives of the programmes. It examines heterosis and heterotic groups in cacao and explores the contrast between 'traditional' and new cacao breeding methods. Finally, the chapter examines the issue of breeding cacao for organoleptic quality.

Part 2 Cultivation techniques

The second part of the volume discusses optimising cultivation techniques to make the most of new varieties. The theme of Chapter 6 is cocoa plant propagation techniques. The availability of high performance planting materials to cocoa farmers is an important part of a package of measures to improve the productivity of cocoa farms and thus the sustainability of the cocoa economy. The chapter reviews the methods, advantages and challenges of techniques of mass propagation, with a focus on seed and conventional vegetative propagation (tissue culture techniques are covered in the following chapter). The chapter discusses key challenges in supplying farmers with improved planting materials. These include availability of source materials, personnel and infrastructure requirements, phytosanitary considerations, costs and levels of demand.

Continuing the theme of propagation, Chapter 7 examines the potential of somatic embryogenesis (SE) for commercial-scale propagation of elite cacao varieties. Plant tissue culture can be used to speed up the development and deployment of genetically-improved genotypes. Research conducted by multiple groups for over 25 years has led to the development of protocols for efficient somatic embryogenesis of cacao. The chapter provides a synthesis of this research on cacao tissue culture methods and field-test evaluations of SE-derived plants. The chapter also reviews current activities by cacao-producing countries in large-scale propagation of important genotypes, with case studies from Africa, Asia and the Americas.

Moving on to agronomic practices, Chapter 8 focuses on the achievement of good agronomic practices in cocoa cultivation and the rehabilitation of cocoa farms. Adoption of good agronomic practices is crucial for the sustainability of cocoa cultivation and this chapter describes good practice at both the pre- and post-planting stages. The chapter focuses on the challenge of rehabilitating cocoa farms, and includes a detailed case study on rehabilitation of farms in Ghana.

The final chapter of the section, Chapter 9, looks specially at the challenge of improving soil and nutrient management in cacao cultivation. The chapter summarises key research on the role and availability of key nutrients affecting cacao growth and health such as phosphorus (P), potassium (K), calcium (Ca) and magnesium (Mg). The variability of yield responses to nutrients has highlighted the ways these nutrients interact both with each other and other factors such as soil, climate, and topography. The chapter outlines a targeted approach to nutrition management based on correcting the soil so that the cacao can find the nutrients it needs in optimal quantities. Through two detailed case studies, the chapter discusses ways of determining an appropriate fertiliser formula for individual plots using physical-chemical analyses of soil samples.

Part 3 Diseases and pests

The third part of the volume summaries the latest research on understanding and combatting the major fungal and viral diseases affecting cocoa. The subject of Chapter 10 is the cocoa disease witches' broom. Caused by the hemibiotrophic basidiomycete *Moniliophthora perniciosa*, witches' broom is one of the most serious cocoa diseases. The pathogen co-evolved with cocoa in the Amazon River basin and is currently restricted to South and Central America but represents a serious risk to other cocoa-producing regions of the world are at risk. In the last few years new technologies such as bioinformatics and the establishment of a robust model of host-plant interaction have enabled significant advances in our knowledge of witches' broom. The chapter reviews the latest research on disease physiology, genomics and transcriptomics, diversity, and management practices to control the disease.

Moving on to another major cocoa disease, Chapter 11 focuses on frosty pod rot, caused by the fungus *Moniliophthora roreri*. The chapter summarises what we know about the pathogen before considering the spread, invasion risk and impact of *moniliophthora roreri* on cocoa. The chapter then examines management options for dealing with the disease, including ways of preventing and minimizing the impact of invasive species which spread disease.

Continuing the theme of cocoa diseases, Chapter 12 looks at vascular-streak dieback (VSD), caused by a previously undescribed tullasneloid basidiomycete, *Oncobasidium theobromae* (now *Ceratobasidium theobromae*). The chapter describes the symptoms of VSD and the biology of the pathogen, the disease's epidemiology and methods of management, as well as including a detailed case study showing the importance of field studies in identifying, tracking and preventing the disease.

Moving away from the subject of diseases, Chapter 13 considers insect pests affecting cacao. Pests have a major impact on cacao production, both from direct damage to crops and as vectors of disease. Estimates of losses due to pest and disease range from 30– 40%. The use of pesticides for pest control can have numerous negative impacts on the environment and human health. An integrated and holistic approach is therefore required to enhance and sustain crop production. The chapter describes the main insect pests attacking cacao in each of the three principal production regions: the Americas, Africa and Asia, and outlines control measures adopted for each of the main insect pests. The chapter also includes detailed case studies that showcase how research can help to develop more sustainable and environmentally-friendly control measures.

The focus of Chapter 14 is on nematode pests of cocoa. Plant-parasitic nematodes cause significant cocoa yield losses, sudden death of trees and retardation of seedling growth in nurseries. The chapter explores the challenges in identifying damage caused by nematodes. It then describes the main nematode pests associated with cocoa and typical damage symptoms. The chapter also addresses available control options in successful nematode management.

The final chapter of the section, Chapter 15, turns to advances in pest and disease-resistant cocoa varieties. Cocoa production is increasingly subject to parasitic pressures. Currently, Cacao Swollen Shoot Virus (CSSV) threatens the production of the largest production basin in West Africa, in Côte d'Ivoire, Ghana, Togo and Nigeria. In Asia, Cocoa Pod Borer and Vascular Streak Dieback (VSD) hamper the extension of cultivation areas and reduce production. In South America, where cocoa originated, Moniliasis and Witches Broom have significantly limited production. Other pests, like mirids or black pod disease due to several species of *Phytophthora*, also affect cacao production. The chapter describes the genetic improvement of resistance, focusing on the widely prevalent black pod disease. The chapter discusses the use of new tools coming from molecular biology, including marker assisted selection (MAS) and genomic selection (GS), and the ways they are being used in breeding for *Phytophthora* disease resistance.

Part 4 Safety and sensory quality

The fourth part of the volume covers safety and quality issues. Echoing themes in Chapter 13, the subject of Chapter 16 is improving best practice with regard to pesticide use in cocoa. Consumer concerns about food safety have been translated into regulations governing minimum pesticide residues in cocoa. Consequently, development and implementation of best practice for pesticide use is critical, but presents major challenges. The chapter includes detailed case studies on establishing baselines, effective monitoring promoting awareness and developing a holistic approach.

Moving from the danger of pesticide traces to toxins left behind by fungi, Chapter 17 considers the causes, detection and control of mycotoxins in cocoa. Mycotoxins are produced as secondary metabolites by various species of filamentous fungi, and may affect many agricultural crops and products. The potential health risks associated with these chemical compounds mean that significant attention has been given to their detection and control. However, most study has so far been dedicated to mycotoxin contamination of agricultural crops such as cereals, with less attention given to cocoa. The chapter presents an overview of the current understanding of mycotoxin contamination of cocoa. The main groups of mycotoxins are discussed, followed by a summary of three methods of detection. The chapter then explains the various methods of controlling mycotoxins in cocoa and discusses attempts to decontaminate infected crops. The chapter concludes that there is a need for more research into the different mycotoxins affecting cocoa and methods of reducing their presence and impact.

The final chapter of the section, Chapter 18, focuses on analysing sensory and processing quality of cocoa. Flavour is a critical aspect of cocoa quality, determining to a large extent the value and end use of traded cocoa beans. The chapter provides a road map towards bridging the knowledge gap that currently exists between industry and cocoa producers by defining approaches for analysing the sensory and processing quality of cocoa. The chapter describes the development of more practical and harmonized quality criteria for cocoa farmers around the world. It also explores research on the genetic expression of flavor, developments in sensory evaluation and the use of both physical and sensory evaluation as a tool to drive improvements in the cocoa value chain.

Part 5 Sustainability

The final part of the book looks at ways of improving the environmental sustainability of cocoa production. The subject of Chapter 19 is the relationship between climate change and cocoa cultivation. The chapter provides a number of case studies which focus on global climate projections for cocoa producing regions, including assessment of regionally-differentiated climate change impacts in Côte d'Ivoire.

Keeping with the theme of the interaction between cocoa production and the environment, Chapter 20 considers critical issues in applying agroforestry science to cocoa cultivation programmes. Cocoa-based agroforestry systems are a conspicuous element of agricultural landscapes worldwide. The chapter concentrates on the analysis and design of the shade canopy in cocoa-based agroforestry systems. Shade canopy analysis and design is a key component of crop husbandry, and requires a good understanding of the interactions, synergies and trade-offs between shade, yield and environmental services. The chapter provides a robust approach to analyze and design an optimal shade canopy that provides a diverse, resilient system which balances differing requirements such as carbon storage and cocoa yields.

Chapter 21 provides an overview of methods for organic cocoa cultivation. As a crop, cocoa can be grown successfully using organic methods, and demand for organic chocolate has risen in line with the overall growth in the organic market. The chapter offers a summary of current issues in the production of organic cocoa. The chapter also considers issues surrounding the certification and pricing of organic cocoa. The chapter summarises a variety of different cultivation methods, such as the differences between agroforestry-based cultivation and full-sun grown cocoa cultivation. Finally, the chapter considers different techniques for controlling pests and diseases without the use of chemical inputs.

Complementing the preceding chapters, Chapter 22 considers cocoa sustainability initiatives. Largely implemented by cocoa farmers and groups, sustainability initiatives are often supported by traders, government agencies, certification organisations and other not-for-profit organisations. The chapter examines the different sustainability initiatives and their social, economic and environmental impacts on cocoa farmers, cocoa farms and cocoa ecosystems. The chapter provides detailed case studies of impacts in Ghana and Côte d'Ivoire.

The volume's final chapter, Chapter 23, looks at supporting smallholders in achieving more sustainable cocoa cultivation. The chapter examines the recent evolution of sustainability in the cocoa and chocolate value chain. Using the case of West Africa, especially the countries of Côte d'Ivoire and Ghana, the chapter examines the key social, economic, and environmental challenges facing smallholders. The chapter provides an overview of initiatives undertaken by public and private actors to address these problems. The chapter focuses on collaborative efforts that bring together various private sector actors in the value chain. The chapter concludes by considering three important developments in the areas of voluntary certification and standards; the discussion of a living income; and the question of price and future supply and demand for cocoa.

Part 1

Genetic resources and breeding

Chapter 1

Taxonomy and classification of cacao

Ranjana Bhattacharjee, International Institute of Tropical Agriculture (IITA), Nigeria; and Malachy Akoroda, Cocoa Research Institute of Nigeria, Nigeria

1 Introduction

Cacao is one of the economically important tree crops in West and Central Africa and is native to the tropical forests of South America. The botanical name of cacao is *Theobroma cacao*, first defined by Carolus Linnaeus, the father of modern-day taxonomic plant classification, and was published in his classic book, *Systema Naturae* in the mid-1700s. *Theobroma* is a Greek word that can be translated as 'Food of Gods': 'theos' meaning 'god' and 'broma' meaning 'food' (Wikipedia, 2013). The word 'cacao' is derived from the Olmec and the subsequent Mayan languages (Kakaw), and the word 'chocolate' is derived from the Nahuatl (the Aztec language) term '*cacahuatl*', which is in turn derived from Olmec/Mayan etymology (Dillinger et al., 2000). The other common names are cocoa (English), cacaoyer (French), cacaoeiro/cacau (Portuguese/Brazil), kakao (Swedish), árbol del cacao/cacaotero/calabacillo/forester (Spanish) and echter Kakaobaum (German). There is archaeological evidence that the Mayans cultivated cacao 2000–4000 years before the Spanish arrived in South America (Bergmann, 1969; Purdy and Schmidt, 1996; Whitlock et al., 2001; Henderson et al., 2007; Powis et al., 2011).

1.1 Distribution of cacao

Whitlock et al. (2001) suggested the region extending from the forests of the Amazon to the Orinoco and Tabasco in Southern Mexico to be the centre of origin of cacao (Van Hall, 1914). Cheesman (1944) believed the Upper Amazon near the Colombian–Ecuadorian border, on the eastern flanks of the Andes, to be its centre of origin. He argued that although cacao cultivation has been practised in Mexico and Central America for over 2000 years, there was no evidence of its wild populations in the region. Later, Schultes

http://dx.doi.org/10.19103/AS.2017.0021.01

(1984) hypothesised that cacao dispersed along two routes from the Amazon Valley, one leading to the north and the other the west. In this way, the domestication of cacao was thought to have occurred in South America and then spread to Central America and Southern Mexico, carried by migrating Indians. It is also estimated that the Mayas and Aztecs domesticated and consumed cocoa, while the Mayas, Olmecs, Toltecs and Aztecs used cocoa beans as the bitter drink, *chocolatl*, as well as a currency for trading, at least 1400 years ago (Purdy and Schmidt, 1996; Rössner, 1997; Nair, 2010; Anon, 2011). Several expeditions and studies have shown that cocoa was first cultivated in Mexico by the Aztecs before it spread to the Caribbean islands. Later, Hernán Cortés, a Spaniard, carried cocoa to Spain in 1520s and used it as a beverage as well as a crop for cultivation. It was the Spaniards who also introduced the crop into Equatorial Guinea in the seventeenth century.

1.1.1 Subspecies and domesticated cacao

Cuatrecasas (1964) suggested the simultaneous origin of cacao separately in South and Central America, a hypothesis supported by most subsequent authors (Cope, 1976; Wood and Lass, 1985; Gómez-Pompa et al, 1990; Laurent et al, 1994; de la Cruz et al, 1995; Whitkus et al, 1998). He also proposed that North and South American cacao populations developed into two forms, separated by the Panama Isthmus, and that this can be attributed to the great morphological diversity observed in Central as well as in South America. These populations evolved independently and are recognised as separate subspecies (*T. cacao* ssp. *cacao* and *T. cacao* ssp. *sphaerocarpum*). Furthermore, Cuatrecasas (1964) hypothesised that wild plants from the Lacandon rainforest from Mexico may possibly be ancestors of domesticated cacao. The subspecies proposed by Cuatrecasas correspond to the two morphogeographic groups namely Criollo, which corresponds to *T. cacao* ssp. *cacao,* and Forastero, which corresponds to *T. cacao* ssp. *sphaerocarpum*, proposed by Cheesman (1944). A third hybrid group was called Trinitario, originating from crosses between Criollo and Forastero. The Forastero group is composed of very diverse populations with different geographic origins such as Upper Amazon, Lower Amazon, Orinoco and the Guianas and can be identified according to the pod morphology, for example, the Amelonado type.

2 Taxonomy and classification

Theobroma is exclusively neotropical in origin and its natural dispersion is in the tropical lowland rainforests which extend from the Amazon basin through Southern Mexico (18°N to 15°S) (Cuatrecasas, 1964). Several studies on its morphological, anatomical, palynological and chemical characteristics as well as its DNA sequences have confirmed the inclusion of the family Sterculiaceae in a broadly defined Malvaceae *sensu latu* (Judd and Manchester, 1997; Alverson et al., 1999; Bayer et al., 1999), tribe Byttnerieae (Bentham and Hooker, 1862) or Theobromeae (Hutchinson, 1967; Takhtajan, 1997). There is minimal evidence on the practice of cacao cultivation by the various indigenous groups of the Amazonian rainforest (Sánchez et al., 1989; Dias, 2001a,b; Bartley, 2005). The wild relatives of cacao thus include two different groups of germplasm. The first is the large spectrum of wild populations which grow simultaneously in the Amazonian rainforest, ranging from French

Guiana to Bolivia. There is no reproductive barrier between cultivated and wild cacao trees. Morphological characteristics of even the ancient cultigens, Criollo cacao from Mesoamerica, are similar to those of their wild counterparts (Iwaro et al., 2005). Wild cacao, therefore, may be used in breeding or commercial production, either as progenitors or as clones (Dias and Resende, 2001; Eskes and Efron, 2006). The second category of wild cacao is that of the 22 related *Theobroma* species (Cuatrecasas, 1964), although these show crossing barrier with *T. cacao*. Three scholarly publications to date have recognised a different number of species of *Theobroma* before Cuatrecasas (1964), which included 18 species (Bernoulli, 1869), 11 species (Schumann, 1886) and 13 species (Chevalier, 1946). Bernoulli (1869) described five categories based on floral morphology and fruit structure. These are:

Cacao: Petal ligule stipitate; staminodes erect, subulate; stamens 2-antheriferous; calyx 5-parted, the laciniae equal. Ovate-oblong fruit included four species: *T. cacao* L.; *T. pentagona*, 'cacao lagarto' from Guatemala; *T. leiocarpa*, 'cumacaco' from Guatemala; and *T. saltzmanniana* from Bahia, Brazil.

Oreanthes: Petal appendix subsessile; staminodes erect, subulate; stamens 3-antheriferous; calyx 5-parted, the laciniae equal, included three species: *T. speciosa* Willd. Ex Spreng.; *T. quinquenervia* and *T. spruceana*, all from Brazil.

Rhytidocarpus: Petal appendix subsessile; staminodes erect, claviform; stamens 2-antheriferous; calyx 5-parted, the laciniae equal; fruit woody, included two species: *T. bicolour* Humb. & Bonpl. (synonym, *T. ovatifolia* DC.) and *T. glauca* Karst.

Telmatocarpus: Petal appendix; staminodes erect, linear-subulate with broad base; stamens 3-antheriferous; calyx 5-parted, the laciniae equal; fruit ovate, lacunose, included only one species: *T. microcarpa* Mart.

Glossopetalum: Petal ligule stipitate; staminodes reflexed, petaloid; stamens 3-antheriferous; calyx irregularly 3-5-fid, foliaceous; fruit sublignose. This is the largest section, including eight species: *T. augustifolia* DC.; *T. subincana* Mart.; *T. sylvestris* Mart.; *T. macrantha* from Brazil; *T. ferruginea* from Peru; *T. obovata* from Brazil; *T. alba* from British Guiana and *T. nitida* from Brazil.

2.1 Classification

Morris (1882) was the first to publish the nomenclature of the cultivars of *Theobroma cacao*. He distinguished two classes, the second of which was divided into the following several varieties:

- Class I: Cacao Criollo (Red)
- Class II: Cacao Forestero
    - Var. a. Cundeamor verugoso Amarillo (rough yellow Cerasee)
    - Var. b. Cundeamor verugoso Colorado (rough red Cerasee)
    - Var. c. Liso amarillo (smooth yellow)
    - Var. d. Liso colorado (smooth red)
    - Var. e. Amelonado amarillo (yellow melon)
    - Var. f. Amelonado colorado (red melon)
    - Var. g. Calabacillo amarillo (yellow calabash)
    - Var. h. Calabacillo colorado (red calabash)

In 1892, Hart modified Morris' classification of the varieties of *T. cacao* and distinguished the following classes:

- Class I: Criollo
  - Trinidad Criollo
    - var. a. Amarillo (yellow, thick-skinned, bottle-necked)
    - var. b. Colorado (red, thin-skinned, bottle-necked)
  - Venezuelan Criollo
    - var. a. Amarillo (yellow, thick-skinned, high-shouldered, sometimes pointed)
    - var. b. Colorado (red, thick-skinned, high-shouldered, sometimes pointed)
  - Nicaraguan Criollo
    - var. a. Amarillo (yellow, thick-skinned, high-shouldered, very large beans with light-coloured interior)
    - var. b. Colorado (red, thick-skinned, high-shouldered, very large beans with light-coloured interior)

- Class II: Forastero
    - var. a. Cundeamor verugoso Amarillo (yellow-warted)
    - var. b. Cundeamor verugoso Colorado (red-warted)
    - var. c. Ordinary amarillo (yellow forastero)
    - var. d. Ordinary colorado (red forastero)
    - var. e. Amelonado amarillo (yellow, melon-shaped)
    - var. f. Amelonado Colorado (red, melon-shaped)

- Class III: Calabacillo
    - var. a. Amarillo (yellow, flat-beaned, smooth, thin- or thick-skinned, small pods)
    - var. b. Colorado (red, flat-beaned, smooth, thin- or thick-skinned, small pods)

Theobroma pentagona

Theobroma pentagona (alligator cacao: yellow, heavily warted pods with five distinctly raised ribs, large beans having white- or light-coloured interior).

2.2 Diversity and varieties

In 1944, Cheesman concluded that the whole assemblage of wild, semi-wild and cultivated cacao constitutes one interbreeding population and supported the main division of cacao into two groups of varieties, Criollo and Forastero. He divided Criollo into Central American and South American Criollos. The Forasteros were divided into Amazonian Forasteros, which can be found wild in Amazonia, and which are widespread in cultivation, and the Trinitarios, possibly originating from the mingling of South American Criollo and Amazonian Forastero stocks. He suggested that *T. pentagonum* was a simple form of *T. cacao*, probably a segregate of the large cross-fertilised population. Cuatrecasas (1964) described 22 species based on the classical method of comparative morphology such as the structure of fruit and vegetative characteristics. He classified the species into six sections: *Andropetalum* (*T. mammosum* Cuatr. & León); *Glossopetalum* [*T. angustifolium* Moçiño & Sessé, *T. canumanense* Pires & Fróes, *T. chocoense* Cuatr., *T. cirmolinae*

Cuatr., *T. grandiflorum* (Willd. ex Spreng.) Schum., *T. hylaeum* Cuatr., *T. nemorale* Cuatr., *T. obovatum* Klotzsch ex Bernoulli, *T. simiarum* Donn. Smith., *T. sinuosum* Pavón ex Hubber, *T. stipulatum* Cuatr., *T. subincanum* Mart.]; *Oreanthes* (*T. bernouillii* Pittier, *T. glaucum* Karst., *T. speciosum* Willd., *T. sylvestre* Mart., *T. velutinum* Benoist); *Rhytidocarpus* (*T. bicolor* Humb. & Bonpl.); *Telmatocarpus* (*T. gileri* Cuatr., *T. microcarpum* Mart.) and *Theobroma* (*T. cacao* L.). Representative species from all the sections can be found in Brazil, except for *Andropetalum* (*T. mammosum*). The species which occur in, and are restricted to, the Amazon basin of Brazil are: *T. grandiflorum*, *T. obovatum*, *T. subincanum*, *T. speciosum*, *T. sylvestre*, *T. microcarpum*, *T. glaucum*, *T. canumanense*, *T. bicolor* and *T. cacao*, (Cuatrecasas 1964; Figueira et al., 1994). The existence of the species *T. canumanense* is based on the description of two exsiccates collected near the river Canumão (municipality of Borba, AM).

2.2.1 Scientific classification/Taxonomic tree

Preferred scientific name: *Theobroma cacao* L.
Preferred common name: Cocoa
Other scientific names: *Theobroma pentagonum* Bernoulli; *Theobroma sativum* (Aubl.) Lign & Le Bey
Domain: Eukaryota
Kingdom: Plantae
Subkingdom: Tracheobionta (Vascular plants)
Superdivision: Spermatophyta (Seed plants)
Division: Magnoliophyta (Flowering plants)
Phylum: Spermatophyta
Subphylum: Angiospermae
Class: Magnoliopsida (Dicotyledons)
Subclass: Dilleniidae
Order: Malvales
Family: Malvaceae/Sterculiaceae
Genus: *Theobroma*
Species: *Theobroma cacao*

Previous phylogenetic studies (Whitlock and Baum, 1999; Silva and Figueira, 2005) have indicated that *Theobroma* is related to *Herrania* Goudot, a genus of about 20 species monographed by Schultes (1958), both representatives of the Theobromeae (Whitlock et al., 2001) along with two other genera, *Glossostemon* Desf., with one species from Arabia, and the Neotropical *Guazuma* Mill., with 2–5 species. The classification of *Theobroma* and *Herrania* into sections was through two clades. The *Herrania* clade was subdivided into the relatively plesiomorphic section *Subcymbicalyx* and the section *Herrania*, with apomorphic characteristics. The *Theobroma* clade was subdivided into a group containing sections *Andropetalum* and *Glossopetalum*, which exhibit many plesiomorphic characteristics, and a second group of the morphologically derived sections *Oreanthes*, *Rhytidocarpus*, *Telmatocarpus* and *Theobroma* (Whitlock and Baum, 1999; Silva and Figueira, 2005). Bernoulli (1869) quoted *Herrania* as a genus differing from *Theobroma* in the habit of the plant and with respect to 5-6-foliolate digitate leaves. However, *Theobroma* and *Herrani* species are both diploid with 2n = 20 chromosomes (Muñoz Ortega, 1948).

2.2.2 Effects of domestication

It is now clear that cacao was brought to Mesoamerica by early humans where its use and domestication may have started around 4000 years ago (Powis et al., 2011). Humans are also responsible for distributions of this species over the Amazon Basin (Clement et al., 2010). In the case of both the *Theobroma* and *Herrania* species, the use of the sweet pulp surrounding the seeds for making a fermented drink is the probable reason for its domestication during the time of European contact (Patiño Rodriguez, 2002), a practice which persists even today (Thomas et al., 2012). Recent discoveries in Honduras showed that the Olmec people in the Ulua valley made an alcoholic drink in this way (Henderson et al., 2007). The diverse nature of cacao's uses as well as its use as a currency led to its wide distribution in Mesoamerica before the Spaniards arrived (Bergmann, 1969; Young, 1994). However, only a very small portion of its diversity has been spread from Amazon to Mesoamerica, and therefore domesticated cacao has a very narrow genetic background (Dias et al., 2001; Bartley, 2005). The order Malvales is large and contains the family Malvaceae to which cacao botanically belongs to.

2.2.3 Species in *Theobroma*

Name	Status	Confidence level	Source	Date supplied	Infra. rec.
Theobroma angustifolium Sessé & Moc. ex DC.	Accepted	★★☆	TRO	18 April 2012	
Theobroma balaense (P.Preuss) De Wild.	Unresolved	★☆☆	WCSP (in review)	26 March 2012	
Theobroma bernoullii Pittier	Accepted	★★☆	TRO	18 April 2012	2
Theobroma bicolor Humb. & Bonpl.	Accepted	★★☆	WCSP (in review)	26 March 2012	
Theobroma cacao L.	Accepted	★★☆	WCSP (in review)	26 March 2012	1
Theobroma calodesme Diels	Unresolved	★☆☆	WCSP (in review)	26 March 2012	
Theobroma canumanense Pires & Fróes ex Cuatrec.	Accepted	★★☆	TRO	18 April 2012	
Theobroma caribaeum Sweet	Unresolved	★☆☆	WCSP (in review)	26 March 2012	
Theobroma celtifolia Salisb.	Unresolved	★☆☆	TRO	18 April 2012	
Theobroma cirmolinae Cuatrec.	Accepted	★★☆	TRO	18 April 2012	
Theobroma duckei Huber	Accepted	★★☆	TRO	18 April 2012	
Theobroma ferrugineum Bernoulli	Unresolved	★☆☆	WCSP (in review)	26 March 2012	

Theobroma fossilium E.W. Berry	Unresolved	★☆☆	TRO	18 April 2012
Theobroma gileri Cuatrec.	Accepted	★★☆	TRO	18 April 2012
Theobroma glaucum H. Karst.	Accepted	★★☆	TRO	18 April 2012
Theobroma grandiflorum (Willd. ex Spreng.) K.Schum.	Accepted	★★☆	WCSP (in review)	26 March 2012
Theobroma guyanense Voigt	Unresolved	★☆☆	WCSP (in review)	26 March 2012
Theobroma hylaeum Cuatrec.	Accepted	★★☆	TRO	18 April 2012
Theobroma macranthum Bernoulli	Unresolved	★☆☆	WCSP (in review)	26 March 2012
Theobroma mammosum Cuatrec. & J. León	Accepted	★★☆	TRO	18 April 2012
Theobroma martianum D.Dietr.	Unresolved	★☆☆	WCSP (in review)	26 March 2012
Theobroma microcarpum Mart.	Accepted	★★☆	WCSP (in review)	26 March 2012
Theobroma montanum Goudot ex Bernoulli	Unresolved	★☆☆	WCSP (in review)	26 March 2012
Theobroma nemorale Cuatrec.	Accepted	★★☆	TRO	18 April 2012
Theobroma obovatum Klotzsch ex Bernoulli	Accepted	★★☆	WCSP (in review)	26 March 2012
Theobroma ovatifolium DC.	Unresolved	★☆☆	WCSP (in review)	26 March 2012
Theobroma pulcherrimum Pittier	Unresolved	★☆☆	WCSP (in review)	26 March 2012
Theobroma quinquenervium Bernoulli	Unresolved	★☆☆	WCSP (in review)	26 March 2012
Theobroma sagittatum Pav. ex A.Chev.	Unresolved	★☆☆	WCSP (in review)	26 March 2012
Theobroma salzmannianum Bernoulli	Unresolved	★☆☆	WCSP (in review)	26 March 2012
Theobroma simiarum Donn. Sm.	Accepted	★★☆	TRO	18 April 2012
Theobroma sinuosum Pav. ex Huber	Accepted	★★☆	TRO	18 April 2012
Theobroma speciosum Willd. ex Spreng.	Accepted	★★☆	WCSP (in review)	26 March 2012

Theobroma stipulatum Cuatrec.	Accepted	★★☆	TRO	18 April 2012
Theobroma subincanum Mart.	Accepted	★★☆	WCSP (in review)	26 March 2012
Theobroma sylvestre Aubl. ex Mart. in Buchner	Accepted	★★☆	TRO	18 April 2012
Theobroma velutinum Benoist	Accepted	★★☆	TRO	18 April 2012

Statistics of the *Plant List* includes 71 scientific plant names of species rank for the genus *Theobroma*. Of these 22 are accepted species names.

Status	Total
Accepted	25
Synonym	46
Unplaced	0
Unassessed	15

2.2.4 Disagreements on classification

The botanical taxonomy of cacao is not straightforward. The high level of synonyms is a hindrance to good classification and there are aspects of disagreement as shown below:

Overall, the taxonomy of *Theobroma cacao* L. is as follows:
Kingdom: *Plantae* (plants)
Infrakingdom: Streptophyta (land plants)
Subkingdom: *Tracheobionta* (vascular plants)
Subkingdom: Viridiplantae
Superdivision: Embryophyta
Superdivision: *Spermatophyta* (seed plants)
Division: Tracheophyta (vascular plants, tracheophytes)
Division: *Magnoliophyta* (flowering plants)
Subdivision: Spermatophytina (spermatophytes, seed plants, phanérogames)
Class: Equisetopsida
Class: *Magnoliopsida* (dicotyledons, e.g. coffee)
Subclass: *Dilleniidae* (e.g. like tea)
Subclass: Magnoliidae
Superorder: Rosanae
Order: Malvales
Family: Malvaceae (formerly *Sterculiaceae*)
Genus: *Theobroma*
Other species of family *Theobroma*
T. grandiflorum
T. bicolor

Since the mid-twentieth century, *T. cacao* varieties have been classified into three general groups: 'Criollo', 'Forastero' and 'Trinitario', based on morphological descriptors and geographic origin (Engels, 1981; Enriquez and Soria, 1967). Recent genetic studies indicate that this broad categorisation is overly simplistic with regard to the term 'Forastero' and fails to acknowledge the significant amount of genetic variation contained within this group. In contrast, 'Criollo' refers to a group of genetically similar trees which produce lightly pigmented seeds and share several other morphological traits. 'Trinitario' refers to hybrids of the 'Criollo' and 'Forastero' strains.

3 The three groups of cocoa varieties

3.1 Criollo

The Criollo group refers to genetically similar trees, believed to have been domesticated by the Mayan civilisation. The seeds (beans) tend to be round in profile and contain white or pale purple cotyledons. Young seedlings can be identified by the presence of green cotyledons and by that of leaf stems (petioles) with a horizontally opposed orientation. This variety displays low vigour, poor productivity and high susceptibility to diseases, insects and stress and is therefore less widely cultivated (Afoakwa, 2010). Until the mid-eighteenth century, Criollo was the most commonly cultivated type of cacao; however, the vast majority of these trees have been replaced with more vigorous and hardy trees of hybrid or 'Forastero' ancestry. This type is very rare and is only found in old plantations in Venezuela, Central America, Madagascar, Sri Lanka and Samoa (Fowler, 2009). The yield of Criollo plantations is lower than that of Forastero. Notable characteristics include a unique growth habit that frequently lacks the usual whorl of five branches (jorquette) and consists entirely of plagiotropic stems with occasional bi- and trifurcating branches. Fruits typically have a soft thin husk with a textured surface and usually have some degree of red pigmentation. Chocolate made from Criollo is light in colour and has a subtle or delicate taste that is low in basic chocolate flavour. Criollo beans are often sold at higher prices, which can somewhat offset their lower yield. Criollo may be translated as 'native' or 'first grown' in Spanish-speaking countries of the Americas. It should be noted that so-called Criollo cacao in many countries may not have the typical genetic profile of pure Criollo.

3.2 Forastero

Forastero means 'foreigner' in Spanish. It refers to any trees that are not Criollo or a hybrid and which usually produce deep purple seeds (Hebbar et al., 2011). The term encompasses a wide range of distinct populations with unique characteristics and is not a meaningful descriptive term. Forastero is native to the Amazon region and largely grown in West Africa and Southeast Asia. It forms 95% of world cocoa production (Afoakwa, 2010; Fowler, 2009; Afoakwa et al., 2012) and is the most widely used due to its higher yield than the Criollo variety. Because of the high genetic variability within this group, Forastero types exhibit greater variability in tree and fruit morphology and are generally more vigorous and less susceptible to disease and pests than are Criollo trees. When ripe, the pods are hard and yellow, have a rounded melon-like shape and contain 30 or more beans ranging in colour from pale to deep purple. Several landraces are particularly well known. Grown mainly in Brazil and Africa, it is hardier, higher yielding and easier to cultivate than Criollo and is used in every blend of chocolate that is produced. The Amelonado type is widely grown throughout Western Africa and produces the majority of 'bulk cacao' beans. Although Forastero beans are said to produce chocolate rich in chocolate flavour but low in complex or fruit flavour notes, there are several well-known exceptions. Forastero may be subdivided into Upper Amazonian [wild or semi-wild cacaos as described by Pound (1938)] and Lower Amazonian, characterised by a rather uniform pod type called Amelonado (non-pigmented, smooth, melon-shaped pods with a blunt end). Lower Amazonian cacaos now constitute the most prevalent cultivated type worldwide and are grown notably in West Africa and Brazil. These types grow wild in the Guyanas and the eastern Brazilian Amazon and may have been partially domesticated by pre-Colombian Amazonian natives for their aromatic pulp (Barrau, 1979).

3.3 Trinitario and other hybrids

In the strict sense, Trinitario refers to clones, or the progenies, of hybrids originally produced in Trinidad between 'Criollo' and 'Forastero' trees which originated in the lower Amazon River basin (Willson, 1999). However, it later spread to Venezuela, Ecuador, Cameroon, Samoa, Sri Lanka, Java and Papua New Guinea (ICCO, 2012). It is believed that it first came into existence on the island of Trinidad, after a hurricane nearly destroyed the local Criollo crops in 1727. Some Trinitario varieties produce cocoa beans with special flavours. They have hard pods and are variable in colour, containing 30 or more beans of variable colour, although white beans are rare. Trinitario is considered to be either an intermediate type between Criollo and Forastero (Lockwood and Gyamfi, 1979) or a group of hybrids displaying characteristics which include the total range of variation (Cheesman, 1944). The Trinitarios combine the best characteristics of the two main varieties: the hardiness and high yield of Forestero and the refined taste of Criollo. Trinitario types have not been found under wild conditions. Because Criollos and Forastero types are so heterogeneous, the resulting hybrids might not be distinct from the parental populations, making Trinitarios impossible to define except by geographical origin (B.G.D. Bartley, unpublished).

3.4 Nacional

The Nacional variety grows in Ecuador and is believed to have originated from the Amazonian area of Ecuador (Fowler, 2009; Afoakwa, 2010). It has distinct characteristics of aroma and flavour. Although less widely cultivated and contributing to only 5% of global cocoa production, it represents more than 50% of the fine cocoa marketed worldwide every year. Until the beginning of twentieth century, Nacional cultivars were the only type of cocoa cultivated in Ecuador (Soria, 1966), with the unique flavour and aroma, known as 'arriba'. The fine flavoured quality chocolate products obtained from Nacional cocoa beans are highly valued by chocolate manufacturers. However, the original Nacional cocoa trees are currently in danger of extinction due to the introduction of an external unrelated germplasm (Solorzano et al., 2012). A large genetic admixture between native Nacional cocoa and foreign germplasm is currently found in modern Ecuador plantations (Solorzano et al., 2009), thus reducing the fine-flavoured cocoa aroma. Currently, pure Nacional cocoa varieties are rare and there is an increasing demand for fine flavour cocoa although Ecuador still represents 6.8% of the supply of 'arriba' flavoured cocoa in the world market (ICCO, 2012). Silva et al. (2004) provided a key to the species and hybrids of genus *Theobroma*.

3.5 Proposed new classification

Motamayor et al. (2008) proposed a new classification into ten major groups within the Forastero cacao, based on genetic diversity in varieties from Latin America. These are Marañon, Curaray, Criollo, Iquitos, Nanay, Contamana, Amelonado, Purús, Nacional and Guiana. This new classification more accurately reflects the genetic diversity of cocoa (Fig. 1).

Since 2008, several scientific studies have been carried out to assess the genetic diversity of cacao (Zhang et al., 2009, 2011; Trognitz et al., 2011; Ji et al., 2013; Santos et al., 2015). However, most of these are limited to only a few geographical locations and to three or four groups (Criollos, Foreseteros, Trinitariors and or Nacional or Cacao Nacional Boliviano types). Thomas et al. (2012) used spatial diversity patterns to explain genetic differentiation in cacao through the possible contribution of climate change, geographical locations, history and culture. However, there are several questions which remain unanswered and a comprehensive study combining application of molecular and modelling approaches is needed to understand the entire genetic diversity and population structure of cacao.

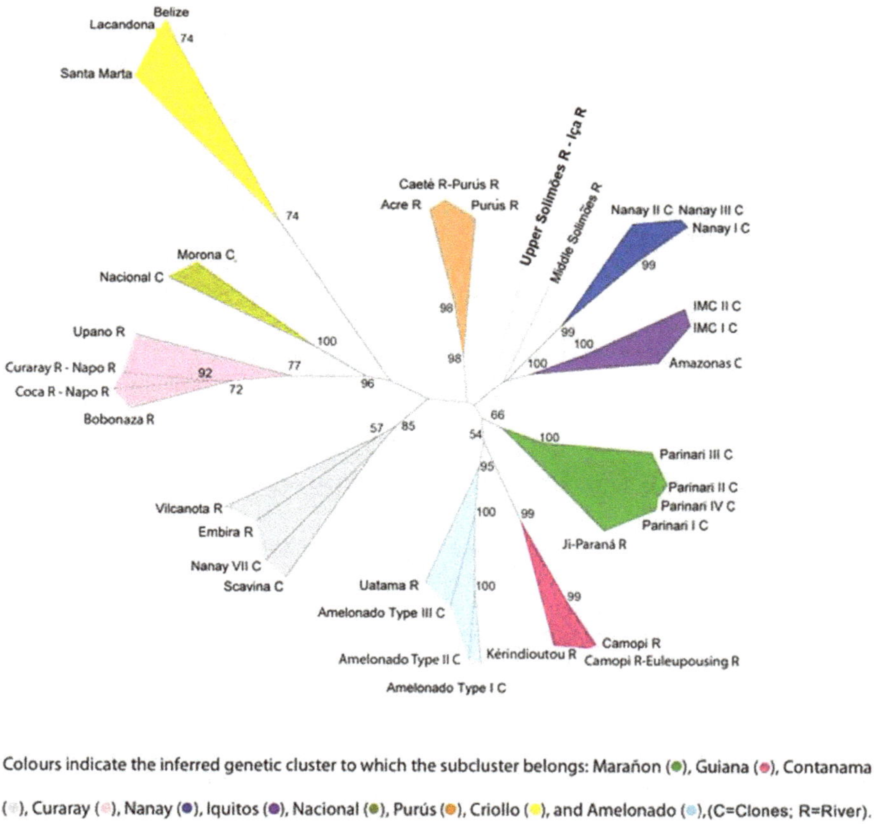

Figure 1 Neighbour joining tree from Cavalli-Sforza and Edwards genetic distance [16] matrix among the 36 subclusters identified using STRUCTURE (559 clones). *Source* Motamayor et al. (2008).

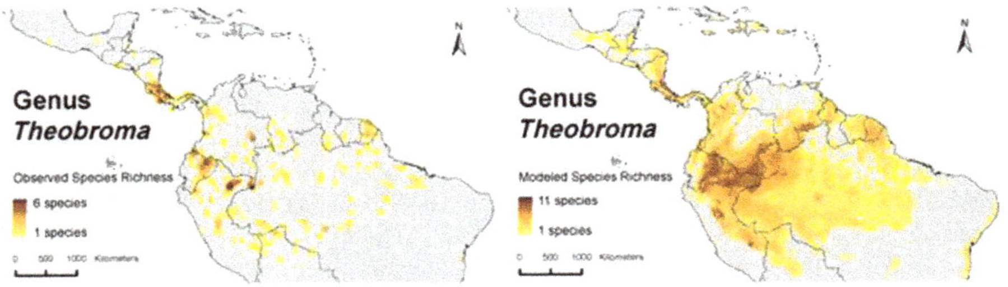

Figure 2 Species richness of the genus *Theobroma*. *Source* Thomas et al. (2012).

4 Conclusions

It has been concluded by many authors that the centre of genetic diversity of cocoa is located in the upper Amazonian regions bordering Peru, Brazil, Colombia and Ecuador, and to a lesser extent, in the southern part of Central America (Motamayor, et al., 2002; Sereno et al., 2006; Zhang, et al., 2006).

A certain level of overlap exists between the areas of highest species richness of the genus *Theobroma* and the area where the highest genetic diversity of cacao is observed. This corroborates the hypothesis that cacao also originated in the Western Upper Amazon. Interpretation of the Pleistocene modelled distribution of cacao, together with present-day genetic observations, suggests that the species was already distributed widely across the Western Amazon. The current distribution of cocoa is likely to have been partly shaped by varying degrees of human interventions. The fact that the pulp of several other species in the genus *Theobroma* was used in a similar fashion as cacao by South-American indigenous groups (Patiño Rodriguez, 2002) might explain why the present-day highest species richness of the genus *Theobroma* is located in this area (Fig. 2).

5 Where to look for further information

The demand for high-quality cocoa is increasing which necessitates the need to broaden the genetic diversity by looking at additional cocoa germplasm from the regions that has not yet been explored. A lot of germplasm characterization research is being undertaken at USDA-ARS, Beltsville in collaboration with the international gene bank, Trinidad and Tobago. Future research should therefore concentrate on the development of cocoa varieties that can give higher yields and produce quality beans in the light of changing climate and associated threats of biotic and abiotic stresses. Although a lot of work and publications are available on defining the origin, taxonomy and classification of cacao, the questions still remains if there are other members of the genus Theobroma that produces cocoa-like products and lot of information can be available in ICCO, WCF and CABI websites.

6 References

Afoakwa, E. O. (2010). *Chocolate Science and Technology*, 1st ed. Wiley-Blackwell Pub, West Sussex, UK.

Afoakwa, E. O., Quao, J., Budu, A. S., Takrama, J. and Saalia, F. K. (2012). Changes in total polyphenols, o-diphenols and anthocyanin concentrations during fermentation of pulp-preconditioned cocoa (*Theobroma cacao* L.) beans. *International Food Research Journal* 19: 1071–7.

Alverson, W. S., Whitlock, B. A., Nyffler, R., Bayer, C. and Baum, D. A. (1999). Phylogeny of the core Malvales: Evidence from *ndh*F sequence data. *American Journal of Botany* 86: 1474–86.

Anon (2011). *Cocoa Atlas – 2010 Edition*. German Cocoa and Chocolate Foundation, Bonn. ISBN: 9-783980-886666.

Barrau, J. (1979). Sur l'origine du cacaoyer, *Theobroma cacao* Linne, Sterculiacées. *Journal D'Agriculture Tropicale Et De Botanique Appliquee* 26: 171–80.

Bartley, B. G. D. (2005). *The Genetic Diversity of Cacao and Its Utilization*. CABI Publishing, Wallingford.

Bayer, C., Fay, M. F., De Bruijn, A. Y., Savolainen, V., Morton, C. M., Kubitzki, K., Alverson, W. S. and Chase, M. W. (1999). Support for an expanded family concept of Malvaceae within a recircumscribed order Malvales: A combined analysis of plastid *atpB* and *rbcL* DNA sequences. *Botanical Journal of Linnaeus Society* 129: 267–303.

Bentham, G. J. and Hooker, D. (1862). Malvaceae. In: G. Bentham and J. D. Hooker (Eds), *Genera Plantarum*. London, England, pp. 195–213.

Bergmann, J. F. (1969). The distribution of cacao cultivation in Pre-Columbian America. *Annual association of American Geography* 59: 85–96.

Bernoulli, G. (1869). Uebersicht der bis jetzt bekannten Arten von *Theobroma*. Neue Denkschriften der Allgemeinin Schweizerischen Gesellschaft für die Gesammten Naturwissenschaften. *Band* 24: 1–15, 7 Taflen, Zürich, 1871.

Cheesman, E. (1944). Notes on the nomenclature, classification and possible relationship of cacao populations. *Tropical Agriculture* 21: 144–59.

Chevalier, A. (1946). Révision du genre Theobroma d'après l'herbier du Muséum national d'histoire naturelle de Paris. *Revue de botanique appliqué* 26: 265–85.

Clement, C. R., de Cristo-Arau´jo, M., d'Eeckenbrugge, G. C., Pereira, A. A. and Picanco-Rodrigues, D. (2010). Origin and domestication of native Amazonian crops. *Diversity* 2: 72–106.

Cope, F. W. (1976) The mechanism of pollen incompatibility in *Theobroma cacao* L. *Heredity* 17: 157–82.

Cuatrecasas, J. (1964). Cacao and its allies. A taxonomic revision of the genus Theobroma. *Contributions to the United States National Herbariuim* 35(6): 379–614.

de la Cruz, M., Whitkus, R., Gomez-Pompa, A. and Mota-Bravo, L. (1995). Origins of cacao cultivation. *Nature London* 375: 542–3.

Dias, L. A. S. (2001a). Chapter 3: Origin and distribution of *Theobroma cacao* L: A new scenario. In: L. A. S. Dias (Ed.), *Genetic Improvement of Cacao* (EcoPort version by P. Griffee). FAO. Available at: at: http://ecoport.org/ep?Search Type=earticleView&earticleId=197&page=-2.

Dias L. A. S. (2001b). Chapter 12: The contributions of breeding. In: L. A. S. Dias (Ed.), *Genetic Improvement of Cacao* (EcoPort version by P. Griffee). FAO. Available at: http://ecoport.org/ep?Search Type=earticleView&earticleId=197&page=-2.

Dias, L. A. S. and Resende, M. D. V. (2001). Chapter 6: Selection strategies and methods. In: L. A. S. Dias (Ed.), *Genetic Improvement of Cacao* (EcoPort version by P. Griffee). FAO.

Dillinger, T. L., Barriga, P., Escárcega, S., Jimenez, M., Lowe, D. S. and Grivetti, L. E. (2000). Food of the gods: cure for humanity? A cultural history of the medicinal and ritual use of chocolate. *The journal of nutrition* 130(8): 20575–725.

Engels, J. M. M. (1981). Genetic resources of cacao: A catalogue of the CATIE collection. Tech. Bul. 7. CATIE Plant Genetic Resources Unit, Turrialba, Costa Rica. p. 196.

Enriquez, G. A. and Soria, J. V. (1967). Cacao Cultivars Register. IICA, Turrialba, Costa Rica.
Eskes, A. B. and Efron, Y. (2006). *Global Approaches to Cocoa Germplasm Utilization and Conservation. Final Report of the CFC/ICCO/IPGRI project on 'Cocoa Germplasm Utilization and Conservation: A Global Approach' (1998–2004)*. ICCO and IPGRI, London and Rome.
Figueira, A., Janick, J., Levy, M. and Goldsbrough, P. (1994). Reexamining the classification of *Theobroma cacao* L. using molecular markers. *Journal of American Society and Horticultural Science* 119: 1073–82
Fowler, M. S. (2009). Cocoa beans: From tree to factory. In: S. T. Beckett (Ed.), *Industrial Chocolate Manufacture and Use*, 4th ed. Blackwell Publishing, Oxford, pp. 10–48.
Gómez-Pompa, A., Flores, J. S. and Fernandez, M. A. (1990). The sacred cacao groves of the Maya. *Latin American Antiquity* 1: 247–57
Hart J. H. (1892). Cocoa*: A Treatise on the Cultivation and Curing of Cacao (Theobroma cacao), Botany and Nomenclature of the Same and Hints on the Selection and Management of Estates*. U.S. Government Printing Office, Port of Spain, Trinidad, pp. 1–77.
Hebbar, P., Bittenbender, H. C. and O'Doherty, D. (2011). Farm and forestry production and marketing profile for cacao (*Theobroma cacao*). In: C. R. Elevitch (Ed.), *Specialty Crops for Pacific Island Agroforestry*. Permanent Agriculture Resources (PAR), Holualoa, Hawaii. Accessed 10 December 2013.
Henderson, J. S., Joyce, R. A., Hall, G. R., Hurst, W. J. and McGovern, P. E. (2007). Chemical and archaeological evidence for the earliest cacao beverages. *Proceedings of the National Academy of Sciences USA*, 104: 18937–40.
Hutchinson, J. (1967). *The Genera of Flowering Plants (Angiospermae), vol. II*. Clarendon Press, Oxford, England. pp.659.
ICCO (2012). *International Cocoa Organization. Quarterly Bulletin of Cocoa Statistics*, vol. XXXVII – no. 3, 26 April 2012.
Iwaro, A. D. S., Bharath, S., Bekele, F. L., Butler, D. R. and Eskes, A. B. (2005). Germplasm enhancement for resistance to black pod disease: strategy and prospects. In: *Proceedings of the 14th International Cocoa Research Conference*, Accra, Ghana, 13–17 October 2003. Cocoa Producers Alliance, Lagos, Nigeria.
Ji, K., Zhang, D. P., Motilal, L., Boccara, M., Lachenaud, P. and Meinhardt, L. W. (2013). Genetic diversity and parentage in farmers' varieties of cacao (*Theobroma cacao* L.) from Honduras and Nicaragua as revealed by single nucleotide polymorphism (SNP) markers. *Genetic Resources and Crop Evolution* 60: 441–53.
Judd, W. S. and Manchester, S. R. (1997). Circumscription of Malvaceae (Malvales) as determined by a preliminary cladistic analysis of morphological, anatomical, palynological, and chemical characters. *Brittonia* 49: 384–405
Laurent, V., Risterucci, A. M. and Lanaud, C. (1994). Genetic diversity in cocoa revealed by cDNA probes. *Theoretical and Applied Genetics* 88: 193–8
Lockwood, G. and Gyamfi, M. M. O. (1979). The CR10 cocoa germplasni collection with notes on codes used in the breeding programme at Tafo and elsewhere. *Technical Bulletin* No. 1: 1–8. Cocoa Research Institute of Ghana, Ghana.
Morris, D. (1882). *Cacao: How to Grow and How to Cure It*. Govt. Printing Establishment, Kingston, Jamaica, pp. 1–45.
Motamayor, J. C., Lachenaud, P., Wallace da Silva e Mota, J., Loor, R., Kuhn, D. N., Brown, S. and Schnell, R. J. (2008). Geographic and genetic population differentiation of the Amazonian chocolate tree (*Theobroma cacao* L). *PloS ONE* 3(10): e3311
Motamayor, J. C., Risterucci, A. M., Lopez, P. A., Ortiz, C. F., Moreno, A. and Lanaud, C. (2002). Cacao domestication I: the origin of the cacao cultivated by the Mayas. *Heredity* 89: 380–6.
Muñoz Ortega, J. M. (1948). Estudios cromosómicos en el género *Theobroma* L. *M. Agr. Diss.*, IICA, Turrialba, Costa Rica.
Nair Prabhakaran, K. P. (2010). Cocoa (*Theobroma cacao* L.). In: Nair Prabhakaran, K. P. (Ed.), *The Agronomy and Economy of Important Tree Crops of the Developing World*. Elsevier, Heidelberg, New York, London, pp. 131–80.

Patiño Rodriguez, V. M. (2002). *Historia y dispersión de los frutales nativos del Neotropico*. Centro Internacional de Agricultura Tropical, Cali, Colombia.
Pound, F. J. (1938). Cacao and witchbroom disease (*Marasmius pemiciosus*) of S. America with notes on other species of Theobroma. *Archives of Cocoa Research* 1: 20–72.
Powis, T. G., Cyphers, A., Gaikwad, N. W., Grivetti, L. and Cheong, K. (2011). Cacao use and the San Lorenzo Olmec. *Proceedings of the National Academy of Sciences USA*, 108, 8595–600.
Purdy, L. H. and Schmidt, R. A. (1996). Status of cacao witches' broom: Biology, epidemiology, and management. *Annual Review of Phytopathology* 34: 573–94.
Rossner, S. (1997). Chocolate- divine food, fattening junk or nutritious supplementation? *European Journal of Clinical Nutrition* 51: 341–5.
Santos, E. S., Cerqueira-Silva, C. B., Mori, G. M., Ahnert, D., Mello, D. L., Pires, J. L., Corrêa, R. X. and de Souza, A. P. (2015). Genetic structure and molecular diversity of cacao plants established as local varieties for more than two centuries: The genetic history of cacao plantations in Bahia, Brazil. *PLoS One* 10(12): e0145276.
Schultes, R. E. (1958). A synopsis of the genus Herrania. *Journal of Arnold Arboretum* 39: 216–78.
Schultes R. E. (1984). Amazonian cultigens and their northward and westward migrations in pre-Columbian times. In: D. Stone (Ed.), *Pre-Columbian Plant Migration, Papers of the Peabody Museum of Archaeology and Ethnology*, Volume 76. Harvard University Press, Cambridge, Massachusetts, pp. 69–83.
Schumann, K. (1886). Family Sterculiaceae genus Theobroma. In: C. P. F. Martius and A. G. Eichler (Eds), *Flora Brasiliensis*, volume XII, part III, pp. 70–77.
Sereno, M. L., Albuquerque, P. S. B., Vencovsky, R. and Figueira, A. (2006). Genetic diversity and natural population structure of cacao (*Theobroma cacao* L.) from the Brazilian Amazon evaluated by microsatellite markers. *Conservation Genetics* 7: 13–24.
Silva, C. R. S. and Figueira, A. (2005). Phylogenetic analysis of Theobroma (Sterculiaceae) based on Kunitz-like trypsin inhibitor sequences. *Plant Systematic and Evolution* 250: 93–104.
Silva, C. R. S., Venturieri, G. A. and Figueira, A. (2004). Description of Amazonian *Theobroma* L.collections, species identification, and characterization of interspecific hybrids. *Acta Botanica Brasilica* 18: 333–41.
Solorzano, R. G. L., Risterucci, A. M., Courtois, B., Fouet, O., Jeanneau, M., Rosenquist, E., Amores, F., Vasco, A., Medina, M. and Lanaud, C. (2009). Tracing the native ancestors of modern *Theobroma cacao* L. population in Ecuador. *Tree Genetics and Genomes* 5(3): 421–33.
Solorzano, R. G. L., Fouet, O., Lemainque, A., Pavek, S., Michel, B., Argout, X., Amores, F., Courtois, B., Risterucci, A. M. and Lanaud, C. (2012). Insight into the wild origin, migration and domestication history of the fine flavour Nacional *Theobroma cacao* L. variety from Ecuador. *PLoS ONE* 7(11): e48438. https://doi.org/10.1371/journal.pone.0048438
Soria J. (1966). Principales variedades de cacao cultivadas en Ame´rica Tropical. *Turrialba* 16(3): 261–5.
Takhtajan, A. (1997). *Diversity and Classification of Flowering Plants*. Columbia University Press, New York, USA.
Thomas, E., van Zonneveld, M., Loo, J., Hodgkin, T., Galluzzi, G. and van Etten, J. (2012). Present spatial diversity patterns of *Theobroma cacao* L. in the neotropics reflect genetic differentiation in Pleistocene refugia followed by human-influenced dispersal. *PLoS ONE* 7: e47676.
Trognitz, B., Scheldeman, X., Hansel-Hohl, K., Kuant, A., Grebe, H. and Hermann, M. (2011). Genetic population structure of cacao plantings within a young production area in Nicaragua. *PloS ONE* 6: e16056.
Van Hall, C. J. J. (1914). *Cocoa*. Macmillan, London, UK.
Whitlock, B. A. and Baum, D. A. (1999). Phylogenetic relationships of *Theobroma* and *Herrania* (Sterculiaceae) based on sequences of the nuclear gene *Vicilin*. *Systematic Botany* 24: 128–38.
Whitlock, B. A., Bayer, C. and Baum, D. A. (2001). Phylogenetic relationships and floral evolution of the Byttnerioideae ('Sterculiaceae' or Malvaceae s.l.) based on sequences of the chloroplast gene *ndh*F. *Systematic Botany* 26: 420–37.
Whitkus, R., De la Cruz, M. and Mota-Bravo, L. (1998). Genetic diversity and relationships of cocoa (*Theobroma cacao* L.) in southern Mexico. *Theoretical and Applied Genetics* 96: 621–7

Willson, K. C. (1999). Coffee, *Cocoa and Tea*. CABI Publishing, New York, pp. 100–10.
Wood, G. A. R. and Lass, R. A. (1985). *Cocoa* 4th Edition. Longman Scientific & Technical, Essex, UK, p. 620
Young A. M. (1994) *The Chocolate Tree: A Natural History of Cacao*. Smithsonian Nature Books, Washington, DC.
Zhang, D., Arevalo-Gardini, E., Mischke, S., Zúñiga-Cernades, L., Barreto-Chavez, A. and Andriazola del Aguila, J. (2006). Genetic diversity and structure of managed and semi-natural populations of cocoa (*Theobroma cacao*) in the Huallaga and Ucayali valleys of Peru. *Annals of Botany* 98(3): 647–55.
Zhang, D., Mischke, B. S., Johnson, E. S., Mora, A., Phillips-Mora, W. and Meinhardt, L. W. (2009). Molecular characterization of an international cacao collection using microsatellite markers. *Tree Genetics and Genomics* 5: 1–10.
Zhang, D., Martinez, W. J., Johnson, E. S., Somarriba, E., Phillips-Mora, W., Astorga, C., Mischke, S. and Meinhardt, L. W. (2011). Genetic diversity and spatial structure in a new distinct *Theobroma cacao* L. population in Bolivia. *Genetic Resources and Crop Evolution* 59: 239–52.

Chapter 2

Conserving and exploiting cocoa genetic resources: the key challenges

Brigitte Laliberté, Bioversity International, Italy; Michelle End, INGENIC (The International Group for Genetic Improvement of Cocoa), UK; Nicholas Cryer, Mondelez International, UK; Andrew Daymond, University of Reading, UK; Jan Engels, Bioversity International, Italy; Albertus Bernardus Eskes, formerly CIRAD and Bioversity International, France; Martin Gilmour, MARS Global Chocolate, UK; Philippe Lachenaud, Centre de coopération internationale en recherche agronomique pour le développement, France; Wilbert Phillips-Mora, Center for Tropical Agriculture Research and Education, Costa Rica; Chris Turnbull, Cocoa Research Association Ltd., UK; Pathmanathan Umaharan, Cocoa Research Centre, The University of the West Indies, Trinidad and Tobago; Dapeng Zhang, USDA-ARS, USA; and Stephan Weise, Bioversity International, Italy

http://dx.doi.org/10.19103/AS.2017.0021.02
Published by Burleigh Dodds Science Publishing Limited, 2018.

1 Introduction

Cacao genetic resources comprise the range of variability that provides the raw material for breeding new and improved varieties to achieve a more economically sustainable cocoa production system, thus contributing to the economies of cacao-producing countries.

Cacao germplasm can be conserved in two ways: (1) *ex situ*, that is comprising all cacao germplasm currently maintained in field genebanks as living trees and/or in *in vitro* collections as tissues and embryos, or (2) *in situ*, that is in farmers' fields for cultivated materials or in protected areas such as nature reserves for wild or semi-wild materials.

Effective conservation and management of cacao genetic resources includes the following routine activities: targeted collecting actions, maintenance of key field collections, effective characterization and identity studies, evaluation for important traits, information management, safe exchange of germplasm and germplasm enhancement (in some cases). All these contribute directly to a sustainable cocoa production (Fig. 1).

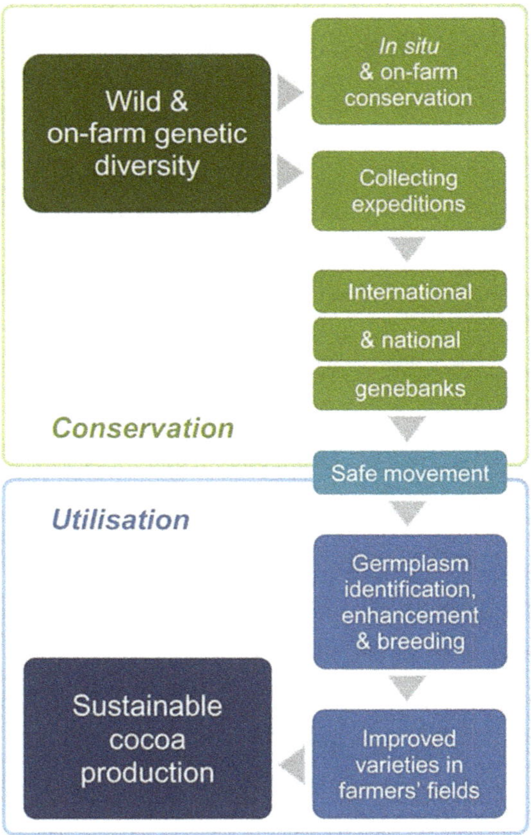

Figure 1 Links between the genetic diversity and sustainable cacao production. Credit: C. Turnbull, Reading University.

Published by Burleigh Dodds Science Publishing Limited, 2018.

The loss of cacao genetic diversity (*in situ*, on-farm and to a lesser extent conserved in collections) is a serious problem, and all its many causes need to be urgently addressed: the destruction of the Amazonian rainforests, changing patterns of land use, the spread of pests and diseases, sudden changes in climate and threats from natural disasters and extreme weather. These factors are resulting in an irreversible loss of the cacao genetic diversity so essential for farmers, breeders and consumers.

The future of the world cocoa economy depends on the availability of genetic diversity and the sustainable use of this broad genetic base to breed improved varieties. Most of the countries involved in the improvement and production of cacao are highly dependent on genes and varieties characterized and conserved in other countries and regions. Effective management of cacao genetic resources can therefore only be carried out through international collaboration.

Since the early part of the twentieth century, numerous missions have been undertaken to collect and conserve cacao *ex situ* in genebanks. The catastrophic impact of cacao diseases led to the expeditions to collect disease-resistant germplasm from the Upper Amazon region in the 1930s, but in recent years, more emphasis has been placed on systematic collections to capture genetic diversity.

The key constraints affecting the current global system are the following:

- Funding for the conservation and sustainable use of cacao genetic resources is well below adequate levels. As a consequence, many collections are conserved in suboptimal standards and their safeguarding is threatened. This seriously impedes the efficient use of available germplasm.
- The lack of sufficient expertise and trained staff for the conservation of collections and the use in breeding is recognized.
- Unnecessary duplication and high frequency of misidentified accessions in collections not only hinder efficient conservation but also hamper the effectiveness of germplasm evaluation and use. Comprehensive assessment of individual accession identity and phylogenetic relationship amongst accessions is therefore a high-priority task.
- Evaluation of germplasm for economically important traits has only been partially achieved, which is a limiting factor in the uptake of accessions in breeding programmes. Lack of reliable evaluation methods in collections also plays a role, especially so for yield and pest and disease resistance, as well as inadequate documentation of the evaluation results.
- There is a lack of sufficient variation for important selection traits in existing collections or in some cases, lack of evaluation for those traits.
- The existing international databases need continued support to improve accessibility and interpretation of the information available. Efforts to install and/or improve local information management systems are needed in some genebanks to facilitate accession management and data sharing.
- There are insufficient, inoperative or inadequately managed seed gardens to reproduce selected hybrid varieties in sufficient quantities to satisfy farmer requirements in many areas.
- The substitution of local varieties (landraces) by clonal varieties in the field can result in irreversible loss of the genetic diversity if not safeguarded in *ex situ* collections. This consequently increases the vulnerability of cacao to sudden changes in climate, and to infection/infestation by new pests and diseases.

- National laws restricting access to and use of plant genetic resources have emerged in many countries. The introduction of intellectual property rights for new varieties and their genetic components in developed countries has been followed by application of national sovereignty and restrictions on access to these resources in developing countries.
- The lack of a long-term global strategy and insufficient regional and international collaboration are reducing the capacity of the global system for the conservation and use of cacao germplasm.

2 How challenges are being addressed: an overview

A considerable portion of the global cacao diversity is *in situ*, in farmers' fields and held in genebanks around the world, including two international collections maintained at the Cocoa Research Centre of the University of the West Indies (CRC/UWI), Trinidad and Tobago, and at the Centro Agronómico Tropical de Investigación y Enseñanza (CATIE), Costa Rica. Unfortunately, much of the genetic resources maintained in national collections is at risk and funding remains insufficient and unstable.

In order to optimize the conservation and maximize the use of cacao genetic resources as the foundation of a sustainable cocoa economy, it is imperative to bring together national and international players in public and private sectors to ensure that the:

- cacao genepool is conserved *in situ* and *ex situ* for the long term by a global network of partners
- global system for the safe exchange of cacao germplasm is strengthened
- use of cacao genetic diversity is optimized
- effectiveness of global efforts to conserve and use cacao genetic resources is assured.

An integrated Global Strategy for the Conservation and Use of Cacao Genetic Resources was developed between 2011 and 2012 by the cacao community including the organization of related information on the materials (origin/donor, taxonomy, basic characters, etc.). The consultation with a wide group of experts in cacao genetic resources research and management was facilitated by the Global Network for Cacao Genetic Resources (CacaoNet) and the Global Strategy was published in 2012. The Global Strategy provides a clear framework to secure funding and collective actions for the most urgent needs to ensure that cacao diversity is conserved and used and provides direct benefits to the millions of small-scale cacao farmers around the world.

Figure 2 illustrates the connections between the different elements of conservation and use of cacao genetic diversity, from the diversity in the wild and on-farms to their conservation *ex situ* and use in evaluation and breeding programmes and development of improved planting materials for farmers contributing to sustainable cocoa production.

The key issues addressed for the conservation and use of cocoa genetic resources are the following (corresponding to the numbers in orange):

1. Securing existing *ex situ* cacao genetic resources and their distribution
2. Developing a Global Strategic Cacao Collection (GSCC)
3. Genetic diversity analysis, gap filling in *ex situ* collections through collecting
4. Ensuring the *in situ* and on-farm conservation of important genetic diversity

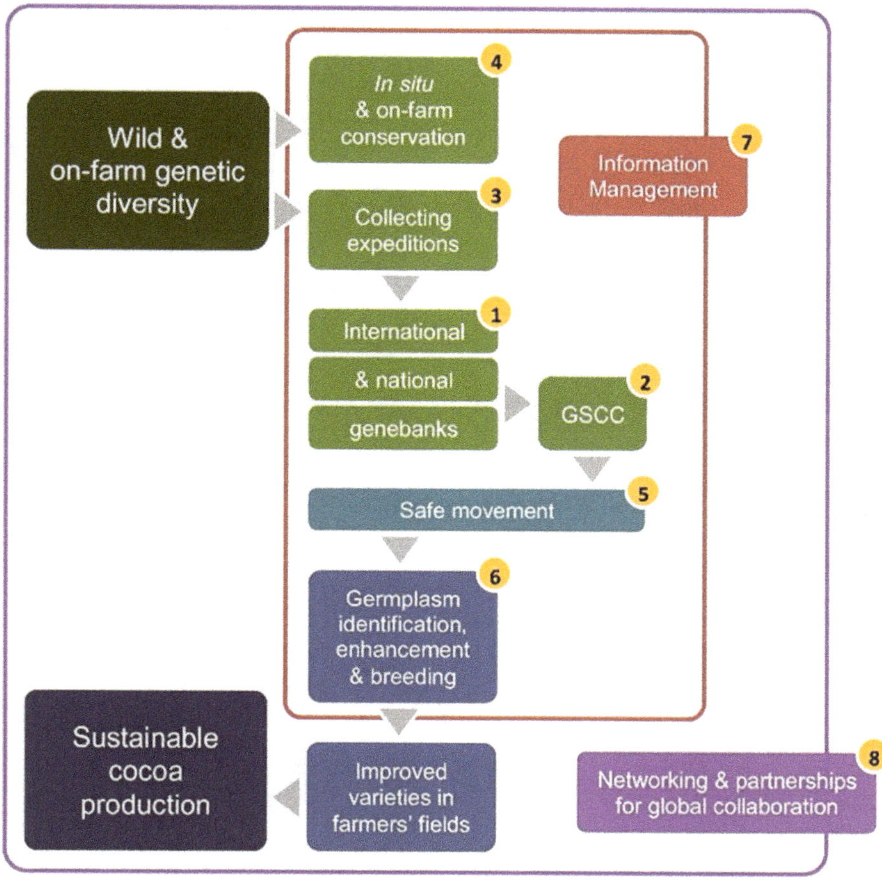

Figure 2 Strategic components from genetic diversity to sustainable cocoa production. Credit: C. Turnbull, Reading University.

5 Strengthening the distribution mechanism and safe movement of germplasm and making the conserved germplasm available to users
6 Strengthening the use of the cacao genetic resources by providing support to breeders and key users through improved characterization, evaluation within collections and supporting population enhancement programmes
7 Improving documentation and sharing of information on germplasm
8 Strengthening the networking and partnerships for global collaboration.

3 Addressing challenges: securing existing *ex situ* cacao genetic resources

Over 40 collections maintain over 24 000 accessions of cacao (see Table 1). Of these, two are international collections managed by the CRC/UWI, Trinidad and Tobago,

Table 1 *Ex situ* collections and acquisition in the past 10 years

Country	Institute	Date of info	No. of accessions today	Accessions acquired since 2000	Accessions lost since 2000
Benin	CRA-SB	March 2012	15	15	0
Brazil	CEPEC-CEPLAC	June 2008	1302	418	10
Brazil	CEPLAC/SUEPA	May 2012	2504	233	96
Brazil	CEPLAC/SUERO	May 2012	773	86	0
Costa Rica	CATIE	February 2012	1146	449	124
Côte d'Ivoire	CNRA	August 2011	1605	1213	168
Cuba	EIC-ECICC	June 2008	127	21	9
Dominican Republic	IDIAF	July 2011	115	36	1
Ecuador	INIAP	March 2012	2332	1354	17
France	CIRAD	February 2012	138	25	30
French Guiana	CIRAD	February 2012	508	30	10
Ghana	CRIG	August 2008	1366	300	200
Guyana	MHOCGA	July 2008	65	0	0
Honduras	FHIA	March 2012	31	0	0
India	CPCRI	July 2008	291	130	9
Indonesia	Bah Lias	March 2012	305	75	30
Indonesia	ICCRI	July 2008	714	15	60
Malaysia	MCB	May 2011	2263	800	50
Nicaragua	UNAN	March 2012	51	51	0
Nigeria	CRIN	August 2011	1100	296	214
Papua New Guinea	CCI	August 2011	1200	1050	150
Peru	CEPICAFE	March 2012	30	0	0
Peru	ICT	March 2012	607	607	8
Peru	UNAS	February 2012	422	262	2
Peru	UNSAAC	March 2012	72	72	0
Thailand	CHRC	March 2012	34	0	0
Togo	CRAF	August 2011	217	189	32
Trinidad and Tobago	CRC/UWI	May 2012	2400	270	25
United Kingdom	ICQC,R	February 2012	395	172	15
United States of America	USDA	August 2011	270	76	10
Venezuela	INIA	February 2012	872	404	0
		Total	23107	8465	1285

Source: CacaoNet survey 2008–12.

and CATIE, Costa Rica. These two institutes have concluded agreements with the International Treaty on Plant Genetic Resources for Food and Agriculture to maintain global collections of cacao genetic resources for the long term and to make this germplasm freely available to any professionally qualified institution or individual (Fig. 3 and 4).

The most urgent priority is to secure the conservation and accessibility to all users of genetic diversity currently in *ex situ* collections, particularly those held in the public domain. This diversity is critical to ensure the future of the world cocoa production to generate planting materials which can improve current productivity. Future improvements in productivity may require very different plant traits in the face of new pest and disease

Figure 3 CATIE International Collection, Costa Rica. Credit: W. Phillips, CATIE.

Figure 4 CRC/UWI International Collection, Trinidad and Tobago. Credit: M. Gilmour, Mars Inc.

Published by Burleigh Dodds Science Publishing Limited, 2018.

challenges, climate change and different agronomic conditions, as well as changing consumer preferences.

The current funding of cacao conservation and use activities is borne by the many national research institutes with the help of industry and international organizations and is below optimal levels. Many national collections are struggling to keep their material alive. Even the funding of the two international collections at CATIE and CRC/UWI and the International Cocoa Quarantine Centre at the University of Reading (ICQC,R) is not guaranteed and their sustainability is not secured over the long term. A stable and long-term source of funding for the major cacao collections is important for their long-term sustainability which must be addressed with urgency.

Securing funding for key *ex situ* collections of distinct and valuable genetic resources will secure the conservation and availability (including its safety duplication); promote participation of all partners through support for documentation systems, characterization and evaluation, collecting to fill gaps and promoting access to and use of materials; and increase efficiency and effectiveness to reduce costs and increase sustainability.

4 Addressing challenges: developing a global strategic cacao collection

No country is self-sufficient when it comes to the range of genetic diversity needed to develop improved planting materials. This diversity is maintained by several research institutes, but only a part of that diversity is in the public domain. Ensuring that the entire cacao genepool is conserved *ex situ*, complementing *in situ* conservation, facilitating its access and serving as a safeguard with greater focus and cost-effectiveness are proposed through the development of a GSCC. The GSCC is a dynamic and geographically dispersed collection of materials that have been identified as unique and interesting, which each of the participating institutes agrees to maintain in the public domain and make readily available to any *bona fide* user.

The development of the GSCC is based on a thorough assessment of the diversity currently conserved in *ex situ* international and national collections and the identification of those unique accessions for use by breeders and researchers. The backbone consists of accessions from the international collections managed by CATIE and CRC/UWI, for which considerable characterization and evaluation data are available in the public domain, complemented with priority accessions from national collections. Unique and valuable material is conserved in these national collections and, thus, collaboration is needed to secure this material and increase the access to and their use in breeding programmes.

A first set of accessions should be selected on the basis of capturing the greatest possible genetic diversity (in the form of allelic richness) held in *ex situ* collections worldwide. Two agreed core selection methods are employed. Both use grouping and selection of an optimum set of accessions followed by a further iteration designed to reduce the redundancy of the core selection (van Raamsdonk and Wijnker, 2000; van Treuren et al., 2008, 2009). The method is illustrated by way of reference to a large public data set of simple sequence repeat (SSR) information (microsatellite markers). The details of the method are described in *Annex 6* of the 2012 Global Strategy.

A further set of accessions will be selected on the basis of key traits of interest to users such as yield, flavour characteristics and disease resistance for which agreed criteria will be developed. Criteria for selection of genotypes may include, in addition to the number of desirable traits present, the genetic diversity amongst the selected types as determined through DNA fingerprints. This part of the GSCC will complement the part selected on allelic diversity and be a dynamic and geographically dispersed collection composed primarily of wild species and populations, landraces, enhanced populations for which characterization and evaluation data are available and used to broaden the basis on which the selection is made.

This material in the public domain is accessible in the collections at CRC/UWI and CATIE for which considerable characterization and evaluation data are already available. Additional materials from national collections will become part of the GSCC if the governments concerned are willing to place them in the public domain.

The formation of the GSCC will require a coordinated effort to characterize and rationalize available cacao genetic resources. A comprehensive assessment of individual tree identity and understanding of the phylogenetic relationship between accessions are of high priority to reduce both levels of misidentification within collections and the redundancy, thereby contributing to the effectiveness of germplasm evaluation and use. With the cost of DNA fingerprinting decreasing with time, verification of accessions at the individual tree level has become feasible.

Since information on accessions is likely to continuously improve over time, the GSCC should be a dynamic, evolving collection, to which individuals may be added or, if really necessary, deleted from time to time. Partners will agree on how to share responsibilities for conserving and distributing material from the GSCC, and long-term funding will be discussed with donors and the private sector.

The accessions within the GSCC should be prioritized for distribution to other international, regional or national genebanks as well as to breeders, via the ICQC,R and intermediate quarantine. The collection managers and breeders around the world will be responsible for comprehensive characterization evaluation and further researching. All related information should be made available to all users through a global information portal.

5 Addressing challenges: collecting and gap filling in *ex situ* collections to reflect genetic diversity

Analysing the status of the cacao genepool in its centre of diversity (Upper Amazon and Mesoamerica) is a priority in order to promote the development of early warning systems for endangered diversity and to understand the threats of genetic erosion, particularly important in view of the rapid rate of destruction of native Amazonian forests. As new technologies are developed, related species and genera of cacao could become important sources of genes for cacao improvement. The collection and conservation of the secondary and tertiary gene pool of cacao are proposed based on an analysis of the diversity currently maintained in *ex situ* collections. Since cacao germplasm moved around the world (see Fig. 5), considerable diversity is present in all cocoa-producing regions of the world, but their conservation and use are very limited.

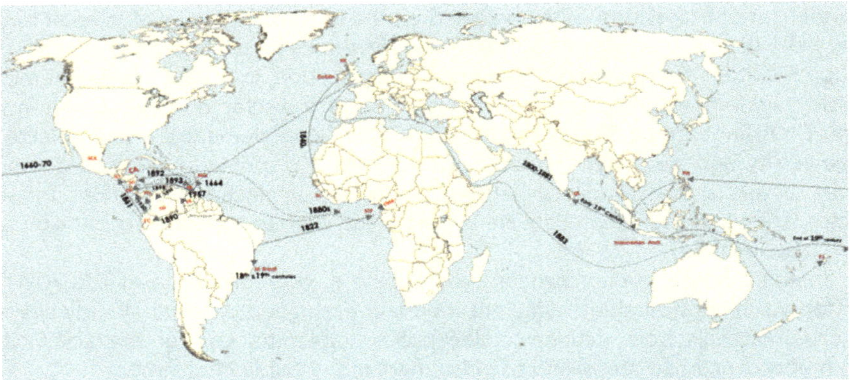

Figure 5 Produced from the map in Bartley (2005). 1660–70 = Mexico to the Philippines; 1664 = Amazon to Martinique; Philippines to Indonesian Archipelago; 1757 = Amazon to Trinidad; Early 19th century = Indonesian Archipelago to Ceylon; 18th and 19th centuries = Amazon to Southeastern Brazil; 1822 = Brazil to Príncipe; 1840s = Dublin to Sierra Leone; 1861 = Ecuador to Guatemala; 1880–1 = Trinidad (via England) to Sri Lanka; 1883 = Trinidad (via England) to Fiji; 1892–3 = Trinidad to Nicaragua; Nicaragua to Trinidad; 1898 = Trinidad to Costa Rica and Colombia; 1890 = Venezuela to Ecuador; 1930s = Ecuador to Costa Rica and Panama; 1880s = Trinidad, Venezuela and Ecuador to Sao Tomé; 1899 = Trinidad, Venezuela, Ecuador and Central America to Cameroon; End of 19th century = Indonesian Archipelago to Samoa.

Geographic information systems should be used to map the spatial distribution of different cacao populations, using additional information about genetic diversity in wild cacao populations to guide future collecting missions. In addition, collecting will be directed to places where one might reasonably expect to find a higher frequency of desirable traits, such as disease resistance.

Gaps in genetic diversity in *ex situ* collections refer to the geographical areas that are under- or not represented in collections and where cacao is expected to occur based on ecological and other factors. Early warning systems should be developed at national and international levels so as to draw attention to immediate threats to *in situ* conservation of wild germplasm or traditional varieties. Gap analysis can be applied to map the actual distribution, agroclimatic preferences and potential distribution of cacao. The degree of variability expected to be found in new collecting areas is another important consideration. Information on allelic diversity in the wild cacao populations will be used as an important criterion to guide future collecting missions.

The priority for collecting to fill gaps in *ex situ* collections should focus on threatened cacao wild relatives and landraces. The main criteria for priority setting to carry out new collecting actions are degrees of threat to the natural habitat of cacao and filling gaps in the existing diversity in collections. Another priority is related to opportunities that may arise to collect materials in areas that have not yet been prospected or where the material collected does not fully represent the diversity present.

Three types of germplasm to be collected should be prioritized: (1) wild populations of *Theobroma cacao* [spontaneous trees (native) and sub-spontaneous trees (not native)]; (2) landraces and traditional varieties (i.e. germplasm with little or even no genetic selection or improvement from farmers). These varieties are generally mass-selected and planted as seeds or seedlings by farmers, but grafting might also be carried out; and (3) Wild *Theobroma* spp. or species from other genera related to *T. cacao*.

The main difference between the wild population and the landraces is that for landraces, selection has reduced the allelic diversity and/or modified allele frequency. This of course depends on how long ago the domestication took place.

Safeguarding the spontaneous wild populations or demes (subpopulations) is paramount and urgent in threatened areas because of intense deforestation, such as in: Bolivia (Ríos Madre de Dios, Beni and Mamoré), Brazil (Rondônia, North Mato Grosso, Pará), Peru (Ríos Ucayali, Huallaga, Madre de Dios, Putumayo) and Venezuela (Río Orinoco). Collecting trips should be planned in areas from which known genetic gaps are known to exist and, hence, not adequately covered so far by collecting actions. These include Guyana (Plateau), Brazil (Amapá, Roraïma), French Guiana, Suriname, Venezuela and Colombia (Amazon region).

Regarding the conservation of landraces, examples are cacao trees with 'amelonado' or 'calabacillo' type pods in Central America ('Indio', 'Matina', 'Ceylán', etc.), West Indies ('Créole') and northern South America. A large-scale collecting mission and comprehensive study of this material should be undertaken in Honduras, Nicaragua, Lesser Antilles, Orinoco Delta, Colombian Pacific coast and so forth. Other examples are the landraces domesticated from wild materials in Peru (e.g. Chuncho) and in Bolivia ('Nacional' types).

With few exceptions, related species including *Theobroma* spp. as well as *Herrania* species are largely under-represented in existing *ex situ* collections. Although there are important barriers in crossing most related species with *T. cacao*, the use of modern tools (protoplasm fusion, embryo rescue, genetic transformation) may make it possible to exploit these species in the longer term for the improvement of cacao. Some protocols should be implemented: (1) the collected germplasm will be planted first (in *ex situ* collections) in the country of origin; and (2) collecting trips should be carried out during the fruiting season.

Many previous collecting expeditions, especially those prior to the 1980s, have focused on collecting germplasm from individual trees showing desired agronomic traits such as disease resistance. Usually, such collections were limited to only those few, or even unique individuals showing the desired traits. This approach has thus reduced the number of samples available for analysis from a given location/population in diversity studies.

Therefore, future collecting missions should have the objective to safeguard alleles in populations and subpopulations (demes). To do so, pods should be collected from an appropriate number of trees (ideally 20), randomly chosen in each deme, with priority given to gaining a representation of the existing phenotypic diversity. If pods are unavailable from sufficient trees, then leaves should be collected for DNA analysis to assess whether the individuals collected from the population are representative of the wild germplasm present. If trees displaying interesting traits, such as a low incidence of disease, are found, pods and budwood should also be collected. For landraces, it will be necessary to collect also numerous historical data and information from old planters and their families.

Because of the Access to Genetic Resources and the Fair and Equitable Sharing of Benefits from their Utilization to the Convention on Biological Diversity in the Nagoya Protocol, it must be stressed that an important step in germplasm collection is to obtain the administrative authorization necessary before any collecting trip is made. When permits to collect germplasm from a given country are obtained, the objectives of the project should be explained to the local populations and the results should be conveyed to them in a meaningful way. Local farmers' associations and non-governmental organizations (NGOs) are important links in facilitating meaningful communication.

6 Addressing challenges: ensuring the *in situ* and on-farm conservation of important diversity

In situ and on-farm conservation is influenced by complex social, political and biological factors. Habitats suitable for wild *Theobroma* species are increasingly degraded and fragmented and the drivers and consequences of these changes are currently not well understood. On-farm conservation depends ultimately on farmers as the final decision makers, choosing particular varieties they wish to use and conserve. The traditional cacao varieties that many farmers prefer often have lower yields, but recently some of these have acquired a reputation for high quality and are increasingly coveted by specialty gourmet markets. Some of these landraces can also be important sources of pest and disease resistance valuable to breeders. It is therefore critical to understand the social and economic factors that influence farmers' decisions to maintain cacao diversity, and to assess the implications of these factors for the design of *in situ* and on-farm conservation strategies.

Relative to *ex situ* conservation, cacao is less studied in terms of *in situ* and on-farm conservation. The latter is influenced by complex social, political and biological factors. There is a threat to the continuity of habitat for *Theobroma* species growing in the wild (see Fig. 6). A greater effort is needed to understand and strategically conserve *in situ* the diversity of cacao and its wild relatives, using a variety of initiatives such as national forest reserves, wildlife refuges and private reserves, all of which can help preserve natural plant communities.

The establishment of protected areas for *in situ* conservation would benefit all countries. Partnerships with governmental agencies with responsibility for forestry and environmental issues and local-level authorities in target countries, plus national and international conservation NGOs, forestry research institutes, farmer communities and civil society organizations, as well as the private sector, are needed for successful *in situ* and on-farm conservation.

Regarding on-farm conservation, farmers make the ultimate decisions about the conservation and use of particular varieties in their community. It is therefore critical to understand the socio-economic determinants that influence their decisions. The traditional cacao varieties often have lower yields, but some of these have acquired a reputation for quality and are increasingly coveted by external gourmet specialty

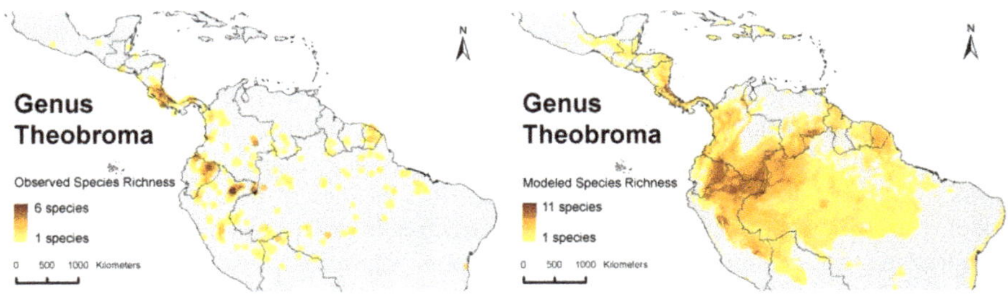

Figure 6 Species richness of genus *Theobroma*. Left: observed species richness in 10-min grid cells and a circular neighbourhood of 1 decimal degree. Right: modelled species richness in 2.5-min grid cells. Source: Adapted from Thomas et al. (2012).

Published by Burleigh Dodds Science Publishing Limited, 2018.

markets. In some cases, these landraces can also be important sources of resistance to pests and diseases. For example, the Bahian Amelonado is more resistant than average to *Ceratocystis*. In the Brazilian Amazon area, there are landraces that have been planted for years (and probably selected for resistance) which tend to be more resistant to witches' broom disease. Production and marketing of differentiated high-value cocoa provides an opportunity for conservation through use of threatened cacao diversity. Higher farm-gate revenues from the premium market, which has been increasing at an annual rate of 16% in the last 5 years, may provide incentives for demand-driven on-farm conservation. On-farm conservation can be strengthened through activities in the farming community, such as participatory variety selection, cacao breeding and farmer field schools. It is critical to understand the socio-economic determinants that influence farmers' decisions about the conservation and use of particular varieties in their community.

In situ and on-farm conservation focus on the generation and management of pertinent socio-economic and biological information. The principal elements identified are the continued use of landraces and traditional varieties for cocoa production, habitat management and alliances and partnership in integrated regional development.

7 Addressing challenges: strengthening the distribution and safe movement of germplasm

Movement of cacao germplasm brings with it the potential risk of transfer of pests and diseases. The risk is particularly acute when germplasm is moved between cacao-growing regions that have different endemic diseases. Risks associated with pests and diseases need to be minimized before cacao diversity can be freely accessible and used by different research institutes around the world.

Several genebanks report that pests/diseases were limiting/preventing them from distributing germplasm. It is essential therefore that all those involved in handling cacao genetic resources have access to information highlighting the risks associated with each particular pest or disease and recommendations on appropriate quarantine measures.

Region-to-region transfer of budwood should take place via a quarantine centre. Currently, the safe global movement of germplasm, including testing for the presence of viruses, is through the ICQC,R at the University of Reading, UK. The USDA/ARS facility in Miami, USA, offers quarantine facilities for regional transfers. Germplasm should be obtained from the safest source possible, for example from a pathogen-tested quarantine collection.

It is essential that users have access to the latest information highlighting the risks associated with pests and diseases and recommendations on appropriate quarantine measures. The 2014 Safe Movement Guidelines for Cacao include descriptions and information on an extensive range of pests and diseases and information on quarantine measures. The guidelines are available in English, French and Spanish and disseminated to relevant institutes and to plant health authorities. The guidelines are updated as new information becomes available (Fig. 7).

It is important to develop effective ways to raise awareness of the importance of safe movement of germplasm to the cacao community, working closely with groups such as INGENIC, INCOSOM, INCOPED, INAFORESTA and CacaoNet.

Figure 7 Front cover images from: Technical guidelines for the safe movement of cacao germplasm. Revised from the FAO/IPGRI Technical Guidelines No. 20 (Second Update, August 2014). Global Cacao Genetic Resources Network (CacaoNet), Bioversity International, Montpellier, France). Source: End et al. (2014).

International standards for phytosanitary measures as published by the Secretariat of the International Plant Protection Convention (IPPC) should be followed. This work is closely linked to the IPPC and its regional organizations to ensure that the safe-movement guidelines are widely available to those responsible for the phytosanitary systems in cocoa-producing countries. The IPPC works with Convention contracting parties, to develop phytosanitary measures that underpin the parties' ability to manage pest risks and the environmental, economic and social impacts of plant pests. The IPPC is governed by the Commission on Phytosanitary Measures, which meets annually to review the state of plant protection, identifies action to control the spread of pests into new areas, develops and adopts international standards and establishes procedures for the sharing of phytosanitary information. The IPPC works with Regional Plant Protection Organizations and international organizations to build phytosanitary capacity and to identify and address risks that cross national borders.

It is important to strengthen the current system of germplasm distribution through the ICQC,R and support the development of new quarantine centres in each of the three regions (Americas, Africa and Asia) to facilitate the safe movement of germplasm within regions. A regional quarantine approach can facilitate the exchange of germplasm between countries where the same cocoa pests and diseases are endemic.

8 Addressing challenges: strengthening the use of cacao genetic resources

The benefits of conserving and utilizing the cacao genetic diversity will only be realized if this diversity is of interest and is made available to researchers engaged in breeding programmes. Strengthening the use of the cacao genetic resources is therefore critical to provide support to breeders and key users through improved characterization, evaluation within collections and supporting population enhancement programmes.

The use of germplasm starts with the evaluation for economically important traits. The main traits assessed are yield determinants, disease resistance, yield components, precocity and tree size (quality and vigour). Although increased yield potential is the main aim, yield potential can only be reliably estimated in collections with uniform planting conditions or by the use of appropriate statistical designs. Drought, heat and CO_2 tolerance and establishment ability in cacao have also become important traits due to the growing concerns about climate change, notably with the dry seasons getting longer, harsher and less predictable in some cocoa-growing regions. Once evaluated, selected accessions that possess the desired traits can be sent to intermediate quarantine for distribution to countries that are interested in acquiring germplasm for use in breeding programmes (Fig. 8).

Cacao breeders and germplasm curators can take advantage of novel molecular tools and genomics to fully utilize the potential of cacao germplasm. The genomics and information revolution has enabled the generation of great amounts of information about the *T. cacao* species. The availability of full sequence information of the cacao genome

Figure 8 Clonal materials for field trials, MMSP nursery, Ghana. Credit: G. Lockwood.

(Criollo and Matina 1–6) represented a critically important step towards unlocking genetic diversity in cacao germplasm with higher precision. There is an urgent need to agree on the composition of an international single-nucleotide polymorphisms set for genetic fingerprinting in the same way that a common set of SSR markers has been internationally agreed upon to facilitate comparison of results from different laboratories.

Utilization of collections and associated science contributes to the delivery of improved varieties only through strong national breeding programmes. The system for practical cocoa breeding is generally weak and underfunded in most countries. Cacao breeding itself is of no use if it is not followed by efficient programmes to multiply selected varieties at the national level. To secure the use of cocoa genetic resources, systems for proliferating and distributing new varieties are of high priority, and capacity building in this area is critically important.

Characterization and evaluation of accessions as well as the systematic documentation and dissemination of the information are imperative to ensure efficient genebank management and use of genetic resources by breeders. This should include the digitization and dissemination of important historical data. Pre-breeding activities are important in selected genebanks so that material in which genes from different sources have been incorporated can be made widely available to accelerate progress in breeding efforts.

It is important to clarify and influence the legal situation regarding access to cacao genetic resources in individual countries, by improving awareness of existing conservation and research/breeding activities in the country and by sharing information regarding international (and sometimes national) policies and legislation regarding access and benefit sharing. This can facilitate the involvement of decision makers in formalizing arrangements for the exchange of cacao genetic resources as well as improve their political understanding of the role of conservation and use of cacao genetic resources towards supporting the development of a sustainable (institutional) cacao industry in the country.

Collaboration between national collections and the International Treaty to promote the placing of cacao germplasm collections in the public domain through designation under the treaty, as was achieved with CATIE and CRC/UWI, is essential. Once accessions have been given a public domain status, they become freely available to *bona fide* users and exchanged legally, transparently and fairly, through the use of the Standard Material Transfer Agreement.

9 Addressing challenges: improving documentation and sharing of information

Exchange of cacao germplasm and access to information on key traits of interest to breeders is an essential condition for plant breeding research and agricultural development. Information on morphology, evaluation, origins and locations of a large number of cacao varieties (genotypes) can be found in the International Cocoa Germplasm Database (ICGD), developed for the cocoa community at the University of Reading, UK, and genetic information is available online through TropGENE, hosted by CIRAD in France. ICGD and TropGENE contain information related to clonal material or varieties. And although they link some information to individual trees or accessions, they are not designed to be germplasm management tools.

A simple yet robust information management system that combines comprehensive and accurate information on the origins, conservation locations, availability and characteristics of individual accessions should be developed to access all relevant information and be a key component in the establishment, management and use of the GSCC. The compilation of characterization and evaluation data from all collections should be coordinated and supported by the molecular verification. A local germplasm management system needs to be in place at each genebank in order to ensure recording and storage of a minimum set of standards of descriptor notes (traits). The local genebank documentation system must be able to link passport data, plus any additional characterization or evaluation information, to specific trees in the field and then make this information more widely available. Technologies such as GPS and barcoding are not widely used in cacao collections at the current time, but they are becoming less expensive and more widely available, and likely to be increasingly important in managing genetic resources in the future. Information on *ex situ* collections should be linked to information and local knowledge on *in situ* and on-farm genetic resources.

10 Strengthening networking and partnerships for global collaboration

An efficient and effective global conservation and use system is based on good planning, coordination and cooperation, to reduce costs and ensure that conservation and management of crop diversity are scientifically sound and financially sustainable. Most nations and regions involved in the improvement and production of cacao are highly dependent on genes and varieties developed and conserved in other countries or regions. The efforts needed to secure this material and increase the access to and their use in breeding programmes can therefore only be carried out through international collaboration.

The global system for conservation and use of cacao genetic resources is based on safeguarding the unique and critical diversity, making it available to breeding programmes and other research institutes via a system for safe movement of germplasm. The system therefore relies on a network of partners including well-functioning and efficiently managed international and national *ex situ* collections collectively conserving and making materials available. National cocoa research institutes and their governments play a key role in ensuring access to a wide range of diverse genetic resources and related information, facilitating the rationalization of collections and building trust amongst partners.

Partnerships and collaboration can ensure that the decisions are advantageous to all countries managing and using cacao diversity. Strengthened regional cooperation facilitates the rationalization of existing and future cacao collections, and reduces duplication of efforts.

Strong networking and partnerships contribute to:

- long-term effective and cost-efficient conservation of important genetic resources
- consolidating and facilitating access to information and knowledge on cacao genetic resources worldwide, on germplasm (characterization and evaluation), on quarantine regulations and on molecular genetics data

- facilitating access to standardized protocols on characterization, evaluation, information analysis and long-term conservation of germplasm
- facilitating access to technologies, procedures and methods to conserve, improve and breed cacao, for example molecular genetics tools; cryopreservation; somatic embryogenesis
- promoting and facilitating the distribution of enhanced populations to users, particularly for resistance to witches' broom, black pod and frosty pod diseases from materials in the two international collections at CRC/UWI and CATIE
- providing pertinent legal information and advice, particularly on access and benefit sharing, and assist partners in implementing regulations
- representing the community through one voice to secure long-term funding of the key strategic elements and research activities.

National cocoa research institutes and governments play a key role in:

- providing access to a wide range of diverse cacao genetic resources and to information on these resources
- building trust between partners by providing information to facilitate rationalization.
- creating awareness of the importance of sharing responsibility for the conservation and use of cacao diversity
- seeking a dialogue with decision-making political/administrative bodies in the country
- working towards the development and implementation of access and benefit-sharing legislation that is conducive to CacaoNet's proposed policy
- facilitating implementation of agreed safe-movement guidelines and cacao quarantine regulations
- seeking funding opportunities, providing advice and developing project proposals with partners
- proactively creating mechanisms and structures that allow and facilitate funding opportunities to be used.

11 Research to strengthen the conservation and use of cacao genetic diversity

Most of the planting material is low yielding, often due to its high susceptibility to prevailing pests and diseases. Cacao has always been plagued by serious losses from pests and diseases, with estimates of losses as high as 30–40% of global production. Compared to many other tree crops, there has been little investment in scientific research to improve cacao production, and the number of breeders is very low.

Scientists worldwide are looking for ways to produce cacao trees that can resist evolving pests and diseases, tolerate droughts, meet manufacturers' needs and produce higher yields. These programmes depend on the availability of substantial genetic diversity together with an understanding of how best to use it, and powerful new technologies, such as molecular genetics, genomics, proteomics and ecogeographical remote-sensing techniques, have increased the value of these genetic resources. The sequencing of the cacao genome is a promising step in advancing breeders' ability to deliver improved trees

Published by Burleigh Dodds Science Publishing Limited, 2018.

to farmers, and advances in informatics have also markedly increased the capacity to use, analyse and communicate related data and information.

The main limiting factors hindering the use of germplasm in breeding are: (1) lack of information and knowledge (particularly evaluation) about the materials, (2) constraints in accessing materials (quarantine and policies), (3) relatively narrow genetic base available in collections, (4) few breeding programmes and breeders and (5) lack of funding for research and breeding programmes.

Research is needed to safeguard the security of cacao diversity, on which the world depends for cocoa production now and in the future, and to ensure its accessibility and sustainable use. Key areas of research are the following:

- Assessing cacao genetic diversity to identify gaps and safeguard the entire genepool. It includes an analysis of the currently conserved diversity in *ex situ* collections, resolving mislabelling problems in the international and national collections, characterization and evaluation of germplasm and identification of duplicates. It also includes assessing the diversity of cacao in the centre of diversity *in situ* and on-farm as a priority (Upper Amazon and Mesoamerica), including understanding the threats to genetic erosion, identifying geographic areas of threatened genetic diversity and collecting priority germplasm. The development of a mechanism to identify and communicate threats to cacao genetic resources (vulnerability and erosion) at national and international levels is also included.
- *Ex situ* conservation and management of cacao germplasm. Development of information systems and a crop ontology to promote the documentation and sharing of information on cacao genetic resources and support breeding. Best practices for cacao collection management and developing genebank management standards and dissemination of germplasm and related information. Coordinating the compilation of characterization and evaluation data from all collections (data to be supported by the molecular verification of genotypes where possible). Stimulating the rescue of historical data collected in genebanks and eventually trials which can provide information useful to breeders.
- Better understanding of the role that *in situ* and on-farm conservation can play. Develop *in situ* and on-farm conservation strategies, including analysis of social economic, market and cultural factors influencing on-farm management and incentives such as Payment for Agrobiodiversity Conservation Services (PACS). Analyse the social, economic, market and cultural factors that influence farmers' maintenance of cacao diversity at the farm level and assess their implications for designing *in situ* and on-farm conservation strategies.
- Field evaluation at multiple sites under controlled and recorded conditions. The use of cacao germplasm can be maximized by providing support to breeders and key users through improved characterization, evaluation within collections and supporting population enhancement programmes. The priority is therefore the evaluation for economically important traits as yield, disease resistance, precocity, tree size (quality and vigour) establishment ability and drought tolerance. Supporting a network of field trials participating in the evaluation of materials at multiple sites. Developing or improving reliable and standardized evaluation methods for major agronomic traits.
- Cacao germplasm safe movement and quarantine procedures. This would include supporting the development of regional quarantine centres to complement the role of securing international movements through the ICQC,R, facilitating regional exchange

of between countries where the same cocoa pest and diseases are endemic; conducting research on tissue culture methods for safe movement of germplasm; and identifying a priority set of accessions for *in vitro* culture to facilitate distribution through quarantine facilities, if the feasibility study validates the use of this technology.

12 Acknowledgements

The 2012 Global Strategy for the Conservation and Use of Cacao Genetic Resources, coordinated by CacaoNet, is the product of expert opinion and detailed discussion amongst diverse stakeholders in the conservation and use of cacao genetic resources. CacaoNet considers that it has provided an informed and realistic foundation for prioritizing cacao research and development. The goal is to use this strategy to invigorate the commercial cacao sector in a sustained manner, while protecting food security, by encouraging partnerships that increase impact of research and adoption of technological innovations. CacaoNet encourages international, regional and national public research organizations, development agencies, NGOs and the private sector to use the priorities set out herein to guide their activities and investment decisions. CacaoNet would like to thank all those organizations and individuals who contributed to the development of the Global Strategy for the Conservation and Use of Cacao Genetic Resources. We particularly thank all the many organizations, institutions and individuals who have assisted us by contributing to the genebank surveys and who have provided constructive feedback during the development of this strategy, which was supported by financial and in-kind contributions from the Cocoa Research Association Ltd., UK (CRA Ltd.); Mars Inc.; the US Department of Agriculture, Agricultural Research Service (USDA/ARS); World Cocoa Foundation (WCF); Bioversity International and the CGIAR Research Programme on Forests, Trees and Agroforestry.

13 Where to look for further information

13.1 Key research institutes

Key research institutes directly involved in the conservation and use of cacao genetic diversity maintaining *ex situ* collections and principal organizations, institutions and networks involved in cacao genetic resources in recent years – in alphabetical order – are as follows:

1. Bah Lias Research Station, Sumatra, Indonesia
2. Bioversity International, Rome, Italy
3. Central Piurana de Cafetaleros (CEPICAFE), Peru
4. Central Plantation Crops Research Institute (CPCRI), India
5. Centre de coopération internationale en recherche agronomique pour le développement (CIRAD), France
6. Centre de coopération internationale en recherche agronomique pour le développement (CIRAD), French Guiana
7. Centre de Recherche Agronomique de la zone Forestière (CRAF), Togo

8 Centre de Recherches Agricoles Sud Bénin (CRA-SB)
9 Centre for Agriculture and Biosciences International – CABI
10 Centre national de recherche agronomique (CNRA), Côte d'Ivoire
11 Centro Agronómico Tropical de Investigación y Enseñanza (CATIE), Costa Rica
12 Chumphon Horticultural Research Centre (CHRC), Thailand
13 Cocoa and Coconut Institute (CCI), Papua New Guinea
14 Cocoa Producers Alliance – COPAL
15 Cocoa Research (UK) Ltd. – CR(UK) Ltd.
16 Cocoa Research Association Ltd. – CRA Ltd.
17 Cocoa Research Centre of the University of the West Indies (CRC/UWI), Trinidad and Tobago
18 Cocoa Research Institute of Ghana (CRIG), Ghana
19 Cocoa Research Institute of Nigeria (CRIN), Nigeria
20 Comissão Executiva do Plano da Lavoura Cacaueira – Ceplac, Superintendência de Desenvolvimento Região Cacaueira no Estado do Pará – Suepa (CEPLAC/SUEPA), Brazil
21 Comissão Executiva do Plano da Lavoura Cacaueira – Ceplac. Superintendência de Desenvolvimento da Região Cacaueira no Estado de Rondônia – Suero (CEPLAC/SUERO), Brazil
22 Comissão Executiva do Plano da Lavoura Cacaueira/Centro de Pesquisas do Cacau (CEPEC-CEPLAC), Brazil
23 Common Fund for Commodities – CFC
24 Corporación Colombiana de Investigación Agropecuaria, Colombia
25 Estación de Investigaciones de Cacao. Estación Central de Investigaciones de Café y Cacao – EIC-ECICC, Cuba
26 European Industry Cocoa Research Outreach Group [including the Association of the Chocolate, Biscuits and Confectionery Industries of Europe (CAOBISCO), the European Cocoa Association (ECA) and the Federation of Cocoa Commerce Limited (FCC)]
27 Fundación Hondureña de Investigación Agrícola (FHIA) Honduras
28 Global Cacao Genetic Resources Network – CacaoNet
29 Global Crop Diversity Trust
30 Government of the Netherlands
31 Indonesian Coffee and Cocoa Research Institute (ICCRI), Indonesia
32 Instituto Agronômico de Campinas (ICA), Brazil
33 Instituto de Cultivos Tropicales (ICT), Peru
34 Instituto Dominicano de Investigaciones Agropecuarias y Forestales (IDIAF), Dominican Republic
35 Instituto Nacional Autónomo de Investigaciones Agropecuarias (INIAP), Ecuador
36 Instituto Nacional de Investigaciones Agrícolas (INIA), Venezuela
37 International Cocoa Organization – ICCO
38 International Cocoa Quarantine Centre – ICQC,R
39 International Cocoa Quarantine Centre, University of Reading (ICQC,R), United Kingdom
40 International Group for Genetic Improvement of Cocoa – INGENIC
41 International Institute for Tropical Agriculture – IITA
42 International Treaty on Plant Genetic Resources for Food and Agriculture – ITPGRFA
43 Laboratorio de BIOciencia, UNAN-Managua, Nicaragua

44 Mabaruma/Hosororo Organic Cocoa Growers Association (MHOCGA), Guyana
45 Malaysian Cocoa Board (MCB), Malaysia
46 Mars Incorporated
47 Mondelez International (formerly Kraft Foods)
48 Nestlé
49 United States Department of Agriculture, Agricultural Research Service, Tropical Agriculture Research Station (USDA/ARS), USA
50 Universidad Nacional Agraria de la Selva (UNAS), Peru
51 Universidad Nacional de San Antonio Abad del Cuzco (UNSAAC), Peru
52 University of Reading, UK
53 World Agroforestry Centre – ICRAF
54 World Cocoa Foundation – WCF

13.2 Useful web links

ACIAR – www.aciar.gov.au
AVRDC – www.avrdc.org
Bioversity – www.bioversityinternational.org
CABI – www.cabi.org
CacaoNet – www.cacaonet.org
Caobisco – www.caobisco.com
CATIE – www.catie.ac.cr
CBD – www.cbd.int
CEPICAFE – www.cepicafe.com.pe
CEPLAC – www.ceplac.gov.br
CFC – www.common-fund.org
CGIAR – www.cgiar.org
CIRAD – www.cirad.fr
CNRA – www.cnra.ci
CPCRI – www.cpcri.gov.in
CPQP – www.idhsustainabletrade.com/CPQP
CRA Ltd. – www.cocoaresearch.org.uk
CRIG – www.crig.org
CRIN – www.crin-ng.org
CRC/UWI – www.sta.uwi.edu/cru
ECA – www.eurococoa.com
FAO – www.fao.org
FCC – www.cocoafederation.com
FEDECACAO – www.fedecacao.com.co
FHIA – www.fhia.org.hn
GPA – www.fao.org/agriculture/crops/core-themes/theme/seeds-pgr/gpa
ICA – www.ica.gov.co
ICCO – www.icco.org
ICCRI – www.iccri.net
ICGD – www.icgd.rdg.ac.uk
ICQC,R – www.icgd.rdg.ac.uk/contact_ICQC.php
ICT – www.ict-peru.org
IDIAF – www.idiaf.gov.do

IDH – www.idhsustainabletrade.com
IICA – www.iica.int
IITA – www.iita.org
INCOCOA – http://www.incocoa.org/
INIA – www.inia.gov.ve
IPPC – www.ippc.int
IRAD – www.irad-cameroon.org
ITIS – www.itis.gov
ITPGRFA – www.planttreaty.org
Mars – www.mars.com/global/brands/cocoa-sustainability-home.aspx
MCB – www.koko.gov.my
Nestlé – www.nestlecocoaplan.com
Trust – www.croptrust.org
UESC – www.uesc.br
UFLA – www.ufla.br
UNAN – www.unan.edu.ni
UNAS – www.unas.edu.pe
UNSAAC – www.unsaac.edu.pe
USAID – www.usaid.gov
USDA/ARS – https://www.ars-grin.gov/
USMARC – www.usm.edu.ph
WCF – www.worldcocoafoundation.org
WIEWS - http://apps3.fao.org/wiews/wiews.jsp

14 References

14.1 References cited in the 2012 Global Strategy

Adu-Gyamfi, R. 2011. Safeguarding cocoa (*Theobroma cacao* L.) germplasm by cryopreservation: The vitrification approach. PhD Thesis, University of Reading, Reading, UK.

Bartley, B. G. D. 1984. Genetic resources programme of the Cocoa Research Centre, University of the West Indies, St. Augustine, Trinidad. Consultant's report. In: *Cocoa Research Centre Report for 1981–1983*. The University of the West Indies, St. Augustine, Trinidad, pp. 13–16.

Bartley, B. G. D. 2005. *The Genetic Diversity of Cacao and Its Utilization*. CABI, Wallingford, UK.

CacaoNet. 2005. Towards a Global Cacao Genetic Resources Network (CacaoNet): A consultation process for its establishment. Proposal Submitted by IPGRI to the Begeleidingscommissie Subsidieregeling Duurzame Ontwikkeling Cacao- en Chocoladesector. December 2005.

CGIAR. 2011. Proposal to the CGIAR Fund Council, submitted by the CGIAR Consortium Board of Trustees, for financial support to the CGIAR Center Genebanks in 2011. Available from URL: http://www.cgiarfund.org/cgiarfund/sites/cgiarfund.org/files/Documents/PDF/fc4_funding_proposal_CGIAR_Genebanks.pdf

Cuatrecasas, J. 1964. Cacao and its allies: A taxonomic revision of the genus *Theobroma*. *Contributions from the United States National Herbarium* 35:379–614.

End, M. J., Daymond, A. J. and Hadley, P. (Eds). 2010. Technical guidelines for the safe movement of cacao germplasm (Revised from the FAO/IPGRI Technical Guidelines No. 20). Global Cacao Genetic Resources Network (CacaoNet), Bioversity International, Montpellier, France.

End, M. J., Daymond, A. J. and Hadley, P. (Eds). 2014. Technical guidelines for the safe movement of cacao germplasm. Revised from the FAO/IPGRI Technical Guidelines No. 20 (Second Update,

August 2014). Global Cacao Genetic Resources Network (CacaoNet), Bioversity International, Montpellier, France.

Escribano, P., Viruel, M. A. and Hormaza, J. I. 2008. Comparison of different methods to construct a core germplasm collection in woody perennial species with simple sequence repeat markers. A case study in Cherimoya (*Annona cherimola, Annonaceae*), an underutilized subtropical fruit tree species. *Annals of Applied Biology* 153:25–32.

Eskes, A. B. and Efron, Y. (Eds). 2006. Global approaches to cocoa germplasm utilization and conservation. Final report of the CFC/ICCO/IPGRI project on 'Cocoa Germplasm Utilization and Conservation: A Global Approach' (1998–2004). CFC, Amsterdam, the Netherlands; ICCO, London, UK; and IPGRI, Rome, Italy.

Fang, J.-Y., Wetten, A. and Hadley, P. 2004. Cryopreservation of cocoa (*Theobroma cacao* L.) somatic embryos for long-term germplasm storage. *Plant Science* 166(3):669–75.

Fang, J.-Y., Wetten, A., Adu-Gyamfi, R., Wilkinson, M. and Rodriguez-Lopez, C. 2009. Use of secondary somatic embryos promotes genetic fidelity in cryopreservation of cocoa (*Theobroma cacao* L.). *Agricultural and Food Science* 18(2):152–9.

FAO. 2010. Chapter 8: The contribution of plant genetic resources for food and agriculture to Food Security and Sustainable Agricultural Development. In: *The Second Report on the State of the World's Plant Genetic Resources for Food and Agriculture*. FAO, Rome, Italy, pp. 184–5.

Frison, E. A., Diekman, M. and Nowell, D. (Eds). 1999. Cacao. FAO/IPGRI technical guidelines for the safe movement of germplasm No. 20 (1st revision). FAO/IPGRI, Rome, Italy. Available from URL: http://www.bioversityinternational.org/nc/publications/publication/issue/cacao.html

Girón, A. and Rita, C. 2010. Evaluación del impacto generado por el uso del germoplasma distribuido en el período del 2003 al 2008, en Costa Rica y otros países y análisis de los costos de conservación en CATIE. Tesis sometida a consideración de la Escuela de Posgrado como requisito para optar por el grado de Magister Scientiae en Socioeconomía Ambiental, Turrialba, Costa Rica.

ICCO. 2010. The world cocoa economy: Past and present. EX/142/6, 30 July 2010. ICCO Executive Committee, 142nd meeting, Bloomsbury House, London, 14–17 September 2010.

ICCO. 2012. May 2012 quarterly bulletin of ICCO statistics. Available from URL: http://www.icco.org/about-us/icco-news/197-may-2012-quarterly-bulletin-of-cocoa-statistics.html

Li, Z., Traore, A., Maximova, S. and Guiltinan, M. J. 1998. Somatic embryogenesis and plant regeneration from floral explants of cacao (*Theobroma cacao* L.) using thidiazuron. *In Vitro Cellular & Developmental Biology – Plant* 34:293–9.

Lockwood, G. and End, M. 1993. History, technique and future needs for cacao collection. In: *Proceedings of International Workshop on Conservation and Utilization of Cocoa Genetic Resources in the 21st Century*, 13–17 September 1992, Trinidad and Tobago. Cocoa Research Centre, The University of the West Indies, Port of Spain, Trinidad and Tobago, pp. 1–14.

Maximova, S. N., Alemanno, L., Young, A., Ferriere, N., Traore, A. and Guiltinan, M. J. 2002. Efficiency, genotypic variability, and cellular origin of primary and secondary somatic embryogenesis of *Theobroma cacao* L. *In Vitro Cellular & Developmental Biology – Plant* 38:252–9.

Motamayor, J. C., Lachenaud, P., da Silva e Mota, J. W., Loor, R., Kuhn, D. N., Brown, J. S. and Schnell, R. J. 2008. Geographic and genetic population differentiation of the Amazonian chocolate tree (*Theobroma cacao* L). *PLoS ONE* 3(10):e3311. e3311.doi:10.1371/journal/pone.0003311

Pence, V. C. 1991. Cryopreservation of immature embryos of *Theobroma cacao*. *Plant Cell Reports* 10:144–7.

Pound, F. J. 1945. A note on the cacao population of South America. In: *Report and Proceedings of Cacao Research Conference*, Colonial Office, May-June 1945. The Colonial Office, His Majesty's Stationery Office, London, UK, pp. 131–3.

Sackville Hamilton, N. R., Engels, J. M. M., van Hintum, Th. J. L., Koo, B. and Smale M. 2002. Accession management: Combining or splitting accessions as a tool to improve germplasm management efficiency. IPGRI Technical Bulletin No. 5. International Plant Genetic Resources Institute, Rome, Italy.

SPC. 2004. Directory of plant genetic resources collections in the Pacific Island countries and territories. Secretariat of the Pacific Community (SPC). Available from URL: www.spc.int/lrd/index.php?option=com_docman&task=doc_download&gid=473&Itemid=298

Thomas, E., van Zonneveld, M., Loo, J., Hodgkin, T., Galluzi, G. and van Etten, J. 2012. Reflections on spatial diversity patterns of *Theobroma cacao* L. in Tropical America. Submitted.

Toxopeus, H. and Kennedy, A. J. 1984. A review of the Cocoa Research Unit Programme. In: *Cocoa Research Unit Annual Report for 1981–1983*. The University of the West Indies, St. Augustine, Trinidad, pp. 8–12.

USDA/ARS. 2010. Sequencing of cacao genome to help chocolate industry, subsistence farmers. 15 September 2010. Available from URL: http://www.sciencedaily.com/releases/2010/09/100915100940.htm

Van Hall, C. J. J. 1914. *Cacao*. Macmillan and Co. Limited, London, UK.

Van Hall, C. J. J. 1932. *Cacao*. 2nd edition. Macmillan and Co. Limited, London, UK.

van Raamsdonk, L. W. D. and Wijnker, J. 2000. The development of a new approach for establishing a core collection using multivariate analyses with tulip as case. *Genetic Resources and Crop Evolution* 47(4):403–16.

van Treuren, R., van Hintum, T. J. L. and de Wiel, C. 2008. Marker-assisted optimization of an expert-based strategy for the acquisition of modern lettuce varieties to improve a genebank collection. *Genetic Resources and Crop Evolution* 55(2):319–30.

van Treuren, R., Engels, J. M. M., Hoekstra, R. and van Hintum T. J. L. 2009. Optimization of the composition of crop collections for *ex situ* conservation. *Plant Genetic Resources Characterization and Utilization* 7(2):185–93.

Wadsworth, R. M., Ford, C. S., Turnbull, C. J. and Hadley P. 2003. *International Cocoa Germplasm Database v. 5.2*. Euronext.life/University of Reading, UK. CD-ROM.

Wood, G. A. R. and Lass, R. A. 1985. *Cocoa*. 4th edition. Longman, London, UK.

14.2 References used for the 2012 Global Strategy but not directly cited

Aikpokpodion, P. O., Motamayor, J. C., Adetimirin, V. O., Adu-Ampomah, Y., Ingelbrecht, I., Eskes, A. B., Schnell, R. J. and Kolesnikova-Allen, M. 2009. Genetic diversity assessment of sub-samples of cacao, *Theobroma cacao* L. collections in West Africa using simple sequence repeats marker. *Tree Genetics & Genomes* 5(4):699–711.

Aime, M. C. and Phillips-Mora, W. 2005. The causal agents of witches' broom and frosty pod rot of cacao (chocolate, *Theobroma cacao*) form a new lineage of *Marasmiaceae*. *Mycologia* 97:1012–22.

Argout, X., Salse, J., Aury, J. M., Droc, G., Gouzy, J., Allegre M., Chaparro, C., Legavre, T., Maximova, S. N., Abrouk, M., Murat, F., Fouet, O., Poulain, J., Ruiz, M., Roguet, Y., Rodier-Goud, M., Barbosa-Neto, J. F., Sabot, F., Kudrna, D., Ammiraju, J. S., Schuster, S. C., Carlson, J. E., Sallet, E., Schiex, T., Dievart, A., Kramer, M., Gelley, L., Shi, Z., Bérard, A., Viot, C., Boccara, M., Risterucci, A. M., Guignon, V., Sabau, X., Axtell, M. J., Ma, Z., Zhang, Y., Brown, S., Bourge, M., Golser, W., Song, X., Clement, D., Rivallan, R., Tahi, M., Akaza, J. M., Pitollat, B., Gramacho, K., D'Hont, A., Brunel, D., Infante, D., Kebe, I., Costet, P., Wing, R., McCombie, W. R., Guiderdoni, E., Quetier, F., Panaud, O., Wincker, P., Bocs, S. and Lanaud, C. 2010. The genome of *Theobroma cacao*. *Nature Genetics* 43:101–8.

Bekele, F., End, M. and Eskes, A. B. (Eds). 2003. *Proceedings of the International Workshop on Cocoa Breeding for Improved Production Systems: INGENIC International Workshop on Cocoa Breeding*, Accra, Ghana, 19–21 October 2003. INGENIC and Ghana Cocoa Board 2005. Available from URL: http://ingenic.cas.psu.edu/proceedings.htm.

Brown, A. H. D. 1989. Core collections – a practical approach to genetic-resources management. *Genome* 31(2):818–24.

CacaoNet. 2009. *Minutes of the Fifth CacaoNet Steering Committee Meeting*. Hyatt Hotel, Port of Spain, Trinidad, 23 March 2009.

CGIAR/SRGP. 2010. The decision-support tool developed as part of the project 'Collective Action for the Rehabilitation of Global Public Goods in the CGIAR Genetic Resources System, Phase 2 (GPG 2)' funded by the World Bank 2007–2010. The Excel tool is available on Crop Genebank Knowledge Base from URL: http://cropgenebank.sgrp.cgiar.org/index.php?option=com_content&view=article&id=45&Itemid=142&lang=english

Cheesman E. 1944. Notes on the nomenclature, classification and possible relationships of cocoa populations. *Tropical Agriculture* 21:144–59

Cryer, N. C., Zhang, D., Lanaud, C., Montamayor, J. C., Wilkinson, M. J. and Hadley, P. 2009. Towards the construction of a global strategic base collection for *Theobroma cacao* L. based on the use of microsatellite markers. Oral presentation, May 2009.

Dias, L. A. S. (Ed.). 2001. Genetic Improvement of Cacao (Melhoramento Genético Do Cacaueiro) FUNAPE-UFG, Brazil. Translated into English by Abreu-Richart, C. E. Available from URL: http://ecoport.org.

Efombagn, I. B. M., Motamayor, J. C., Sounigo, O., Eskes, A. B., Nyasse, S., Cilas, C., Schnell, R., Manzanares-Dauleux, M. J. and Kolesnikova-Allen, M. 2008. Genetic diversity and structure of farm and genebank accessions of cacao (*Theobroma cacao* L.) in Cameroon revealed by microsatellite markers. *Tree Genetics & Genomes* 4(4):821–31.

End, M. J., Engels, J. M. M. and Zhang, D. 2008. Report of a meeting on the establishment and composition of the Global Strategic Cacao Collection held at The University of Reading, March 2008. Available from URL: http://www.cacaonet.org/images/stories/document/Expert consultation draft report (13 05 08 final).doc.

Engels, J. M. M. 2007. Generating benefits of conserving and distributing cacao genetic diversity: What are the roles of CacaoNet and its partners? A discussion paper. Rome, Italy.

Engels, J. M. M., End, M. J., Turnbull, C. J., Daymond, A. J., Cryer, N., Ruas, M. and Rouard, M. 2009. A global strategy for the conservation and utilization of cacao genetic resources. Paper presented at the 16th International Cocoa Research Conference, Bali, Indonesia, November 2009.

Esquinas-Alcázar, J. 2005. Protecting crop genetic diversity for food security: political, ethical and technical challenges. *Nature Reviews Genetics* 6:946–53. Abstract available from URL: www.nature.com/reviews/genetics.

FAO. 2009. International treaty on plant genetic resources for food and agriculture. Available from URL: http://www.fao.org/AG/cgrfa/itpgr.htm.

FAO/IPGRI. 1994. *Genebank Standards*. FAO, Rome, Italy and IPGRI, Rome, Italy, 13p. Available from URL: http://www.bioversityinternational.org/index.php?id=19&user_bioversitypublications_pi1%5BshowUid%5D=2399.

Figueira, A., Albuquerque, P. and Leal-Jr., G. 2006. Genetic mapping and differential gene expression of Brazilian alternative resistance sources to witches' broom (causal agent *Crinipellis perniciosa*). In: *Proceedings 15th International Cocoa Research Conference, organized by the Cocoa Producers' Alliance*, 9–14 October 2006. San José, Costa Rica.

Gepts, P. 2006. Plant genetic resources conservation and utilization: The accomplishments and future of a societal insurance policy. *Crop Science* 46(5):2278–92.

Henderson, J. S., Joyce, R. A., Hall, G. R., Hurst, W. J. and McGovern, P. E. 2007. Chemical and archaeological evidence for the earliest cacao beverages. *Proceedings of the National Academy of Sciences of USA* 104:18937–40.

Hu, J., Zhu, J. and Xu, H. M. 2000. Methods of constructing core collections by stepwise clustering with three sampling strategies based on the genotypic values of crops. *Theoretical and Applied Genetics* 101(1–2):264–8.

Iwaro, A. D., Bekele, F. L. and Butler, D. R. 2003. Evaluation and utilization of cacao (*Theobroma cacao* L.) germplasm at the International Cocoa Genebank, Trinidad. *Euphytica* 130:207–21.

Iwaro, A. D., Bharath, S., Bekele, F. L., Butler, D. R. and Eskes, A. B. 2006. Germplasm enhancement for resistance to black pod disease. In: *Global Approaches to Cocoa Germplasm Utilization and*

Conservation. Final Report of the CFC/ICCO/IPGRI Project on 'Cocoa germplasm utilization and conservation: a global approach' (1998–2004). CFC Technical Paper No. 50. CFC/ICCO/IPGRI, pp. 47–57.

Lanaud, C., Risterucci, A. M., Pieretti, I., Falque, M., Bouet, A. and Lagoda, P. J. 1999. Isolation and characterization of microsatellites in *Theobroma cacao* L. *Molecular Ecology* 8:2141–3.

Lockwood, G. and Gyamfi, M. M. O. 1979. *The CRIG Cocoa Germplasm Collection with Notes on Codes Used in the Breeding Programme at Tafo and Elsewhere*. Cocoa Research Institute, Ghana.

Motamayor, J. C., Risterucci, A. M., Lopez, P. A., Ortiz, C. F., Moreno, A. and Lanaud, C. 2002. Cacao domestication I: The origin of the cacao cultivated by the Mayas. *Heredity* 89:380–6.

Motilal, L. and Butler, D. 2003). Verification of identities in global cacao germplasm collections. *Genetic Resources and Crop Evolution* 50:799–807.

Penn State. 2010. Finest chocolate may get better: Cacao tree genome sequenced. *ScienceDaily*, 28 December 2010. Retrieved 4 September 2012 from: http://www.sciencedaily.com/releases/2010/12/101226131600.htm

Phillips-Mora, W., Mora, A., Johnson, E. and Astorga, C. 2006. Recent efforts to improve the genetic and physical conditions of the international cacao collection at CATIE. *Proceedings of the 14th International Cocoa Research Conference*, San José, Costa Rica. Vol. I, pp. 611–23.

Phillips-Mora, W., Aime, M. C. and Wilkinson, M. J. 2007. Biodiversity and biogeography of the cacao (*Theobroma cacao*) pathogen *Moniliophthora roreri* in tropical America. *Plant Pathology* 56:911–22.

Posnette, A. F. 1986. *Fifty Years of Cocoa Research in Trinidad and Tobago*. Cocoa Research Centre, University of the West Indies, St. Augustine, Trinidad.

Pound, F. J. 1938. *Cacao and Witches' Broom Disease (Marasmius perniciosus) of South America with Notes on Other Species of Theobroma*. Yuille's Printery, Port of Spain, Trinidad and Tobago.

Silva, C. R. S. and Figueira, A. 2005. Phylogenetic analysis of *Theobroma* (Sterculiaceae) based on Kunitz-like trypsin inhibitor sequences. *Plant Systematics and Evolution* 250:93–104.

Sonwa, D. J., Nkongmeneck, B. A., Weise, S. F., Tchatat, M. and Janssens, M. J. J. 2008. Chocolate Forest and the Peri-Urban Landscape: Tree diversity of the cocoa agroforests around Yaoundé (southern Cameroon). The Sustainable Tree Crops Program (STCP) Working Paper Series 5. January 2008.

Soria, V. J. 1970. The latest cocoa expeditions to the Amazon basin. *Cacao* 15:5–15.

Thevenin, J. M., Umaharan, R., Surujdeo-Maharaj, S., Latchman, B., Cilas, C. and Butler, D. R. 2005. Relationships between black pod and witches'-broom diseases in *Theobroma cacao*. *Phytopathology* 95:1301–7.

Toxopeus, H. 1985. Botany, types and populations. In: Wood, G. A. R. and Lass, R. A. (Eds), *Cocoa*. 4th edition. Longman, London, UK, pp. 11–37.

Turnbull, C. J. and Hadley P. 2012. International Cocoa Germplasm Database, ICGD. Online Database. CRA Ltd./NYSE Liffe/University of Reading, UK. Available from URL: http://www.icgd.reading.ac.uk.

Turnbull, C. J., Butler, D. R., Cryer, N. C., Zhang, D., Lanaud, C., Daymond, A. J., Ford, C. S., Wilkinson, M. J. and Hadley, P. 2004. Tackling mislabelling in cocoa germplasm collections. *Ingenic Newsletter* 9:8–11.

Turnbull, C. J., Rouard, M., Ruas, M., Argout, X., End, M. J., Roux, N., Arnaud, E., Johnson, V., Hadley, P. and Engels, J. M. M. 2009. An Information Management System for CacaoNet. Poster Presentation at the 16th International Cocoa Research Conference, Bali, Indonesia, 16–21 November 2009.

van de Wouw, M., van Hintum, T., Kik, C., van Treuren, R. and Visser, B. 2010. Genetic diversity trends in twentieth century crop cultivars: A meta analysis. *Theoretical and Applied Genetics* 120(6):1241–52.

Wood, G. A. R. and Lass, R. A. (Eds). 2001. *Cocoa*, Tropical Agricultural Series, 4th edition. Longman Scientific and Technical and John Wiley & Sons Inc., London, UK and New York, NY.

Young, A. M. 1994. *The Chocolate Tree: A Natural History of Cacao*. Smithsonian Nature Books, Washington DC.

Zhang, D. P., Boccara, M., Motilal, L., Mischke, S., Johnson, E. S., Butler, D. R., Bailey, B. and Meinhardt, L. W. 2009a. Molecular characterization of an earliest cacao (*Theobroma cacao* L.) collection from Peruvian Amazon using microsatellite DNA markers. *Tree Genetics & Genomes* 5:595–607.

Zhang, D., Mischke, S., Johnson, E. S., Phillips-Mora, W. and Meinhardt, L. 2009b. Molecular characterization of an international cacao collection using microsatellite markers. *Tree Genetics and Genomes* 5:1–10.

Zhang, D. P., Figueira, A., Motilal, L. A., Lachenaud, P and Meinhardt, L. W. 2011. Wild populations and related species of *Theobroma*. In: Kole C. (Ed.), *Wealth of Crop Relatives: Genomic and Breeding Resources. Plantation and Ornamental Crops*. Springer-Verlag Berlin Heidelberg, pp. 277–96. doi:10.1007/978-3-642-21201-7_13.

Chapter 3

The role of gene banks in preserving the genetic diversity of cacao

Lambert A. Motilal, The University of the West Indies, Trinidad and Tobago

1 Introduction

Gene banks are repositories created by human involvement as an investment to safeguard against extinction in the worst-case scenario and loss of diversity in the best-case scenario. The conservation of plant genetic resources is seen as a societal insurance policy (Gepts 2006). Cacao (*Theobroma cacao* L.) is an industrial tree crop whose main economic returns are derived from the fermented and dried seeds. Cacao is important to many developing countries in Central America, South America, the Caribbean, Africa, Asia, Indonesia and the Pacific. Cacao was therefore designated as a priority crop for conservation (IBPGR 1981). The centre of origin and diversity of cacao is in the Upper Amazon of South America within the

http://dx.doi.org/10.19103/AS.2018.0021.03

present-day countries of Peru, Ecuador, Colombia and Brazil (Bartley 2005; Cheesman 1944). The natural flora in this area is under threat by various human activities. About 70% of the Amazonian rainforest in Latin America was converted into grazing lands (FAO 2006). Gene banks are therefore important to safeguard cacao diversity and to act as an expedient source for developing improved cultivars in the best interest of farmers and the industry. Zhang and Motilal (2016) reviewed germplasm groups, collections and diversity of farmed and wild cacao. They emphasised the low on-farm diversity in South East Asia and West Africa and the small number of wild clones that have generated the cacao industry today. Yet, with the exception of Morera and Paredes (1990), CacaoNet (2012) and Laliberté et al. (2012), there is an absence of recent work detailing the role of gene banks in preserving the genetic diversity of cacao.

2 History of cacao collections

The Royal Botanic Gardens at Kew, England, were used for cacao germplasm from as early as 1768 and for distributing cacao throughout the British Empire (Dring 1976). The first major collecting expedition was in response to the spread of witches' broom disease in Trinidad. Germplasm with resistance to this fungal disease was sought by F.J. Pound in the Upper Amazon in 1937 in Ecuador and in 1938 in Peru (Kennedy et al. 1987; Kennedy 1985; Pound 1938, 1943, 1945). A cacao collection was created in Ghana in the early 1940s from germplasm on farms and agricultural stations which were derived from germplasm transfer during 1905–09 from Kew (Glendinning 1967).

The importance of wild populations was recognised for breeding purposes (Allen 1987; Chalmers 1972) and early expeditions focused on wild populations with desirable traits especially disease resistance. The 1950s Anglo-Colombian expedition specifically collected related *Theobroma* and *Herrania* species (Baker et al. 1954). The London Cocoa Trade Amazon Project had a systematic but random sampling of wild *T. cacao*, with fruits rather than budsticks being preferentially collected (Allen 1984; Allen and Lass 1983). Similarly, a population approach to collect the phenotypic variability in Brazil was made during 1976–85 (Almeida et al. 1987).

From 1930 to 1988, 28 collecting missions were undertaken in the countries of Bolivia, Brazil, Colombia, Ecuador, French Guiana, Peru, Suriname and Venezuela (Lockwood and End 1993). Collection expeditions in Belize (in 1994), Guyana (in 1998), Peru (in 1987–89, 2008 and 2009) and Ecuador (in 2010) were undertaken (CacaoNet 2012). Other studies described the genetic background of new germplasm sampled either from the wild or farmers' fields from Bolivia (Zhang et al. 2012), Cuba (Martínez et al. 2015), Mexico (Vázquez-Ovando et al. 2014; Whitkus et al. 1998) and Venezuela (Sánchez and Jaffé 1989) but did not indicate whether the material was conserved either as *in situ* or *ex situ* collections.

3 Gene bank concepts

Vavilov (1951) was probably the first to formally understand genetic diversity for crop improvement. General overviews of types, methods and roles of gene banks can be found (Engels and Fassil 2004; Engelmann and Engels 2002; Jaramillo and Baena 2002;

Scarascia-Mugnozza and Perrino 2002; Swaminathan 2002). Gene banks can be broadly considered to be *in situ* or *ex situ*. Gene banks are *in situ* if the conservation is at the same place that the plants inhabited. *In situ* gene banks include on-farm gene banks in which landraces or farmer varieties are identified and left under the care of the farmer or landowner. *Ex situ* gene banks maintain the plant germplasm at a site distant from the original area of the germplasm. Safety duplication of each accession should be undertaken in at least one more site or with an alternative conservation measure (FAO 2014). Internal safety duplication in cacao gene banks may be considered as having multiple clonal copies of each accession. Most accessions in global cacao gene banks are not safety-duplicated as the former case among gene banks but are duplicated as the latter case within gene banks. Reed et al. (2004) recommended four clonal copies for heterozygotic material and eight clonal copies for homozygotic material of *Coffea* with wild genotypes being represented by two clonal copies. Internal safety duplication is an insurance against the risks of genetic erosion, and in tropical countries in which most cacao gene banks are located, there must be adequate clonal copies to protect against natural disasters and disease. Under low input maintenance and management, the same number of clonal copies regardless of accession is a simpler strategy. A set of eight clonal cacao plants offers a suitable sample size for the collection of phenotypic data and genetic erosion thresholds of 25% or 50% can be easily set for management decisions.

Complementary strategies for the conservation of genetic resources exist and are partially dependent on the biology of the organism and the facilities available (Khoury et al. 2010; Engelmann and Dulloo 2007; Maunder et al. 2004; Engelmann and Engels 2002; Jaramillo and Baena 2002; Tao 2001; Maxted et al. 1997; Damania 1996; Cohen et al. 1991). Gene banks contain plants from locales that may be difficult to access or undergoing threats from natural disasters (e.g. severe weather, fire, pests and disease) and man's activities. The latter includes the replacement of traditional varieties (Upadhyaya et al. 2010) likely due to the adoption of elite germplasm (Jaramillo and Baena 2002). The general reduction in germplasm richness even in native biodiversity by monoculture systems using one variety is recognised (Committee on Germplasm Resources 1978). In Mexico, cacao types with highly desirable flavour attributes were replaced with more productive cacao and less flavour attributes (Ogata 2003).

About 75% of the genetic diversity of crop plants has been lost since the 1900s (Upadhyaya et al. 2010). Although the loss of genetic diversity in cacao is not accurately quantified, it is deemed severe and partially attributed to the destruction of the Amazon rainforest (CacaoNet 2012; Laliberté et al. 2012). Unlike other crop gene banks, the majority control of cacao germplasm is in the South rather than the North (Mooney 1983). Genetic resources may either have an uncertain value (Ramanatha Rao 2001) or an expected value for commercial exploitation (Smale 2005). Nass et al. (2012) suggested a cost analysis approach to raise efficiency of *ex situ* conservation.

The collected germplasm ideally represents the entire genetic diversity of the plant of interest. This would include cultivated types, land races, breeding material, species in the wild and relatives of the species of interest (Ramanatha Rao 2001). These genetic resources can be collected from the country in which the collection is based, as well as from other countries, from the centres of diversity, other gene banks, farmers' fields, semi-cultivated and abandoned areas. The germplasm contains a repertoire of alleles which may be used to improve yield, disease resistance and stress resistance, as well as for changing market needs and climatic conditions (Upadhyaya et al. 2010). The lack of long-term and reliable funding resulted in a less-than-satisfactory evaluation of the primary genetic resources in universal collections to provide reliable information on heritable traits (Bloomfield and Lass 1992). These workers suggested that this is the primary reason for many problems in the world cacao industry and the inability to match the increase in productivity seen in other tropical and temperate crops. Few trees from few collecting expeditions provided planting material in the cacao industry (Schultes 1984; Cope 1976) with much less than 5% of the genetic variability of cacao being used (Drenth and Guest 2016).

4 Crux of conservation

A gene bank collection is a dynamic time capsule of plant genetic resources as the individuals were collected at particular periods along the evolutionary timeline, domestication history and human interest. We are therefore conserving a complement of genes that represent the population diversity at the time of collection. Since we cannot accurately predict the genetic drift or selection pressures, the change in gene or allele frequency and therefore the types of cultivars or species that would be obtained is unknown. The challenge is therefore to collect the best representation of the genetic diversity at the time of collection which can be used for breeding purposes to protect the species or ultimately to replenish the species as needed. Ensuring sufficient representation provides for maximum exploitation for any possible scenario. It is recommended to have at least one copy of 95% of the alleles at a frequency greater than 0.05 in the collected sample which translates to a sample of 50 individuals from each population (Brown and Marshall 1995; Marshall and Brown 1975). Collecting a smaller number would effectively lead to a bottleneck and a reduced genetic diversity is captured (Nordin and Saad 2001). However, the constraint is appreciable for very small sample sizes, as ten individuals could contain 95% of the genetic variance of the population (Frankel and Soulé 1981).

If seeds are collected, the allelic compositions of the populations are effectively stored. However, if the budwood is collected, the resultant plant is a clone of the genotype of the tree from which the budwood was collected at the time of collection. This represents just a part of the mix that contributed to the evolutionary route. Dias and Almeida (2002) reported that the collection of mother trees was inefficient due to the low heritability of yield and disease resistance. A random sampling of fruits from 50 to 80 mother trees, with eight or five seeds selected per fruit, to obtain 400 seeds was recommended for an effective size of 150 for safe conservation of a cacao population (Dias and Almeida 2002).

The minimum viable population (MVP), defined as the number of individuals necessary to ensure the long-term survival of a species (Shaffer 1981), has not been applied to cacao, probably because over 3000 million cacao trees are under cultivation. Nevertheless, cacao is still conserved because the full range of the genetic diversity surpasses that of cultivated cacao. The number of plants will influence the economic cost of conservation. The MVP may serve as a guide to the lower limit of the size of a cacao collection. Reed (2005) estimated that at least 2000 plants would maintain population fitness. Although forbs were used in that study, the same may apply for trees and a minimum of 2000 genetically unique accessions should be maintained over cacao gene banks or in the universal gene banks.

5 Gene pool coverage

Gene bank holdings insure against genetic erosion by including accessions that represent the gene pool (Fig. 1). Gene pools, by definition, include wild, semi-wild and cultivated germplasm. Wild germplasm refers to the species of interest, related species within the same genus and related genera within the same family (see Chapter 1 for further details). The primary gene pool contains the ten populations of Motamayor et al. (2008) described mainly from gene bank holdings. This gene pool also contains three other populations which originate from Bolivia along the Beni River (Zhang et al. 2012), from Hulluaga in Peru and the Nanay VII subgroup which became newly delineated (Motamayor et al. 2010). As more collections from South and Central America occur, the actual number of cacao populations should increase. Relic varieties as natural hybrids would encompass the Refractario and Trinitario germplasm. Refractario accessions were selected from trees which survived in a mass seedling nursery maintained under trees in Ecuador that were heavily infected with witches' broom disease (Bartley 2001; Pound 1943, 1938). Refractarios may be hybrids between Nacional types of Ecuador and Trinitario types (Zhang et al. 2008). The latter most likely originated from Venezuela or Trinidad (Bartley 2005). Repeated movements of cacao between Trinidad and Venezuela occurred in the early colonial period (Bartley 2005) even prior to 1727 (Ciferi and Ciferi 1957; Aspinall 1928). Furthermore, the identification of about 18 types of cacao in Trinidad sometime before 1880, and the introduction of

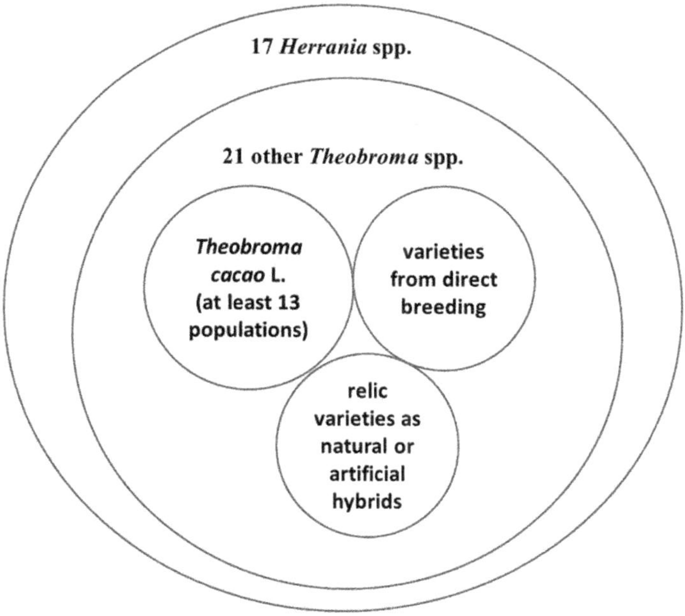

Figure 1 Gene pools of *Theobroma* germplasm. Primary gene pool is composed of wild *T. cacao*, cultivars from breeding programmes and relic varieties arising either naturally or artificially. Secondary and tertiary gene pools are the circular areas moving outwards from the centre. The genus *Gauzuma* may be considered as a quaternary gene pool.

cacao germplasm into Ecuador about 1890 (Bartley 2005), would suggest that the material moved into Ecuador could be from germplasm that was admixed between Trinidad and Venezuela which constituted the same Spanish territory. The Trinitario germplasm is now closely linked to the natural hybrids that developed in Trinidad (Bartley 2005) as a result of a combination of climatic, genetic and agronomic factors in 1725 (Motilal and Sreenivasan 2013, 2012). Although the name 'Trinitario' was reportedly originally applied to the type of Amelonado cacao from Trinidad (Bartley 2005), it was said to be Venezuelan in origin (Cheesman 1932). Reintroduction of the material from the same region of Venezuela into Trinidad in 1757 resulted in the Trinitario name being associated with Trinidad cacao (Cheesman 1932; Van Hall 1914; Preuss 1901).

Semi-wild includes those genetic resources that may have been directly influenced by man's domestication or farming activities and would encompass Refractario and Trinitario material. Cultivated germplasm includes farmers' varieties and landraces. Genes from wild types and landraces have been bred into modern crop varieties to improve yield, resistance to pest and diseases and stress tolerance, among others (Committee on Managing Global Genetic Resources 1991). Maxted et al. (2012) explored the value and conservation of wild relatives.

Cacao has not been truly domesticated as in other crops like coconut, corn, potato, rice or tomato, where the plant parts harvested for consumption have changed drastically either in form or content. The fruits of cultivated cacao are similar to those of wild *T. cacao*. There are possibly three groups of cacao (Criollo, Nacional and Amelonado) that have undergone domestication events to some extent. Criollo, as a genetic group term, came to be applied to the plants favoured by the Aztecs and Mayans which had white beans. Descendants and hybrids of these plants, including Trinitario material from Venezuela, Trinidad and Indonesia, among others, have become an integral part of the fine flavour cacao industry. Germplasm from Honduras, Belize, Nicaragua and Mexico contains substantial Criollo ancestry especially when relic trees are sampled. The Nacional material from Ecuador is also fine flavour cacao. The Refractario germplasm is a mixed assemblage similar to the Trinitario group but showing more defined partitioning into genetic subclusters. The Amelonado material coming from the Brazilian Amazonas have had far-reaching influence in the cacao industry, forming part of the genetic ancestry of the ICS accessions in Trinidad and the subsequent TSH cultivars which were derived in part from ICS 1 (Gonsalves 1996). The Amelonado group was also the primary genesis of the West African cacao industry, the bulk cacao that forms the majority of exported cacao.

The 22 related species within the *Theobroma* genus show limited, if any, natural cross-pollination or cross-fertilisation events with *T. cacao*. However, artificial hybrids from manual pollinations have been documented from intergeneric cross of *T. cacao* with *Theobroma* spp. (Lanaud 1986) and *T. cacao* and the related genus *Herrania* (Doodnath 1996; Lanaud 1986). Frosty pod has been recorded on native hosts of *T. bicolor* and *Herrania* (Rorer 1918 cited in Evans (2016)), as well as *T. gileri* (Holliday 1957; Baker et al. 1954). Fifteen of the 22 *Theobroma* species are edible with *T. angustifolium*, *T. bicolor* and *T. grandiflorum* being cultivated for their pulp (CacaoNet 2012). The contribution of genes from the secondary and tertiary gene pools to cacao breeding is at present non-existent. Cacao collections should still seek to maintain several of these species, as potential sources for crop improvement in the universal collections, as well as in the national collections of the South and Central American countries. Uchola (2016) indicated that cacao diversity was represented by several cultivars or clones. However, an estimate of the number of cultivars was not given and only examples from LCT EEN, PB and RIM were provided. The

Amelonado which makes the bulk of West African industry or the ICS and TSH cultivars from Trinidad which have spread throughout cacao-growing countries were not listed.

Wild, semi-wild and related species of *Theobroma* are considered to be diploids with ten pairs of chromosomes (Davie 1935; Simmonds 1954; Cope 1976; Toxopeus 1985). Germplasm with different ploidy levels may be possible. Zven and Zhukovsky (1975) (cited in Engels (1984)) have reported 16 and 26 instead of 20 chromosomes in cacao. Haploid plants were obtained from the embryo rescue of mature flat beans (Zuñiga 1988). Natural haploids from polyembryonic seeds (Dublin 1972) were used to produce double haploids and the latter were recommended as tools to create diversity and as pollen donors (Sounigo et al. 2003). Lanaud (1988) showed that haploids could come from either parent, be chimaeric haploids or be chimaeric diploid/haploids. Opeke and Jacob (1969), on the basis of cytological irregularities, suggested that cacao is an autotetraploid which is evolving into a secondary diploid as a result of continuous outbreeding. Evidence for tetraploidy was reportedly obtained from the segregation ratios of five morphological and qualitative traits (Engels 1984). Artificial tetraploids were induced in cacao (Martinson 1986). There is a scarcity or indirect records of non-diploid accessions in cacao collections. The DE accessions were reportedly obtained from double or polyembryonic seeds (Turnbull and Hadley 2017) and their ploidy status should be examined. Eighty-nine H accessions held in the Centre National de Recherche Agronomique collection of Côte d'Ivoire have the identifier (HAP) and include the double haploids listed in Lanaud (1988) and Sounigo et al. (2003). Non-diploid accessions should be sent for safety conservation to other cacao gene banks. If this is not possible, then safety duplication within and among the national collections should be undertaken to reduce the risk of loss. Landrace varieties are often termed local and in the case of cacao history, the word form 'creole' is often applied. The importance of the local, creole or landrace germplasm was recognised for various crops in Brazil (Albuquerque and Nass 2009). Aikpokpodion (2010), using agromorphological traits, showed that Nigerian farms were less reliant on the traditional local Amelonado landrace variety and recommended it for conservation. The much-distributed ICS varieties, as well as the TRD varieties, may be landrace varieties in Trinidad. These varieties arose out of a natural disaster in 1725 that destroyed the majority of the Criollo trees on the island and elsewhere (Motilal and Sreenivasan 2012, 2013). The valuation of landrace varieties is not restricted to being cultivars which are best adapted to particular local conditions. Landrace varieties need to be comprehensively evaluated as they represent a selection of the evolutionary path of the crop that can be more readily accessed than that occurring in the wild.

Frosty pod is an important cacao disease in Central and South America, but the disease has not been encountered in the Caribbean, except for Jamaica, where the disease was reported in late 2016 (S. Webster 2016, pers. comm.). The accession ICS 95 is tolerant to frosty pod rot (Ramirez et al. 2013; Jaimes et al. 2011). It was fortuitous that this accession had a genetic background that allowed it to combat a pathogen to which it was not exposed. The other 99 of the 100 selected ICS accessions should be evaluated for their resistance to this disease. Ramirez et al. (2013) found that only four other genotypes (UF 273, UF 712, EET 75 and PA 169 [PER]) out of over 600 were tolerant to frosty pod rot. These accessions were present in variable numbers of gene banks – four (UF 712), seven (EET 75 [ECU]), eight (UF 273), 19 (PA 169 [PER]) and 31 (ICS 95) according to the records in Turnbull and Hadley (2017). It may be prudent for cacao gene banks to request the required frosty-pod-tolerant germplasm to enhance their holdings. Since frosty pod resistance may be a rare trait (Ramirez et al. 2013), broadening the search to include gene banks other than that in CATIE should be considered. A molecular screen to determine the common factors indicative of frosty pod resistance should be undertaken. This would facilitate the screening for frosty pod resistance in gene banks where the pathogen is absent.

Augmentation and replenishment of gene bank holdings is a continual activity. Some curators may decide to maintain a set limit of accessions within their holdings. If accession holdings are to be increased, gap analysis may be undertaken to systematically determine the deficiencies in gene pool coverage. Maxted et al. (2008) proposed a four-step gap analysis for the genetic conservation inclusive of assessing diversity through a review of intrinsic, taxonomic, genetic and ecogeographical diversities combined with threat assessment. Spatial gaps in geographic and environmental areas have been used to improve conservation efforts of crop wild relatives (Garcia et al. 2017a,b), *Lupinus* (Parra-Quijano et al. 2012), pearl millet (Upadhyaya et al. 2012) and *Phaseolus* beans (Ramirez-Villegas et al. (2010). Lack of or under-representation of accession groups, genetic clusters, populations and wild relatives of cacao are considered in gap analysis.

6 Techniques for conserving cacao germplasm

Several techniques are available to conserve plant genetic resources due to the totipotency of plant parts. We can therefore consider the 'conservation propagule', which is the unit item being conserved from the plant. The conservation propagule can be the entire plant, seeds or explants. The type of conservation propagule influences the storage conditions. In all cases, cellular integrity needs to be maintained and the conservation propagule needs to be protected from mechanical damage. Techniques for *in situ* (protected area, on-farm and home garden) and *ex situ* (seed storage, *in vitro* storage, DNA storage, pollen storage, field gene bank and botanical garden collections) are detailed in the literature (Khoury et al. 2010; Maunder et al. 2004; Engelmann and Engels 2002; Maxted et al. 1997). An *in vitro* collecting technique for cacao has also been described (Yidana 1988; Yidana et al. 1987).

Explants as cuttings offer a convenient means for storing plant species that root easily or can be induced to root easily. In tissue culture, the tissue used can come from any part of the plant and because small portions are used, the storage vessel can contain several portions of an individual. This reduces the chance of loss and increases the availability of the plant to the users of the collection. However, there is a poor success rate of cacao plants derived from tissue culture (Bloomfield and Lass 1992; Palma and Villalobos 1989). This was likely due to a low rate of, and variably efficient, primary somatic embryogenesis (Li et al. 1998). The carbon source for embryo induction was dependent on the genotype (da Silva et al. 2008). Silva Pupo et al. (2005) recorded a 20% success rate in converting and maturing embryos to a stage at which they could become plants. Embryogenic cell lines of cacao germplasm were reportedly conserved on a large scale by Nestlé (Florin et al. 1999). Chaudhury and Malik (2004) and Cruz-Cruz et al. (2013) reviewed the cryopreservation methods including those for recalcitrant species like cacao. They recorded success at the secondary embryogenesis stage but the regeneration of adult plants was not addressed. In contrast, Paulin et al. (2007) showed that plantlets could be produced but that the yield of suitable plants was limited by the inefficiency of the primary embryogenesis phase and the presence of substantial abnormal plantlets and large losses during the acclimatisation phase. Maximova et al. (2002) developed a secondary somatic embryogenesis process which was used in protocols by Fang et al. (2004) and later by Adu-Gyamfi (2011) to improve the production efficiency and stability (Fang et al. 2009) of regenerated plants. Plants derived from tissue culture were supplied to farmers in Brazil, Ecuador, Indonesia and Côte d'Ivoire by Nestlé (CacaoNet 2012). The inefficiency and genotype-dependent nature of the *in vitro* techniques limit the adoption of *in vitro* conservation and consequently, *in vitro* storage of cacao is uncommon.

Pollen grains and certain seeds are naturally packaged for prolonged storage by having a tough wall and minimal water content to prevent damage from ice crystals during freeze–thaw cycles. The small size facilitates economy of storage space and their low water content enables them to withstand prolonged periods of storage, whether natural or artificial, until the regeneration of the plant under suitable conditions. Low temperatures at subzero conditions add an extra level of safety. Tissue culture techniques can be used to regenerate plants from the pollen grains. Since the pollen grains were formed from meiotic events, sterile haploid plants will be regenerated. The use of colchicine will enable the formation of fertile double haploids.

Seeds offer an advantage over pollen grains in being able to generate a fertile plant without resorting to toxic chemicals. They are, however, a natural food source and must therefore be protected from pest and microbial attack. Storage under sterile conditions should be mandatory. Some plants, however, possess seeds that are not conveniently stored as they have a short dormancy period, possess high water content and are damaged when dried and cooled. These seeds are termed recalcitrant but are better described as desiccation sensitive (Committee on Managing Global Genetic Resources 1991). Such recalcitrant seeds can be found in avocado, citrus, rubber and cacao. There is as yet no suitable method for storage of cacao seeds although some early work using polyethylene glycol and charcoal showed promise (Sánchez and Velázquez 1989). Notwithstanding the advances made in tissue culture and seed storage, cacao is currently conserved as living trees in field gene banks.

7 *In situ* and *ex situ* gene banks

In situ gene banks avoid the logistics in sample collection, transportation, nursery establishment, quarantine and relocation to final planting site. Conservation of 'Globally Important Agricultural Heritage Systems' is a recognised facet of *in situ* conservation which can protect agricultural biodiversity, resilient ecosystems, cultural heritage, food security and livelihood security for millions of poor and small-scale farmers covering about five million ha (Koohafkan and Altieri 2010). Yet, Ahtiainen and Pouta (2011) found that knowledge of the value of plant genetic resources was an important gap that negatively impacted policy formulation. Approaches to *in situ* conservation are available (de Boef et al. 2013; Engels et al. 2002; Jarvis et al. 2000 and references therein) but have not been applied or studied in cacao (CacaoNet 2012).

However, *in situ* gene banks can suffer from vagaries of land tenureship. Management issues arise when a clear policy for the responsibility of maintenance of the health and care of the trees is lacking. *In situ* conservation in the form of on-farm conservation relies heavily on farmer goodwill. If the farmer views conservation as unimportant, the danger of genetic erosion becomes high. Furthermore, if ownership of the land changes, there is no guarantee that the new owner(s) would be willing to undertake the conservation activity. CacaoNet (2012) and Laliberté et al. (2012) recommended that the social and economic factors that influence farmers' decisions to conserve cacao diversity be understood in order to better design *in situ* and on-farm conservation approaches. Ideally, *in situ* conservation can be within national parks or forest reserves or other protected areas sanctioned by the government. In Bolivia, the Parque Nacional y Territorio Indígena Isiboro-Secure, a 1.372 million ha protected area established in 1965, was designed in part to protect wild relatives of cacao (Hunter and Heywood 2011). This offers some measure of relief against human threats to genetic erosion.

However, these areas may still be under threat if the legal sanctions are not enforced and if squatting or illegal logging activities are present. In addition, multiple sites would be required to safeguard genetic diversity of cacao whether as national sanctuaries or as farm sites. The management of multiple discrete units becomes more complex especially as the number of *in situ* conservation sites increases and as the number of responsible parties increases. The cost of *in situ* conservation is also expected to be higher when habitats are conserved rather than when the genetic resources of a species are maintained in a dedicated site as an *ex situ* collection. However, unlike *ex situ* conservation, *in situ* conservation protects ecosystems, species and genetic diversity along with the evolutionary processes and pressures (Jarvis et al. 2000) which may lead to the development of new germplasm (Hayward and Sackville Hamilton 1997; Brush 1995). Hayward and Sackville Hamilton (1997) stress that efficient *in situ* conservation areas must have a suitable size, heterogeneity and structure that can allow evolutionary processes to maximise genetic variability. Efficient *in situ* conservation is therefore dependent on accurate knowledge of spatial and temporal genetic diversity and the factors that control such diversity (Hayward and Sackville Hamilton 1997). Samuel et al. (2013) demonstrated that *in situ* conservation could be prioritised based in part on cacao population clustering and risk assessment.

On-farm conservation forms a part of the participatory conservation approach in which non-governmental organisations (NGOs), community-based organisations (CBOs), cooperatives and interested citizens play an active role in safeguarding the genetic diversity. When the interested party is a farmer or a farmer's cooperative, then the plants are maintained in the farmers' fields, making on-farm conservation also participatory conservation. Conservation of germplasm in farmers' fields is a major part of *in situ* conservation (Sastrapradja and Balakhrishna 2002; Scarascia-Mugnozza and Perrino 2002). The other groups involved in participatory conservation may manage the genetic diversity in an arboretum or botanic garden. The NGOs and CBOs may be the better partners for working alongside gene banks in safeguarding genetic diversity as they are usually already interested in germplasm conservation and have a regulatory framework in place for partnership activities.

The trend for niche and ultra-niche markets for fine or flavour cacao beans may be advantageous for *in situ* conservation. In these markets, the location and genetic background of the chocolate source become important selling features. On-farm varieties, single estate, single fields and wild areas for sustainable harvesting would become more attractive for conservation where the cacao is found.

There are over 7.4 million *ex situ* plant germplasm accessions conserved within approximately 1750 gene banks (Upadhyaya et al. 2010). *Ex situ* collections may contain the same wealth of diversity as *in situ* collections but would be likely dependent on the species in question. Rice et al. (2006) using 22 SSRs showed that *ex situ* and *in situ* collections of Jala maize had similar overall genetic diversity with a minimal differentiation between them. In general, *ex situ* conservation includes (a) plant or field gene bank, (b) seed bank, (c) *in vitro* banks of cells or organs and (d) DNA banks. Currently, cacao germplasm collections are primarily *ex situ* field gene banks. Cacao is primarily an outcrossing species and this combined with the presence of recalcitrant seeds and the lack of an adopted tissue culture propagation protocol has led to the conservation of genetic resources as living trees in field gene banks. An advantage of field gene banks for vegetatively propagated plants, outcrossing plants or plants with recalcitrant seeds, is that the genetic composition of the accession is always intact and is not subject to drift due to the sampling of seed or inconsistent numbers of progeny produced. Thus, an accession in a cacao field gene bank is a defined genotype, whereas an accession from a seed gene bank is a composite representation of several genotypes. Hayward and Sackville Hamilton (1997) have explored this issue. The advantage of *ex situ* field gene banks in providing budwood, seed and pollen from long-lived trees must be balanced against the risk of natural disasters and maintenance costs (CacaoNet 2012).

The intent of the germplasm collection separates gene banks into active and base collections. Active collections are tasked with multiplying the material and are the primary sites for the distribution, characterisation and evaluation of the genetic resources of the species. Base collections are backup reserves of the active collections and are held under the conditions of long-term storage (Committee on Managing Global Genetic Resources 1991). These definitions are, however, unsuitable for field gene banks, where seeds are not conserved and an *in vitro* collection is absent for the species of interest (Sackville Hamilton et al. 2003). In a working collection, the material is stored for a short term and is regularly used for crop-improvement programmes. Since most cacao gene banks evaluate their genetic resources towards crop improvement, many cacao gene banks may also be considered as working collections. Working collections are those that are accessible to the cacao community and contain accessions which are not held in trust. A listing of some working collections for different tropical plant species is provided in Table 1.

Quarantine centres are the focal point for germplasm distribution and this has been used to support quarantine centres as working collections (Anon 2007; End et al. 2007). However, a quarantine centre does not own the germplasm material but exists as a transit point among donors and recipients. Instead, the donor collection, from which the germplasm was obtained, and which should be identified in the material transfer agreement (MTA), has acted as the working collection.

IBPGR (1981) designated the major collections in Brazil (Centro de Pesquisos do Cacau, Itabuna; Comissão Executiva de Planejamento da Lavoura Cacaueira, Belém; CEPEC/CEPLAC), Costa Rica (International Cocoa Collection of CATIE; Centro Agronómico Tropical de Investigacioñ y Enseñanza (CATIE), Turrialba; IC3), Ecuador (Estacíon Experimental Tropical INIAP, Pichilingue), Trinidad (International Cocoa Genebank Trinidad; ICGT) and Venezuela (Estacíon Experimental Caucagua; EEC) as primary collections. All other collections in Africa and Asia were designated as secondary collections and three quarantine transfer collections were also identified. The ICGT and IC3 were designated as universal collections to duplicate their germplasm material and all other germplasm from primary collections, as well as to be involved with all aspects of genetic resources.

National collections are those that fall within the regulatory framework of a political state and to which the access is restricted according to the governmental policy. Details of these collections may be treated as confidential and an accurate count of trees, accessions and genetic groups may not be available to the wider cacao community. If the budwood is desired from a particular variety within a national collection, there is no guarantee that the

Table 1 Examples of working collections of tropical plant germplasm

Common name	Scientific name	Description of tropical plant	Holdings	Reference
Bananas and related	*Musa* spp.	monocot; largest herbaceous flowering plant with edible fruits	9051 accessions in 52 collections worldwide	MusaNet (2016)
Cashew	*Anacardium occidentale* L.	dicot evergreen tree with edible one-seeded hypocarpium	419 clonal accessions (National Cashew Gene Bank in Puttur, Karnataka India)	Swamy et al. (2002)
Cassava	*Manihot esculenta*	dicot woody shrub with edible starchy roots	over 6500 accessions in Centro Internacional de Agricultura Tropical (CIAT), Colombia	Robinson and Srinivasan (2013)
Cotton	*Gossypium* spp.	dicot woody shrub with cellulose fibres surrounding seeds	18 971 (Uzbekistan); 10 469 (India); 10 318 (USA); 8837 (China); 6276 (Russia), 4296 (Brazil); 3070 (CIRAD, France); 1711 (Australia)	Abdurakhmonov et al. (2012)
Mango	*Mangifera indica*	dicot evergreen tree with edible drupe fruit	150 accessions (MGCollection, Brazil)	Ribeiro et al. (2013)
Pepper	*Capsicum* spp.	dicot herb with edible berry-like fruit	8165 accessions (Asian Vegetable Research and Development Center)	Lin et al. (2013)
Pineapple	*Ananas comosus*	monocot herb with short stem, tough spiny leaves and an edible, multiple fruit	55 genotypes (Cuba)	Yanes et al. (2012)
Potato	*Solanum tuberosum*	dicot perennial herb with edible stem tuber	~98 000 accessions (Centro Internacional de la Papa, Spain)	Muzzalupo et al. (2012)
Rice	*Oryza* spp.	monocot perennial grass with edible starchy seeds	24 673 accessions (Rural Development Administration Gene Bank, Korea)	Khaing et al. (2014)
Rubber	*Hevea brasiliensis*	dicot evergreen tree used for latex production	3542 accessions (Vietnam)	Lam et al. (2012)
Sugarcane	*Saccharum* spp.	monocot perennial grass with high sucrose content in stems	1002 accessions (USDA/ARS Genebank Repository, Miami, USA)	Nayak et al. (2014)

Common name	Scientific name	Description of tropical plant	Holdings	Reference
Sweet potato	*Ipomoea batatas*	dicot vining plant with edible tuberous roots	1303 accessions (National Crops Resources Research Institute, Uganda); >6400 accessions (Centro Internacional de la Papa, Spain)	Yada et al. (2010); Muzzalupo et al. (2012)
Taro	*Colocasia* spp.	monocot perennial herb with edible starchy stem corm and edible leaves	2298 accessions in eight countries: 172 (Philippines); 350 (Vietnam); 300 (Thailand); 135 (Malaysia); 685 (Indonesia); 278 (Papua New Guinea); 378 (Vanuatu)	Lebot et al. (2010)
Tomato	*Solanum lycopersicum*	dicot perennial herb with edible berry fruits	82 400 accessions in *ex situ* collections; 8259 accessions (Asian Vegetable Research and Development Center)	Ebert and Chou (2015)
Turmeric	*Curcuma longa*	monocot rhizomatous perennial herb	39 accessions (Department of Genetics of the University of São Paulo)	Sigrist et al. (2011)
Yam	*Dioscorea* spp.	monocot perennial with vining habit and starchy root tubers	53 landraces & seven improved cultivars of four species	Sartie et al. (2012)

budwood will be provided, especially if the requisition is from another country or a gene bank is from another country. From their survey of cacao collections, CacaoNet (2012) and Laliberté et al. (2012) stated that access to gene bank holdings and information of both national and international collections was restricted by a 'lack of a clear legal and policy framework at the institutional, national and regional levels'. Mooney (1983), however, treated the issue as rights of the source countries to have control, prevent gene drain and derive economic benefit from patents and varieties from the germplasm drawn from their own countries. Farmers' rights as part of intellectual property rights were also advanced by Scarascia-Mugnozza and Perrino (2002).

The Committee on Managing Global Genetic Resources (1991) stated that many nations which are rich in genetic diversity are reluctant to allow foreigners to collect and exchange germplasm. The Convention on Biological Diversity allows the source countries to exercise sovereign rights over their genetic resources. The issue of restricted access (CacaoNet 2012; Laliberté et al. 2012) can be reconciled with the latter authorities provided that MTAs are implemented and enforced. Barton and Siebeck (1994) argued for this position and proposed that germplasm distribution centres should 'facilitate the realisation of source country rights to compensation and technology transfer without disrupting or jeopardising the operation of their genetic resource collections' since free access to genetic resources from developing countries would become untenable to the developing countries. CacaoNet (2012) has indicated that MTAs are employed by the International Cocoa Quarantine Centre at the University of Reading (ICQC,R). Whether there is an evidentiary support for the routine application of MTAs in the cacao community; between quarantine centres and the gene banks from which they source their material; among quarantine centres, the entities to which they release material and the donor of the germplasm; or when genetic resources are sampled in the source country and the research is performed by an entity outside the source country, is unknown.

8 Status of cacao field gene banks

An accurate account of the cacao genetic resources held in gene banks worldwide is presently unavailable from a single source document, except perhaps for CacaoNet (2012). However, the International Cocoa Germplasm Database (ICGD; Turnbull and Hadley 2017) contains information on at least 40 major gene banks distributed internationally as quarantine centres, national and universal collections (Table 2). This was about 14 less than that reported by Motilal and Butler (2003) which was obtained from the then version of the ICGD. Cacao germplasm in The Royal Botanic Gardens, Kew, is presently limited to six cacao plants of unknown accession or population name, one Amelonado seedling from Ghana and one plant of *T. bicolor* (C. Brough 2016, pers. comm.; S. Stark 2016, pers. comm.). Other wild *Theobroma* spp. conserved in gene banks are provided in Table 3.

A single entity with a global mandate to protect cacao germplasm was found lacking by Brenner (1992) and remains so to this day. Fifty cacao germplasm collections worldwide were surveyed during 2008–12 and over 24 000 cacao accessions were distributed over 40 collections (CacaoNet 2012; Laliberté et al. 2012). The accessions per 'gene bank' in the ICGD, however, reflect only those accessions for which information is available at the time at which the database was created. There may be cases where the gene bank is no longer operational (e.g. Jamaica), has not been listed (e.g. Cuba, Dominica) or for which the current holdings have not been accurately listed. The latter is probably the case for the majority of the collections listed in Table 1 and 2. The Brazilian germplasm collection, CEPLAC, for instance, by the end of 1985, reportedly contained 1749 accessions (984 clones and 765 progenies) representing 17 251 genotypes (Almeida et al. 1987) and by 2016 reportedly 1766 accessions. This suggests that there is a difference in the application of the word 'accession' since most cacao gene banks use an accession to indicate a particular genotype, whereas in CEPLAC's case, it appeared that 'accession' was used as a group name to cover multiple different genotypes. This would be similar to a seed gene bank where the accession is maintained as a collective pool of seeds, but with each seed as an individual genotype.

The two universal cacao collections of IC3, in Costa Rica under management by CATIE, and the ICGT under management by Cocoa Research Centre of The University of the West Indies, supposedly contain the majority of the genetic diversity of cacao present in field gene banks. The five main repositories by the number of accessions in decreasing order are ICGT, CEPLAC, EET, CEPEC and IC3. If the number of holdings is adjusted for CEPLAC's use of the term 'accession' and 'genotype', a census of 17 251 distinct

Table 2 Status of *Theobroma cacao* L. holdings in global cacao gene banks

	Country	Collection	Accessions recorded in ICGD[1]	Total accession count[2]	Latest confirmed holdings
1	Benin	IRCC, Niaouli[3]	14	15	
2	Brazil	MARS Centre for Cocoa Science	392		
3	Brazil	CEPEC	1240		
4	Brazil	CEPLAC	1766		
5	Brazil	CEPEC/CEPLAC		1302	
6	Brazil	CEPLAC/SUEPA		2504	
7	Brazil	CEPLAC/SUERO-Rondonia		773	
8	Brazil	ICA		130	
9	Cameroon	IRAD	224		330 (O. Sounigo 2016, pers. comm.)
10	Cameroon	N'koemvone	6		
11	Colombia	ICA, Palmira	400		
12	Colombia	CORPOICA La Selva		745	
13	Costa Rica	IC3, CATIE	1100	1146	1217 (W. Philips-Mora 2016, pers. comm.)
14	Cote d'Ivoire	CNRA, Bingerville	744	1605	
15	Cuba	EIC/ECICC		127	
16	Dominican Republic	Comision del Cacao	60		
17	Dominican Republic	IDIAF	116	115	109: 64 local clones & 45 introduced clones (R. Villar 2016, pers. comm.; M. Ventural-Lopez 2016, pers. comm.)
18	Ecuador	EET Napo	298		
19	Ecuador	EET Pichilingue	1462		
20	Ecuador	INIAP		2332	
21	Fiji	Naduruloulou Research Centre	40		
22	Fiji	Dobuilevu Research Station		115	
23	France	CIRAD-CP Montpellier	136	138	
24	French Guiana	CRB-PPG cacao	550	508	545 (P. Lachenaud 2016, pers. comm.)
25	Gabon	SOCAGAB	35		

(*Continued*)

Table 2 (*Continued*)

	Country	Collection	Accessions recorded in ICGD[1]	Total accession count[2]	Latest confirmed holdings
26	Ghana	CRIG	994	1366	
27	Grenada	Ministry of Agriculture, Lands, Forestry and Fisheries	93		
28	Guatemala	Los Brillantes	113		
29	Guyana	NARI	15		
30	Guyana	MHOCGA		65+	
31	Honduras	CEDEC/FHIA	144	31	
32	India	CPCRI, Karnataka	200	291	430 (E. Apshara 2017, pers. comm.)
33	India	Kerala Agricultural University	235		564 (E. Apshara 2017, pers. comm.)
34	Indonesia	Bah Lias Research Station, LONSUM	225	305	
35	Indonesia	ICCRI	564	714	400 (A. Susilo 2016, pers. comm.)
36	Jamaica	Cocoa Industry Board	11		8 at Orange River Research Station (Pryce et al. 2008)
37	Madagascar	FOFIFA, Ambanja	33		
38	Malaysia	ARC, Tuaran	707		514 (H. Bin Ramba 2017, pers. comm.)
39	Malaysia	MCB Kota Kinabalu	156	2263	1950 (H. Bin Ramba 2017, pers. comm.)
40	Mexico	Rosario Izapa	191		
41	Nicaragua	UNAN		51	
42	Nigeria	CRIN	261	1100	
43	Papua New Guinea	Cocoa and Coconut Research Institute	75	1200	
44	Peru	CEPICAFE		30	
45	Peru	ICT		607	
46	Peru	UNAS, Tinga Maria	72	422	310 clones & 89 hybrid populations (D. Guarda Sotelo 2016, pers. comm.)
47	Peru	UNSAAC		72	
48	Solomon Islands	Solomon Islands Plantation	29		

(*Continued*)

Table 2 (*Continued*)

	Country	Collection	Accessions recorded in ICGD[1]	Total accession count[2]	Latest confirmed holdings
49	Solomon Islands	Black Post Cocoa Unit		95	
50	Sri Lanka	CARI	13		
51	Suriname	Agricultural Experiment Station	9		
52	Thailand	CHRC, Chumphon	34	34	34 (P. Nopchinwong 2017, pers. comm.)
53	Togo	CRAF (previously IRCC)	152	217	261 (E. Ametefe 2017, pers. comm.)
54	Trinidad and Tobago	ICGT	2263	2400	2024 (A. Holder-John 2016, pers. comm.)
55	United Kingdom	ICQC,R	378	395	396 (A. Daymond 2016, pers. comm.)
56	Puerto Rico, USA	TARS, USDA Mayaguez	230	200	251 (R. Goenaga 2017, pers. comm.)
57	United States of America	USDA Miami	80		
58	Vanuatu	Valeteruru Station	66	85	
59	Venezuela	INIA		872	
60	Venezuela	MAC Caucagua	180		
61	Venezuela	Ocumare	210		

[1] Online records in International Cocoa Germplasm Database (Turnbull and Hadley 2017).
[2] As provided in Table 2 of CacaoNet (2012) and from http://www.cacaonet.org/(information last updated March 2012).
[3] CRA-SB; Centre des Recherches Agricoles Sud Benin (Niaouli) given in CacaoNet (2012).

individuals would place CEPLAC as the gene bank with the highest number of distinct cacao varieties. Pires et al. (2000) considered CEPEC/CEPLAC to contain nearly all the variabilities of cacao. The Malaysian Cocoa Germplasm Collection and INIAP in Ecuador contain over 2200 (Johnsiul and Awang 2016; CacaoNet 2012) and 2332 (CacaoNet 2012) accessions, respectively. Growth, loss and deliberate reduction of holdings of field gene banks were recognised (CacaoNet 2012).

However, the number of accessions cannot be considered in isolation. The relatedness of the accessions and the incidence of mislabelling in the gene bank are two facets of the information matrix needed to objectively and accurately determine whether a gene bank is representative of the diversity of the species and whether a gene bank should be considered as a top conservatory of cacao genetic diversity (Fig. 2). The extent of representation of the genetic diversity was said to be more important than the size of the collection (Committee on Managing Global Genetic Resources 1991).

Table 3 Wild related species in the secondary (*Theobroma*) and tertiary (*Herrania*) gene pools of cacao conserved in gene banks

Country	Collection	*Theobroma* and *Herrania* spp.	Reference
England	Kew Gardens	*T. bicolor*	C. Brough 2016, pers. comm.; S. Stark 2016, pers. comm.
Costa Rica	IC3, CATIE	*T. angustifolium, T. bicolor, T. grandiflorum, T. mammosum, T. microcarpum, T. simiarum, T. speciosum, T. subincanum, H. albiflora, H. baloensis, H. cuatrecasana, H. nitida, H. nycterodendron, H. purpurea, H. umbratica, T. mammosum × T. simiarum, T. simiarum × T. mammosum*	W. Philips-Mora 2016, pers. comm.
Cameroon	IRAD	*T. grandiflora, T. microcarpum;*	O. Soungo 2016, pers. comm.
French Guiana	CRB-PPG cacao	*T. grandiflorum, T. speciosum, T. subincanum, H. kanukuensis, Herrania* spp.;	P. Lachenaud 2016, pers. comm.
India	CPCRI, Karnataka	*T. grandiflorum*	E. Apshara 2017, pers. comm.
Peru	UNAS, Tinga Maria	*T. bicolor, T. grandiflorum;*	D. G. Sotelo 2016, pers. comm.
Puerto Rico	TARS, USDA Mayaguez	*T. angustifolium, T. bicolor, T. gileri, T. grandiflorum, Theobroma spp., H. pulcherrima, H. purpurea, Herrania spp., H. umbratica*	R. Goenaga 2017, pers. comm.
Trinidad	ICGT	*T. grandiflorum, T. microcarpum, T. obovatum, T. speciosum, T. spp., H. albiflora, H. carmargoana, H. kanukuensis, H. nycterendon, H. pulcherrima, Herrania spp.* crosses	R. Umaharan 2016, pers. comm.

Enlarging the number of conserved cacao accessions in *ex situ* gene banks will occur largely as a result of new collecting expeditions, sharing of germplasm between source country and gene banks, sharing of germplasm from national collections to other gene banks and the identification of landrace varieties that could be deposited in national, working or universal collections. Brenner (1992) documented IBPGR (now Bioversity International) as recommending that the germplasm in CATIE be duplicated in the ICGT and vice versa. However, while substantial movement of germplasm from ICGT has occurred over the years, the redistribution of Criollo germplasm which is better represented in CATIE has been limited. Conservation of plant genetic resources, especially for tropical fruit trees that are maintained in field gene banks, would likely be negatively impacted by climate change (Ramanatha Rao and Sthapit 2012). Cacao gene banks should therefore be evaluated for their risk of genetic erosion in the face of climate change and to implement strategies that would minimise the loss of accessions.

Collecting plant genetic resources involves collecting the maximum amount of genetic variation in the smallest number of samples (Marshall and Brown 1975) from ecologically diverse environments (Brown and Marshall 1995) and geographical extremes (Allard 1970).

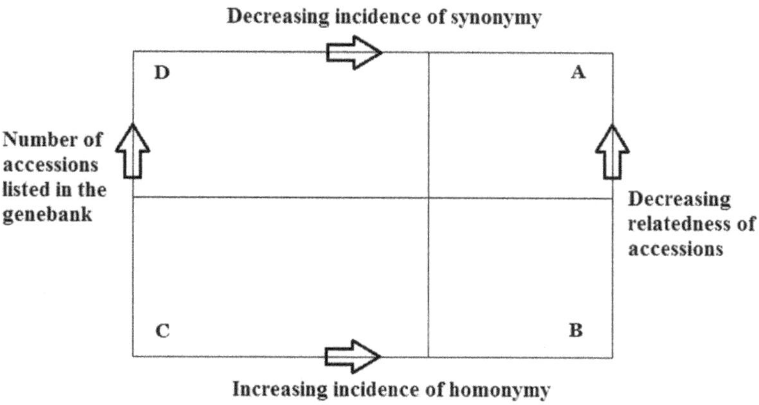

Figure 2 Assessment matrix for accurately identifying gene banks with greatest representation of cacao genetic diversity. Gene banks at A contain maximal diversity with maximal number of truly distinct accessions. Universal collections should occupy this position. Gene banks at D contain many accessions but have a high degree of repetition in the gene bank as duplicated accessions but have a substantial amount of genetically dissimilar accessions. Gene banks at B contain a fair number of accessions that are uniquely identified but have a substantial amount of genetically similar accessions. Gene banks at C contain the least number of distinct accessions and contain the maximal amount of genetically similar accessions. Gene banks occupying this position need to be critically evaluated for continuity. Homonymy: Accessions with the same name but are different genotypically. Synonymy: Accessions with the same genotype but different names.

Hence, with the collection of farmers' material, relic varieties and landraces, in addition to the mislabelling issues (Section 10), it is even more imperative that the genetic diversity of cacao collections be accurately described even if the sole intent is to preserve the genetic diversity of cacao. Cacao varieties in gene banks are termed accessions and the accessions may be grouped into families which belong to an accession series or an accession group. One or more accession groups may comprise a genetic population. Some accessions are hybrids of two or more ancestral lines. Previously, agromorphological groups were used to describe cacao types based on the fruit morphology and the location where the cacao trees were first grown (Toxopeus 1985; Soria 1970). The ten population clusters of Motamayor et al. (2008) are variously represented in the different global collections. Details on the cacao accession nomenclature can be found online at the ICGD (Turnbull and Hadley 2017) and elsewhere (Bartley 1984; Lockwood and Gyamfi 1979; Mooleedhar et al. n.d.). Fingerprinting of the collections followed by ancestry analysis is needed to obtain an accurate picture of the population distribution within and among collections.

9 Assessing genetic diversity

Hunter (1990) made five recommendations that are still valid today: document existing collections before acquiring new material or starting new expeditions, obtain voucher specimens, obtain purebred lines, test cultivars in multilocational trials and conduct studies on stock/scion relationships. A genetic diversity study is a facet of the first three recommendations. Genetic variation is considered to be the lowest level that underpins all biodiversities (Culver et al. 2011). Karp et al. (1997) and Spooner et al. (2005) described how the genetic information could be used in conserving genetic diversity. This included the detection of gaps, augmentation, guiding new collection, obtaining accurate genetic relationships, identification of valuable genotypes, identification of duplicates, devising core collections and monitoring of genetic stability and integrity of accessions. Traditional *ex situ* gene banks can have several genotypes that represent an accession and in such cases gene flow that can cause introgression of genes must be minimised (Visser and Rao 2005). In the case of cacao, this is not an issue as an accession is treated as a single genotype.

Genetic diversity can be revealed using differentiation patterns of morphological, protein or DNA markers. Morphological data for germplasm characterisation has been reviewed (Idris and Saad 2001) and descriptors for cacao are available (Bekele et al. 2006; Engels et al. 1980). Aikpokpodion (2012) provided a brief review with the application to the cacao industry in West and Central Africa. In general, molecular data is better than fruit character data in tracing origin and pedigree (Sorkheh et al. 2007) and in differentiating cultivars and wild types (Nybom and Weising 2010; de Andres et al. 2007; Sorkheh et al. 2007). Trait data are dependent on, and confounded by, environmental conditions although molecular data may underestimate the variation of traits (e.g. vigour) that are highly affected by evolutionary pressure (Shukor 2001). The general history of molecular markers in DNA fingerprinting and the applications of fingerprinting of plants have been treated (Nybom et al. 2014; Weising et al. 2005), as well as, in the context of gene bank management (Spooner et al. 2005; Shukor 2001; Karp et al. 1997; Strauss et al. 1988). The advantages and disadvantages of different fingerprinting methods have been discussed (Weising et al. 2005; Nybom 2004). In cacao, the majority of DNA fingerprinting and diversity studies have currently relied on SSR and recently on SNP genotyping. Variation from SNPs, rather than SSRs, was recommended for estimating genetic diversity within, and population differentiation of, *Arabidopsis* (Fischer et al. 2017).

Aside from differences among accessions, the differences among and within accession groups, populations, regions and countries are the facets of genetic diversity which cacao gene bank curators can use to improve management of the collections. A unique multilocus fingerprint profile must first be assigned as the signature reference profile for a given accession. The assigned signature profile should be allocated to the most original tree of the earliest known collection or germplasm identified. The signature profiles can then be compared to the nomenclature of the trees and accessions in the gene bank to determine the extent of homonymy (same accession name but different signature profile) and synonymy (different accession names but same signature profile).

Both SSR and SNP profile data are available online for various cacao accessions (Turnbull and Hadley 2017). However, unlike SNP data, SSR data are not easily compared across genotyping platforms. Furthermore, the attributed signature profile for an accession is not always indicated in the ICGD. Collaborative research with gene banks that hold the correct signature accessions is needed. The genetic diversity of relic plants from introduced areas would facilitate the detection of unique allelic combinations, which allow for adaption to the local ecological conditions. Reference accessions can be used to decipher the population structure following the grouping of Motamayor et al. (2008). Genetic parameters that should be obtained by gene bank curators should include the probability of identity and expected heterozygosity (H_e) for each individual; the number of effective alleles (N_E), allelic richness, H_e and polymorphism information content (Botstein et al. 1980) for each marker and diversity and differentiation measures of individual accessions and groups. The estimation of these parameters would obviously be affected by the number of markers used. Polymorphisms from SNPs rather than SSRs are now more easily determined from medium- to high-throughput systems and SNPs can be selected to cover a wider and better distribution of the genome than other molecular markers. Curators are advised to employ a combination of measures of genetic diversity. In particular, H_e may not be a good estimator of genetic diversity (Jost 2008; Greenbaum et al. 2014) or fitness (Savolainen and Hedrick 1995). The preservation of heterozygosity may not be sufficient to preserve allelic richness (Greenbaum et al. 2014).

The correct accession name that is to be assigned to each tree should be decided after DNA fingerprinting using a suitable set of molecular markers. However, since only a limited number of markers are employed, the proof of being different rather than being identical is obtained (Spooner et al. 2005). The term redundancy rather than duplicate was recommended (Spooner et al. 2005) but this was applied to gene banks where a batch lot of samples represent an accession rather than in the case of cacao where an accession is a single genotype. Evett and Weir (1998) and Waits et al. (2001) pointed out that many plant populations have related individuals which would influence the multilocus match probability of declaring a unique individual. A conservative measure, the probability of identity among siblings $P_{(ID)sib}$, was developed by these workers and was first implemented in cacao by Zhang et al. (2006a,b). The choice of detection system, the type of molecular marker, the actual markers used and the number of markers affect the accuracy in resolving identity issues (Motilal et al. 2009). The genetic dissimilarity of individuals can be assessed with various statistics including, but not limited to, Nei's genetic distance (Nei et al. 1983).

Phylogenetic analysis, employing the best-suited algorithms and principal coordinate analysis (PCoA) of the genetic dissimilarity matrix, allows for a graphic representation of the distribution of accessions. This enables curators to decide which, and by how much, accessions are related to each other, as well as the grouping or hierarchical system under which they may be arrayed. Estimating the extent of differentiation among hierarchical groups may be achieved using F-statistics (Wright 1965). Two other group measures, the estimator of actual differentiation (D_{est}; Jost 2008) and the G'_{ST} measure of Hedrick (2005), can be employed. The former is reportedly the most accurate measure of population differentiation (Jost 2009, 2008). The Shannon Information Index (Lewontin 1972) and the Shannon Mutual Information Index (Sherwin et al. 2006) may also prove worthwhile. Estimating the genetic differentiation of groups followed by PCoA to reveal graphical genetic relationships among groups was recommended for cacao (Motilal et al. 2013). The allocation of ancestral contribution can be undertaken using STRUCTURE (Pritchard et al. 2000), BAPS (Corander et al. 2008), AWclust (Gao and Starmer 2008), TESS (Caye et al. 2016) or other applicable programme.

Exploration of the spatial genetic pattern of genetic diversity using geographical information including distance, elevation and GIS coordinates remains an understudied area. Thomas et al. (2012) suggested that glaciation events created bottlenecks, which with human-mediated processes affected the original wild populations of cacao in the Amazon basin. The northeastern region of Peru in the Amazonian region was recommended for *in situ conservation* based on the high likelihood that cacao would survive in this region in the future (Thomas et al. 2012). Spatial genetic structure was also detected in the Cacao Nacional Boliviano from microsatellite data indicative of short-distance gene flow (Zhang et al. 2012). Spatial genetic structure was recommended as being useful for *in situ* conservation decisions (Zhang et al. 2012).

10 Curating field gene banks

Management of on-farm or agrobiodiversity *in situ* conservation has been variously described (Ramanatha Rao 2001; Jarvis et al. 2000; Frankel et al. 1995; Altieri and Merrick 1987; Oldfield and Alcorn 1987; Brush 1986; Simmonds 1962) and will not be treated here. Curation of field gene banks involves several issues which intermingle with each other. However, safeguarding the true genetic identity from loss or confusion by human error or natural causes, as well as preventing genetic erosion remains common to all *ex situ* gene banks, including cacao gene banks. As a field gene bank, picket maps and field layout must be available and continually updated as soft and hard copies. Subsumed within this is the need to have each tree uniquely identified since an accession is represented by more than one clonal tree within a gene bank. The extent of the safety duplication is a subjective one and is dependent on the amount of land available for the establishment and expansion of the gene bank. A minimum of eight trees can be maintained with picket positions as two rows of four with wider spacing between accession rows than within accession rows. By using eight trees, a reduction due to unforeseen events (such as fire, floods, malicious damage and severe pest or pathogen attack) to four trees would represent a 50.0% genetic erosion level. This can be used as a critical maximum threshold to trigger a repropagation and resupply exercise.

Each tree must have its own position tag attached in a non-lethal manner to the tree. Trees must be continually assessed for the presence of tags and any fallen or illegible tags should be replaced with the correct tag. A census of the gene bank should also be undertaken at regular intervals, the cycling of which would be influenced by the resource personnel available and the funds needed to support this activity. Nevertheless, a census should be undertaken at least every three years to have a reliable accession inventory.

Box 1 Twenty questions for curators of cacao gene banks

1. Has each tree been fingerprinted using a suitable set of molecular markers?
2. Does each tree have its correct unique tree identifier tag?
3. What is the level of homonymy and synonymy in the gene bank?
4. What is the sibship structure in the gene bank?
5. How many accessions are there in each accession group?
6. How many accession groups are represented in the gene bank?
7. How many populations are present in the gene bank?
8. How many representatives of each population are present in the gene bank?
9. How many landrace varieties are present in the gene bank?
10. Have phenotypic data been collected and organised in a hierarchical manner?
11. Is there a good representation of the morphological diversity in the gene bank?
12. What is the safety duplication of each accession?
13. What risks will affect, or are expected to affect, the gene bank?
14. What measures can be taken to minimise risks?
15. What is the risk of genetic erosion for each accession?
16. Which accessions are more prone to genetic erosion?
17. Is there a ready supply of restock for accessions at risk of genetic erosion?
18. Which areas of the gene bank collection are more prone to genetic erosion?
19. How many other *Theobroma* species and *Herrania* species are present?
20. Can data on the gene bank accessions be obtained efficiently by end users?

The curator can use the census to answer particular questions for gene bank curation and management (Box 1).

Mislabelling in plant germplasm collections is recognised as a major hindrance to the efficient conservation, use and dissemination of crop germplasm (Hurka et al. 2004). Various mislabelling problems have been identified (de Andres et al. 2007; Spooner et al. 2005) and documented in several germplasm collections (Van Treuren et al. 2010; Leão et al. 2009). Morphological information deposited in the ICGD at the time was used to obtain an estimated error rate of 43.9% in *ex situ* cacao collections worldwide (Motilal and Butler 2003). Cacao accessions can be identified from the phenotypic evaluation (Johnson et al. 2007; Bekele et al. 2006; Engels et al. 1980). However, the identification of mislabelling is more accurate with molecular data than with phenotypic data (Nybom and Weising 2010; de Andres et al. 2007).

$P_{(ID)sib}$ (Waits et al. 2001) from SSR or SNP markers must be stringent enough to discriminate amongst cacao accessions. Although a panel of 15 SSR loci was recommended for fingerprinting purposes (Saunders et al. 2004), it was unable to unambiguously resolve closely related accessions like those in the NA group (Motilal et al. 2009). The number of markers and the composition of the marker set are essential facets in determining a DNA fingerprinting panel (Motilal et al. 2009). $P_{(ID)sib}$ of 1.48×10^{-13} could be used as a maximum threshold in allocating the accession identity (Motilal et al. 2009). The advent of SNP markers has resulted in the cacao community having access to much more markers and as yet a defined panel composition has not been adopted. However, 125 SNPs were useful in discriminating among 338 accessions from the ICGT (Motilal 2017). Two SNP panels have been recommended for use (Motilal et al. 2017). McCouch et al. (2012) similarly found that the number of markers and the degree of similarity to be used for determining duplicates in rice were as yet undetermined. In the case of cacao gene banks, both redundancy (closely related) and duplicates (genetically equivalent) can be employed. However, to safeguard against the potential error in allocating duplication, phenotypic data when available should be used before a final decision is made.

Removal of duplicate accessions can reduce management costs (Chang 1989). Van Hintum et al. (2002) advocated for lumping or splitting accessions to improve the efficiency of gene bank management. Although the recommendation was predicated on an accession being a mix of genotypes, which is not the case with cacao gene banks, the general idea is still applicable. Should we lump synonyms and split homonyms in cacao gene banks? Similar to Van Hintum et al. (2002), Motilal (2017) recommended that the decision should also use phenotypic variation to help determine whether accessions are similar or different.

Mislabelling in cacao field gene banks has been documented using randomly amplified polymorphic DNA (Faleiro et al. 2002; Sounigo et al. 2001; Christopher et al. 1999) and SSRs (Cryer et al. 2006; Zhang et al. 2006b; Saunders et al. 2004; Risterucci et al. 2001; Figueira 1998; see Table 4). Although SSRs have been preferred for identity issues in germplasm collections (Cipriani et al. 2008), SNPs are becoming cheaper and more accessible, even if outsourcing needs to be implemented. SNP profiling is poised to become the preferred method for high-throughput assessment of cacao diversity.

Fingerprinting with DNA markers across all the accessions in all the cacao gene banks held worldwide (see Table 2) is yet to be completed or recorded in the literature. Risterucci et al. (2001) found a 30.0% mislabelling rate on assessing 28 accessions from nine germplasm collections. Cryer et al. (2006) found 2% homonymy and 6.6% synonymy with an overall 5.2% error rate in the ICQC,R. Complete profiling for the cacao collection of the USDA-ARS Tropical Agricultural Research Station (TARS) (Irish et al. 2010) and for

Table 4 Comparison of error rates in selected *ex situ* collections of cacao germplasm

Location	Sample size	Error estimate[a]	Reference
Global cacao genebanks	335 accessions	43.9%	Motilal and Butler (2003)
CRIN, Nigeria	250 accessions	46.4%	Aikpokpodion et al. (2010)
IC3, Costa Rica	688 accessions	14.4% synonymy	Zhang et al. (2009b)
ICGT, Trinidad	387 accessions	39.4%; *27% homonymy; 7.5% synonymy*	Motilal et al. (2012)
IDIAF Mata Larga, Dominican Republic	803 accessions	40.9%	Boza et al. (2013)
Ghana	2424 trees (64 SNPs)	*0–100% homonymy according to farm*	Padi et al. (2015)
Malaysian Cocoa Germplasm Collection		37.3% homonymy; *10.9% synonymy*	Johnsiul and Awang (2016)
TARS, Puerto Rico	154 accessions; 924 trees	19.5%; *12.3% homonymy; 20.1% synonymy*	Irish et al. (2010)
Average		34.5%	

[a]Error estimates are as quoted in reference or examination of reported data. Estimates in italics were not used in determining the average.

the international IC3 collection in CATIE, Costa Rica (Zhang et al. 2009b), was completed. Error rates in these and other collections have been reported (Motilal et al. 2011) and a revised mislabelling rate of 34.5% is suggested from the literature (Table 4).

11 Genetic diversity in cacao collections

Zhang et al. (2011a) compared ten collecting missions and showed that the POUND collection present in the ICGT had the highest allelic richness. CacaoNet (2012) stated that there was 'a considerable range of cacao genetic diversity' in *ex situ* field gene banks, but that this was not fully representative of the known range of diversity. Increasing the genetic diversity in collections may be achieved by exploring farmers' fields and the wild regions of Bolivia, Brazil, Colombia, Ecuador, Peru and Venezuela (CacaoNet 2012). New germplasm, not only in phenotypic trait combination, but in genetic composition, especially for the presence of new or rare alleles, has a high chance of being obtained from unexplored areas within the centre of origin and in bottlenecked populations. Since wild *T. cacao* germplasm is stratified according to river basins (Thomas et al. 2012), a collection strategy in these countries should incorporate stratified sampling of river basins. The Contamana, Curaray, Nacional and Purus clusters are most likely the closest related to the original wild population (Thomas et al. 2012) and should be prioritised.

Farmed cacao originated from few populations and few accessions from within a given population (Bartley 2005; Warren and Kennedy 1991; Kennedy et al. 1987; Toxopeus 1985; and references therein). Limited genetic diversity may therefore be present on farms, especially in areas where cacao was introduced as limited genotypes and in areas that are away from the centre of diversity. For instance, Efombagn et al. (2009), using 12 SSRs, showed that the parentage of farmed cacao varieties in Cameroon was principally derived from the Amelonado group and 24 parental accessions. Opoku et al. (2007) found a high level of similarity using 17 SSRs among the 11 groups of 377 accessions obtained from parental types, breeders' collections and farmers' collection in Ghana. Dinarti et al. (2015) and Lukman et al. (2014) found that Trinitario, Marañon and Nanay made the greatest ancestral contributions to 53 Sulawesi and 80 Aceh farmer selections in Indonesia, respectively. Martínez et al. (2015) in a study with 15 SSRs on 537 cacao samples from Central and East Cuba found that the majority were various hybrids among Amelonado, Criollo and Marañón groups with few pure representatives from Contamana, Iquitos, Marañón, Nacional and Nanay groups.

Therefore, although the combination of alleles may be unique, the alleles themselves may be already present in accessions in *ex situ* collections. Zhang et al. (2011b) showed that 220 farmer selections in Peru contained only 25% Trinitario hybrids and were comparable to the wild Pound collection in the ICGT. Wild and cultivated germplasm in Bolivia had similar gene diversities (0.57 and 0.55, respectively) and allelic richness (5.84 and 5.68, respectively) although they were both surpassed by the Ucayali population (gene diversity = 0.74; allelic richness = 8.93) in Peru (Zhang et al. 2012). Private alleles present in the *ex situ* collections may not be represented in farmers' fields as were found in the Dominican Republic (Boza et al. 2013). In contrast, de Schawe et al. (2013) found higher gene diversity, higher allelic richness and higher numbers of alleles in cultivated rather than wild cacao from Bolivia. Sereno et al. (2006), using nine SSRs, estimated gene diversity of 0.524–0.630 and 3.0–5.1 alleles per locus for four regions in Brazil. Ji et al. (2013) partitioned 84 farmer selections from Honduras and Nicaragua using 70 SNPs into five clusters and showed that the Trinitario types from Honduras and Nicaragua were different from each other.

Several POUND accessions were distributed amongst the Contamana (SCA accessions), Iquitos (IMC accessions), Marañon (PA accessions) and Nanay (NA accessions) from the Upper Amazon (Zhang et al. 2009a). Maternal contributions of POUND 27 [POU] to the Iquitos group and seven POUND accessions (POUND 1 [POU], POUND 2 [POU], POUND 5 [POU], POUND 7 [POU], POUND 8 [POU], POUND 10 [POU] and POUND 15 [POU]) to the Nanay group exist (Zhang et al. 2009a). The Nacional group was probably descended in part from the wild cacao accessions of LCT EEN 85, LCT EEN 86 and LCT EEN 91 (Loor Solorzano et al. 2012). Lukman et al. (2014) and Dinarti et al. (2015) working with Indonesian farmer selections found 13–15 known parental clones (using 53 SNPs) in Aceh and six parental clones (using 15 SSRs) in Sulawesi. Conserving the parental types of breeding programmes or accession groups should therefore be an essential component of a cacao collection.

Estimation of genetic diversity is influenced by sample size, sample composition of study, category of marker (e.g. SSR vs. SNP), number of markers employed, informativeness of marker used and scoring system employed (Table 5 and 6). Comparison of Cuba and Indonesia on-farm diversity (Table 5) shows that Indonesia had a higher gene diversity as H_e even though the total number of alleles and the average number of alleles per locus were higher in farmed cacao from Cuba. This indicated that areas with less admixed plants may still be a valuable source of alleles and that the types of germplasm moved into these areas may have been different. Farmed cacao from Cuba, Indonesia, Nicaragua and Peru, which was assessed with the same number and loci of SSRs, yielded higher but similar H_e for Indonesia and Peru, whereas Cuba and Nicaragua had lower but similar H_e estimates. Thus, proximity to the centre of diversity is not a guarantee of on-farm genetic diversity.

In general, on-farm diversity (as the total number of alleles and the number of alleles per locus) was less than that contained in *ex situ* gene banks, illustrating the usefulness of the

Table 5 On-farm cacao diversity in selected countries worldwide

Sampled area	Number of samples analysed	Number & type of marker system[1]	Range of alleles	Total alleles (average per locus)	mean H_e	Reference
Bolivia	72	9 a	n.a.[2]	n.a. (4.8)	0.68	de Schawe et al. (2013)
Cameroon	400	12 a	n.a.	113 (9.4)	0.50	Efombagn et al. (2009, 2008)
Cuba	537	15 a	6–17	139 (9.3)	0.42	Martínez et al. (2015)
Dominican Republic	115	14 a	3–11	113 (8.1)	0.62	Boza et al. (2013)
Ghana	235	17 a	n.a.	107 (6.3)	0.73	Opoku et al. (2007)
Honduras & Nicaragua	84	70 b	n.a.	n.a. (n.a.)	0.37	Ji et al. (2013)
India (Tamil Nadu)	27	10 a	1–3	23 (2.3)	n.a.	Thondaiman et al. (2013)
Indonesia (Aceh)	80	53 b	n.a.	n.a. (n.a.)	0.37	Lukman et al. (2014)
Indonesia (Sulawesi)	53	15 a	n.a.	130 (8.7)	0.67	Dinarti et al. (2015)
Soconusco, Mexico	108	12 a	n.a.	n.a. (3.54)	0.54	Vázquez-Ovando et al. (2014)
Nicaragua	70	10 a	7–15	95 (9.5)	0.60	Ruiz et al. (2011)
Nicaragua (Waslala)	317	15 a	n.a.	116 (7.7)	0.48	Trognitz et al. (2011)
Peru (Huallaga)	50	15 a	n.a.	n.a. (3.7)	0.61	Zhang et al. (2006a)[3]
West Africa	111 (rainforest vegetation zone Nigeria)	12 a	n.a.	99 (8.25)	0.62	Aikpokpodion et al. (2009)
West Africa	156 (Nigeria-Cameroon border)	12 a	n.a.	108 (9.00)	0.56	Aikpokpodion et al. (2009)

[1] a = SSRs, simple sequence repeats, microsatellites; b = SNPs, single nucleotide polymorphisms.
[2] Not available.
[3] Allelic richness from a rarefaction approach was used by Zhang et al. (2006a).

latter in conserving the genetic resources of cacao. Furthermore, the larger gene banks closer to the centre of diversity (IC3, ICGT and TARS) generally had the highest diversity measures (Table 6) except for H_e, which may be expected as the germplasm in outlying countries contains a larger amount of hybrid material. An exception was the Malaysian gene bank (Johnsiul and Awang 2016). Examination of their results revealed a very narrow bin size for the SSR alleles across all loci which would have resulted in the inflated estimates.

Table 6 Genetic diversity of selected cacao gene banks worldwide

Sampled area	Number of samples analysed	Number of SSRs[1]	Range of alleles	Total alleles (mean)	avgH_e[3]	Reference
CRIG, Ghana	142 Upper Amazon, Trinitario, Amelonado and hybrids	17	n.a.[2]	114 (6.7)	0.76	Opoku et al. (2007)
ERJOH, Pará, Brazil	94 CAB accessions	11	1–8	49 (4.5)	0.50	Sereno et al. (2006)
IC3, CATIE, Costa Rica	548 unique accessions	15	n.a.	231 (14.2)	0.51	Zhang et al. (2009b)
ICGT, Trinidad	345–8 Refractarios	15	2–7	63 (4.2)	0.56	Zhang et al. (2008)
ICGT, Trinidad	316 Upper Amazon from Pound collection	15	5–18	180 (12)	0.63	Zhang et al. (2009a)
ICGT + Criollo and Amelonado references	263 from Amelonado, Criollo, IMC, NA, PA, Refractario	24	5–14	204 (8.5)	0.67	Motilal et al. (2013)
IDIAF Mata Larga, Dominican Republic	66	14	4–12	117 (8.4)	0.71	Boza et al. (2013)
Indonesian National germplasm collection	84 accessions	15	2–15	114 (7.6)	0.77	Susilo et al. (2011)
MCGC, Malaysia	1000 accessions (includes synonyms)	15	15–23	273 (18.2)	0.73	Johnsiul and Awang (2016)
CRIN, Nigeria	250 accessions	12	14–24	218 (18.2)	0.77	Aikpokpodion et al. (2010)
TARS, Puerto Rico	139 unique accessions	15	5–12	132 (8.8)	0.66	Irish et al. (2010)
Tingo Maria, Peru	36 Ucayali accessions	15	n.a.	n.a. (5.7)	0.74	Zhang et al. (2006a)
WACRI, Ghana	108 1st generation progenies of crosses of Upper Amazon	12	n.a.	110 (9.2)	0.70	Aikpokpodion et al. (2009)
WACRI, Ghana	73 WACRI selections of 2nd generation progenies of Upper Amazon crosses	12	n.a.	96 (8.0)	0.60	Aikpokpodion et al. (2009)

[1] SSRs, simple sequence repeats, microsatellites.
[2] Not available.
[3] Average expected heterozygosity.

The molecular diversity of TARS cacao gene bank was less than that of the IC3 collection although the major alleles were similar between these two collections (Irish et al. 2010). The IC3 collection contained more alleles and had a greater number of alleles per locus than the ICGT, although the ICGT had more heterozygous individuals (Table 6). This may be due to the presence of more Criollo germplasm in the IC3 collection and the complete characterisation of the IC3 collection, whereas only a subset of the ICGT was fingerprinted. Similarly, Susilo et al. (2011) found that a sample of 84 accessions from the National Germplasm collection in Kaliwining Garden, Java, Indonesia, captured the major alleles of 15 SSRs as that of 546 accessions in the IC3 collection and the POUND accessions of the ICGT.

12 Core and minicore

As collection and acquisition efforts continue, the size of most cacao gene banks is expected to increase. Chang (1989) supported large numbers of accessions as valuable resources. The number of cacao accessions in *ex situ* gene banks does not surpass that of cotton, potato, rice or tomato (Table 1). It is also much less than the 41 813 accessions of barley held at Plant Gene Resources of Canada (Diederichsen 2016) or of the 28 000 maize and 150 000 wheat collections held in the gene bank of the International Maize and Wheat Improvement Center (CIMMYT 2016). However, for some of the cacao gene banks listed in Table 2, the number of accessions and hence the number of trees after taking the internal safety duplication into account may make the management of these gene banks problematic. This is especially so when resources are low, funding is challenged and there is limited land space to conserve the germplasm. Notwithstanding the lower number of cacao accessions in global field gene banks, the reduction in the size of germplasm holdings would be beneficial due to the reduction in annual costs for equipment maintenance and resupply, weed management, pruning, fertilising and labour. Management of a reduced subset becomes more efficient and accurate as DNA fingerprinting to verify genetic identities can be more quickly determined and there is a reduced cost to attaching metal name and location tags or barcode labels to the trees. A reduced subset also increases the opportunity for the replication of the core collection at multiple sites which would be advantageous for genotype × environment studies and reduces the risk of genetic erosion. The lowered acreage needed for the core set is also attractive to small island states in which land is in heavy demand for housing and agriculture. Resource allocation becomes more efficient in targeting a particular group of plants in a minimal spatial area. Finally, evaluation and characterisation of a core set of cacao accessions may become accelerated as data acquisition by trained technicians can be done for most or all of the accessions within a growing season.

The core collection concept was introduced by Frankel (1984) to represent the genetic diversity of a species and its wild relatives with minimal redundancy (Brown 1989a,b; Chang 1989; Frankel 1984). Brown (1995) defined a core as representing the entire collection and including as much of the diversity as possible. Core collections are therefore planned for users to have as much diversity as possible in a subset of the entire collection (Frison et al. 2003). Notwithstanding this, Brown and Spillane (1999) pointed out that, in a survey of 1346 various gene banks and institutions involved with conservation, there was considerable confusion on core collections. Core collections capture most of the available diversity in about 10% of the entire collection (Brown 1989a,b) without needing to capture the maximum absolute diversity (Upadhyaya et al. 2010). Sizes of 5–10% were optimal in capturing 75–90% of the diversity (Bisht et al. 1998; Charmet and Balfourier 1995). However, this means that up to 30% of the allelic diversity is unrepresented (Brown 1989a,b; Strauss et al. 1988) and the entire collection should therefore be conserved (Committee on Managing Plant Genetic Resources 1991). Brown (1989b) suggested that a core should not exceed 2000 entries, which agrees with the MVP (see Section 4). Van Hintum et al. (2000) found that most core collections were 5–20% of the target collection and that core size could be unique to the collection under analysis. Noirot et al. (1996) suggested that cores of 20–30% of the collection would be needed to capture the genetic diversity of traits of a quantitative nature. Genetic redundancy in the collection would however affect the actual optimum core size (Chavarriaga-Aguirre et al. 1999; Yonezawa et al. 1995). Cacao collections should therefore be assessed for genetic redundancy and reframed into its minimal actual holdings, before compiling a core collection.

Core collections, however, do not provide information for eliminating, bulking or combining accessions (Committee on Managing Global Genetic Resources 1991). Anthony and Dussert (2007) argued that core collections are intended to facilitate the use and distribution of conserved material. The Convention on Biological Diversity was open for signature in 1992 and, after several ratifications, entered into force on 29th December 1993. Van Hintum et al. (2000) stated that germplasm collected after the latter date is governed by different procedures, with regard to its distribution, than material collected before said date. Creation of a core collection would therefore be based not on really capturing the genetic diversity of the species (and its wild relatives) but on capturing the genetic diversity of the accessions in the public domain. Accessions that are not in the public domain, but which may be interesting or valuable to include in a core, would need to have the necessary support from the relevant parties and if agreed upon, suitable MTAs can be made for the distribution of these core germplasm accessions.

Although core collections reduce the number of accessions to a more manageable size, further reduction is often needed when a gene bank contains thousands of accessions. A 10% subset of the core forms the minicore (Upadhyaya and Ortiz 2001). The accessions in a minicore can then be intensely evaluated. The MVP may represent an upper threshold for a core collection.

Methods for constructing core collections have been described (Van Hintum et al. 2000; Schoen and Brown 1995; Brown and Schoen 1994; Brown 1989a) and have traditionally been constructed using multiple criteria (Spooner et al. 2005). The combined use of quantitative and qualitative data adequately constructed core collections of various plants (Kiers et al. 2000; Hamon et al. 1998). Comparing the core with the remainder of the collection (Shashidhara et al. 2003; Grenier et al. 2000; Huamán et al. 2000) or the entire collection with molecular markers should reveal no significant differences, indicating that the core is representative of the collection. Spooner et al. (2005) suggested that molecular markers would be unsuitable to form a core that maximised functional diversity. The inability of a core constructed solely on molecular markers, whether SSRs or SNPs, to capture phenotypic diversity in cacao was recently demonstrated (Motilal 2017).

Eskes (1995) proposed the establishment of a 'users' core collection containing disease resistance, pest resistance, quality and special physiological features. Based on this, Lockwood (1995) identified 50 accessions that could be used in a multilocational trial. Core collections for cacao have been developed using RAPDs (Marita et al. 2000). Ronning and Schnell (1994) suggested that a combination of eight polymorphic isozyme loci, geographical origin and morphological characters was the best for developing a core collection of cacao. Irish et al. (2010) found that 37 accessions could capture 90% of the SSR genetic diversity (measured by Shannon's Index) of the 139 unique accessions in the TARS cacao collection. However, the identities of the 37 accessions were not presented.

A set of 110 accessions was selected as part of the CFC/ICCO/Bioversity project collection (Soungio et al. 2006) and was transferred to ICQC,R between 2005 and 2010 (CacaoNet 2012). International evaluation of these accessions across multiple traits and their subsequent use would increase the visibility of this collection. At present, there is little evidence that this collection has become a core collection for cacao germplasm although it was supposed to represent 'a high level of diversity which would also be a rich source of favourable agronomic traits such as resistance to diseases and good bean and pod characteristics'. In addition, concerns as to the impact of mislabelling on the data collection were raised (Motilal et al. 2013). As an outcrosser, the majority of the genetic variation of cacao is contained within groups rather than among groups (Motilal et al. 2013; Zhang et al. 2008, 2009b, 2012; Motamayor et al. 2008; Sereno et al. 2006; Whitkus et al. 1998). The set of 110 could therefore have captured a suitable amount of the genetic diversity, even though the selection was done prior to the identification of the ten populations of cacao (Motamayor et al. 2008). Unfortunately, the extent of genetic capture compared to the ICGT or other collections was not determined and the set of 110 accessions therefore should not be considered as a core collection. Ten accessions were common to the sets identified by Lockwood (1995) and Sounigo et al. (2006): COCA 3348/44 [CHA], CCN 51, ICS 95, LCT EEN 241, LCT EEN 46, MA 12, MO 20 (or its misidentified counterpart in the ICGT), NA 33, PA 13 [PER] and POUND 7/B [POU] and should be kept in a core collection of cacao.

Motilal et al. (2013) identified a core set of 59 accessions that captured all the alleles of 24 SSR loci as a set of 414 accessions, and was genetically similar based on R_{ST} and D_{est} measures. In this core, five accessions (IMC 20; IMC 47_mis Field 6B Section F Plot 401 T3; IMC 94; MATINA 1/7 Field 6B D236 T15 and MOQ 6/95_mis Field 5B, Section C, Plot 221 T3) were also selected in Sounigo et al. (2006). Some trees selected by Motilal et al. (2013) for their allelic composition were different from their expected group. This was used as support for the retention of mislabelled trees in a field gene bank. The allelic composition can be such that it retained unique alleles or was able to combine in one individual several alleles that would otherwise have needed more than one accession represented in the core. Hybrid plants and misidentified plants offer the added advantage of potentially reducing the size of a core collection, thereby improving the versatility and utility of the collection. If we are interested in truly conserving the genetic diversity of cacao, then we should not be limited by whether population groups are captured or even adequately represented. The primary focus is whether the alleles, no matter where they are found, are well represented within the core collection. This focus is maintained whilst searching for the minimum number of plants to contain the alleles. If this approach is taken, then the evaluation of hybrid material, misidentified plants, relic varieties and breeding lines become a priority.

Motilal (2017) found that a subset of 123 accessions adequately represented a set of 400 accessions assessed with 95 SSR loci and that a subset of 24 accessions adequately represented a set of 338 accessions assessed with 775 SNP loci. However, these subsets did not adequately capture the phenotypic diversity leading to the suggestion of thematic core collections as has been earlier advocated (Odong et al. 2013; Glaszmann et al. 2010; Pessoa-Filho et al. 2010; and references therein). Since the number and type of marker used to assess genetic diversity may influence core selection (Motilal 2017), it is recommended that a universal set of DNA markers be identified and adopted by the international community. The panel of SNP markers of Motilal et al. (2017) could be a good starting point. The evaluation of all public domain germplasm or accessions that will be placed in the public domain as agreed upon from private or national collections with this panel would then have to be undertaken internationally. The dedication and consensus of parties sufficiently engaged in the conservation of cacao genetic resources are essential for this process. The introduction of elite lines or landraces that are deemed essential due to breeding and conservation efforts can then be iteratively integrated to avoid redundancy. If the core collection is viewed as a means to conserve the maximal genetic diversity whilst being useful for breeding activities, the process becomes efficient in conserving the least number of accessions with maximal genetic diversity and maximal utility.

The core collection so identified could then be placed as part of the universal collections, as well as any other collection with the requisite resources. This would allow for a safety duplication of the core collection. The core collection should be maintained as a discrete set planted in a dedicated site. Management decisions can then be made for the core set independent of the remaining collection. Although population stratification would need to be controlled, a core collection of adequate size could be used in a genome-wide association study to identify favourable alleles. A breeding programme involving genomic selection methods can then be used to identify the individuals with the best genotypic combination (Spindel et al. 2016).

13 Future trends and conclusion

In the early 2000s, the concept of a virtual collection (CacaoNet's Global Strategic Cacao Collection (GSCC); End et al. 2007) was raised. This virtual collection was proposed to consist of *T. cacao*, other *Theobroma* species and *Herrania* so as to cover the entire *Theobroma* gene pool. Emphasis was placed on cacao accessions in the public domain for long-term conservation, accessibility and use of the germplasm material to all current and future users. Priority accessions were to be selected from *ex situ* collections for allelic richness and a further dynamic set based on the traits of interest to breeders and farmers (CacaoNet 2012; Laliberté et al. 2012). A core of 261 accessions across the ten population groups was identified that reportedly captured the majority of the genetic and geographic diversity within worldwide *ex situ* collections (Table 7). However, the absence of Refractario accessions but the inclusion of Trinitario accessions appeared to be contradictory. Comparison of the accession names with the sample set of Motamayor et al. (2008) revealed that this dataset contained all but the ICS accessions, VILLANO 2 [CHA] and WILD of the GSCC core set. The last two accessions may have been VIL 2 (ARBOL 3) in the Curaray population and WILD which was in the Criollo population according to Motamayor et al. (2008). Since the latter did not include Refractario accessions, it seemed remiss to say that the core of the GSCC contained the majority of the genetic diversity held in *ex situ* collections especially since the Refractario accessions form a substantial part of the ICGT which was part of the GSCC study.

Storage of DNA is an alternative conservation strategy (Dulloo et al. 2006; Ebert et al. 2006; Andersson 2004; Callow et al. 1997; Ramanatha Rao and Riley 1994; Adams and Adams 1992). Accessions may not always be readily accessible especially from *in situ* conservation sites or be present on revisits due to genetic erosion in gene banks. In addition, the inefficiency and waste of resources in having to recollect and re-extract DNA

Table 7 Core collection from the Global Strategic Cacao Collection

Population	Number of accessions
Amelonado	15
Contamana	27
Criollo	13
Curaray	38
Guiana	10
Iquitos	23
Marañon	37
Nacional	17
Nanay	23
Purús	46
Trinitario	11
Unknown (likely Criollo)	1 (name given as 'WILD')
Total	261

Information obtained from Annex 7 in CacaoNet (2012) A Global Strategy for the Conservation and Use of Cacao Genetic Resources, as the Foundation for a Sustainable Cocoa Economy (B. Laliberté, compiler). Bioversity International, Montpellier, France.

are adequate reasons to have a DNA bank maintained alongside a gene bank collection. Ebert et al. (2006) outlined the opportunities of DNA banks which included the prospect of association genetics and improved gene bank management.

Cell cultures, tissues, RNA, nuclear DNA, organelle DNA, amplified DNA and genomic libraries (de Vicente et al. 2006; Dulloo et al. 2006; Ebert et al. 2006) are maintained. DNA can be stored suspended in a liquid medium, as a precipitate under molecular grade alcohol, in buffered solution at −80°C, in a lyophilised state at −80°C or flash frozen and maintained under liquid nitrogen. DNA in aqueous solution will degrade quickly at ambient temperature, so loss of electrical power or too low levels of liquid nitrogen in cryotanks is detrimental to DNA banks. Avoiding liquids and storing dry DNA on nitrocellulose can minimise degradation. DNA banks may be present in research centres that are closely aligned with cacao gene banks, as a consequence of DNA being extracted from accessions in the gene bank and stored for research purposes.

Even if stored dry or frozen, the DNA macromolecule can still be used as a voucher specimen for the accession from which it was extracted, once the plant is accepted to be the reference type for that accession name. DNA samples can then be sent out to other institutes who may need to confirm accession identities within their germplasm collection. The shipment of plant genomic DNA circumvents phytosanitary and quarantine issues (Ebert et al. 2006), including survival during shipment. National policies that inform on the use of DNA would however need to be considered. This is particularly relevant since international treaties for ownership, use and transfer of DNA may be variably interpreted (Graner et al. 2006; Gepts 2004; Beachy 2003; Feyt 2003; Resnik 2001).

DNA banks can be used as a repository for genomic work, as the DNA of many different accessions become readily available. After diversity analysis, differential genomes may be selected to undergo deep sequencing or whole genome sequencing. Whole genome sequencing is presently becoming more available either as an outsourced or autonomous event. Genotyping by sequencing is currently available and is expected to become more popular as costs decrease. McCouch et al. (2012) suggested that new samples should be sequenced to determine if new alleles or new combinations of alleles are present. Curators would then acquire only those samples that add novelty and thereby increase the molecular diversity of the collection. The systematic large-scale and deep sequencing of accessions in gene banks was recommended (McCouch et al. 2012). Fine-scale genotyping and the sharing of these genomic data would facilitate the identification of gene markers using association mapping approaches in different countries.

Fully sequenced plant genomes (Bevan and Uauy 2013; https://www.ncbi.nlm.nih.gov/genome/) are available. At present, two genomes of cacao have been published, one from MATINA 1/6 (Saski et al. 2011) and other from a Criollo accession from Belize (Argout et al. 2011). Recently, 14 Colombian cacao genomes were sequenced by CasaLuker (www.lukeringredients.com) and 250 cacao genomes, most from the public domain germplasm from the two universal collections, are underway (P. Umaharan 2016, pers. comm.). These genomes would become the reference-type genomes for these accessions. Identity and genetic diversity studies seeking to use a whole genome approach would then need to compare the genome of the accession of interest with that of the reference genomes. However, it would appear to be more practical to use a consensus SNP reference panel for identity and genetic diversity studies while using genotyping by sequencing or whole genome sequencing for dissecting genetic control of traits.

In DNA barcoding, sequenced DNA from a stipulated genomic area is compared. The *trnH-psbA* and *ITS* spacer regions (Kress et al. 2005) and the *rbcL* and *matK* regions (Fazekas et al. 2012) have been used for species identification of land plants. Motilal et al. (2017) raised the issue of discriminant intra-accession variation needed for identity resolution. Six segregating sites in the *trnH-psbA* intergenic spacer (Gutiérrez-López et al. 2016) were able to detect several haplotypes among a mixed cacao germplasm dataset but were unable to resolve all accessions. The use of additional genes may be able to differentiate within haplotype groups of the *trnH-psbA* region. If an agreed panel of SNPs is used, the ordered allelic states may also be used as the input data for a barcoding system. In this way, an SNP fingerprinting panel may be used in worldwide *ex situ* collections of cacao germplasm.

Cacao diversity is currently predominantly conserved in *ex situ* collections as field gene banks. A brief history of cacao collections which led to the existing germplasm collections is provided. General gene bank and conservation concepts are explained and applied based on the way that an accession is defined in cacao collections. Although *in situ* and *ex situ* conservation strategies are available, current cacao collections are predominantly *ex situ* field gene banks. Cacao genetic resources were partitioned into gene pools and the close relationship between the current cultivars and the wild *T. cacao* was highlighted. The importance of hybrids, natural as Trinitario and Refractario assemblages, or artificial as commercial hybrids, semi-wild or relic material in containing the genetic diversity and being vital to preserving unique combination of alleles, is discussed. The status of existing collections is compared and the issue of fidelity management to obtain an accurate count of accessions is raised. The extent of mislabelling was estimated at 34.5% for cacao collections. The relatedness of accessions and the number of unique accessions need to be determined for an accurate representation of the genetic diversity contained in a collection. $P_{ID(SIB)}$ of 1.48×10^{-13} was suggested as the maximum threshold to declare an identity match.

A proper comparison of the genetic diversity of collections was restricted because different types and numbers of markers were used by various researchers. However, although on-farm diversity was usually less than that of *ex situ* collections, and the larger collections contained a larger wealth of diversity, the presence of unique alleles in collections removed from the centre of diversity or in on-farm varieties was recognised. Collecting to augment existing diversity was recommended to be in wild populations stratified by river basins as well as from relic and on-farm varieties. The latter two were suggested to be good choices to obtain in few individuals a maximal allele set. A set of 2000 accessions was suggested as the minimum size to be maintained either within or among cacao collections if the genetic diversity of cacao is to be preserved. A core set of germplasm is dependent on the collection, and the core selection needs to be based on both molecular and phenotypic information. Conserving a set of germplasm for genetic diversity is recommended to be based on the allelic information with minimum individuals. The contribution of landraces, relic varieties and hybrid material in bringing together these alleles is highlighted. The definition of a core with a focus on its evaluation and use needs to be recognised and adopted so that the accessions from which the core is drawn from can be agreed upon.

14 Where to look for further information

Cacao and cocoa online

Alliance of Cocoa Producing Countries (COPAL; http://www.copal-cpa.org/)
Cocoa Genome Database (http://www.cacaogenomedb.org/)
CocoaGenDB (http://cocoagendb.cirad.fr/)
European Cocoa Association (http://www.eurococoa.com/)
Global Network for Cacao Genetic Resources (http://www.cacaonet.org/)
Heirloom Cacao Preservation (http://www.finechocolateindustry.org/hcp)
International Cocoa Germplasm Database (www.icgd.reading.ac.uk)
International Cocoa Organization (https://www.icco.org/)
Trinitario Cacao Resource (http://www.trinitario-cacao.org/)
TropGENE (http://tropgenedb.cirad.fr/tropgene/)
World Cocoa Foundation (http://www.worldcocoafoundation.org/)

Cacao books

Bartley, B. G. D. (2005), *The genetic diversity of cacao and its utilization*. CAB International, Wallingford, Oxfordshire.

CacaoNet. (2012), *A global strategy for the conservation and use of cacao genetic resources, as the foundation for a sustainable cocoa economy* (B. Laliberté, compiler). Bioversity International, Montpellier, France.

Eskes, A. B. (Ed.) (2010), *Collaborative and participatory approaches to cocoa variety improvement. Final report of the CFC/ICCO/Bioversity project on 'Cocoa Productivity and Quality Improvement: A Participatory Approach' (2004–2010)*. CFC, Amsterdam, The Netherlands/ICCO, London, UK/Bioversity International, Rome, Italy.

Dias, L. A. S. (2001), *Genetic improvement of cacao*. Translation by Abreu-Reichart, C. E., Viçosa, M. G., Brazil; aided by the Editor and FAO EcoPort version by Peter Griffee, FAO. http://ecoport.org/ep?SearchType=earticleView&earticleId=197

Wood, G. A. R. and Lass, R. A. (1985), *Cocoa*. 4th ed., Longman, London.

Cacao book chapters

Bhattacharjee, R. and Kumar, P. L. (2007), 'Cacao', in C. Kole (Ed.), *Genome mapping and molecular breeding in plants. Volume 6. Technical crops*, pp. 127–42. Springer Verlaag, Berlin. doi: 10.1007/978-3-540-34538-1_7.

Guiltinan, M. J., Verica, J., Zhang D. and Figueira, A. (2008), 'Genomics of *Theobroma cacao*, 'the Food of the Gods'', in P. H. Moore and R. Ming (Eds), *Genomics of Tropical Crop Plants*, pp. 145–70. Springer Verlag, New York.

Monteiro, W. R., Lopes, U. V. and Clement, D. (2009), 'Genetic improvement in cocoa', in S. M. Jain and P. M. Priyadarshan (eds), *Breeding Plantation Tree Crops: Tropical Species*, pp. 589–626. Springer Verlaag, New York.

Motamayor, J. C. and Lanaud C. (2002), 'Molecular analysis of the origin and domestication of *Theobroma cacao* L', in J. M. M. Engels, V. Ramanatha Rao, A. H. D. Brown and M. T. Jackson, (Eds), pp. 77–87. *Managing plant genetic diversity*. CABI Publishing, UK.

Zhang, D., Figueira, A., Motilal, L., Lachenaud, P. and Meinhardt, L. W. (2011), 'Theobroma', in C. Kole (Ed.), *Wild crop relatives: genomic and breeding resources. Plantation and ornamental crops*, pp. 277–96. Springer-Verlag, Berlin.

Zhang, D. and Motilal, L. (2016), 'Origin, dispersal, and current global distribution of cacao genetic diversity', in B. A. Bailey and L. W. Meinhardt (Eds), *Cacao Diseases*, pp. 3–30. Springer International Publishing, Switzerland. doi: 10.1007/978-3-319-24789-2_1

Cacao germplasm centres and research centres

IC3@CentroAgronómicoTropicaldeInvestigacióñyEnseñanza(https://www.catie.ac.cr/en/products-and-services/collections-and-germplasm- banks/international-cocoa-collection.html)

ICGT @ Cocoa Research Centre, The University of the West Indies (http://sta.uwi.edu/cru/GeneticResources.asp; www.cocoacentre.com)

International Cocoa Quarantine Centre, Reading (www.icgd.reading.ac.uk/quarantine.php)

Centre de coopération internationale en recherche agronomique pour le développement (CIRAD; http://www.cirad.fr/en)

United States Department of Agriculture. Agricultural Research Service (https://www.ars.usda.gov/)

DNA banks

Australian Plant DNA Bank (http://www.dnabank.com.au/)

National Botanical Institute (http://www.nbi.ac.za/research/dnabank.htm)

National Institute of Agrobiological Sciences (http://www.dna.affrc.go.jp/bank/index.html)

New York Botanical Garden (http://www.nybg.org/science/dna-bank.php)

The Royal Botanic Gardens, Kew (UK; http://www.kew.org/data/dnaBank/homepage.html), and US Missouri Botanical Gardens (http://www.mobot.org/MOBOT/research/applied_research/).

General germplasm information and research

Bioversity International (formerly IPGRI; https://www.bioversityinternational.org/e-library/)

Crop Trust (https://www.croptrust.org/)

Food and Agriculture Organization (http://www.fao.org/publications/en/)

CGIAR (http://library.cgiar.org/)

Journal articles

Cuatrecasas, J. (1964), 'Cacao and its allies: A taxonomic revision of the genus Theobroma', *Contri. U.S. Natl. Herb.*, 35, 379–614.

Kim, K.-W., Chung, H.-K., Cho, G.-T., Ma, K.-H., Chandrabalan, D., Gwag, J.-G., Kim, T.-S., Cho, E.-G. and Park, Y.-J. (2007), 'PowerCore: A program applying the advanced M strategy with a heuristic search for establishing core sets', *Bioinformatics*, 23, 2155–62. doi:10.1093/bioinformatics/btm313.

Krishnan, S., Ranker, T. A., Davis, A. P. and Rakotomalala, J. J. (2013), 'An assessment of the genetic integrity of *ex situ* germplasm collections of three endangered species of *Coffea* from Madagascar: Implications for the management of field germplasm collections', *Genet. Resour. Crop Evol.*, 60, 1021–36. doi:10.1007/s10722-012-9898-3.

Micheli, F., Guiltinan, M., Gramacho, K. P., Wilkinson, M. J., Figueira, A. V. O., Cascardo, J. C. M., Maximova, S. and Lanaud, C. (2010), 'Functional genomics of cacao', *Adv. Bot. Res.*, 55, 119–77.

Rice, R. A. and Greenberg, R. (2000), 'Cacao cultivation and the conservation of biological diversity', *Ambio*, 29(3), 167–73.

Van Hintum, T. J. L. (2000), 'Duplication within and between germplasm collections. III. A quantitativemodel',*Genet.Resour.CropEvol.*,47,507–13.doi:10.1023/A:1008703031415.

Van Hintum, T. J. L., and Van Treuren, R. (2002), 'Molecular markers: tools to improve genebank efficiency', *Cell. Mol. Biol. Lett.*, 7, 737–44.

Plant DNA barcoding

Smithsonian National Museum of Natural History Plant DNA Barcode Project (http://botany.si.edu/projects/dnabarcode/pubs.htm)

Plant genetic resources

Evenson, R. E, Gollin, D. and Santaniello, V. (Eds) (1998), *Agricultural values of plant genetic resources*, CABI Publishing, UK.

Guarino, L., Ramanatha Rao, V. and Reid, R. (Eds) (1995), *Collecting plant genetic diversity. Technical guidelines.* CAB International, UK.

Walton, M. (2016), *Germplasm: Characteristics, diversity and preservation.* Nova Science Pub. Inc.

15 Acknowledgements

The author thanks Ms. Rena K. Kalloo, Ms. Kadine David, Ms. Saila Ramkissoon and Mr Mitra Benny for suggestions that improved this document. Ms. Amrita Mahabir is thanked for sourcing and compiling the information framework, as well as editing the chapter.

16 References

Abdurakhmonov, I. Y., Buriev, Z. T., Shermatov, S. E., Abdullaev, A. A., Urmonov, K., Kushanov, F., Egamberdiev, S. S., Shapulatov, U., Abdukarimov, A., Saha, S., Jenkins, J. N., Kohel, J. R., Yu, J. Z., Pepper, A. E., Kumpatla, S. P. and Ulloa, M. (2012), 'Genetic diversity in *Gossypium* genus', In M. Caliskan (Ed.), *Genetic Diversity in Plants*, InTech, pp. 313–38. http://www.intechopen.com/books/genetic-diversity-inplants/genetic-diversity-in-gossypium-genus. doi:10.5772/35384.

Adams, R. P. and Adams, J. E. (1992), *Conservation of plant genes: DNA banking and in vitro biotechnology*. Academic Press, San Diego, USA.

Adu-Gyamfi, R. (2011), 'Safeguarding cocoa (*Theobroma cacao* L.) germplasm by cryopreservation: The vitrification approach', PhD Thesis, University of Reading, UK.

Ahtiainen, H. and Pouta, E. (2011), 'The value of genetic resources in agriculture: A meta-analysis assessing existing knowledge and future research needs', *Int. J. Biodivers. Sci., Ecosys. Ser. Manag.*, 7(1), 27–38, doi:10.1080/21513732.2011.593557.

Aikpokpodion, P. O. (2010), 'Variation in agro-morphological characteristics of cacao, *Theobroma cacao* L., in farmers' fields in Nigeria', *N. Z. J. Crop Hortic. Sci.*, 38(2), 157–70, doi:10.1080/0028825X.2010.488786.

Aikpokpodion, P. O. (2012), 'Defining genetic diversity in the chocolate tree, *Theobroma cacao* L. grown in West and Central Africa', In M. Caliskan (Ed.), *Genetic Diversity in Plants*, InTech, pp. 185–212, http://www.intechopen.com/books/genetic-diversity-in-plants/defining-geneticdiversity-in-the-chocolate-tree-theobroma-cacao-l-grown-in-west-and-central-africa. doi:10.5772/33101.

Aikpokpodion, P. O., Kolesnikova-Allen, M., Adetmirin, V. O., Guiltinan, M. J., Eskes, A. B., Motamayor, J. C. and Schnell, R. C. (2010), 'Population structure and molecular characterization of Nigerian field genebank collections of cacao, *Theobroma cacao* L.', *Silvae Genet.*, 59, 273–85.

Aikpokpodion, P. O., Motamayor, J. C., Adetimirin, V. O., Adu-Ampomah, Y., Ingelbrecht, I., Eskes, A. B., Schnell, R. J., and Kolesnikova-Allen, M. (2009), 'Genetic diversity assessment of sub-samples of cacao, *Theobroma cacao* L. collections in West Africa using simple sequence repeats marker', *Tree Genet. Genomes*, 5, 699–711, doi:10.1007/s11295-009-0221-1.

Albuquerque, A. C. and Nass, L. L. (2009), 'The state of use', In Da Silva Mariante A., Amstalden Sampaio M. J. and Valadares Inglis M.C. (Eds.), *State of the Brazil's Plant Genetic Resources. Second National Report*. EMBRAPA and FAO: Brasilia, DF, Brazil and Rome, Italy, pp. 54–89, Available online: www.fao.org/docrep/013/i1500e/Brazil.pdf.

Allard, R. W. (1970), 'Population structure and sampling methods', In O. H. Frankel and E. Bennett (Eds), *Genetic Resources in Plants: Their Exploration and Conservation*, F.A. Davis Company, Philadelphia, USA, pp. 97–107.

Allen, J. B. (1984), 'Strategies and methods for collecting *Theobroma*', *Plant Genet. Resour. Newsl.*, 57, 8–14.

Allen, J. B. (1987), 'London Cocoa Trade Amazon Project. Final report Phase 2', *Cocoa Growers' Bull.*, 39, 1–94.

Allen, J. B. and Lass, R. A. (1983), 'London Cocoa Trade Amazon Project. Final report phase 1', *Cocoa Growers' Bull.*, 34, 1–71.

Almeida, C. M. V. C. de, Barriga, J. P., Machado, P. F. R. and Bartley, B. G. D. (1987), 'Evolução do programa de conservação dos recursos genéticos de cacao na Amazônia Brasileira', *Boletim Técnico*, 5, Comissão Executiva do Plano da Lavoura Cacaueira, Belém, Brasil.

Altieri, M. A. and Merrick, L. C. (1987), '*In situ* conservation of crop genetic resources through maintenance of traditional farming systems', *Econ. Bot.*, 41(1), 86–96, doi:10.1007/BF02859354.

Andersson, M. S. (2004), *Feasibility Study for DNA Storage and Use: DNA Banks – Providing Novel Options for Genebanks? A Worldwide Survey*, IPGRI Report No. 6900/D10A, International Plant Genetic Resources Institute (IPGRI), Rome, Italy, 38p.

Anon. (2007), 'INGENIC survey on the development of CacaoNet conservation strategy summary of replies of the opinions received 1st June 2007', *INGENIC Newsl.*, 11, 32–6.

Anthony, F. and Dussert, S. (2007), 'Construction of coffee core collections', In F. Engelmann, M. E. Dulloo, C. Astorga, S. Dussert and F. Anthony (Eds), *Conserving Coffee Genetic Resources. Complementary Strategies for ex situ Conservation of Coffee (Coffea arabica L.) Genetic Resources. A Case Study in CATIE, Costa Rica*, Topical reviews in Agricultural Biodiversity. Bioversity International, Rome, Italy, pp. 45–8, http://www.bioversityinternational.org/e-library/publications/detail/conserving-coffee-genetic-resources/.

Argout, X., Salse, J., Aury, J. M., Guiltinan, M. J., Droc, G., Gouzy, J., Allegre, M., Chaparro, C., Legavre, T., Maximova, S. N., Abrouk, M., et al. (2011), 'The genome of *Theobroma cacao*', *Nat. Genet.*, 43(2), 101-8, doi:10.1038/ng.736.

Aspinall, A. (1928), *A wayfarer in the West Indies*, Methuen & Co., London.
Baker, R. E. D., Cope, F. W., Holliday, P. C., Bartley, B. G. and Taylor, D. J. (1954), 'The Anglo-Colombian cacao-collecting expedition', In *A Report on Cacao Research 1953*, Imperial College of Tropical Agriculture, Trinidad, pp. 8–29.
Bartley, B. G. D. (1984), 'Genetic resources programme of the Cocoa Research Unit, University of the West Indies, St Augustine, Trinidad. Consultant's Report', In *Report for 1981–1983*, Cocoa Research Unit, The University of the West Indies, St. Augustine, Trinidad, pp. 13–16.
Bartley, B. G. D. (2001), 'Refractario – an explanation of the meaning of the term and its relationship to the introductions from Ecuador in 1937', *INGENIC Newsl.*, 5, 4–6.
Bartley, B. G. D. (2005), *The Genetic Diversity of Cacao and Its Utilization*, CAB International, Wallingford, Oxfordshire.
Barton, J. H. and Siebeck, W. E. (1994), 'Material transfer agreements in genetic resources exchange - the case of the International Agricultural Research Centres', In *Issues in Genetic Resources No. 1*, International Plant Genetic Resources Institute, Rome, Italy, http://www.bioversityinternational.org/e-library/publications/detail/material-transfer-agreements-in-genetic-resources-exchange/.
Beachy, R. N. (2003), 'IP policies and serving the public', *Science*, 299, 473, doi:10.1126/science.299.5606.473.
Bekele, F. L., Bekele, I., Butler, D. R. and Bidaisee, G. G. (2006), 'Patterns of morphological variation in a sample of cacao (*Theobroma cacao* L.) germplasm from the International Cocoa Genebank, Trinidad', *Genet. Resour. Crop Evol.*, 53(5), 933–48.
Bevan, M. W. and Uauy, C. (2013), 'Genomics reveals new landscapes for crop improvement', *Genome Biol.*, 14, 206, doi:10.1186/gb-2013-14-6-206.
Bisht, I. S., Mahajan, R. K., Lokknathan, T. R. and Agrawal, R. C. (1998), 'Diversity in Indian sesame collection and stratification of germplasm accessions in different diversity groups', *Genet. Resour. Crop Evol.*, 45, 325–35, doi:10.1007/s10722-004-6692-x.
Bloomfield, E. M. and Lass, R. A. (1992), *Impact of Structural Adjustment and Adoption of Technology on Competitiveness of Major Cocoa Producing Countries*, OECD DEVELOPMENT CENTRE Working Paper No. 69 (Formerly Technical Paper No. 69) OCDE/GD(92)120, OECD Publishing, Paris, http://www.oecd-ilibrary.org/development/impact-of-structural-adjustment-and-adoption-of-technology-on-competitiveness-of-major-cocoa-producing-countries_275313772128, doi:10.1787/275313772128.
Botstein, D., White, R. L., Skolnick, M., and Davis, R. W. (1980), 'Construction of a genetic linkage map in man using restriction fragment length polymorphisms', *Am. J. Hum. Genet.*, 32, 314–31.
Boza, E. J., Irish, B. M., Meerow, A. W., Tondo, C. L., Rodríguez, O. A., Ventura-López, M., Gómez, J. A., et al. (2013), 'Genetic diversity, conservation, and utilization of *Theobroma cacao* L.: Genetic resources in the Dominican Republic', *Genet. Resour. Crop Evol.*, 60(2), 605-19, doi:10.1007/s10722-012-9860-4.
Brenner, C. (1992), *Biotechnology and the Changing Public/Private Sector Balance: Developments in Rice and Cocoa*, OECD Development Centre Working Paper No. 72, OECD Publishing, Paris, doi:10.1787/324280502022.
Brown, A. H. D. (1989a), 'Core collections: A practical approach to genetic resources management', *Genome*, 31(2), 818–24, doi:10.1139/g89-144.
Brown, A. H. D. (1989b), 'The case for core collections', In A. H. D. Brown, O. H. Frankel, D. R. Marshall and J. T. Williams (Eds), *The Use of Plant Genetic Resources*, Cambridge Univ. Press, Cambridge, pp. 136–56.
Brown, A. H. D. (1995), 'The core collection at the crossroads', In T. Hodgkin, A. H. D. Brown, Th. J. L. Van Hintum and E. A. V. Morales (Eds), *Core Collections of Plant Genetic Resources*, John Wiley and Sons, UK, pp. 3–19.
Brown, A. H. D. and Marshall, D. R. (1995), 'A basic sampling strategy: Theory and practice', In L. Guarino, V. R. Rao and R. Reid (Eds), *Collecting Plant Genetic Diversity, Technical Guidelines*, CAB International, Wallingford, UK, pp. 75–91, http://www.bioversityinternational.org/e-library/publications/detail/collecting-plant-genetic-diversity/.

Brown, A. H. D. and Schoen, D. J. (1994), 'Optimal sampling strategies for core collections of plant genetic resources', In V. Loeschke, J. Tomiuk and S. K. Jain (Eds), *Conservation Genetics*, Birkhäuser Verlag Basel, Switzerland, pp. 357–69.

Brown, A. H. D. and Spillane, C. (1999), 'Implementing core collections – principles, procedures, progress, problems and promise', In R. C. Johnson and T. Hodgkin (Eds), *Core Collections for Today and Tomorrow*, IPGRI, Rome, Italy, pp. 1–9, http://www.bioversityinternational.org/e-library/publications/detail/core-collections-for-today-and-tomorrow/.

Brush, S. B. (1986), 'Genetic diversity and conservation in traditional farming systems', *J. Ethnobiol.*, 6(1), 151–67.

Brush, S. B. (1995), '*In situ* conservation of landraces in centres of crop diversity', *Crop Sci.*, 35(2), 346–54, doi:10.2135/cropsci1995.0011183X003500020009x.

CacaoNet. (2012), *A Global Strategy for the Conservation and Use of Cacao Genetic Resources, as the Foundation for a Sustainable Cocoa Economy* (B. Laliberté, compiler), Bioversity International, Montpellier, France, http://cacaonet.org/.

Callow, J. A., Ford-Lloyd, B. V. and Newbury, H. J. (Eds), (1997), *Biotechnology and Plant Genetic Resources: Conservation and Use*. Biotechnology in Agriculture Series 19, CAB International, Wallingford, UK.

Caye, K., Deist, T. M., Martins, H., Michel, O. and Francois, O. (2016), 'TESS3: Fast inference of spatial population structure and genome scans for selection', *Mol. Ecol. Resour.*, 16(2), 540–8, doi:10.1111/1755-0998.

Chalmers, W. S. (1972), 'The conservation of wild cacao populations: The plant breeders most urgent task', In *Proc. Fourth Intl. Cocoa Res. Conf.*, Government of Trinidad and Tobago, West Indies, pp. 114–18.

Chang, T. T. (1989), 'The case for large collections', In A. D. H. Brown, O. H. Frankel, D. R. Marshall and J. T. Williams (Eds), *The Use of Plant Genetic Resources*, Cambridge University Press, New York, pp. 123–35.

Charmet, G. and Balfourier, F. (1995), 'The use of geostatistics for sampling a core collection of perennial ryegrass populations', *Genet. Resour. Crop Evol.*, 42(4), 303–9, doi:10.1007/BF02432134.

Chaudhury, R. and Malik, S. K. (2004), 'Genetic conservation of plantation crops and spices using cryopreservation', *Indian J. Biotechnol.*, 3, 348–58, http://nopr.niscair.res.in/handle/123456789/5866.

Chavarriaga-Aguirre, P., Maya, M. M., Tohme, J., Duque, M. C., Iglesias, C., Bonierbale, M. W., Kresovich, S. and Kochert, G. (1999), 'Using microsatellites, isozymes and AFLPs to evaluate genetic diversity and redundancy in the cassava core collection and to assess the usefulness of DNA-based markers to maintain germplasm collections', *Mol. Breed.*, 5(3), 263–73, doi:10.1023/A:1009627231450.

Cheesman, E. E. (1932), *The Economic Botany of Cacao. A Critical Survey of the Literature to the End of 1930. Supplement to Tropical Agriculture*. Government Printing Office, Port-of-Spain, Trinidad.

Cheesman, E. E. (1944), 'Notes on the nomenclature, classification and possible relationships of cocoa populations', *Trop. Agric.*, 21, 144–59.

Christopher, Y., Mooleedhar, V., Bekele, F. and Hosein F. (1999), 'Verification of accession in the ICG,T using botanical descriptors and RAPD analysis', In *Annual Report 1998*, Cocoa Research Unit, The University of the West Indies, St. Augustine, Trinidad and Tobago, pp. 15–18.

Ciferri, R., and Ciferri, F. (1957), 'The evolution of cultivated cacao', *Evolution*, 11(4), 381–97, doi:10.2307/2406059.

CIMMYT. (2016), *Building Resilience to Risk. CIMMYT Annual Report 2015*. CIMMYT, CDMX, Mexico.

Cipriani, G., Marrazzo, M. T., Di Gaspero, G., Pfeiffer, A., Morgante, M. and Testolin, R. (2008), 'A set of microsatellite markers with long core repeat optimized for grape (*Vitis* spp.) genotyping', *BMC Plant Biol.*, 8, 127, doi:10.1186/1471-2229-8-127.

Cohen, J. I., Williams, J. T., Plucknett, D. L. and Shands, H. (1991), 'Ex situ conservation of plant genetic resources: Global development and environmental concerns', *Science*, 253(5022), 866–72, doi:10.1126/science.253.5022.866.

Committee on Germplasm Resources. (1978), *Conservation of Germplasm Resources. An Imperative*, National Academy of Sciences, Washington, DC, https://www.nap.edu/catalog/19943/conservation-of-germplasm-resources-an-imperative.

Committee on Managing Global Genetic Resources. (1991), *Managing Global Genetic Resources. The U.S. National Plant Germplasm System*. National Academy Press, Washington, DC, https://www.nap.edu/catalog/1583/the-us-national-plant-germplasm-system.

Cope, F. W. (1976), 'Cacao, *Theobroma cacao* (Sterculiaceae)', In N. W. Simmonds (Ed.), *Evolution of Crop Plants*, Longman, London, pp. 285–89.

Corander, J., Marttinen, P., Sirén, J. and Tang, J. (2008), 'Enhanced Bayesian modelling in BAPS software for learning genetic structures of populations', *BMC Bioinformatics*, 9, 539, doi:10.1186/1471-2105-9-539.

Cruz-Cruz, C. A., González-Arnao, M. T. and Engelmann, F. (2013), 'Biotechnology and conservation of plant biodiversity', *Resources*, 2, 73–95, doi:10.3390/resources2020073.

Cryer, N. C., Fenn, M. G. E., Turnbull, C. J. and Wilkinson, M. J. (2006), 'Allelic size standards and reference genotypes to unify international cocoa (*Theobroma cacao* L.) microsatellite data', *Genet. Resour. Crop Evol.*, 53(8), 1643–52, doi:10.1007/s10722-005-1286-9.

Culver, M., Fitak, R. and Herrmann, H. (2011), 'Genetic methods for biodiversity assessment', In A. E. Magurran and B. J. McGill (Eds), *Biological Diversity: Frontiers in Measurement and Assessment*, Oxford University Press, Oxford, pp. 175–93.

Damania, A. B. (1996), 'Biodiversity conservation: A review of options complementary to standard *ex situ* methods', *Plant Genet. Resour. Newsl.*, 107, 1–18.

da Silva, T. E. R., Cidade, L. C., Alvim, F. C., Cascardo, J. C. M. and Costa, M. G. C. (2008), 'Somatic embryogenesis and plant regeneration in elite clones of *Theobroma cacao*', *Pesq. Agropec. Bras., Brasília*, 43(10), 1433–6.,

Davie, J. H. (1935), 'Chromosome studies in the malvaceae and certain related families. II', *Genetica*, 17, 487–98.

de Andres, M. T., Cabezas, J. A., Cervera, M. T., Borrego, J., Martínez-Zapater, J. M. and Jouve, N. (2007), 'Molecular characterization of grapevine rootstocks maintained in germplasm collections', *Am. J. Enol. Vitic.*, 58, 75–86, http://www.ajevonline.org/content/58/1/75.

de Boef, W. S., Subedi, A., Peroni, N., Thijssen, M., and O'Keeffe, E. (Eds). (2013), *Community Biodiversity Management: Promoting Resilience and the Conservation of Plant Genetic Resources*, Routledge, New York, USA. https://www.bioversityinternational.org/e-library/publications/detail/community-biodiversity-management.

de Schawe, C. C., Durka, W., Tscharntke, T., Hensen, I. and Kessler, M. (2013), 'Gene flow and genetic diversity in cultivated and wild cacao (*Theobroma cacao*) in Bolivia', *Am. J. Bot.*, 100(11), 2271–9, doi:10.3732/ajb.1300025.

de Vicente, M. C., Andersson, M. S. and Engels, J. (2006), 'Introduction and overview', In M. C. de Vicente and M. S. Andersson (Eds), *DNA Banks – Providing Novel Options for Genebanks?*, Topical Reviews in Agricultural Biodiversity, International Plant Genetic Resources Institute, Rome, Italy, pp. 1–5, http://www.bioversityinternational.org/e-library/publications/detail/dna-banks-providing-novel-options-for-genebanks/.

Dias, L. A. S. and Almeida, C. M. V. C. (2002), 'Cacao germplasm: How and how much to collect?', *INGENIC Newsl.*, 7, 8–11.

Diederichsen, A. (2016), '*Ex situ Germplasm Conservation by Plant Gene Resources of Canada as a Complement to On-farm Conservation in Indigenous Agriculture*'. Available online 9 July 2017. https://chaire-diversite-alimentaire.ulaval.ca/wp-content/uploads/2016/11/diederichsen-quebec_workshop_2016.pdf.

Dinarti, D., Susilo, A. W., Meinhardt, L. W., Ji, K., Motilal, L. A., Mischke S. and Zhang D. (2015), 'Genetic diversity and parentage in farmer selections of cacao from Southern Sulawesi, Indonesia revealed by microsatellite markers', *Breed. Sci.*, 65(5), 438–46, doi:10.1270/jsbbs.65.438.

Doodnath, R. (1996), 'A study of intergeneric hybridization between *Theobroma cacao* L. and *Herrania species*', MPhil thesis, The University of the West Indies, St. Augustine, Trinidad.
Drenth, A. and Guest, D. I. (2016), 'Fungal and oomycete diseases of tropical tree fruit crops', *Annu. Rev. Phytopathol.*, 54, 373–95, doi:10.1146/annurev-phyto-080615-095944.
Dring, D. M. (1976), 'The recent role of the Royal Botanic Gardens, Kew, in distribution of cocoa', *Cocoa Growers' Bull.*, 24, 12–14.
Dublin, P. (1972), 'Polyembryonie et haploidie chez *Theobroma cacao*', *Café Cacao Thé*, 26(4), 295–311.
Dulloo, E., Nagamura, Y. and Ryder, O. (2006), 'DNA storage as a complementary conservation strategy', In M. C. de Vicente and M. S. Andersson (Eds), *DNA Banks – Providing Novel Options for Genebanks?*, Topical Reviews in Agricultural Biodiversity, International Plant Genetic Resources Institute, Rome, Italy, pp. 11–24, http://www.bioversityinternational.org/e-library/publications/detail/dna-banks-providing-novel-options-for-genebanks/.
Ebert, A. W. and Chou, Y. Y. (2015), 'The tomato collection maintained by AVRDC-The World Vegetable Center: Composition, germplasm dissemination and use in breeding', In *XXIX International Horticultural Congress on Horticulture: Sustaining Lives, Livelihoods and Landscapes (IHC2014): IV 1101*, pp. 169–76, http://www.actahort.org/books/1101/1101_26.htm.
Ebert, A. W., Karihaloo, J. L. and Ferreira, M. E. (2006), 'Opportunities, limitations and needs for DNA banks', In M. C. de Vicente, and M. S. Andersson (Eds), *DNA Banks – Providing Novel Options for Genebanks?*, Topical Reviews in Agricultural Biodiversity, International Plant Genetic Resources Institute, Rome, Italy, pp. 61–8, http://www.bioversityinternational.org/e-library/publications/detail/dna-banks-providing-novel-options-for-genebanks/.
Efombagn, I. B. M., Motamayor, J. C., Sounigo, O., Eskes, A. B., Nyassé, S. Cilas, C., Schnell, R., Manzanares-Dauleux, M. J. and Kolesnikova-Allen, M. (2008), 'Genetic diversity and structure of farm and GenBank accessions of cacao (*Theobroma cacao* L.) in Cameroon revealed by microsatellite markers', *Tree Genet. Genom.*, 4(4), 821–3, doi:10.1007/s11295-008-0155-z.
Efombagn, M. I. B, Sounigo, O., Eskes, A. B., Motamayor, J. C., Manzanares-Dauleux, M. J., Schnell, R. and Nyassé, S. (2009), 'Parentage analysis and outcrossing patterns in cacao (*Theobroma cacao* L.) farms in Cameroon', *Heredity*, 103(1), 46–53, doi:10.1038/hdy.2009.30.
End, M., Engels, J. and Eskes, B. (2007), 'Discussion paper on germplasm to be considered in setting up a long-term global cacao conservation strategy by CacaoNet', *INGENIC Newsl.*, 11, 37–9.
Engelmann, F. and Dulloo, M. E. (2007), 'Introduction', In F. Engelman, M. E. Dulloo, C. Astorga, S. Dussert and F. Anthony (Eds), *Complementary Strategies for ex situ Conservation of Coffee (Coffea arabica L.) Genetic Resources. A Case study in CATIE, Costa Rica*, Topical reviews in Agricultural Biodiverstity, Bioversity International, Rome, Italy, pp. 1–11, http://www.bioversityinternational.org/e-library/publications/detail/conserving-coffee-genetic-resources/.
Engelmann, F. and Engels, J. M. M. (2002), 'Technologies and strategies for *ex situ* conservation', In J. M. M. Engels, V. Ramanatha Rao, A. H. D. Brown and M. T. Jackson (Eds), *Managing Plant Genetic Diversity*, IPGRI, CABI Publishing, UK, pp. 89–103, http://www.bioversityinternational.org/e-library/publications/detail/managing-plant-genetic-diversity/.
Engels, J. M. M. (1984), 'A systematic description of cacao clones IV. Some evidences of tetraploid inheritance', *Café Cacao Thé*, 28(2), 51–7.
Engels, J. M. M. and Fassil, H. (2004), 'Plant and animal genebanks', In V. R. Squires (Ed.), *The Role of Food, Agriculture, Forestry and Fisheries in Human Nutrition*, Encyclopedia of Life Support Systems (EOLSS), Developed under the Auspices of the UNESCO, Eolss Publishers, Oxford, UK, http://www.eolss.net.
Engels, J. M. M., Bartley, B. G. D. and Enriquez, G. A. (1980), 'Cacao descriptors, their states and modus operandi', *Turrialba*, 30, 209–18.
Engels, J. M. M., Ramanatha Rao, V., Brown, A. H. D. and Jackson, M. T. (Eds) (2002), *Managing Plant Genetic Diversity*. IPGRI, CABI Publishing, UK, http://www.bioversityinternational.org/e-library/publications/detail/managing-plant-genetic-diversity/.
Eskes, A. (1995), 'International cooperation in cocoa breeding', *INGENIC Newsl.*, 1, 4.

Evans, H. C. (2016), 'Frosty pod rot (*Moniliophthora roreri*)', In B. A. Bailey and L. W. Meinhardt (Eds), *Cacao Diseases: A History of Old Enemies and New Encounters*, Springer International Publishing, Switzerland, pp. 63–96, doi:10.1007/978-3-319-24789-2_3.

Evett, I. W. and Weir, B. S. (1998), *Interpreting DNA evidence: Statistical Genetics for Forensic Scientists*. Sinauer, Sunderland, MA.

Faleiro, F. G., Yamada, M. M., Lopes, U. V., Faleiro, A. S. G., Bahia, R. C. S., Gomes, L. M. C., Santos, R. C. and dos Santos, R. F. (2002), 'Genetic similarity of *Theobroma cacao* L. accessions maintained in duplicate at the cacao research centre germplasm collection based on RAPD markers', *Crop Breed. Appl. Biotechnol.*, 2(3), 439–44, doi:10.12702/1984-7033.v02n03a15.

Fang, J.-Y., Wetten, A. and Hadley, P. (2004), 'Cryopreservation of cocoa (*Theobroma cacao* L.) somatic embryos for long-term germplasm storage', *Plant Sci.*, 166(3), 669–75, doi:10.1016/j.plantsci.2003.11.002.

Fang, J.-Y., Wetten A., Adu-Gyamfi, R., Wilkinson, M. and Rodriguez-Lopez, C. (2009), 'Use of secondary somatic embryos promotes genetic fidelity in cryopreservation of cocoa (*Theobroma cacao* L.)', *Agric. Food Sci.*, 18(2), 152–9, doi:10.2137/145960609789267579.

FAO. (2006), *Livestock's Long Shadow: Environmental Issues and Options*, Food and Agriculture Organization of the United Nations, Rome, http://www.fao.org/docrep/010/a0701e/a0701e00.HTM.

FAO. (2014), *Genebank Standards for Plant Genetic Resources for Food and Agriculture*. Rev. ed. FAO, Rome, http://www.fao.org/3/a-i3704e.pdf.

Fazekas, A. J., Kuzmina, M. L., Newmaster, S. G. and Hollingsworth, P. M. (2012), 'DNA barcoding methods for land plants', In W. J. Kress and D. L. Erickson (Eds), *DNA Barcodes: Methods and Protocols, Methods in Molecular Biology (Methods and Protocols)*, 858, 223–52, Humana Press, Totowa, NJ, doi:10.1007/978-1-61779-591-6_11.

Feyt, H. (2003), 'Intellectual property rights on varieties and access to plant genetic resources: Towards a coherent and an ethically based approach', *Plant Genet. Res.*, 1, 97–102, doi:10.1079/pgr200316.

Figueira, A. (1998), 'Homonymous genotypes and misidentification in germplasm collections of Brazil and Malaysia', *INGENIC Newslett.*, 4, 4–8. http://ingenic.cas.psu.edu/newsletters.htm.

Fischer, M. C., Rellstab, C., Leuzinger, M., Roumet, M., Gugerli, F., Shimizu, K. K., Holderegger, R. and Widmer, A. (2017), 'Estimating genomic diversity and population differentiation – an empirical comparison of microsatellite and SNP variation in *Arabidopsis halleri*', *BMC Genomics*, 18, 69, doi:10.1186/s12864-016-3459-7.

Florin, B., Brulard, E. and Lepage, B. (1999), 'Establishment of a cryopreserved coffee germplasm bank', *Cryobiology*, 39, 309.

Frankel, O. H. (1984), 'Genetic perspectives of germplasm conservation', In W. K. Arber, K. Llimensee, W. J. Peacock and P. Starlinger (Eds), *Genetic Manipulation: Impact on Man and Society Part III, Paper No. 15*, Cambridge University Press, Cambridge, pp. 161–70.

Frankel, O. H. and Soulé, M. E. (1981), *Conservation and Evolution*. Cambridge University Press, Cambridge.

Frankel, O. H., Brown, A. H. D. and Burdon, J. J. (1995), *The Conservation of Plant Biodiversity*. Cambridge University Press, Cambridge, UK.

Frison, E., Mitteau, M., Sharrock, S. and Visser, B. (2003), 'Sharing responsibilities', In J. M. M. Engels. and L. Visser, (Eds), *A Guide to Effective Management of Germplasm Collections, IPGRI Handbooks for Genebanks No. 6*, IPGRI, Rome, Italy, pp. 107–21, http://www.bioversityinternational.org/e-library/publications/detail/a-guide-to-effective-management-of-germplasm-collections/.

Gao, X. and Starmer, J. D. (2008), 'AWclust: Point-and-click software for non-parmetric population structure analysis', *BMC Bioinform.* 9, 77, doi:10.1186/1471-2105-9-77.

Garcia, R. M., Parra-Quijano, M. and Iriondo, J. M. (2017a), 'A multispecies collecting strategy for crop wild relatives based on complementary areas with a high density of ecogeographical gaps', *Crop Sci.*, 57, 1059–69, doi:10.2135/cropsci2016.10.0860.

Garcia, R. M., Parra-Quijano, M. and Iriondo, J. M. (2017b), 'Identification of ecogeographical gaps in the Spanish *Aegilops* collections with potential tolerance to drought and salinity', *Peer J.*, 5, e3494, doi:10.7717/peerj.3494.

Gepts, P. (2004), 'Who owns biodiversity, and how should the owners be compensated?', *Plant Physiol.*, 134(4), 1295–307, doi:10.1104/pp.103.038885.

Gepts, P. (2006), 'Plant genetic resources conservation and utilization: The accomplishments and future of a societal insurance policy', *Crop Sci.*, 46(5), 2278–92, doi:10.2135/cropsci2006.03.0169gas.

Glaszmann, J. C., Kilian, B., Upadhyaya, H. D. and Varshney, R. K. (2010), 'Accessing genetic diversity for crop improvement', *Curr. Opin. Plant Biol.*, 13(2), 1–7, doi:10.1016/j.pbi.2010.01.004.

Glendinning, D. R. (1967), 'New cocoa varieties in Ghana', *Cocoa Growers' Bull.*, 8, 19–24.

Gonsalves, C. (1996), 'History of cocoa breeding in the Ministry of Agriculture, Trinidad', *Cocoa Research Unit Newsl.*, 3, 4–6.

Graner, A., Andersson, M. S. and de Vicente, M. C. (2006), 'A model for DNA banking to enhance the management, distribution and use of *ex situ* stored PGR', In M. C. de Vicente and M. S. Andersson, (Eds), *DNA Banks – Providing Novel Options for Genebanks?*, Topical Reviews in Agricultural Biodiversity, International Plant Genetic Resources Institute, Rome, Italy, pp. 69–75, https://www.bioversityinternational.org/e-library/publications/detail/dna-banks-providing-novel-options-for-genebanks/.

Greenbaum, G., Templeton, A. R., Zarmi, Y. and Bar-David, S. (2014), 'Allelic richness following population founding events – a stochastic modeling framework incorporating gene flow and genetic drift', *PLoS ONE* 9(12), e115203, doi:10.1371/journal.pone.0115203.

Grenier, C., Deu, M., Kresovich, S., Bramel-Cox, P. J. and Hamon, P. (2000), 'Assessment of genetic diversity in three subsets constituted from the ICRISAT Sorghum collection using random vs. non-random sampling procedures. B. Using molecular markers', *Theor. Appl. Genet.*, 101(1–2), 197–202, doi:10.1007/s001220051469.

Gutiérrez-López N., Ovando-Medina, I., Figueroa, M. S., Molina-Freaner, F., Avendaño-Arrazate, C. H. and Vázquez-Ovando, A. (2016), 'Unique haplotypes of cacao trees as revealed by trnH-psbA chloroplast DNA', *Peer J.*, 4, e1855, doi:10.7717/peerj.1855.

Hamon, S., Dussert, S., Deu, M., Hamon, P., Seguin, M., Glaszmann, J. C., Grivet, L., Chantereau, J., Chevallier, M. H., Flori, A., Lashermes, P., Legnate, H. and Noirot, M. (1998), 'Methodologies de gestion et de conservation des ressources genetiques', *Genet. Sel. Evol.*, 30, S237–S258 (suppl.).

Hayward, M. D. and Sackville Hamilton, N. R. (1997), 'Genetic diversity – population structure and conservation', In J. A. Callow, B. V. Ford-Lloyd and H. J. Newbury (Eds), *Biotechnology and Plant Genetic Resources: Conservation and Use*, Biotechnology in Agriculture Series 19, CAB International, Wallingford, UK, pp. 49–76.

Hedrick, P. (2005), 'A standardized genetic differentiation measure', *Evolution*, 59(8), 1633–8, doi:10.1111/j.0014-3820.2005.tb01814.x.

Holliday, P. (1957), 'Spread of pod rot of cacao', *Commonwealth Phytopathol. News*, 3(1), 12.

Huamán, Z., Ortiz, R. and Gomez, R. (2000), 'Selecting a *Solanum tuberosum* subsp. *andigena* core collection using morphological, geographical, disease and pest descriptors', *Am. J. Potato Res.*, 77, 183–90, doi:10.1007/bf02853943.

Hunter, J. R. (1990), 'The status of cacao (*Theobroma cacao*, Sterculiaceae) in the Western Hemisphere', *Econ. Bot.*, 44(4), 425–39, doi:10.1007/BF02859775.

Hunter, D. and Heywood, V. (Eds) (2011), *Crop Wild Relatives. A Manual of in situ Conservation.* Earthscan, Washington, DC, http://www.bioversityinternational.org/e-library/publications/detail/crop-wild-relatives/.

Hurka, H., Neuffer, B. and Friesen, N. (2004), 'Plant genetic resources in botanical gardens', In G. Forkmann and S. Michaelis (Eds), *Proceedings of the 21st International Symposium on Classical vs. Molecular Breeding of Ornamentals*, Part II, 25–29 August 2003, *Acta Horticulturae*, 651, pp. 35–44, doi:10.17660/ActaHortic.2004.651.2.

IBPGR (1981), *Genetic Resources of Cocoa (AGP: IBPGR/80/56)*, Report of the working group on genetic resources of cocoa, International Board for Plant Genetic Resources, Rome, Italy.

Idris, S. and Saad, M. S. (2001), 'Characterization of plant genetic resources', In M. S. Saad and V. Ramanatha Rao (Eds), *Establishment and Management of Field Genebank. A Training Manual*, IPGRI-APO, Serdang, pp. 81–6, http://www.bioversityinternational.org/e-library/publications/detail/establishment-and-management-of-field-genebank/.

Irish, B. M., Goenaga, R., Zhang, D., Schnell, R., Brown, J. S. and Motamayor, J. C. (2010), 'Microsatellite fingerprinting of the USDA-ARS Tropical Agriculture Research Station cacao (*Theobroma cacao* L.) germplasm collection', *Crop Sci.*, 50(2), 656–67, doi:10.2135/cropsci2009.06.0299.

Jaimes, Y., Aránzazu, F., Rodriguez, E. and Martínez, N. (2011), 'Behaviour of introduced regional clones of *Theobroma cacao* toward the infection *Moniliophthora roreri* in three different regions of Colombia', *Agronomía Colombiana*, 29(2), 361–71, http://www.scielo.org.co/pdf/agc/v29n2/v29n2a2.pdf.

Jaramillo, S. and Baena, M. (2002), *Ex situ Conservation of Plant Genetic Resources: Training Module*. International Plant Genetic Resources Institute, Cali, Colombia, http://www.bioversityinternational.org/e-library/publications/detail/ex-situ-conservation-of-plant-genetic-resources/.

Jarvis, D. I., Myer, L., Klemick, H., Guarino, L., Smale, M., Brown, A. H. D., Sadiki, M., Sthapit, B. and Hodgkin, T. (2000), *A Training Guide for In Situ Conservation On-farm. Version 1*. IPGRI, Rome, Italy, https://www.bioversityinternational.org/e-library/publications/detail/a-training-guide-for-in-situ-conservation-on-farm/.

Ji, K., Zhang, D., Motilal, L. A., Boccara, M., Lachenaud, P. and Meinhardt, L. W. (2013), 'Genetic diversity and parentage in farmer varieties of cacao (*Theobroma cacao* L.) from Honduras and Nicaragua as revealed by single nucleotide polymorphism (SNP) markers', *Genet. Resour. Crop Evol.*, 60(2), 441–53, doi:10.1007/s10722-012-9847-1.

Johnsiul, J. and Awang, A. (2016), 'Utilization of molecular markers to detect the authenticity of cocoa clones', *Intl. J. Agricult. For. Plant.*, 3, 101–4.

Johnson, E. S., Mora, A. and Schnell, R. J. (2007), 'Field guide efficacy in the identification of reallocated clonally propagated accessions of cacao (*Theobroma cacao* L.)', *Genet. Resour. Crop Evol.*, 54(6), 1301–13, doi:10.1007/s10722-006-9111-7.

Jost, L. (2008), 'G_{ST} and its relatives do not measure differentiation', *Mol. Ecol.*, 17(18), 4015–26, doi:10.1111/j.1365-294x.2008.03887.x.

Jost, L. (2009), 'D vs. G_{ST}: Response to Heller and Siegismund (2009) and Ryman and Leimar (2009)', *Mol. Ecol.*, 18(10), 2088–91, doi:10.1111/j.1365-294x.2009.04186.x.

Karp, A., Kresovich, S., Bhat, K. V., Ayad, W. G. and Hodgkin, T. (1997), *Molecular Tools in Plant Genetic Resources Conservation: A Guide to the Technologies*. IPGRI Technical Bulletin No. 2, International Plant Genetic Resources Institute, Rome, Italy, https://www.bioversityinternational.org/e-library/publications/detail/molecular-tools-in-plant-genetic-resources-conservation/.

Kennedy, A. J. (1985), 'The International Cocoa Genebank, Trinidad: Its role in future cocoa research and development', *Cocoa Growers' Bull.*, 36, 5–10.

Kennedy, A. J., Lockwood, G., Mossu, G., Simmonds, N. W. and Tan, G. Y. (1987), 'Cocoa breeding: Past, present and future', *Cocoa Growers' Bull.*, 38, 5–22.

Khaing, A. A., Li, G., Wang, X. Q., Yoon, M. Y., Kwon, S. W., Lee, C. Y., Park, B. S. and Park, Y. J. (2014), 'Rice germplasm in Korea and association mapping', In W. Yan and J. Bao (Eds.), *Rice - Germplasm, Genetics and Improvement*, InTech, pp. 316, http://www.intechopen.com/books/rice-germplasm-genetics-andimprovement. doi:10.5772/51100.

Khoury, C., Laliberté, B. and Guarino, L. (2010), 'Trends in *ex situ* conservation of plant genetic resources: A review of global crop and regional conservation strategies', *Genet. Resour. Crop Evol.*, 57(4), 625–39, doi:10.1007/s10722-010-9534-z.

Kiers, A. M., Mes, T. H. M., van der Meijden, R. and Bachmann, K. (2000), 'A search for diagnostic AFLP markers in *Cichorium* species with emphasis on endive and chicory cultivar groups', *Genome*, 43(3), 470–6, doi:10.1139/g00–024.

Koohafkan, P. and Altieri, M. A. (2010), *Globally Important Agricultural Heritage Systems. A Legacy for the Future*. Food and Agriculture Organization of the United Nations, Rome, Italy, www.fao.org/docrep/013/i1979e/i1979e00.htm.

Kress, W. J., Wurdack, K. J., Zimmer, E. A., Weigt, L. A. and Janzen, D. H. (2005), 'Use of DNA barcodes to identify flowering plants', *Proc. Natl. Acad. Sci. (USA)*, 102(23), 8389–74, doi:10.1073/pnas.0503123102.

Laliberté, B., Cryer, N., Daymond, A. J., End, M. J., Engels, J., Eskes, B., Gilmour, M., Lachenaud, P., Phillips-Mora, W., Turnbull, C. J., Umaharan, P., Zhang, D. and Wiese, S. (2012), 'A global strategy for the conservation and use of cacao genetic resources, as the foundation for a sustainable cocoa economy', In *Proc. 17th Intl. Cocoa Res. Conf.*, Yaoundé, Cameroon, 15–20 October 2012.

Lam, L. V., Thanh, T., Trang, L. T. T., Truong, V. V., Lam, H. B. and Tuy, L. M. (2012), '*Hevea* germplasm in Vietnam: Conservation, characterization, evaluation and utilization', In M. Caliskan (Ed.), *Genetic Diversity in Plants*, InTech, pp. 433–56, doi:10.5772/35086, http://www.intechopen.com/books/genetic-diversity-in-plants/hevea-germplasm-in-vietnam-conservationcharacterization-evaluation-and-utilization.

Lanaud, C. (1986), 'Origine genetiques des plantes a phenotype maternal issues de croisements intra ou interspecific, de feve plates ou de grains polyembryonees chez *Theobroma cacao* L.', *Café Cacao Thé*, 3191, 3–12.

Lanaud, C. (1988), 'Origin of haploids and semigamy in *Theobroma cacao* L.', *Euphytica*, 38, 221–8.

Leão, P. C. S., Riaz, S., Graziani, R., Dangl, G. S., Motoike, S. Y. and Walker, M. A. (2009), 'Characterization of a Brazilian grape germplasm collection using microsatellite markers', *Am. J. Enol. Vitic.*, 60(4), 517–24.

Lebot, V., Hartati, S., Hue, N. T., Viet, N. V., Nghia, N. H., Okpul, T., Pardales, J., Prana, M. S., Prana, T. K., Thongjiem, M., Krieke, C. M., VanEck, H., Yap, T. C. and Ivancic, A. (2010), 'Characterizing taro using isozymes and morpho-agronomic descriptors', In V. Ramanatha Rao, P. J. Matthews, P. B. Eyzaguirre and D. Hunter (Eds), *The Global Diversity of Taro: Ethnobotany and Conservation*, Bioversity International, Rome, Italy, pp. 39–55, http://www.bioversityinternational.org/e-library/publications/detail/the-global-diversity-of-taro-ethnobotany-and-conservation/.

Lewontin, R. C. (1972), 'The apportionment of human diversity', *Evol. Biol.*, 6, 381–98, doi:10.1007/978-1-4684-9063-3_14.

Li, Z., Traore, A., Maximova, S. and Guiltinan, M. J. (1998), 'Somatic embryogenesis and plant regeneration from floral explants of cacao (*Theobroma cacao* (L.) using thidiazuron', *In Vitro Cell. Dev. Biol. Plant*, 34(4), 293–9, doi:10.1007/bf02822737.

Lin, S.-W., Chou, Y.-Y., Shieh, H.-C., Ebert, A. W., Kumar, S., Mavlyanova, R., Rouamba, A., Tenkouano, A., Afari-Sefa, V. and Gniffke, P. A. (2013), 'Pepper (*Capsicum* spp.) germplasm dissemination by AVRDC – The World Vegetable Center: An overview and introspection', *Chronica Hortic.*, 53(3), 21–7.

Lockwood, G. (1995), 'Core collection of cocoa germplasm', *INGENIC Newsl.*, 1, 4–5.

Lockwood, G. and End, M. (1993), 'History, techniques and future needs for cacao collection', In *Proc. International Workshop on Conservation, Characterization and Utilization of Cocoa Genetic Resources in the 21st Century*, Cocoa Research Unit, The University of the West Indies, St. Augustine, Trinidad, Trinidad and Tobago, pp. 1–14.

Lockwood, G. and Gyamfi, M. M. O. (1979), *The CRIG Cocoa Germplasm Collection with Notes on Codes Used in the Breeding Programme at Tafo and Elsewhere*. Technical Bulletin 10, Cocoa Research Institute, Tafo, Ghana.

Loor Solorzano, R. G., Fouet, O., Lemainque, A., Pavek, S., Boccara, M., Argout, X., Amores, F., Courtois, B., Risterucci, A. M. and Lanaud, C. (2012), 'Insight into the wild origin, migration and domestication history of the fine flavour Nacional *Theobroma cacao* L. variety from Ecuador', *PLoS ONE*, 7(11), e48438, doi:10.1371/journal.pone.0048438.

Lukman, Zhang, D., Susilo, A. W., Dinarti, D., Bailey, B., Mischke, S. and Meinhardt, L. W. (2014), 'Genetic identity, ancestry and parentage in farmer selections of cacao from Aceh, Indonesia revealed by single nucleotide polymorphism (SNP) markers', *Trop. Plant Biol.*, 7(3), 133–43, doi:10.1007/s12042-014-9144-6.

Marita, J. M., Rodríguez, J. M. and Nienhuis, J. M. (2000), 'Development of an algorithm identifying maximally diverse core collections', *Genet. Resour. Crop Evol.*, 47(5), 515–26, doi:10.1023/A:1008784610962.

Marshall, D. R. and Brown, A. H. D. (1975), 'Optimum sampling strategies in genetic conservation', In O. H. Frankel and J. G. Hawkes (Eds), *Crop Genetic Resources for Today and Tomorrow*, Cambridge University Press, Cambridge, UK, pp. 53–80.

Martínez, I. B., Nelson, M. R., Flamand, M.-C. and Bertin, P. (2015), 'Genetic diversity and population structure of anciently introduced Cuban cacao *Theobroma cacao* plants', *Genet. Resour. Crop Evol.*, 62(1), 67–84, doi:10.1007/s10722-014-0136-z.

Martinson, V. A. (1986), 'Fertility improvement in induced tetraploid Theobroma cacao', *J. Hered.*, 77(4), 279–81, https://doi.org/10.1093/oxfordjournals.jhered.a110238.

Maunder, M., Havens, K., Guerrant Jr., E. O. and Falk, D. A. (2004), '*Ex situ* methods a vital but underused set of conservation resources', In E. O. Guerrant, K. Havens, Jr. and M. Maunder (Eds), *Ex situ Plant Conservation Supporting Species Survival in the Wild*, Island Press, Washington DC, USA, pp. 3–20.

Maximova, S. N., Alemanno, L., Young, A., Ferriere, N., Traore, A. and Guiltinan, M. (2002), 'Efficiency, genotypic variability, and cellular origin of primary and secondary somatic embryogenesis of *Theobroma cacao* L.', *In Vitro Cell. Dev. Biol. Plant*, 38(3), 252–9, doi:10.1079/ivp2001257.

Maxted, N., Dulloo, E., Ford-Lloyd, B. V., Iriondo, J. M. and Jarvis, A. (2008), 'Gap analysis: A tool for complementary genetic conservation assessment', *Divers. Distrib.*, 14, 1018–30, doi:10.1111/j.1472-4642.2008.00512.x.

Maxted, N., Ford-Lloyd, B. V. and Hawkes, J. G. (1997), *Plant genetic conservation: The in situ approach*. Chapman and Hall, London, UK, doi:10.1007/978-94-009-1437-7.

Maxted, N., Kell, S., Ford-Lloyd, B., Dulloo, E. and Toledo, A. (2012), 'Toward the systematic conservation of global crop wild relative diversity', *Crop Sci.*, 52(2), 774–85, doi:10.2135/cropsci2011.08.0415.

McCouch, S. R., McNally, K. L., Wang, W. and Sackville Hamilton, R. (2012), 'Genomics of gene banks: A case study in rice', *Am. J. Bot.*, 99(2), 407–23, doi:10.3732/ajb.1100385.

Mooleedhar, V., Bekele, F., Cassie, T., Warren, J. and Spence, J. (n.d.), 'Accessions held in the, International Cocoa Genebank, Trinidad (ICG, T)', In *Report for 1990*, Cocoa Research Unit, The University of the West Indies, St. Augustine, pp. 30–45.

Mooney, P. R. (1983), *The Law of the Seed – Another Development and Plant Genetic Resources*. Dag Hammarskjold Report, Uppasala, http://www.daghammarskjold.se/publication/law-seed-another-development-plant-geneticresources/.

Morera, J. A. and Paredes, A. (Eds) (1990), *Memoria del Seminario Manejo de Germoplasma de Cacao*, 19–20 October 1989, Turrialba, Costa Rica. Serie de ponencias, resultados y recomendaciones de eventos tecnicos, San José, Costa Rica.

Motamayor, J. C., Lachenaud, P., da Silva e Mota, J. W., Loor, R., Kuhn, D. N., Brown, J. S., and Schnell, R. J. (2008), 'Geographic and genetic population differentiation of the Amazonian chocolate tree (*Theobroma cacao* L)', *PLoS ONE*, 3(10), e3311, doi:10.1371/journal.pone.0003311.

Motamayor, J. C., Lachenaud, P., da Silva e Mota, J. W., Loor, R., Martinez, W. J., Graham, J., Kuhn, D. N., Brown, S. and Schnell, R. J. (2010), 'No mas 'Forastero': A new protocol for meaningful cacao germplasm classification', In *Proceedings of the 16th International Cacao Research Conference*, COPAL, Nigeria, pp. 179–85.

Motilal, L. A. (2017), 'A molecular genetic study of the International Cocoa Genebank, Trinidad towards efficient conservation and utilisation', PhD thesis, The University of the West Indies, St. Augustine, Trinidad.

Motilal, L. A. and Butler, D. (2003), 'Verification of identities in global cacao germplasm collections', *Genet. Resour. Crop Evol.*, 50(8), 799–807, doi:10.1023/A:1025950902827.

Motilal, L. A. and Sreenivasan, T. N. (2012), 'Revisiting 1727: Crop failure leads to the birth of Trinitario cacao', *J. Crop Improv.*, 26(5), 599–626, doi:10.1080/15427528.2012.663734.

Motilal, L. A. and Sreenivasan, T. N. (2013), 'Birth of Trinitario cacao: History intertwined with myths and edaphic and climatic factors', In M. S. Kang and S. S. Banga (eds) *Combating Climate Change: An Agricultural Perspective*, CRC Press, Taylor & Francis Group, pp. 305–24, doi:10.1201/b14056-17.

Motilal, L. A., Sankar, A., Gopaulchan, D. and Umaharan, P. (2017), 'Cocoa', In P. Chowdappa, A. Karun, M. K. Rajesh and S. V. Ramesh (Eds), *Biotechnology of Plantation Crops*, Daya Publishing House, New Delhi, India, pp. 313–54.

Motilal, L. A., Zhang, D., Mischke, S., Meinhardt, L. W. and Umaharan, P. (2013), 'Microsatellite-aided detection of genetic redundancy improves management of the International Cocoa Genebank, Trinidad', *Tree Genet. Genomes*, 9(6), 1395–411, doi:10.1007/s11295-013-0645-5.

Motilal, L. A., Zhang, D., Umaharan, P., Boccara, M., Mischke, S., Sankar, A. and Meinhardt, L. W. (2012), 'Elucidation of genetic identity and population structure of cacao germplasm within an international cacao genebank', *Plant Genet. Resour.*, 10(3), 232–41, doi:10.1017/S1479262112000305.

Motilal, L. A., Zhang, D., Umaharan, P., Mischke, S., Boccara, M. and Pinney, S. (2009), 'Increasing accuracy and throughput in large-scale microsatellite fingerprinting of cacao field germplasm collections', *Trop. Plant Biol.*, 2(1), 23–37, doi:10.1007/s12042-008-9016-z.

Motilal, L. A., Zhang, D., Umaharan, P., Mischke, S., Pinney, S. and Meinhardt, L. W. (2011), 'Microsatellite fingerprinting in the International Cocoa Genebank, Trinidad: Accession and plot homogeneity information for germplasm management', *Plant Genet. Resour.*, 9(3), 430–8, doi:10.1017//S1479/26211100058X.

MusaNet. (2016), *Global Strategy for the Conservation and Use of Musa Genetic Resources* (B. Laliberté, compiler). Bioversity International, Montpellier, France, http://www.bioversityinternational.org/e-library/publications/detail/global-strategy-for-the-conservation-and-use-of-musa-genetic-resources/.

Muzzalupo, I., Perri, E. and Chiappetta, A. (2012), 'Fruit germplasm characterization: Genomics approaches for the valorisation of genetic diversity', In M. Caliskan (Ed.), *Genetic Diversity in Plants*, InTech, pp. 55–86, doi:10.5772/32906, http://www.intechopen.com/books/geneticdiversity-in-plants/fruit-germplasm-characterization-genomics-approaches-for-the-valorisation-of-geneticdiversity.

Nass, L. L., Sigrist, M. S., Ribeiro, C. S. C. and Reifschneider, F. J. B. (2012), 'Genetic resources: The basis for sustainable and competitive plant breeding', *Crop Breed. Appl. Biot.*, 12, 75–86, doi:10.1590/s1984-70332012000500009.

Nayak, S. N., Song, J., Villa, A., Pathak, B., Ayala-Silva, T., Yang, X., Todd, J., Glynn, N. C., Kuhn, D. N., Glaz, B., Gilbert, R. A., Comstock, J. C. and Wang, J. (2014), 'Promoting utilization of *Saccharum* spp. genetic resources through genetic diversity analysis and core collection construction', *PLoS ONE*, 9(10), e110856, doi:10.1371/journal.pone.0110856.

Nei, M., Tajima, F. and Tateno, Y. (1983), 'Accuracy of estimated phylogenetic trees from molecular data. II. Gene frequency data', *J. Mol. Evol.*, 19(2), 153–70, doi:10.1007/bf02300753.

Noirot, M., Hamon, S. and Anthony, F. (1996), 'The principal component scoring: A new method of constituting a core collection using quantitative data', *Genet. Resour. Crop Evol.*, 43(1),1–6, doi:10.1007/BF00126934.

Nordin, M. S. and Saad, M. S. (2001), 'Genetic considerations in field genebank conservation', In M. S. Saad and V. Ramanatha Rao (Eds), *Establishment and Management of Field Genebank. A Training Manual.* IPGRI-APO, Serdang, pp. 66–72, http://www.bioversityinternational.org/e-library/publications/detail/establishment-and-management-of-field-genebank/.

Nybom, H. (2004), 'Comparison of different nuclear DNA markers for estimating intraspecific genetic diversity in plants', *Mol. Ecol.*, 13(5), 1143–55, doi:10.1111/j.1365-294x.2004.02141.x.

Nybom, H. and Weising, K. (2010), 'DNA-based identification of clonally propagated cultivars', *Plant Breed. Rev.*, 34(1), 221–95, doi:10.1002/9780470880579.ch6.

Nybom, H., Weising, K. and Rotter, B. (2014), 'DNA fingerprinting in botany: Past, present, future', *Investig. Genet.*, 5(1), 1, doi:10.1186/2041-2223-5-1.

Odong, T. L., Jansen, J., van Eeuwijk, F. A. and Van Hintum, T. J. L. (2013), 'Quality of core collections for effective utilisation of genetic resources review, discussion and interpretation', *Theor. Appl. Genet.*, 126(2), 289–305, doi:10.1007/s00122-012-1971-y.

Ogata, N. (2003), 'Domestication and distribution of the chocolate tree (*Theobroma cacao* L.) in Mesoamerica', In A. Gómez-Pompa, M. F. Allen, S. L. Fedick and J. J. Jiménez-Osorio (Eds), *The Lowland Maya Area. Three Millennia at the Human-wildland Interface*, The Haworth Press Inc., New York, pp. 415–38.

Oldfield, M. L. and Alcorn, J. B. (1987), 'Conservation of traditional agroecosystems: Can age-old farming practices conserve crop genetic resources?', *BioScience*, 37(3), 199–208, doi:10.2307/1310519.

Opeke, L. K. and Jacob, V. J. (1969), 'Cytological irregularities in *Theobroma cacao* L.', In *Proceedings of the Second International Cacao Research Conference*, 19–26 November 1967, CEPLAC, Bahia, Brasil, pp. 114–16.

Opoku, S. Y., Bhattacharjee, R., Kolesnikova-Allen, M., Motamayor, J. C., Schnell, R., Ingelbrecht, I., Enu-Kwesi, L. and Adu-Ampomah, Y. (2007), 'Genetic diversity in cocoa (*Theobroma cacao* L.) germplasm collection from Ghana', *J. Crop Improv.*, 20(1/2), 73–87.

Padi, F. K., Ofori, A., Takrama, J., Djan, E., Stephen, Y., Opoku, A. M., Dadzie, R. B., Motamayor, J. C. and Zhang, D. (2015), 'The impact of SNP fingerprinting and parentage analysis on the effectiveness of variety recommendations in cacao', *Tree Genet. Genomes* 11, 44, doi:10.1007/s11295-015-0875-9.

Palma, T. and Villalobos, V. M. (1989), 'Rescate in vitro de embriones provenientes de semillas aplanadas de cacao (*Theobroma cacao*)', *Turrialba*, 39(4), 525–9.

Parra-Quijano, M., Iriondo, J. M. and Torres, E. (2012), 'Improving representativeness of genebank collections through species distribution models, gap analysis and ecogeographical maps', *Biodivers. Conserv.* 21, 79–96, doi:10.1007/s10531-011-0167-0.

Paulin, D., Garzon, I. and Alemanno, L. (2007), 'Clonal propagation by somatic embryogenesis: Preliminary results', *INGENIC Newsl.*, 11, 16–19.

Pessoa-Filho, M., Rangel, P. H. N., and Ferreira, M. E. (2010), 'Extracting samples of high diversity from thematic collections of large gene banks using a genetic-distance based approach', *BMC Plant Biol.*, 10(1), 127, doi:10.1186/1471-2229-10-127.

Pires, J. L., Marita, J. M., Lopes, U. V., Yamada, M. M., Aitken, W. M., Melo, G. P., Monteiro, W. R. and Ahnert, D. (2000), 'Diversity for phenotypic traits and molecular markers in CEPEC's germplasm collection in Bahia, Brazil', In *Proceedings of the International Workshop on New Technologies and Cocoa Breeding*, INGENIC, Malaysia, pp. 72–8.

Pound, F. J. (1938), 'Cacao and witches' broom disease (*Marasmius perniciosus*) of South America', *Arch. Cacao Res.*, 1, 20–72.

Pound, F. J. (1943), *Cacao and Witches' Broom Disease (Marasmius perniciosus). Report on a Recent Visit to the Amazon Territory of Peru, September, 1942–February, 1943.* Yuille's Printery, Port of Spain, Trinidad and Tobago.

Pound, F. J. (1945), 'A note on the cocoa population of South America', In *Report and Proceedings of the Cocoa Research Conference*, May-June 1945, London, Colonial No. 192, The Colonial Office, London, pp. 131–3.

Preuss, P. (1901), 'Chapter 13: Expedition nach Central-und Süd Amerika', In *Der Kakao, seine Kultur und Aufbereitung [Cocoa, Its Cultivation and Preparation]*, International Office of Cocoa and Chocolate, Brussels, pp. 167–277. Reprint, *Arch. Cocoa Res.*, 3, 15–115.

Pritchard, J. K., Stephens, M. and Donnelly, P. (2000), 'Inference of population structure from multilocus genotype data', *Genetics*, 155(2), 945–59.

Pryce, M., Mitchell, S., Burke, A., McKenzie, C., Stirling, S., Ryan, J., Simpson, W. and McGlashan, D. (2008), *Jamaica: Country Report to the FAO International Technical Conference on Plant Genetic Resources for Food and Agriculture.* Kingston, Jamaica, www.fao.org/docrep/013/i1500e/Jamaica.pdf.

Ramanatha Rao, V. (2001), 'Principles and concepts in plant genetic resources conservation and use', In M. S. Saad and V. Ramanatha Rao (Eds), *Establishment and Management of Field Genebank. A Training Manual*, IPGRI-APO, Serdang, pp. 1–16, http://www.bioversityinternational.org/e-library/publications/detail/establishment-and-management-of-field-genebank/.

Ramanatha Rao, V. and Riley, K. W. (1994), 'The use of biotechnology for conservation and utilization of plant genetic resources', *Plant Genet. Resour. Newsl.*, 97, 3–19.

Ramanatha Rao, V. and Sthapit, B. R. (2012), 'Tropical fruit tree genetic resources: Status and effect of climate change', In B. R. Sthapit, V. Ramanatha Rao and S. R. Sthapit (Eds), *Tropical Fruit Tree Species and Climate Change*, Bioversity International, New Delhi, India, pp. 97–128.

Ramirez, M., Ortiz, R., Taba, S., Sebastián, L., Peralta, E., Williams, D. E., Ebert, A. W. and Vézina, A. (2013), 'Demonstrating interdependence on plant genetic resources for food and agriculture', In M. Halewood, I. L. Noriega and S. Louafi (Eds), *Crop Genetic Resources as a Global Commons. Challenges in International Law and Governance*, Routledge, New York, pp. 39–61, http://www.bioversityinternational.org/e-library/publications/detail/crop-genetic-resources-as-a-global-commons/.

Ramirez-Villegas, J., Khoury, C., Jarvis, A., Debouck, D. G. and Guarino, L, (2010), 'A gap analysis methodology for collecting crop genepools: A case study with *Phaseolus* beans', *PLoS ONE* 5(10), e13497, doi:10.1371/journal.pone.0013497.

Reed, D. H. (2005), 'Relationship between population size and fitness', *Conserv. Biol.*, 19(2): 563–8, doi:10.1111/j.1523-1739.2005.00444.x.

Reed, B. M., Engelmann, F., Dulloo, M. E. and Engels, J. M. M. (2004), *Technical Guidelines for the Management of Field and in vitro Germplasm Collections*, IPGRI Handbooks for Genebanks No. 7, International Plant Genetic Resources Institute, Rome, Italy.

Resnik, D. B. (2001), 'DNA patents and scientific discovery and innovation: Assessing benefits and risks', *Sci. Eng. Ethics*, 7(1), 29–62, doi:10.1007/s11948-001-0023-9.

Ribeiro, I. C. N. S, Santos, C. A. F., and Neto, F. P. L. (2013), 'Morphological characterization of mango (*Mangifera indica*) accessions based on Brazilian adapted descriptors', *J. Agric. Sci. Technol.*, B11, 798–806, doi: 10.17265/2161-6264/2013.11B.005.

Rice, E. B., Smith, M. E., Mitchell, S. E. and Kresovich, S. (2006), 'Conservation and change: A comparison of *in situ* and *ex situ* conservation of Jala maize germplasm', *Crop Sci.*, 46, 428–36, doi:10.2135/cropsci2005.06-0116.

Risterucci, A. M., Eskes, B., Fargeas, D., Motamayor, J. C. and Lanaud, C. (2001), 'Use of microsatellite markers for germplasm identity analysis in cocoa', In *Proceedings of the Third International Group for Genetic Improvement of Cocoa (INGENIC) International Workshop on the New Technologies and Cocoa Breeding*, INGENIC, UK, pp. 25–33.

Robinson, J. and Srinivasan, C. S. (2013), *Case-studies on the Impact of Germplasm Collection, Conservation, Characterization and Evaluation (GCCCE) in the CGIAR*. Standing Panel on Impact Assessment, CGIAR. http://impact.cgiar.org/publications/GCCCE-in-the-cgiar.

Ronning, C. M. and Schnell, R. J. S. (1994), 'Allozyme diversity in a germplasm collection of *Theobroma cacao* L.', *J. Hered.*, 85(4), 291–5.

Ruiz, J. C., Gamboa, O. R. and Arguello, I. M. (2011), 'Molecular ecology of genetic diversity of cacao cultivated in the south-east region of Nicaragua', *Int. Res. J. Agric. Sci.*, 1(1), 6–13.

Sackville Hamilton, R., Engels, J. and van Hintum, T. (2003), 'Considerations for improved conservation and utilization concepts and strategies', In J. M. M. Engels and L. Visser (Eds), *A Guide to Effective Management of Germplasm Collections*, IPGRI Handbooks for Genebanks No. 6, IPGRI, Rome, Italy, pp. 43–59, https://www.bioversityinternational.org/e-library/publications/detail/a-guide-to-effective-management-of-germplasm-collections.

Samuel, A. F., Drucker, A. G., Andersen, S. B., Simianer, H. and van Zonneveld, M. (2013), 'Development of a cost-effective diversity-maximising decision-support tool for *in situ* crop genetic resources conservation: The case of cacao', *Ecol. Econ.*, 96, 155–64, http://dx.doi.org/10.1016/j.ecolecon.2013.10.018.

Sánchez, J. E. and Veláquez, A. (1989), 'Conservacíon de las semillas del cacao (*Theobroma cacao*)', *Turrialba*, 39(4), 483–7.

Sánchez, P. and Jaffé, K. (1989), 'El género *Theobroma* en el Térritorio Federal Amazonas (Venezuela). II. Distribución geográfica', *Turrialba*, 39(4), 446–54.

Sartie, A., Asiedu, R. and Franco, J. (2012), 'Genetic and phenotypic diversity in a germplasm working collection of cultivated tropical yams (*Dioscorea* spp.)', *Genet. Resour. Crop Evol.*, 59(8), 1753–65, doi:10.1007/s10722-012-9797-7.

Saski, C. A., Feltus, F. A., Staton, M. E., Blackmon, B. P., Ficklin, S. P., Kuhn, D. N., Schnell, R. J., Shapiro, H. and Motamayor, J. C. (2011), 'A genetically anchored physical framework for *Theobroma cacao* cv. Matina 1-6', *BMC Genomics*, 12: 413, doi:10.1186/1471-2164-12-413.

Sastrapradja, S. D. and Balakhrishna, P. (2002), 'The deployment and management of genetic diversity in agroecosystems', In J. M. M. Engels, V. Ramanatha Rao, A. H. D. Brown and M. T. Jackson (Eds), *Managing Plant Genetic Diversity*, IPGRI, CABI Publishing, UK, pp. 121–32, http://www.bioversityinternational.org/e-library/publications/detail/managing-plant-genetic-diversity/.

Saunders, J. A., Mischke, S., Leamy, E. A. and Hemeida, A. A. (2004), 'Selection of international molecular standard for DNA fingerprinting of *Theobroma cacao*', *Theor. Appl. Genet.*, 110(1), 41–7, doi:10.1007/s00122-004-1762-1.

Savolainen, O. and Hedrick, P. (1995), 'Heterozygosity and fitness: No association in Scots Pine', *Genetics*, 140, 755–66.

Scarascia-Mugnozza, G. T. and Perrino, P. (2002), 'The history of *ex situ* conservation and use of plant genetic resources', In J. M. M. Engels, V. Ramanatha Rao, A. H. D. Brown and M. T. Jackson (Eds), *Managing Plant Genetic Diversity*, IPGRI, CABI Publishing, UK, pp. 1–22, http://www.bioversityinternational.org/e-library/publications/detail/managing-plant-genetic-diversity/.

Schoen, D. J. and Brown, A. H. D. (1995), 'Maximising genetic diversity in core collections of wild relatives of crop species', In T. Hodgkin, A. H. D. Brown, Th. J. L. van Hintum and E. A. V. Morales (Eds), *Core Collections of Plant Genetic Resources*, John Wiley and Sons, UK, pp. 55–76.

Schultes, R. E. (1984), 'Amazonian cultigens and their northward and westward migration in pre-Columbian times', In D. Stone (Ed.), *Pre-Columbian Plant Migration*, Volume 76, Harvard Univ. Press, Cambridge, MA, 19–37.

Sereno, M. L., Albuquerque, P. S. B., Vencovsky, R. and Figueira, A. (2006), 'Genetic diversity and natural population structure of cacao (*Theobroma cacao* L.) from the Brazilian Amazon evaluated by microsatellite markers', *Conserv. Genet.*, 7(1), 13–24, doi:10.1007/s10592-005-7568-0.

Shaffer, M. L. (1981), 'Minimum population sizes for species conservation', *BioScience*, 31, 131–4, doi:10.2307/1308256.

Shashidhara G., Hema, M. V., Koshy, B. and Farooqi A. A. (2003), 'Assessment of genetic diversity and identification of core collection in sandalwood germplasm using RAPDs', *J. Hortic. Sci. Biotechnol.*, 78(4), 528–36, doi:10.1080/14620316.2003.11511659.

Sherwin, W., Jabot, F., Rush, R. and Rossetto, M. (2006), 'Measurement of biological information with applications from genes to landscapes', *Mol. Ecol.*, 15(10), 2857–69, doi:10.1111/j.1365-294x.2006.02992.x.

Shukor, N. A. A. (2001), 'Biochemical markers in plant genetic resources characterization', In M. S. Saad and V. Ramanatha Rao (Eds), *Establishment and Management of Field Genebank, a Training Manual*, IPGRI-APO, Serdang, pp. 95–106, http://www.bioversityinternational.org/e-library/publications/detail/establishment-and-management-of-field-genebank/.

Sigrist, M. S., Pinheiro, J. B., Azevedo Filho, J. A. and Zucchi, M. I. (2011), 'Genetic diversity of turmeric germplasm (*Curcuma longa*; Zingiberaceae) identified by microsatellite markers', *Genet. Mol. Res.*, 10(1), 419–28, doi:10.4238/vol10-1gmr1047.

Silva Pupo, J. J., Cruz, S. M., Pompa, L. A., Arias, E. and García, A. (2005), 'Embriogénesis somática: Una alternativa para la propagación, mejoramiento y conservación de germoplasma en cacao', *Cuadernos de Biodiversidad, [S.l.]*, 16, 9–12, doi:https://doi.org/10.14198/cdbio.2005.16.

Simmonds, N. W. (1954), 'Chromosome behaviour in some tropical plants', *Heredity*, 8, 139–46, doi:10.1038/hdy.1954.11.

Simmonds, N. W. (1962), 'Variability in crop plants, its use and conservation', *Biol. Rev.*, 37(3), 422–65, doi:10.1111/j.1469-185x.1962.tb01620.x.

Smale, M. (2005), 'Economics literature about crop biodiversity: Findings, methods, and limitations', In A. G. Drucker, M. Smale and P. Zambrano (Eds), *Valuation and Sustainable Management of Crop and Livestock Biodiversity: A Review of Applied Economics Literature*, CGIAR System-wide Genetic Resources Programme, IFPRI, IPGRI and ILRI, pp. 22–41, http://www.bioversityinternational.org/e-library/publications/detail/valuation-and-sustainable-management-of-crop-and-livestock-biodiversity/.

Soria, V. J. (1970), 'Principal varieties of cocoa cultivated in tropical America', *Cocoa Growers' Bull.*, 15, 13–21.

Sorkheh, K., Shiran, B., Gradziel, T. M., Epperson, B. K., Martínez-Gomez, P. and Asadi, E. (2007), 'Amplified fragment length polymorphism as a tool for molecular characterization of almond germplasm: Genetic diversity among cultivated genotypes and related wild species of almond, and its relationship with agronomic traits', *Euphytica*, 156(3), 327–44, doi:10.1007/s10681-007-9382-x.

Sounigo, O., Lachenaud, P., Bastide, P., Cilas, C., N'Goran, J. and Lanaud, C. (2003), 'Assessment of the value of double haploids as progenitors in cocoa (*Theobroma cacao* L.) breeding', *J. Appl. Genet.*, 44(3), 339–53.

Sounigo, O., Bekele, F. L., Iwaro, A. D., Thévenin, J-M., Bidaisee, G., Umaharan, R., Sankar, A., Sukha, D., Motilal, L., Butler, D. R. and Eskes, A. B. (2006), 'Description of cocoa clones proposed for the 'CFC/ICCO/IPGRI project collection'', In A. B. Eskes and Y. Efron (Eds), *Global Approaches to Cocoa Germplasm Utilization and Conservation. Final Report of the CFC/ICCO/IPGRI Project on 'Cocoa Germplasm Utilization and Conservation: A Global Approach' (1998–2004)* , CFC, Amsterdam, The Netherlands; ICCO, London, UK; IPGRI, Rome, Italy, pp. 67–81, https://www.bioversityinternational.org/e-library/publications/detail/global-approaches-to-cocoa-germplasm-utilization-and-conservation/.

Sounigo, O., Christopher, Y., Bekele, F., Mooleedhar, V., and Hosein, F. (2001), 'The detection of mislabelled trees in the International Cocoa Genebank, Trinidad (ICG,T)', In *Proceedings of the Third International Group for Genetic Improvement of Cocoa (INGENIC) International Workshop on the New Technologies and Cocoa Breeding*, INGENIC, UK, pp. 34–9.

Spindel, J. E., Begum, H., Akdemir, D., Collard, B., Redoña, E., Jannink, J.-L. and McCouch, S. (2016), 'Genome-wide prediction models that incorporate *de novo* GWAS are a powerful new tool for tropical rice improvement', *Heredity*, 116, 395–408, doi:10.1038/hdy.2015.113.

Spooner D., van Treuren, R. and de Vicente, M. C. (2005), *Molecular Markers for Genebank Management*, IPGRI Technical Bulletin No. 10. International Plant Genetic Resources Institute, Rome, Italy, http://www.bioversityinternational.org/e-library/publications/detail/molecular-markers-for-genebank-management/.

Strauss, M. S., Pino, J. A., and Cohen, J. I. (1988), 'Quantification of diversity in *ex-situ* plant collections', *Diversity*, 16, 30–2.

Susilo, A. W., Zhang, D., Motilal, L. A., Mischke, S., and Meinhardt, L. W. (2011), 'Assessing genetic diversity in Java fine-flavor cocoa (*Theobroma cacao* L.) germplasm by using simple sequence repeat (SSR) markers', *Trop. Agric. Dev.*, 55(2), 84–92, doi:10.11248/jsta.55.84.

Swaminathan, M. S. (2002), 'The past, present and future contributions of farmers to the conservation and development of genetic diversity', In J. M. M. Engels, V. Ramanatha Rao, A. H. D. Brown and M. T. Jackson (Eds), *Managing Plant Genetic Diversity*, IPGRI, CABI Publishing, UK, pp. 23–63, http://www.bioversityinternational.org/e-library/publications/detail/managing-plant-genetic-diversity/.

Swamy, K. R. M., Thoyajaksha, and Rao, E. V. V. B. (2002), 'Collecting cashew (*Anacardium occidentale* L.) germplasm from forest plantations in Karnataka, India', *Plant Genet. Resour. Newsl.*, 132, 35–8.

Tao, K.-L. (2001), 'Complementary conservation strategy for plant genetic resources', In M. S. Saad and V. Ramanatha Rao (Eds), *Establishment and Management of Field Genebank. A Training Manual*, IPGRI-APO, Serdang, pp. 46–53, http://www.bioversityinternational.org/e-library/publications/detail/establishment-and-management-of-field-genebank/.

Thomas, E., van Zonneveld, M., Loo, J., Hodgkin, T., Galluzzi, G. and van Etten, J. (2012), 'Present spatial diversity patterns of *Theobroma cacao* L. in the Neotropics reflect genetic differentiation in Pleistocene refugia followed by human-influenced dispersal', *PLOS ONE*, 7(10), e47676, doi:10.1371/journal.pone.0047676.

Thondaiman, V., Rajamani, K., Senthil, N., Shoba, N. and John Joel, A. (2013), 'Genetic diversity in cocoa (*Theobroma cacao* L.) plus trees in Tamil Nadu by simple sequence repeat (SSR) markers', *Afr. J. Biotechnol.*, 12(30), 4747–53, doi:10.5897/AJB2013.12423.

Toxopeus, H. (1985), 'Botany, types and populations', In G. A. R Wood and R. A. Lass (Eds), *Cocoa*, 4th ed., Longman, London, pp. 11–37.
Trognitz, B., Scheldeman, X., Hansel-Hohl, K., Kuant, A., Grebe, H. and Hermann, M. (2011), 'Genetic population structure of cacao plantings within a young production area in Nicaragua', *PLOS ONE*, 6(1), e16056, doi:10.1371/journal.pone.0016056.
Turnbull, C. J. and Hadley, P. (2017), *International Cocoa Germplasm Database (ICGD)*, [Online Database], CRA Ltd./ICE Futures Europe/University of Reading, UK, http://www.icgd.reading.ac.uk (17 July 2017).
Uchola, B. E. (2016), 'A concept of wild flora and biodiversity based on tree resource domestication in agriculture', *Int. J. Agric. For.*, 6(1), 1–7, doi:10.5923/j.ijaf.20160601.01.
Upadhyaya, H. D. and Oritz, R. (2001), 'A mini core subset for capturing diversity and promoting utilization of chickpea genetic resources in crop improvement', *Theor. Appl. Genet.*, 102(8), 1292–8, doi:10.1007/s00122-001-0556-y98.
Upadhyaya, H. D., Reddy, K. N., Ahmed, M. I. and Gowda, C. L. L. (2012), 'Identification of gaps in pearl millet germplasm from East and Southern Africa conserved at the ICRISAT genebank', *Plant Genet. Resour.*, 10(3), 202–13, doi:https://doi.org/10.1017/S1479262112000275.
Upadhyaya, H. D., Yadav, D., Dronavalli, N., Gowda, C. L. L. and Singh, S. (2010), 'Mini core germplasm collections for infusing genetic diversity in plant breeding programs', *Electr. J. Plant Breeding*, 1(4), 1294–309.
Van Hall, C. J. J. (1914), *Cocoa*. Macmillan, London.
Van Hintum, Th. J. L., Brown, A. H. D., Spillane, C. and Hodgkin, T. (2000), *Core Collections of Plant Genetic Resources*. IPGRI Technical Bulletin No. 3, International Plant Genetic Resources Institute, Rome, Italy, http://www.bioversityinternational.org/e-library/publications/detail/core-collections-of-plant-genetic-resources/.
Van Hintum, Th.J. L., Sackville Hamilton, N. R., Engels, J. M. M. and Van Treuren, R. (2002), 'Accession management strategies: Splitting and lumping', In J. M. M. Engels, V. Ramanatha Rao, A. H. D. Brown and M. T. Jackson (Eds), *Managing Plant Genetic Diversity*, IPGRI, CABI Publishing, UK, pp. 113–20.
Van Treuren, R., de Groot, E. C., Boukema, I. W., Van de Wiel, C. C. M. and Van Hintum, Th.J. L. (2010), 'Marker-assisted reduction of redundancy in a genebank collection of cultivated lettuce', *Plant Genet. Resour.*, 8(2), 95–105, doi:10.1017/S1479262109990220.
Vavilov, N. I. (1951), 'The origin, variation, immunity and breeding of cultivated plants', (translated from Russian by K. Starr Chester) *Chronica Botanica*, 13, 1–366.
Vázquez-Ovando, J. A., Molina-Freaner, F., Nuñez-Farfán, J., Ovando-Medina, I. and Salvador-Figueroa, M. (2014), 'Genetic identification of *Theobroma cacao* L. trees with high Criollo ancestry in Soconusco, Chiapas, Mexico', *Genet. Mol. Res.*, 13(4), 10404–14, doi:10.4238/2014.december.12.2.
Visser, B. and Rao, V. R. (2005), 'Gene flow and the management of *ex situ* collections', In M. C. de Vicente (Ed.), *Gene Flow and Germplasm Management*, Topical Reviews in Agricultural Biodiversity, International Plant Genetic Resources Institute, Rome, Italy, pp. 30–4, http://www.bioversityinternational.org/e-library/publications/detail/issues-on-gene-flow-and-germplasm-management/.
Waits, L. P., Luikart, G. and Taberlet, P. (2001), 'Estimating the probability of identity among genotypes in natural populations: Cautions and guidelines', *Mol. Ecol.*, 10(1), 249–56, doi:10.1046/j.1365-294x.2001.01185.x.
Warren, J. M. and Kennedy, A. J. (1991), 'Cocoa breeding revisited', *Cocoa Growers' Bull.*, 44, 18–24.
Weising, K., Nybom, H., Wolff, K. and Kahl, G. (2005), *DNA Fingerprinting in Plants: Principles, Methods and Applications*. Taylor & Francis, CRC Press, Boca Raton.
Whitkus, R., de la Cruz, M., Mota-Bravo, L. and Gómez-Pompa, A. (1998), 'Genetic diversity and relationships of cacao (*Theobroma cacao* L.) in southern Mexico', *Theor. Appl. Genet.*, 96(5), 621–7, doi:10.1007/s001220050780.

Wright, S. (1965), 'The interpretation of population structure by F-statistics with special regard to systems of mating', *Evolution*, 19(3), 395–420, doi:10.2307/2406450.
Yada, B., Tukamuhabwa, P., Villordon, A., Alajo, A., and Mwanga R. O. M. (2010), 'An online database of sweet potato germplasm collection in Uganda', *Hortsci.*, 45(1), 153.
Yanes, E., Gil, K., Rebolledo, L., Rebolledo, A., Uriza, D., Martínez, O., Isidrón, M., Díaz, L., Lorenzo, J. C. and Simpson, J. (2012), 'Genetic diversity of Cuban pineapple germplasm assessed by AFLP markers', *Crop Breed. Appl. Biot.*, 12, 104–10, doi:10.1590/s1984-70332012000200002.
Yidana, J. A. (1988), 'The development of *in vitro* collecting and isozyme characterization of cocoa germplasm', Ph.D. Thesis, University Nottingham, Nottingham, UK.
Yidana, J. A., Withers, L. A. and Ivins, J. (1987), 'Development of a simple method for collecting and propagating cocoa germplasm *in vitro*', *Acta Hortic.*, 212, 95–8, doi:10.17660/actahortic.1987.212.13.
Yonezawa, K., Nomura, T. and Morishima, H. (1995), 'Sampling strategies for use in stratified germplasm collections', In T. Hodgkin, A. H. D. Brown, Th. J. L. Van Hintum and E. A. V. Morales (Eds), *Core Collections of Plant Genetic Resources*, John Wiley and Sons, UK, pp. 35–54.
Zhang, D. and Motilal, L. (2016), 'Origin, dispersal, and current global distribution of cacao genetic diversity', In B. A. Bailey and L. W. Meinhardt (Eds), *Cacao Diseases*, Springer International Publishing, Switzerland, pp. 3–30, doi:10.1007/978-3-319-24789-2_1.
Zhang, D., Arevalo-Gardini, E., Mischke, S., Zúñiga-Cernades, L., Barreto-Chavez, A., Adriazola del Aguila, J. (2006a), 'Genetic diversity and structure of managed and semi-natural populations of cocoa (*Theobroma cacao*) in the Huallaga and Ucayali Valleys of Peru', *Ann. Bot.*, 98, 647–55, doi:10.1093/aob/mcl146.
Zhang, D., Mischke, S., Goenaga, R., Hemeida, A. A., and Saunders, J. A. (2006b), 'Accuracy and reliability of high-throughput microsatellite genotyping for cacao clone identification', *Crop Sci.*, 46(5), 2084–92, doi:10.2135/cropsci2006.01.0004.
Zhang, D., Boccara, M., Motilal, L., Butler, D. R., Umaharan, P., Mischke, S. and Meinhardt, L. (2008), 'Microsatellite variation and population structure in the 'Refractario' cacao of Ecuador', *Conserv. Genet.*, 9(2), 327–37, doi:10.1007/s10592-007-9345-8.
Zhang, D., Boccara, M., Motilal, L., Mischke, S., Johnson, E. S., Butler, D. R., Bailey, B., and Meinhardt, L. (2009a), 'Molecular characterization of an earliest cacao (*Theobroma cacao* L.) collection from Upper Amazon using microsatellite DNA markers', *Tree Genet Genomes*, 5(4), 595–607, doi:10.1007/s11295-009-0212-2.
Zhang, D., Mischke, S., Johnson, E. S., Phillips-Mora, W., and Meinhardt, L. (2009b), 'Molecular characterization of an international cacao collection using microsatellite markers', *Tree Genet Genomes*, 5(1), 1–10, doi:10.1007/s11295-008-0163-z.
Zhang, D., Figueira, A., Motilal, L., Lachenaud, P. and Meinhardt, L. W. (2011a), 'Theobroma', In C. Kole (Ed.), *Wild Crop Relatives: Genomic and Breeding Resources. Plantation and Ornamental Crops*, Springer-Verlag, Berlin, Heidelberg, pp. 277–96.
Zhang, D., Gardini, E. A., Motilal, L. A., Baligar, V., Bailey, B., Zuñiga-Cernades, L., Arevalo-Arevalo, C. E. and Meinhardt, L. (2011b), 'Dissecting genetic structure in farmer selections of *Theobroma cacao* in the Peruvian Amazon: Implications for on farm conservation and rehabilitation', *Trop. Plant Biol.*, 4(2), 106–16, doi:10.1007/s12042-010-9064-z.
Zhang, D., Martínez, W. J., Johnson, E. S., Somarriba, E., Phillips-Mora, W., Astorga, C., Mischke, S. and Meinhardt, L. W. (2012), 'Genetic diversity and spatial structure in a new distinct *Theobroma cacao* L. population in Bolivia', *Genet. Resour. Crop Evol.*, 59, 239–52, doi:10.1007/s10722-011-9680-y.
Zuñiga, T. P. (1988), 'Estudio moreogenico de cacao (*Theobroma cacao* L.) empleando embriones haploides', MSc thesis, CATIE, Costa Rica.

Chapter 4

Safe handling and movement of cocoa germplasm for breeding

Andrew Daymond, University of Reading, UK

1 Introduction

Whilst the centre of diversity of cocoa (*Theobroma cacao* L.) is in Central and South America, figures from ICCO (2016) indicate that 84% of current cultivation takes place in West Africa and Southeast Asia. Since the majority of cocoa is cultivated away from its centre of diversity, movement of germplasm between countries and from the international cocoa genebanks has been integral to the expansion of production.

Cocoa farmers frequently face challenges in the production stage in the form of low potential yields of existing cultivated varieties or losses associated with pest and disease pressures. Furthermore, abiotic stress (such as periods of drought) can influence both crop establishment and yield development of cocoa. With a changing climate, such stresses are likely to become more acute in some growing regions in the future. A route to improving on-farm yields is through breeding varieties that have a higher yield potential and a greater tolerance to pests, diseases and abiotic stresses. Access to a broad range of genetic diversity is an important prerequisite for breeding programmes (Padi et al., 2016). This is particularly important since most of the cultivated varieties in West Africa have been bred from a relatively narrow genetic base (Zhang and Motilal, 2016). Given that much of the production of cocoa is away from the centre of diversity, there is a need for movement from the global genebanks, such as those in Trinidad and Tobago (Iwaro et al., 2003) and Costa Rica (CacaoNet, 2012) to research institutes engaged in evaluation and breeding of cocoa. However, such movement brings with it the risk of transfer of pests and diseases. Therefore, it is critical that movement of germplasm is conducted within a quarantine framework.

http://dx.doi.org/10.19103/AS.2017.0021.05

2 Overview of risks associated with plant movement

Potential losses to pests and diseases are difficult to quantify with some having a much greater impact than others. However, a global figure of a loss in excess of 30% is often quoted (Hebbar, 2007). Some pests and diseases of cocoa are found throughout cocoa-growing regions, while others have a non-uniform geographical distribution. An example of a widespread pathogen is *Phytophthora palmivora* (one of the causal agents of black pod disease). Pathogens confined to particular regions include frosty pod rot (causal agent: *Moniliophthora roreri*), which is encountered in parts of South and Central America and more recently in one Caribbean island (IPPC, 2016); cacao swollen shoot virus (CSSV) disease, which is encountered in West Africa (Ghana, Côte d'Ivoire, Togo and Nigeria; Muller, 2016) and vascular streak dieback, which is found in several countries in Southeast Asia (McMahon and Purwantara, 2016). A number of pests of cocoa are also confined to particular countries or regions. An example is cocoa pod borer which is an important pest in a number of countries within Southeast Asia, including Malaysia, Indonesia, the Philippines and Papua New Guinea (Awang and Lamin, 2014). The geographical spread of major pests and diseases is summarised in Table 1.

A notable example of the devastating effects of the introduction of a new disease into a new area is that of witches' broom disease in Brazil. Historically, in Brazil witches' broom disease was only found within the Amazon region where it is endemic (Evans, 2016). In 1989, it was first observed in the state of Bahia, where the majority of cocoa production takes place (Pereira et al., 1989). Since there was little genetic resistance amongst cultivated varieties, the introduction of the disease had a devastating impact on yields and consequently the economy and employment within the region (Pereira et al., 1996). Other examples of disease and pest movement include the spread of frosty pod rot within Central America (Phillips-Mora et al., 2006) and the spread of the cocoa pod borer within Southeast Asia (Yen et al., 2010). These examples serve to illustrate the dangers of new pest and disease encounters in a region or country and hence the vital need for quarantine measures.

3 Levels of risk

3.1 Assessing and managing risks

Movement of cocoa germplasm for breeding may comprise new material being imported into the country in question, typically in the form of budwood, which is then grafted onto the rootstocks at the destination. Movement may also take place within a given country as part of a breeding programme, for example, the setting up of clonal trials in a region to expose cocoa clones to given stresses or that of progeny trials at distant locations, which involves the movement of seed. Hence, both international and within-country movement need to be considered regarding safe handling of cocoa germplasm for breeding.

The level of risk involved in moving plant material will depend upon a combination of the survivorship of pests and diseases that are present at the source, whether the pests and diseases present at source are also present at the destination and the form in which the plant material is transported. As a general rule, international movement of cocoa germplasm should be via an intermediate quarantine facility. When movement of material

Table 1 Distribution of the main pests and diseases of cocoa by country (adapted from End et al. (2017); information is based on published information at the time of writing). Species of *Phytophthora*, the causal agent of black pod are widespread and so are not listed, with the exception of *Phytophthora megakarya*, which is a particularly virulent strain. *Rosellina* root rot and mirid species are also widespread

Geographical region	Country	Pest/Disease risk
Central America and Caribbean	Belize	*Moniliophthora* pod rot
	Costa Rica	*Moniliophthora* pod rot *Ceratocystis* wilt
	El Salvador	*Moniliophthora* pod rot
	Grenada	Witches' broom disease
	Guatemala	*Moniliophthora* pod rot
	Jamaica	*Moniliophthora* pod rot
	Mexico	*Moniliophthora* pod rot
	Nicaragua	*Moniliophthora* pod rot
	Panama	*Moniliophthora* pod rot Witches' broom disease
	St Vincent	Witches' broom disease
South America	Bolivia	Witches' broom disease *Moniliophthora* pod rot
	Brazil	*Ceratocystis* wilt *Verticillium* wilt of cacao Witches' broom disease
	Colombia	*Ceratocystis* wilt *Moniliophthora* pod rot *Verticillium* wilt of cacao Witches' broom disease
	Ecuador	*Ceratocystis* wilt *Moniliophthora* pod rot Witches' broom disease
	French Guiana	Witches' broom disease
	Guyana	Witches' broom disease
	Honduras	*Moniliophthora* pod rot
	Peru	Witches' broom disease *Moniliophthora* pod rot
	Suriname	Witches' broom disease
	Trinidad and Tobago	Witches' broom disease *Ceratocystis* wilt
	Venezuela	Witches' broom disease *Moniliophthora* pod rot *Ceratocystis* wilt
Africa	Benin	*Cacao swollen shoot virus* (CSSV)

(*Continued*)

Table 1 (*Continued*)

Geographical region	Country	Pest/Disease risk
	Bioko (Fernando Po)	*Phytophthora megakarya*
	Cameroon	*Phytophthora megakarya*
	Côte d'Ivoire	*Cacao swollen shoot virus* (CSSV) *Phytophthora megakarya*
	Ghana	*Cacao necrosis virus* (CNV) *Cacao swollen shoot virus* (CSSV) *Phytophthora megakarya*
	Liberia	*Cacao swollen shoot virus* (CSSV)
	Nigeria	*Cacao necrosis virus* (CNV) *Cacao swollen shoot virus* (CSSV) *Phytophthora megakarya*
	Sierra Leone	*Cacao swollen shoot virus* (CSSV) *Cacao yellow mosaic virus* (CYMV)
	Sri Lanka	Virus (unknown type reported)
	Togo	*Cacao swollen shoot virus* (CSSV) *Phytophthora megakarya*
	Uganda	*Verticillium* root rot
South and Southeast Asia	India	Vascular streak dieback
	Indonesia	Cocoa pod borer Vascular streak dieback
	Malaysia	Cocoa pod borer Vascular streak dieback
	Papua New Guinea	Cocoa pod borer Vascular streak dieback
	Philippines	Vascular streak dieback Cocoa pod borer
	Thailand	Vascular streak dieback
	Vietnam	Vascular streak dieback

takes place from a country or region where a given pest or disease is present to one where it is not present, it *should always* be via an intermediate quarantine facility. For example, movement of germplasm from any Central or South American country to Africa and Southeast Asia or movement within South/Central America from a country where frosty pod rot (*M. roreri*) is present to one where it is not present.

It is important that scientists who are active in research are aware of the risks of moving pests or diseases on their clothes or footwear. In this respect, when travelling from one cocoa-growing region to another, research scientists need to employ a procedure that includes a change of footwear and clothing. Particular care needs to be taken when travelling from a region where frosty pod rot (*M. roreri*) is present as the spores of this disease are very resilient (Evans, 1981). Hence, it is not recommended to travel directly from such regions to cocoa-growing areas where this disease is not present.

3.2 Risks associated with the movement of different plant materials

Seed

In terms of plant parts, movement of seed, which may be required, for example, in the setting up of progeny trials, represents one of the safer means of moving cocoa germplasm (End et al., 2017). However, since pod husks can be infected with fungi or house insect pests, movement of whole pods is not recommended. Furthermore, care should be taken to select pods that do not exhibit visual symptoms of diseases, and a treatment with fungicide should be considered before opening the pods to prevent surface contamination of seed. In areas where CSSV is present, precautions are needed to prevent transport of mealybug vectors of the disease. Campbell (2017) highlighted the fact that mealybug nymphs can feed on the cotyledons of cocoa seeds that might be damaged during pod-splitting and therefore suggested dipping pods in an insecticide solution before seeds are extracted and transported. Awang and Lamin (2017) highlighted the importance of selecting pods with no signs of insect boring and recommended washing the seed in an insecticide solution.

Budwood

Movement of budwood is commonly practised for the purpose of plant breeding in order to maintain the genetic integrity of the material. As budwood may be infected with a number of viruses, international movement of budwood should only take place via an intermediate quarantine facility where virus indexing is conducted. Quarantine measures may be needed when budwood is moved within a country where viruses, such as CSSV, are present (see sections 4.1 and 4.2).

Whole plants

Whole plants represent a high-risk route for cocoa germplasm movement due to the risk of transfer of insect pests and soil-borne organisms such as nematodes and therefore plants should not be moved internationally in soil. International movement of bare-rooted cocoa plants is not recommended, unless it takes place via an intermediate quarantine facility.

In vitro movement

Movement of plant material *in vitro*, for example, somatic embryos, should be in sealed containers containing sterile growth media. The material should be indexed for the presence of pathogens within a quarantine facility.

4 Risk management governance and procedures

4.1 International governance of plant movement

The International Plant Protection Convention (IPPC) was established to regulate the movement of plant material in order to reduce the spread of pests and diseases (IPPC, 2017). In particular, the convention provides a template for phytosanitary certificates. It also requires that signatories to the agreement should have a specifically assigned health authority to deal with imports and exports of plant material. Thus, any international movement of cocoa germplasm, regardless of the form in which it is moved (e.g. budwood, somatic embryos, seed), must comply with IPPC regulations. Specifically, the movement of germplasm should take place in consultation with the relevant plant health authorities in both the exporting and importing countries. This involves the plant health authority in the importing country providing permission in the form of an import permit for germplasm to enter the country from a stated source. The import permit will also state whether any specific tests should have taken place on the material being exported and whether any treatments are required, such as a pesticide/fungicide dip, before export. The plant health authority in the exporting country will provide a phytosanitary certificate which lists any treatments that the plant material was subjected to.

4.2 Within-country quarantine procedures

Within-country movement of cocoa germplasm sometimes takes places as part of a breeding programme, for example, the establishment of clonal trials at a research substation. Here, a quarantine procedure may be needed, particularly if a pest or disease is present at the source location which is not present at the destination. A notable example is the within-country movement of cocoa budwood from sources where CSSV is present. Here, a procedure should be adopted to index donor plants before budwood is transported.

5 Case study: International Cocoa Quarantine Centre, Reading

The International Cocoa Quarantine Centre at the University of Reading (ICQC,R), UK, is currently the main hub for international movement of cocoa germplasm (Daymond et al., 2006; Turnbull et al, 2010; see Fig. 1). It has been in operation since 1985, when it acquired a number of cocoa accessions from the Royal Botanic Gardens at Kew, London, which previously had a responsibility for intermediate quarantine of cocoa. It is currently funded by the Cocoa Research Association and the United States Department of Agriculture. The location of ICQC,R in a temperate country in which cocoa is not cultivated, is significant in that there is no risk of specific pests and diseases of cocoa entering the facility. Indeed, it is a requirement of some cocoa-producing countries that when they import material from abroad it should have undergone quarantine in a non-producing country. It is an intermediate quarantine facility in that further post-entry quarantine procedures may take place after the material has been exported. The facilities comprise 1000 m^2 of compartmentalised greenhouses that are heated to provide tropical conditions. Plants are grown in an inert medium and are regularly watered via an automated system with a nutrient solution. This setup minimises the chance of establishment of invertebrate pests in the medium and also eases plant management. The greenhouses are insect-proofed and access to them is strictly controlled to prevent entry of pests on the clothes or footwear of individuals. Specifically, a standard operational procedure is maintained which prohibits the entry of anybody that has recently been in the field.

As of 2017, the International Cocoa Quarantine Centre contains 400 cocoa accessions, 369 of which are available for international export. These accessions have been received

Figure 1 The International Cocoa Quarantine Centre at the University of Reading, UK.

from the two international cocoa genebanks in Trinidad and Tobago and Costa Rica as well as from a number of national genebanks. Material that has particularly desirable traits for breeders such as high yield potential, large bean size and a level of resistance/tolerance to economically important pests and diseases is brought into quarantine.

The procedures in place at the ICQC,R for receipt and quarantine of cocoa germplasm, which serve as an exemplar of a quarantine procedure, are summarised as follows:-

1 At the source (donor country providing germplasm to ICQC,R), budwood is selected from stock trees that show no visible signs of pests or diseases. After cutting the budwood it is then dipped in a mixture of pesticide and fungicide.
2 The material is inspected by the relevant authority and a phytosanitary certificate is provided.
3 On receipt at ICQC,R, the budwood is inspected in the laboratory under a microscope for any signs of the presence of insects, eggs or fungal spores. In the event that any insects, eggs or fungi should be observed, the material in question is autoclaved and destroyed.
4 In the case of cocoa budwood that is received from a country in which vascular streak dieback is present, samples of budwood are dissected to examine for the presence of characteristic streaking symptoms.
5 Imported material is then grafted onto rootstocks in a greenhouse compartment that is maintained specifically to establish new material from abroad. After grafting, the budwood along with the packaging material is autoclaved.
6 The grafted plants are initially kept in insect-proof cages and are carefully observed after grafting for signs of the presence of any insects that may have survived the insecticide treatment.

7 Received material is subjected to virus indexing using two methodologies: a laboratory-based polymerase chain reaction (PCR) technique and *in situ* screening. Leaf samples are taken at an early stage after the establishment of the imported grafted plant from an actively growing flush. These are then tested in the laboratory using a suite of PCR probes. Since PCR-based methods do not currently detect all viruses of cocoa, a visual indexing test is also utilised (Thresh, 1960), which involves the following steps:
   a Budwood is taken from an accession that has recently been established from abroad ('mother plant') and buds are grafted onto seedlings of West African Amelonado, a variety that shows clear symptoms in the flush leaves when it is infected with viruses such as CSSV. A minimum of three such indicator plants are established.
   b Once the buds have formed a union with the seedlings, the mother plant and indicator plants are moved to a greenhouse compartment specifically for the purpose of virus indexing. Here, indicator plants are inspected on a weekly basis for the characteristic leaf symptoms and stem swellings.
   c In the event that virus symptoms are observed, the mother plant, along with the test plants, is destroyed by autoclaving or incineration.
8 After the two-year quarantine period has been completed for a given accession and if it has a clean quarantine record, the mother plant is moved to the greenhouse compartment that houses the post-quarantine collection and the test plants are destroyed.

The inventory of cocoa accessions ('clones') held at the ICQC,R is maintained within an online database that can be accessed at www.icgd.reading.ac.uk/icqc/.

In response to requests from institutes engaged in cocoa breeding and research, plant material in the form of budwood is periodically exported from the ICQC,R post-quarantine collection. To facilitate this, the recipient institute provides an import permit that has been issued by the relevant authority (Ministry of Agriculture or Plant Health body), along with the requested clone list. On the designated day of export, budwood is cut from actively growing branches and the leaves are removed. The material is then inspected by an officer of the UK Animal and Plant Health Authority who provides the phytosanitary certificate. The budwood is treated by dipping it in a mixture of pesticide and fungicide if this is a stated requirement of the import permit. It is then wrapped in damp paper towel and placed in a sealed polythene bag to maintain viability after it has been transported. A material transfer agreement is also issued which ensures that the germplasm remains in the public domain.

A series of quality control measures are in place within the ICQC,R facilities. These include twice-yearly inspections by expert consultants in pathology, virology and entomology, who report to a Quarantine Advisory Board. The board also reviews the activities of the ICQC,R and considers any necessary amendments to the *modus operandi* of the facility. An annual inspection of the facilities by an officer of the UK Animal and Plant Health authority is also conducted. As of 2017, germplasm has been provided from the International Cocoa Quarantine Centre to institutes in over thirty different countries.

6 Concluding remarks

Breeding new varieties of cocoa is a major component of improving on-farm yields to meet expected future demand for cocoa products. Quarantine procedures and facilities provide the mechanism to enable researchers to have access to the genetic diversity required for such breeding programmes. Continued research is needed to increase the range of tools available to quarantine centres, such as laboratory screens for a broader range of diseases. Furthermore, in an increasingly globalised world, there is an ongoing requirement to raise awareness of the risks associated with cocoa germplasm movement and the need for quarantine measures (both intermediate quarantine and within-country measures) to negate these risks.

7 Where to look for further information

The Technical Guidelines for the Safe Movement of Cacao Germplasm are published online at http://www.cacaonet.org/and are currently available in English, French and Spanish. The document provides general guidance for the safe movement of cocoa germplasm and also specific information on different pests and diseases, particular risks associated with them and quarantine measures required. Each section on a particular pest or disease is authored by an expert in that area. The guidelines are periodically reviewed and updated to take into account new information, for example, on the spread of a particular disease or new disease detection methods.

Information on the International Cocoa Quarantine Centre may be found at www.icgd.reading.ac.uk/icqc/. The website includes a continually updated list of cocoa germplasm and a link to the International Cocoa Germplasm Database (ICGD), which provides information on specific accessions as well as their genetic fingerprint.

Further information about the International Plant Protection Convention can be found at: https://www.ippc.int/en/.

8 References

Awang, A. and Lamin, K. 2017. Cocoa pod borer. In: *Technical Guidelines for the Safe Movement of Cacao Germplasm. Revised from the FAO/IPGRI Technical Guidelines No. 20* (Third update, October, 2017) (M. J. End, A. J. Daymond and P. Hadley (Eds)). Global Cacao Genetic Resources Network (CacaoNet), Bioversity International, Rome, Italy, pp. 66–69.

CacaoNet. 2012. A global strategy for the conservation and use of cacao genetic resources, as the foundation for a sustainable cocoa economy (B Laliberté complier). Bioversity International, Montpellier, France.

Campbell, C. A. M. 2017. Mealybugs and other insects. In: *Technical Guidelines for the Safe Movement of Cacao Germplasm. Revised from the FAO/IPGRI Technical Guidelines No. 20* (Third update, October, 2017) (M. J. End, A. J., Daymond and P. Hadley (Eds)). Global Cacao Genetic Resources Network (CacaoNet), Bioversity International, Rome, Italy, pp. 76–8.

Daymond, A. J., Fenn, M. G. E., Binns, H. and Hadley, P. 2006. The university of reading intermediate cocoa quarantine facility. In: *Proceedings of the Malaysian International Cocoa Conference, Kuala Lumpur, Malaysia*. July 2005, pp. 362–6.

End, M. J., Daymond, A. J. and Hadley, P. (Eds). 2017. *Technical Guidelines for the Safe Movement of Cacao Germplasm. Revised from the FAO/IPGRI Technical Guidelines No. 20* (Third update, October 2017). Global Cacao Genetic Resources Network (CacaoNet), Bioversity International, Rome, Italy.

Evans, H. C. 1981. Pod rot of cacao caused by *Moniliophthora* (*Monilia*) *roreri*. Phytopathological Papers, 24, 44.

Evans, H. C. 2016. Witches' broom disease (*Moniliophthora perniciosa*): History and biology. In: *Cacao Diseases: A History of Old Enemies and New Encounters* (B. A. Bailey and L. W. Meinhardt(Eds)). Springer International Publishing, New York, Dordrecht, London, pp. 137–78.

Hebbar, P. K. 2007. Cacao diseases: A global perspective from an industry point of view. *Phytopathology*, 97, 1658–63.

ICCO. 2016. *Quarterly Bulletin of Cocoa Statistics*, Vol. XLII, No. 3, Cocoa Year 2015/16. http://www.icco.org/about-us/international-cocoa-agreements/cat_view/30-related-documents/46-statistics-production.html. Accessed November 2016.

IPPC. 2016. *Detection of Frosty Pod Rot in Jamaica*. Pest Report, September 2016.

IPPC. 2017. https://www.ippc.int/en/structure/. Accessed June 2017.

Iwaro, A. D., Bekele, F. L. and Butler, D. R. 2003. Evaluation and utilisation of cacao (*Theobroma cacao* L.) genotypes at the International Cocoa Genebank, Trinidad. *Euphytica*, 130, 207–21.

McMahon, P. and Purwantara, A. 2016. Vascular streak dieback (*Ceratobasidium theobromae*): History and biology. In: *Cacao Diseases: A History of Old Enemies and New Encounters* (B. A. Bailey and L. W. Meinhardt (Eds)). Springer International Publishing, pp. 307–35.

Muller, E. 2016. Cacao swollen shoot virus (CSSV): History, biology and genome. In: *Cacao Diseases: A History of Old Enemies and New Encounters* (B. A. Bailey and L. W. Meinhardt (Eds)). Springer International Publishing, pp. 337–58.

Padi, F. W., Ofori, A. and Akpertey, A. 2016. Genetic base-broadening of cacao for precocity and cropping efficiency. *Plant Genetic Resources*. https://doi.org/10.1017/S1479262116000277

Pereira, J. L., Ram, A., Figueredo, J. M. and Almeida, L. C. C. 1989. Primeira ocorrência de vassoura-de-bruxa na principal região produtora de cacao de Brasil. *Agrotrópica*, 1, 79–81.

Pereira, J. L., Almeida, L. C. C. and Santos, S. M. 1996. Witches' broom disease of cocoa in Bahia: Attempts at eradication and containment. *Crop Protection*, 15, 743–52.

Phillips-Mora, W., Coutiño, A., Ortiz, C. F., Lopez, A. P., Hernández, J. and Aime, M. C. 2006. First report of *Moniliophthora roreri* causing frosty pod rot (Moniliasis disease) of cocoa in Mexico. *Plant Pathology*, 55, 584.

Thresh, J. M. 1960. Quarantine arrangements for intercepting cocoa material infected with West African viruses. FAO Plant Protection Bulletin, 8, pp. 89–92.

Turnbull, C. J., Daymond, A. J., Lake, H., Main, N. E., Radha, B., Cryer, N. C, End, M. J. and Hadley, P. 2010. The role of the international cocoa germplasm database and the international cocoa quarantine centre in information management and distribution of cocoa genetic resources. In: *Proceeding of the 16th International Cocoa Research Conference*. Bali, November 2009, pp. 33–9.

Yen, D. L., Water, E. K. and Hamilton, A. J. 2010. Cocoa pod borer (*Conopomorpha cramerella* Snellen) in Papua New Guinea: Biosecurity models for New Ireland and the Autonomous Region of Bougainville. *Risk Analysis*, 30, 293– 309.

Zhang, D. and Motilal, L. 2016. Origin, dispersal, and current global distribution of cacao genetic diversity. In: *Cacao Diseases: A History of Old Enemies and New Encounters* (B. A. Bailey and L. W. Meinhardt (Eds)). Springer International Publishing, pp. 3–31.

Chapter 5

Developments in cacao breeding programmes in Africa and the Americas

Dário Ahnert, Universidade Estadual de Santa Cruz, Brazil; and
Albertus Bernardus Eskes, formerly CIRAD and Bioversity International, France

1 Introduction

Cacao (*Theobroma cacao* L.) breeding is a long-term process that requires large quantities of good-quality germplasm, large experimental areas, testing of a large quantity of progenies and clones, appropriate schemes of selection and a team of experienced breeders focused on clear breeding objectives that meet the needs of farmers and the industry. In this chapter we will describe the main developments in the cacao breeding programmes in Trinidad, Brazil, Ecuador and Costa Rica in the Americas, and in Ghana, Ivory Coast, Nigeria and Cameroon in Africa. These programmes began between the 1930s and 1950s by applying mass selection to local populations and by establishing the first germplasm collections.

Initially, Trinidad selected the series ICS (Imperial College Selection), Brazil the series SIC and SIAL ('Seleção Instituto do Cacau' and 'Seleção Instituto Agronomico do Leste', respectively), Ecuador the series EET ('Estación Experimental Tropical'), Costa Rica the series UF and CC (United Fruit and 'Centro de Cacao', respectively) and the West African producing countries the series Amelonado and local Trinitario (LT). After evaluation of these initial selections, the programmes released clones or open-pollinated cultivars as planting material (Toxopeus, 1985). However, no genetic gains were obtained because local selections with low genetic variability for agronomic traits were used (Bartley, 1994).

http://dx.doi.org/10.19103/AS.2017.0021.06

Between 1938 and 1943, Trinidad's breeding programme collected Upper Amazon (UA) germplasm with the series Scavina (Sca), Iquitos Mixed Calabacillo (IMC), Nanay (Na), Morona (Mo), Parinari (Pa) and Pound (P) (Kennedy et al., 1987). This germplasm was made available to all breeding programmes worldwide from the 1940s onwards. It soon became widely used because of its resistance to almost all biotic and abiotic stresses, high yield and many other important agronomic traits (Bartley, 2005). The quality of this germplasm showed that the wild cacao populations in the Amazon were an important source of genetic variability that could be used to overcome the threats faced by growers worldwide. This has resulted in many botanical expeditions to the Amazon to collect more germplasm (Chalmers, 1972; Allen and Lass, 1983; Almeida et al., 1987; Monteiro et al., 2010).

The utilisation of heterosis in cacao can be considered the most important cacao breeding achievement in the twentieth century, boosting the use of hybrid cultivars. This phenomenon was discovered in cacao between the 1940s and 1960s (Posnette, 1943; Russell, 1952; Montserin et al., 1957) and it was adopted by all breeding programmes to produce hybrid cultivars from the 1960s onwards, with the exception of the Caribbean countries that used clones. The hybrids were produced by crossing UA × UA, UA × Lower Amazon (LA), UA × Trinitario (T) or local selections and T × local selections. Seed gardens were established in the producing countries and hybrid seeds were produced and made available to farmers.

Genetic gains for agronomic traits were reported for these hybrids, showing their superiority over local selections (Soria and Esquivel, 1967; Lockwood, 1976; Bartley et al., 1982; Paulin et al., 1993; Adomako et al., 1999a). Yields of up to 3000 kg/ha/year were reported at experimental stations, depending on the region and the hybrid. The hybrids were also more resistant to biotic and abiotic stresses than local cultivars but, in general, their long-term genetic gains proved to be low in subsequent hybrid cacao releases (Eskes et al., 1994; Monteiro and Ahnert, 2012). This has been attributed to the use of unimproved parents (Lockwood and Pang, 1994; Pires et al., 1999a; Eskes et al., 1994). To overcome this problem, breeders in the Americas and Africa started using population breeding and recurrent selection from the 1990s onwards. This decision was supported by quantitative genetics, which showed that agronomic traits have strong additive effects in cacao (Monteiro et al., 1985; Lockwood and Pang, 1994; Eskes et al., 1994; Adomako et al., 1999b; Tan, 1990; dos Santos et al., 2016). This allowed complementary populations with a high frequency of favourable alleles for parental selection to be developed.

All the cacao breeding programmes described here (except those in Cameroon and Nigeria) have been using population breeding to improve parents or obtain superior clones since the beginning of the twenty-first century. Farmers in Brazil and Ecuador have been directly involved in selecting breeding materials with increased disease resistance. Ivory Coast (Clement et al., 1994) and Brazil (Pires et al., 1999a) have been using reciprocal recurrent selection schemes.

Clonal cultivars are currently the most important type of planting material in the Americas and Asia. In future, they will become the most important cultivars in the world, because they are easier to breed and have better agronomic, resistance and quality traits than traditional cacao hybrid cultivars. High-tech cacao plantations using clones, fertilisers, irrigation and mechanisation have been established, mainly growing the clone CCN51 (Cacao Castro Naranjal), especially in drier regions with less than 1200 mm of annual rainfall. These modern industrial cacao plantations have had a tremendously positive impact on breeding, encouraging the production of high-yielding cultivars adapted to this cultivation system. All American cacao-producing countries now deliver resistant and highly productive clonal cultivars to plant in new areas and rehabilitate old plantations for large-, medium- and small-scale farmers.

There is currently great demand for new planting material in all producing countries due to the increasing impact of pests, diseases and drought and the need to reduce production costs. To accelerate breeding, more support is required for large-scale testing of progenies and clones from the current population breeding programmes. More of the genetic variability available in germplasm banks needs to be used, to incorporate new alleles into the populations that are being improved.

Breeding for rootstocks should also become a priority once the synergy between rootstock and scion has been explored. Rootstocks with important agronomic attributes can be multiplied by tissue culture to produce genetically uniform plants. The use of uniform rootstocks, grafted with clonal cultivars, could possibly reduce variation between trees, which is very important for higher yields/ha. Rootstocks resistant to biotic and abiotic stresses can also be obtained in this way, to resist soil diseases, drought and waterlogging.

Ideotype plants should be developed for each cultivation system, to meet demands for ease of cultivation, high planting density and mechanisation. To produce ideal clonal cultivars, plants with low to moderate vigour and an upright growth habit should be selected, to facilitate pruning and increase planting density. These plants should have a low pod index to reduce the cost of harvesting and opening the pods, and better partitioning of photo-assimilates between pods and vegetative growth (yield efficiency). In this respect, further knowledge of the physiological aspects of the cacao plants being selected and utilisation of this knowledge in the breeding programme are very important.

However, adopting an agronomically superior cultivar is not enough to increase yields/ha. The full genetic potential of a cultivar can only be expressed if farmers also adopt good agricultural practices, such as correcting soil pH, managing shade, fertilisation, cleaning, sanitation, pruning, irrigation and so forth to maintain good tree health, with enough branches and leaves to produce the necessary photo-assimilates for high fruit load and high yields of beans/ha.

With the adoption of clonal cultivars and rootstocks by most countries in America, Asia and, we hope, Africa before very long, breeding can be accelerated and can provide solutions to current and future constraints. The use of improved clones and good agricultural practices could help farmers produce more kg/ha, making cacao cultivation more profitable and producing enough beans to meet industry demand.

These considerations only relate to breeding for productivity and disease or pest resistance. Specific cacao breeding for sensory quality traits has been limited to the Trinidad TSH (Trinidad selected hybrids) breeding programme started in the 1970s (Maharaj et al., 2011). In 1995, cacao breeders considered that breeding for quality traits was being hampered by a lack of uniformity in demands from the industry, and also by a lack of efficient single-tree selection methods (Lockwood and Eskes, 1995). This situation was transformed by the discovery of a direct relationship between fresh pulp flavour and aroma traits and fine flavour chocolate sensory traits in 2006, facilitating rapid single-tree selection based on sensory pulp evaluations. There is a brief discussion at the end of this chapter on recent developments and perspectives in selection for organoleptic quality, including the discovery of 64 unique aroma profiles in the Peruvian native Chuncho variety in 2015.

2 Types of commercial cacao cultivars

Three types of commercial cacao cultivars are grown worldwide: hybrids, clones and local or landrace cultivars. Hybrids and clones are usually produced by breeding programmes, while local cultivars are selected by farmers. In West Africa, around 80% of the cacao plants cultivated originate from local farmers' selections and the remaining 20% are hybrids selected by breeders. In the Americas, around 55% are local cultivars, 40% hybrids and 5% clones. In Asia, 65% are local cultivars, 30% hybrids and the remaining 5% are clones (Paulin and Eskes, 1995). These figures may have changed since 1995 because cacao planting is increasing by around 3% per year worldwide (ICCO, 2016) and new plantations mainly use hybrids and clonal cultivars. However, this does demonstrate that plantations in some parts of the world are still based on unscientifically selected breeding material, and replacement of these cultivars by new, more productive and resistant plants is a priority. Farmers do not generally replace existing trees, except in Brazil, Asia and some Latin American countries, but they may plant new trees.

A good commercial cacao cultivar needs to possess the agronomic traits required by farmers, the physical, chemical and organoleptic traits needed by industry and the ability to adapt and produce well when challenged with the biotic and abiotic stresses where it is cultivated (Toxopeus, 1985). Farmers prefer fast-growing cultivars (vigorous plants) to facilitate establishment in the field, early yielding for rapid returns on income, production and stress resistance to increase income and reduce production costs, with easy husbandry to facilitate management, and low pod index to reduce the cost of harvesting and opening the pods. Industry requires dried beans with an average weight above 1.0 g, around 55% fat and 11–12% shell, well-fermented, free of chemical and biological contaminants and with the appropriate organoleptic characteristics. Each growing area has its typical soil, climate, pests and diseases, requiring specific planting material which is adapted to each site and capable of maintaining stable production over the years. These general features have guided breeding for cultivar development.

Planting material to establish commercial plantations can be reproduced by farmers, private nurseries or the government. Local cultivars are produced by farmers themselves, by selecting the parents to produce seeds or buds, and preparing seedlings. This choice gives farmers more autonomy as they do not have to pay for the seeds, buds or seedlings, but they really should be choosing hybrids or clonal cultivars developed by breeding programmes for improved planting material.

Hybrid seeds are produced by the governments of the producing countries by first establishing a seed production garden, usually consisting of self-compatible or cross-compatible pollen donor parents and self-incompatible female parents from different genetic groups (Lockwood, 2015). Seeds are produced by natural or artificial (hand) pollination of fresh flowers. Hand pollination is preferable because it reduces pollen contamination by selfing or undesirable crosses (Lanaud et al., 1987). Farmers usually receive a mixture of fruits, containing hybrid seeds, or seedlings, which originate from a minimum of five different mother clones because hybrid seeds give rise to a certain proportion of self-incompatible trees. Therefore, mixing seeds to produce a plantation with a mixture of hybrid plants increases the chance of intercrossing. The mixture of different genotypes is also believed to reduce the incidence of pests and diseases, and improve adaptation to the local environment, making the plantation more resilient. However, the mixture of hybrids produces a heterogeneous plantation made up of plants with different heights and vigour, making management very difficult for the farmer.

Clonal cultivars can be produced by farmers, local governments or private nurseries and give farmers more autonomy as they can either produce or purchase their own planting material. In the case of hybrid seeds, farmers have no choice and depend on the government to release a mixture of seeds or seedlings. Clones are easily produced by rooting cuttings, budding or grafting onto rootstocks or chupons of old plants. Self-compatible clones can be planted in monoclonal blocks, creating homogeneous plantations that are easier to manage. The technological barriers to reproducing cacao by clonal propagation have now been removed and the technology can be easily transferred to small farmers (Sena Gomes and Sodré, 2015).

3 Breeding objectives

The constraints on cacao bean production in all producing countries have increased in the last few years for various reasons, such as spread of diseases and insect pests throughout the production regions, extreme drought and increased temperatures caused by worldwide climate change, soil exhaustion due to overuse and a lack of nutrient replacement, a shortage of labour in rural areas and increased production costs. This means that the list of objectives to breed for is also increasing, and that requires more effort, investment and new breeding strategies.

Breeding for improved yield, as well as resistance to pests and diseases, has been, and continues to be, the most important objective for all programmes. However, programmes are also breeding for improving other traits such as drought resistance, establishment ability, husbandry and other attributes, depending on the importance of each in the particular production region. Yield is a very complex trait that depends on intrinsic plant factors (genetic, physiological and biochemical), and also on biotic and abiotic factors where the plant is grown. The most important factor is that there should be genetic diversity for intrinsic yield factors once plants with high yielding capacity have been selected. For example, CCN51 is a well-known highly productive clone, but there are others with the same potential in the producing countries. Breeder experience and field results indicate that there is great potential to improve cacao for higher yields, but the cultivation system also needs to be improved as has occurred in high-tech agriculture. There is also great genetic diversity to be explored for yield component variables, such as the total number of fruits produced by a plant, number and weight of beans per fruit, shell and placenta weight per fruit, water content in beans and pulp and number of ovules per ovary (Bartley, 2005). In conclusion, the new breeding schemes adopted by some programmes to improve parents and exploit heterosis should allow selection of new cultivars with higher yield potential to continue.

Diseases and insect pests are different in each growing region, requiring specific breeding solutions for each region. In West Africa, efforts are mainly concentrated on resistance to swollen shoot virus disease (CSSV), *Phytophthora palmivora*, *P. megakarya* and mirids (*Distantiella theobroma* and *Sahlbergella singularis*). In the Americas, these are mainly aimed at witches' broom (*Moniliophthora perniciosa*), moniliasis (*M. roreri*), black pod (*Phytophthora spp*) and mal de machete (*Ceratocystis cacaofunesta*). Some producing countries now have two or more important diseases, such as moniliasis, witches' broom and pod rot (*P. palmivora*) in the Americas, and CSSV and pod rot (*P. megakarya* and *P. palmivora)* in Africa, increasing the drive to produce new resistant and productive cultivars.

In Asia, vascular streak dieback (VSD, *Oncobasidium theobromae*) and *P. palmivora* are the main foci of breeding for disease resistance, and there is also some breeding for resistance to cocoa pod borer (*Conopomorpha cramerella*), a devastating insect pest restricted to Asia and parts of the Pacific.

Various sources of resistance to insect pests and diseases have been identified in the Sca, P, Pa, Na, IMC, UF, EET and ICS gene pools, among others (Bartley, 2005; Thresh et al., 1988; Pires et al., 1999b; Phillips-Mora, 1999). These sources of resistance were initially used to find immune plants with one or more sources of resistance. In the Americas, for example, witches' broom-resistant hybrid cultivars were obtained by crossing resistant parent clones, such as Sca6 and 12, with clones ICS1, 6, 8 and others. The hybrids of Sca crossed with these ICS clones have one major resistant gene that worked in Trinidad but has failed in other regions in Latin America, probably because of different strains of the fungus (Bartley, 1981; Monteiro and Ahnert, 2012). In Ghana, a great effort was made to produce CSSV- and insect pest-resistant hybrids, with interesting results (Thresh et al., 1988), but breeding concepts failed because the resistant plants did not remain resistant for more than a few years (Eskes et al., 1994).

Field experience and disease inheritance studies have recently demonstrated that resistance in cacao is a quantitative trait (Bartley, 1994; Djiekpor' et al., 1994; Pires et al., 1999b; Phillips-Mora, 1999; Lopes et al., 2011). New breeding strategies incorporate this pattern of inheritance are integrated with breeding for yield (Eskes et al., 1994; Lopes et al., 2011; Monteiro and Ahnert, 2012). Some practical results of these new breeding strategies are clonal cultivars tolerant to witches' broom in Brazil (Lopes et al., 2011), Trinidad (Maharaj et al., 2011) and Ecuador (Amores et al., 2011), and moniliasis in Costa Rica (Phillips-Mora et al., 2013). Tolerant cultivars in these countries have resisted well so far, especially when their cultivation is based on an integrated system involving chemical fungicides, resistance and good agronomic management.

Breeding for good establishment ability and drought resistance has become an important objective in all producing countries due to the effects of worldwide climate change, and hybrid cultivars with these traits have been selected in both Ghana (Abdul-karimu et al., 2003) and Nigeria (Jacob et al., 1971). Two studies in greenhouses in Brazil identified drought-resistant plantlets that were then selected for rootstocks (Santos et al., 2014; dos Santos et al., 2016).

Another objective that is becoming more important is the development of agronomically more manageable cultivars, suitable for different cultivation systems. The ideal plants should be high yielding, be resistant to biotic and abiotic stresses, have a low pod index, high seed index and erect growing habit, and easily be managed even at higher plant densities (Ahnert, 2006; Lockwood and Pang, 1994).

4 Heterosis and heterotic groups in cacao

Heterosis and the correct allocation of parents to their heterotic groups are very important in cacao breeding, to produce hybrid and clonal cultivars. Heterosis occurs in cacao when genotypes from different geographic origins, such as UA × UA (Upper Amazon from different geographic locations), UA × LA, UA × T, T × LA, etc., are crossed, and the progeny are superior to the average of the parents, or to the best parent, for a certain trait. Heterosis is demonstrated by plant vigour, establishment ability, precocity, yield and yield component variables, fat content and other important traits (Posnette, 1943; Russell, 1952; Montserin et al., 1957; Eskes et al., 1994). Breeders have used heterosis to produce hybrids and clones since the 1960s, so they are aware of its importance in improving cacao. However, there has not yet been enough investment of time or financial resources to classify and separate parents into their different hetorotic groups. In maize, this was considered a priority and started in the early days of breeding to facilitate the choice of parents and hybrid production (Hallauer and Miranda, 1988). Anyone wishing to start this work in cacao could refer to the genetic groups recently proposed: Maranon, Curaray, Crioulo, Iquitus, Nanay, Contanama, Amelonado, Purús, Nacional and Guiana (Motamayor et al., 2008). Genotypes with promising features could be tested in appropriate mating designs, and combined with genetic markers to classify and allocate them to heterotic groups. This would require time and financial resources, but would be studied and understood more effectively provided it is considered a priority to be developed internationally.

5 'Traditional' and new cacao breeding methods

'Traditional' breeding to develop hybrids is based on bi-parental crossings between clones to explore heterosis and combine each parent's traits to improve the progeny. The parents used in this scheme are accessions collected in the Amazon, and local selections, a limited number of which have been improved by some degree of selection and/or breeding. Breeders have tried to develop more homozygous parents by self-pollination or using double haploids to produce more uniform hybrid progenies, but this work was terminated due to the scarcity of haploid plants and the duration required. In other words, parent improvement has not been systematic in most of the programmes described here. Instead, they have focused on producing cultivars from existing germplasm to meet more immediate demands, especially for resistant and productive cultivars (Lockwood and Pang, 1994).

However, it became clear in the 1990s that it was difficult to progress genetically with the 'traditional breeding' approach and develop cultivars with all the necessary traits, both for the present and the future, so some breeding programmes in Africa and America started population breeding. This approach is supported by quantitative genetic studies in cacao which demonstrate that inheritance of agronomic traits is preferentially controlled by additive genes (Monteiro et al., 1985; Lockwood and Pang, 1994; Adomako et al., 1999b; dos Santos et al., 2016), facilitating selection of parents with an increased frequency of favourable alleles. Variability of important agronomic traits in cacao is also spread across different populations and the best way of combining is by successive cycles of crossing, evaluation, selection and crossing in a population breeding scheme.

Trinidad's breeding programme included recurrent selection since it started in the 1950s, generating the TSH (Trinidad selected hybrids) and TSA (Trinidad selected Amazon) hybrids and some very interesting clones, demonstrating how useful this breeding strategy is in cacao. Two interesting reciprocal recurrent selection programmes were then reported; one in Ivory Coast (Clement et al., 1994) and one in Brazil (Pires et al., 1999a). These schemes use intra- and inter-population breeding to improve two populations, each belonging to a specific heterotic group, simultaneously. Parents are improved for general combining ability (GCA) and specific combining ability (SCA) at the same time. It is also possible to integrate tools, such as molecular markers, candidate genes and wide genome selection, into these schemes to accelerate cacao breeding and improve precision. More details of these schemes are given in the description of breeding programmes in various countries in the following sections.

6 Cacao breeding programmes and outputs

6.1 Trinidad

Trinidad's breeding programme began in 1930, laying the foundations for cacao breeding and collecting, developing and releasing the most important germplasm used worldwide (Bartley, 1994, 2005). Pound (1932) and Cheesman and Pound (1933) defined the methodologies for individual plant selection and, at that same time, agronomists in Trinidad understood that selecting individual trees and using them as clones would fix important traits and produce homogeneous plantations (Bartley, 1994). Genetic resources were collected from the UA region under Trinidad's programme and made available for worldwide breeding in the 1940–50s and samples of ICS, TSA and TSH clones were also made available. Population breeding, or successive cycles of crossing, evaluation, selection and crossing, started in Trinidad in 1949, and the breeding programme has been delivering clonal cultivars to farmers in the Caribbean region as commercial planting material from the beginning of the programme until today (Kennedy et al., 1987). The programme is currently delivering 20 new clonal cultivars to farmers (Maharaj et al., 2011) and it also maintains an International Cacao Collection of 2400 accessions (CacaoNet, 2016).

6.1.1 Mass and clone selection

Mass selection was conducted on plantations from 1931 to 1935 to select individual plants to start a germplasm collection and a programme of clone selection. Individual plants were initially selected on a large scale (50 000 candidate plants), based on visual traits and information provided by farmers with regard to their yield capacity. These candidate plants were then reselected using negative selection, eliminating any plants with small pods and thick pod walls, reducing the number to 1000 plants. These plants were further reselected for the production of at least 50 pods per year during the two years of evaluation, and a pod index of 16.5, reducing the total to 100 plants. These 100 plants were then propagated as clones at River State in Diego Martin in 1937 and coded ICS1 to ICS100 (Cope, 1952; Toxopeus, 1985).

The ICS clones were evaluated by testing in replicated blocks in one location, and then in many locations in Trinidad and the results showed average yields of 800–1000 kg/ha/year, with the best being ICS1, 6 8, 39, 40, 43, 48, 60 and 95. The ICS95 clone was the most tolerant to witches' broom at that time. Some clones performed better in one environment than in another, demonstrating a strong interaction between genotype and environment. The superior clones were released as cultivars for commercial cultivation in Trinidad and some were used as parents in the local breeding programme. Samples of ICS clones, especially ICS1, 6, 45, 60 and 95, were introduced to almost all producing countries and have been used as parents for crossing and hybrid seed production (Toxopeus, 1985).

In 1928, an outbreak of witches' broom disease was registered in Trinidad (Bekele, 2004). The disease affected local farmers and the industry very seriously, as it had in Ecuador in 1920 (Amores et al., 2011) and, more recently, in Brazil in 1989 (Pereira et al., 1996). In Ecuador, some resistant plants were found after infection with the disease and these were named 'refractarios'. Pound visited Ecuador and noticed that 'refractario' trees did not belong to the 'Nacional' population or other types grown locally for commercial production, but to another population (Pound, 1943). He hypothesised that the 'refractario' germplasm could belong to populations from the headwaters of the Amazon River.

To confirm this hypothesis, Pound organised an expedition to collect cacao in the UA in 1938. He navigated the Amazon River from Brazil to Peru and Ecuador, and described the cacao population along the river and its tributaries, discovering disease-free plants upstream on the River Iquitos in Peru/Ecuador. A total of 250 pods were collected from the UA region and sent to Barbados for quarantine, generating 2500 seedlings. These plants were clonally propagated and established in an area infected with high levels of witches' broom at Marper Farm, eastern Trinidad and this germplasm formed the series IMC, Mo, Na, Pa and Sca. Pound returned to the UA region in 1943 and collected budwood from eight parental plants which had previously provided pods for the pod collection, and from 24 other local trees, generating the 'P' series. He also collected 320 pods from 80 'refractario' trees in Ecuador, quarantined them and established the seedlings in Trinidad (Kennedy et al., 1987).

Field evaluation of these accessions revealed that only two, Sca6 and Sca12, were highly resistant to witches' broom (Bartley, 1994). According to Laker et al. (1988), these clones were released as cultivars in Trinidad for some years, but were later dropped because of the small size of their beans. These two Scavina accessions are very important sources of resistance to witches' broom and other traits, so they are widely used as parents. It is worth mentioning that some UA accessions collected by Pound or derived from the material collected are the most important parents for hybrid seed production in the world. The success of Pound's expedition and the possibility of finding new sources of resistance and other agronomic traits have encouraged many other expeditions, such as those reported by Chalmers (1972) and Allen and Lass (1983), and those developed by Brazil in the Amazon Basin (Almeida et al., 1995; Bartley, 2005).

6.1.2 Population breeding and clone selection

Breeders in Trinidad soon realised that ICS and the UA clones had not only very interesting traits but also weaknesses. They used population breeding to combine the important features of each group, developing and improving a segregating population to select agronomically superior parents and hybrids to produce clones. Population breeding started in 1949 and continues until today (Maharaj et al., 2011), delivering the well-known TSH and TSA clones. Breeding objectives at the beginning of the programme were as follows: improve yield and yield component variables, such as bean size and pod index; increase resistance to *Ceratocystis* and witches' broom diseases; and maintain the organoleptic quality of fine flavour from Trinitario plants. They used ICS1 (large bean size and low pod index), Sca6 (resistance to witches' broom, and low vigour), P18 (low pod index, precocious, vigorous and resistant to *Ceratocystis*) and IMC67 (*Ceratocystis* resistance, low pod index and a large number of beans per pod) as parents. Others were used in addition to this core germplasm, such as ICS95 (high yielding), P7 and some Pa clones (Maharaj et al., 2011).

In total, 20 TSH clonal cultivars were selected by this breeding programme and used as planting material. Early selections were the clones TSH730 (IMC67 × ICS95); TSH919 [IMC67 × TSH538 (ICS1 × Sca6)]; TSH973; TSH1076 [P18 × TSH758 (TSA641 × OPEN)]; TSH1095 (P18 × TSH75) and TSH1102. Later selections were TSH1102 [P18 × TSH792 (TSA644 × OPEN)]; TSH1104; TSH1188 (P18 × TSH753); and TSH1220 [TSH565 (Sca6 × ICS1) × IMC67]. The newest selections are TSH1313; TSH1315 (TSH938 × TSH1095); TSH1330 (IMC67 × TSH932); TSH1344 (IMC67 × TSH932), TSH1347; TSH1350 [IMC67 × TSH728 (ICS95 × IMC67)], TSH1352; TSH1356; TSH1362; TSH1364 (TSH1105 × TSH1103) and TSH1380 (Table 1). Simple, double and complex crosses were created, but the number of parents was limited to the original selection of parents. These cultivars have the potential to produce in excess of 2000 kg/ha and the range of some of their yield production variables is as follows: bean number 42.2 ± 2.42 to 61.4 ± 2.89, cotyledon dry weight 0.74 g ± 0.02 to 1.49 g ± 0.06 and pod index 12.0 ± 0.67 to 26.1 ± 0.7 (Maharaj et al., 2011).

Table 1 Yield components of 20 clonal cultivars released by Trinidad's breeding programme

Cultivar	Pod weight (g)	Total wet bean weight (g)	Number of beans	Dry bean weight (g)	Pod index
TSH730	844.54	250.51	52.1	1.5	13.77
TSH919	847.62	225.31	52.4	0.78	26.16
TSH973	970.2	199.6	49.9	1.10	19.72
TSH1076	877.65	216.03	52.2	1.19	17.50
TSH1095	831.73	230.33	52.1	1.58	13.18
TSH1102	988.23	274.95	61.4	1.48	12.00
TSH1104	802.91	263.9	51.0	1.44	14.84
TSH1188	839.36	208.27	48.1	1.24	18.36
TSH1220	938.77	210.15	48.4	0.87	25.76
TSH1313	569.4	201.8	46.4	1.35	17.52
TSH1315	465.92	154.7	42.2	1.27	20.03
TSH1330	903.19	218.5	50.1	1.32	16.34
TSH1344	722.03	187.13	46.3	1.37	17.40
TSH1347	813.79	193.11	49.6	1.34	16.31
TSH1350	810.99	222.66	52.2	1.81	16.79
TSH1352	758.46	220.27	51.7	1.34	15.47
TSH1356	654.1	209.36	49.4	1.37	16.14
TSH1362	812.3	286.29	50.9	1.51	14.14
TSH1364	886.9	194.21	50.0	1.25	17.51
TSH1380	852.33	210.51	48.7	1.28	17.10

Source: Maharaj et al. (2011).

6.2 Brazil

In Brazil, the breeding programme was started in Bahia in 1930 by the now-defunct 'Instituto de Cacau da Bahia' which selected local cultivars and used them to release clones and open-pollinated seeds to farmers. The 'Comissão Executiva do Plano da Lavoura Cacaueira' (CEPLAC) was created in 1956 and took over the breeding programme. From 1960 to 1997, CEPLAC developed and released bi-clonal hybrid cultivars and produced hybrid seeds as commercial planting material for farmers. From 1989 onwards, with the outbreak of witches' broom disease in Bahia, the programme was completely reorganised in terms of concepts and breeding methods. Population breeding started in 1996 and the concept of participatory breeding was adopted. Advanced materials are evaluated in farmers' fields, using farmer's participation to increase testing capacity and improve the evaluation system. Clonal cultivars started to be released in Bahia and Espirito Santo in 1997; ten are currently available and others will soon be delivered. In the Brazilian Amazon, CEPLAC continues to release bi-clonal hybrid seeds to farmers and is also starting to release clones. Brazil began to collect germplasm in the Amazon Basin in 1965 and the last expedition took place in 2007 (Monteiro et al., 2010). Brazil currently has one of the largest wild germplasm collections in the world, with 2000 accessions (Lopes et al., 2011), and an inventory describing the wild accessions established in the field until 1985 can be found in Almeida et al. (1987). A total of 3300 accessions are conserved in germplasm banks maintained by CEPLAC (Lopes et al., 2011).

6.2.1 Mass and clone selection

Mass selection was conducted from 1930 to the 1950s on plantations made up of the local cultivars 'Cacau Comum', 'Pará', 'Parazinho' and 'Maranhão' (LA types). The main objectives at that time were to improve yield and yield component variables (low pod index and seed index above 1.0 g) and increase resistance to pod rot (*P. palmivora*) (Vello et al., 1969). The SIC series and the SIAL series in Bahia and EEG ('Estação Experimental de Goitacazes') in the Espirito Santo State were selected during this period.

A total of 300 clones were selected and established in local experimental stations for evaluation and conservation, and agronomically superior clones were released as clonal cultivars. The government started to produce and release clonal seedlings, but this initiative soon stopped because of transportation difficulties and poor acceptance of this type of seedling by farmers. The clone selections were then planted in blocks and used to produce open-pollinated seeds as planting material for farmers (Vello et al., 1969).

There are now around 100 SIC, SIAL and EEG germplasm accessions in CEPLAC/CEPEC's cacao germplasm collection. They are very important sources of genes for breeding in Brazil, because of their adaptation to the cacao region of Bahia and their interesting agronomic traits. These accessions have been used as parents for hybrid seed production and other breeding purposes, and they have also been released to international cacao germplasm banks for use elsewhere. They are currently used as rootstocks in Bahia because of their resistance to *Ceratocystis*.

Mass selection was conducted on local Bahia cacao plantations for the second time in the 1990s, to select resistant clones after the witches' broom disease outbreak, and farmers were involved to improve the selection process (Monteiro and Ahnert, 2012). The on-farm cacao population in the 1990s was composed of local cultivars and hybrids delivered by CEPLAC since the 1960s. CEPLAC and the local government of Bahia launched a campaign to select resistant plants on farmers' fields and at CEPLAC's experimental stations. Thousands of resistant plants were initially identified and, after two years of field evaluations for yield, resistance (cushion, fruit and vegetative brooms), seed index and shell thickness, superior plants were selected and cloned. These clones were evaluated in experimental stations and field trials in many locations.

In 1997, CEPLAC initially released five resistant self-incompatible clones (TSH1188, TSH516, TSH565, EET397 and CEPEC42), then other clones (TSH774, TSA792, TSA654 and TSA656.), without large-scale testing in the field because there was no time to do so; something had to be done to alleviate farmers' anxiety. Unfortunately, these clones did not perform well and the plants were soon severely attacked by the fungus. Their resistance was based only on the Scavina alleles and it had not withstood attack, as had occurred in Ecuador. They also did not produce well because of their self-incompatibility and high incidence of cushion brooms: it is a well-known fact that self-incompatible plants produce more flowers than self-compatible plants, increasing points of infection (Monteiro and Ahnert, 2012).

Resistant and self-compatible clones were then selected from farmers' fields and these proved to be superior and more resistant. Evidence from genetic studies and molecular markers suggests that self-compatible clones have resistant alleles from more than one source (Pires, pers. comm.; Monterio and Ahnert, 2012). The first group of self-compatible clones was released in 2007 and others are being tested in a participatory system (Lopes et al., 2011; Monteiro and Ahnert, 2012). The ten most planted self-compatible and resistant clones are: PS1319, PS1030, SJ02, CEPEC2002, PH15, PH16, BN34, FA13, CP49 and CCN51. The yield potential of these ten clones varies from 2000 to 3000 kg/ha and they are highly resistant/tolerant to witches' broom. The yield and yield production components of some clones are presented in Table 2.

The participatory method adopted in Bahia provided very important information and experience to breeders and institutions working in cacao research and extension services. Advanced cacao selections are now tested in farmers' fields, so cultivars are tested under

Table 2 Yield and its components in nine clonal cultivars in Brazil

Cultivar	Yield* kg/ha	Pod weight (g)	Total wet bean weight (g)	Number of beans	Dry bean weight (g)	Pod index
SJ02	1774	1217.4	194	46.5	1.5	14.1
BN34	–	1133.5	192.3	40.8	1.5	15.9
CCN51	1614	1362.5	245.3	46.1	1.8	12
PH16	1359	1054.3	198.1	43.7	1.4	16.2
FA13	–	893.3	137.9	45.3	1.1	21.9
PS1319	1427	998.8	210.1	45	1.5	14.4
CEPEC2002	1681	808.2	136.9	38.9	1.1	23.4
PS1030	–	729.7	165.6	46.5	1.2	18.7
PH15	–	920.1	199.7	47.7	1.5	14.4

* Yield data from the fourth year of field evaluation according to Lopes et al. (2011).
– Data not available. Other data were collected by the authors.

farm conditions and in different environments. All cacao clonal cultivars delivered now are self-compatible to increase the chances of pollination and reduce the points of entry for witches' broom on the excess flowers produced by self-incompatible plants (Monteiro and Ahnert, 2012).

6.2.2 Hybrid selection

A total of 600 hybrid combinations were tested in the producing regions of Bahia and Amazon from the 1960s to the 1980s to select cultivars and study the inheritance of agronomically important traits (Lopes et al., 2011). These hybrids were mainly based on crosses between UA or T and local selections, to explore heterosis among these groups and combine the important agronomic traits in each group.

The potential of a representative sample of parents tested in the state of Bahia was analysed and published by Mariano and Bartley (1981) and Bartley et al. (1982). They used field data collected from five experiments conducted at CEPEC experimental stations from 1965 to 1969, using a randomised block design with three to five repetitions, plant spacing of 3 × 3 m and shade provided by *Erythrina glauca*. The following variables were used in the analysis: (i) production of wet bean (g/plant/year) and (ii) percentage of fruits infected by *P. palmivora* (pod rot). The following parents were tested: local selections ('Cacau Comum' from the series SIC, EEG, CEPEC and 'Catongo'); T (ICS1, 8, UF168, 613, 667) and UA (Sca6, 12, IMC67 and P12). The Catongo parent was used as a control.

These experiments demonstrated very variable behaviour by the UA and T parents in terms of combining ability for yield (wet bean/plant/year) and percentage of fruits infected by pod rot. On the other hand, when the locally selected parents were crossed with the other parents, there was almost no difference between them in terms of combining ability for these variables because of their low genetic diversity (Mariano and Bartley, 1981).

When the ICS1 and 8 parents were crossed with local selections, there was high GCA for wet bean yield and, although the hybrids started producing later, production was more stable over the years. They also performed better, producing more wet beans per plant than the other parents. The UF168 and 667 parents had high GCA for yield, but a higher percentage of fruits were infected by pod rot due to their increased susceptibility to *P. palmivora*. The UF613 parent had low GCA for yield, but was more resistant to pod rot. Their yields were the same as the local selections, which did not demonstrate heterosis. Behaviour was similar between the Sca6, 12 and P12 parents, with high GCA for yield and yield precocity, producing more yield in the first years in the field but decreasing from the sixth year onwards. They had the smallest average wet bean/plant and a relatively high percentage pod rot infection compared to the other parents. IMC67 displayed a certain combining ability, precocity and intermediate yield, and a high percentage pod rot infection (Bartley et al., 1982).

The adult phase of the most productive hybrids, such as SIC923 × ICS1 and SIC1104 × UF168, yielded around 2500 kg/ha/year dry bean, while the less productive ones, such as IMC67 × Catongo and SIC841 × UF613, yielded 1622 and 1718 kg/ha/year dry bean, respectively, in experimental stations. The Trinitario hybrids had the highest average wet beans per pod, such as UF168 × Catongo with 128 g, while the UA parents had the lowest, such as EEG64 × P12 with 79 g. The Catongo control produced around 1900 kg/ha/year of dry bean (Bartley et al., 1982). Based on the results of the test of hybrid combinations the following parents were used to produce hybrid seeds in Bahia:

ICS1, ICS6, ICS8, UF613, UF168, UF667 (T), IMC67, Pa30, Pa150, Sca6, Sca12, TSA644 (UA), SIAL70, SIAL169, SIAL325, SIAL505, SIC17, SIC19, SIC328, SIC329 and SIC813 (LA) (Yamada et al., 2001). It is interesting that hybrids have not been accepted well by famers in Bahia, as they prefer local cultivars. Hybrids were replaced by clonal cultivars resistant to witches' broom in the 1990s.

A new generation of hybrids was developed for the Brazilian Amazon in the 1970s. The results of one trial which tested 18 hybrid combinations in Pará were published (Machado et al., 1983, 1986; Ahnert et al., 1983) and will be used in this chapter to illustrate the more recent hybrids selected in Brazil. The trial was conducted at the 'Estação Experimental de Altamira', Pará, in 1978, in a randomised block design with seven repetitions, plant spacing of 3 × 3 m and with sparse permanent shading by native trees. The hybrids were evaluated for yield when they started producing and for seed weight per fruit during one harvesting season. The net annual yield data for the four years of evaluation (1983–87) are presented in Table 3. The following parents were tested in this trial: local LA selections from Bahia (SIC864, 831, 644, 328, 17, Catongo, CEPEC16 and SIAL505);

Table 3 Annual net yield of cacao hybrids at the 'Estação Experimental de Altamira, Pará' from 1983 to 1987, with planting density of 3 × 3 m, 12 plants per plot, seven repetitions and sparse native shading

	Net dry bean yield (kg/ha/year)					Wet bean weight/fruit (g)
Hybrid	1983	1984	1986	1987	Average	1987
P7 × SIC864	2275	2715	2443	2334	2442	130
P7 × Ma15	2275	2677	2818	2478	2562	155
P7 × SIAL505	2204	2748	2508	2625	2521	130
Sca6 × Mocorongo1	2160	2676	2667	2412	2479	100
Sca6 × ICS1	2017	2578	2868	2615	2519	120
Sca6 × Be10	1893	2050	1806	1978	1932	100
P7 × SIC644	1853	2210	2238	2184	2121	125
P7 × Be10	1760	2139	1837	1879	1904	135
Pa121 × CEPEC16	1689	2348	2096	2162	2074	120
P12 × SIC831	1662	2237	2128	2105	2033	105
IMC67 × Catongo	1644	2209	1789	1872	1878	130
IMC67 × Be9	1566	1959	1792	1706	1756	135
IMC67 × Mocorongo1	1444	1895	1845	1656	1710	140
IMC67 × Ca4	1418	1688	1756	1613	1619	145
P12 × Be10	1386	1639	1679	1750	1613	110
Pa150 × SIC17	1253	1824	1893	1586	1639	120
Pa150 × SIC328	1191	2033	2173	1703	1775	135
Pa150 × SIC864	1138	1924	2090	1704	1714	125

Source: Ahnert et al. (1983), Machado et al. (1983, 1984, 1986).

local LA selections from Amazon [Ma15, Be9 and Be10, Mocorongo1 and Ca4; T (ICS1) and UA (Sca6, IMC67, Pa121, Pa150, P7 and P12]. The highest yielding hybrids were P7 × Ma15, P7 × SIAL505 and Sca6 × ICS1, with average production of 2562, 2521 and 2519 kg/ha/year dry bean, respectively. The least productive hybrids were P12 × Be10, IMC67 × Ca4 and Pa150 × SIC17, with 1613, 1619 and 1639 kg/ha/year dry bean, respectively. The hybrid P7 × Ma15 yielded the highest wet seed weight/fruit with 155 g, while Sca6 × Mocorongo1 and Sca6 × Be10 yielded the lowest with 100 g. The hybrids selected and delivered by CEPLAC in the Brazilian Amazon are based on UA × LA crosses and they have high potential productivity and are relatively tolerant to witches' broom disease. Farmers in Pará use these hybrids to obtain an average production of 846 kg/ha/year (IBGE, 2014), one of the highest in Brazil.

Based on the results of hybrid testing in the Amazon, the following parents are used by CEPLAC to produce hybrid seeds: IMC67, P7, Pa121, Pa150, Sca6 (UA) Ma15, Ca6, Mocorongo1, Be8, Be10, SIC644, SIAL505, SIC17, CAB17, CAB24, CAB28 (LA) and ICS1 (T) (Francisco Neto, 2008).

6.2.3 Population breeding and clone selection

A very interesting recurrent selection population breeding scheme using the North Carolina II mating design to form the population has been under way at CEPLAC/CEPEC since 1996 (Pires et al., 1999a; Lopes et al., 2011). A basic scheme showing the reasoning used in the programme is presented in Fig. 1. Eight parents from each genetic (heterotic) group (UA and T) are crossed to form three types of populations: (i) AA – progenies of UA × UA; (ii) TT – progenies of T × T and (iii) AT – progenies of UA × T. Other heterotic groups can also be included in each basic scheme. Parents can be chosen according to genotypic and phenotypic features, and on genetic distance based on marker studies. The most important aspect of this scheme is the simultaneous improvement of two populations, each belonging to a specific heterotic group, and the exploitation of heterosis by crossing the two populations. This crossing scheme can be used to investigate combining ability and eventually produce cultivars in each generation. Based on the performance of the AA, TT and AT families, individuals/families are selected from AA and TT, to be crossed to form the next generation, and from AT to be tested in replicated trials for eventual release as clonal or hybrid cultivars (Pires et al., 1999a).

Parents		A – Upper Amazon 1 2 3 4	T – Trinitario 5 6 7 8
A	9 10 11 12	Progenies AA: intra-population breeding	Progenies AT: explore heterosis to produce cultivars
T	13 14 15 16	Progenies TA: explore heterosis to produce clones	Progenies TT: intra-population breeding

According to Pires *et al*, (1999a).

Figure 1 Basic scheme of recurrent selection at CEPLAC. Source: Pires et al. (1999a).

6.3 Ecuador

Ecuador was the largest cacao producer in the Americas in the 1900s, but the outbreak of two diseases, moniliasis and witches' broom, in the 1920s reduced production drastically, from 40 000 to 10 000 t, causing a major crisis in the cacao industry and the economy in general. The Experimental Station of Pichilingue was set up in 1934 and INIAP ('Instituto Nacional de Investigaciones Agropecuarias') was created in 1959 to accommodate ongoing research activities. In the 1940s and 1950s, EET ('Estación Experimental Tropical') clones were selected from local farm material. In the 1960s and 1970s, hybrid cultivars were delivered to farmers as planting material and the breeding programme has been delivering EET clonal cultivars since the 1970s (Amores et al., 2011). Ecuador is recognised worldwide for 'Cacao Nacional' with 'Arriba' flavour, an important source for *gourmet* chocolate, but this landrace started to be replaced by other cultivars, especially CCN51, in the 1990s. Currently, 60% of the total cultivated area is planted with CCN51, 35% with hybrids and 5% Nacional (INIAP breeding programme slide presentation at the Third Annual Breeders Group Meeting, Guayaquil, Ecuador, September 2016). A great deal of effort is currently being dedicated to the selection of new clones as productive and resistant as CCN51, but with the Arriba flavour, for release to farmers and INIAP is releasing 14 EET clonal cultivars for commercial cultivation. Ecuador's breeding programme has also introduced germplasm from other countries and collected wild accessions to widen genetic diversity for breeding purposes, and it maintains 2332 accessions in its germplasm collection (CacaoNet, 2016).

6.3.1 Mass and clone selection

Mass selection was conducted on farm plantations in the late 1940s and early 1950s, generating the first series of EET clones. These accessions are maintained by INIAP at the Experimental Station of Pichilingue at Quevedo (Amores et al., 2011) and 597 EET accessions are currently registered in the International Cacao Germplasm Database (Turnbull and Hadley, 2016). Further mass selection occurred from 2004 to 2010, to select a new set of resistant plants, especially in areas where plantations had higher concentrations of the Nacional and Criollo types. The farmers themselves assisted in the selection, to improve the process, and a total of 84 plants were selected, reproduced vegetatively, planted at the Experimental Station of Pichilingue and are currently being evaluated (Amores et al., 2011).

Eight clonal trials were conducted in the 1950s and 60s to test 294 clones, and six of these were selected and distributed to farmers as planting material in the late 1970s. This was followed by the release of a new series of eight clones in 2009 (Table 4). Three of the first generation of clones (EET103, EET96 and EET95) display higher yielding capacity, especially EET103, which adapts well to several local growing environments.

The famous CCN51 was developed in the 1960s, and it originated from the crosses (IMC67 × ICS95) × O-1, where O-1 is an accession from Valle de los Canelos, Eastern Ecuador. This clone is widely planted in the Americas today and 60% of the plantations in Ecuador are made up of this unique clone because of its high yield performance, resistance to local disease and upright growing habit that facilitates husbandry and mechanisation. It can produce 2000 kg/ha/year or more under high-tech irrigation conditions in Ecuador and other growing regions. Experiments comparing the yield potential of EET103 and CCN51 clones demonstrated that the latter yields much more (twice as much) under non-irrigated

Table 4 Clonal cultivars released in Ecuador in the 1970s and in 2009, and their sexual compatibility

Clones delivered in 1970s	Sexual compatibility	Clones delivered in 2009	Sexual compatibility
EET103	SC	EET544	SC
EET96	SC	EET558	SC
EET95	SI	EET575	SC
EET64	SC	EET559	SC
EET48	SI	EET576	SC
EET19	SC	EET577	SC
		EET450	–
CCN51*	SC	EET454	SC

* CCN51 was developed in the 1960s. According to a slide presentation at the Third Annual American Cocoa Breeders Group Meeting, Guayaquil, Ecuador, September 2016; and Amores et al. (2011).

conditions, while the difference reduces to 20% under irrigation. However, yields from the clones released in 2009 are comparable to CCN51 when irrigated (Amores et al., 2011).

6.3.2 Hybrid selection

Efforts to develop hybrid cultivars started in the 1960s by testing hybrid progeny mainly originating from crosses of UA accessions, especially witches' broom-resistant cultivars such as Sca6, Sca12, IMC67, Silecia1 and Silecia5, with local EET selections. There were also crosses between Trinitario accessions, such as ICS1 and ICS6, and EETs. These experiments showed that the UA clones Sca6 and 12, as well as Silecia, when crossed with EETs and ICS1 and 6, had the highest production and resistance to witches' broom. Crossing Trinitarios with EETs also produced good agronomic traits. In the 1960s and 70s, as a result of these findings, INIAP distributed hybrid seed obtained by crossing parents including EETs, ICS1, ICS6, Sca6, Sca12, Silecia1 and IMC67 (Amores et al., 2011). According to the information available, it seems that Sca hybrids have quickly lost their resistance.

6.3.3 Population breeding and clone selection

Population breeding is currently aiming to develop new clones that are as good as CCN51, but with the Nacional Arriba flavour. Ecuador is concerned about the use of a single clone, CCN51, and the loss of the Arriba flavour. Different sources of germplasm are being combined in several crossing schemes to generate improved segregating populations with an increased frequency of alleles associated with resistance, high yield and Arriba flavour, to develop and release new clones and increase the clonal cultivars available to farmers in the country (Amores et al., 2011).

6.4 Costa Rica

Costa Rica's cacao breeding programme was started by the Inter-American Institute for Cooperation on Agriculture (IICA) in 1944. The institute established an International

Cacao Collection at CATIE, the Tropical Agricultural Research and Higher Education Center in Turrialba (Phillips-Mora et al., 2013). In the 1940s and 50s, breeding activity was also conducted by the United Fruit Company, selecting UF clones and providing genetic material for CATIE, and by STICA (Ministry of Agriculture with the cooperation of AID), which undertook large-scale mass selection of the local population (Hunter, 1961). Costa Rica's breeding programme released hybrid cultivars as planting material for local farmers and for neighbouring Spanish-speaking countries until the 1980s (Phillips-Mora, pers. comm.). Six clonal cultivars resistant to moniliasis were released in 2007 and these are being tested in many Latin America countries. CATIE maintains an international Cacao Collection of 1146 accessions (CacaoNet, 2016).

6.4.1 Clone selection

The first clonal trials were established in 1960, comparing local and introduced clones. The study was more academic than practical because farmers preferred seed-propagated planting material at that time. The most productive of the 25 clones tested in plantings spaced at 4 × 4 m with permanent shade and no fertiliser were UF29, UF221 and CC10, yielding around 1500 kg/ha in the sixth year (Soria and Esquivel, 1967). In 1985, another trial tested 12 clones, but breeders were already considering the possibility of delivering clonal cultivars to farmers. The main aim of this experiment was to evaluate resistance to *M. roreri* and *P. palmivora*, and it was set up in a completely randomised design, with six plants per plot, plant spacing of 3 × 4 m and permanent shade. Clones were chosen based on previous indications of resistance and were sourced from different origins: five local selections (CC137, CC210, CC266, UF11 and UF296); five from Ecuador (EET59, EET67, EET75, EET183 and EET 76); one from Brazil (RB41) and one from Panama (UF273). After five years of evaluation, no significant statistical differences were found between them with regard to pod rot resistance, but differences were found for yield and resistance to moniliasis. The clones with the highest dry bean yields were CC137 (811 kg/ha); EET67 (741 kg/ha); EET183 (684 kg/ha) and EET59 (683 kg/ha). UF296 and UF273 had the lowest incidence of moniliasis, while EET376 and RB41 had the highest. Clone EET183 had a low incidence of both diseases and one of the higher yields. Pod index varied from 15.5 (UF11) to 37.5 (RB41) and seed index from 0.99 (CC266) to 2.5 (UF 11) (Morera and Mora, 1996).

6.4.2 Hybrid selection

In 1967, around 200 cacao hybrid combinations were being evaluated at experimental stations in Costa Rica, to produce hybrid cultivars and develop basic studies on the inheritance of important agronomic traits (Soria and Esquivel, 1967). Crosses were mainly based on UA × local selections (UF and CC), T × local selections, UA × Amazon and UA × T. These experiments revealed the following general trends: (i) the most productive and precocious hybrid combinations originated from crosses of clones from different geographic and genetic sources, usually UA × local selections, UA × T, UA × Amazon (from different geographic areas) or quasi-Criollo clones and (ii) hybrid combinations were less productive than clones from similar geographic and genetic origins. The most productive hybrid combinations included P12 × Catongo, P12 × Sca12, UF29 × IMC67, UF29 × UF667 and Sca6 × ICS1 which started to produce in the second year in the field, and productivity ranged from 1000 to 1500 kg/ha in the third and fourth years. The least productive hybrid combinations were UF296 × CC9 and UF296 × CC18 (Soria and Esquivel, 1967).

Based on the results of these and other experiments, the following parents were selected for seed production: IMC67, P12, P7, Sca6, Sca12, SPA9 (UA), Catongo (LA), UF12, UF29, UF296, UF613, UF654, UF667, UF668, UF676, UF677, UF12 (T) and CC9, CC18, CC276 (local selections). In 1966, Soria described the situation of local plantations as follows: 'in some areas new plantations of cacao are done with open pollinated or controlled pollinated seeds originated from Trinitario clones', which suggests that the programme was already delivering seeds to farmers at that time. CATIE has also delivered a mixture of hybrid seeds to Central America and Spanish-speaking countries in the Caribbean since 1976 (Toxopeus, 1985). In general, the hybrids were poorly accepted by farmers, so the programme was terminated in the 1980s and a new programme to produce clonal cultivars was started (Phillips-Mora, pers, comm,).

6.4.3 Population breeding and clone selection

CATIE started a population breeding programme in 1996 to identify genotypes with resistance to moniliasis and black pod and submit them to successive cycles of crossing, evaluation and selection to develop high-yielding cultivars resistant to pests and diseases. The programme consisted of the following stages: (i) crossing selected parents and evaluating the progeny in replicated trials to form populations; (ii) selecting the best families and evaluating them in regional trials; (iii) selecting individual plants from the best families; (iv) propagating clones of the best individuals for conservation, breeding and testing in regional trials and (v) planting clones approved during regional trials in observation plots and clonal gardens. At the end of the cycle, agronomically superior clones are delivered to farmers. When one cycle is under way, a new one starts by crossing the selected clones and evaluating the progeny in replicated trials. The other steps proceed in the same way, in a continuous cycle of population improvement and cultivar delivery (Phillips-Mora et al., 2013).

The programme is currently evaluating 17 clonal trials and ten progeny trials. Two clone trials established in 1998 and 1999 tested 42 clones chosen based on information on their yield potential and tolerance to pod rot, moniliasis or witches' broom disease obtained from previous clone trials and from the germplasm collection. Trials were conducted in one location, using a randomised block design, with four replications and eight plants per plot, at a plant spacing of 3 × 3 m, with permanent shade and regular fertilisation. Data on yield and losses due to pod rot and moniliasis were collected for 11 years. The average yield over the 11-year period ranged from 61 (GU33 – N) to 1485 kg/ha (CATIE-R6), and pod losses from moniliasis ranged from 5% (CATIER6) to 78% (RB41, Sca12) and from pod rot from 0% (many clones) to a maximum of 9% (UF273T2).

Based on the results of these trials and other regional trials, six clones tolerant to moniliasis and pod rot were released in 2007 (Table 5). These cultivars are: CATIER1 (UF273T1 × CATIE1000) (CATIE1000 originated from P12 × Catongo); CATIER4 (UF273T1 × Pa169); CATIER6 (UF273T1 × Pa169); CC137 (open-pollination of UF12); ICS95T1 (Trinitario) and PMCT58 (Trinitario hybrid of unknown parentage) (Phillips-Mora et al., 2013).

Table 5 Cacao clones tolerant to moniliasis and pod rot, released by CATIE in 2007

	Yield (kg/ha/year)	Yield components				Infection (%)		
Cultivar	11-year data	Fresh weight/fruit (g)	Fruit/tree/year	Seeds/fruit	Sexual compatibility	Monilia	Pod rot	Fat (%)
CATIER1	1066	93.4 ± 3.82	29	29 ± 1.08	SC	12	7	52.3
CATIER4	1336	144.7 ± 5.7	18	35 ± 1.30	SI	9	1	56.2
CATIER6	1485	127.2 ± 4.42	24	31 ± 1.11	SI	5	0	55.7
CC 137	990	117.3 ± 3.87	24	27 ± 0.78	SC	32	1	50.6
ICS95T1	636	102.0 ± 2.93	22	33 ± 0.76	SC	26	6	–
PMCT58	789	93.1 ± 4.48	27	37 ± 1.32	SI	26	4	59.1

Source: Phillips-Mora et al. (2013).

6.5 Ghana

Ghana's cacao breeding programme began in 1937 by selecting local populations and introducing germplasm from South America and Trinidad to establish a germplasm collection and produce planting material for farmers (Rogers and Knight, 1953). A list of these initial selections can be found in Lockwood and Gyamfi (1979). CRIG (Cacao Research Institute of Ghana) was created in 1938, and the now-defunct WACRI (West African Cacao Research Institute) in 1944, to support research, especially into resistance to CSSV. WACRI conducted research in Ghana, Nigeria and Sierra Leone. The West African Amelonado cacao was the predominant type cultivated in the region at that time (Lockwood and Gyamfi, 1979), but around 5% of the plants in Ghana were Local-Trinitarios (LT) (Thresh et al., 1988).

In the 1930s, cacao plantations in Ghana became infected by CSSV and the Amelonado plants were highly susceptible to the disease. In an attempt to find new sources of resistance, WACRI introduced one of the most important UA germplasm collections from Trinidad into Ghana in 1944. This germplasm helped revolutionise breeding in Africa and other countries because of its variability for agronomic traits, such as resistance to pests and diseases, yield and its variables and resistance to abiotic stresses (Thresh et al., 1988). The introduction process continues, and Ghana currently maintains a germplasm collection of 1366 accessions (CacaoNet, 2016), facilitating the choice of parents.

Ghana's cacao breeding programme initially delivered open-pollinated cultivars from local selections as planting material for farmers (Toxopeus, 1985) and CRIG produced and released mixed F_2/F_3 UA open-pollinated populations from 1954 to 1970. Following the development and selection of Series II bi-clonal hybrids, CRIG started to deliver these as planting material in the 1960s, and it now delivers seeds from inter-Amazon bi-clonal hybrids, considered to be more tolerant to CSSV and pod rot (Thresh et al., 1988; Adomako et al., 1999a; Lockwood, 2015).

6.5.1 Mass- and open-pollinated selection

When the breeding programme began, Amelonado clones were selected from experimental stations and LT clones from farm fields, after individual trees were evaluated for yield, and pod and bean size (Posnette, 1943). The selected plants were vegetatively propagated and planted on experimental stations for evaluation. Agronomically superior plants were selfed and the selfed progeny were used to plant blocks of seed garden to produce open-pollinated seeds from local selections as planting material. The seeds released to farmers from the first seed gardens were produced by natural selfing or intercrossing of the plants (Toxopeus, 1985).

Some of the Amelonado selfed progeny were evaluated in the 1st progeny trial area (PTA) (1939–41) and LT progeny in the 2nd PTA established at CRIG (Posnette, 1943). These trials showed that Amelonado progeny had little genetic variation for economic traits (Rogers and Knight, 1953). The Amelonado cacao has some agronomic advantages, such as uniformity, good organoleptic flavour and suitability for growing at reduced spacing, and yields can be high in a suitable environment (Toxopeus, 1985; Lockwood, 1976). However, it does also have disadvantages, such as susceptibility to CSSV, poor establishment ability, bearing pods late and slow initial growth (Lockwood, 1976). Although there was more genetic diversity in the LT selection than the Amelonado, the quality of the diversity was insufficient for breeding progress (Rogers and Knight, 1953). These genetic materials had some level of resistance to CSSV, but less than that of the UA germplasm (Thresh et al., 1988).

The UA germplasm brought to Ghana from Trinidad in 1944 originated from IMC, Na, Pa and Sca genetic material collected in Peru. Around 100 pods were introduced, giving rise to 3000 seedlings that were planted at CRIG for evaluation, using the Amelonado as control. Each pod was allocated a number with the prefix 'T' (Trinidad) and those numbered T60–T102 were obtained from hand pollination between UA types and open-pollinated T1–T59. It soon became evident that the UA trees were more vigorous and could grow more rapidly, start to produce pods earlier and yield more than the Amelonado, LT and other genetic material on the experimental stations (Rogers and Knight, 1953). The following were selected from the trials: T60 (Pa7 × Na32); T63 (Pa35 × Na32); T65 (Pa7 × IMC 47); T72 (Na32 × IMC60); T73 (Na33 × IMC60); T76 (Pa35 × Na31); T79 (Na32 × Pa7); T82 (Na32 × Pa35); T85 (IMC60 × Na34) and T87 (IMC60 × Na34). T12, which originated from Sca12 open-pollination, was selected later. These clones were selected for their superior precocity, vigour, yield, fruit and bean sizes and physical and organoleptic qualities (Rogers, 1955).

At that time, there was great demand for seed to replant areas cut out during CSSV eradication campaigns. Open-pollinated F_2 seeds from these selections were planted in observation plots and evaluated with the aim of producing planting material. After inspection, they were approved by the local government for seed production and the open-pollinated F_2 and F_3 seeds (pods), referred to as F_2/F_3 Amazon, were widely released in Ghana and on a smaller scale in Nigeria and Sierra Leone, between 1954 and the 1970s. To clarify, T selections are the F_1 (generation 1) plants that originated from the original F_0 clones in Trinidad, their seeds generated the F_2 plants and these were used to produce the F_3 seeds that were released to farmers. The mixed F_2/F_3 Amazon plants were more productive and vigorous and easier to establish in the field than the Amelonado (Rogers,

1955). However, they were also more susceptible to *Phytophthora* pod rot, but recovered more rapidly from capsid attack (Toxopeus, 1985). The next step was to develop hybrids that could yield more than the mixed F_2/F_3 Amazon cultivars.

6.5.2 Hybrid selection

Different hybrid combinations have been tested in Ghana since 1952 to develop bi-clonal hybrids as planting material. Crosses have been made for heterosis and complementary agronomic features, by combining the UA × UA (from separate geographic origins), UA × Amelonado, UA × LT and UA × LA genomes and other sources. The UA germplasm was mostly from series T (F_1 generation) and also the original clones P, Pa, Na and IMC. The hybrid progenies were allocated a series number based on the type and agronomic traits of the parents chosen for hybrid combination. Series I mainly had UA × UA parent crosses, Series II and III mainly UA × Amelonado and UA × LT, Series IV consisted of crosses between parents with resistance to CSSV and Series V consisted of parents with black pod resistance (Glendinning, 1966). The parents of the most important hybrid series, Series II (SII), were: SIIA (P4/9 × T63/971 = Amelonado × UA); SIIB (EI × T63/967 = LT × UA); SIIC (E1 × T63/971 = LT × UA); SIID (E1 × T16/613 = LT × UA); SIIE (T79/416 × EI = UA × LT); SIIF (T79/501 × T85/799 = UA × UA); SIIG (T60/887 × T60/888 = UA × UA); SIIH (S84 × T85/779 = Amelonado × UA); SIIJ (T85/779 × S84 = UA × Amelonado); SIIK (W4I × T63/967 = LT × UA) SIIL (T63/967 × W4I = UA × LT); SIIM (T79/467 × S72 = UA × Amelonado) and SIIN (T79/467 × E9 = UA × LT) (Thresh et al., 1988). Breeders considered 12 attributes when choosing parents for crossing: growth rate, commencement of bearing, yield, bean size, flavour, fat content, resistance to CSSV, black pod, rodent attack, drought, vigour and growth habit (Glendinning, 1966). SII progenies were assessed in the 8th and 9th PTAs, planted in 1952 and 1954, respectively. The results of these trials were published by Lockwood (1976) and will be used as an example of the potential of this important series. Series A, B, C, D and G and controls (WA1 and WA2) were tested in the 8th PTA, and Series E, F, H, J, K, L, M and N and control TF1 in the 9th. Other progenies were also tested in these trials, but they will not be discussed here. Plant spacing was 2.44 m^2 in the 8th PTA and 3.05 m^2 in the 9th PTA, and plants were shaded by local forest trees in both cases. Both experiments used a randomised block design with 36 tree plots, and there were four incomplete replications in the 8th PTA and five complete replications in the 9th (Lockwood, 1976).

The results of the hybrid combinations in Series II, grouped according to PTA, are presented in Table 6. Tree sizes, based on trunk circumference, varied from 440.3 mm in Series G to 579 mm in Series A in the 8th PTA, and from 524.3 mm in Series E to 631.5 mm in Series K in the 9th PTA. UA × UA Series G (T60/887 × T60/888) progeny were the smallest plants, smaller than the controls, possibly because T60/887 only grows to a small size. From a practical point of view, these results indicate that variation in plant size among Series II crosses must have been the main cause of heterogeneity in the plantation. To minimise heterogeneity in commercial plantations, plants with similar sizes should ideally be mixed to avoid competition and excessive pruning to equilibrate the sizes of the canopy. Of course, hybrids are segregating, so even if plants with similar sizes are mixed, there will still be variation, but at lower levels and more easily manageable.

Bean number in crosses varied from 31.0 in Series G to 40.1 in Series D. The two UA × UA progenies of Series G (T60/887 × T60/888) and Series F (T79/501 × T85/799) had

Table 6 Results from analysis of Series II in the 8th and 9th PTAs in Ghana

Hybrids: 8th PTA	Series II	Circumference, 1974 (mm)	Number of beans, 1974	Diseased pods, 1952–73 (%)	Gross yield, 1964–74 (kg/ha)
P4/9 × T63/971	A	579.0	38.9	18.21	23145
EI × T63/967	B	517.2	38.3	22.89	13345
E1 × T63/971	C	501.3	37.7	24.49	16937
E1 × T16/613	D	493.0	40.1	20.56	16616
T60/887 × T60/888	G	440.3	31.0	16.87	11958
WAA (WA 1)	Contr.	500.0	37.9	8.95	13742
WAA (WA 2)	Contr.	492.9	37.0	9.08	11885
Hybrids: 9th PTA				1959–73 (%)	
S84 × T85/779	H	570.8	36.5	18.46	11671
T85/779 × S84	J	550.5	36.9	19.16	12885
T79/467 × S72	M	570.8	35.5	16.44	14479
T79/416 × EI	E	524.3	35.9	23.26	13678
T79/467 × E9	N	580.7	34.2	21.07	9019
T63/967 × W4I	L	619.0	37.8	26.27	11148
W4I × T63/967	K	631.5	36.7	26.63	11156
T79/501 × T85/799	F	545.8	31.8	11.70	10206
L-T (TF1)	Contr.	504.9	36.2	21.97	7536

Source: Lockwood (1976).

the smallest number of beans per pod and all parents involved in these crosses are self-incompatible. Series D and A, with the highest number of beans per pod, had only one self-compatible parent each: T16/613 and P4/9, respectively (Thresh et al., 1988). In general, the number of seeds per pod depends on the number of ovules per ovary, pollinators, the environment and compatible pollen available for pollination. Breeders know that self-incompatible plants may have fewer seeds per pod under some conditions, but a higher number of beans per pod is important for farmers, to reduce production costs, because there will be fewer pods to harvest and break to remove the seeds. Sexual incompatibility and its effects in a mixture of hybrids have not been adequately studied and this must therefore be considered. In the case of plantations made up of clones, it is very clear that incompatibility should be avoided because the use of self-incompatible clones in Brazil was disastrous for farmers (Monteiro and Ahnert, 2012). All breeding textbooks also recommend not using self-incompatible plants if the aim is to produce seeds commercially (Borem, 1997).

The percentage of diseased pods in Series L varied from 9% (controls WA1 and WA2) to 27% and most of the hybrids lost around 20% of the harvest, except T79/501 × T85/799 which lost 11.7%. Losing 20% of the harvest is a great economic challenge for farmers, but this is what happens when a new disease, or a new disease strain, spreads to a plantation, such as the spread of *P. megakarya* in West Africa. Integrated control, the use of resistant plants, application of fungicides and good agronomic practices help reduce losses.

Accumulated yield from mature plants from 1964 to 1974 varied from 11 885 (WA2) to 23 145 kg/ha for Series A in the 8th PTA, and from 7536 (LT) to 14 479 kg/ha for Series M in the 9th PTA. All hybrids in both trials produced more than the Amelonado and LT controls. The hybrid P4/9 × T63/971, a cross originating from Amelonado × UA from Series A, had the highest yields, far superior to all the other hybrids. The hybrids from the Series II had better yields, and other features, than the local Amelonado, and LT selections had better establishment ability. Some also had a certain level of tolerance to CSSV and pod rot, and good organoleptic flavour, which made them suitable for further breeding (Lockwood, 1976; Thresh et al., 1988).

To replace the CSSV-susceptible parents from Series II, a new group of UA parents that was more resistant and had desirable agronomic traits was selected. Female parents from Series II T63/967, T63/971, T79/467 and T85/799 that were already established in field gardens in the Eastern Region of Ghana, a highly infected area, were used to accelerate the breeding process. These were crossed with UA parents and the progenies tested in laboratory and field trials to select the best pollen parents. Once these parents were selected, they could replace Amelonado and LT in field gardens, keeping the females already planted in the fields, accelerating the process and reducing costs. After years of field testing, the following crosses were selected and recommended for seed production: T63/967 × (T17/524, T60/887, T65/238 and IMC60); T63/971 × (T60/887, T65/238 and IMC60); T79/467 × (T17/524, T87/1312, IMC47 and IMC60) and T85/799 × (T65/238, T65/326, T79/501 and Pa7). These hybrids are considered to be as productive as the SII series, with good industry acceptance, similar or lower losses due to black pod and some resistance to CSSV (Thresh et al., 1988).

Results from hybrids obtained by crossing two widely used females (T85/887 and T60/887) with different Amazon clones were published by Adomako et al. (1999a). The hybrids with the best yield, yield-vigour relationship, vegetative traits and resistance to black pod and CSSV were: T60/887 × (Be8, P10, T17/524 and Eqx3338) in trial 25 and T85/887 × (Pa150, Pa7 and Ma12) in trial 26. The net yields (kg/ha/year) of the hybrids with the female parent T60/887 from year 1 to 10 of evaluation were, on average, 905.8 (Be8); 912.5 (P7); 904.3 (Eqx3338) and 857.6 (T17/524). Net yields of the hybrids with the female parent T 85/887 were, on average, 704.9 (Pa150); 646.9 (Pa7) and 594.1 (Ma12). The incidence of black pod disease was low in these hybrids, at around 5%. Despite the selection and use of superior hybrids by farmers, new breeding challenges are arising from the introduction of new diseases, such as *P. megakarya*, and climate change with prolonged droughts (Adu-Ampomah, 1994)

6.5.3 Parents used to produce hybrid cultivars

Based on the results of the breeding programme, the following parents are currently used to produce hybrid seeds for farmers: Pa7, Pa150, P7 and T60/887. These were previously T85/799, T63/967, T63/971, T79/467 and T79/501 (Lockwood, 2015).

6.5.4 Population breeding

CRIG has started a population breeding programme to target two important diseases, CSSV and black pod, yield and its variables, and abiotic problems such as drought and exhausted soils (Adomako et al., 1999a). The traditional breeding scheme, using one single generation of bi-clonal hybrids, has limitations, and studies have shown that most of the important agronomic traits, including resistance, are controlled by many genes. Therefore, cacao breeders have started to implement population improvement schemes to advance by more than one breeding cycle, as implemented with great success in annual crops like corn (Hallauer and Miranda 1988). Ghana's population breeding scheme is selfing and crossing inter- and intra-groups of UA germplasm from series Pa, Na, IMC and T clones, and also using participatory breeding (Opoku et al., 2010).

6.6 Nigeria

Nigeria's cacao breeding programme began in 1931 by conducting mass selection in local populations to establish a germplasm collection and select parents to produce seeds for distribution to farmers. The programme has four distinct phases, and each has contributed to the establishment of a large collection of germplasm and the production of hybrid cultivars. The four phases are as follows: i) First Nigerian Programme; ii) WACRI Programme; iii) Second Nigerian Programme and iv) Third Nigerian Programme. The Cacao Research Institute of Nigeria (CRIN) was created in 1964 after WACRI ended. In terms of planting material, the breeding programme began releasing NT38 hybrid seeds and open-pollinated cultivars from local selections in the 1940s. It then started to release open-pollinated cultivars originating from F_2/F_3 UA selections in 1954, and Series I and II hybrids, and CRIN Establishment Elites from the 1960s. CRIN started to release the CRINTC1 to CRINTC8 hybrid series in 2010.

6.6.1 The first Nigerian breeding programme

The first Nigerian breeding programme started in 1931 and ended in 1956. It initially employed mass selection in local plantations, and these selected plants were selfed and their progenies were evaluated for further use. Agronomically superior progenies were planted in blocks and used to produce open-pollinated seeds for farmers in the 1940s (Williams and Akinwale, 1994). NT38, a self-incompatible Nigerian Trinitario clone with high bean weight, was selected and two progeny trials were run in 1942 and 1945, respectively, to test hybrid combinations of NT38 and introduced Trinitario germplasm. Hybrid vigour or heterosis was observed in the progenies of the crosses, and seed gardens were set up to produce seeds for farmers (Russell, 1952). One clone trial was also run, and 11 clones selected: NT39, 114, 164, 215, 216, 284, 310, 655, HH268, S36 and T38 (Atanda, 1977).

6.6.2 WACRI programme

WACRI had its headquarters in Ghana and a substation in Nigeria, and the programme ran in Nigeria from 1938 to 1960. UA germplasm (T1–T120), originating in Trinidad and selected in Ghana, was introduced into Nigeria in 1946 (Omalaja et al., 2006). 2000 F_2 progeny originating from the F_1 population of T germplasm in Ghana were also introduced into Nigeria (Jacob et al., 1971) and open-pollinated F_2 and F_3 seeds, referred to as F_2/F_3 Amazon, started to be released in the country in 1954 (Thresh et al., 1988). Four Series I and Series II progeny trials were planted in western Nigeria from 1956 to 1957. These trials demonstrated that progenies from these series generally produced more than the F_3 Amazon and local selections, and some, such as Series II K, E, A and D, were better adapted to drier areas. Series II F, K and L had both good establishment ability and yields (Jacob et al., 1971; Atanda, 1977). This programme has been very important for Nigeria because most of the hybrids cultivated in the country originated from the F_3 Amazon and the UA germplasm introduced is still very important for the ongoing programme (Omalaja et al., 2006).

6.6.3 The second Nigerian breeding programme

The second Nigerian breeding programme ran from 1961 to 1970, so it only lasted ten years. When the programme started, germplasm was available from the first two programmes and from seed gardens to supply local demand. This programme mainly focused on selecting hybrids with good establishment ability/drought resistance, CSSV tolerance, pod rot resistance or escape and quantitative genetic studies (Jacob et al., 1971). A total of twelve hybrids with good establishment ability were selected, CRIN Establishment Ability Elites belonging to CRIN Hybrid Series I: Na173 × Na257; Na173 × Na251; Na26 × Na387; Na342 × Na 387; C 40 × C64; C42 × C43; C42 × C85; C42 × C68; C67 × C14; C43 × C22; and WACRI Series IIF and Series IIK (Atanda, 1977).

A total of 15 drought-tolerant hybrids were selected, CRIN Establishment Ability Elites, also belonging to CRIN Hybrid Series I: WACRI Series II D, K/L and F; and C77 × C64, T88/45 × Na32, T85/5 × Pa35, T65/7 × Na32, T30/10 × NA32, T20/21 × Na32, C77 × C23, ICS7 × Na32, ICS1 × Na32, T17/11 × Na32, T12/5 × Pa35 and T7/12 × Na32. Three hybrid progenies obtained by hand-pollinating WACRI Series II and Series K, H and E gave much better hybrids than the mixed Series II hybrids. These hybrids were designated CRIN Hybrid Series II (Omalaja et al., 2006).

Evaluation for tolerance to CSSV allowed promising genetic material to be selected, but only D (E1 × T16/613) out of WACRI Series II demonstrated tolerance. Out of the other available germplasm, the following demonstrated tolerance: Na and IMC genotypes; clones C77, T9/15, T65/10, T73/55, T101/15, T101/42 and T72/20; and progenies C23 × C77 and C77 × Na32 (Jacob et al., 1971).

6.6.4 The third Nigerian breeding programme

The third Nigerian breeding programme started in 1971 and focused on selecting cultivars with high-yielding potential and tolerance to the main constraints for cacao production in Nigeria: black pod (*P. megakarya*) and CSSV; mirids (*S. singularis*) and drought (good establishment ability and tolerance to water deficiency in the soil).

The programme generated the CRIN Synthetic series of cultivars: CRIN Synthetic II, resistant to pod rot and consisting of parents T0/15, T12/11, T19/9, T24/12, T50/13 and T86/2. CRIN Synthetic III, resistant to CSSV and consisting of hybrids T22 × T9/15; T90/22 × T9/18; C57 × C42; C77 × C23; C15 × C14; C68 × C26; T72/20 × C77; C73 × C25; C65 × C77; C64 × C77; C73 × T9/15; C75 × C14; T65 × C77; C67 × C77 and WACRI Series IIH (Omolaja et al., 2006).

Finally, a breeding programme started at the end of the 1990s aiming to combine pod rot resistance, tolerance to CSSV and good yield capacity. In 2010, the programme released eight hybrid cultivars with moderate to high levels of pod rot resistance, and production potential ranging from 1200 to 2300 kg/ha/year (Table 7). The parents are UA, except N38, which is a local selection (CRIN, 2016).

Table 7 Cacao hybrid cultivars released by CRIN in 2010

Hybrid	Pedigree	Resistance	Yield potential (kg/ha/year)
CRINTC1	T65/7 × N38	Moderate pod rot	1900–2200
CRINTC2	T101/15 × N38	Moderate pod rot	1940–2300
CRINTC3	P7 × Pa 150	High pod rot/mirids	1700–2000
CRINTC4	T65/7 × T57/22	Moderate pod rot/mirids	1500–1800
CRINTC5	T82/27 × T12/11	Mirids	1500–1850
CRINTC6	Pa 150 × T60/887	High pod rot	1400–1650
CRINTC7	T82/27 × T16/17	Moderate pod rot and high mirids	1600–1900
CRINTC8	T65/7 × T9/15	Pod rot and mirids	1200–1500

Available at: http://www.crin-ng.org/index.php/14-crin-monthly-seminar/26-the-new-cocoa-hybrids. Accessed date: December 2016.

6.7 Ivory Coast

The breeding programme in Ivory Coast started in 1946 with mass selection of local cultivars and introduction of germplasm from other countries to establish an initial collection (Besse, 1964). A few clones were selected from the local population and IFC5 (Institut Français du Café, Cacao) is considered the best, with outstanding yield and vigour (Eskes et al., 1994). There are now a total of 238 IFC clones registered at the Turnbull and Hadley (2016). Clonal cultivars were not released in Ivory Coast, even though the programme tested many clones and selected some with outstanding performance. The first generation of hybrid cultivars was released in 1975, with productivity in experimental stations ranging from 1800 to 2400 kg/ha/year. The second generation of hybrid cultivars was released in 1995, with productivity ranging between 2000 and 3100 kg/ha/year. The programme currently maintains a germplasm collection with 1605 accessions from different origins (CacaoNet, 2016).

It is very interesting to note that the Ivory Coast programme is very well structured in terms of breeding strategies, and it has had excellent results. According to the literature, highly productive clones and hybrids are available, and a reciprocal recurrent selection programme has been under way since 1991/92 and is now entering the second phase. It is also worth mentioning that the best hybrids in Ivory Coast originate from UA × Amelonado crosses, while UA × UA crosses give the best hybrids in other West African countries.

6.7.1 Clone selection

The programme has tested many clones and we will report the results of five trials established between 1970 and 1983, published by Sounigo et al. (1994), to illustrate the programme of clonal selection in Ivory Coast. A total of 59 clones from the following genetic groups were tested: five LA; 27 UA; 14 T; two Criollos (C); two descendants of C; and nine UA and T or LA hybrids. Four trials were evaluated for yield: D11/2 (three years of data, four replicates and 12 plants/replicate); D13/4 (five years of data, four reps and 12 plants/rep); G8 (four years of data, four reps and 12 plants/rep) and A18/2 (five years of data, four reps and 12 plants/rep). The same experiments were also evaluated for resistance to pod rot (data from rotted pods). Experiments D11/2, D13/4, G8 and 117 (62 reps, and one plant/rep) were evaluated for mirid resistance (accumulated data of many years). The LA clone IFC5 was used as the control throughout.

Of the 59 clones evaluated, 25 with productivity over 2000 kg/ha/year are described to illustrate the potential of some clones (Table 8). The UA clones were the most productive, with two exceptional clones: T79/501, with productivity of 4708 kg/ha/year; and Na 32,

Table 8 Clonal productivity and resistance to pod rot and mirids in Ivory Coast

Clone/group	Productivity (kg/ha/year)	Rotten pods/total pods	Mirid attack (notes 0–4)	Trial
T79/501 (UA)	4708	6.65	1.77	D11/2
IFC5 (LA)	4536	10.02	2.36	D11/2
Na32 (UA)	4393	19.40	1.65	D11/2
Mo 81 (UA)	3924	19.45	–	A18/2
UPA620 (UA)	3551	22.65	–	A18/2
Pa4 (UA)	3487	18.40	–	A18/2
Pa13 (UA)	3348	14.11	–	A18/2
Pa 121 (UA)	3189	18.17	–	A18/2
A1–214-9 (UA)	3177	12.78	1.86	D11/2
IFC25 (Hyb)	3169	15.23	1.91	D11/2
MO 9 (UA)	3157	14.69	–	A18/2
P19A (UA)	3050	14.39	–	A18/2
IFC27 (Hyb)	3047	11.65	1.93	D11/2
IMC57 (UA)	2869	14.11	–	A18/2
UPA402 (UA)	2560	13.14	1.90	D11/2
T60/887 (UA)	2548	8.56	1.85	D13/4
GS36 (T)	2488	11.45	2.71	D11/2
ACU85 (T)	2388	12.78	2.54	D11/2
UPA413 (UA)	2367	10.22	1.55	D13/4
ICS95 (T)	2196	13.41	3.38	D11/2
ICS39 (C)	2189	11.46	2.80	D11/2
MOQ413 (UA)	2122	11.50	–	A18/2
T63/967 (UA)	2095	10.46	1.73	D13/4
Wa40 (T)	2086	16.07	2.38	G8
IMC5 (UA)	2005	22.23	–	A18/2

Source: Sounigo et al. (1994).

with productivity of 4393 kg/ha/year. The clone T79/501 was the most resistant to pod rot and mirids. IFC5, the control LA clone, produced 4536 kg/ha/year, was moderately resistant to pod rot and moderately susceptible to mirids. Ten clones had productivity ranging between 3000 and 3900 kg/ha/year and 12 between 2000 and 2900 kg/ha/year. They had varying degrees of pod rot and mirid resistance. Although the programme produced these exceptional clones, no clonal cultivars have been released in Ivory Coast because of the difficulty of producing and distributing them to farmers on a large scale (Besse, 1964).

6.7.2 Hybrid selection

The development of bi-clonal hybrid cultivars started in 1958 and has been based on combinations of germplasm from different genetic groups: UA × LA and UA × T (Paulin, 1993). The UA germplasm was introduced to Ivory Coast from Ghana in the form of F_2 seeds, originating from the F_1 generation in Ghana. The selected F_2 plants in Ivory Coast were coded UPA (Eskes et al., 1994). Starting in 1975, the first phase of the programme tested a total of 510 hybrid combinations and selected 12 hybrids, based mainly on yield and yield variables, which were released to farmers (Besse, 1975). According to this author, the yield potential of these 12 hybrids varied from 1800 to 2400 kg/ha/year of commercial beans and the seven with the best performance are listed (Table 9). These seven hybrids or generation (I) hybrids have fat percentages above 55%, except H153 which is still close, and bean weight close to 1.0 g.

In the 1970s, the breeding programme started to develop a second generation (II) of hybrids, testing more than 120 hybrid combinations after the introduction of new germplasm (Eskes et al., 1994). In 1978, testing started on 62 hybrid combinations of UA × Amelonado clones in multiple locations (Paulin et al., 1993) and testing on a series of new UA × UA, UA × C and LA × T combinations for hybrid selection was started in 1979 (Eskes et al., 1994).

The results of four trials established in 1978 and published by Paulin et al. (1993) will be described in this chapter to illustrate the potential of generation II hybrids in Ivory Coast. The hybrid combinations tested in these trials were developed by crossing 16 UA parents with five Amelonado parents. The UA parents were P7; Na79; Sca6; IMC67; Pa150, Na32; Pa35; Pa7; IMC78; T60/887; T79/501; T79/467; T79/416; T63/967;

Table 9 Generation I cacao hybrids distributed in Ivory Coast since 1975

Hybrid/pedigree	Yield (t/ha/year)	Number of beans/100 g	Fat (%)
H109 (UPA402 × UF676)	2.2	120	57.0
H125 (UPA40 × C1)	2.0	107	58.5
H117 (UPA 405 × C412)	2.1	117	58.0
H140 (UPA419 × C1)	2.0	113	58.7
H153 (UPA608 × C412)	2.1	130	54.4
RB9 (UPA603 × UF667)	2.4	117	57.8
H435 (T79/416 × E1)	2.3	105	56.5

Source: Tahi et al.'s (2012) slide presentation in 7th INGENIC Workshop.

T63/971; T85/799 and UPA 402. The T series originates from Ghana and has the following parents: T60/887 = (Pa7 × Na32); T79/501 = (Na32 × Pa7); T79/467 = (Na32 × Pa7); T79/416 = (Na32 × Pa7); T63/967 = (Pa35 × Na32); T63/971 = (Pa35 × Na32); T85/799 = (IMC60 × Na34) and UPA402 = (T87/1329 × T87/1211). The Amelonado parents were IFC1, IFC2, IFC3, IFC4 and IFC5, and the trials were established in four locations: Abengourou, Bingerville, Divo and San Pedro. Each trial was set up in a completely randomised design, with 20 plants per treatment and one plant per plot. Plant spacing was 2.5 × 3 m or 1333 plants/ha, temporary shade was provided by *Gliricidia maculata* for the first two years, and plants were fertilised and insects controlled regularly.

Table 10 Cumulative production (t/ha of commercial beans) from the 20 most productive cacao hybrid combinations in four locations in Ivory Coast from 1983 to 1987

Hybrid	B	D	A	S	Average
Sca6 × IFC1	8.43	9.21	11.78	–	9.81
Sca6 × IFC15	9.85	–	11.53	5.36	8.91
Sca6 × IFC5	11.41	–	9.79	5.23	8.81
Sca6 × IFC2	11.28	8.34	10.25	4.97	8.71
T79/501 × IFC5	6.55	8.30	10.2	6.13	7.80
T60/887 × IFC5	8.86	8.32	8.15	4.92	7.56
Na32 × IFC15	8.53	8.29	8.66	4.72	7.55
Pa7 × IFC15	8.15	7.95	8.84	4.68	7.41
Na32 × IFC5	7.57	8.76	7.68	5.18	7.30
Na32 × IFC1	5.65	8.23	10.16	5.08	7.28
T79/501 × IFC1	5.25	10.11	7.30	5.98	7.16
Pa150 × IFC15	–	7.06	–	–	7.06
IMC67 × IFC1	6.40	8.67	8.50	4.22	6.95
P7 × IFC15	6.68	8.09	8.92	4.09	6.95
T60/887 × IFC1	–	–	8.75	4.92	6.84
Pa150 × IFC5	4.89	9.59	7.85	4.93	6.82
T79/501 × IFC2	5.92	7.44	7.09	6.78	6.81
T79/501 × IFC15	4.72	8.54	7.50	6.26	6.76
IMC67 × IFC5	6.82	9.37	6.06	4.47	6.68
T60/887 × IFC2	6.97	8.11	6.14	5.49	6.68
T60/887 × IFC15	5.94	6.68	9.05	4.87	6.64
T63/971 × IFC15	6.73	7.46	7.39	4.31	6.47
T85/799 × IFC15	6.50	8.23	7.14	3.99	6.47
Pa7 × IFC5	7.13	7.74	6.19	4.67	6.43
T63/967 × IFC5	6.81	7.46	6.48	4.95	6.43

B: Bingerville; D: Divo; A: Abengourou; S: San Pedro. Source: Paulin et al. (1993).

The cumulative production from the 20 most productive hybrids per location and the average over the locations from 1983 to 1987 are presented (Table 10). The Sca6 parents generated the most productive hybrids, but the seeds were very small so they could not be released for commercial production. The T79/501, T60/887, Na32, Pa7 and Pa150 parents also gave rise to highly productive hybrids. According to Paulin et al. (1993), the most productive hybrids in the years 1985/1986 and 1986/1987 yielded over 3000 kg/ha/year. Some productive hybrids also presented fewer pods that had been attacked by *Phytophthora* sp., suggesting that it is possible to select hybrids for both tolerance and yield. The authors also found a high correlation between girth increments in the juvenile plant phase and cumulative yield. The correlation coefficients were r = 0.66 in Abengourou and r = 0.85 in Bingerville, with more favourable conditions for the plants, and r = 0.22 in Divo and r = 0.31 in San Pedro, where plant growth was slow. This suggests that early vigour in good growing conditions can be a good trait for preselection (Eskes et al., 1994)

Early vigour and adult vigour had higher correlations in Bingerville (r = 0.64) and Abengourou (r = 0.49) than in Divo (r = 0.06) and San Pedro (r = 0.32). Adult stem girth was correlated with cumulated yield (r = 0.56–0.64) and canopy size (r = 0.60–0.71). Canopy size was also correlated to yield (r = 0.40–0.60), but the authors identified plants with a twofold difference in canopy size that had the same production potential (Paulin et al., 1993). This indicates that these relationships depend on how the genotypes partition photo-assimilates for yield and canopy size.

Based on the results of the hybrid trials and on complementary information on resistance to pod rot, mirid attack, yield/vigour relationships and other traits, (Eskes et al., 1994), the best hybrid combinations were released in 1995 (Table 11). Recommendations may change as new information from the hybrid trials becomes available.

6.7.3 Population breeding

Reciprocal recurrent selection started in Ivory Coast in 1991/92 to improve two populations from different genetic or heterotic groups: UA and LA (Clément et al., 1994). Around 25–35 parents with known phenotypic or genotypic values were chosen to start the breeding process following the scheme presented in Fig. 2. Parents within each genetic group were initially crossed and evaluated in replicated trials, and the best progeny were selected. Another selection cycle followed, and the selected genotypes were crossed and evaluated, and agronomically superior individuals selected. The best individuals from

Table 11 Generation II cacao hybrids distributed in Ivory Coast since 1995

Hybrid	Yield (t/ha/year)	Number of beans/100 g	Fat (%)	Phythophthora pod rot (%)
H490	3.1	70	51.5	7.8
H155	2.0	105	50.5	6.4
H795	2.9	87	51.6	6.7
H608	2.7	86	51.3	8.4
H321	3.0	84	52.5	8.9
H6	3.1	120	49.3	9.4
H720	2.3	77	51.2	7.8

Source: Tahi et al.'s (2012) slide presentation in 7th INGENIC Workshop.

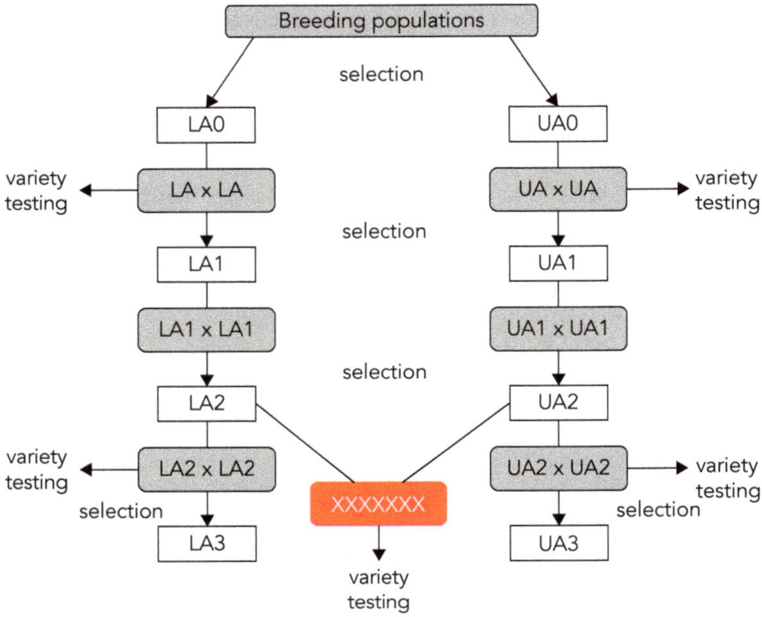

Figure 2 Basic scheme of recurrent selection in Ivory Coast. Source: Eskes et al. (1994).

the two genetic groups are crossed with each other to explore heterosis. Agronomically superior families can be used as hybrids, or the best individuals can be cloned and tested as clones. Indeed, the scheme is very flexible and allows superior families or individuals in each cycle of selection to be selected and new accessions to be introduced to each genetic group (Eskes et al., 1994).

6.8 Cameroon

In Cameroon, the local cacao population mainly consists of Trinitario-type introductions from São Tomé, so they are genetically different from the other West African producers (Nya Ngatchou, 1981). Cameroon's breeding programme started in 1949, with the mass selection of local populations and release of clones as planting material. Bi-clonal hybrids were later released, but most were very susceptible to *P. megakarya*. More recently, there has been a great effort to introduce resistant germplasm and select local material with farmers' participation.

6.8.1 Mass and clone selection

In Cameroon, an initial survey on farmers' fields allowed a collection of 350 seed-growing progenies to be set up in 1950, with a total of 15 000 plants, mainly LT. Those progenies were mass-selected and agronomically superior plants were propagated by rooted cuttings. After precocity, productivity and other traits were evaluated, 35 clones were released to farmers between 1957 and 1968. Twenty-one were locally selected clones, designated SNK (Nkoemyone Selections): SNK 10, 12, 13, 16, 30, 37, 48, 64, 109, 111,

203, 213, 332, 343, 348, 413, 415, 416, 450, 456 and 460. Three were locally selected Amazon clones: UPA 134, 143 and 337. Eleven clones were introduced from Trinidad: ICS 1, 6, 16, 39, 43, 46, 60, 61, 84 and 95 (Liabeuf, 1967). These clones were replaced by hybrids because of difficulties with large-scale plantings and low establishment ability in small farmers' environments (Eskes et al., 1994).

6.8.2 Hybrid selection

Hybrid development in Cameroon has been based on crossing parents from different heterotic groups, generally UA clones or their first or second generations, with parents from local selections belonging to the LA or T clones, as is the case in other countries. Trinitario parents have also been crossed with locally selected parents to produce hybrids, increasing the number of T parents. Liabeuf (1967) listed 11 bi-clonal hybrids used to produce seeds for farmers in 1967: SNK37 × SNK37; SNK13 × ICS95; UPA143 × SNK64; ICS43 × SNK37; SNK109 × SNK37; UPA143 × ICS40; SNK16 × ICS6; ICS60 × SNK456; ICS46 × SNK460; UPA134 × SNK12 and UPA143 × ICS95. These hybrids are mainly based on crosses between LT and introduced T. An extended list of hybrids was presented by Nya N'gatchou (1981): UPA143 × SNK64; SNK13 × ICS95; UPA143 × SNK64; T60/887 × ICS46; UPA143 × ICS95; IMC67 × SNK10; ICS43 × UPA143; UPA143 × ICS40; ICS40 × UPA134; SNK109 × IMC67; T79/501 × SNK109; SNK10 × IMC67; T79/501 × SNK13; T79/467 × SNK13; T79/467 × SNK64; SNK48 × IMC67 and T60/1774 × SNK64.

These hybrids are mainly based on UA × LT or introduced T and they generally proved highly susceptible to *P. megakarya*. The Cameroon research team therefore turned their attention to identifying and introducing more resistant clones. This was accomplished successfully and included accessions in germplasm collections as well as material identified through participatory selection of resistant trees in farmers' fields (Efombagn et al., 2007, 2013).

7 Selection of cacao for organoleptic quality

7.1 Fine flavour and bulk cacao

The world cacao market distinguishes between 'bulk' and 'fine flavour' cacao. Bulk cacao represents 95% of the world cacao market and fine flavour cacao currently corresponds to 220000 t annually. The bonus paid for fine flavour cacao at the farm gate varies from +15% to +300% (Pipitone, 2016), corresponding to a total bonus of 260 million USD annually for fine flavour cacao producers. Bulk cacao varieties do not generally express fine flavour traits, at least not at a commercial level. Fine flavour cacao has historically been provided mainly by the Criollo, Trinitario and Nacional varieties. These express the caramel and nutty, floral/herbal and fresh fruit sensory profiles, respectively. Caramel is not considered to be a true varietal sensory trait here, because it is derived from commonly occurring non-aromatic sugars. Over the past century, Criollo and Nacional hybridised with Trinitario in commercial plantations express additional fine flavour traits: 'raisin', 'floral' and 'spicy' (Afoakwa et al., 2008). The Amazon Sca6 genotypes that have been widely used in cacao variety development, such as the TSH varieties, also express a 'raisin or brown fruit' flavour and a 'heavy floral' aroma. In summary, the six varietal sensory traits that are associated with recognised fine flavour cacao varieties are fresh fruit, raisin or brown fruit, nutty, floral, heavy floral and spicy.

7.2 Limitations in cacao selection flavour quality traits

Cacao variety improvement since the 1930s has mainly addressed constraints such as yield and biotic stresses. Specific breeding for fine flavour quality traits has probably been limited to the TSH varieties. Selection for yield and resistance has occasionally been associated with lower than average bulk cacao quality, as is the case with the CCN51 clone selected in Ecuador in the mid-1960s. Lockwood and Eskes (1995) concluded that selection for cacao quality was hampered by differences in quality objectives among chocolate manufacturers and by inefficient methods of selection at individual tree level. They proposed that breeders should restrict themselves to verifying that commercial qualities in advanced breeding material are similar to the qualities of cacao produced traditionally in their country. This rather poor perspective on cacao quality selection is now understood to be due to a lack of knowledge on the expression of fine flavour cacao fruit pulp traits in chocolates.

7.3 Lack of cacao pulp sensory trait studies

For a long time, the main role attributed to cacao pulp was as a substrate for the various microorganisms essential to the development of chocolate flavour precursors during fermentation that react to form 'cacao' aroma components during the roasting process. The effect of cacao pulp sensory traits on chocolate qualities has only recently received attention. Regular visitors to cacao plantations might be familiar with the wide variety of cacao pulp flavours and aromas, and acidity and sweetness vary the most. For example, Criollo and Scavina genotypes have very sweet pulps, whereas the pulp of Trinitario and Forastero is often rather acid. While pulp flavour and aroma intensity used to be low and unspecific in the vast majority of genotypes, a few rare clones, such as Sca6 and EET62, are known to have very flavourful and/or aromatic pulp.

7.4 Cacao pulp versus fine chocolate sensory trait relationships

Farmers' field tasting of cacao pulp in 11 cacao-producing countries led to the conclusion in 2005 that 'countries that produce fine chocolates (e.g., Trinidad, Ecuador, Venezuela and Peru) grow varieties that have nice cacao pulps'. Cacao pulp in fine flavour cacao-producing countries was generally perceived to be sweet, with a variety of flavour and/or aroma expressions and low bitterness in the fresh cacao bean. However, countries that produce bulk cacao grow varieties with pulp that was quite acidic, with generally unexpressive flavours or aromas and rather bitter beans, which are known to be characteristics of bulk cacao chocolate. These empirical observations strongly suggested that cacao pulp sensory traits can be positively correlated with fine flavour chocolate traits.

7.5 Pulp sensory traits are discriminative varietal traits

In 2008, six panellists assessed seven pulp sensory trait intensities in nine cacao genotypes in Tingo Maria, Peru (Eskes et al., 2010). The level of discrimination between varieties was high for all traits ($p = 0.05$), with mean significance groups varying from a to d except for bean bitterness (a–c). EET62 had the highest average preference score (4.3) and CCN51 (2.0) and H56 (1.8) the lowest (Table 12). Panellists' scores for sweetness,

Table 12 Sensory preference and associated traits for the pulp of nine cacao genotypes assessed in UNAS, Tingo Maria, Peru, by six panellists in 2008

Clone	Pulp description	Preference
EET62	Aroma, fruity and sweet	4.3 a
Pandora1	Sweet, soursop flavour	3.2 b
ICS1	Sweet/acid, citric (lemon)	2.7 bcd
U43, 45, 53	Sweet and intermediate flavour	2.6 bcd
H56, 60	Astringent, acid, low fruity	2.1 bcd/d
CCN51	Astringent, acid, low flavour, very bitter bean	2.0 cd

acidity and overall preference did not vary significantly, suggesting uniformity of human perception of these traits. The sensory traits of CCN51 and EET62 generally represented the opposite extremes among the nine clones. The Pandora genotype presented a surprising 'soursop' pulp flavour, ICS1 a strong 'fresh fruit' flavour and EET62 had a 'floral' aroma as well as a fruity flavour. The ICS1 and EET62 pulp traits are in full agreement with the Trinitario and Nacional chocolate sensory traits of fresh fruit and floral, respectively. Overall preference scores were positively related to pulp odour, flavour and sweetness and negatively, but not significantly, to acidity and astringency. Pulp acidity, astringency and bean bitterness were positively correlated with each other and negatively correlated with pulp sweetness. The varieties with the most contrasting pulp sensory traits are EET62 (sweet pulp, strong flavour and odour, high preference), Ucayali clones (sweet pulp, low acidity and astringency, medium preference), CCN51 (acid and astringent pulp, very bitter beans and low preference) and the Huallaga clones (acid and quite astringent pulp, low preference).

7.6 New pulp and chocolate sensory traits discovered between 2007 and 2012

Between 2007 and 2012, ten trees with five hitherto unknown cacao pulp flavour traits were discovered as a result of limited pulp assessment exercises carried out in Trinidad, Peru, Ecuador and Brazil (Table 13). The most commonly detected new flavour was soursop, identified in four different locations in genetically very different cacao varieties. A banana flavour was first found in Trinidad in 2008 in an old Trinitario plantation with predominating pulp sensory variation being only for sweetness and acidity. A mandarin/jasmine sensory profile was first found in Chuncho cacao in Quillabamba, Peru, in 2011 and in local White Piura in 2012. Beans from the Trinitario 'banana' cacao tree were transformed into chocolates by Ed Seguine (ex-Guittard and Mars quality specialist) and three cacao quality specialists described the chocolate fruit flavour as follows: 'The banana flavour was detectable from the mid-session' (K. Maharaj, pers. comm.), 'The chocolate comes across with banana skins/strings astringency' (Ed Seguine) and 'The long-lasting aftertaste remembers that of a banana jam' (Albertus Eskes). These limited pulp surveys and chocolate tasting exercises suggest that interesting new pulp flavours and chocolates can be found in many different genetic varieties of cacao and many different countries.

Table 13 Five unknown fruit flavour traits found by the main author in cacao pulp from seven locations between 2007 and 2012

Year	Flavour/aroma trait	Cacao population	Country
2007	Soursop	Amazon collection, INIAP	Quevedo, Ecuador
2007	Soursop	Pandora clone from Colombia	Tingo Maria, Peru
2008	Mango/rose	EET397 from Ecuador	Bahia, Brazil
2008	Banana	Old Trinitario	Trinidad
2009	Soursop	French Guiana collection	Bahia, Brazil
2010	Soursop*	Old Trinitario	Trinidad
2011	Mandarin/jasmine	Chuncho variety	Cusco, Peru
2012	Mandarin, soursop and grape	Piura 'White' or 'Criollo' variety	Piura, Peru

* K. Maharaj, pers. comm.

7.7 The native Chuncho aromatic pulp and bean survey in Peru (2012–15)

'Chuncho' ('from the jungle') cacao is native to La Convención province (Urubamba valley) in the Cusco region of Peru. It was already cultivated in the seventeenth century, but also occurred spontaneously in association with the indigenous Matsigenkas ('human beings'). This tribe has always had a special interest in consuming Chuncho fruit pulp and slightly fermented and roasted raw beans (Carlos Rodriguez, Pers. Obs.). Paz Soldan (1852), cited by Gade (1975), labelled Chuncho cacao from the Echerate Estate as 'the best cacao in the world'. Chuncho cultivation culminated in the 1980s with 14000 ha and a production of 10000 t, but has declined to the current approximately 11000 ha and 2600 t. This is due to low productivity (200–300 kg/ha) and ageing trees, and especially to competition from CCN51. Chuncho is still mainly used as a low-value bulk cacao bean source for butter and powder extraction. However, farmers also consume pulp juices made from individually selected Chuncho trees, which is a practice that is not known to occur anywhere else in the world. This implies that some Chuncho trees produce pulp with special sensory qualities so our pulp and bean survey focused on trees used for juice production. Commercial Chuncho beans already express a noticeable level of unspecific flavour and aroma, and even when they are unfermented or slightly fermented, Chuncho beans are neither acidic and bitter nor astringent but may become so when fermented for longer than four days.

The Chuncho variety is one of the ten currently recognised major *T. cacao* genetic groups, and it belongs to the 'Contamana' Forastero group, being very closely related to Beni and Madre de Dios accessions. Recent DNA diversity studies are expected to define the indigenous origins of Chuncho more precisely (Dapeng Zhang, pers. comm.). Our brief report on an aroma survey of 226 Chuncho trees growing in a 200-km stretch along the Urubamba River in the La Convención province in Peru can be found in the following section.

7.7.1 Pulp survey

The pulp flavours and aromas of 100 Chuncho trees in La Convención were evaluated in 2012. The choice of trees with flavourful pulps for the survey was mainly guided by observations by farmers. Eighty-seven of the 100 trees presented qualitative expressions of pulp flavours and/or aromas representing 29 distinct pulp sensory profiles. Fifteen were represented by just one tree and the rest by 2–16 trees, showing the rare occurrence of these individual sensory profiles. Profiles included 16 fruit flavours and eight aromas or spices. The 17 fruit flavours are: annona, soursop, citrus, mandarin, tangelo, banana, grape, fruity, guava, rose apple, green apple, mango, custard apple (cherimoya), inga, banana var. 'Manzana', 'Isla' and 'Cavendish', 16 of which mimic known tropical or temperate fruit species. The eight floral or spice aromas are: floral, mint, jasmine, heavy floral, rose, vanilla, lily and yeast, with five being related to known flower or spice species. The most frequently encountered sensory profiles were soursop/jasmine (16 trees) and floral (14 trees). Fifteen (54%) of the twenty-nine profiles were made up of more than one flavour and/or aroma, and fourteen (46%) by one flavour and/or aroma.

7.7.2 Bean survey

In April 2015, a panel of three cacao specialists and two fine chocolate manufacturers from Peru identified sensory traits and assigned preferences to 126 slightly fermented raw bean samples. The beans were again obtained from individual Chuncho trees in La Convención valley used by farmers to prepare cacao pulp juices. Of the 126 bean samples, 92 received overall preference scores of three and above, and a total of 40 different sensory profiles were identified, involving 16 fruit flavours and 9 aromas or spices. The fruit flavours identified were soursop, citrus, mandarin, cranberry, nuts, peach, jackfruit, dried apple, ripe red fruit, fresh red fruit, ripe yellow fruit, ripe plum, raisin, jackfruit, mango and custard apple (cherimoya), and the most frequently encountered sensory profiles were mandarin/jasmine, citrus, cranberry and malt, found in six or more trees. The aromas and spices identified were jasmine, floral, rose, lily, cinnamon, mint, malt and spices. Of the 16 bean flavours 5 were also found among the 16 pulp flavours identified, resulting in a total of 26 Chuncho pulp and bean fruit flavours. Five of the eight bean aromas were also found among the eight pulp aromas, resulting in a total of eleven Chuncho pulp and bean aromas/spices. Of the 40 bean sensory profiles 30 (75%) consist of more than one sensory trait, with 20 (50%) of these consisting of flavour × aroma combinations.

These pulp and bean evaluations resulted in the identification of 62 unique and highly diverse sensory profiles among the 226 Chuncho trees studied. Only 16 (26%) of the 62 unique sensory profiles are made up of one flavour or aroma trait. The remaining 46 profiles (74%) are based on combinations of two or more sensory traits. Thirty-two (52%) contain exclusively fruit flavour and flower or spice aroma combinations. Six profiles (10%) are made up of three flavour and/or aroma combinations. Of the 64 unique sensory profiles 16 (26%) were identified in only one tree and should therefore be considered extremely rare at present.

7.8 Conclusions

7.8.1 Sensory studies of cacao pulp

No scientific references to cacao pulp sensory traits had been published until recently. Pulp sensory trait studies have arisen from the conclusion that countries that produce fine chocolates have cacao varieties with flavourful pulp, which suggests that pulp and chocolate sensory traits may be linked and studies of nine cacao genotypes have revealed that these traits have a high discriminative varietal value. Cacao pulp studies advanced significantly between 2007 and 2011, through the discovery of single-aroma pulp trees in four countries which presented hitherto unknown sensory profiles: banana, soursop, mango/rose and mandarin/jasmine. Chocolates made with beans from trees with banana profiles have strong banana traits, and cacao pulp science is exploding with the discovery of 64 diverse and expressive pulp and bean sensory profiles in the Chuncho cacao variety. The Chuncho cacao producers themselves are responsible for this amazing find as they know exactly which trees have the most interesting sensory pulp traits to make juice. Of the 126 evaluated as raw beans in the study, 92 (73%) gained medium to high overall panellists' preferences, with 40 different sensory profiles based on 23 flavours and aromas or spices. This surprisingly large diversity of fine flavour sensory traits and profiles suggests that farmers select their trees effectively for aromatic juice making. Aroma identification carried out by our research team confirms that of the farmers, validating the enormous sensory diversity of Chuncho cacao.

7.8.2 Chuncho sensory diversity compared to fine flavour cacao varieties

Traditional fine flavour cacao varieties present six flavour or aroma traits that are identical or very similar to six of the thirty-four Chuncho pulp and bean sensory traits. This indicates that 26 of the 34 Chuncho sensory traits are new with regard to the traditional fine flavour varieties. However, more sensory trait variation might exist in fine flavour varieties than have been recognised so far, as suggested by the discovery of two trees in old Trinitario plantations in Trinidad that express banana and soursop pulp flavours, both of which are completely new for Trinitario trees.

7.8.3 Chuncho sensory diversity compared to other cacao varieties

Between 2007 and 2012, four different fruit flavours and two aromas were detected in the pulp of ten trees from very different varieties and locations. Commercial chocolates made from native cacao plantations in Peru (Amazon and San Martin regions) and Ecuador reveal the presence of citrus flavour, and vanilla and rose aromas, respectively. These observations suggest that Chuncho-like flavours and aromas can also be found in very different places and *T. cacao* varieties. It is likely that systematic single-tree pulp assessments will detect many of the known Chuncho flavours and aromas, and possibly others, in varieties that are not currently known to possess any fine flavour traits.

In 2017, pulp flavours and aromas were identified in 25 commercial Brazilian clonal varieties including also Amelonado. Sixteen varieties showed to be fine-flavour varieties expressing a total of six fruity flavours and four aromas. Amelonado, traditionally considered as a bulk variety expresses a strong jasmine aroma! These surprising results suggest that fine-flavour cocoa genotypes can be found in many varieties possibly worldwide, even in the so-called bulk varieties.

7.9 Further selection, multiplication and breeding of aromatic cacao varieties

Further selection, multiplication and breeding of rare aromatic genotypes, such as those found in Chuncho cacao, can best be conducted as follows:

1 Worldwide surveys to identify more aromatic genotypes, especially among germplasm collections, cultivated varieties and native cacao populations;
2 Clonal multiplication to conserve rare aromatic genotypes;
3 Open-pollinated seed multiplication to reproduce the same sensory traits as the mother tree through expected pleiotropic dominant inheritance, to establish small commercial plots, followed by large-scale clonal multiplication of superior progeny;
4 Somatic embryogenesis to establish large-scale commercial plots of Chuncho selections;
5 Crossing aromatic genotypes with resistant heterotic genotypes (e.g., Pa121 and Pa169) that have mild bean characteristics, to create productive, resistant and high-quality hybrid varieties and finally
6 Creation of hybrids with very complex sensory traits by crossing genotypes with complementary sensory profiles, for example, combining one or more fruity flavours with an aroma and/or spice.

8 Future trends and conclusion

With almost 100 years of cacao breeding in the producing countries of the world many lessons were learned. Accumulated experience is now available to indicate the most appropriate methods for the present and the future of cacao breeding. In the beginning (1930s), the programmes released clonal and open pollinated seed cultivars as planting material for farmers, but almost no genetic gains were obtained because selection was done in populations with low genetic diversity for agronomic traits available at that time. In the 1960s with the distribution of germplasm collected by Pound in Peru to nearly all producing countries, hybrid cultivars were developed and released to farmers as planting material. There were genetic gains in the first generations of hybrids but the gains were not continuous because of the lack of improved progenitors. In the 1990s, many breeding programmes started population breeding to improve parents for developing hybrids and clonal cultivars. The exploitation of heterosis should be considered in all breeding programmes. More recently, with the selection and good results of some clonal cultivars in farmers' fields, they are becoming the main type o planting material in the Americas and Asia. Selection of rootstocks resistant to biotic and abiotic stresses is also important with the adoption of clonal cultivars. Continuous genetic gains are now expected with the use of clonal and population breeding programmes. But more resources and the participation of farmers are required because the challenges to breed for have increased with the spread of disease and pests and with the climate changes in producing countries. Besides high yield and resistance to pests and diseases, the development of shorter and more erect plants with better partitioning of assimilates and more easy husbandry are required to increase planting density. Adoption of molecular biology tools such as wide genome selection and CRISPR by the ongoing population and clonal breeding programmes might help to speed up breeding. The recent identification of a large number of fresh fruit flavours and aromas present in *Theobroma cacao* warrants significant and quick selection progress for fine-flavour varieties. Once adopted, this may well represent a revolution for the fine-flavour cacao market. Finally, we have to consider that the genetics of cultivars is considered to be responsible only for 50% of the actual yield while the production system responsible for the other 50%. Therefore good genetics needs to be going side by side with good agronomical practices.

9 References

Abdul-Karimu, A., Adu-Ampomah, Y., Frimpong, E. B. and Tafo, A. (2003), Field evaluation of agronomic characters of some selected cocoa hybrids in a marginal area of Ghana. In: *Proceedings of the 14th International Cocoa Research Conference*, Volume 3. Cocoa Producers' Alliance (COPAL), Lagos, Nigeria, pp. 111–20.

Adomako, B., Allen, R. C. and Adu-Ampomah, Y. (1999a), Evaluation of hybrids between Upper Amazon cocoa selections in Ghana. *Plantations, Recherche, Développement*, 6(6), 455–65.

Adomako, B., Allen, R. C. and Adu-Ampomah, Y. (1999b), Combining abilities for yield and vegetative traits of Upper Amazon cocoa selections in Ghana. *Plantations, Recherché, Développement*, 6(3), 183–92.

Adu-Ampomah, Y. (1994), The cocoa breeding programme in Ghana: Achievements and prospects for the future. In: End, M. J. (Ed.), *Proceedings of the International Workshop on Cocoa Breeding Strategies*. INGENIC, Kuala Lumpur, Malaysia, pp. 29–32.

Afoakwa, E. O., Paterson, A., Fowler, M., Ryan, A. (2008), Flavor formation and character in cocoa and chocolate: A critical review. *Crit Rev Food Sci Nutr*, 48:839–64.

Ahnert, D. (2006), Ideotype Breeding in Cocoa. In: *Proceedings of the International Workshop on Cocoa Breeding for Farmers' Needs*. INGENIC, San Jose, Costa Rica, pp. 157–66.

Ahnert, D., Machado, P. H., Bartley, B. G. D. and Neto, F. E. (1983), Desenvolvimento e avaliação de cultivares de cacau. In: Informe de Pesquisa. CEPLAC/DEPEA, Belém, p. 134.

Allen, J. B. and Lass, R. A. (1983), *London Cacao Trade Amazon Project. Final Report Phase I*. Cocoa Growers' Bulletin 34. Cadbury Schweppes, Birmingham, UK.

Almeida, C. M. V. C., Barriga, J. P., Machado, P. F. R. and Bartley, B. G. D. (1987), *Evolução do Programa de Conservação de Recursos Genéticos de Cacau da Amazonia Brasileira*, Boletin Técnico no. 5. Comissão Executiva do Plano da Lavoura Cacaueira, DEPEA, Brazil, p. 108

Almeida, C. M. V. C., Machado, P. F. R., Barriga, J. P. and Silva, F. C. O. (1995), *Coleta de Cacau (Theobroma cacao L.) da Amazônia Brasileira: Uma Abordagem Histórica e Analítica*. Comissão Executiva do Plano da Lavoura Cacaueira, SUPOC, Brazil, 83p.

Amores, F. M., Vasco, S. A., Eskes, A. B., Suarez, C., Quiroz, J. G., Loor, R. G., Jimenez, J. C., Zambrano, J., Bolanos, M. J., Reynel, V. H., Teran, M. M. and Quijano, G. C. (2011), On-farm and on-station selection of new cocoa varieties in Ecuador. In: A. B. Eskes (Ed.), *Collaborative and Participatory Approaches to Cocoa Variety Improvement: Final Report of the CFC/ICCO/Biodiversity International Project on 'Cocoa productivity and quality improvement: a participatory approach' (2004–2010)*. CFC, Amsterdam, pp. 59–72.

Atanda, O. A. (1977), The third Nigerian cocoa breeding programme: A progress report. In: *Proceedings of the 5th International Cocoa Research Conference, 1993*. CRIN, Ibadan, Nigeria, pp. 85–90.

Bartley, B. G. D. (1981), The status of genetic resistance in cacao to *Crinipellis perniciosa* (Stahel) Singer [*Theobroma cacao*]. In: *Proceedings of the 6th Internacional Cocoa Research Conference*. Cocoa Producers' Alliance, Lagos, pp. 56–69.

Bartley, B. G. D. (1994), A review of cacao improvement: Fundamentals, methods and results. In: End, M. J. (Ed.), *Proceedings of the International Workshop on Cocoa Breeding Strategies*. INGENIC, Kuala Lumpur, Malaysia, pp. 3–17.

Bartley, B. G. D. (2005), *The Genetic Diversity of Cacao and Its Utilization*. CABI Publishing, Wallingford, UK, 341p.

Bartley, B. G. D., Monteiro, W. R. and Carletto, G. A. (1982), Comportamento dos clones introduzidos como progenitores de híbridos na Bahia. In: *8th International Cacao Research Conference. Cartagena, Colombia*. Cocoa Producers' Alliance, London, pp. 703–12.

Bekele, F. (2004), The history of cocoa production in Trinidad and Tobago. In: *APASTT Seminar Series on the Re-vitalization of the Trinidad & Tobago Cocoa Industry*. Univ. West Indies, St. Augustine, Trinidad, pp. 4–12.

Besse, J. (1964), L'amélioration du cacaoyer en Côte d'Ivoire. *Café, Cacao, Thé (Francia)*, 8(4), 245–62.

Besse, J. (1975), La selection generative du cacaoyer en Côte d'Ivoire: Bilan et orientation des recherches en 1975. In: *5th International Cacao Research Conference*. Cocoa Producers' Alliance, Ibadan, Nigeria, pp. 95–103.

Borém, A. (1997), *Melhoramento de Plantas*. UFV, Viçosa, 547p.

CacaoNet (2016). Available: http://www.cacaonet.org/. Accessed December 2016.

Chalmers, W. S. (1972), The conservation of wild cacao populations: The plant breeder's most urgent work. In: *Proceedings of the 4th International Cacao Research Conference*, Port-of-Spain, Trinidad, pp. 114–18.

Cheesman, E. E. and Pound, F. J. (1933), *Further Notes on Criteria of Selection in Cacao. Annual Report on Cacao Research*, Volume 3. Imperial College of Tropical Agriculture, Trinidad, pp. 21–4.

Clément, D., Eskes, A., Sounigo, O. and N'Goran, J. A. (1994), Amélioration génétique du cacaoyer en Côte d'Ivoire. Présentation d'un nouveau schéma de sélection. In: *Proceedings of the 11th International Cocoa Research Conference, 1993*, Yamoussoukro, Côte d'Ivoire. Cocoa Producers' Alliance, Lagos, Nigeria, pp. 451–5.

Cocoa Research Institute of Nigeria (CRIN) (2016), The eight new cocoa hybrids. In: *CRIN Monthly Seminar*. Available: http://www.crin-ng.org/index.php/14-crin-monthly-seminar/26-the-new-cocoa-hybrids.

Cope, F. W. (1952), *Some Results of the Cacao Clonal Trials at River State. Annual Reporto n Cacao Research 1945–1951*. ICTA, Trinidad, pp. 12–24.

Djekpor, E. J., Cilas, C. and Paulin, D. (1994), Spotlight on cacao breeding for resistance to swollen shoot. In: *Proceedings of the International Workshop on Cocoa Breeding Strategies*. INGENIC, Kuala Lumpur, Malaysia, pp. 18–19.

dos Santos, E. A., de Almeida, A. A. F., Ahnert, D., da Silva Branco, M. C., Valle, R. R. and Baligar, V. C. (2016), Diallel analysis and growth parameters as selection tools for drought tolerance in young *Theobroma cacao* plants. *PLoS One*, 11(8), e0160647.

Efombagn, M. I. B., Nyassé, S., Sounigo, O., Kolesnikova-Allen, M. and Eskes, A. B. (2007), Participatory cacao (*Theobroma cacao*) selection in Cameroon: *Phytophthora* pod rot resistant accessions identified in farmers' fields. *Crop Protection*, 26(10), 1467–73.

Efombagn, M. I. B., Nyassé, S., Bieysse, D. and Sounigo, O. (2013), Analysis of the resistance to *Phytophthora* pod rot within local selections of cacao (*Theobroma cacao* L.) for breeding purpose in Cameroon. *Journal of Plant Breeding Crop Science*, 5, 111–19.

Eskes, A. B., Paulin, D., Clement, D., N'Goran, J. A. K., Sounigo, P. L., Cilas, C. and Yagmpam, A. (1994), Selection methods applied and genetic knowledge generated in cocoa breeding in Côte d'Ivoire and Cameroon. In: *Proceedings of the International Workshop on Cocoa Breeding Strategies*. INGENIC, Kuala Lumpur, Malaysia, pp. 41–55.

Francisco Neto, E. (2008), *Parâmetros genéticos seleção genotípica de cacaueiro na Amazonia brasileira*. Tese (Doutorado em Genética e Bioquímica)-Universidade Estadual de Uberlândia, Uberlândia, 138p.

Gade, D. W. (1975), *Plants Man and the Land in the Vilcanota Valley*. Biogeographical volume 6, Dr. W. Junk B.V. Publishers, the Hague, The Netherlands.

Glendinning, D. R. (1966), Further developments in the breeding programme at the Cacao Research Institute, Tafo. *Ghana Journal of Science*, 6, 52–62.

Hallauer, A. R. and Miranda, F. J. B. (1988), *Quantitative Genetics in Plant Breeding*. Iowa State. Univ. Press, Ames, pp. 337–48.

Hunter, J. R. (1961), The status of cacao variety improvement work in the western hemisphere. *Cacao*, Interamerican Institute of Agriculture, Turrialba, Costa Rica, 6(4), 1–16.

Instituto Brasileiro de Geografia e Estatistica (2014), Available: http://www.ibge.gov.br/home/. Accessed December 2016.

International Cocoa Organization (2016), Available: https://www.icco.org/. Accessed December 2016.

Jacob, V. J., Atanda, O. A. and Opeke, L. K. (1971), Cacao breeding. In: *Annual Report of Cocoa Research Institute of Nigeria*. CRIN, Ibadan, pp. 28–31.

Kennedy, A. J., Lockwood, G., Mossu, G., Simmonds, N. W. and Tan, G. Y. (1987), Cacao breeding: past, present and future. *Cacao Growers Bulletin*, 38, pp. 5–22.

Laker, H. A., Sreenivasan, T. N. and Kumar, D. R. (1988), The resistance of some cacao clones to *Crinipellis perniciosa* in Trinidad. In: *Proceedings of the 10th International Cacao Research Conference*. Cocoa Producers' Alliance, Lagos, Nigeria, pp. 637–41.

Lanaud, C. S., Sounigo, O., Amefia, Y. K., Paulin, D., Lachenaud, P. and Clement, D. (1987), Nouvelles données sur le fonctionnement du système d'incompatibilité du cacaoyer et ses conséquences pour la sélection. *Café, Cacao, Thé (Francia)*, 31(4), 267–82.

Liabeuf, J. (1967), Etat des travaux effectués pour la sélection du cacaoyer au Cameroun et perspectives d'avenir. In: *2nd International Cacao Research Conference*, Salvador and Itabuna, Bahia. Centro de Pesquisas do Cacau, Itabuna, pp. 86–93.

Lockwood, G. (1976), A comparison of the growth and yield during a 20 year period of Amelonado and Upper Amazon hybrid cacao in Ghana. *Euphytica*, 25(1), 647–58.

Lockwood, G. (2015), Propagation by seeds. In: Laliberté, B. and End, M. (Eds), *Supplying New Cocoa Planting Material to Farmers: A Review of Propagation Methodologies*. Bioversity International, Rome, Italy, pp. 12–17.

Lockwood, G. and Gyamfi, M. M. O. (1979), *The CRIG Cacao Germplasm Collection with Notes on Codes Used in the Breeding Programme at Tafo and Elsewhere*. Ghana Technical Bulletin, pp. 1–62, 10, Cacao Research Institute, Ghana.

Lockwood, G. and Pang Thai Yii, J. (1994), Additive inheritance of yield in cacao. In: *Proceedings of the 11th International Cacao Research Conference, 1993*, Yamoussoukro, Côte d'Ivoire. CIRAD, Montpellier, France, pp. 415–24.

Lockwood, G. and Eskes, A. B. (1995), Relationship between cocoa variety and quality. In: *Seminar Proceedings, Cocoa Meetings 'The Various Aspects of Quality'*, 30 June 1995, AFCC, CIRAD, Montpellier, France, pp. 159–67.

Lopes, U. V., Monteiro, W. R., Pires, J. L., Clement, D., Yamada, M. M. and Gramacho, K. P. (2011), Cacao breeding in Bahia, Brazil: Strategies and results. *Crop Breeding and Applied Biotechnology*, 1, 73–81.

Machado, P. H., Ahnert, D., Bartley, B. G. D. and Neto, F. E. (1983), Desenvolvimento e avaliação de cultivares de cacau. In: *Informe de Pesquisa*. CEPLAC/DEPEA, Belém, pp. 79–103.

Machado, P. F. R., Bartley, B. G. D., Ahnert, D. and Barriga, J. P. (1984), *Desenvolvimento e Avaliação de Cultivares de Cacau*. CEPLAC/DEPEA, Belém, pp. 90–103.

Machado, P. H., Ahnert, D., Bartley, B. G. D. and Neto, F. E. (1986), Desenvolvimento e avaliação de cultivares de cacau. In: *Informe de Pesquisa*. CEPLAC/DEPEA, Belém, pp. 88–104.

Maharaj, K., Maharaj, P., Bekele, F. L., Ramnath, D., Bidaisee, G. G., Bekele, I. and Sankar, R. (2011), Trinidad selected hybrids: An investigation of the phenotypic and agro-economic traits of 20 selected cacao cultivars. *Tropical Agriculture*, 88(4), 175–85.

Mariano, A. H. and Bartley, B. G. (1981), Comportamento das seleções baianas na produção de híbridos de cacaueiros [*Theobroma cacao*; Brasil]. In: *7th International Cacao Research Conference*, London. Actes. J. de Lafforest and Transla-Inter, London, pp. 527–33.

Monteiro, W. R., Carletto, G. A. and Bartley, B. G. D. (1985), Avaliação da capacidade combinatória de clones de cacaueiro. In: *Proceedings of the* 9th *International Cacao Research Conference*, Lomé, Togo. Cocoa Producers' Alliance (COPAL), Great Britain, p. 11.

Monteiro, W. R. and Ahnert. D. (2012), Melhoramento genético do cacaueiro. In: Valle, R. R. (Ed.), *Ciência, tecnologia e manejo do cacaueiro*. CEPLAC/CEPEC, Brasília/Itabuna, p. 688

Monteiro, W. R., Vanderlei Lopes, U., Pires, J. L. Newman Luz, E. D. M., Midlej Silva, S. D. V., Silva Reis, D. and Silva de Araujo, D. C. (2010), On-farm and on-station selection of new cacao varieties in the State of Bahia, Brazil. In: A. B. Eskes (Ed.), *Collaborative and Participatory Approaches to Cacao Variety Improvement*. CFC, Amsterdam, The Netherlands; ICCO, London, UK; Bioversity International, Rome, Italy, pp. 8–30.

Montserin, B. G., de Verteuil, L. L. and Freeman, W. E. (1957), A note on cocoa hybridization in Trinidad with reference to clonal selections and hybrid seed. *Caribbean Commission Public Exchange Service*, 33, 156–9.

Morera, J. and Mora, A. (1996), *Comportamiento agronómico de 12 clones de cacao bajo las condiciones de Turrialba, Costa Rica*. CATIE M843co. CATIE, Turrialba (Costa Rica).

Motamayor, J. C., Lachenaud, P., e Mota, J. W. D. S., Loor, R., Kuhn, D. N., Brown, J. S. and Schnell, R. J. (2008), Geographic and genetic population differentiation of the Amazonian chocolate tree (*Theobroma cacao* L). *PLoS One*, 3(10), e3311.

Nya Ngatchou, J. (1981), Etat d'avancement des travaux de génétique et d'amélioration du cacaoyer au Cameroun Proceedings. In *Proceedings of the 7th International Cacao Research Conference*. Douala (Camerún), Cacao Producers' Alliance, Lagos (Nigeria), pp. 507–11.

Omolaja, S. S., Aikpokpodion, P. and Iremiren, G. O. (2006), Cocoa Genetic Resources in Nigeria. In: *15th International Cocoa Research Conference*, San Jose, Costa Rica. Cocoa Producers' Alliance (COPAL), Lagos, Nigeria, pp. 1805–10.

Opoku, S. Y., Dadzie, M. A. Padi, F. K., Opoku, I. Y., Adu-Ampomah and Adomako, B. (2010), Selection of new varieties on-farm and on-station in Ghana. In: *Collaborative and Participatory Approaches to Cocoa Variety Improvement*, pp. 73–9.

Paulin, D. and Eskes A. B. (1995), Le cacaoyer: Strategies de selection. *Plantations, Recherche, Développement* 2, 5–18.

Paulin, D., Mossu, G., Lachenaud, P. and Cilas, C. (1993), La sélection du cacaoyer en Côte d'Ivoire: analyse du comportement de soixante-deux hybrides dans quatre localités. *Café, Cacao, Thé*, 7, 3–20.

Pereira, J. L., De Almeida, L. C. C. and Santos, S. M. (1996), Witches' broom disease of cacao in Bahia: Attempts at eradication and containment. *Crop Protection*, 15(8), 743–52.

Phillips Mora, W. (1999), Studies at CATIE on moniliasis resistance (*Moniliophthora roreri* (Cif. And Par.) Evans et al.). In: *Proceedings of the Workshop on the Contribution of Disease Resistance to Cacao Variety Improvement*. INGENIC, Reading, pp. 11–117.

Phillips-Mora, W., Arciniegas-Leal, A., Mata-Quiros, A. and Motamayor-Arias, J. C. (2013), *Catalogue of cacao clones selected by CATIE for commercial plantings*. Technical series. Technical manual no. 105. CATIE, Turrialba.

Pipitone, L. (2016), Global Situation and Prospects of the Cocoa and Chocolate Markets – Current Challenges and Opportunities, 2nd Cocoa Revolution, Ho Chi Minh City, 09–11 March 2016.

Pires, J. L., Monteiro, W. R., Pinto, L. R. M., Figueira, A., Yamada, M. M. and Ahnert, D. (1999a), A proposal for cacao breeding. In: *Proceedings of the 12th International Cacao Research Conference*. Cocoa Producers' Alliance, Lagos, pp. 287–92.

Pires, J. L., Monteiro, W. R., Pinto, L. R. M. and Luz, E. D. M. N. (1999b), Resistance to witches' broom evaluation of genotypes from different origins. In: *Proceedings of the 12th International Cacao Research Conference*. Cocoa Producers' Alliance, Lagos, pp. 389–98.

Posnette, A. F. (1943), Cacao selection on the gold coast. *Tropical Agriculture (Trinidad)*, 22, 184–7.

Pound, F. J. (1932), The principles of cacao selection. *Agricultural Society of Trinidad and Tobago, Proceedings*, 32, 122–7.

Pound, F. J. (1943), *Cacao and Witches' Broom Disease (Marasmius perniciosus). Report on a Recent Visit to the Amazon Territory of Peru, September 1942 – February 1943*. A. L. Rhodes, Port-of-Spain, Trinidad and Tobago, 14p.

Rogers, H. H. (1955), New varieties of cacao in West Africa. In: *Report of the Cacao Conference held at Grosvenor House, London, W.1., 13th and 14th September 1950*. Cocoa, Choc. Confect. Alliance, London, pp. 76–9.

Rogers, H. H. and Knight, R. (1953), Plant breeding and selection. In: *Report of West African Cocoa Research Institute*. The Alliance, London, pp. 55–79.

Russell, T. A. (1952), The vigour of some cacao hybrids. *Tropical Agriculture (Trinidad)*, 29, 102–6.

Santos, I. C., de Almeida, A. A. F., Ahnert, D., da Conceição, A. S., Pirovani, C. P., Pires, J. L. and Baligar, V. C. (2014), Molecular, physiological and biochemical responses of *Theobroma cacao* L. genotypes to soil water deficit. *PloS one*, 9(12), e115746.

Sena Gomes, A. R. and Sodré, J. A. (2015), Conventional vegetative propagation. In: Laliberté, B. and End, M. (Eds), *Supplying New Cocoa Planting Material to Farmers: A Review of Propagation Methodologies*. Bioversity International, Rome, Italy, pp. 34–66.

Soria, J. (1966), Principales variedades de cacao cultivadas en América Tropical. *Turrialba (IICA)*, 16(3), 261–6.

Soria, J. and Esquivel, O. (1967), Estudio preliminar sobre el periodo mínimo y confiabile de producción en cacao para su uso en experimentos de evaluación de cultivars. *Cacao*, 12, 9–14.

Sounigo, O., N'Goran, J. A., Coulibaly, N., Clément, D. and Lachenaud, P. (1994), Evaluation de clones de cacaoyers pour la productivité, la résistance aux mirides et la résistance à la pourriture des cabosses. In: *Proceedings of the 11th International Cacao Research Conference*, Yamoussoukro, Côte d'Ivoire. Cocoa Producers' Alliance (COPAL), Lagos, pp. 437–42.

Tan, G. Y. (1990), Combining ability analysis of yield and its components in cacao. *Journal of the American Society for Horticultural Science*, 115(3), 509–12.

Thresh, J. M., Owusu, G. K., Boamah, A. and Lockwood, G. (1988), Ghanaian cacao varieties and swollen shoot virus. *Crop Protection*, 7(4), 219–31.

Toxopeus, H. (1985), Planting material (4 ed.). In: Wood, G. A. R. and Lass, R. A. (Eds), *Cocoa*. Longman, London, pp. 81–92.

Turnbull, C. J. and Hadley, P. (2016), *International Cacao Germplasm Database* [Online Database]. CRA Ltd./ICE Futures Europe/University of Reading, UK. Available: http://www.icgd.reading.ac.uk. Accessed December 2016.

Vello, F., Mariano, A. H., Garcia, J. R., Nascimento, I. F. and Magalhães, W. S. (1969), O Programa de melhoramento genético do cacau na Bahia. In: *Proceedings of the 2th International Cacao Research Conference*. CEPLAC, Salvador/Itabuna, pp. 43–56.

Williams, J. A. and Akinwale, S. A. (1994), Breeding strategies adopted in Nigerian cacao collections. In: End, M. J. (Ed.), *Proceedings of the International Workshop on Cocoa Breeding Strategies*. INGENIC, Kuala Lumpur, Malaysia, pp. 33–7.

Yamada, M. M., Faleiro, F. G. and Lopes, U. V. (2001), Genetic variability in cultivated cacao populations in Bahia, Brazil, detected by isozymes and RAPD markers. *Crop Breeding and Applied Biotechnology*, 1, 377–84.

Part 2

Cultivation techniques

Chapter 6

Cocoa plant propagation techniques to supply farmers with improved planting materials

Michelle End, INGENIC (The International Group for Genetic Improvement of Cocoa), UK; Brigitte Laliberté, Bioversity International, Italy; Rob Lockwood, Consultant, UK; Augusto Roberto Sena Gomes, Consultant, Brazil; George Andrade Sodré, CEPLAC/CEPEC, Brazil; and Mark Guiltinan and Siela Maximova, The Pennsylvania State University, USA

1 Introduction

The availability of high-performance planting materials to cocoa farmers is an important component in a package of measures to improve the productivity of cocoa farms and thus the sustainability of the cocoa economy. It has been estimated that currently only about 25% of all cacao plantings consist of improved varieties. If farmers cannot access improved materials, or cannot afford to do so, then they will continue to use their own materials, or those of their neighbours, and be unable to benefit from the breeding work that is delivering varieties with improved yield, quality and resilience to biotic and abiotic stresses.

The challenge is to integrate breeding programmes with systems to ensure that recommended varieties can be delivered to the farmers in sufficient quantities to satisfy demand in a cost-effective and timely way. As described elsewhere in this book, there are three types of commercial cacao cultivars: local or landrace cultivars, hybrid seedling

http://dx.doi.org/10.19103/AS.2017.0021.11
Published by Burleigh Dodds Science Publishing Limited, 2018.

varieties (more correctly bi-parental crosses) and clones. The techniques used to propagate them are:

- Seed propagation: Although many farmers plant seeds from pods harvested from their own trees or obtained from neighbouring farmers, improved seedling varieties have been developed and are being produced and distributed in large numbers from national and/or commercial seed gardens. Some farmers prefer to plant the seed directly in their fields (also known as 'planting at stake'), whilst others sow the seed in a nursery, or obtain young seedlings for planting when they are 3–6 months old, often depending on weather conditions.
- Conventional vegetative (clonal) propagation: Various techniques may be used to bud or graft materials from improved varieties onto seedling rootstocks in the nursery, or onto plants already growing in the field. The latter can include side grafting onto the trunk, lateral branches or upright suckers ('chupons') of mature trees. Once the bud or graft has established ('takes'), the aerial part of the rootstock/older part of the tree is cut away. Another technique involves producing rooted cuttings of improved materials. Usually these operations are carried out professionally, with the establishment of a clonal garden to supply budwood/graftwood or cuttings of the improved types, and nursery facilities in order to supply farmers with young container-grown plants. Graftwood can be obtained from clonal gardens for use in professional field-grafting operations, and, occasionally, for use by the farmers themselves. It is important to note that regardless of the technique used, the choice of plagiotropic or orthotropic material for the scion or cutting, as well as the breeder's selection criteria, will determine the architecture of the resulting plant.
- Tissue culture (clonal): This technique involves the regeneration of somatic embryos from floral tissues and the maturation of these embryos, *in vitro* through tissue culture, into plantlets. Laboratory-based procedures are used to multiply large numbers of selected, improved varieties, which are supplied to farmers as young plants.

Whilst some of these techniques are well known, there is still scope to improve the efficiency and economics of the production methods and the quality of the materials produced. Since the methods involved and prospects for tissue culture propagation are fully described in an accompanying chapter (Maximova and Guiltinan), this chapter will focus on the current status and prospects for the seed and conventional vegetative propagation methods. However, a summary will be provided of the cross-cutting issues that should be considered when deciding on the most appropriate propagation method, or indeed combination of methods, to be used to supply farmers in any given situation.

The material presented in this chapter is adapted from 'Supplying new cocoa planting material to farmers: A review of propagation methodologies': a review coordinated by Bioversity International and published in 2015 (Sena et al., 2015) and also including reference to some more recent publications. This review is based on three technical reports that were developed by Rob Lockwood (Propagation by Seeds), Augusto R. Sena Gomes and George Sodre (Conventional Vegetative Propagation) and Siela Maximova and Mark Guiltinan (Tissue Culture). The technical reports were consolidated into one multi-authored expert report which was reviewed by a broader group of experts, who met at a workshop held in May 2013 to discuss the technical findings and provide input into the final report on which this chapter is based. Since an accompanying chapter (Maximova and Guiltinan) in this book covers the potential of somatic embryogenesis (tissue culture) methods for mass propagation, we will focus on the methodologies used to supply planting materials as seed and by conventional vegetative propagation techniques and discuss the cross-cutting issues in supplying improved planting materials to farmers.

2 Propagation by seeds

Lockwood's review of use of seeds to supply improved planting material to farmers covers the history of seed garden development and current status in the most important cocoa-producing countries (Lockwood, 2015a,b). He also set out the factors that need to be taken into account when modelling requirements for seed garden capacity and for estimating the costs for establishing and maintaining seed garden operations. Lockwood concluded that a seed garden system can be an efficient and economical way to provide high-quality planting materials in the large quantities needed by the farmers. However, despite the development of high-performance seedling varieties through the breeding programmes of many cocoa-producing countries, there is still limited seed garden capacity in many areas and especially those which are at the frontier of cocoa development.

2.1 Availability of seedling varieties

Many traditional cocoa varieties including West African Amelonado, Brazilian Comum, Ecuadorian Nacional and Venezuelan Criollo types are largely homozygous and thus will breed true if self-pollinated or cross-pollinated with a tree of the same variety. Farmers have traditionally propagated these varieties from their own seed, or that of their neighbours. Though the resulting plantings are reasonably genetically uniform, susceptibility to pests, diseases and abiotic stress of some of these varieties can result in poor yields, especially where husbandry/local growing conditions are not uniform and/or suboptimal. Many other varieties of cocoa, including the Trinitario complex and improved bi-parental crosses, are highly heterozygous and farmers are likely to experience problems with greater variability in the performance of their tree stocks if they attempt to propagate using seed from their own, or their neighbours', trees of these types.

High-performance seedling varieties have been developed, and proven suitable for a particular region, by various breeding programmes. These are generally bi-parental crosses (often referred to as 'hybrid') which have been selected on the basis that they generate a uniformly productive progeny. A summary of the bi-parental crosses that have been recommended and/or are in current production in the seed gardens of most cocoa producing countries is provided in Lockwood (2015b).

2.2 Seed garden design and operation

Historically, many seed garden designs have sought to take advantage of the self-incompatibility of at least one of the parents and natural pollination in order to produce bi-parental seed varieties. However, Lanaud and colleagues (Lanaud et al., 1987) showed that 20–97% of the output from this type of seed garden in West Africa was actually produced by self-fertilization and not the expected bi-parental cross. Although the genetic and physiological basis of the incompatibility mechanism is not yet fully understood, it seems that the mechanism can break down under some circumstances. A number of studies have found that the yield performance and uniformity of plantings made from seeds obtained from naturally pollinated seed gardens may not match those of plantings from seed resulting from manual pollinations and/or performance in breeders' trials (N'Goran et al., 2000; Ooi, 1981; Padi et al., 2015). Carefully controlled manual pollination, together with verification of the identity of parental clones in the seed gardens, is therefore strongly recommended to ensure the quality of the seed garden output.

Manual pollination is carried out by touching the anther anywhere on the stigma or style of a freshly opened flower. Suitable flowers from the 'male' or pollen donor tree can easily be identified since their anthers are creamy-white when viewed through the translucent petal hoods (the anthers turn brown within 24 h of anthesis). Freshly opened flowers are normally collected early morning from the 'male' or pollen donor tree and transported to the 'female' or pod-bearing trees in plastic pots for immediate use. Each anther is generally used only once, though Edwards (1972) has shown that individual anthers can be used to pollinate up to five separate flowers which can be useful if flowers on the pollen donor trees are scarce. The pod-bearing trees should be prepared by stripping all natural sets on a frequent basis in advance of the planned manual pollination date both to stimulate flowering and to minimize the risk of harvesting naturally set pods. The aim is then to complete all the pollinations on an individual tree early in the morning of a single day. Depending on factors such as the variety, age and condition of the tree, the aim would be to pollinate approximately 75 flowers. This synchronous mass pollination results in cherelles of a uniform age and appears to minimize losses through the normal cherelle wilt thinning mechanism.

Based on experience in seed gardens in Ghana (Edwards, 1972) and Côte d'Ivoire (Bastide and Sounigo, 1993) and his own experience, Lockwood suggests that pollinators should be trained to reach a standard of a minimum of 125 pollinations/h with 75% setting rate (successful pollinations as measured 10 days after pollination) and levels of contamination of extraneous pollen below 5%.

Lockwood modelled the costs involved in establishing a seed garden using budded materials and the costs of maintaining the seed gardens from a zero-base (i.e. built up from first principles rather than by modification of an existing budget) using estimates for labour, productivity and materials gained from various sources in West Africa. He estimated that the cost of producing seed destined to be planted at stake, or raised in a farmer's nursery, from an established seed garden using manual pollination would be US$0.002, or US$0.088 if the costs of establishing and maintaining the seed garden are taken into account. He estimated that the additional cost of using manual pollination to produce a seed, compared to natural pollination, from a seed garden would be US cents 0.088–0.786 depending on the cost of the labour [estimates based on West African (Ghana) conditions and costs at 2015]. At the lowest labour cost (US$5/day), this would equate to a marginal difference of US$2.08 more to establish a hectare of cocoa of known parentage over the cost of establishing cocoa from open-pollinated pods (based on the following assumptions: planting at stake at 1600 trees/ha, 90% germination, 75% nursery outturn, with similar costs involved in harvesting and transporting the pods, and with similar depreciation and management costs for the seed garden) and at the highest labour cost modelled (US$45/day), this difference would be US$18.71/ha.

2.3 Challenges and research needs

Though seed gardens can take time to establish and reach maximum productivity, they can remain productive for over 40 years. During this period, their output can be changed by introducing new pollen parents for existing 'mother' trees to take advantage of new recommendations from breeding programmes and to increase the genetic diversity of the material supplied to farmers in the area in order to lower the risks associated with monocultures. Effective mechanisms are needed to establish the parental trees in the quantities needed with good management to bring them to early flowering (pollen donor trees) and fruiting. The availability of stocks of the parental types has sometimes been a limiting factor in establishing seed gardens, especially in new growing areas or following new recommendations from breeding programmes. A combination of tissue culture and traditional vegetative propagation techniques can be useful in bulking up supplies of parental types (see sections below). Full coordination with breeding programmes is particularly important to ensure that supplies of promising parents can be bulked up as the promising candidate varieties progress through their evaluation trials. This can ensure that a new recommendation can be implemented at the seed gardens as soon as it is confirmed.

Seed gardens are a highly cost-effective way to supply new planting materials to farmers. Accessibility could be further improved if a cost-effective way can be found to ensure its availability at the time of the year when the farmer can make best use of it. Rain-fed seed gardens tend to produce most ripe pods at the same time as the farmers are heavily involved in harvesting their own crop and thus have little time for nursery preparations. Moreover, weather patterns in some of the major growing regions are not conducive to planting at this time of year. Keeping seedlings in the nursery until the next rainy season incurs additional costs and these older seedlings may not establish so well when they are eventually planted in the field. Optimising the timing of pod production in seed gardens to better match farmer requirements might involve the use of irrigation and/or continued stripping of flowers/cherelles to prolong or induce flowering, and thus sustain pod production, outside of the usual season or application of seed storage technology. Cocoa seeds are recalcitrant, and are killed when desiccated below 20% moisture or when cooled to 4°C (Mumford and Brett, 1982). However, studies have shown that if seed can be kept free from microbial contamination, it is possible to store them for up to 175 days (Mumford and Brett, 1982) or even 210 days (Figueiredo, 1986) with good germination rates, using solutions of polyethylene glycol and methyl cellulose, respectively. Lockwood recommends further research to develop such a method for large-scale use at seed gardens and notes that if seed could be stored and transported without the weight of the surrounding pod husk and mucilage, it could reduce the costs as well as the phytosanitary risk of supplying seed to remote areas.

3 Conventional vegetative propagation

The review undertaken by A. R. Sena Gomes and G. A. Sodre of the use of conventional vegetative propagation techniques to supply improved planting materials to farmers covers the use of rooted cuttings and marcotting as well as techniques to produce composite plants where the improved material is budded or grafted onto a rootstock (Sena-Gomes and Sodre, 2015). Vegetative propagation produces planting materials that are genetically identical to the source material, though there may be variation in vigour and architecture, depending on the propagation method used and on agronomic and environmental factors. Conventional vegetative propagation has been used in many cocoa-growing countries to produce clonal material for use in seed gardens or to supply directly to farmers, though the latter is far more common in the Americas and in some parts of South-east Asia than in West Africa. The supply of vegetatively propagated material has been particularly important in Brazil where there has been a high demand for clonal planting materials to rehabilitate cocoa farms following the outbreak of witches' broom disease in the main cocoa-growing state of Bahia in the early 1990s. Details of the facilities and inputs required and the protocols used in various countries to mass-produce vegetatively propagated materials for farmers are provided in the appendices of the review together with cost estimates based on the Brazilian experience.

3.1 Availability of proven clones

Clones have been selected and evaluated using conventionally vegetative propagation techniques in many countries in South America and some parts of South-east Asia (Sena Gomes and Sodre, 2015). Brazil and Ecuador, in particular, have large-scale programmes using rooted cuttings and other techniques to supply these clones to farmers for new plantings and to rehabilitate existing fields. There are several recent initiatives to develop clones suitable for use in West Africa, though it will be many years before the first outputs from these initiatives will have been thoroughly proven.

3.2 Techniques and operation of clonal propagation facilities

Techniques to vegetatively propagate cocoa have been under development since the late 1890s and have been refined for use in mass propagation in each cocoa-growing region. Some conventional propagation techniques, such as grafting and budding, generate a composite plant consisting of a rootstock and the clonal scion material, whereas other techniques, such as cuttings and marcotting, generate a plant consisting entirely of the clonal scion material. Supplies of clonal material are often generated through the use of a clonal garden.

3.2.1 Rootstocks

Rootstocks are required for budding or grafting techniques and can include seedlings raised specifically for this purpose, or trunks, branches and chupons of mature field plantings where the intention is to replace part of the existing canopy with improved varieties. Seeds for use as rootstocks are generally raised in plastic bags in a greenhouse facility for periods of just a few days to up to 8 months depending on the technique they are to be used for. The seed can be obtained from pods produced by manual pollination to ensure a known bi-parental cross, or open-pollinated pods from a certain clone or mixture of maternal trees. The costs of producing seedling rootstocks will depend largely on labour costs and the length of time that the plant is maintained in the nursery, with estimates ranging from US$0.07–0.17 for 2.5- to 3-month old seedlings in Ecuador to US$0.50–0.7 for 3- to 6-month old seedlings in Indonesia.

Although strong rootstock effects on vigour, yield and fruit quality are known in other fruit tree crops, a review of the literature by Lockwood and Nsiah (in prep.) finds little evidence as yet that the rootstock selected will improve yield efficiency in cocoa, though effects on resistance to certain diseases, on overall vigour and on establishment have been reported (e.g. Ayegboyin et al., n.d.; Ribeiro et al., 2016). The genotype of the rootstock has also been found to affect success rates for grafting and budding, though the method used and other factors such as age of the seedling and environmental conditions will also have an effect (e.g. González et al., 2006). Whilst there have been some limited experimental studies on the use of clonal rootstocks (e.g. Murray and Cope, 1959), results from research on the potential value of clonal rootstocks for commercial production of planting materials are still awaited (Sodre, pers. comm.). Moreover, there is scope for further investigation of the rootstock/scion interaction and its potential impact on cocoa productivity with a view to selecting rootstocks that optimize the performance of the scions whilst still offering other advantages such as disease resistance.

3.2.2 Grafting

Various grafting techniques have been developed for cocoa which involve the insertion of a scion-stick containing several buds into a top split (top grafting) or laterally under the bark (splice grafting) of a rootstock. This rootstock can be a nursery- or field-planted seedling, or a basal chupon, stem or high branch of an established tree.

Top or split grafting: Generally, a semi-hardwood scion about 15 cm long is prepared with a long v-shape gradually tapering to a wedge at the base. This can be inserted into a vertical cut made in the stem of a seedling (4–8 months old depending on whether it has been nursery or field grown) or a basal chupon of a tree in the field. Two or more scions may be needed if the diameter of the chupon is greater than 2.5 cm. The graft is taped and covered with a plastic bag, or a special biodegradable tape can be used, to protect it over the two to three weeks needed to form a union. This technique is simple and quick and can achieve success rates of at least 80% under optimal conditions.

Side grafting: The semi-hardwood scion is cut obliquely to form a slender wedge which can be inserted under the bark of stock so that the cambial layers are in tight contact. The graft can be made on seedling rootstocks or on the stems or higher branches of field established trees.

Crown-grafting or canopy replacement: This technique involves grafting of scionwood onto the main branches, above the jorquette, of field grown trees. It is a popular technique in Brazil since it is seen as the fastest way to increase production on farms with unproductive trees less than 20 years old. It is particularly useful where the trees do not produce chupons, and therefore cannot be side grafted, and where there is potential to replace a diseased canopy with material offering more disease resilience.

Cleft-grafting or nurse-seed grafting: As noted above, the costs of producing rootstocks are directly related to the time that the material is kept in the nursery. There has been interest therefore in using very young seedlings (less than five weeks old) as rootstocks for green (softwood) scions with a view to reducing nursery costs, though a mist humification is generally needed for a month after grafting, and to take advantage of the vigour of the green scionwood. Methods developed in Malaysia (Ramadasan and Ahmad, 1986; Yow and Lim, 1994) gave promising results, though it is difficult to implement on a commercial scale in some regions. Recently, a modification of a green grafting method has been developed which uses sections of plastic drinking straws instead of grafting tape to secure the graft union when side grafting onto 12- to 14-day old seedlings (Bhola, 2017).

3.2.3 Budding

Techniques have been developed to transfer a single bud from a stock plant to a rootstock which may be either a seedling raised specifically for this purpose or a chupon, stem or branch of an existing field-grown tree. The techniques vary according to the age of the scion and stocks but include green patch budding where young seedlings are budded (Giesberger and Coester, 1977) and brown budding for use on older material. Although the technique has the advantage that only one bud is required for each plant propagated, which is especially important when quantities of scion wood are limiting, it requires more skill and higher levels of sanitation than some of the other techniques. In Bahia, Brazil, initial 'take' success rates can be 60–70% though the rate at which the bud grows out can be rather slow depending on factors including the nutritional status of the scion and the rootstock.

3.2.4 Rooted cuttings

Pyke (1933) first demonstrated a method for rooting cuttings using sand as the potting medium and a closed bin system in 1933. It is possible to take cuttings from both plagiotropic (fan branch) and orthotropic (chupon) material. Various systems have been developed using different potting media and methods of maintaining the high humidity needed to establish a rooted cutting. These range from simple non-misting propagation chambers constructed using polythene sheeting and suitable for use in rural areas/on farms (Tee and Lamin, 2015) to a commercial system using automated, intermittent fertigation with rooting taking place in plastic tubes filled with a mixture of compost and shredded coconut fibre (Sena Gomes et al., 2000). The costs of producing cuttings will vary widely depending on the techniques used but details of protocols for the production of rooted cuttings under different commercial systems as used in Brazil and Ecuador, and projected budgets are presented in the appendix of Sena Gomes and Sodre's review.

There have been a number of studies on the factors affecting rooting success which are reviewed by Amoah (2006a,b) and Sena-Gomes and Sodre (2015), and include varietal effects, use of different phytohormones and environmental conditions. The nutritional status of the stock plant, and in particular the nodal position of the cutting, has been found to be important with highest success rates obtained from cuttings taken at nodal positions 4–6 from the apex of the shoot.

The choice of plagiotropic or orthotropic material for the cuttings affects the architecture of the resulting plant. Cuttings taken from fan branches of some, but by no means all, cacao varieties develop a prostrate, bush-like canopy which requires repeated pruning or staking for easier management and harvest. However, clones selected for their performance when propagated as plagiotropic cuttings can offer a satisfactory growth habit and a good balance between crop, continuing vegetative growth and planting density that does not necessitate continuous pruning. Cuttings derived from orthotropic material develop the same dimorphic framework as trees grown from seed. Some farmers, particularly those in West Africa, are more familiar with this tree architecture, but regular pruning is needed to prevent the trees from becoming too tall.

There has been some discussion on the rooting characteristics of cuttings derived from plagiotropic and orthotropic materials and the impact on the performance of the resulting trees. There have been some reports that trees derived from plagiotropic cuttings succumb more easily to lodging and are susceptible to periods of soil moisture stress during establishment than plants derived from seed (Lee, 2000; Mooleedhar, 2000). This has been linked by some to the inconsistent development of an orthogeotropic adventitious root framework which would provide the anchorage associated with a seedling tree's taproot. Orthotropic cuttings, as produced using the methods developed at Pennsylvania State University (see accompanying chapter in this book by Maximova and Guiltinan), often develop taproot-like structures. However, it is possible that the lack of a good orthogeotropic framework is related more to poor rooting technology than to whether the source material is plagiotropic or orthotropic. Indeed, Goenaga and co-authors (Goenaga et al., 2015) found that whilst some trees derived from orthotropic cuttings still had taproot-like structures when they were excavated 9 years after planting, others did not. The authors suggested that factors affecting the growth (or loss) of the taproot could include aspects of tree management such as length of time the cuttings were kept in plastic bags in the nursery and transplanting damage, and soil factors such as waterlogging and aluminium toxicity and that further research under controlled conditions is needed to determine the frequency of the development of taproot-like structures in orthotropic cuttings.

3.2.5 Marcotting

Cocoa can be propagated by marcotting, or air-layering, whereby rooting is induced by wounding a stem whilst it remains attached to the parent tree. Though the technique requires relatively little skill and may be used on farm, it is not used commercially since other more efficient methods are available.

3.2.6 Clonal gardens

Clonal gardens are maintained under an intensive management regime to ensure that stock plants of the desired clones are kept in a juvenile growth phase and generate plentiful supplies of scion material. Procedures will vary according to local conditions but usually involve careful pest, disease and weed control, pruning and fertilizer application. In Bahia, Brazil, clonal gardens for cuttings are planted at 3 × 3 m and generate between 50 to over 100 cuttings per plant per year depending on the clone. Since the quality of the propagating material declines with the age of the stock tree, the clonal gardens are replanted every 6 years.

3.3 Challenges and research needs

Conventional vegetative propagation systems are being successfully used in countries such as Brazil and Ecuador for the large-scale supply of planting materials to farmers, whilst the technology has not been so widely adopted in other regions, particularly in West Africa. This may largely be due to the lack of clones that have been proven as suitable for farmer use and cost. Sena Gomes and Sodre recommend further research aimed at introducing techniques developed by the forestry industry such as container-grown stock plants, and other ways to reduce production costs, particularly through shortening the time of maintenance of young plants under nursery conditions. They also advocate research to improve understanding of the rootstock-scion interaction and its implications for the selection of rootstocks.

4 Cross-cutting issues

The decision to adopt a specific propagation method, or combination of methods, for the production of improved planting materials will depend on a number of factors. These will include the availability of source materials and other resources and the timeframe involved, as well as the farmer's needs and ability to invest in new planting materials and associated inputs. It will also depend on whether the farmer aims to plan a new area, or rehabilitate existing plantings where productivity is low due to factors such as poor planting materials, low tree density, pest/disease problems and tree age.

4.1 The availability of source materials

The availability of improved types, proven suitable for a particular region, is of overarching importance when considering the most appropriate propagation method for supplying farmers with new planting materials. In Latin America and the Caribbean, much of the recent breeding work has been targeted at the identification of clones that have resistance to the prevalent diseases (mainly witches' broom and/or frosty pod and black pod diseases), and that can produce cocoa with the desired quality characteristics (often fine flavour profiles). Most farmers in West Africa currently use seedling planting materials. The breeding programmes in the region have largely focused on the development of high-yielding seedling varieties with resistance to black pod disease and, in some cases, cocoa swollen shoot virus (CSSV) disease, and on the quality characteristics required for mainstream cocoa (good cocoa flavour and high butter fat content). However, several recent initiatives focus on developing clones suitable for use in West Africa, though it will be many years before the first outputs from these initiatives will have been thoroughly proven. In South-east Asia, breeding for both seedling varieties and clones has been carried out, though most of the large breeding programmes set up by plantation companies in Malaysia in the 1970s/1980s have lapsed. A few clones identified in these programmes are in use in the region today.

The performance of a variety as grown in a farmer's field will be affected by a combination of genetic, environmental and agronomic factors, but the impact of the method used to propagate the material should not be underestimated. It is therefore important to ensure that these materials have been evaluated using the same propagation method that will be eventually used to mass-propagate them. For example, the uniformity and yields from trials established using bi-parental seed from carefully controlled manual pollination could be quite different to the results obtained by farmers supplied by seed gardens relying on the self-incompatibility reaction to favour, but not guarantee, the production of bi-parental seed. For clonally propagated varieties, the method used may have substantial effects on vigour and tree architecture, such that a clone selected following trials established using budded plagiotropic material, for example, may not achieve the same performance when propagated by somatic embryogenesis, particularly if no adjustments are made to planting density and management practices. The process of developing and registering new varieties, and establishing selection criteria and requirements for trial data, will vary between countries, but it is essential that the materials that are recommended and propagated for distribution have been thoroughly evaluated to reduce the risk to the farmers who will depend on the performance of this material for their livelihoods. Attention should be paid to the legal conditions associated with the propagation and distribution of planting materials to ensure that there are no infringements to plant breeders' rights (for registered varieties) or benefit-sharing (for materials distributed from the international cocoa genebanks using the Standard Material Transfer Agreement of the International Treaty on Plant Genetic Resources for Food and Agriculture).

Although farmers may seek uniformity in tree architecture and production within their individual fields, extensive use of very closely related or clonal material across a wide area increases the vulnerability of the crop to changes in local growing conditions and to threats of pests and disease – whether due to the evolution of more devastating forms of native types, or to the accidental or natural introduction of new species into cacao-growing areas. It is therefore important to ensure that farmers are provided with an appropriate mixture of clones or seed varieties that have been confirmed as suitable for planting together, taking into account agronomic and genetic factors such as tree architecture, vigour and nutritional requirements. Moreover, the period of flowering, and most importantly, pollination cross-compatibility must be taken into account given cocoa's complex self-incompatibility system. It is also essential that the planting materials provided produce cocoa of the quality required by intended buyers and that these quality characteristics are stable across different growing environments.

4.2 Technologies and facilities for propagation

The facilities required for propagation vary from simple field nurseries suitable for raising seedlings or cuttings to the well-equipped laboratories and misting units which might be needed to raise plantlets through somatic embryogenesis systems or mass production of cuttings. Whatever the scale, the establishment and operation of any facilities for producing planting materials require careful management to ensure that the materials are 'true to type' and free from pests and diseases. This can be accomplished by providing training and materials for operatives so that off-types and phytosanitary issues can be quickly recognized using visual or molecular techniques such as genetic fingerprinting. Clean stock schemes are especially important in keeping the materials free from graft-transmissible diseases, such as those caused by CSSV and other viruses (see 'Phytosanitary considerations').

Well-managed nurseries are a key part of large-scale propagation operations, whether the objective is to supply farmers with seedlings, rooted cuttings, grafted/budded plants or weaned plantlets propagated through somatic embryogenesis. Since materials can be kept in the nursery for periods of up to several months, they offer the possibility of supplying planting materials to farmers at the most convenient time and when weather conditions are favourable for planting. However, such nursery operations can be expensive, thus increasing the costs of supplying the materials to farmers. This is especially the case in areas where labour costs are high, electric power is either unavailable or unreliable and automatic irrigation/fertigation is not available. There is considerable scope to improve nursery operations by optimizing the time that materials remain in the nursery. As discussed in the previous sections, this might involve improving the availability of seed through manipulating the cropping period or seed storage, and/or using improved vegetative propagation techniques. Moreover, there is also scope for developing and utilizing potting materials which would improve water and nutrient-use efficiency to facilitate nursery and distribution operations whilst also ensuring the development of a good root system to aid field establishment and long-term performance.

4.3 Labour, skill requirements and farmer training

The availability of local labour, existing skills and training needs are important factors when considering the establishment of new propagation facilities. The development of technologies to reduce labour requirements, such as irrigation, may become increasingly important in areas where there are likely to be inadequate rainfall, labour shortages and/or where labour is expensive.

Farmers may require training in order to be able to manage their new planting materials appropriately. They will need to have an understanding of the planting density, harvesting frequency, crop protection, shading and fertilizer required to reach the reported yields for those materials. Even if the materials underperform due to poor agronomy, rather than the quality of the materials themselves, farmer dissatisfaction may lead to rejection of future opportunities to obtain new planting materials. Training will be of even greater importance in situations where the new planting materials are clonally propagated from plagiotropic material, whereas previous experience has been with seed-raised materials, to ensure that the plants are well pruned and rootstock outgrowths removed appropriately.

5 Phytosanitary considerations

Care must be taken at all stages of propagation and dissemination of planting materials to minimize the risk of spread of pests and diseases. Viruses pose a particular risk when vegetatively propagating cacao by conventional techniques, since they can be spread through graft transmission and cannot be eliminated by pesticide treatments. Depending on the strain of the virus and the cacao variety, symptoms may appear within weeks of infection, but in some cases, the plants may remain symptomless for several years. Therefore, a clean stock scheme is needed to ensure that the stock plants have been confirmed to be free of virus, through frequent inspection by trained disease spotters, and ideally using molecular probes or virus indexing. Although CSSV DNA has been detected in pollen from CSSV-infected trees using some molecular probes, there is no evidence of pollen transmission of the virus through cross-pollination from infected cocoa parents to healthy cocoa trees (Ameyaw et al., 2013). Therefore, seed is often considered to be the safest way to distribute planting materials in the areas of West Africa affected by CSSV, in terms of avoiding disease transmission.

Particular care must be taken to avoid the spread of fungal spores and insect pests associated with pod husks, poorly cleaned seeds, young plants and packaging materials, especially when moving materials between regions that do not share the same species/strains of pests and diseases. In such cases, the movement of whole pods is not recommended and some countries have placed their own restrictions on the movement of seeds/whole pods within their country or region. In Indonesia and Malaysia, for example, there is a perceived risk that larvae of boring insects such as the cacao pod borer (*Conopomorpha cramerella*) could be carried with poorly cleaned and treated seeds, while in Brazil and Ghana, different species and/or strains of diseases are endemic to particular regions. Moving plants in containers, especially when soil has been used in the potting medium, pose a high risk of spreading soil-borne pests and diseases, which can include *Rosellinia* spp., *Phytophthora* spp., nematodes and soil-borne mealy bugs.

Contamination from microorganisms and pests is generally eliminated at an early stage in tissue culture, and somatic embryos and plantlets in sterile media are considered to be of low risk from a phytosanitary perspective. Research results suggest that the somatic embryogenesis process considerably reduces the risk of CSSV transmission; however, since the possibility cannot be absolutely ruled out, it would be prudent to confirm that the source materials used are virus-free.

6 Predicting demand for planting materials

Estimating demand for planting materials for new plantings and rehabilitation is complex and should take into account factors such as planting density, area to be planted, the efficiency of the utilization/establishment of planting materials, farmer demand and national policies, if any, towards diversification of perennial cash crops. However, efforts are underway in several countries to estimate demand as part of the national cocoa plans being developed under the Global Cocoa Agenda.[1] Lockwood (2015a) discussed the challenges in modelling the seed garden capacity required to match actual demand. Whilst a better understanding of some of the factors involved is still needed to refine the models, he concluded that there is a considerable shortfall in seed garden capacity in West Africa compared to predicted demand.

Experience gained from initiatives such as Project Lembrance in Brazil (Sena-Gomes and Sodre, 2015), where rooted cuttings were supplied to farmers for intercropping with papaya, and initiatives involving the side grafting of young tree trunks or high branches with improved materials, may well be useful in developing propagation strategies for other areas where there is a need for the rehabilitation of existing cacao farms to rapidly increase their productivity.

Somatic embryogenesis, possibly in conjunction with conventional vegetative propagation techniques, may also offer unique opportunities, particularly in the initial scale up of new varieties, by saving up to 2 or 3 years in cases where there are very limited supplies of the source clone (Maximova and Guiltinan, 2015).

7 Considerations on estimating costs

Cost considerations for large-scale plant propagation will usually include selecting one or more propagation methods specifically suitable for a given region or country, and describe the resources needed; capital for investment in infrastructure (such as propagation laboratories, greenhouses, nurseries, water for irrigation, shade and electricity); plant materials; equipment and supplies; labour; technical assistance; and administration. These costs will vary according to location, and will be particularly influenced by local labour costs. However, in general terms, improved seed that is destined to be planted at stake or raised in a farmer's nursery is by far the cheapest option, with somatic-embryogenesis-derived plants the most expensive. Though costs for various propagation options have been estimated, for example for seed production in West Africa (Lockwood, 2015a), rooted cuttings and grafted seedlings in Brazil (Sena-Gomes and Sodre, 2015) and somatic-embryo-derived plantlets in Ecuador (Maximova and Guiltinan, 2015), it is difficult to compare them or use them to estimate costs in other countries or situations since:

- The production objectives may vary. The relative cost of establishing new plantings vs. rehabilitating old, low-performing trees on an existing farm depends on different sets of criteria and assumptions. There will clearly be more possibilities to invest in large infrastructures and expensive equipment if the production objective is on a large scale.
- A large proportion of the costs of propagation, particularly for the techniques involving nursery care, are dependent on local labour rates, though other factors

1. The Global Cocoa Agenda was agreed at the First World Cocoa Conference, Abidjan, Côte d'Ivoire, in November 2012. It provides the roadmap towards achieving a sustainable world cocoa economy; see www.icco.org.

such as local availability/cost of materials and distribution costs will also have an impact.

- Although the cost of nursery management and distribution of a 'ready-to-plant' propagule might be similar for a given region or country, there may be considerable variation in the costs of producing the materials entering the nursery. This will depend on the scale of the operation and the starting point of comparison, that is, whether or not the costs of establishing the laboratories, seed gardens, budwood gardens and nursery facilities should be taken into account, and, if so, how the increase in productivity/efficiency over time can be taken into account – for example, as procedures are optimized, source plants mature and management and labour become more skilled.

8 Future trends and conclusion

Most cocoa-producing countries already have some capacity to produce improved seed at very low cost, though in many cases considerable further investment would be needed to increase this capacity, and to develop seed storage or other mechanisms, to ensure sufficient seed is available to farmers wherever they are located and whenever they need it. Some countries, particularly those in Latin America and the Caribbean and South-east Asia, have successfully used traditional vegetative propagation techniques to supply clonal material to farmers either as ready-to-plant propagules or for *in situ* grafting to rehabilitate existing plantings. There is scope to expand the clonal gardens and optimize large-scale propagating facilities, or develop low-cost systems for community-level use, to make these materials more widely available. Once suitable clones have been selected for use in West Africa, this technology can be introduced for large-scale use in that region too. Somatic embryogenesis has the advantage that, once suitable clones have been identified, it can provide a rapid means to produce clonal plantlets with the seedling growth habit familiar to most farmers. Although the technique requires substantial investment to set up and run the laboratories/nursery facilities, and consequently results in higher costs per propagule, it can be used as a means to provide materials directly to farmers (such as the Gernas Kakao programme in Indonesia, referred to elsewhere in this book). There are also possibilities for its use in combination with orthotropic cuttings to increase the output of clonal materials with a seedling habit.

It is clear that the availability of recommended source materials that have been thoroughly evaluated using the same propagation technique as that which will be used to mass-produce the materials for the farmers, will be a key factor when considering the most appropriate technology or technologies to include in a propagation strategy. The production of improved types is a very long and expensive process and the breeding programmes must be fully integrated with propagation systems so that materials are tested for their ease of propagation, and stocks of candidate materials should be built up during the later stages of evaluation so that adequate quantities of planting materials are available for distribution as soon as new recommendations are made.

The success factors of any effort to supply new planting materials to farmers will depend not only on the characteristics of the material to be propagated but also on the effectiveness of quality control during the propagation procedure and distribution, and the care taken after delivery of the plants, which may also be influenced by the levels of farmer training and access to inputs such as fertilizers. Ultimately, the benefits of the new

Published by Burleigh Dodds Science Publishing Limited, 2018.

materials will have to be seen by the farmer to outweigh the costs and risks involved in replanting if they are to be widely adopted.

9 Acknowledgements

The authors would like to express their appreciation to Drs. Lockwood, Sena Gomes, Sodre, Maximova and Guiltinan and indeed to all those individuals and organizations who participated in the review process and the contributed information presented in this chapter. A full list is given in the Acknowledgements section of the review document [Laliberté and End (Eds), 2015] available at https://www.bioversityinternational.org/e-library/publications/detail/supplying-new-cocoa-planting-material-to-farmers-a-review-of-propagation-methodologies/

10 References

Ameyaw, G. A., Wetten, A., Dzahini-Obiatey, H., Domfeh, O. and Allainguillaume, J. (2013). Investigation on Cacao swollen shoot virus (CSSV) pollen transmission through cross-pollination. *Plant Pathology*, 62(2), 421–7. https://doi.org/10.1111/j.1365-3059.2012.02640.x

Amoah, F. M. (2006a). Review of vegetative propagation of cocoa (*Theobroma cacao*) by rooted cuttings. I Physiological considerations. *Ghana Journal of Agricultural Science*, 39(2), 209–16.

Amoah, F. M. (2006b). Review of vegetative propagation of cocoa (*Theobroma cacao* L.) by rooted cuttings. 2. Environmental and technical considerations. *Ghana Journal of Agricultural Science*, 39(2), 217–26.

Ayegboyin, K. O., Hadley, P. and Daymond, A. J. (n.d.). The effect of rootstock on the establishment ability of *Theobroma cacao* L. in Nigeria. *Proceedings of the 17th International Cocoa Research Conference*, 15–20 October 2012, Yaoundé, Cameroon. Lagos, Nigeria: COPAL.

Bastide, P. and Sounigo, O. (1993). Production of cocoa hybrid seeds in the Côte d'Ivoire – Various hand pollination trials. *Cafe Cacao The*, 37(4), 295–302.

Bhola, J. (2017). Modification of the Bhola method. *Tropical Agriculture*, 94(2), 209–12.

Edwards, D. F. (1972). The production of hybrid seed in Ghana. *IV International Cocoa Research Conference*, 8–18 January, 1972, St. Augustine, Trinidad. St. Augustine (Trinidad and Tobago): Government of Trinidad and Tobago, pp. 69–72.

Figueiredo, S. F. L. (1986). Conservação da viabilidade de semente de cacao. III. Solução de éteres celulósicos. *Revista Theobroma (Brazil)*, 16(3), 115–26.

Giesberger, G. and Coester, W. A. (1977). Green budding and grafting in cacao nurseries. *Acta-Horticulturae (Netherlands)*, 53, 263–71.

Goenaga, R., Guiltinan, M., Maximova, S., Seguine, E. and Irizarry, H. (2015). Yield performance and bean quality traits of cacao propagated by grafting and somatic embryo-derived cuttings. *HortScience*, 50(3), 358–62.

González, A., González, J. and Ventura, M. (2006). Evaluation of rootstocks for the grafting of plantlets of cocoa (*Theobroma cacao* L.) in the nursery. *Proceedings of the 15th International Cocoa Research Conference*, 9–14 October 2006, San José, Costa Rica. COPAL.

Lanaud, C., Sounigo, O., Amefia, Y. K., Paulin, D., Lachenaud, P. and Clement, D. (1987). New data on the mechanisms of incompatibility in cocoa and its consequences on breeding. *Cafe Cacao The*, 31(4), 267–77.

Lee, M. T. (2000). Recent experiences in field use of cocoa clones for large scale commercial planting in Malaysia: Pros and cons. *Proceedings of the Technical Meeting: State of Knowledge on Mass Production of Clonal Cocoa Plants*, 19–23 October 1998, Ilhéus, Bahia, pp. 117–21.

Lockwood, R. (2015a). Propagation by seeds. In: B. Laliberté and M. J. End (Eds), *Supplying New Cocoa Planting Material to Farmers: A Review of Propagation Methodologies*. Rome: Bioversity International, pp. 12–33.

Lockwood, R. (2015b). Report on cacao planting materials from various cocoa producing countries. In: B. Laliberté and M. J. End (Eds), *Supplying New Cocoa Planting Material to Farmers: A Review of Propagation Methodologies*. Rome: Bioversity International, pp. 121–50.

Maximova, S. and Guiltinan, M. J. (2015). Tissue culture. In: B. Laliberté and M. End (Eds), *Supplying New Cocoa Planting Material to Farmers: A Review of Propagation Methodologies*. Rome: Bioversity International, pp. 67–88.

Mooleedhar, V. (2000). A review of vegetative propagation methods in cocoa in Trinidad and the implications for mass production of clonal plants. *Proceedings of the Technical Meeting: State of Knowledge on Mass Production of Clonal Cocoa Plants*, 19–23 October 1998, Ihéus, Bahia. CEPLAC, pp. 122–5.

Mumford, P. M. and Brett, A. C. (1982). Conservation of cacao seed. *Tropical Agriculture*, 59(4), 306–10.

Murray, D. B. and Cope, F. W. (1959). A stock-scion experiment with cacao III. A report on cacao research, 1957–1958, pp. 29–35.

N'Goran, J. A. K., Clement, D. and Sounigo, O. (2000). Performance of self-progenies among a set of hybrids released in Côte d'Ivoire. *Proceedings of the 14th International Cocoa Research Conference*, Accra, Ghana, 9–14 October 2000. COPAL, pp. 297–301.

Ooi, L. H. (1981). Highlands research unit. *Paper Presented during the Mini-Seminar: Review of Cocoa Breeding and Forward Thinking in Malaysia*. United Planters' Association of Malaysia.

Padi, F. K., Ofori, A., Takrama, J., Djan, E., Opoku, S. Y., Dadzie, A. M., Bhattacharjee, R., Motamayor, J. C. and Zhang, D. P. (2015). The impact of SNP fingerprinting and parentage analysis on the effectiveness of variety recommendations in cacao. *Tree Genetics & Genomes*, 11(3), 1–14. https://doi.org/ARTN 4410.1007/s11295-015-0875-9

Pyke, E. E. (1933). The vegetative propagation of cacao. II softwood cuttings, Second annual report on cacao research, Trinidad, 2, 3–9.

Ramadasan, K. and Ahmad, M. Y. (1986). Recent advances in vegetative propagation techniques of *Theobroma cacao* L. under Malaysian conditions. *Proceedings of the International Conference on Cocoa and Coconuts*, 15–17 October 1984, Kuala Lumpur, Malaysia, pp. 133–46.

Ribeiro, M. A. Q., de Almeida, A. A. F., Alves, T. F. O., Gramacho, K. P., Pirovani, C. P. and Valle, R. R. (2016). Rootstock x scion interactions on *Theobroma cacao* resistance to witches' broom: Photosynthetic, nutritional and antioxidant metabolism responses. *Acta Physiologiae Plantarum*, 38(3), 73. https://doi.org/10.1007/s11738-016-2095-9

Sena, R., Sodré, G. A., Guiltinan, M., Lockwood, R. and Maximova, S. (2015). *Supplying New Cocoa Planting Material to Farmers: A Review of Propagation Methodologies*. B. Laliberté and M. J. End (Eds). Rome: Bioversity International, 189p.

Sena-Gomes, A. R. and Sodre, G. A. (2015). Conventional vegetative propagation. In: B. Laliberté and M. J. End (Eds), *Supplying New Cocoa Planting Material to Farmers: A Review of Propagation Methodologies*. Rome: Bioversity International, pp. 34–76.

Sena Gomes, A. R., Castro, G. C. de., Moreno-Ruiz, M. A. and Almeida, H. A. de. (2000). Avancos na propagacao clonal do cacaueiro no Sudeste da Bahia. *Proceedings of the Technical Meeting on the State of Knowledge on Mass Production of Genetically Improved Propagules of Cocoa*, 19–23 October 1998, Ilhéus, Bahia, Brazil, pp. 85–90.

Tee, Y. K. and Lamin, K. (2015). Vegetative propagation in cocoa (*Theobroma cacao*): Effects of propagation environment and rooting substrates on rooting behaviour of cocoa stem cuttings. *Proceedings of the International Conference on Plant Physiology 2014 (ICPP 2014)*, 26–28 August 2014, Sanur Paradise Plaza Hotel and Suites, Bali, Indonesia.

Yow, S. T. K. and Lim, D. H. K. (1994). Green-patch budding in very young cocoa rootstock and side-grafting of mature trees. *Cocoa Growers' Bulletin*, 47, 27–41.

Chapter 7

The potential of somatic embryogenesis for commercial-scale propagation of elite cacao varieties

Siela N. Maximova and Mark J. Guiltinan, The Pennsylvania State University, USA

1 Introduction

A high degree of yield variation is a general characteristic of cacao plantations worldwide, due in part to the predominant use of seed propagation in this mostly self-incompatible and highly heterozygous tree crop (Figueira and Janick 1995). Some of the earliest studies on individual tree yield in cacao-producing fields and experimental plots demonstrated that only 7–20% of the trees propagated from seed produce the majority of the yield measured in dry weight of cocoa beans per ha (Freeman 1929; Cheesman and Pound 1932; Anonymous 1933, 1934, 1937; Shephard 1936; Pound 1945). Additionally, a study of the early yield of five high-producing cacao families grown in full sun in Puerto Rico reported that only 2–3% of the trees in the population produced more than 60% of the yield (Irizarry and Rivera 1999). Genetic improvements made by plant breeders are often rapidly lost as farmers propagate plants through seeds, and segregation results in highly

http://dx.doi.org/10.19103/AS.2017.0021.09

heterozygous populations of plants. Vegetative propagation systems provide a means to capture genetic gain and are commonly used for cocoa (grafting and rooted cutting techniques); however, these methods suffer from low multiplication rates and require several years to establish a nursery with sufficient numbers of individuals to support a large-scale propagation programme.

Cacao somatic embryogenesis (SE) has been the topic of research of many scientists and laboratories around the world for the last 30 years, and perhaps is one of the most well-documented methods for the vegetative propagation of cacao. Efficient methods for multiplying cacao by SE have been developed as a result of research conducted at DNA Plant Technology Corporation, Cinnaminson, New Jersey, USA (Sondahl et al. 1993); de Coopération Internationale en Recherche Agronomique Pour le Développement (CIRAD), in Montpellier, France (Alemanno et al. 1996a,b, 1997); Nestlé Research and Development Center in Tours, France (Masseret et al. 2005, 2008; Masseret 2009; Florin 1997; Florin et al. 2000, 2001); and the Pennsylvania State University (Guiltinan et al. 1997, 2000; Lambert et al. 2000a,b; Li et al. 1998; Maximova and Guiltinan 2000; Maximova et al. 2002, 2005, 2008; Traoré et al. 2003). Protocols for SE have been published by Maximova et al (2005).

The main advantages of cacao tissue culture methods like SE include a very high multiplication rate and scalability, which speed up the development and deployment of genetically improved genotypes. It helps in rapid generation of large numbers of genetically uniform plants, in the production of orthotropic plants with normal dimorphic architecture (jorquette and taproot formation) and in the production and testing of disease-free materials; it also works as a tool for germplasm conservation via cryopreservation.

This review aims to synthesize publically available information and literature describing the research and development activities on cacao tissue culture and on field-test evaluations of SE-derived plants. We also review current activities in cacao-producing countries involving adoption and scale-up for large-scale propagation of important genotypes.

2 The somatic embryogenesis method for cacao

Zygotic embryos (ZEs) result from the fusion of female and male gametes formed by meiosis during the sexual phase of plant development within flowers. Gametic fusion (fertilization) is followed by cell division and the differentiation of the dividing cells into tissues and organs, comprising ZEs. These embryos develop within the protective and nourishing environment of the ovule, and at maturity they are found within the confines of a fully developed seed (the cacao bean), with 40 or so of such seeds developing within each cacao fruit (pod).

SE is a process that does not involve meiosis or fertilization. The newly propagated plants or embryos are derived from a single parental somatic cell (any cell in the plant other than a gamete or germ cell), or from a group of somatic cells. This process occurs in nature in certain plants that have evolved the ability to propagate asexually. One example is the propagation mechanism of *Kalanchoe* (commonly known as 'Mother of Thousands'). In these plants, new individuals develop vegetatively as plantlets along the leaves of the mother plant, and the young plants eventually drop off and develop their own roots (Garces et al. 2007).

Plant tissue culture offers a host of new approaches to speed up the development and deployment of genetically improved genotypes because of its potentially very high multiplication rate and scalability. The main advantages of cacao tissue culture methods include the rapid generation of large numbers of genetically uniform plants, for the production of orthotropic plants with normal dimorphic architecture (jorquette and taproot formation); for the production and testing of disease-free materials; and as a tool for germplasm conservation via cryopreservation. Currently, these methods are only beginning to be applied in cocoa-producing countries.

The SE process begins with immature flowers, which are collected from trees, surface sterilized, dissected to excise the petals and staminodes and introduced into the tissue culture media. The cultures are grown in sterile laboratory conditions using either solid or liquid nutrient media, which contain various mixtures of plant hormones. Hormonal treatments trigger a reprogramming of the somatic cells to de-differentiate, and then initiate embryogenesis development. During SE, a single somatic cell or group of somatic cells begins to divide and form pro-embryonic structures, then embryos. Plants derived from SE differ from those that have developed from ZEs, mainly in the lack of development of the cotyledons and the absence of maternal seed and fruit tissues.

The process of SE involves five main stages, as illustrated in Fig. 1.

- *Primary somatic embryogenesis*, where flower explants are cultured to produce embryogenic calli, which further develop into primary somatic embryos.
- *Secondary somatic embryogenesis*, in which cotyledons from primary embryos are re-cultured to produce secondary embryos.
- *Conversion*, where mature embryos are converted into plantlets.
- *Acclimatization*, where plantlets are acclimated to greenhouse or field conditions (conversion and acclimation).
- *Plant development,* where plants grow through plantlet stage and then transition through the juvenile to adult stages of development.

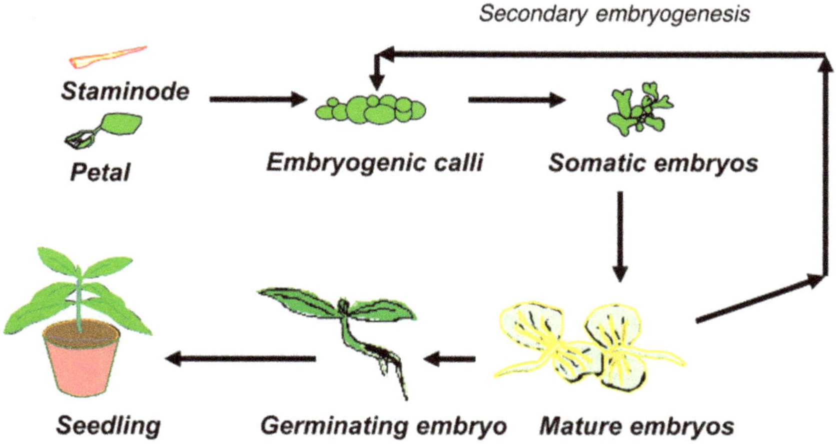

Figure 1 Graphic representation of the cacao somatic embryogenesis process. Source: Maximova and Guiltinan (2014).

3 Field evaluation of somatic embryo trees

Over the past decade, scientific groups in cocoa-producing countries have begun to integrate the cacao SE technology into their research and commercial programmes. Training has been provided in many countries through intensive workshops and visiting scientist and graduate student programmes. Today, more than 150 cacao genotypes worldwide have been introduced into sterile culture for the production of somatic embryos, and since 1999 more than 20 individual genotypes propagated in tissue culture have been planted in long-term field trials in Saint Lucia, Puerto Rico, Ecuador, Brazil and Indonesia. During the past 18 years, the growth of SE plants in the field has been compared to the growth of plants propagated by seedlings, rooted cuttings and grafted plants, by measuring parameters including stem diameter, stem height, longest jorquette branch, number of jorquette branches, occurrence of multiple stems, flowering and fruiting and so forth. The results of all these studies demonstrated that there were no major differences in growth parameters among the different genotypes and the propagation methods evaluated, based on which it was concluded that SE plants demonstrate normal phenotypes in field conditions and have growth parameters similar to plants propagated by traditional methods. Details of the major field trials of SE plants are given below.

3.1 Saint Lucia

To compare the growth of the SE plants to plants propagated by traditional methods, the Penn State conducted a five-year field test at Union Vale Estate, Saint Lucia (2001–06) (Maximova et al. 2008). The results of the study demonstrated that during the first few years of growth there were no significant differences in growth parameters among the different genotypes and propagation methods evaluated, with the exception of the orthotropic rooted cuttings (ORCs).

Trees propagated via ORCs appeared to be smaller in size, with smaller average stem diameters, shorter stems to the jorquette and shorter jorquette branches. It was concluded that SE plants demonstrated normal phenotypes in field conditions and have growth parameters similar to plants propagated by traditional methods (Fig. 2).

3.2 Ecuador (I)

Field evaluation was also carried out in Ecuador, in 1998, on the first 110 SE plants generated in the laboratory at the Nestlé's Research and Development Center in Tours, France. The trees demonstrated normal development and bean production similar to the control trees in the test (Fontanel et al. 2002). An extensive evaluation of SE plants was initiated in 2000 in two locations in Ecuador, with the objective of evaluating the agronomic performance and conformity of SE plants in comparison with other methods of cacao tree propagation. The test included SE plants, plants from orthotropic and plagiotropic grafts of EET95 on CCN51, rooted cuttings and seedlings of genotype EET95 grown in Ecuador (on Nestlé's El Chollo farm). In Ecuador, various parameters including measurements per tree of: trunk diameter, jorquette height (except for plagiotropic grafts and cuttings that do not develop jorquettes), the length of time required to produce the first pod (in months), the total number of pods produced and the average weight of 100 dry beans (Masseret et al. 2005), have been collected since 2001. The study reported that no differences could be

Figure 2 SE plants growing in the field at Union Vale Estate, St. Lucia. (a) Plantlets shortly after planting with drip irrigation, inter-planted with banana and coconut shade; (b) Secondary SE plant UV1 genotype; (c) Secondary SE plant UV1 genotype, November 2005 and (d) Primary SE plant, UV56 genotype, November 2005.

detected in the average trunk diameters and jorquette formation between SE trees and trees propagated by seeds (Fig. 3).

It was observed that cacao trees from orthotropic cuttings developed jorquette at a lower height, similar to the report from the Penn State test in Saint Lucia (Masseret et al. 2005; Maximova et al. 2008). Interestingly, SE plants on average produced the first pods four months earlier than those produced through the other types of propagation. In addition to earlier fruit production, the SE plants displayed good physiological vigour compared with the other trees, which is of particular importance during their early establishment in the field. In the third year after planting, SE trees produced significantly more pods compared with all other propagation methods and reached a consistent production rate. In 2008, yields of 1.7 tons per hectare (t/ha) were recorded for graft orthotropic trees and

Figure 3 Trees propagated by SE at the Nestlé farm in Ecuador five years after planting. Source: Maximova and Guiltinan (2014).

rooted cuttings, 1.2 t/ha for plants propagated by seeds and 2.5 t/ha for SE plants. Later trials, some of which were conducted at a commercial scale, confirmed the initial findings in terms of early fruiting and morphological characteristics.

However, while the productivity of SE trees was also better in the first few years after planting, no significant differences were observed during the later years in comparison with trees propagated using other methods (Fig. 4) (Masseret et al. 2008).

Field tests were conducted at the Nestlé farm in Ecuador between 2000 and 2009 to compare the field performance of SE plants from genotypes EET95, EET96 and EET103 generated in two different locations – the Nestlé research laboratory in France and the research laboratory at Penn State in the United States – using published protocols (Fontanel et al. 2002; Li et al. 1998). The SE plants were transported to Ecuador in 1999 and data were collected from all trees for a minimum of five years. Similar to the previous reports, all trees displayed normal morphology and growth with respect to the other propagation methods. Primary SE plants from all genotypes generated at Penn State produced pods after two years of planting, which is six months earlier than the rest of the trees in the test. Trees derived from secondary somatic embryos were more productive than those derived from primary embryos. This difference is expected, considering previous reports that secondary embryogenesis results in higher quality (faster growing, more robust) and more uniform plants (Maximova et al. 2002).

In 2006, Nestlé began planting a collection of cacao genotypes in Ecuador, consisting of SE trees. To date, 70 different accessions have been planted and an evaluation of the agronomic performance of the accessions has been initiated. A replication of this collection is being established in Côte d'Ivoire. This work is ongoing, with the integration of new accessions. It offers a unique opportunity to evaluate the performance of clones presenting all the characteristics of the original selected tree. So far, all the quantitative parameters recorded show a clear homogeneity between trees and indicate that the trees

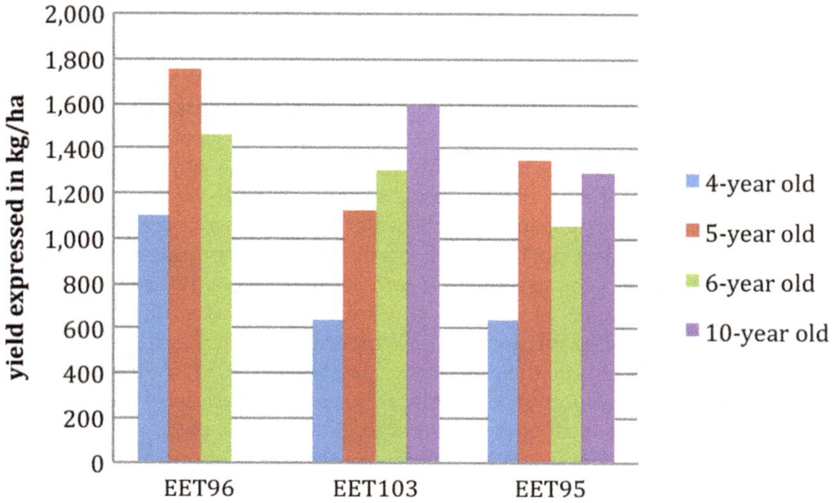

Figure 4 Productivity of SE trees planted at the Nestlé farm in Ecuador. Data from years 4 to 6 are from large-scale trials (at least 1000 trees); data from year 10 are from the 2000 small-scale trial. Source: Pierre Broun, Nestlé.

are developing normally. In order to ensure the long-term survival of the collection, SE multiplication of each clone is coupled with the cryopreservation of the embryogenic callus. The observations from this collection suggest that SE clones are 'true-to-type', regardless of genotype.

3.3 Ecuador (II)

A collaborative project between CIRAD in France and the Instituto Nacional Autónomo de Investigaciones Agropecuarias (INIAP) in Ecuador, initiated in 2002, focused on the adoption and evaluation of the SE technology. The project involved INIAP's national cocoa programme (Programa Nacional de Cacao del INIAP) and its tissue culture and biotechnology laboratory in Pichilingue. The project goals were to introduce the SE technology into the country and optimize it for important local Nacional varieties. The project objectives included an evaluation of the embryogenic capacity of multiple cacao genotypes and the establishment of a field trial to compare the performance of plants propagated by the different methods, including SE, rooted cuttings, grafting and seeds (Garzón et al. 2011b).

Using protocols developed by the Penn State, the embryogenic capacity of 44 different Nacional cacao genotypes was evaluated. The first 30 months of the project included building the infrastructure for tissue culture in Pichilingue, training personnel and conducting experiments aimed at optimizing the protocols for local conditions. Large numbers of embryos were successfully converted to plants. Based on the results obtained from the tissue culture experiments, 12 promising genotypes were selected for field testing and for comparing SE plants to conventional methods of propagation (i.e. grafting, rooted cuttings and seeds). In addition to monitoring the SE field trials, INIAP is currently working on optimizing SE protocols for Nacional clones and on adapting protocols for the low-tech propagation of

ORCs. During the first four years of the field trial (2005–09), data on agronomic, morphological and physiological performances were recorded. Yield data collection is still in progress. The analysis of data recorded during the first three years demonstrated that the SE plants have a high degree of morphological conformity compared with plants propagated by seeds. It was observed that the SE plants had better growth rates and faster jorquette formation, when compared with the other types of plants (Fig. 4). In general, the physiological parameters and pathological status of the SE plants were the same as those of plants propagated by other methods. The SE plants produced pods earlier than the rest of the plants (20 months after planting in the field), and in all cases SE plants had higher production than the plants obtained by other forms of propagation. The cumulative yield per hectare of plants during the first three years of production reached 3.0 t/ha (fresh cacao) and was 2.4 t/ha (fresh cacao) for the fourth year alone. SE plants yielded 43% more than rooted cuttings, 44–56% more than grafted plants and 49% more than seedlings (Garzón et al. 2011). At the INIAP station, it was observed that some SE plants produce more chupons at the base of the tree compared with plants produced from seeds, grafts and rooted cuttings. This could potentially increase the labour requirements for SE plants to periodically remove chupons. Potentially, the new chupons could be used as orthotropic material.

3.4 General conclusions of field studies

1. SE trees had the same architecture as seed-propagated trees.
2. Compared with the other vegetative propagation methods, like grafting or rooted cuttings, the SE trees needed approximately two times less pruning during the first three years.
3. The SE trees were more vigorous (larger trunk diameters) and more homogeneous, they developed pods earlier and their first crops produced higher yields.

4 Applying cacao somatic embryogenesis for long-term storage of germplasm

Live cacao seeds cannot be stored longer than a few weeks due to their sensitivity to temperature variation and dehydration. Thus, the world cacao germplasm is maintained as living collections in tropical countries or greenhouses. Preserving genetic resources as living trees is risky and quite expensive due to the changing climatic, environmental, funding and political conditions. One alternative or a backup option is the storage of living cells/tissues and/or organs in a frozen state (cryopreservation), which could decrease the risks and reduce the maintenance costs of germplasm conservation considerably. Additionally, SE-induced cryopreserved tissues potentially could be used as a starter material for *in vitro* propagation, an approach that does not require culture initiation with fresh floral tissues and that potentially could shorten the overall time for plant propagation; however, currently there is no published information providing data to support this application in cacao. Methods for cryopreservation of embryogenic tissues from cacao have been developed by Nestlé (European patent EP1119246B1) (Florin 1997; Florin et al. 2000). Plants regenerated from cryopreserved tissue have been evaluated by the company in field trials in Ecuador.

The research group in the School of Biological Sciences, University of Reading, Reading, UK, has developed an encapsulation–dehydration system for somatic embryos regenerated using the regeneration protocol developed at Penn State (Fang et al. 2004, 2008). The results have indicated that somatic embryos are currently the most suitable targets for clonal cryopreservation. The encapsulation–dehydration of cacao somatic embryos allows for appropriate management of the water content of the propagules during cryopreservation (Fang et al. 2009) such that tissue disruption is minimized (Fang and Wetten 2011) allowing for recovery of up to 72% of evaluated genotypes (Fang et al. 2004). The somatic embryo regeneration protocol developed at Penn State utilized in this approach has the added benefit of pathogen elimination with PCR-based detection showing that *Cacao Swollen Shoot Virus* (CSSV)-infected floral tissue can give rise to virus-free SEs (Quainoo et al. 2008). Key clones in the cacao collection at the University of Reading, UK, are now being routinely cryopreserved using a 'Plant Vitrification Solution' (PVS2)–based approach, which is becoming the method of choice for most tropical crop cryopreservation methods (Adu-Gyamfi and Wetten 2012). This vitrification methodology that involves incubation of somatic embryos in a simple cryoprotectant mixture followed by rapid cooling in liquid nitrogen produces recovery rates comparable to encapsulation–dehydration but is significantly less labour-intensive. According to the Reading scientists... 'As long as suitable precautions are taken to manage somaclonal variation risks associated with the use of propagules that arise from callus (Fang et al. 2009), cacao somatic embryogenesis represents a practical methodology for the establishment of compact, cryopreserved back-up germplasm collections with the potential to benefit from automated handling (Niemenak et al. 2008)'.

5 Using somatic embryo plants for propagation by orthotropic rooted cuttings

In addition to using SE to produce a large number of clones quickly for distribution to farmers, this technology can also be used for the rapid establishment of seed and clonal gardens. It has been shown that somatic embryo-derived trees (SE trees) managed as a bentwood clonal garden can produce orthotropic shoots from which orthotropic trees can be derived (Maximova et al. 2005; Miller, 2009; Miller and Guiltinan 2003). Thus, ORCs with single or multiple nodes can be produced using SE plants as a source of stem cuttings (Guiltinan et al. 2000; Maximova et al. 2005; Miller 2009). ORC plantlets produced with this method grow in a similar way to seedlings, with an orthotropic growth phase followed by jorquetting, and the transition to an adult growth phase characterized by the development of plagiotropic branches followed by flower and fruit development.

To test the field performance of rooted orthotropic cuttings collected from SE plants in bentwood gardens, a large-scale clonal garden was established on Nestlé's experimental farm (El Chollo) near Quevado, Ecuador (Fig. 5). Ivan Garzón and colleagues at the Pichilingue experimental station of INIAP have implemented this technology and are now conducting field tests for the production of rooted cuttings under high-density planting. Results of the first year of the INIAP trial indicated that orthotropic trees can be produced at a competitive cost, and that their agronomic performance is comparable to that of trees produced by SE or traditional methods. An account of the use of these methods in projects for the mass propagation of cacao was recently reviewed by Maximova and Guiltinan (2012).

Figure 5 Production of orthotropic rooted cuttings from SE trees using a bentwood clonal garden. Source: Pierre Broun, Nestlé.

These results are similar to those reported in a publication by Miller (2009), describing the use of bentwood stock plants as a source for orthotropic branches, from which single-node cuttings can be harvested for producing plants with the characteristic jorquette branching pattern. Miller (2009) further noted that occasionally the plants propagated by this method produced jorquettes at heights lower than normal. This limitation can be easily overcome with a single heading cut applied to the first jorquette, which will induce a subtending orthotropic shoot to grow to a normal jorquette height. Complete field evaluation (including yield) of plants produced by this method will require several more years of monitoring. This technique can be used with SE in the large-scale production of orthotropic cacao trees.

6 Field evaluation of orthotropic rooted cuttings

6.1 Puerto Rico

In a collaborative project between the USDA in Puerto Rico and Penn State, genotypes from USDA cacao field trials were propagated via SE at the Penn State laboratory between 2001 and 2002. At Penn State, some of the first primary SE plants were used as stock plants to produce ORCs. Twenty different genotypes were propagated and approximately

Figure 6 Field test in Puerto Rico. (a) Plants propagated by orthotropic rooted cuttings (ORC) and grafting, 2001; (b) ORC and grafted plants; (c and d) ORC plants during harvest. Source: Ricardo Goenaga, USDA-ARS, Puerto Rico.

350 plants (ORCs) were transported as bare-rooted plantlets back to Puerto Rico. The field site for this test was located directly adjacent to the original two sites from which the evaluated accessions were selected by Irizarry, Rivera and Goenaga (Irizarry and Rivera 1999; Irizarry and Goenaga 2000) (Fig. 6).

In Puerto Rico, plants of the same genotypes were generated by plagiotropic rooted cuttings and by grafting plagiotropic and orthotropic cuttings onto EET400 rootstock. The test plots contain approximately ten genotypes propagated by four propagation methods, planted in a split-plot design. The test includes approximately 900 trees. Physiological and yield data were continuously collected from 2001 to 2012 and have indicated that all the plants are growing normally and, independent of the propagation method, are producing on average high yields of 2007 kg of dry beans/ha per year, with the top yielder averaging 2538 kg of dry beans/ha/year (Goenaga et al. 2015). The results from this study have suggested that SE followed by ORC propagation is a viable and reliable propagation method for cacao.

6.2 Ecuador (III)

In 2006, INIAP initiated a project to adopt and evaluate the ORC method for Nacional varieties (Fig. 7). SE plants from six Nacional genotypes were established in an experimental plot as bentwood stock plants for rooted cuttings, as described by the Penn State group (see Maximova et al. 2005; Miller 2009). The bentwood stock plants produced numerous

Figure 7 ORC plants in INIAP, 2012. Source: Maximova and Guiltinan.

axillary shoots and, although there were differences between genotypes, on average a single stock plant could produce between four and eleven shoots within a period of three months. The average length of the shoots was 42–65 cm and each produced five to eight quality cuttings (Garzón et al. 2011b). Thus, an annual multiplication rate of over 350 rooted cuttings per stock plant was achieved.

In 2007, an ambitious project was initiated by INIAP, together with the Secretaría Nacional de Planificación y Desarrollo, and the Secretaría Nacional de Ciencia y Tecnología, aimed at developing a commercial system for the production of planting material from cacao plants of the Nacional cacao. The main goal was to scale up the multiplication process for the production of SE plantlets and ORCs obtained from SE stock plants. SE cultures were initiated with explants from seven of the ten cacao genotypes of Nacional cacao currently recommended by INIAP for the main cocoa-producing areas of Ecuador, and two promising genotypes recommended for the Amazon region.

Cultures have been continuously initiated since 2008, with more than 20000 flowers dissected. The resulting SE plants have been planted in the first bentwood clonal gardens established at INIAP's tropical experimental station in Pichilingue, and at central experimental stations in Santo Domingo and Amazonia. Data are being collected on the number of shoots produced per SE stock plant, shoot lengths and the total number of cuttings generated per shoot. Experiments were performed to optimize the light levels, type of shade and fertilization regimes. The highest yields (31 shoots/plant; average shoot length of 69 cm) were obtained at a planting density of 2963 plants/ha, under shade (60%

plastic mesh or 1111 plantain plants/ha), with 120 mm irrigation/month during the dry season and 1000 kg of NPK micronutrients/ha per year. Shoots are harvested throughout the year at 132-day cycles. The average number of cuttings per shoot recorded so far was 5.8, yielding approximately 180 cuttings per plant per year, and a potential total yield of approximately 534 060 cuttings/ha per year, at a density of 2967 stock plants/ha; or potentially 1 000 000 cuttings/ha per year, at a density of 6667 stock plants/ha. Additional experiments were conducted to optimize the protocols for rooting the orthotropic cuttings and for plant establishment, and to test different environmental conditions, substrates and hormone types and concentrations. The best results obtained were 80% successful in producing rooted plants in a period of 120 days from rooting initiation.

Plants propagated by this technology are currently under evaluation in four one-hectare demonstration plots in different regions of the country. The main objective of these tests is to confirm the field performance of plants in different environmental conditions. Based on preliminary analysis of data collected 18 months after planting, it was concluded that the orthotropic plants demonstrated normal growth, but in general the plants were smaller and developed shorter jorquettes in comparison with SE plants, which is consistent with previous reports (Maximova et al. 2008; Miller 2009).

7 Integrated system for vegetative propagation of cacao

A paper on the integration of multiple propagation methods to attain efficient large-scale mass propagation of orthotropic cacao plants was published in 2005 (Maximova et al. 2005). An extension of this idea is presented in Fig. 8. The protocol combines low- and high-tech approaches, including somatic embryo-derived and orthogonal rooted cuttings in non-sterile greenhouse or field conditions using SE plants as stock plants as described above. This step can be performed at a low cost, greatly reducing the overall expense of a rapid, integrated, orthotropic, clonal multiplication system starting from somatic embryos. Plants produced through these methods could be used for planting in the fields, or as self-rooted stock for budwood and/or cuttings, as required.

Depending on the responsiveness of the individual genotypes to the SE protocols, and the availability of flowers to be used as tissue culture explants, the production of SE plants via primary and secondary embryogenesis requires 12–18 months. When the plants are established in the nursery and converted to stock plants for cuttings, the process for producing rooted plants ready for planting in the field requires an additional eight to ten months.

8 Technology transfer

8.1 West Africa

As part of a long-term goal, and with initial funding from the Sustainable Tree Crops Program, Penn State and Mars Incorporated, the team from Penn State has been providing training and education in SE technology in Africa since 2000. Technology transfer and

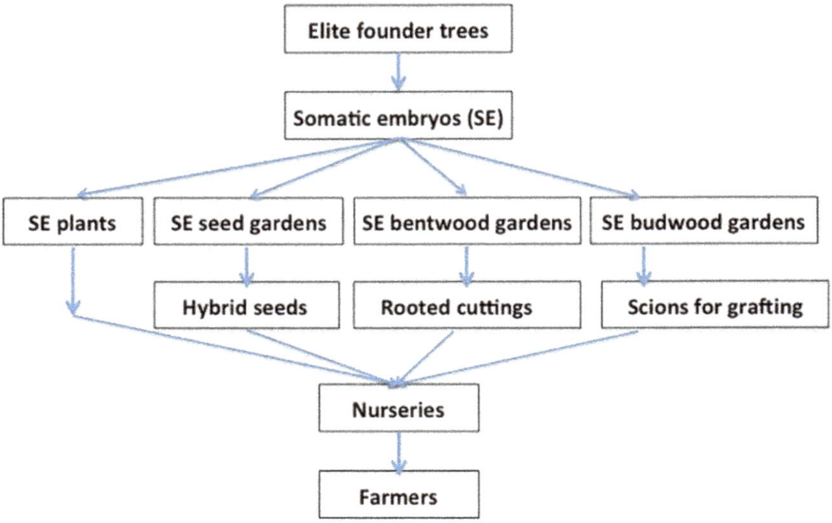

Figure 8 Integrated system for the vegetative propagation of orthotropic SE cacao plants. Source: Maximova and Guiltinan.

training have been carried out at CRIG in Ghana, National Center for Agricultural Research (CNRA) in Côte d'Ivoire; CRIN in Nigeria; and at the Institute of Agricultural Research for Development of Cameroon.

The genotypes introduced into tissue culture in Ghana and Côte d'Ivoire between 2000 and 2001 were cacao clones selected in the national breeding programmes. In Ghana, these included clones resistant to CSSV, which were previously selected through breeding. Cacao floral explants from 17 clones in Ghana and from 31 clones in Côte d'Ivoire were cultured, and somatic embryos from 14 clones in Ghana and 24 clones in Côte d'Ivoire were produced. In Ghana, four out of seven CSSV-tolerant clones produced somatic embryos. In Côte d'Ivoire, over 8000 somatic embryos were produced and maintained in tissue culture, and more than 200 SE plantlets were successfully acclimated. The CNRA group has continued their efforts to optimize the system for local varieties and has described the results of their work in two publications (Issali et al. 2008, 2010).

In 2009, CNRA in collaboration with Nestlé reactivated SE activities at the new research and development centre in Abidjan, Côte d'Ivoire. The goal of this project is to produce about one million new, disease-resistant cacao trees per year to help with the rehabilitation of the ageing cocoa farms in Côte d'Ivoire (Masseret 2009).

Research and adaptation of the cacao SE methods are ongoing in Cameroon since 1997, mainly involving a group of scientists from the University of Yaoundé. This group has performed research in collaboration with the University of Bamenda in Cameroon, CNRA and the Training and Research Unit (UFR) Biosciences in Côte d'Ivoire; and with laboratories at the University of Hamburg, Germany; CIRAD, France; the University of Perpignan, France; as well as Penn State, USA. The results of these efforts were reported in a number of publications (Ndoumou et al. 1997; Verdeil et al. 2007; Niemenak et al. 2008, 2012; Alemanno et al. 2008; Minyaka et al. 2008).

8.2 Asia

8.2.1 Centre for Cocoa Biotechnology Research, Malaysian Cocoa Board, Sabah, Malaysia

In a study to explore the possibility of using SE technology for producing planting materials in Malaysia, several commercial cacao clones were evaluated to assess their potential for SE. Similar to published results, the best regeneration success was achieved using flower staminode tissues (Tan and Furtek 2003). The success of embryo production was influenced by the genotype of cacao and also the time of year in which cultures were initiated (Tan Chia Lock, unpublished data). The protocols used for this research are detailed in Maximova et al. (2008).

As part of this study, molecular fingerprinting and morphological characterization were performed to test for mutations and somaclonal variation resulting from SE. Two- to three-year-old SE plants and their mother plants were fingerprinted using 39 different microsatellite markers (Tan Chia Lock, unpublished data), and no differences were observed between the SE plants and the flower source plants. Cytological analysis of the root tips of the SE plants and the source plants revealed an equal chromosome count of 2n = 20 (Tan Chia Lock, unpublished data). Preliminary morphological examination of leaves and flowers also showed no differences from the mother plants.

8.2.2 Indonesian Coffee and Cocoa Research Institute (ICCRI)

To rehabilitate and intensify the cacao-growing sector, in 2009 the Indonesian government initiated a national programme aimed at improving Indonesian cocoa production and quality (Gernas Kakao). The programme's goals were to rehabilitate a total area of 450000 ha by side grafting (245000 ha), intensifying or applying better horticultural practices for existing trees (135000 ha) and replanting with new planting materials (70000 ha). The 70 000 ha were replanted with clonal material multiplied by SE. To satisfy the demand a tissue culture laboratory with production capacity of up to 50 million plants per year was established by ICCRI in Jember. Under the direction of scientists from Nestlé R&D Tours, France, the SE technology was transferred to ICCRI over the course of two years (2006–07). About 260 people worked at the laboratory up until 2012, with more than 2000 working in the nearby acclimation facility.

SE was initiated for five clones selected for high yield and resistance or tolerance to major diseases of cacao, in particular vascular streak dieback disease (VSD): ICCRI03, ICCRI04, Sca6, Sulawesi1 and Sulawesi2 (Agung Susilo, personal communication, 2012; see also http://www.iccri.net). The ICCRI laboratory generated and delivered more than 74.5 million SE clones to cocoa-developing areas, mainly in Sulawesi, during the period of 2009–11. ICCRI reported that SE cacao clones were found to be normal (true-to-type) and similar to plants propagated by other methods of clonal propagation. Acclimated plantlets (transferred from sterile tissue culture vessels and adapted to nursery conditions) were shipped bare rooted to the cacao-growing regions for distribution by selected private nurseries contracted by the Indonesian government. To maintain quality standards, the plants distributed by the nurseries were certified and workers at distribution nurseries in the growing regions were required to complete training in the handling and management of the SE plants and to obtain certification from ICCRI (A. Susilo, ICCRI pers. comm.).

Today, ICCRI is the most successful institution in scaling up SE technology for cacao for commercial purposes.

In 2016, during the International Symposium 'Frontiers in Science and Technology for Cacao Quality, Productivity and Sustainability' at University Park, PA, Agung Susilo reported the results of an assessment of the field performance of the SE trees. ICCRI researchers evaluated the performance of SE trees in five Provinces (Central Sulawesi, West Sulawesi, South Sulawesi, South-east Sulawesi and Bali). The results indicated that the proportion of SE plants lost during establishment was normal for cacao plants propagated by any method (5–10%) and overall the SE plants performed better than seedlings during the first two years. SE plants were distributed to the five provinces between 2009 and 2011. A total of 5900 SE trees that had replaced old or unproductive trees on 295 cacao farms were sampled and evaluated with stratified random sampling. Six teams of four people each were deployed to the selected areas to interview farmers and personally collect tree performance data. Characterization of the individual trees was carried out based on plant vigour, growth parameters, number of flowering trees, number of fruits per tree, incidents of VSD and angle of the main trunk in relation to the ground. Overall, 73% of the surveyed farms belonged to the 'good' and 'fair' categories, while 27% were categorized as 'bad'. It was observed that the conditions of SE farms in Central Sulawesi were better than the other locations surveyed. More specifically, the percentages of 'bad' SE farms planted between 2009 and 2010 were on average the lowest recorded: 4.74% and 11.52%, respectively. Like seed-grown control plants, the cloned SE trees displayed normal dimorphic structure

Figure 9 Young SE cocoa trees established in the field in Sulawesi. (a, b) Images of SE trees on farms; (c) Image of a taproot of SE trees similar to the taproot of seedlings; normal development was observed in 95% of the SE plants. Source: Agung Susilo, ICCRI.

of seed-grown plants with the formation of jorquette branches and taproots (Fig. 9). The mean jorquette height of the SE trees was 106.5 cm with few SE trees (0.04%) forming jorquettes higher than two metres. The majority of the SE trees (95%) were in the upright position, and few had medium (3.45%) to severe (1.26%) inclines. Observations of the root systems of some of the inclined trees revealed twisted roots or very short taproots. Further enquiries revealed that these symptoms resulted from improper management of the SE plants in the distribution nurseries where the bare-rooted plants received from ICCRI were transplanted into plastic bags. In some cases, where the plants were too large, the taproots were bent, twisted or cut to fit into the bags during transplantation. In some cases, plants were also grown with bad soil drainage, likely resulting in waterlogging and/or poor root development. In 2012, the SE trees exhibited good flower precocity. An average of 56% of the SE trees planted in 2009 (two-and-a-half years old) were producing flowers. The average number of pods recorded at the time of the evaluation (July–August 2012) for the 2009 plants was 3.97 pods/tree (Fig. 10). This indicated the flowering precocity of SE trees compared with traditional cacao trees, which normally reach this level of productivity at four years after planting. The number of pods was also greatly affected by the application

Figure 10 SE cocoa trees producing fruit in the field in Sulawesi, 2.5 years after planting. Source: Agung Susilo, ICCRI.

of agricultural practices by the farmers. Because of this it is impossible to determine how much of the precocity reported is due to SE technology and how much is the result of better agronomic practices.

9 Advantages and constraints of tissue culture

The main advantages of using the SE technology for propagating cacao are:

- SE can produce genetically uniform plants and capture high-value genetic traits.
- The SE technology is applicable to many different genotypes; although variation in efficiency has been observed between different genotypes, most genotypes can be propagated by this method.
- The SE process can be carried out year-round, the work takes place in a laboratory setting and it is not strongly dependent on the time of year. However, plant propagation activities should be timed to ensure the plants will be ready for the planting seasons of individual countries.
- SE plants have morphological characteristics similar to plants grown from seeds. Cacao seedlings have characteristic dimorphic growth that is preferred by farmers because the young plants establish better owing to a single main taproot and jorquette formation that requires less formative pruning. SE plantlets have a similar morphology to seed plants.
- SE plants develop their own roots, unlike grafted plantlets that are commonly grown on rootstock obtained from seedlings. It is possible that genetic segregation for root traits may cause a variation in the yield potential of the resulting grafted populations.
- SE can be used to scale up from one or a few mother plant(s) more rapidly than other clonal propagation methods due to the multiplication rate of the SE process and the large number of floral explants possible per mother plant (rates of up to 50 embryos per flower are common). Since a healthy cacao tree can produce thousands of flowers per year, a potential multiplication rate of 50 000 fold is therefore possible. The use of embryogenic suspension cultures can further increase the multiplication potential.
- Cacao SE can be used for the production and testing of disease-free materials (see Quainoo et al. 2008).

The main disadvantages of using the SE tissue culture method for cacao propagation are:

- Some genotypes exhibit very low SE potential. For these, optimization of media and hormone concentrations may be necessary.
- Risk of somaclonal variation; this has been observed with other plant species and is a result of mutations occurring during the tissue culture process. While field testing has shown that this is not a major problem for SE cacao plants, there is a potential for problems if the protocol is modified, if cultures are carried out for too long or, possibly, for certain genotypes. Careful monitoring of plantlet production and performance in the field can mitigate this potential problem.
- Cacao SE is more expensive than traditional propagation methods. In fact, the cost of building and operating a tissue culture facility is much higher than that of a traditional

propagation facility (see previous section on costs of applying the tissue culture technology).

- Embryogenesis has to be conducted in sterile laboratories that are expensive to build and operate, and are often located far from the cacao farms. Whilst SE cannot be carried out by the farmers themselves, clonal propagation by cuttings and grafting/ micro-cutting can be conducted on farm by the farmers.

10 Future trends and conclusion

Tissue culture has been commercialized for more than four decades. It is estimated that the current market of tissue-cultured products is 500 million to 1 billion plants annually. Perennial species produced commercially through SE include teak; shade trees like maple, birch and hawthorn; roses; fruit trees; shrubs; lilacs; sugar cane; banana; coconut; bamboo; date palms and forest trees (http://www.monfori.com; http://www.agriforestbiotech.com; http://www.indvitrolab.com; http://www.davaomusat.com; http://www.lapandybio.com; http://www.vitrobiovalen.com; http://www.marionnet.com; http://www.dbtmicropropagation.nic.in). Additionally, plant tissue culture has been used for the commercial propagation of a large number of coffee trees (http://www.ecomtrading.com/) and plantain (http://www.galiltec.com). Initially, abnormalities related to clonal oil palm were reported in 1986 (Corley and Tinker, 2003). However, over the course of the two decades that followed, SE protocols were optimized and by 2010 twelve tissue culture laboratories in Malaysia were producing clonal oil palm ramets (Kushairi et al. 2010). In 2011, oil palm clone producers in Malaysia produced 3.5 million ramets (http://www.myjurnal.my/public/article-view.php?id=48496).

The success of these ventures encourages us to consider the use of tissue culture for cacao and has driven the investments in research and field testing, as reviewed above. The effectiveness of this technology has been proved in many field tests, which have positively demonstrated the successful and reproducible growth of test-tube plantlets into fully producing mature trees. The recent experience in Indonesia demonstrates that SE could be a viable commercial technology for cacao.

The demand for clonally propagated cacao plantlets is expected to increase dramatically over the coming years, due to the intersection of several synergistic drivers: increased cocoa demand; the release of new, high-value, disease-resistant varieties; increased demand for special high-flavour cocoa varieties; the need to replace a large number of ageing or diseased cacao trees currently in production; demand for the supply of specialty cocoa varieties and demand for planting new land with cacao. We conservatively estimate that there are about nine billion cacao trees in production worldwide and if these are replanted at a rate of once every 25 years, about 450 million new cacao plantlets will be required each year just to keep up with current replanting demands. As new varieties are developed, demand for new planting materials will increase dramatically because more farmers will want to grow the new varieties.

Based on the promising results presented in this review, we conclude that cacao tissue culture holds a very strong potential to help address the future demand for cacao planting material. Future research should be focused on protocol optimization for new elite varieties according. Furthermore, with the rapid pace of scientific advancement in the fields of cacao genetics and breeding, it will become even more important to implement fast clonal propagation systems for cacao to provide more efficient and rapid testing and deployment of new varieties. SE technology may have its greatest impact in the initial scale up of new varieties by saving two or three years when only one or a few mother trees exist. Our analysis suggests that cacao SE technology should be utilized not as a stand-alone technology, but in conjunction with other methods of propagation such as rooted cuttings and grafting.

11 Acknowledgements

The authors would like to express their appreciation to all the organizations and individuals that contributed to the information presented in this chapter, including: Pierre Broun, Anne Buchwalder, Bernard Masseret and Vincent Petiard from Nestlé France; Ivan Garzón and Freddy Amores from INIAP, Ecuador; the Indonesian Coffee and Cocoa Research Institute (ICCRI) team, and especially Agung Susilo; Ricardo Goenaga from United States Department of Agriculture (USDA), Puerto Rico; and Andrew Wetten from the University of Reading, UK.

This work was supported by The Pennsylvania State University's College of Agricultural Sciences, The Huck Institutes of the Life Sciences, the Penn State Endowed Program in Molecular Biology of Cacao, by the USDA National Institute of Food and Agriculture Hatch project 1003147 and Bioversity International.

The material presented in this chapter is adopted from Chapter 4 of 'Supplying new cocoa planting material to farmers: a review of propagation methodologies' written by Dr. Maximova and Guiltinan and published by Bioversity International in 2015 (Laliberté and End 2015). This chapter includes additional information and updated literature review.

12 References

Adu-Gyamfi, R. and Wetten, A. (2012). Cryopreservation of cocoa (*Theobroma cacao* L.) somatic embryos by vitrification. *CryoLetters*, 33(6), pp. 494–505.

Alemanno, L., Berthouly, M. and Michaux-Ferrière, N. (1996a). Embryogenèse somatique du cacaoyer à partir de pièces florales. *Plantations, Recherche, Développement*, 3, pp. 225–37.

Alemanno, L., Berthouly, M. and Michaux-Ferrière, N. (1996b). Histology of somatic embryogenesis from floral tissues cocoa. *Plant Cell, Tissue and Organ Culture*, 46, pp. 187–94.

Alemanno, L., Berthouly, M. and Michaux-Ferrière, N. (1997). A comparison between *Theobroma cacao* L. zygotic embryogenesis and somatic embryogenesis from floral explants. *In vitro Cellular and Developmental Biology-Plant*, 33(3), pp. 163–72.

Alemanno, L., Devic, M., Niemenak, N., Sanier, C., Guilleminot, J., Rio, M., Verdeil, J. L. and Montoro, P. (2008). Characterization of leafy cotyledon1-like during embryogenesis in *Theobroma cacao* L. *Planta*, 227, pp. 853–66.

Anonymous. (1933). Cacao research. *Tropical Agriculture*, 10(8), pp. 209–11.

Anonymous. (1934). Cacao research. *Tropical Agriculture*, 11(6), pp. 129–31.

Anonymous. (1937). Efficient cacao production. *Tropical Agriculture*, 14(6), pp. 161–2.

Cheesman, E. E. and Pound, F. J. (1932). Uniformity trials on cacao. *Tropical Agriculture*, 9(9), pp. 277–87.

Corley, R. H. V. and Tinker, P. B. H. (2008). *The Oil Palm*. Fourth edition. Wiley-Blackwell. ISBN: 978-0-470-75036-0 (e-book).

Fang, J. Y. and Wetten, A. (2011). Importance of structural integrity of somatic embryos for long-term cryopreservation of cocoa (*Theobroma cacao* L.) germplasm. *African Journal of Agricultural Research*, 6, pp. 3954–61. Available at: http://bit.ly/1yNjsVs (Accessed: 10 April 2014).
Fang, J. Y., Wetten, A. and Hadley, P. (2004). Cryopreservation of cocoa (*Theobroma cacao L.)* somatic embryos for long-term germplasm storage. *Plant Science*, 166(3), pp. 669–75.
Fang, J. Y., Wetten, A. and Johnston, J. (2008). Headspace volatile markers for sensitivity of cocoa (*Theobroma cacao* L.) somatic embryos to cryopreservation. *Plant Cell Reports*, 27(3), pp. 453–61.
Fang, J. Y., Sacande, M., Pritchard, H. and Wetten, A. (2009). Influence of freezable/non-freezable water and sucrose on the viability of *Theobroma cacao* somatic embryos following desiccation and freezing. *Plant Cell Reports*, 28(6), pp. 883–9.
Figueira, A. and Janick, J. (1995). Somatic embryogenesis in cacao (*Theobroma cacao* L.). In: Newton, R. (Ed.), *Somatic Embryogenesis in Woody Plants*, Volume 2, Klumer Academic Publishers, Netherlands, pp. 291–310.
Florin, B. (1997). *In vitro* cryopreservation of cocoa germplasm. CRN-TOURS activity report, 3, pp. 1–14.
Florin, B., Brulard, E. and Pétiard, V. (2000). *In vitro* cryopreservation of cacao genetic resources. In: Engelman, F. and Tagaki, H. (Eds), *Cryopreservation of Tropical Plant Germplasm*, pp. 344–7. Japanese International Research Center for Agricultural Sciences, Tsukuba; The International Plant Genetic Resources Institute, Rome. Available at: http://bit.ly/1FS8rED (Accessed: 10 April 2014).
Florin, L., Tsokoglou, A. and Happe, T. (2001). A novel type of iron hydrogenase in the green alga *Scenedesmus obliquus* is linked to the photosynthetic electron transport chain. *Journal of Biological Chemistry*, 276, pp. 6125–32. Available at: http://www.jbc.org/content/276/9/6125.full (Accessed: 10 April 2014).
Fontanel, A., Gire-Bobin, S., Labbé, G., Favereau, P., Álvarez, M., Rutte, S. and Pétiard, V. (2002). *In vitro* multiplication and plant regeneration of *Theobroma cacao* L. via stable embryogenic calli, paper presented at *IAPTC Congress,* Plant Biotechnology, Orlando, Florida, USA.
Freeman, W. G. (1929). Results of cacao research at River Estate. *Tropical Agriculture*, 6, pp. 127–33.
Garces, H. M., Champagne, C. E., Townsley, B. T., Park, S., Malho, R., Pedroso, M. C., Harada, J. J. and Sinha, N. R. (2007). Evolution of asexual reproduction in leaves of the genus Kalanchoe, *Proceedings of the National Academy of Sciences of the United States of America*, 104(39), pp. 15578–83. Available at: http://www.pnas.org/content/104/39/15578.full (Accessed: 10 April 2014).
Garzón, A., Castillo, B., Eguez, E., Guevara, J. and Osorio, E. (2011b). Estado de desarrollo de la tecnología de embriogénesis somática y nuevas tecnologías para la propagación de plantas clonales de cacao National. EET Pichilingue, INIAP, Ecuador.
Goenaga, R., Guiltinan M. J., Maximova, S. N., Seguine, E. and Irizarry, H. (2015). Yield performance and bean quality traits of cacao propagated by grafting and somatic embryo-derived cuttings. *HortScience*, 50(3), pp. 358–62.
Guiltinan, M. J., Zhijan, L., Traoré, A. and Maximova, S. N. (1997). High efficiency somatic embryogenesis and genetic transformation of cacao. *INGENIC Newsletter*, 3, pp. 7–8.
Guiltinan M. J., Miller, C. R., Traoré, A. and Maximova, S. N. (2000). Greenhouse and field evaluation of orthotropic cocoa plants produced via somatic embryogenesis, micro and macropropagation. In: *Proceedings of the 13th International Cocoa Research Conference*, 9–14 October 2000, Kota Kinabalu, Sabah, Malaysia, Cocoa Producers' Alliance, Sabah, Malaysia, pp. 323–30.
Irizarry, H. and Goenaga, R. (2000). Clonal selection in cacao based on early yield performance of grafted trees. *Journal of Agriculture of the University of Puerto Rico*, 84, pp. 153–63. Available at: http://1.usa.gov/1DwSTzl (Accessed: 10 April 2014).
Irizarry H. and Rivera, E. (1999). Early yield of five cacao families at three locations in Puerto Rico. *Journal of Agriculture of the University of Puerto Rico*, 82(3–4), pp. 167–76.

Issali, A. E., Traoré, A., N'Goran, J. A. K., Koffi, E. K. and Sangaré, A. (2008). Relationship between some phenological parameters and somatic embryogenesis in cocoa (*Theobroma cacao* L.). *Journal of Crop Science and Biotechnology*, 11, pp. 23–30. Available at: http://bit.ly/1Hqvw07 (Accessed: 10 April 2014).

Issali, A. E., Traoré, A., Konan, J. L., Mpika, J., Minyaka, E., N'Goran, J. A. K. and Sangaré, A. (2010). Relationship between five climatic parameters and somatic embryogenesis from sporophytic floral explants of *Theobroma cacao* L.. *African Journal of Biotechnology*, 9(40), pp. 6614–25. Available at: http://bit.ly/1O9UyWL (Accessed: 10 April 2014).

Kushairi, A., Tarmizi, A. H., Samsuri, I., Ong-Abdullah, M., Samsul Kamal, R., Ooi, S. E. and Rajanaidu, N. (2010). Production, performance and advances in oil palm tissue culture, paper presented at the *International Seminar on Advances in Oil Palm Tissue Culture*, organized by the International Society for Oil Palm Breeders (ISOPB), Yogyakarta, Indonesia, 29 May 2010, pp. 1–25.

Laliberté, B. and End, M. (Eds) 2015. *Supplying New Cocoa Planting Material to Farmers: A Review of Propagation Methodologies*. Bioversity International, Rome, Italy, pp. 67–87.

Lambert, S. V., Guiltinan, M. J. and Maximova, S. N. (2000a). *Ex vitro* propagation of acclimated somatic embryo derived cocoa plants. In: *Proceedings of the 13th International Cocoa Research Conference*, 9–14 October 2000, Kota Kinabalu, Sabah, Malaysia, Cocoa Producers' Alliance, Sabah, Malaysia, pp. 5–10.

Lambert S. V., Guiltinan, M. J., Maximova, S. N. and Aitken, W. M. (2000b). *Ex vitro* propagation via mini-cuttings of acclimatized somatic embryo derived cocoa plants. In: *Proceedings of the 13th International Cocoa Research Conference*, 9–14 October 2000, Kota Kinabalu, Sabah, Malaysia, Cocoa Producers' Alliance, Sabah, Malaysia, pp. 237–45.

Li, Z., Traoré, A., Maximova, S. N. and Guiltinan, M. J. (1998). Somatic embryogenesis and plant regeneration from floral explants of cacao (*Theobroma cacao* L.) using thidiazuron. *In vitro Cellular and Developmental Biology-Plant*, 34(4), pp. 293–9.

Manurung, H. (2010). *Cocoa, A Passionate Food for Your Daily Start*. Export News, Indonesia, BPEN/MJL/XX/04/2010, pp. 3–10.

Masseret, B. (2009). Implementation of somatic embryogenesis as a propagation technique for cacao, paper presented at the *INGENIC workshop*, October 2009, Bali, Indonesia.

Masseret, B., Alvarez, M., Fontanel, A. and Petiard, V. (2006). Conformity of cocoa trees produced via somatic embryogenesis. In: *Proceedings of the Malaysian International Cocoa Research Conference*, 18–19 July 2005, Kuala Lumpur, Malaysia, Malaysian Cocoa Board, Malaysia, pp. 122–30.

Masseret, B., Vachet, C., Florin, B., Gianforcaro, M., Fillodeau, A. Brulard, E., Bouquet, J. F., Alvarez, M. and Broun, P. (2008). Propagation of *cacao (Theobroma cacao* L.) by somatic embryogenesis and field performance of the trees. In: *Proceedings of the Indonesian Cocoa Symposium*, October 2008. Indonesian Coffee and Cocoa Research Institute, Bali.

Maximova, S. N. and Guiltinan, M. J. (2000). Greenhouse and field evaluation of orthotropic cacao plants produced via somatic embryogenesis, micro and macro-propagation. In: *Proceedings of the 13th International Cocoa Research Conference*, Kota Kinabalu, Sabah, Malaysia, 9–14 October 2000, Cocoa Producers' Alliance, Sabah, Malaysia, pp. 10–15.

Maximova, S. N. and Guiltinan, M. J. (2012). Potentials for application and utilization of cacao somatic embryogenesis for commercial plant propagation, *paper presented at the INGENIC workshop*, 15–20 October 2012, Yaoundé, Cameroon.

Maximova, S. N., Alemanno, L., Young, A., Ferrière, N., Traoré, A. and Guiltinan, M. J. (2002). Efficiency, genotypic variability and cellular origin of primary and secondary somatic embryogenesis of *Theobroma cacao* L. *In vitro Cellular and Developmental Biology-Plant*, 38(3), pp. 252–9.

Maximova, S. N., Young, A., Pishak, S., Miller, C., Traoré, A. and Guiltinan, M. J. (2005). Integrated system for propagation of *Theobroma cacao* L. In: Jain, S. M. and Gupta, P. K. (Eds) *Protocol for Somatic Embryogenesis in Woody Plants* (*Forestry Sciences*, 11), pp. 209–27.

Maximova, S. N., Young, A., Pishak, S. and Guiltinan, M. J. (2008). Field performance of *Theobroma cacao* L. plants propagated via somatic embryogenesis. *In vitro Cellular and Developmental Biology-Plant*, 44(6), pp. 487–93.

Miller, C. R. (2009). An integrated in vitro and greenhouse orthotropic clonal propagation system for *Theobroma cacao* L. Ph.D. thesis, The Pennsylvania State University, USA.

Miller, C. R. and Guiltinan, M. J. (2005). Perspective on rapid vegetative multiplication for orthotropic scion and rootstock varieties of cocoa. In: *Proceedings of the International Workshop on Cocoa Breeding for Improved Production Systems (INGENIC)*, 19–21 October 2003, Accra, Ghana, INGENIC and COCOBOD, pp. 189–94.

Minyaka, E., Niemenak, N., Fotso, Sangare, A. and Omokolo, D. N. (2008). Effect of MgSO4 and K2SO4 on somatic embryo differentiation in *Theobroma cacao* L.. *Plant Cell, Tissue and Organ Culture*, 94(2), pp. 149–60.

Ndoumou, D. O., Ndzomo, G. T. and Niemenak, N. (1997). Phenol content, acidic peroxidase and IAA-oxidase during somatic embryogenesis in *Theobroma cacao* L. *Biologia Plantarum*, 39(3), pp. 337–47.

Niemenak, N., Saare-Surminski, K., Rohsius, C., Ndoumou, D. O. and Lieberei, R. (2008). Regeneration of somatic embryos in *Theobroma cacao* L. in temporary immersion bioreactor and analyses of free amino acids in different tissues. *Plant Cell Reports*, 27(4), pp. 667–76.

Niemenak, N., Awah, T. M. and Lieberei, R. (2012). Establishment of suspension culture in *Theobroma cacao* and polyamines associated with cacao embryogenesis. *Plant Growth Regulation*, 67(1), pp. 1–8.

Pardomuan, L., Taylor, M. (2012). Indonesia's 'Frankentrees' turn cocoa dream into nightmare. *Reuters*, 15 October. Available at: http://reut.rs/1FSo0we (Accessed: 10 April 2014).

Pound, F. J. (1945). The selection of cocoa trees on the basis of productive efficiency and the requirements of the manufacturers of chocolate. In: *Cocoa Research Conference, London.* Imprint: H.M.S.O., London.

Quainoo, A. K., Wetten, A. C. and Allainguillaume, J. (2008). The effectiveness of somatic embryogenesis in eliminating the cocoa swollen shoot virus from infected cocoa trees. *Journal of Virological Methods*, 149, pp. 91–6. Available at: http://worldcocoafoundation.org/wp-content/files_mf/quainoo200883.pdf (Accessed: 10 April 2014).

Shephard, C. Y. (1936). Some economic aspects of cacao production in Trinidad with special reference to the Montserrat district. *Tropical Agriculture*, 13(4), pp. 85–90.

Sondahl, M. R., Liu, S., Bellato, C. and Bragin, A. (1993). Cacao somatic embryogenesis, *Acta Horticulturae*, 336, pp. 245–8.

Tan, C. L. and Furtek, D. B. (2003). Development of an in vitro regeneration system for *Theobroma cacao* from mature tissues. *Plant Science*, 164, pp. 407–12.

Traoré, A., Maximova, S. N. and Guiltinan, M. J. (2003). Micropropagation of *Theobroma cacao* using somatic embryo-derived plants. *In vitro Cellular and Developmental Biology-Plant*, 39(3), pp. 332–7.

Verdeil, J. L., Alemanno, L., Niemenak, N. and Tranbarger, T. J. (2007). Pluripotent versus totipotent plant stem cells: dependence versus autonomy? *Trends in Plant Science*, 12(6), pp. 245–52. Available at: http://bit.ly/1DI8WPj (Accessed: 10 April 2014).

Chapter 8

Good agronomic practices in cocoa cultivation: rehabilitating cocoa farms

Richard Asare, International Institute of Tropical Agriculture (IITA), Ghana; Victor Afari-Sefa, World Vegetable Center, Benin; Sander Muilerman, Wageningen University, The Netherlands; and Gilbert J. Anim-Kwapong, Cocoa Research Institute of Ghana, Ghana

1 Introduction

Globally, over 2.9 million households depend on cocoa production as the major economic activity for their livelihoods. Cocoa cultivation covers a total area of about 5.9 million ha with around 73% of this area found in the four large cocoa-producing countries in West Africa, namely Cameroon, Côte d'Ivoire, Ghana and Nigeria.

Average cocoa yields recorded in West and Central Africa is between 200 and 700 kg per ha (Asare et al., 2010) as against a potential yield of about 2000 kg or more reported on research stations. This huge yield gap has been attributed to several factors including ageing trees, poor planting materials, poor soil fertility management, losses due to diseases and pests and old age of farmers coupled with an overall lack of knowledge on good agricultural practices (Asare et al., 2016; Gockowski et al., 2013). The average age of cocoa farmers in West Africa is between 45 and 50 years, which is higher than in most agricultural enterprises in the developing world. Interest among the younger generation is low and dwindling due to perceived lower incomes from the sector as a result of less productive ageing trees, uncompetitive farm-gate prices, insecurities of land tenure as well as poverty, food and nutritional insecurity and poor infrastructure in rural areas,

http://dx.doi.org/10.19103/AS.2017.0021.12

among others. Limited innovation capacity along the value chain underlies the poor productivity performance of the global cocoa sector over the last 20 years. Improved high-yielding cocoa varieties developed by the national cocoa research institutes, especially in West Africa have seen limited adoption (Asare, 2010). Even though high-yielding cocoa varieties have been hailed as the silver bullet, rapid senescence owing to the physiological stresses of higher yields requires more frequent replanting (Ruf and Burger 2001, World Bank, 2011). Low yields on farms have resulted in decades of extensive migratory cocoa farming that have contributed immensely to deforestation and biodiversity loss in the Upper Guinean rainforest, of which today less than 20% remains in West Africa.

In response to these challenges, West African governments over the past decades have embarked on various interventions to help cocoa farmers to rehabilitate their old cocoa farms or establish new ones. In countries such as Ghana, this includes governmental subsidies for hybrid cocoa planting materials and fertilizers on the assumption that potential productivity gains from the adoption of hybrid cocoa and fertilization could potentially double current yields in much the same way as improved cultivars and fertilizer use had contributed to improved yields in other crops during the last decade.

For instance, under the current policy initiative of the Ghana Cocoa Board (Cocobod)[1], 50 million improved hybrid cocoa seedlings raised at 87 nursery sites were distributed gratis to farmers during the 2014/2015 season. In 2015/2016 season, nursery sites were increased to 331 which allowed 60 million seedlings to be distributed across the cocoa growing regions in Ghana. The Cocobod also distributed 2.4 million 50 kg bags of granular fertilizers and 1.6 million litres of foliar fertilizers to farmers free of charge during 2015/16. In addition, 1500 tons of fungicides and 1.7 million litres of insecticides were supplied free of charge. Similarly, in Cameroon through the Cocoa and Coffee Seed Project, an additional 25 ha of cocoa seed garden was established in 2011. It is estimated that the new seed garden at its full production capacity will distribute 500,000 pods to supplement what is currently being produced in the other seed gardens. With support from the Federal Government of Nigeria, the Cocoa Research Institute of Nigeria established two-hectare seed gardens in each of the 14 cocoa growing states of Nigeria during the period 2007–2011 to give farmers access to cocoa varieties with yield potentials in excess of 2.5 to 3.5 tons/ha (Asare, 2010).

However, in terms of fertilizers, even though governmental policies in West African countries aim at improving accessibility and affordability through provision of subsidies, the implementation process has proved inefficient and inequitable leading to only between 13 and 30% of farmers applying mineral fertilizers (Gockowski et al., 2013).

There is an urgent demand to adopt a more comprehensive approach within the framework of a sustainable cocoa-based farming system. A sustainable cocoa farm is a long-term family business with intergenerational opportunities for men, women and youth. These are well-planned and -managed diversified farm enterprises in which cocoa is part of a farming system that includes food and cash crops as well as trees that yield both timber and non-timber forest products. The cash and food crops from the diversified farm will improve family health, welfare and well-being and create diverse livelihood opportunities within the community. Cocoa trees within the system are periodically rejuvenated with

1. Cocobod is a Ghanaian government-regulatory institution that controls and manages produce marketing and fixes the buying price for cocoa in Ghana.

quality planting materials and are nurtured using good agricultural practices to realistically attain up to 1,200–2,000 kg/ha. Challenges with regard to soil fertility or pests and diseases are managed in an integrated way. This will result in a profitable, integrated and diversified farm that will contribute to a gender-equitable, non-predatory and sedentary cocoa farming system.

The main objective of this chapter is to review good agricultural practices and cost/returns associated with cocoa farm rehabilitation under three regimes, viz. under planting, gradual replanting and complete replanting, using Ghana as a case study. This will provide a broad range of investment choices to cocoa farmers while ensuring that the practices promote optimal environmental stewardship and opportunities for improved ecosystem services. The aim is to provide a comprehensive framework that helps farmers to obtain the necessary knowledge required to embark on a successful cocoa farm rehabilitation and/or to establish a new cocoa farm.

2 Good pre-planting practices in cocoa cultivation

Good agricultural practices in cocoa cultivation constitute a portfolio of management practices that ensure increased productivity while at the same time maintain and protect the environmental integrity of the landscape. These practices include, but are not limited to, the use of improved planting materials and the implementation of sound cocoa agronomic practices. The combination ensures maximum production at a reduced financial and environmental costs.

2.1 Planting materials for cocoa cultivation

Across West Africa, hybrid pods (improved planting materials) from government-managed seed gardens are the main source of planting material to farmers, although other sources of propagules also exist. These hybrids, which are mostly crosses between the Amazon type of cocoa and Amelonado are early bearing, high yielding and possess some level of tolerance to diseases and insect pests such as black pod and mirids. The availability of these hybrids, however, remain a bottleneck even though countries such as Ghana have raised and distributed a significant quantity of free improved hybrid seedlings to farmers since 2014. The challenge has been that sowing time is out of phase with the peak period of hybrid pod production from these seed gardens, and hence planting materials of questionable quality are sourced by farmers from informal channels. The poor inherent genetic and physical qualities of the seeds and susceptibility of such materials to severe pests and disease infestation (Asare et al., 2010) at least partly is responsible for low productivity of cocoa farms. The use of poor quality seeds has also been attributed to high seedling mortality in the field (Padi, 2013). To overcome this challenge, Asare et al. (2016) advocated a collective purchase agreement between farmers and seed gardens in the form of a seed brokerage system that will ensure timely delivery of improved seeds to farmers. The study reports that timely receipt of hybrid pods by farmers along with a complementary 3–6 months' nursery management phase, results in an almost 80% establishment rate of transplanted seedlings in the field irrespective of the method of rehabilitation and the land-use type (Fig. 1).

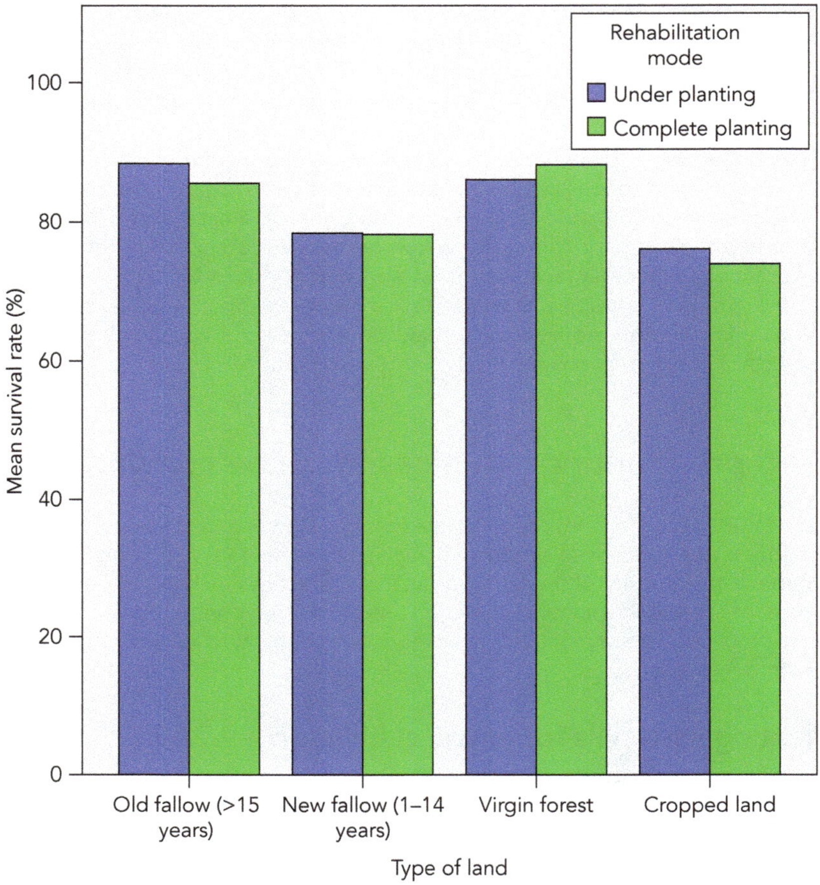

Figure 1 Impact of rehabilitation method on survival rate of hybrid cocoa seedlings in Ghana. Adapted from Asare et al. (2016).

2.2 Site selection

In cocoa cultivation, the selection of a suitable site for plantation establishment is key. Cocoa has been found to be quite/rather demanding in terms of its soil requirements (Smyth, 1975). Forest areas with deep, well-drained soils that vary from loamy sands to friable clays that are red or reddish brown in colour are preferred to enhance crop growth (Charter, 1953; cited in Asare et al., 2016). On the contrary, lands filled with rocks or lands subjected to excessive water accumulation are detrimental for cocoa cultivation. That said, other more practical considerations like distance from the residence and access to water also play a major part in farmers' decisions. Cocoa can typically be grown within latitude 20° S and 20° N of the Equator. However, most of the crop is produced within latitude 10° N and 10° S. Rainfall and temperature are the most critical climatic factors necessary for growth and development. According to Wood and Lass (2008), cocoa is best suited in areas with annual rainfall ranging between 1250 and 3000 mm and a dry period of not more than three months. Temperature requirements vary between 30°–32°C mean maximum and 18°–2°C mean minimum and for optimum production there should not be more than one month when the average daily maximum temperature exceeds 33.5°C.

2.3 Land preparation

Most cocoa plantations worldwide are established either on land committed to different types of land use, ranging from forest to cropped lands. In order to plant cocoa, farmers either thin or completely eliminate the over-story, after which they clear and/or burn the understory vegetation and trees, to make growing space for their cocoa and food crops (Ruf and Zadi, 1998). However, there have been mixed reviews on burning after clear felling as a means of land preparation. Burning destroys a great deal of potential humus and releases large amounts of ash, causing imbalance of nutrients and exposure of the soil to sun and rain leading to severe erosion and losses of organic matter (Wood and Lass, 1985; Asare and David, 2011). In the same vein, burning after land clearing can result in increased early growth of cocoa (Ling and Mainstone, 1983; Odegboro, 1977).

2.4 Establishing shade cover in cocoa

During land preparation some farmers choose to preserve certain mature trees for shade, instead of removing them from the system. In the initial phase of cocoa establishment, food crops such as plantain, cocoyam, maize and cassava are planted, followed by cocoa, to provide key initial shade to the cocoa seedlings (Asare and Asare, 2008). This is also an important source of food and income to the farm family over the next growing season (Osei-Bonsu et al., 1998). Subsequent planting of native and/or exotic trees following clear felling, or by thinning the forest before planting cocoa is practised to provide permanent shade for cocoa. In the process about 35–45 dominant intermediate trees should be preserved per ha at a forest site as permanent shade (Hammond, 1962). Farmers in West Africa have also learned from experience that not all forest trees provide suitable shade for cocoa. Asare (2005) presented a review of shade tree species preferred by farmers and recommended by researchers in West African cocoa growing areas. Thinned forest is used almost in West Africa and Brazil where the forest trees provide suitable shade and its advantages are of paramount importance to small farmers. However, the use of thinned forest trees for shade in cocoa is suboptimal as forest shade is hardly uniform across the landscape in addition to being more difficult to control and adjust than planted shade. It is a challenge to use farming equipment such as tractor and spraying machines due to the uneven distribution of different plants including cocoa in the field, especially when the forest has been thinned.

2.5 Lining, pegging and spacing in cocoa systems

It has been argued that lining and pegging are one of the most important operations during cocoa establishment, as recommended spacing will give the greatest economic return of cocoa yield per unit area (Urquhart, 1961). Lining and pegging leads to the creation of holes for seedling planting or seed sowing. For seedlings from the nursery, it is recommended the holes should be large and deep enough to provide sufficient space for accommodating the poly bags containing seedlings. These holes need to be created well in advance (at least two weeks) before transplanting seedlings. This will allow the holes to settle (siltation) properly to ensure that seedlings fit properly in the holes. After placing seedlings in the holes each hole should be firmly covered with the same soil dug from it while ensuring that the soil level is the same as under nursery conditions. Do not heap soil around the stem of the seedling above the stem collar to prevent moisture getting in contact with the stem which will otherwise lead to rotting of seedlings in the field. In the situation where seeds are planted at stake, the hole should be twice the size of the fresh cocoa beans. Sow one or two cocoa beans in each hole and cover with soil. Remove one of the seedlings at the four-leaf stage and plant in a different hole at the recommended spacing.

There are many different opinions concerning the spacing of cocoa depending on the country where cocoa is grown, as different countries have adopted spacing in relation to their climate, type of planting material, shade conditions and soil type (Van Hall, 1932). Wood (1964) argues that the optimum plant spacing is influenced by factors such as the type of cocoa planted, its vigour and branching habit. For example, a vigorous type, besides growing more quickly than others, will probably grow to a larger size at maturity, and could therefore be planted at a wider spacing. Cocoa types with branching habits vary from type to type: Amelonado-seeded cocoa tends to grow upwards and bear the bulk of its crop on the trunk; the proportion of the crop borne on the trunk varies with spacing, but at a given spacing this proportion is higher with Amelonado than with seeded Amazonian cocoa types. Shade conditions will influence the optimum spacing; cocoa grown with little or no shade has much denser foliage and does not spread to the same extent as shaded cocoa (Wood and Lass, 1985). Spacing plays a significant role on the effect of certain pests and diseases. There is evidence from Ghana and Nigeria that capsid attack is more severe on widely spaced trees. In Nigeria, only 6% of trees were affected by capsid attack when planted at 1.5 m × 1.5 m. However, 100% of the trees were affected when the planting density was increased to 4.5 m × 4.5 m (Wood, 1964). The magnitude of black pod incidence is usually greater at close spacing, presumably owing to higher humidity under the dense canopy (Wood, 1964). According to Squibbs (1943) the wide variations in the planting distances of different varieties of cocoa even in one locality or on one soil type showed that there is no fixed planting distance except what is appropriate for a particular location.

Despite the wide divergence of opinion about the spacing to adopt, a number of experiments on spacing have been conducted in Trinidad, Ghana, Nigeria and Papua New Guinea. Most of the experimental evidence points to close spacing for attaining the highest yields. Work by Van Hall (1932) at River Estate, Trinidad, showed that 3.7 m planting distance gave the highest yield per hectare for the first six years of production but was then surpassed by the 4.3 m as the trees matured. Russell (1953) reported on a spacing trial in Nigeria showed that under the prevailing conditions close planting of trees at 2 m to 2.4 m apart was advantageous for high early yields. Knowal (1959) further noted that the average weight of wet beans and the number of pods per tree increased with decreasing stand density, but this was not enough to increase the yield per hectare at lower densities.

Van der Knapp (1958) in Indonesia reported higher yields, in the first three years of cropping, from cocoa growing at 4 m × 2 m. In Papua New Guinea, Charles (1961) found that earlier plantings of both sole cocoa (Trinitario and Amazon types) and cocoa interplanted with coconuts at a 3.7 m triangular spacing (850 trees ha^{-1}) resulted in greater yields per unit area than the wider spacing of 4.6 m, particularly in the earlier years. Byrne (1972) reported on comparisons made on five closer spacing: 1.5 m, 2.4 m, 3 m and 3.7 m triangular and 4.6 m × 4.6 m and 5.5 m × 5.5 m hedge spacing in Papua New Guinea, and indicated that even at age 21 months, inter-tree competition was obvious in the 1.5 m triangular spacing and under the country's conditions would have limited yields as it matured. Optimum spacing for Papua New Guinea conditions would be 3.7 m triangular or less but greater than 2.4 m triangular spacing.

Current spacing trials in Ghana using Inter-Amazons and Series II hybrids[2] indicate that spacing ranging from 2.0 m × 3.0 m, 2.5 m × 2.5 m, 2.0 m × 3.5 m and 2.5 m × 3.5 m gave better yields than plantings at 3.0 m × 3.0 m and 3.0 m × 3.5 m (Osei-Bonsu et al., 2002a). Also, cocoa spaced at 2.5 m triangular spacing under coconuts spaced at 9.8 m × 9.8 m spacing was more profitable than widely spaced coconut intercropped with cocoa spaced at 3 m × 3 m (Osei-Bonsu et al., 2002b). In their review of the planting distance and stand design in cocoa systems in Ghana, Anim-Kwapong and Anim-Kwapong (2012) suggested that superior yields are associated with spacing ranging between 2.3 m × 2.3 m and 3 m × 3 m for Amelonado, 3.5 m × 3.5 m for vigorous Amazons and 3 m × 3 m as best compromise for estate cocoa. High-density planting of cocoa has recently generated much interest due to its tremendous potential in increasing yield per unit area. High-density planting requires among other things a suitable planting design and density and the use of appropriate cultivars. In Malaysia, Kahar et al. (1991), testing 7 clones, three planting designs and densities ranging from 643 to 5000 plants ha^{-1} over three locations, observed that high-density planting could increase early tree yields significantly. However, any increase in density above 3330 plants ha^{-1} did not give significant yield increases. The results further indicated that site conditions had a profound influence on yields.

3 Good post-planting practices in cocoa cultivation

3.1 Planting of cocoa

In West Africa, cocoa farms are normally established from seedlings raised in polythene bags in a nursery (Wood and Lass, 1985) or on nursery beds and transplanted as bare-rooted plants. Transplanting of seedlings grown in polythene bags is cumbersome and highly labour intensive especially when seedlings have to be transported over long distances from the nursery to the field due to the bulkiness of the bagged seedlings (Amoah et al., 2000). In many cocoa communities, labour is in short supply and otherwise an expensive resource. The use of bare-root seedlings transplanted from nursery beds cuts down on the labour and input costs during the establishment of cocoa farm (Amoah et al., 2000; Alvim et al., 1981; De Souza et al., 1981a,b), particularly in sloping terrains.

Cocoa farms may also be established by planting seeds at stake. This method is popular with farmers who lack the resources to buy inputs for raising seedlings in the nursery. The initial performance of such seedlings is generally lower than the nursed seedlings since they suffer from competition from weeds and rodent damage (Amoah et al., 2000). It is important to provide approximately the same shade conditions at the time of transplanting cocoa seedlings from a nursery to the field (Asare and David, 2011). If there is a great

2. These are cocoa varieties released by the Cocoa Research Institute of Ghana (CRIG) in the 1970s and 1980s, respectively, through its cocoa breeding programme.

difference then shade should be reduced in the nursery several days before the seedlings are transplanted (Wood and Lass, 1985). In seasonally wet and dry conditions, as occurs in West Africa, the timing of transplanting cocoa seedlings or sowing seeds at stake in the field is very crucial for the survival of seedlings. Thus, as a rule in West Africa, it is better to plant at the beginning of the wet season. Other countries are fortunate to have more suitable conditions during most times of the year, thus allowing for a continuous planting regime.

3.2 Importance of shade in cocoa

The use of shade in cocoa is an ancient practice, reportedly dating to the domestication of the crop. As the use of leguminous shade trees such as *Gliricidia sepium* for cocoa occurs at what are generally considered to be the points of domestication, it was generally assumed that shade trees, and more especially leguminous shade trees, were essential for the cultivation of the crop (Sanchez et al., 1990). The origin of the use of shade is usually attributed to early cultivators mimicking the natural sub-canopy environment of wild cocoa trees in the forest (Murray, 1955) of the upper Amazon and Orinoco River basin (Simpson and Ogorzaly, 1986). This agro-ecosystem is apparently ideal for the cultivation of the crop since the leaves of cocoa have a low light saturation point of 400 μ E m^{-2} S^{-1} and a low maximum photosynthetic rate (7 mg dm^{-1} h^{-1}) at light saturation (Hutcheon, 1981). The photosynthetic rate of the crop is known to decrease if the photosynthetic apparatus is exposed to light intensities exceeding 60% full sunlight, while prolonged exposure to high light intensities damages the photosynthetic mechanism of the leaves (Raja Harun and Hardwick, 1987). Shade or low light intensities, however, decrease or suppress flower production. All other factors set aside, light levels under 1,800 hours per year is known to have a considerable depressing effect on production (Asomaning et al., 1971; Gerritsma and Wessel, 1994).

Although shade limits cocoa yield, it provides several agro-ecological benefits. Beer et al. (1998) summarized the major physiological benefits that cocoa receives from shade trees. Despite these good attributes there is also evidence by Ahenkorah et al. (1987) on the depressing effect of shade on yields. The shade requirement of cocoa has been questioned and the practice has been subject to wide experimentation. A principal line of inquiry has been to uncover whether shade is an innate requirement of the cocoa tree itself or whether it serves a secondary role by maintaining appropriate soil, insect population and other condition for the cocoa plants, conditions that could be potentially maintained with the application of suitable chemical inputs. However, in a recent publication Asare et al. (2016) provided evidence that in shaded systems, yield of cocoa increases with increases in canopy cover at plot level in four cocoa growing districts in Ghana (Fig. 2).

The main physiological drawback is competition, especially when excessive shade is used. In seasonally drier zones, root competition for water may limit the use of shade trees. Pest and disease problems may also have been shown to increase with increased shade (Dakwa, 1976; Smith, 1981; Ahenkorah et al., 1974). The perceived negative effect of shade trees on the yields of cocoa has led to ideals of compatibility of shade trees in cocoa systems. According to Asare (2005) while farmers are concerned about exploiting all the necessary components in the system and their interactions to maximize income and reduce risks, research has focused heavily on parts of the system, which in most cases led to an improvement in yield of the cocoa tree. Asare (2005) provided lists of farmer-preferred and research-recommended shade tree species across West Africa. These include desirable trees such as *Terminalia* sp., *Milicia* (*Chlorophora*) *excelsa*, *Albizia*

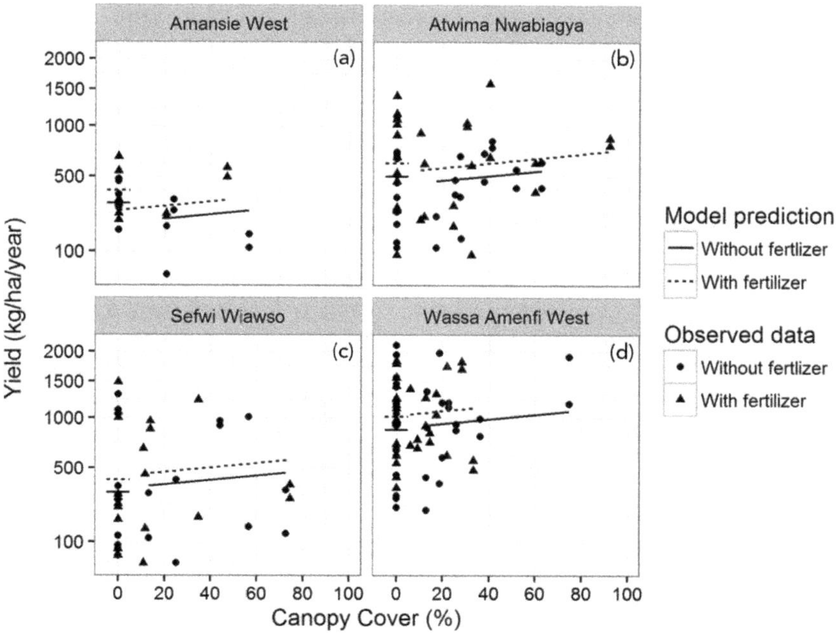

Figure 2 Canopy cover and yield of cocoa (triangles = with fertilizer, crosses = without fertilizer) in Atwima Nwabiagya, Amansie West, Wassa Amenfi West and Sefwi Wiawso. The lines show the model predictions (full lines = without fertilizer, dashed lines = with fertilizer). These model lines are drawn inside the range of the canopy cover in the four districts and two fertilizer regimes, but for visualization extended around cover=0% for the subplots without shade. Adapted from Asare et al. (2016).

coriaria, Entandrophragma angolense, Funtumia elastica, Alstonia boonei and *Pycnanthus angolensis.* The undesirable forest trees are *Myrianthus arboreus, Lecaniodiscus cupanoides, Carapa procera, Blighia welwitschii, Ceiba pentandra, Canthium glabriflorum, Cola gigantea, Adansonia digitata, Cola chlamydantha* and *Musanga cecropioides.*

3.3 Intercropping in young cocoa farms

Traditionally, as part of an establishment method, cocoa is grown in association with food crops up to the stage of canopy closure (Hammond, 1962; La Anyane, 1963). Intercropping in cocoa also referred to as plant diversification is defined as the strategic integration in time (at various stages in the establishment and management of cocoa farms) and space (the three-dimensional arrangement of trees on the ground and into the canopy) of suitable and valuable non-cocoa tree species and other plants. Species may include as mentioned earlier indigenous forest species, fruit trees, common agroforestry species, food crops, shrubs and so on (Asare, 2006). The benefits of such a practice may include food sustainability for the household, income generation to partially offset the cost of establishment, provision of temporary shade for the young cocoa, weed control and reduction in insect pest damage (Hammond, 1962; Adeyemi, 1987; Egbe and Adenikinju, 1990; Idowu, 1996). In West Africa, food crops grown in association with cocoa vary according to the country and mainly includes plantain, cocoyam, yam, pepper, maize, cassava, cowpea, melon and okra. Research into the suitability of some food crops for cocoa intercropping has received considerable attention in the major cocoa growing countries in West Africa. In Nigeria, Iweama (1963) and Adeyemi (1988) reported significantly better growth of cocoa seedlings grown in association with crops such as yam, maize, cowpea, melon, okra and papaya than those in pure stands. A similar trend was observed in the early yield of cocoa beans, which Adeyemi (1988) attributed to favourable environmental conditions created as a result of effective weed suppression from the intercrops.

In Ghana, researchers including Bonaparte and Toseafa (1979), Osei-Bonsu et al. (1998) and Opoku-Ameyaw et al. (2003) have provided evidence on the biological and economic benefits of intercropping cocoa with plantains, maize and cassava at the establishment phase. Contrary to the findings of Bonaparte and Toseafa (1979) which showed a depressive effect of cassava on the growth of cocoa seedlings and the generally held view that it is a strong feeder and may impoverish the soil, the work of Osei-Bonsu et al. (1998) and Opoku-Ameyaw et al. (2003) indicated its suitability as an intercrop in the cocoa culture even in the absence of fertilizer application. As the range of staples grown by the peasant farmer is so wide, current research efforts are geared towards investigating their suitability for intercropping with cocoa. The findings of the current studies will provide the farmer with many options to improve the household needs in terms of food and income.

3.4 Intercropping in mature cocoa farms

Cocoa at the mature stage is intercropped with other tree crops in addition to timber trees for various reasons. Mature cocoa farms can incorporate tree crops such as avocado, oil palm, coconut, mango, citrus, cola and breadfruit for economic benefit. Some of these tree crops may even have a higher farm-gate value than cocoa. The tree crops studied include coconut (Fagan and Topper, 1988), cola (Egbe and Adenikinju, 1990), oil palm (Kolade, 1986; Egbe and Adenikinju, 1990), rubber (Wood and Lass, 1985), citrus (Famaye et al., 2003) and peach palm (Silva and Dias, 1987). Some of these studies have shown that cocoa can be successfully intercropped with some tree crops without any detrimental effect but in some cases the success depended on the environment, density and spatial arrangement which may influence the biological relationships existing within the system. Studies carried out on coconut in Malaysia and on oil palm in Nigeria have indicated that the presence of the crops did not significantly affect the performance of cocoa (Ramadasan et al., 1978; Egbe and Adenikinju, 1990). This led Egbe and Adenikinju (1990) to speculate that synergistic interactions existed between cocoa and oil palm in the system. Studies carried out in Jamaica, however, indicated that interspecific competition occurred between coconut and cocoa (Fagan and Topper, 1988). The divergent observation could be due to notable environmental differences between countries. Famaye et al. (2003) also showed that in Nigeria, the growth performance of polycultured cocoa, kola and citrus was comparable to that of their corresponding pure stands indicating that there were no deleterious or allelopathic effects on any of the component crops.

In Ghana, the need to find additional trees to solve the problem of shade for cocoa, which is becoming critical, has prompted studies into intercropping cocoa with coconut and oil palm. Results from these studies indicated that cocoa could be grown under the shade of these tree crops when planted at wider spacing with the additional income from the latter serving as a 'bonus' (Osei-Bonsu, et al., 2002b; Amoah et al., 1995). The usefulness of the cocoa–coconut intercropping system is particularly evident in coconut plantations along the coastal regions of Ghana where it has helped in managing the devastating effect of the Cape St Paul wilt disease of coconut. Research is in progress to introduce the technology to farmers in this environment to diversify sources of farm income. Thresh (1958) and Thresh et al. (1988) suggested that the spread of the cocoa swollen shoot virus (CSSV) disease can be reduced by interplanting cocoa with associated immune crops. This prompted a barrier crop experiment at the Cocoa Research Institute of Ghana (CRIG) using cola, oil palm and citrus to slow down the movement of the mealy bug, which is the vector of the CSSV. Preliminary results indicate that both citrus and oil palm barriers were effective in isolating the test cocoa trees from CSSV infection. Furthermore, the economic analysis of the three-barrier crop system suggested that the citrus and oil palm systems adequately compensated for the cocoa space they occupied (Ollennu et al., 2003).

3.5 Weed control in cocoa cultivation

Weed management is critical in newly established cocoa plantations and this can amount to 70% of all costs during the first two to three years of a cocoa plantation (Corven, 1993). Weeds in cocoa farm can be managed manually or with the use of herbicides. In young cocoa, the frequency of manual weeding ranges between 4 and 6 times per year. As the cocoa canopy closes the frequency reduces to 2–3 times a year. The high frequencies of weeding render manual weed control expensive and time-consuming. Investigations aimed at developing labour-saving methods to help farmers maintain their cocoa farms started in the mid-1960s. Bonaparte (1966) tested the effectiveness of the herbicides: atrazine, simazine, dalapon and a mixture of amitrole and ammonium thiocyanate in controlling weeds in old cocoa plantation and concluded that the amitrole and ammonium thiocyanate mixture satisfactorily controlled most of the weed. The search for more cost-effective herbicides continued and Oppong et al. (1995) reported that low volume application of the herbicide glyphosate at 1–2 L/ha in 100 L of water resulted in better growth of cocoa seedlings and substantially higher yield of plantains used as temporary shade. Based on these studies, glyphosate herbicides were recommended for weed control in cocoa production in Ghana.

However, since the adoption of chemical weed control strategies involves initial injection of capital, it is still beyond the reach of most peasant cocoa farmers. Agrochemical application, unlike fertilizer application, generally must be perceived as a direct economic loss, not as an investment. Further, there is evidence that many weed species are evolving herbicide resistance, thus making their control less effective and costlier (Lebaron and Gressel, 1982). Under such a situation, the need for an alternative method of weed control in cocoa becomes imperative.

The use of selected plant species to control weeds was investigated by Opoku (1967) who compared the effects of several leguminous and non-leguminous cover plants on weed control in young cocoa. He reported higher yield increases of 30% or above in plots with *Flemingia congesta*, *Indigofera sumatrana*, *Tephrosia* spp., *Desmodium asperum* and *Pennisetum purpureum* than in plots where plant cover was made up of natural regeneration. Oppong et al. (1998) tested the potential of *Flemingia macrophylla* as mulch for managing weeds in young cocoa. When combined with three times clean weeding, it was observed that *F. macrophylla* contributed to the suppression of weeds leading to the rapid growth of cocoa seedlings at lower cost than the traditional method of manual weeding four times per year. Another study by Oppong et al. (2003) indicated that controlling weeds with glyphosate (low volume application) at 1.5 L/ha three times a year promoted rapid growth and early yield of cocoa and were more profitable and cost-effective in the long-term.

3.6 Pruning of cocoa

Pruning is a common management practice on tree crops, which involves the selective removal of plant parts. Reasons for undertaking pruning on cocoa have been given to include facilitation of harvesting and spraying, general movement through the farm and the production of a well-balanced plant. When undisturbed, cocoa seedlings will form their first jorquette at a height of about 1.5 m two years after sowing the seed and subsequently produce chupons from below the fixed jorquette to form another story of fan branches from a second jorquette, a process that may be repeated. For effective pruning it can be detrimental to cut more than 25% of the tree canopy since this will only serve to starve the tree, causing it to develop fast growing, weakly attached sucker growths, which will increase maintenance cost and break off easily in strong winds (Asare and David, 2011). In terms of management, there are three kinds of pruning in cocoa, namely, shape pruning, maintenance pruning and rejuvenation pruning (Bonaparte, 1966; cited from Van der Knaap, 1955). However, in relating the art of pruning to the developmental stages of the cocoa plant Wood and Lass (1985) described two kinds of pruning: (i) formation/structural pruning and (ii) sanitary pruning. Adoption of proper pruning practices can be hampered by farmers' misconceptions about hurting the tree or eliminating branches that may potentially bear pods, even if only few.

3.6.1 Formation/structural pruning

This is done during the juvenile stage of the plant and also in mature plants. The objective is to adapt the plant's habit of growth so as to create a productive and efficient cocoa field that allows easy farm operations. According to Wood and Lass (1985) there are two major aspects of formation/structural pruning: (i) height adjustment of the first jorquette and (ii) control of vertical growth. Studies by Bonaparte (1966) in Ghana showed that limiting vertical growth to the first jorquette of the cocoa tree gave slightly higher yields than with unpruned trees in the early years, but from about the tenth year after planting the unpruned commenced to out yield the uniformly pruned. He argues that the second-tier and higher-order chupons compete with the branch system of the lower jorquettes for nutrients, which would tend to lower the yields obtained from the unpruned trees in the early years of fruit bearing. In addition, the apical dominance effects on the lower fan branches led to the early ageing of those branches which became senescent and fell off. Pod production on the lower fan branches of unpruned cocoa trees thus increased to a maximum followed by a decline to zero. As a result, Bonaparte (1966) suggested that it can be advantageous to remove the lower fan branches from the unpruned trees as soon as a drop in their pod production due to ageing processes. Hall (1961) suggested that the higher yields that were being obtained from the pruned trees in the early years of bearing were due to the fact that vegetative growth of the pruned cocoa was limited, whereas vegetative growth continued on the unpruned cocoa at the expense of pod production.

In many instances, the immediate effect of pruning is to reduce fruit production, because carbohydrate reserves are lost in the wood is removed and the area of leaf surface is reduced. Severe pruning should not be practiced, because it substantially reduces the hardened wood of the tree and thus the carbohydrate reserves, which would have been utilized in pod production in subsequent months. Light pruning is likely to be beneficial, but further research is still required on its effects (Wood and Lass, 1985). It has been established that among other factors, the productivity of a stand of trees is more influenced by competition for light between leaves in a crown (mutual shading) than between the crowns of plants (Kellomaki et al., 1984; Oker-Blom and Kellomaki, 1983). Light is not redistributed as are nutrients or water. If a leaf is heavily shaded, it will respire more than it photosynthesizes and hence will not be able to maintain itself above the compensation point and will consequently die. Unlike nutrients and water, where a period of plentiful supply will reduce or eliminate competition, periods of bright sunlight will not reduce competition. This makes the efficiency by which light is distributed through the canopy an important factor (Hadley and Yapp, 1992). The same authors observed that where cocoa canopies were still expanding, total canopy radiation interception was the main determinant of productivity. However, in continuous canopies it was demonstrated that yield was not so much associated with total solar radiation interception or leaf area index as radiation attenuation through the canopy. Consequently, these observations lend support to the idea that light pruning could be beneficial.

3.6.2 Sanitary pruning

Sanitary pruning is mostly done on mature cocoa plants to maintain the health and vigour of the tree. Hammond (1962) observed that sanitary pruning is remove unnecessary chupons, dead branches, mistletoes, epiphytes, climbing plants, ant nests, old and diseased, and rodent-damaged pods and over-ripe pods. This helps to reduce humidity and pest and disease load within and below the canopy.

3.7 Mulching in cocoa

A great deal can be done to restore cocoa soils by the use of mulches where these are readily available. Among the mulching materials that are suitable are cut-bush, cocoa husks, straw, chopped leaves and grass (Asare and David, 2011). Conservation of soil moisture by mulching and weed control is likely to have beneficial effects on the growth of young cocoa where the canopy is still open (Oppong et al., 1998). Protection of the soil, the maintenance of the right atmospheric conditions near its surface and the conservation of the crumb soil layer with its attendant cover of decomposing litter, should be the essential aims in attempting to rehabilitate old cocoa lands. Urquhart (1961) reported that among the dry vegetable waste products that might be used as mulch for cocoa are sawdust, wood shavings, stover and sugar cane bagasse. Where these may be obtained readily and cheaply, they are highly beneficial when applied in layers, three to six inches thick and allowed to rot down, and when appropriate and timely replenishment is made so as to maintain a constant covering. Pen manure and other organic manures, when available and affordable, may also be used, not only as mulch but also as nutrient suppliers. Manuring of cocoa as a general practice is only common in a few parts of the world. Ceylon (Sri Lanka) uses both farmyard and artificial manure whereas plantations in Grenada and Trinidad apply mulches.

Urquhart (1961) noted that the use of mulches in cocoa plantations is mainly to provide a suitable medium for the development of the cocoa root system by maintaining a porous layer, which is well aerated and retentive of water. In addition, mulches supply variable but often considerable amounts of plant nutrients, namely nitrogen, phosphate and potassium as well as other essential elements including zinc and copper. Moreover, these nutrients are liberated slowly but continuously by mulches, which, in contrast with fertilizers supply large quantities of nutrient at one time. When mulches and fertilizers are applied adequately, eroded and deteriorated cocoa soils may be rapidly rejuvenated and the growth and vigour of the trees increased so that the normal cycle of nutrient circulation involving the accumulation and decomposition of litter is soon reinstated. Eventually, the amount of mulch dressings might deliberately be diminished and the quantities of fertilizers reduced and adjusted so as to make good the losses brought about by the removal of the crop, thus reducing the cost of mulching which is an expensive operation. Despite the numerous benefits of mulching, the practice has achieved limited application in cocoa establishment. This is mainly due to the high cost and drudgery and labour requirements of establishing, transporting and applying large quantities of organic mulching materials.

4 Cocoa farm rehabilitation

4.1 Stages of the cocoa farm

A cocoa tree when given the right conditions can be productive for more than eight decades. However, studies have shown that annual yields can greatly reduce after 25 years of growth (Mahrizal et al., 2013). The gradual nature of annual yield losses over time makes it difficult for smallholder farmers to time and implement rehabilitation methods that will improve the value of their investment (Mahrizal et al., 2013). In their description of cocoa farm rehabilitation, Mahrizal et al. (2013) and Asare and David (2011) described cocoa farms as having a production life cycle that spans through four stages: (1) the young age, which is an early period of no yield, normally occurring between years 1 and 3. During this period the cocoa is planted among a combination of food crops, fruit trees and other shade tree species. The food crops provide a source of income for the farmer and also serve as vegetation cover protecting the soil and also acting as temporary shade cover for the young cocoa seedlings, while the fruit and timber trees provide multipurpose functions including serving as permanent shade cover; (2) the mature stage that is a period of increasing yield at an increasing rate, which occurs from year 4 to year 13. During this period the closed canopy (mutual shading) of the cocoa trees makes it impossible for food crops to thrive underneath the cocoa canopy. This period is critical in terms of economic pod production and so farmers tend to reduce both above- and below-ground competition from associated plant species in the system to allow for maximum production; (3) the adult stage, which is a period of increasing yield at a decreasing rate, which occurs from 14 to 25 years of the plantation. During this stage there will be the need for a lot of investment into soil fertility and disease and pest management; and (4) old age, which is a period of decreasing yields, which occurs beyond 25 years of establishment and involves cocoa trees that are past their yield prime.

4.2 What is cocoa farm rehabilitation?

Cocoa farm rehabilitation or rejuvenation can refer to any method or scheme designed to breathe new life into an unhealthy stand of cocoa, which is potentially productive but has been hindered due to the use of poor planting materials, pests and diseases attack, decline in soil fertility and lack of appropriate vegetation cover (Ahenkorah et al., 1979). In a broader context, rehabilitation can also refer to the planting of improved planting materials in the form of hybrid seedlings or grafted materials, the application of fertilizer and other agrochemicals at the recommended dosages, good weeding regime and the planting of compatible associated plant species as shade cover and for soil fertility improvement. Hence, cocoa rehabilitation encompasses both improvement of cocoa trees and also land recycling due to soil fertility restoration and appropriate vegetation cover. In order to rehabilitate a cocoa plantation there are three methods, which are discussed elaborately in the following sections.

5 Methods for cocoa farm rehabilitation

Rehabilitation begins when a farmer is able to quantify annual losses due to attributable factors as mentioned above. It has been demonstrated by researchers that maximum cocoa yields are attained between the ages of 15 and 25 years of cocoa tree establishment (Obiri et al., 2007; Montgomery, 1981), after which yield declines gradually then rapidly until the trees no longer bear fruit even though there are cases where cocoa trees have been productive beyond 25 years (Mahriz et al., 2013). While it is true that some cocoa trees can be profitable up to 50 years, Ahenkorah et al. (1987) provides evidence that yield declines after 25 years in no shade, medium shade and high shade scenarios (Fig. 3). It has been established that keeping cocoa trees after the adult age into the old age will lead to drastic reduction of yields and a consequent income loss. Hence, farmers need to know when and how to decide what percentage of trees to replace to maximize the present value of their annual income.

To determine the productivity of a cocoa tree Asare and David (2010) suggested that if a cocoa tree produces less than 25 pods per year (i.e. less than 1 kg dry cocoa beans

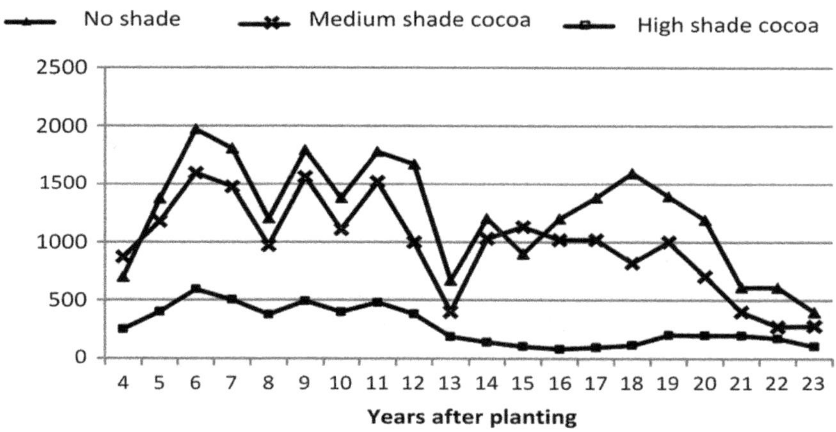

Figure 3 Twenty years' results from a shade and fertilizer trial on Amazon cocoa (*Theobroma cacao*) in Ghana. Ahenkorah, et al. (1987).

per year per tree), then there is the need for the farmer to consider rehabilitation. Yields per tree in a year can be classified as good (25 or more pods), average (15–24 pods) and poor (14 or less pods). The authors suggest that farmers who have unproductive trees (producing 10 or less pods per year) should urgently rethink what they can do to improve their productivity. Unproductive trees can be identified by selecting at least 50 trees at random and counting the number of pods per annum. In the process the individual trees could be grouped according to the number of pods observed as above 25 pods, between 15 and 24 pods, 9–14 pods, 1–8 pods and 0 pod. Per cent of trees observed in each pod count group could be estimated by calculating the number of trees observed as a percentage of the number of total number of trees counted as below: For example, if a farmer observed 10 trees out of 50 trees with pods between 9 and 14, divide 10 by 50 and multiply by 100.

Percentage of observed trees with poor pod production= (trees observed with average production/Total number of trees surveyed) × 100

$$= (10/50) \times 100 = 20\%$$

Extrapolating this to the entire field could mean replacing 20% of the old trees with new improved planting materials.

Based on this, Asare and David (2011) proposed three rehabilitation methods, which include but are not limited to under planting between old cocoa trees, gradual replanting and complete replanting of old cocoa trees. These three methods have also been suggested by Lass (2001).

5.1 Under planting

This method of planting is the best when trees are old (over 26 years). Under this approach, old and unproductive trees are carefully removed from the farm and new improved cocoa seedlings are planted under the cocoa trees. In this way the remaining cocoa trees in the field provide shade to the newly planted trees. This method is not recommended where CSSV is endemic. In order to implement this, there is the need to identify and remove all diseased/old and unproductive cocoa trees based on annual yields obtained. By so doing farmers still get income from existing trees. There is no requirement for additional land area, which makes it less expensive for smallholder farmers with limited cash resources.

In the short-term, under planting seems attractive to the farmer because a certain level of income would be maintained for the duration of the rehabilitation process. Are (1971) reported that under planting produced the best result in Nigeria in terms of yield. However, this was not confirmed in Ghana (Ampofo and Osei-Bonsu, 1988). In cocoa swollen shoot disease endemic areas, any rehabilitation in which old cocoa trees are left in the field is not advisable since rapid loss of trees through CSSV attack makes this method of rehabilitation unprofitable in the long-term (Ampofo and Osei-Bonsu, 1988).

5.2 Gradual replanting

Gradual or partial replanting is a rotation process in which the field is compartmentalized to allow for a certain proportion of the field to be gradually replanted annually (Asare and David, 2011; Lass, 2001). Partial planting can be done in two modifications. The first modification enables the inclusion of food crop cultivation in a system of strip cultivation known as a 'box hedge' (Vaast, 1988; Decazy et al., 1988). With this method, old cocoa is completely removed and improved planting material planted in double row hedges. The spaces between the hedges are used for growing food crops. The cultivation of food crops in association with cocoa performs both service functions (temporary shade) and production (food) functions during the establishment phase. This is a good technique for replanting cocoa (Decazy et al., 1988; Vaast, 1988; Osei-Bonsu et al., 1996; Frimpong et al., 1996).

However, studies in the Côte d'Ivoire indicated significant depletion of phosphorus under food crops, requiring compensatory fertilizer for food crop stabilization (Lechenaud, 1988). In the second modification, partial felling is done by taking out between 6 m and 12 m strips of old cocoa trees and leaving the intervening strips between them standing. This method removes half the old cocoa stock over the whole plantation. The retained strips are cut down 4 years later as the young replanted trees start to come into production. This has the advantage of maintaining an income for the farmer for the duration of the rehabilitation process (Vaast, 1988). For this method to be successful in cocoa swollen shoot disease endemic areas, it is recommended that a CSSV immune crop, such as *Cola nitida*, oil palm or citrus, is planted as a barrier between the strips of young and old cocoa (Ollennu et al., 1997).

Partial replanting is often more realistic in the context of a smallholder farmer. It allows for more efficient use of existing labour and other resources over a longer period, generally four years, although it requires better planning and multitasking across the various portions of the farm by the farmer. Other advantages include a continuous stream of income to the farmer from existing trees and no additional land requirement. Nevertheless, accumulated diseases and pests could affect newly planted materials and pose a problem. Gradual replanting is not recommended where CSSV disease is a problem. Furthermore, since soil fertility may be a problem in old farms, there may be a need for fertility augmentation.

5.3 Complete replanting

This third method involves clearing all plant species including cocoa from the field and re-establishing cocoa. However, re-establishing cocoa on impoverished lands that had previously carried the crop has proven to be difficult due to diminished soil fertility and inadequate vegetation cover to provide shade for young cocoa (Adams, 1962; Ayanlaja, 1983). The soil fertility problem can be addressed through the use of organic and inorganic fertilizers or through the use of a managed tree fallow system, which uses trees with the potential to restore degraded lands. In this regard the deliberate planting of fast growing trees, shrub and herbaceous legume species to improve soil fertility has been proposed by Anim-Kwapong (2003). A number of shrubs and tree legumes including *Flemingia macrophylla* and *Senna siamea* have been evaluated for use as temporary shade and for the amelioration of degraded cocoa soils in Ghana (Oppong et al., 1996).

Since the use of thinned secondary forest is the common practice and may be preferred to planted shade for economic reasons or for the type of shade it provides (Sanchez et al., 1988), a strategy based on the manipulation of native tree species in a successional sequence has been proposed for the rehabilitation of degraded cocoa lands and ultimately for the provision of shade for replanting cocoa (Anim-Kwapong and Teklehaimanot, 1995).

In areas where cocoa swollen shoot disease is prevalent, it helps to disrupt the cycle of disease transmission by removing old trees to prevent re-infection of young cocoa. However, this is costly, laborious and entails a loss of income to the farmer (Asare and

David, 2011). To avoid re-infection of newly planted cocoa seedlings with the CSSV and to improve soil fertility, the field can be fallowed for at least three years before replanting. This method is best applied in farms where most cocoa trees are unproductive due to old age or pest and diseases and there is not much the farmer can do to make the farm more productive. Although complete removal of derelict cocoa and replanting *in situ* allows for the introduction of improved planting material and may prove to be economically advantageous in the long-term, it will result in a complete loss of income for at least three years (Ampofo and Osei-Bonsu, 1988).

6 Case study: farmers' preferred rehabilitation methods in Ghana

6.1 Background

Cocoa production is the major economic activity for over 800,000 households in Ghana, where around 6.3 million Ghanaians (around 30% of the total population) depend upon cocoa production for their livelihood. Limited innovation underlies the poor productivity performance of the Ghanaian cocoa sector over the past 20 years. For instance, there has been limited adoption of the hybrid cocoa varieties developed by the CRIG (Padi and Owusu, 1998; Asare, 2005). Furthermore, migratory extensive cocoa farming has resulted in deforestation and biodiversity loss in the country. It is for this reason that the government together with other development partners are exploring good rehabilitation approaches to restore productivity and income of old cocoa farms. It is in this context that a project entitled, 'Cocoa-Eco Project', was initiated as a collaboration between the International Institute of Tropical Agriculture and SNV and Kuapa Kokoo (Farmer Associations) to assess three alternative cocoa rehabilitation regimes (i.e. under planting, gradual replanting and complete replanting) and their potential costs and benefits to enterprises within ten business development centers (BDC) towards providing realistic recommendations and investment strategies for farmers in these clusters.

6.2 Methodological approach

The study used cross-sectional survey data from four cocoa-producing regions, namely Ashanti, Western, Brong Ahafo and Central regions of Ghana, where a total of 10 BDC of Kuapa Kokoo cooperative farmers are located. The four regions constitute over 90% of Ghana's cocoa production and exports. The corresponding 10 BDCs were located in New Edubiase, Antoakrom, Offinso, Agona, Asumura, Goaso, Dadieso, Enchi, Wasa Akropong and Twifo Praso. A stratified random sampling of 245 farm households was drawn from the 10 BDCs for the purpose of this study. Data was collected on the three key rehabilitation methods discussed in this chapter. This was used to estimate farm enterprise budgets for the 2014/2015 production season, which served as the reference year for the survey.

The survey was implemented from October to November 2015 and involved administering a semi-structured questionnaire to a stratified random sample of 222 targeted beneficiary farmers of the project, who had received hybrid cocoa seedlings for rehabilitating their ageing cocoa farms (based on their declared available land sizes for the purpose) at the commencement of the project. Targeted households were randomly sampled, proportionate to their size in each of the 10 BDC villages. A total of 222 out of the 245 farmers successfully participated in the survey. The questionnaire elicited information on the socio-economic characteristics of sampled households, cocoa seedling planting methods employed and the corresponding field characteristics of the farms along with their corresponding plot-level input–output data for both the 2014–15 and 2015–2016 production seasons as well as an assessment of seedling survival rates of seedlings planted or replanted in the field. The one-on-one survey was complemented with field observations, selected focus group discussions in the various BDCs as well as secondary data to develop and validate representative farm data for each of the three planting/replanting regimes. Cross-sectional data for the reference season was then discounted over a 20-year economic productive life span of cocoa trees in Ghana.

6.3 Data analysis

Quantitative data from the field survey were encoded into Microsoft Excel and loaded into IBM SPSS Statistics Version 21 for statistical analysis. Descriptive and/or inferential statistics were used to characterize the socio-economic parameters of the sampled households and the cocoa planting and replanting methods adopted by the various project beneficiaries. The one-way ANOVA and chi-square test (χ^2) were, respectively used to gain insights into major differences in scale-variable household characteristics and categorical variable scores between the four regions and three planting/replanting methods adopted by respondents for the purpose of this study. Clustered results of sample mean for the three rehabilitation methods of the sampled households showed that only 30% (Fig. 4) of them adopted either under planting, gradual replanting or complete replanting.

The data from the relatively small sample size of adopters of the three rehabilitation regimes was used to develop a representative farm household typology using a standard cost–benefit analysis based on farm enterprise budgets and secondary data of alternative farm production activities to estimate the opportunity costs of the three rehabilitation regimes using three economic decision criteria, namely net present value (NPV), benefit–cost ratio (BCR) and internal rate of return (IRR) (Gockowski et al., 2013). Input–output data at the plot level included information on individual farm characteristics; labour quantities; and agrochemical application levels, crop yield parameters and quantities sold.

For evaluating accurately the NPV returns to the hypothetical cocoa planting/replanting regimes, extant counterfactual production systems and empirical data were needed on the agronomic relationships between age of the tree stock, yields, shade, fertilizer response, and pests and diseases. Farmers for both the hypothetical and counterfactual scenarios were assumed to follow best practices introduced as part of the project intervention and as a result the assumed yields were above national averages. Best practices in the context of this chapter included recommended levels of fertilizer, use of integrated crop and pest management practices, and minimum shade levels for cocoa. Moreover, modelling the returns to a perennial crop such as cocoa is complicated by the dynamics of yield over time as the tree stock ages. On the basis of raw data from shade-yield curves estimated

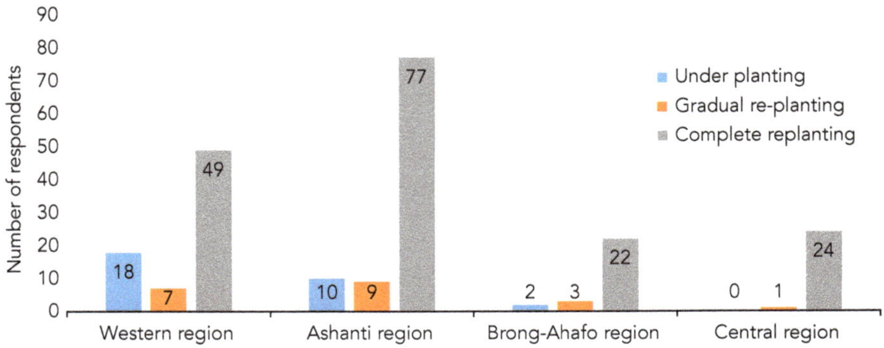

Figure 4 Type of planting for new cocoa seedlings by respondents per region.

from average annual yields of longitudinal field age-yield trial data spanning over two decades by Ahenkorah et al. (1987) and modified and adapted by Gockowski et al. (2013).

According to the authors, cocoa trees over 20 years old on farms in Ghana may have long passed their economic bearing age. Moreover, Ofori-Bah and Asafu-Adjaye (2011) confirmed that the ageing cocoa farming population could reduce farm technical efficiency. A 20-year age-yield profile (Fig. 3) was assumed as the cut-off point, particularly for full-sun cocoa production systems that will be typical for the complete replanting scenario, which are expected to produce suboptimal yields after two decades and would require replanting. Based on the three economic decision criteria of projected costs and revenues earlier highlighted, a conservative baseline scenario of cocoa production systems assuming no planting/replanting (though hardly is the case in practice) is then compared with the three alternative planting/replanting regimes scenarios being implemented in this project. Analysis was conducted in MS Excel spreadsheet scenario analysis.

6.4 Results and discussions

The field survey conducted with the 222 sampled farm households to assess three alternative cocoa rehabilitation regimes, viz. under planting, gradual replanting and complete replanting, with respect to potential costs and benefits to their enterprises showed that complete planting was the preferred planting/replanting method adopted by majority of farmers, as represented by 78% of the total sampled households (Fig. 4). This could be attributed to the ease and uniformity with which farm management practices can be adopted under such a system. Another consideration for this choice was that most cocoa farms may have had overaged trees yielding below the economic optimum. The second preferred option was under planting as represented by 13% of the sampled households with gradual planting being the least preferred accounting only for 9% of the sampled households. At the plot level, however, only 30% of sampled households adopted one of the three major planting/replanting regimes for the 2014/15 reference production season (Fig. 4, 5).

An assessment of the survival characteristics of planted cocoa trees in the 2014/2015 production season is presented in Table 1 based on a one-way ANOVA analysis that allows for a comparison of selected parameters across the three rehabilitation regimes.

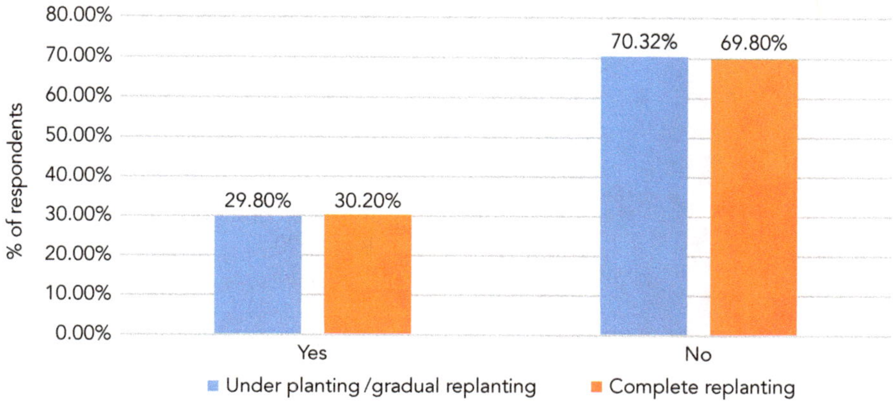

Figure 5 Proportion (%) of respondents engaged in under planting and gradual replanting.

Table 1 Farm characteristics and seedlings condition of sampled households by planting method (one-way ANOVA)

	Parameters	Number of cocoa seedlings available to farmer	Land area of hybrid cocoa seedlings established with seeds (ha)	Number of hybrid cocoa seedlings transplanted in 2014	Number of hybrid cocoa seedlings survived	Number of dead seedlings	Survival rate (per unit fraction)
Under planting (*N* = 30)	Mean	460.03	0.48	48.75	33.47	16.87	0.59
	SD [a]	289.66	0.33	58.49	58.04	15.33	0.17
Gradual replanting (*N* = 20)	Mean	422.50	0.41	39.67	25.64	16.73	0.70
	SD	180.99	0.18	53.26	27.74	28.33	0.16
Complete replanting (*N* = 172)	Mean	449.01	0.42	33.33	19.91	12.12	0.61
	SD	159.40	0.18	26.28	19.92	13.04	0.22
Total (*N* = 222)	Mean	448.10	0.42	35.43	21.84	13.00	0.61
	SD	183.31	0.21	33.58	27.22	14.96	0.21
	F-value	0.259	1.138	1.613	1.777	1.039	0.383
	P-value	0.772	0.322	0.203	0.173	0.357	0.683

[a] SD denotes standard deviation of corresponding variable.

Neither the number of seedlings planted nor seedling survival rates were statistically significant between the rehabilitation schemes. The former is expected, since the computation of number of seedlings planted is based on the land availability and farmers' choice of planting density, both of which are typically predetermined by farmers prior to rehabilitation interventions.

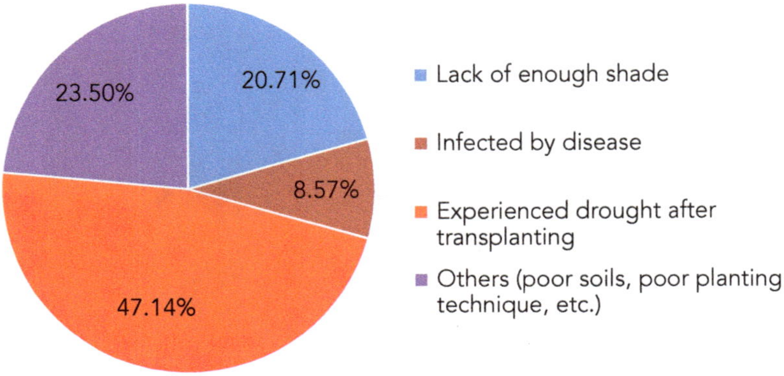

Figure 6 Reasons mentioned by respondents as being the cause of death of planted cocoa seedlings.

Nonetheless, complete replanting experienced relatively the lowest number of seedling deaths, amounting to an average of 12.12 seedlings per 0.4 ha (1 acre). Seedling survival rates consequently ranged from 0.59 (with a standard deviation of 0.17) for under planting to 0.70 (with a standard deviation of 0.16) for gradual planting, *albeit* not statistically significant among the three rehabilitation methods. The reasons outlined by farmers for the observed death of seedlings transplanted were differences in competition for nutrients, environmental factors such as precipitation (Fig. 6) that are essential for seedling establishment and uneven shade. The pie chart illustration shows that drought was the major reason leading to seedling deaths as indicated by 44% of the sampled households. This is followed by poor planting techniques/soils (23.5%), lack of adequate shade for cocoa at initial stages of tree establishment (20.71%) and finally, disease infestation (8.57%).

6.4.1 Cost–benefit analysis of three planting methods at plot level

Based on the scenario analysis under varying assumptions of cultural management practices applied for the cost–benefit analysis at plot level, complete replanting offers higher returns over time as depicted by a cost–benefit ratio of 1.24 and an IRR of 33% which is well above the bank lending rate for agriculture (Table 2). The standard cocoa farming practice with a conservative no-planting regime is clearly not profitable and results in high farm production costs for management practices.

In terms of rehabilitation regime choices, the second best enterprise to complete replanting is gradual replanting. Given, the use of a representative farm for varying agro-ecologies in the modelling analysis, however, the actual rates/time lag for gradual replanting needs to be determined to test the robustness of the cultural/farm management assumptions made for much specific recommendations.

Table 2 Computed economic indicators of scenarios on premium price and assumed environmental benefits per annum

Crop/farming system	Yield of main crop in kg ha^{-1} year^{-1}	Total production cost+ (¢GHC ha^{-1})	Total revenue+ (¢GHC ha^{-1})	Economic profitability indicators		
				NPV (¢GHC ha^{-1})	BCR	IRR (%)
Standard cocoa farm (no planting/ replanting regime)	602.34	1,329.15	1,747.33	725.12	1.04	19.0
Under planting	501.28	722.15	1,219.36	666.19	1.09	24.0
Gradual replanting	598.28	716.73	1,243.61	681.13	1.12	27.0
Complete replanting	673.23	782.34	1578.45		1.24	33%

+ Production and revenue figures are annualized assuming two production cycles per year.

7 Conclusion and future trends

This chapter has focused on good agronomic practices in cocoa systems. Cocoa is a crop that can be established from seeds sown at stake, transplanted from seedlings raised in polythene bags in a nursery or from bare-root seedling raised on nursery beds. Temporary and permanent shade are planted following clear felling, or by thinning the forest. The cultivation of food crops in association with cocoa performs both service (temporary shade) and production (food) functions during the establishment phase and this is reported to be a promising technique for replanting cocoa.

Planting cocoa at the recommended spacing will give the best economic return and optimal cocoa yield per unit area. However, there are many different opinions concerning the spacing of cocoa and this is conditioned by local climatic and edaphic factors. The shade requirement of cocoa has also been questioned and the practice has been subject to wide experimentation. Although shade limits cocoa yield, it provides several agro-ecological and ecosystem benefits. Yield potential, competition for water and pest/disease incidence depends on site conditions and are fundamental issues to be considered in determining the shade regime to be used in cocoa. Despite the controversy, shade is invariably recommended for the establishment of cocoa. Weed management is critical in newly established cocoa plantations and the proper selection and management of shade tree species can reduce labour input and weeding cost considerably.

To sustain the cocoa industry in the face of dwindling natural forest and the general decline in production, there is the need to explore alternatives such as rehabilitating abandoned cocoa farms by replanting cocoa in the old growing areas. In an assessment of three rehabilitations methods, results show that complete replanting is the most common planting regime adopted by farmers within the 10 BDCs in Ghana as it tends to offer the highest economic benefit in relation to farmer investment. This is followed by gradual replanting with under planting yielding the lowest returns to investment, albeit it provided a much better cost–benefit ratio when compared with a hypothetical scenario of no replanting of old cocoa farms. However, complete replanting using new and improved high-yielding planting materials is the only viable means of re-establishing cocoa in the CSSV endemic areas.

New studies are required to determine the optimal cultural and/or good agricultural practices under varying agro-climatic conditions to ascertain which methods would work best under specific conditions while allowing for testing the robustness of the current assumptions made in the analysis thereof.

8 Where to look for further information

Literature cited for this work is documented in the next section. However, in the likely event that readers will want to have further access to cocoa literature for academic or other purposes, the following repositories could be helpful: the annual reports on cocoa activities and proceedings from the International Cocoa Conferences housed in resident libraries at the Cocoa Research Institutes of Ghana and Nigeria, National Agricultural Research Centre (CNRA) in Côte d'Ivoire, and the Institute of Agricultural Research for Development (IRAD) in Cameroon. More information could also be found on the on-line libraries of the International Cocoa Organization (ICCO), the World Cocoa Foundation, the International Institute of Tropical Agriculture (IITA) and the World Agroforestry Center (ICRAF) just to name a few.

9 References

Adams, S. N. (1962). Cocoa: Soils and Manuring. In: Wills, J. B. (ed.), *Agriculture and Land Use in Ghana*, pp. 266–273. Oxford University Press.

Adeyemi, A. A. (1987). Improvement of farming systems for tree crops (1) Cocoa/arable crops intercropping experiment. Annual Report of Cocoa Research Institute of Nigeria 1987, pp. 67–8.

Ahenkorah, Y., Halm, B. J., Appiah, M. R.; Akrofi, G. S. and Yirenkyi, J. E. K. (1987). Twenty years' results from shade and fertilizer trial on Amazon cocoa (Theobroma cacao L.). *Experimental Agriculture* 23(1): 31–9.

Ahenkorah, Y., Halm, B. J. and Akrofi, G. S. (1979). Some agronomic factors affecting cocoa rehabilitation in Ghana. Proceedings of the 5th International Cocoa Research Conference, Ibadan, Nigeria, pp. 199–203.

Ahenkorah, Y., Akrofi, G. S. and Adri, A. K., (1974). The end of the first shade and Manurial experiment at the Cocoa Research Institute of Ghana. *Journal of Horticultural Science* 49: 43–51.

Alvim, R., Lima, F. J. M. P. and Alfonso, E. C. A. (1981). Possibilidade de transplante do cacaueiro com raizes nuas. Proceedings of the 8th International Cocoa Research Conference, Cartegena, Colombia, pp. 21–6.

Amoah, F. F., Nuertey, B. N., Baidoo-Addo, K., Oppong, F. K., Osei-Bonsu, K. and Asamoah, T. E. O (1995). Underplanting oil palm with cacao in Ghana. *Agroforestry Systems*, 30: 3–15.

Amoah, F. M., Opoku-Ameyaw. K., Osei-Bonsu, K. and Oppong, F. K. (2000). The effect of seedling age on the survival rate of transplanted bare rooted cocoa seedling. Proceedings of the 13th International Cocoa Research Conference, October Kota Kinabalu, Malaysia, pp. 14–18.

Ampofo, S. T. and Osei-Bonsu, K. (1988). Models for rehabilitating small scale cocoa farms in Ghana. Proceedings of the 10th International Cocoa Research Conference, Santo Domingo, Dominican Republic, 17–23 May 1987: 51–5.

Anim-Kwapong, G. J. and Teklehaimanot, Z. (1995). Reclamation of degraded cocoa lands using Albizia zygia. *Land Degradation and Rehabilitation* 6: 109–23.

Anim-Kwapong, G. J. (2003). Potential of some Neotropical Albizia species as shade trees when replanting cacao in Ghana. *Agroforestry Systems* 58: 185–93.

Are, L. A. (1971). Cocoa Rehabilitation in Nigeria. Proceedings of the 3rd International Cocoa Research Conference. Accra, Ghana, 1969: 29–36.

Asare, R., Asare, R. A., Asante, A. W., Markussen, B. and Ræbild, A. (2016). Influences of shading and fertilization on on-farm yields of cocoa in Ghana. *Experimental Agriculture*, pp. 1–16. doi:10.1017/S0014479716000466.

Asare, R., Afari-Sefa, V. and Muilerman, S. (2016). Access to improved hybrid seeds in Ghana: im-plications for establishment and rehabilitation of cocoa farms. *Experimental Agriculture*, pp. 1–13. doi:10.1017/S0014479716000247.

Asare, R. and David, S. (2011). Good agricultural practices for sustainable cocoa production: a guide for farmer training, manual no. 1: planting, replanting and tree diversification in cocoa systems. Development and Environment Series 13–2010. Forest and Landscape Denmark.

Asare, R., Afari-Sefa, V., Gyamfi, I., Okafor, C. and Mva, M. J. (2010). Cocoa seed multiplication: an assessment of seed gardens in Cameroon, Ghana and Nigeria. STCP Working Paper Series No. 11.

Asare, R., and Asare R. A. (2008). A participatory approach for tree diversification in cocoa farms: Ghanaian farmers' experience. STCP Working Paper Series No. 9. International Institute of Tropical Agriculture, Accra, Ghana.

Asare, R. (2006). Learning about neighbour trees in cocoa growing systems – a manual for farmer trainers. Development and Environment Series 4–2006. Forest and Landscape Denmark.

Asare, R. (2005). Cocoa Agroforests in West Africa: a look at activities in the farming systems. Forest and Landscape Working Papers No. 6. Forest and Landscape, Denmark.

Asomaning, E. J. A., Kwakwa, R. S. and Hutcheon, W. V. (1971). Physiological studies on an amazon shade and fertilizer trial at the Cocoa Research Institute of Ghana. *Journal of Agricultural Science* 4: 47–64.

Ayanlaja S. A. (1983). Rehabilitation of cocoa (Theobroma cacao L) in Nigeria. Major problems and possible solution. I. Causes of difficulty of seedling establishment. *Plant and Soil* 73 (3): 403–9.

Beer, J., Muschler, R., Kass, D. and Somarriba, E. (1998). Shade management in coffee and cacao plantations. *Agroforestry Systems* 38: 139–64.

Bonaparte, E. E. N. A. (1966). Pruning studies on Amazon and Amelonado Cocoa in Ghana. *Tropical Agriculture* 43: 25–34. Published by Butterworths, London.

Bonaparte, E. E. N. A. (1966). Preliminary trials on chemical weed control in cocoa. *Ghana Journal of Science*, 6: 92–6.

Bonaparte, E. E. N. A. and Toseafa, D. W. K. (1979). Cocoa/Food crop experiment. Rep. Cocoa Res. Ghana 1979/80–1980/82, 15–16.

Byrne, P. M. (1972). Cocoa Shade, spacing and fertilizer trials in Papua New Guinea. Proceedings of the. IV International Cocoa Research Conference, pp. 275–85.

Charter, C. (1953). Cocoa soils. Good and bad. WACRI Report, Tafo, Ghana.

Corven, J. M. (1993). Asociacion de cultivos en cacao: aspectos economicos. In: Phillips-Mora, W. (ed.), *Sombras Y Cultivos Asociados con Cacao*. Series Tecnica Informe Tecnico No. 206. CATIE, Turrialba, Coata Rica, pp. 11–18.

Dakwa, J. T. (1976). The effects of shade and NPK fertilizers on the incidence of cocoa blackpod disease in Ghana. *Ghana Journal of Agricultural Science* 9(3): 179–84.

Decazy, B., Delsol, E., Paulin, D. and Coulibaly, N. (1988). La regeneration d'une vieille cacaoyere par le systeme en 'Haies Fruitieres'. Proceedings Of the 10th International Cocoa Research Conference. Santo Domingo, Dominican Republic, 17–23 May 1987: 71–6.

De Souza L., Garcia, J. R., Muller, M. W. and Midlei, P. R. M. (1981a). Transplanto de mudas de cacaueiro com raizes nuas. Proceedings of the 8th International Cocoa Research Conference, Cartagena, pp. 65–8.

De Souza, L., Carcia, J. R., Muller, M. W. and Midlei, P. R. M. (1981b). Influencia de idade de muda no desenvolvimento do cacaueiro aposo transplante. Proceedings of the 8th International Cocoa Research Conference, Cartagena.

Egbe, N. E. and Adenikinju, S. A. (1990). Effects of intercropping on potential yield of cacao in south western Nigeria. *Café, Cacao The* 34(3): 142–6.

Fagan, H. J. and Topper, B. F. (1988). An agronomic research on cocoa in Jamaica. 1950–1980 and current research trends. *Trop. Agric. (Trin.)*, 65: 291–4.

Famaye, A. O., Adeyemi, E. A. and Olaiya, A. O. (2003). Spacing trials in cocoa/kola/citrus interccrop. Paper presented at the 14th International Cocoa Research Conference, Accra, Ghana. 13–18 October.

Frimpong, E. B., Osei-Bonsu, K. and Akuoko, S. (1996). Cassava as temporary shade tree in cocoa establishment in degraded areas of Ghana. Rep. Cocoa Res. Inst. Ghana, 1994/95, pp. 24.

Gerritsma, W. and Wessel, M. (1994). Calculated yield trends of cocoa in different countries. In: Malaysian International Cocoa Conference, Kuala Lumpur. Malaysia, 20–21 October 1994.

Gockowski, J., Afari-Sefa, V., Sarpong, D. B., Osei-Asare, Y. B. and Agyeman, N. F. (2013). Improving the productivity and income of Ghanaian cocoa farmers while maintaining environmental services: what role for certification? *International Journal of Agricultural Sustainability* 11(4): 331–46.

Gordon, J. (1976). Pests and diseases. In: Simmons (ed.), *Cocoa Production: Economic and Botanical Perspective*. Praeger Publishers, New York, pp. 30–43.

Hadley, P. and Yapp. J. H. H. (1992). Measurement of physiological parameters with respect to yield. International Workshop on Conservation, Characterization and Utilization of cocoa Genetic Resources in the 21st century, CRU, Trinidad, 1992, pp. 121–36.

Hall, T. H. R. (1961). 'A note on the pruning of Amazon cocoa in Ghana' *Trop. Agriculture, Trin*. 38: 333–6.

Hammond, P. S. (1962). Cocoa Agronomy. In Wills, J. B. (ed.), *Agriculture and Land Use in Ghana*. Oxford Univ. Press: Oxford, pp. 252–6.

Hutcheon, W. V. (1981). The swollen shoot research project at the Cocoa Research Institute of Ghana. Section V. Physiological studies, Tafo, Ghana, pp. 67–115.

Idowu, O. L. (1996). Using food crop/cacao intercropping as insect IPM tool for cocoa production in Nigeria. Proceedings of the 12th International Cocoa Research Conference. Salvador, Bahia, Brazil, 17–23 Nov. 1996.

Iweama V. O. (1963). A cocoa establishment trial using shade plants of economic value. Report of the West African Cocoa Research Institute (Nigeria) 1962/63.

Kahar, S., A. B., Abdullah, O., Denamay, G.and Jelani, B. M. (1991). Early performance of high density cocoa plantings in different locations in Peninsular Malaysia. Proceedings of the International Cocoa Conference: Challenges of the 90s. 25–8 September 1991. Kualar Lumpur, Malaysia.

Kellomaki, S., Oker-Blom, P. and Kuulvainen, T. (1984). The effect of crown and canopy structure on light interception and distribution in a tree stand. In: Tigerstedt, P. M. A., Puttonen P. and Koski, V. (eds), *Crop Physiology of Forest Trees*. Proceedings of an International Conference on Managing Forest Trees as Cultivated Plants. Helsinki, Finland, July 23–28, 1984, pp. 107–15.

Knowal, J. N. L. (1959). The effect of spacing on the environment and performance of cocoa under Nigerian conditions. I. Agronomy. Emp. *Journal of Experimental Agriculture*, 27: 27–34.

Kolade, J. A. (1986). Influence of different densities of cocoa and oil palm on yield performance of cocoa. *Turrialba*, 36(3): 345–53.

La Anyane, S. (1963). *Ghana Agriculture: Economic development from early times to the middle of the 20th century*. Oxford University Press, London.

Lachenaud, P. (1988). La plantation en haies fruitiers une techique permettant l'association avec les cultures vivrieres et leur stabilization. Proceedings of the 10th International Cocoa Research Conference. Santo Domingo, Dominican Republic, 17–23 May 1987: 45–50.

Lass, R. A., (2001). Replanting and rehabilitation of old cocoa farms. In: Wood, G. A. R. and Lass, R. A. (eds), *Cocoa*. Blackwell Science, Iowa, pp. 210–33.

Lebaron, H. M. and Gressel, J. (1982). *Herbicide Resistance in Plants*. John Wiley and Sons, New York.

Ling, A. H. and Mainstone, B. J. (1983). Effects of burning of rubber timber during land preparation on soil fertility and growth of Theobroma cacao and Gliricia maculata. *Planter* 59: 52–9.

Mahrizal, L., Nalley, L., Dixon, L. B. and Popp, S. J. (2013). An optimal phased replanting approach for cocoa trees with application to Ghana. *Agricultural Economics* 45: 1–12.
Montgomery, P. J., (1981). Some thoughts on the life span of cocoa. *Planter* 57: 604–9.
Murray, D. B. (1955). A shade and fertilizer experiment with cacao III. In: A report on cacao research, 1954. Imperial College of Tropical Agriculture, St. Augustine, Trinidad, pp. 30–6.
Obiri, D. B., Geoff, A. B., Morag, A. M., Anglaare, L. C. N. and Cobbina, J. (2007). Financial analysis of shaded cocoa in Ghana. *Agroforestry Systems* 71: 139–49.
Odegbaro, O. A. (1977). prospects of rehabilitation Amelonado cocoa with improved varieties in Nigeria without completely replanting. Proceedings of the 5th International Cocoa Research Conference. Ibadan, Nigeria, 1975: 259–64.
Ofori-Bah, A. and Asafu-Adjaye, J. (2011). Scope economies and technical efficiency of cocoa agroforesty systems in Ghana. *Ecological Economics* 70(8): 1508–18.
Oker-Blom, P. and Kellomaki, S. (1983). Effect of grouping of foliage on the within-stand and within-crown light regime: comparison of random and grouping canopy models. *Agricultural Meteorology* 28: 143–55.
Ollennu, L. A. A., Osei-Bonsu, K., Acheampong, K. and Aneani, F. (2003). Preliminary studies of the control of cocoa swollen shoot disease by the use of immune crops as barrier. Paper presented at the 14th International Cocoa Research Conference, Accra, Ghana. 13–18 October.
Ollennu, L. A. A., Osei-Bonsu, K. and Acheampong, K. (1997). Epidemiology and Control of CSSV-Barrier cropping trial. Rep. Cocoa Res. Inst., Ghana, 1995/96, 156.
Opoku, A. A. (1967). Effects of various cover plants on perennial crops. University of Ghana Agricultural Research Station Annual Report 1966/67.
Opoku-Ameyaw K., Osei-Bonsu, K., Ofori-Frimpong, K. and Acheampong K. (2003). Biological and economic evaluation of some cocoa food crop intercropping packages at establishment. Proceedings of the 14th International Cocoa Research Conference, Accra, Ghana. 13–18 October, 2003 (in press).
Oppong, F. K., Osei-Bonsu, K., Amoah, F. M. and Acheampong, K. (1995). Effectiveness of glyphosate in controlling weeds during cocoa establishment. Proceedings of the 1st International Seminar on Cocoa Pests and Diseases, Accra, 6–10 November, 1995, 280–5.
Oppong, F. K., Osei-Bonsu, K., Amoah, F. M. and Acheampong, K. (1998). Potential use of Flemingia macrophylla as mulch for managing weeds in young cocoa in Ghana. Ghana *Journal of Agricultural Science* 31:67–72.
Oppong, F. K., Asante, E. G., Osei-Bonsu, K., Acheampong, K. and Amoah, F. M. (2003). Agronomic and economic evaluation of different weed management options in young cocoa. Proceedings of the 14th International Cocoa Research Conference, Accra, Ghana. 13–18 October, 2003.
Osei-Bonsu, K., Amoah, F. M., Oppong, F. K. and Acheampong, K. (1996). Cocoa-Food intercropping. Rep. Cocoa Res. Inst. Ghana. 1994/95, p. 25.
Osei-Bonsu, K., Amoah, F. M. and Oppong, F. K. (1998). The establishment and early yield of cocoa intercropped with food crops in Ghana. Ghana *Journal of Agricultural Science* 31: 45–53.
Osei-Bonsu, K., Acheampong, K., Amoah, F. M. Oppong, F. K. and Opoku-Ameyaw, K. (2002a). Cocoa shade variety and spacing trial, D2, Bunso. Rep. Cocoa Res. Inst., Ghana. 1999/2000, p. 38–9.
Osei-Bonsu, K., Opoku-Ameyaw, K., Amoah, F. M. and Oppong, F. K. (2002b). Cocoa-Coconut intercropping in Ghana: agronomic perspectives. *Agroforestry Systems* 55(1): 1–8.
Padi, B. and Owusu, G. K. (1998). Towards an Integrated Pest Management for Sustainable Cocoa Production in Ghana. Paper from workshop held in Panama, 3/30–4/2, 1998. Smithsonian institution. Washington, D. C.
Padi, F., Adu-Gyamfi, P., Akpertey, A., Arthur, A. and Ofori, A. (2013). Differential response of cocoa (Theobroma cacao) families to field establishment stress. *Plant Breeding* 132(2): 229–36.
Raja Harun, R. M. and Hardwick, K. (1988). The effects of prolonged exposure to different light intensities on the photosynthesis of cocoa leaves. Proceedings of the 10th International Cocoa Research Conference, 1987, Santo Domingo, pp. 205–9.

Ramadasan, K., Abdullah, I., Yeoh, K. C., Vanialingam, T. and Chan, E. (1978). Intercropping coconuts with cocoa in Malaysia. *Planter*, 54, 329–42.
Ruf, F. and Burger, K. (2001). Planting and replanting tree crops, smallholders' investment decisions[online]. Research Report of the Faculty of Economic and Business Administration, University of Amsterdam, Amsterdam, Netherlands. Available from: http://www.feweb.vu.nl/esi/bin/pdf/876.pdf [Assessed 27 January 2009].
Ruf, F. and Zadi, H. (1998). Cocoa: From deforestation to reforestation. Paper from workshop on shade Grown Cocoa held in Panama, 3/30–4/2, 1998. Smithsonian Institution. Washington DC.
Russell, T. A. (1953). The spacing of Nigerian cocoa. *Emp. J. exp. Agric.* 21: 145–53.
Sanchez, G., Kass, D., Borel, R., Bonneman, A. and Beer, J. (1990). Agroforestry Land-use systems. Pages 75–82. In: Proceedings of a special session on Agroforestry land-use systems in International Agronomy. American Society of Agronomy annual meeting. November 28–9, 1988, Anaheim, California.
Silva, I. C. and Dias, A. C. and da, C. P. (1987). Intercropping of peach palm and cacao in the Brazilian Amazon: partial results. *Revista Theobroma*, 17(2): 93–100.
Simpson, B. B. and Ogorzaly, M. C. (1986). *Economic Botany: Plants of our World*. MacGraw-Hill, New York.
Smith, E. S. C. (1981). The interrelationships between shade types and cocoa pest and disease problems in Papua New Guinea. In: Proceedings of the 7th International Cocoa Research Conference, Cameroon, 1970, pp. 37–43.
Smyth, A. J. (1975). Soils. In G. A. R. Wood cocoa (3rd edu.). Longman: London, pp. 34–47.
Squibbs, F. L. (1943). Uniformity trial on established cocoa. Report on central cocoa research station, Tafo, 1938–42, pp. 6–9.
Thresh, J. M. (1958). The spread of virus disease in cacao. Techn. Bull. 5, West African Cocoa Res. Inst., p. 36.
Thresh, J. M., Owusu, G. K. and Ollennu, L. A. A. (1988). Cocoa swollen shoot virus: An archetypal crowd disease. *Journal of Plant Diseases and Protection*, 95(4): 428–46.
Urquhart, D. H. (1961). Cocoa, 2nd ed. London: Longmans, Green.
Vaast, P. (1988). Presentation des systemes de regeneration cacaoyere en milieu paysan au Togo. Proceedings of the 10th International Cocoa Research Conference. Santo Domingo, Dominican Republic, 17–23 May 1987: 63–9.
Van Der Knaap, W. P. (1955). 'Snoei Van cacaobomen'. *Bergcultures* 24: 517–21.
Van der Knapp, W. P. (1958). Tree density and spacing in cocoa. Menara Perkebunan, Indonesia, 27: 39–47.
Van Hall, C. J. J. (1932). *Cacao*. 2nd Edition. Macmillan and Co. London.
Wood, G. A. R. (1964). Cocoa Growers' Bulletin, No. 2. Bourneville, England, pp. 16–18.
Wood, G. A. R. (1985). Establishment: temporary shade. In *Cocoa*. Wood, G. A. R. and Lass, R. A. (eds), 131. Longman Scientific and Technical, U. K. 620p.
Wood, G. A. R. and Lass, R. A. (1985). Cocoa. (4th edn.) Longman Scientific and Technical, pp. 119–20.
World Bank, 2011: Supply chain risk assessment of cocoa in Ghana [online]. World Bank Agricultural Risk Management Team Report. Washington DC, USA. Available from: http://www.agriskmanagementforum.org/sites/agriskmanagementforum.org/files/Documents/Ghana%20Cocoa%20SCRA%20Report%202011%20ARMT.pdf [Accessed 2 July 2011].

Chapter 9

Improving soil and nutrient management for cacao cultivation

Didier Snoeck and Bernard Dubos, CIRAD, UR Systèmes de pérennes, France

1 Introduction

The need to fertilise cacao (*Theobroma cacao* L.) emerged at the very beginning of the establishment of high-yielding commercial cacao plantations. Results of fertiliser trials were first published in a study conducted in Cameroon in 1910, quickly followed by Java, Trinidad and Ghana, among others (Verlière, 1981). The objective of these studies was to evaluate the effect of each nutrient supplied separately on cocoa yields. The results showed that phosphorus (P) and potassium (K) have always had positive effects on cacao yields. Although calcium (Ca) and magnesium (Mg) have also had positive effects on yields, the responses were much more variable and the quantities were dependent on the environmental condition. The variability of yield responses to nutrients pointed to the need to link the nutrients with each other or with other factors, in particular with soil, climate and topography. In the case of nitrogen (N), a significant effect on yields was rarely observed, except when the intensity of shading is reduced (Evans and Murray, 1953). There are three reasons for the contrasting effects of N (positive, neutral or negative) and its relationship to shading.

First, due to the high vigour of the hybrid cacaos, the supply of nitrogen promotes vegetative growth at the expense of the production of pods (Alvim, 1977; Bastide et al., 2003). Consequently, the positive effect of N on yields is only visible when the yield of pods is high and nutrient requirements are consequently high, which is more frequently the case when cacao is grown without shade.

http://dx.doi.org/10.19103/AS.2017.0021.13

Second, since cacao is cauliflorous, the fruits grow directly on the trunk and primary branches (at the base of the tree), while leaf production is concentrated in the apical branches (at the top of the tree). As N promotes vegetative growth, cacao uses the products of photosynthesis to produce more branches at the expense of flowering and fruiting. By increasing the amount of shade over the fruiting area, N can thus have a rather negative effect on production unless shading is controlled (Wessel, 1985).

Third, when nitrogen deficiency occurs in the soil, the need for N may emerge and it becomes a limiting factor for other nutrients, which is quite a common phenomenon in soils with low organic matter content (Snoeck et al., 2016). However, our study also showed that when cacao is grown under shade trees, recycling of organic matter increases soil N content, which improves soil quality with age, thus reducing the need for N in older plots.

Nitrogen management should therefore aim for a balance between soil poverty and shade intensity.

Based on the findings that P and K are always important and most often the only nutrients required, many governments recommend only one P–K–Ca–Mg-based fertiliser formula for their countries. The same formula may even be recommended in several other countries (e.g. Côte d'Ivoire, Ghana, Togo and Cameroon). Only a few countries (particularly, Brazil and Malaysia) have more specific recommendations, where fertiliser formulas are calculated based on critical N, P and K levels in the soil (Chepote et al., 2013; Ling, 1984; Malavolta, 1997). This simplification, which consists in using very few formulations or even only one for a whole country, goes against the principle of nutrient management of the soil–plant system, which is based on two observations:

- Each species has unique nutritional requirements and its own mechanisms underlying soil nutrient absorption. The specific nutrient requirements of cacao were deduced from the results of many trials conducted in many countries (Jadin and Snoeck, 1985).
- Each soil can only supply what it contains, and the availability of each nutrient in the soil varies with the type of soil in interaction and with the properties of the soil–root interface. In particular, pH may be a determining factor for P (retrograded forms) content and cations.

Thus, the objective of nutrition management is to correct the soil so that the cacao plant can find the nutrients it needs in optimal quantities and ratios. The fertiliser formula is determined by physical-chemical analyses of a sample of soil taken from each plot to be corrected. A plot corresponds to a map unit with the same soil and climatic conditions and homogeneous farming practices.

2 Case study: the need for one formula per plot

Because each soil has a different origin, composition and environment and each soil is subjected to different cultivation practices, the current recommendation to use a single formula for all cacao plantations is not successful. Hence, it would be ideal to have one fertiliser formula for each cultivated plot. The aim being that the soil in each plot contains optimal nutrient levels to allow the cacaos to achieve their optimal vegetative state and sustainable cocoa production.

Soils with optimum characteristics in terms of high cocoa production tend to have an average pH of 5.6–7.0, organic carbon not less than 3%, available P greater than 20 ppm, cation exchange capacity of 10–25 cmol/kg (Metson method), exchangeable K not less than 0.25 cmol/kg and (Ca + Mg) about 8–13 cmol/kg (Snoeck et al., 2016). However, there are many other factors that can modify the nutrient status of soils under cacao. The observation of soil nutrient chemical status of some selected cacao-growing soils compared to the neighbouring forest soils highlights the need for fertiliser applications (Table 1).

The variability of cacao reactions in relation to the variety of cultivated soils has been confirmed by many authors. For example, in a trial in Ghana, Appiah et al. (2000) showed that the use of the same fertiliser in a large number of plantations resulted in an overall average positive increase in cocoa yields. This average positive result enabled these authors to confirm that the choice of the formula recommended at national level is reasonably satisfactory. However, the results of a real-time detailed analysis at the level of cacao plantation level were much more diversified as only 75% of the cacao plots had actually benefited from the fertiliser, while in the remaining 25% of cacao plots, the use of the same fertiliser resulted in yield losses (Fig. 1).

Moreover, among the plots that responded positively to fertiliser, only 26% doubled their yield compared to the control plot (no fertiliser). In the other plots, the increase in yield was average to low. When the formula was not appropriate for the soil conditions, the use of fertiliser had an adverse effect, resulting in yield losses. This reinforces the idea that all soil parameters should be analysed together because, according to the principle of the limiting factor, when one nutrient is added in large quantities while others are only present in small quantities, this can counteract the effect of the added nutrient. The application of this principle to cacao was confirmed by the studies of Loué (1961) and Verlière (1981) in Côte d'Ivoire and that of Malavolta (1997) in Brazil.

To calculate the fertiliser formulation, the physical–chemical analysis of the soil must be correlated with cacao requirements by simultaneously using nutrient thresholds and nutrient balances to avoid limiting factors. To facilitate the analysis of a large number of

Table 1 Soil nutrient chemical status of some selected cacao plots compared to the neighbouring forest

Country	Nigeria		Cameroon		Ghana	
Crop	Forest	Cacao	Forest	Cacao	Forest	Cacao
pH	6.8	5.9	5.1	5.1	5.1	6.3
C org (%)	2.60	1.90	1.50	1.10	4.00	1.70
N tot (%)	0.17	0.13	0.11	0.07	0.46	0.19
P avail (ppm)	14	12	4	4	24	10
CEC (cmol·kg)	14.4	9.5	8.5	7.5	11.5	8.5
Ca (cmol·kg)	9.00	5.30	2.50	2.09	5.75	3.70
Mg (cmol·kg)	3.80	2.00	0.56	0.33	2.18	1.18
K (cmol·kg)	0.70	0.40	0.22	0.20	0.20	0.10

Sources: Nigeria: Adejuwon and Ekanade (1988); Cameroon: Snoeck et al. (2010a) and Ghana: Owusu-Sekyere et al. (2006).

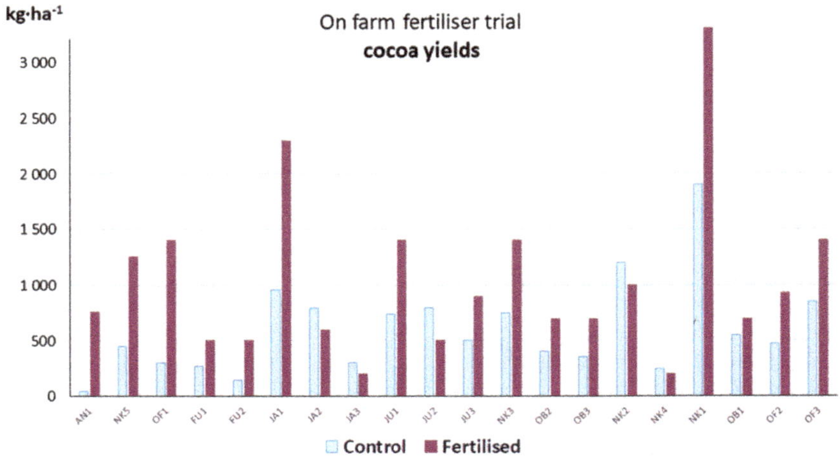

Figure 1 Evaluation of fertiliser application on some peasant cocoa farms in Ghana (after Appiah et al. (2000)).

soil samples, a software tool was developed to compute the fertiliser formulas and doses (Jadin and Snoeck, 1985). This tool was created from the results of the many long-term fertiliser trials conducted on different types of soils and in different environments in Côte d'Ivoire (Jadin, 1975). They were supplemented by greenhouse trials to better understand how major nutrients and trace elements are correlated in a given soil (Verlière, 1981). The nutrient requirements identified in these studies were interpreted from the nutrient expert's point of view to create a decision flowchart and convert it into a computer tool to aid in the diagnosis of nutrient requirements using soil analyses (Jadin and Snoeck, 1985). The thresholds of each nutrient and balances were then further adjusted using the results obtained by other authors in other countries to obtain a tool which could be used worldwide (Snoeck and Jadin, 1992). In particular, the thresholds defined in Brazil (Cabala-Rosand et al., 1975) and Malaysia (Wyrley Birch, 1987) do not differ significantly from those observed in West Africa. The tool was checked in trials in other countries: São Tomé (Jadin, 1988), Togo (Jadin and Vaast, 1990) and Ghana (Afrifa et al., 2009). The tool uses values determined and interpreted at pH close to neutral (6–7), which is appropriate for cacao species. Indeed, in the field, the first step is to correct the soil pH to ensure the nutrient steady state is also correct.

This step of soil correction makes it possible to achieve the right soil balances. It constitutes the basal application that must be distinguished from the maintenance fertilisation which consists of replacing the nutrients exported by harvesting, leaching, volatilisation and using them for the growth of the cacao itself. Indeed, if maintenance fertilisation is not used, it can lead to excessive depletion of soil reserves and thus to a marked decline of cocoa yields over the long term (Appiah et al., 1997).

Among the factors which influence soil fertility, the pedological origin of the soil has been shown to be an important parameter because it is largely responsible for the physical and chemical characteristics of soils. As early as the 1950s, Charter (1953) proposed soil suitability classifications based on the soil pedological origin for cacao cultivation in

Ghana. In Côte d'Ivoire, Verlière (1965) showed that the soils derived from tertiary sands are often deficient in N and K, soils which originate from granitic rocks are poor in P and sometimes in N and soils which originate from schist rocks are deficient in both P and K. Similarly, in Brazil, a study on the influence of the soil origin on its fertility under cacao showed a close correlation between soil classes and some nutrients. In particular, it was shown that nitisols are much richer in N, P and trace elements than latosols (Santana, 1972). Furthermore, the texture, structure and soil depth, which are important to ensure good root development of cacao, also depend on the pedological origin of the soils (Kummerow and Alvim, 1981).

To illustrate the importance of the hypothesis in the first case study stating that 'We need one formula per soil unit', we analysed the diversity of fertiliser requirements with regard to the soil types in two countries, one in Ghana (Snoeck et al., 2010b) and one in Côte d'Ivoire (N'guessan et al., 2017), which enabled us to build two thematic fertiliser maps. The thematic maps were produced by linking each map unit to one fertiliser recommendation. The map units were obtained by linking each pedological unit to the maps of optimal rainfall and temperatures for cacao. The average fertiliser recommendation per map unit was computed from the average of five fertiliser recommendations obtained from five soil samples taken in adult cacao plots of the corresponding map unit based on their GPS coordinates. The number of map units was 120 in Ghana and 130 in Côte d'Ivoire, resulting in two thematic maps with, respectively, 120 and 130 fertiliser formulas. However, after grouping formulas with similar nutrient balances and which differed only in the amount of fertiliser to be applied, the number of formulas was reduced to 33 for Ghana and 23 for Côte d'Ivoire (Fig. 2 and 3).

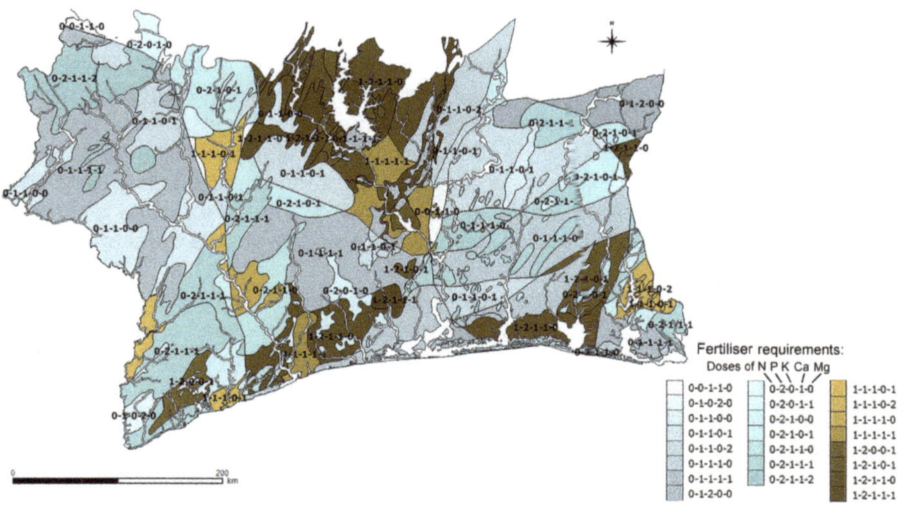

Figure 2 Map showing the areas covered by the recommended fertiliser formulas in Côte d'Ivoire (based on N, P, K, Ca and Mg nutrients).

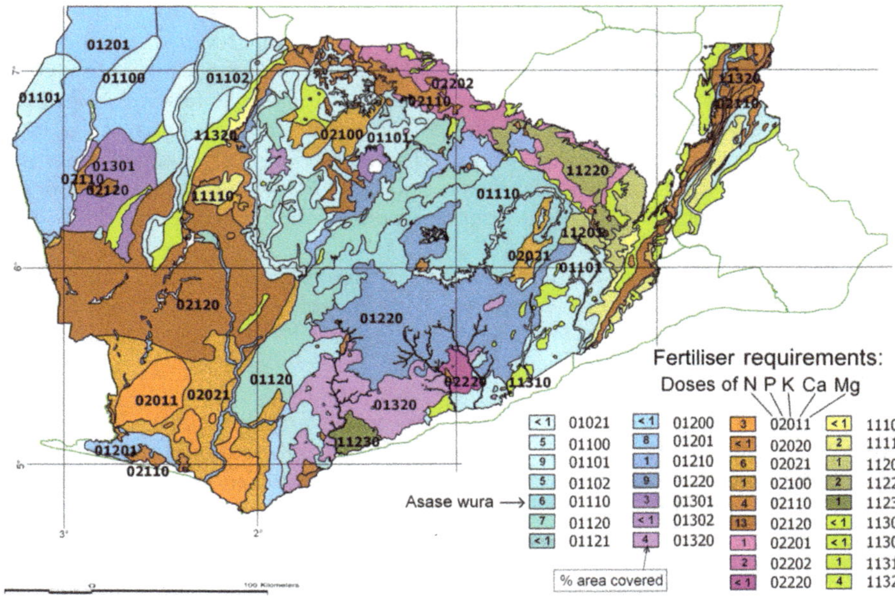

Figure 3 Map showing the areas covered by the recommended fertiliser formulas in Ghana (based on N, P, K, Ca and Mg nutrients). (Figures in the rectangles are the % of cocoa area covered by the recommendation.)

3 Case study: homogeneity in diversity

Our original hypothesis was that, because of the variability of the soil analyses, each situation would require an individual fertiliser formula. In practice, the number of fertiliser formulas turned out to be much smaller than the number of map units, that is, about thirty per country and less than five for a given area, suggesting that hypothesis 1 is not rational. Altogether, this simplification extends the area of application of soil diagnosis and shows that it can be operational even in the case of small farms. There are several reasons why a process of homogenisation is possible despite the diversity of soils that emerged from the original hypothesis.

On the one hand, the effect of cultivating cacao in an agroforestry system, even a simple one, leads to homogenisation of the soils due to repeated supplies of organic matter and the more or less rapid recycling of the nutrients in this type of cropping system (Arévalo-Gardini et al., 2015). This hypothesis is consistent with the observations made by Araujo et al. (2013), who showed that soils in cacao systems associated with different crops or in agroforestry have equivalent levels of C irrespective of the pedological origin of the soil. Similarly, Saj et al. (2013) observed a gradient in the sequestration of carbon linked to different levels of cacao-based agroforests in Cameroon. The type of trees grown in association with cacao is also important as it influences the nutrient composition of the organic matter, which can modify soil fertility, soil pH and the balance between nutrients (Somarriba and Beer, 2011). This is consistent with the results obtained by Alpízar et al.

(1986), who showed that legumes (e.g. Erythrina) have a more rapid nutrient cycle than forest trees (e.g. Laurel) and supply nitrogen through symbiotic fixation of atmospheric N_2. In addition to the planting density of the legumes, soil acidity may limit the capacity of legumes to fix the atmospheric nitrogen and hence the quantity of fixed nitrogen (Snoeck and Domenach, 1999). This was confirmed by the work of Nygren and Leblanc (2009) showing that the association with legumes can cover cacao nitrogen requirements, which are low. This is even more true when the number of shade trees is increased as this reduces photosynthetic activity and hence cacao yields (Jagoret et al., 2012).

On the other hand, the repeated use of the same type of cultivation practices over a period of many years leads to a predominance of nutrient levels and regulates the soil quality. Several authors have published studies illustrating the processes involved. Snoeck et al. (2010a) observed a marked influence of the age of the cacao plot on nutrient levels and particularly on the level of N which is closely linked to modifications in soil biological activity. Somarriba and Beer (2011) showed that association with trees can facilitate nutrition, thanks to a better nutrient transfer via the recycling of leaves after pruning. Jagoret et al. (2011) showed that a few years of cultivating annual or shrub crops were sufficient to favourably improve soil properties (in particular C and N) before starting a new cacao cropping cycle. The positive effects of agroforestry play a leading role in the homogenisation process through their role in C and N stocks (Hartemink, 2005; Schroth et al., 2001).

4 Soil diagnosis in practice

In addition, it is clear that a soil diagnosis is needed to enhance soil properties but that certain steps need to be followed. The first step should be to adjust the pH of the surface soil because other chemical properties, including the availability of most soil nutrients, depend on it (Snoeck et al., 2016). Since soil acidity is easy to measure *in situ*, rapid and low-cost results can be envisaged. In particular, it is possible to simplify the number of fertiliser formulations needed to maintain the balance of soil nutrients in the long term according to sustainability criteria (near-maximum yields, conservation of soil reserves). Over time, the adoption of cultural practices adapted to each environment will take precedence over the origin of soils due to homogenisation processes linked to the recycling of organic matter. The choice of species (fruit trees, timber) associated with cacao is important not only for economic reasons but also for the management of C, N and pH (shade tree species, legumes). The planting density of each associated species and pruning practices are also important. This shows that when cacao is grown in agroforestry systems, the type of association can be a criterion for assessing fertiliser requirements.

5 Future trends and conclusion

Research is ongoing around two cultural approaches: on the one hand, smallholdings in which cacao is not at all or only a little pruned and is often grown in an agroforestry system, which explains the limited effects of nitrogen; on the other hand, industrial cropping systems which increasingly use high-yielding and heavily pruned clonal material to minimise vegetative growth and increase the fruit:leaf ratio.

There is a strong demand for a method to establish simple recommendations for small-scale farmers' plantations using parameters that can be assessed from field observations (texture, pH, C) linked to the history of the plot and cultural practices, such as tree associations. In Africa, for example, the adaptation of this approach to smallholder farmers who practise agroforestry would benefit 70% (Côte d'Ivoire) to 97% (Nigeria) of cacao growers (Gockowski and Sonwa, 2011). The choice of useful variables for the development of the tool should be guided by the following facts: (i) Yields are low and the need for nutrients is thus also low as shown by Hartemink (2005); (ii) The plot history and cultivation choices are good keys for the aggregation of the diagnosis of a production area; (iii) The simplified tool must be easy to use and the cost of analyses must be affordable by the smallholders.

In the case of industrial plantations, the objective is to improve the production: vigour ratio through genetic improvement and better management of farming practices in monoculture or in association with shade trees (that must necessarily be simple). The choice of cultural practices suggests that nutrient requirements will be higher and will affect soil reserves in a short time. This will therefore require taking all soil variables into account and possibly increasing the density of control points per unit area following the principle of precision farming. Two practices appear to be promising to refine the interpretation of soil variables: (i) pruning, which makes it possible to modify the ratio of the production of vegetative biomass to reproductive biomass, and which will have an impact on nutrition management and (ii) the use of clonal material, which enables the introduction of controlled variability for the production of the same biomass while increasing the fruit:leaf ratio.

Today, only a few industrial plantations use clonal material and prune their cacaos with the aim of reducing vegetative development to promote fruiting (Wibaux et al., 2017). Consequently, few results are available. However, the future of cacao in industrial plantations appears to be moving in this direction.

Although cacao has been cultivated for a long time, only a few studies are available on nutrition management when plantations in which cacao was also the previous crop are replanted. This is because traditionally, when a plantation becomes old, the farmer abandons it and establishes a new plot elsewhere rather than replanting it. In this way, the farmer benefits from the fertility of the soil inherited from the forest after clearing. To describe this strategy Ruf and Siswoputranto (1995) used the term 'differential forest income'. But over the last thirty years, this type of shifting cultivation, which was made possible through forest clearing, has become impossible due to land pressure, and cacao farmers are forced to create new plantations in less favourable soils, either after a previous cacao cycle or in marginal areas (Ruf, 2015). It has therefore become essential to consider the consequences of a first cacao cycle for soil properties in relation to its duration and to the previous practices used and what is left for the following crop. This approach will help anticipate the need for soil correction through amendments and fertilisers. This would help validating the soil diagnosis over long periods of time. The sustainability of cacao farming is also a legitimate concern and should be of interest to socio-economic decision-makers.

There is a strong demand for simple recommendations for small-scale cacao farmers to enable them to benefit from adapting to the context of their land using easily measurable field parameters, for example, by observing visible leaf deficiencies. The possibility of using foliar diagnosis has been studied for this purpose (Nelson et al., 2010; Verlière, 1981). Although the results showed that leaf diagnosis is a useful indicator for the health of the cacao tree, the method is imprecise and is too late because a symptom of leaf deficiency only becomes visible when there is already a very strong deficiency in the soil or the plant (Chepote et al., 2013; Loué, 1961; Malavolta, 1997). The authors agreed that foliar diagnosis can help in the assessment of fertiliser only qualitatively. But this can be acceptable for use in industrial plantations when soil was previously corrected. As mentioned earlier, soil acidity is an important indicator because it affects the availability of nutrients and is easy to measure. In particular, the pH of the soil will influence the composition of the phosphate fertiliser.

Analysis of macrofauna (especially earthworms) as an indicator of soil fertility and health is a recent and as yet little documented approach in cacao plantations. The first studies using this approach were used to show that cacao-based agroforestry systems do not differ significantly from the macrofaunal systems in neighbouring forest soils, but do significantly differ from those in savannah soils or heavily cultivated soils (Rousseau et al., 2012). However, these types of studies are worth being improved because the microbiological activity can play a role in soil fertility and nutrient availability (Moço et al., 2010). Such studies on the soil biodiversity in cacao plantations can learn from those used for oil palm that are a little more advanced (Dubos and Snoeck, 2017).

Future research may consider how climate change can affect soil fertility and cocoa sustainability. One consequence is that the production area will change, either by moving or shrinking, which will lead to the use of new soil types (Schroth, 2016). In addition, dry seasons will be longer and more irregular, with the result that, as soils are warmer, the organic matter turnover will be faster and soil moisture regime will be different, probably reduced in some seasons and increased in others. This will have a consequence on the soil nutrient management (Karmakar, 2016). The authors conclude that more studies on the cultivation of cacao in association with trees should be promoted, in particular the effect on soil humidity and litter turnover as this could make the system more resilient to climate change.

6 Where to look for further information

General information and soil and mapping data can be found on the International Soil Reference and Information Centre (ISRIC) website (http://www.isric.org). ISRIC is an independent, science-based foundation created on the initiative of the International Soil Science Society (ISSS) and of the United Nations Educational, Scientific and Cultural Organization (UNESCO).

Useful data on cacao cultivation, its exploitation and trade are available on the World Cocoa Foundation (http://www.worldcocoafoundation.org) and the International Cocoa Organization (https://www.icco.org) websites.

General information and pedagogical material on soil health concepts (assessment and management) can be found on the United States Department of Agriculture website (http://www.nrcs.usda.gov/wps/portal/nrcs/main/soils/health).

7 References

Adejuwon, J. O. and Ekanade, O. 1988. A comparison of soil properties under different land use types in a part of the Nigerian cocoa belt. *Catena*, 15, 319–31.

Afrifa, A. A., Ofori Frimpong, K., Acquaye, S., Snoeck, D. and Abekoe, M. K. 2009. Soil nutrient management strategy required for sustainable and competitive cocoa production in Ghana. In: *16th International Cocoa Research Conference*, Bali (Indonesia), 16–21 November 2009. Lagos (Nigeria): COPAL.

Alpízar, L., Fassbender, H. W., Heuveldop, J., Fölster, H. and Enríquez, G. 1986. Modelling agroforestry systems of cacao (*Theobroma cacao*) with laurel (*Cordia alliodora*) and poro (*Erythrina poeppigiana*) in Costa Rica. *Agroforestry Systems*, 4, 175–89.

Alvim, P. D. T. 1977. Cacao. In: Alvim, P. T. and Kozlowski, T. T. (Eds), *Ecophysiology of Tropical Crops*. London: Academic Press, pp. 279–313.

Appiah, M. R., Sackey, S. T., Ofori-Frimpong, K. and Afrifa, A. A. 1997. The consequences of cocoa production on soil fertility in Ghana: A review. *Ghana Journal of Agricultural Science*, 30, 183–90.

Appiah, M. R., Ofori-Frimpong, K. and Afrifa, A. A. 2000. Evaluation of fertilizer application on some peasant cocoa farms in Ghana. *Ghana Journal of Agricultural Science*, 33, 183–90.

Araujo, Q. R., Loureiro, G. A. H. A., Santana, S. O. and Baligar, V. C. 2013. Soil classification and carbon storage in cacao agroforestry farming systems of Bahia, Brazil. *Journal of Sustainable Forestry*, 32, 625–47.

Arévalo-Gardini, E., Canto, M., Alegre, J., Loli, O., Julca, A. and Baligar, V. 2015. Changes in soil physical and chemical properties in long term improved natural and traditional agroforestry management systems of cacao genotypes in Peruvian Amazon. *PLoS ONE*, 10, e0132147.

Bastide, P., Cilas, C. and Jimmy, I. 2003. Variability of leaf parameters in the foliage of cocoa trees in Vanuatu. In: *14th International Cocoa Research Conference*. Lagos: COPAL, Lagos.

Cabala-Rosand, F. P., Ruy De Miranda, E., Santana, M. B. M. and Santana, C. J. L. 1975. Exigências nutricionais e fertilização do cacaueiro. *Boletim Técnico* n°30, p. 59.

Charter, C. 1953. The need for manuring cocoa in the Gold Coast in order to maintain and augment the level of production. *Appendix II. Report of the 1953 London Cocoa Conference*. London: Cocoa, Chocolate and Confectionery Alliance.

Chepote, R. E., Sodré, G. A., Reis, E. L., Pacheco, R. G., Marrocos, P. C. and Valle, R. R. 2013. Recomendações de corretivos e fertilizantes na cultura do cacaueiro no sul da Bahia. *Boletim Técnico* n°203, 1–43.

Dubos, B. and Snoeck, D. 2017. Maintaining soil health in oil palm cultivation. In: Rival, A. (Ed.), *Achieving Sustainable Cultivation of Oil Palm*. Cambridge CB22 3HJ, UK: Burleigh Dodds Science Publishing Limited.

Evans, H. and Murray, D. 1953. A shade and fertilizer experiment on young cacao. In: *Report on Cacao Research 1945–51*, pp. 67–76.

Gockowski, J. and Sonwa, D. 2011. Cocoa intensification scenarios and their predicted impact on CO2 emissions, biodiversity conservation, and rural livelihoods in the Guinea rain forest of West Africa. *Environmental Management*, 48, 307–21.

Hartemink, A. E. 2005. Nutrient stocks, nutrient cycling, and soil changes in cocoa ecosystems: A review. *Advanced in Agronomy*, 86, 227–53.

Jadin, P. 1975. L'utilisation du diagnostic sol pour l'estimation des besoins en engrais des cacaoyers ivoiriens. *Café Cacao Thé*, 19, 203–20.

Jadin, P. 1988. Fertilité et fertilisation des sols à vocation cacaoyère de Sâo Tomé. *Café Cacao Thé*, 32, 111–26.

Jadin, P. and Snoeck, J. 1985. La méthode du diagnosis sol pour calculer les besoins en engrais des cacaoyers. *Café Cacao Thé*, 29, 255–72.

Jadin, P. and Vaast, P. 1990. Estimation des besoins en engrais des sols à vocation cacaoyère dans le Litimé Togo. *Café Cacao Thé* 34, 179–88.

Jagoret, P., Michel-Dounias, I. and Malézieux, E. 2011. Long-term dynamics of cocoa agroforests: A case study in central Cameroon. *Agroforestry Systems*, 81, 267–78.

Jagoret, P., Michel-Dounias, I., Snoeck, D., Ngnogué, H. and Malézieux, E. 2012. Afforestation of savannah with cocoa agroforestry systems: A small-farmer innovation in central Cameroon. *Agroforest Systems*, 86, 493–504.

Karmakar, R., Das, I., Dutta, D. and Rakshit, A. 2016. Potential effects of climate change on soil properties: A review. *Science International*, 4, 51–73.

Kummerow, J. K. and Alvim, A. 1981. Root biomass in a mature cacao (*Theobroma cacao* L.) plantation. *Revista Theobroma (Brasil)*, 11(1), 77–85.

Ling, A. H. 1984. Cocoa nutrition and manuring on inland soils in Peninsular Malaysia. *Planter*, 60, 12–24.

Loué, A. 1961. *Etude des carences et des déficiences minérales sur le cacaoyer.* Bingerville, Côte d'Ivoire: IRCC.

Malavolta, E. 1997. Nutrição e adubação do cacaueiro. In: Abeas, E. U. (Ed.), *Curso de nutrição mineral de plantas. nutrição mineral do cacaueiro e do cafeeiro.* Piracicaba: módulo 06.

Moço, M. K. S., Gama-Rodrigues, E. F., Gama-Rodrigues, A. C., Machado, R. C. R. and Baligar, V. C. 2010. Relationships between invertebrate communities, litter quality and soil attributes under different cacao agroforestry systems in the south of Bahia, Brazil. *Applied Soil Ecology*, 46, 347–54.

Nelson, P. N., Fidelis, C., Webb, M. J., Berthelsen, S. and Yinil, D. 2010. Nutrient status of cocoa (*Theobromae cacao*) in Papua New Guinea: results from a survey. In: *Proceedings of the 19th World Congress of Soil Science: Soil Solutions for a Changing World*, Brisbane, Australia, 1–6 August 2010. Symposium 4.3. 1 Impacts of land use change in unsustainable ecosystems, 2010. International Union of Soil Sciences (IUSS), c/o Institut für Bodenforschung, Universität für Bodenkultur, 71–4.

N'guessan, K. J.-C., Traore, M. J., Snoeck, D., Kassin, K. E., Koko, L., Camara, M. and Yao Kouamé, A. 2017. Mapping cacao fertiliser requirements in Côte d'Ivoire. *Imperial Journal of Interdisciplinary Research*, 3, 504–15.

Nygren, P. and Leblanc, H. A. 2009. Natural abundance of 15N in two cacao plantations with legume and non-legume shade trees. *Agroforestry Systems*, 76, 303–15.

Owusu-Sekyere, E., Cobbina, J. and Wakatsuki, T. 2006. Nutrient cycling in primary, secondary forests and cocoa plantation in the Ashanti Region, Ghana. *West African Journal of Applied Ecology*, 9, 1–9.

Rousseau, G. X., Deheuvels, O., Rodriguez Arias, I. and Somarriba, E. 2012. Indicating soil quality in cacao-based agroforestry systems and old-growth forests: The potential of soil macrofauna assemblage. *Ecological Indicators*, 23, 535–43.

Ruf, F. 2015. La fertilisation des cacaoyères en Côte d'Ivoire, 35 ans d'innovations villageoises. *Agronomie, Environnement et Sociétés*, 5, 65–74.

Ruf, F. and Siswoputranto, P. 1995. *Cocoa Cycles: The Economics of Cocoa Supply.* Cambridge, England: Woodhead Publishing, Ltd., p. 385.

Saj, S., Jagoret, P. and Ngogue, H. T. 2013. Carbon storage and density dynamics of associated trees in three contrasting *Theobroma cacao* agroforests of Central Cameroon. *Agroforestry Systems*, 87, 1309–20.

Santana, C. D. 1972. Formas de micronutrientes em solos da região cacaueira da Bahia. *Turrialba (IICA)*, 22(1), 73–80.

Schroth, G., Lehmann, J., Rodrigues, M. R. L., Barros, E. and Macêdo, J. L. V. 2001. Plant-soil interactions in multistrata agroforestry in the humid tropicsa. *Agroforestry Systems*, 53, 85–102.

Schroth, G., Läderach, P., Martinez-Valle, A. I., Bunn, C. and Jassogne, L. 2016. Vulnerability to climate change of cocoa in West Africa: Patterns, opportunities and limits to adaptation. *Science of the Total Environment*, 556, 231–41.

Snoeck, J. and Jadin, P. 1992. Cacao (*Theobroma cacao* L.). In: Halliday, D. J. and Trenkel, M. E. W. W. (Eds), *IFA World Fertilizer Use Manual.* Paris, France: IFA.

Snoeck, D. and Domenach, A. 1999. Mise en Evidence du Transfert d'Azote des Légumineuses aux Caféiers par l'Utilisation des Méthodes Isotopiques. In: *Colloque Scientifique International sur le Café*, 18, Helsinki (Finlandia), 2–6 Agosto 1999.

Snoeck, D., Abolo, D. and Jagoret, P. 2010a. Temporal changes in VAM fungi in the cocoa agroforestry systems of central Cameroon. *Agroforestry Systems*, 78, 323–8.

Snoeck, D., Afrifa, A. A., Ofori Frimpong, K., Boateng, E. and Abekoe, M. K. 2010b. Mapping fertilizer recommendations for cocoa production in Ghana using soil diagnosis and GIS tools. *West African Journal of Applied Ecology*, 17, 97–107.

Snoeck, D., Koko, L., Joffre, J., Bastide, P. and Jagoret, P. 2016. Cacao nutrition and fertilization. In: Lichtfouse, E. (Ed.), *Sustainable Agriculture Reviews: Volume 19*. Chap. 4: Springer International Publishing, 155–202.

Somarriba, E. and Beer, J. 2011. Productivity of *Theobroma cacao* agroforestry systems with timber or legume service shade trees. *Agroforestry Systems*, 81, 1`09–21.

Verlière, G. 1965. Un essai d'engrais sur cacaoyers en Côte d'Ivoire. Relations entre les rendements et les teneurs des feuilles en azote, phosphore, potassium, calcium et magnésium. In: *Les recherches agronomiques cacaoyères*. Paris: Jouve, pp. 45–56.

Verlière, G. 1981. Etude par la méthode du diagnosis foliaire de la fertilization et de la nutrition minérale du cacaoyer (*Theobroma cacao* L.) en Côte d'Ivoire. Travaux et documents de l'ORSTOM n° 131. Paris (France): Office de la Recherche Scientifique et Technique Outre-Mer (Ed), p. 273.

Wessel, M. 1985. Shade and nutrition of cocoa. In: Wood, G. A. R. and Lass, R. A. (Eds), *Cocoa*. Essex: Longman Scientific and Technical, pp. 166–94.

Wibaux, T., Konan, D.-C., Snoeck, D., Jagoret, P. and Bastide, P. 2017. Study of tree-to-tree yield variability among seedling-based cacao populations in an industrial plantation in Côte d'Ivoire. *Experimental Agriculture*, 1–12, published online (First view): 25 July 2017, p. 1–12.

Wyrley Birch, E. 1987. *Results of Cocoa Agronomy Trials in Sabah, Malaysia from 1964 to 1977*. Surrey, London: Land Resources Development Centre.

Part 3

Diseases and pests

Chapter 10

Cocoa diseases: witches' broom

Jorge Teodoro De Souza, Federal University of Lavras, Brazil; Fernando Pereira Monteiro, Federal University of Lavras and UNIVAG Centro Universitário, Brazil; Maria Alves Ferreira, Federal University of Lavras, Brazil; and Karina Peres Gramacho and Edna Dora Martins Newman Luz, Comissão Executiva do Plano da Lavoura Cacaueira (CEPLAC), Brazil

1 Introduction

Cocoa witches' broom, caused by the fungus *Moniliophthora perniciosa*, is economically considered the second most damaging disease that affects the cocoa tree (*Theobroma cacao* L.). It is only surpassed by black pod rot caused by several *Phytophthora* species with a more widespread occurrence. The witches' broom disease is known by different names in different countries: 'krulloten' in Suriname, 'balai de sorcière' in French-speaking countries, 'escoba de bruja' in Spanish-speaking countries and 'vassoura de bruxa' in Portuguese. The phylogenetically related fungus *M. roreri* causes a disease known as frosty pod rot, which is also a cocoa disease of economic importance (Bowers et al., 2001).

Witches' broom is restricted to South America and parts of Central America and the Caribbean (Purdy and Schmidt, 1996; End et al., 2014). Fortunately it does not occur in Africa and Asia, from where approximately 85% of the world's cocoa production comes from (FAOSTAT, 2014).

http://dx.doi.org/10.19103/AS.2017.0021.14

The witches' broom pathogen is thought to have co-evolved with cocoa in the Amazon River basin (Purdy and Schmidt, 1996). It has dispersed to commercial cocoa-producing areas of South America and parts of Central America and the Caribbean causing severe losses (Evans, 2002). The introduction of the disease has been responsible for socio-economical decline and environmental disasters in many countries (Alger and Caldas, 1994; Trevizan and Marques, 2002), and poses a risk to countries still free of the pathogen.

In this chapter, we review the current knowledge on witches' broom, including information on disease physiology, genomics and transcriptomics, diversity and management practices. In some sections of this chapter examples of Bahia State, Brazil, will be highlighted reflecting the experience of the authors.

2 Geographical distribution and historical perspective

The disease was first encountered in the Brazilian Amazon by A.R. Ferreira in his explorations of the lower Amazon and Rio Negro between 1785 and 1787 (Silva, 1987). Its destructive potential was identified by the Portuguese colonists of the upper Rio Negro who called the disease 'lagartão' due to the lizard-like shape of the brooms hanging on the infected trees (Silva, 1987).

The first report of the economical impact of the witches' broom was in Suriname in 1895, where serious losses were noted in the municipality of Saramacca. The disease was known by the Dutch name 'krulloten' due to the distortions of infected branches. The production of cocoa beans in Suriname decreased from 4.5 million kilograms in 1895 to 850,000 kilograms in 1904 and as a result cocoa was mostly replaced by banana and coffee (Holliday, 1952).

By 1915, Ecuador was the largest cocoa producer, exporting approximately 40,000 tons per year, but these exports fell to less than 15,000 tons by 1925, due to the combined effect of frosty pod rot introduced in the 1850s and witches' broom in 1920 (Evans, 2002). The witches' broom disease was detected in Trinidad in 1928 (Stell, 1928) where cocoa exports dropped from a high of 30,000 tons to 8,000 tons per year, a decade later (Padwick, 1956).

During most of the twentieth century, Brazil occupied the second position in cocoa production thanks to the yields obtained in Bahia State, where cocoa is grown under the shade of the thinned Atlantic forest in a traditional cultivation system known as 'cabruca'. This agro-ecosystem helps protect the fauna and flora of this region, considered endangered to be one of the world's most diverse tropical forests (Alger and Caldas, 1994). An expansion of the cocoa plantations was attempted in the early 1970s with the construction of the Trans-Amazonian Highway and the establishment of farmers' settlements, especially in Rondônia State, Brazil, which is located more than 3,000 km away from Bahia State. However, Rondônia State was covered by the Amazon rainforest, where the pathogen is endemic. The national agency Instituto Nacional de Colonização e Reforma Agrária (INCRA) coordinated the settlements and Comissão Executiva do Plano da Lavoura Cacaueira (CEPLAC), established in 1957, supplied witches' broom-resistant genetic materials introduced from Trinidad, especially the Scavinas from the Pound collection (Pound, 1938). Unfortunately, witches' broom ravaged the plantations in Rondônia State and pod losses reached 90% leading to economical, ecological and social catastrophe in the 1970s (Evans and Prior, 1987).

Many years later and despite the phytosanitary isolation imposed by the government and coordinated by CEPLAC to avoid the introduction of witches' broom, the pathogen was reported in Bahia State in 1989 (Pereira et al., 1989). In a period of 10 years after its arrival, production fell from 400,000 to 150,000 tons per year (Pereira et al., 1990). Brazil became an importer of cocoa beans and now occupies the fifth position in cocoa production (FAOSTAT, 2014). Many of the plantations were abandoned, with serious socio-economical and ecological consequences, including job losses, increased crime rates and clearing of the Atlantic forest remnants to plant pasture (Alger and Caldas, 1994; Trevizan and Marques, 2002). The pathogen also spread to Espírito Santo State in 2001, but owing to a well-defined dry season, losses were not as great as in Bahia (Silva et al., 2002). Most of the large cocoa plantations of Bahia State no longer exist, although cocoa cultivation has continued on a much smaller scale. With production levels being stable, there has been a renewed interest in cocoa cultivation thanks to the integrated management of witches' broom introduced by CEPLAC and its sustained rehabilitation campaign (Medeiros et al., 2010).

Moniliophthora perniciosa is thought to have evolved with cocoa in the Amazon basin and dispersed to other areas of the Americas when its host was transferred (Griffith et al., 2003; Purdy and Schmidt, 1996). It occurs in most cocoa-producing countries of South America, but is delimited by the Panama Canal in Central America and parts of the Caribbean (Table 1). The danger of its introduction to other cocoa-producing areas outside of the Americas underscores the importance of quarantine and safe movement of germplasm around the world (End et al., 2014).

Table 1 Geographic distribution of *Moniliophthora perniciosa* by country

Country	First report	Reference
Brazil – Lower Amazon	1785–1787	Silva et al., 1987
Suriname	1895	Ritzema-Bos, 1901
Guyana	1906	Holliday, 1952
Colombia	1917	Pound, 1938
Ecuador	1920	Rorer, 1926
Trinidad	1928	Stell, 1928
Tobago	1939	Baker and Holliday, 1957
Peru	1943	Pound, 1943
Grenada	1948	Baker and Holliday, 1957
Panama	1978/1988	Espino, 1989
Brazil – Bahia	1989	Pereira et al., 1989
Dominican Republic	2002	Evans, 2002
Saint Vincent	2002	Evans, 2002
Grenadines	2002	Evans, 2002
Saint Lucia	2009	Kelly et al., 2009
French Guiana	Unknown*	
Venezuela	Unknown*	

* Probably occurred in these countries before 1928.

3 Pathogen taxonomy, biology and disease cycle

3.1 Taxonomy

The taxonomy of the witches' broom agent has been the subject of many scientific disputes throughout history. The disease was first noted in the Brazilian Amazon (Silva, 1987) and subsequently in Suriname, but the identity of the pathogen remained unknown for many years. The first taxonomical studies were done in the Netherlands with samples received from Suriname and the pathogen was named *Exoascus theobromae* (Ritzema-Bos, 1901). Other scientists, who also received samples from Suriname, reported it to be *Fusarium* (Howard, 1901) or *Lasiodiplodia* (Charles, 1906). Studies done in Suriname lead to the isolation and description of the pathogen as *Colletotrichum luxicum* (Van Hall and Drost, 1907), although Koch's postulates were not carried out. Because of the lack of pathogenicity proof this binomial was viewed with suspicion by scientists.

Rorer (1913) isolated the fungus from infected brooms and pods, and based on the clamp connections formed in cultures classified it as a basidiomycete. Stahel (1915) studied the disease in Suriname, completed Koch's postulates for the first time and named the pathogen *Marasmius perniciosus* Stahel. Later studies done in Ecuador by the same author lead to the description of a new variety viz. *M. perniciosus* var. *ecuadoriensis* (Stahel, 1924), based on the darker red mushrooms produced. The name *M. perniciosus* became established and was used by phytopathologists for more than three decades. *Marasmius perniciosus* was transferred to the genus *Crinipellis*, as *C. perniciosa* (Stahel) Singer, based on the presence of amyloid hairs on the cuticle of the pileus (Singer, 1942), although some authors considered *C. perniciosa* as unique, since it was the only pathogenic species among others in the genus that were litter fungi (Holliday, 1971; Aime and Phillips-Mora, 2005).

Cytological investigations on the causal agent of another cocoa disease, frosty pod rot, led to the conclusion that it was caused by a basidiomycete closely related to *C. perniciosa*, which was named *M. roreri* (Evans et al., 1978). The authors then transferred *M. roreri* to the genus *Crinipellis*, as *C. roreri* (Evans et al., 2002), although they recognized that no analysis of the type species of *Crinipellis* was done. Further studies, this time involving a detailed phylogenetic analysis of different genetic regions and the type species of *Crinipellis*, *C. stipitaria*, revealed that witches' broom and frosty pod rot pathogens, although closely related to each other, did not belong in the genus *Crinipellis* (Aime and Phillips-Mora, 2005). The authors proposed the transference of *C. perniciosa* to the genus *Moniliophthora* as *M. perniciosa* (Stahel) Aime and Phillips-Mora. Nowadays, the names *M. perniciosa* and *M. roreri* are widely accepted as comprising sister species that form a distinct lineage in the family Marasmiaceae. The most closely related genera to *Moniliophthora* include *Chaetocalathus*, *Crinipellis* and *Marasmius*, which are all mainly saprophytic litter fungi (Aime and Phillips-Mora, 2005).

3.2 Biology and disease cycle

Moniliophthora perniciosa is a species with primary homothallism, where a single basidiospore is able to reproduce sexually without the need of cross-fertilization. This condition is observed in only 1% of the known basidiomycetes (Lemke, 1969). The fungus has a hemibiotrophic life cycle with marked biotrophic and necrotrophic phases.

The basidiospores (9–12.5 x 4.5–6 μm) are produced on basidia located at the lower side of caps of pink/reddish mushroom-like basidiomata or basidiocarps (Fig. 1). Basidiospores are single-celled, haploid and asymmetric in shape varying from oblong to ellipsoid with an apical germ pore (Wheeler and Mepsted, 1988). The basidiocarps are naturally formed on previously infected plant tissue, such as pods and vegetative tissues (Fig. 1). Basidiospores suitable for experimentation may also be produced on natural culture media under special laboratory conditions (Niella et al., 1999). A single basidiocarp is estimated to produce 80-90 million basidiospores (Bowers et al., 2001).

Basidiospores attach to the host surface by means of a secreted mucilage. Germination usually starts 2 h after deposition and is followed by the emergence of a germ tube with a 'club-shaped tip' that probably allows the fungus to penetrate the shoot apex cuticle within 6–9 h, establishing a successful biotrophic infection of the host (Sena et al., 2014; Fig. 1). Other experiments showed that the minimum time for infection and establishment of the pathogen in the host was 4–5 h. However, the maximum number of infected seedlings was obtained after 12–15 h under 100% relative humidity at 25°C (Frias et al., 1991, 1995). Penetration occurs through the trichome base, natural openings on the cuticle surface, stomata and directly through the cuticle (Sreenivasan and Dabydeen, 1989; Sena et al., 2014). In plant tissue, the biotrophic phase is characterized by an intercellular monokaryotic primary mycelium that measures 5–20 μm in diameter originating from germinating basidiospores (Stahel, 1915). Appressoria and haustoria were not observed

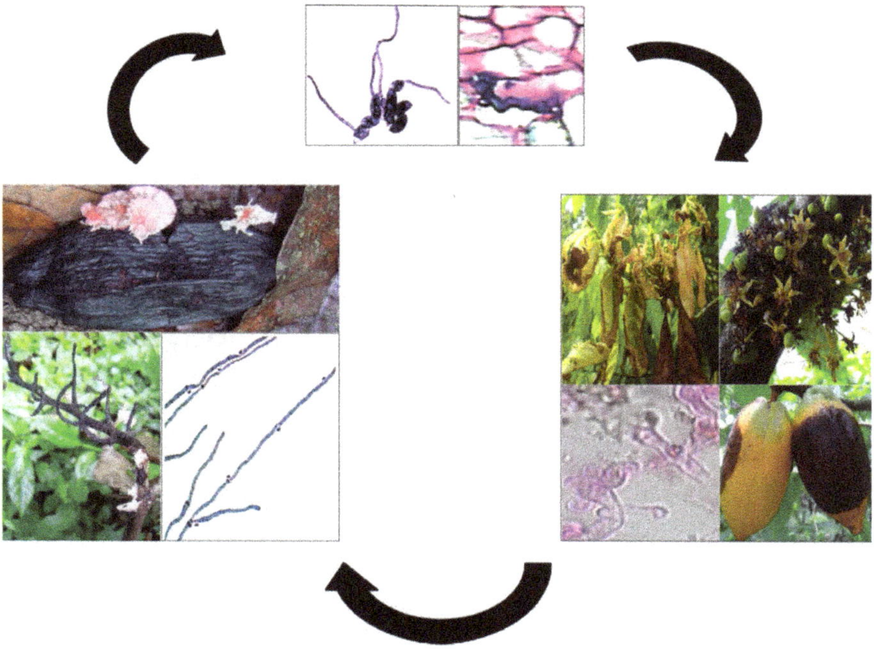

Figure 1 Witches' broom disease cycle. Basidiospores germinate and infect all actively growing tissues (shoots, flower cushions, pods). The primary convoluted monokaryotic mycelium infects intercellularly and induces the typical symptoms – green brooms, flower cushion brooms, abortive pods and pod lesions. Infected tissues suffer accelerated senescence and necrosis upon transition to the secondary dikaryotic mycelium with clamp connections, producing dead brooms and pods where basidiocarps are formed. Basidiospores are produced at the lower side of basidiocarp caps and re-initiate the cycle.

(Sena et al., 2014), but a hyphal matrix in intimate contact with the host cells (Silva and Matsuoka, 1999) likely functions as a haustorium to absorb nutrients from the apoplast.

On artificial growth media, uninucleate hyphae occur at the edges of colonies and these were confirmed to have one nucleus and lack clamp connections and therefore were thought to be primary mycelium (Kilaru and Hasenstein, 2005). This uninucleate mycelium may be maintained for up to five days on potato dextrose agar (PDA) and up to two months on glycerol as a sole carbon source, but in none of the cases it was able to infect cocoa plants in a manner similar to the infection caused by basidiospores, the only infective propagules of the pathogen (Kilaru and Hasenstein, 2005; Meinhardt et al., 2006). It is thus hypothesised that transition to the necrotrophic phase is dependent on the type and availability of nutrients to the mycelium as well as host cues (Scarpari et al., 2005). The factors that govern the transition from biotrophic to necrotrophic phase involve an alternative oxidase chain (AOX) that regulates this process together with starvation and senescence stresses (Thomazella et al., 2012).

The necrotrophic phase is characterized by the secondary dikaryotic mycelium that appears three to nine weeks after host infection (Calle et al., 1982). The necrotrophic mycelium measures 1–3 μm in diameter and has clamp connections (Fig. 1) responsible for dikaryotization of the cells.

Infected tissues develop symptoms known as brooms (Fig. 1), which are initially called green brooms. During the transition to dry brooms both biotrophic and necrotrophic mycelium are present in infected branches (Stahel, 1915). Dry brooms remain hanging on the tree canopy. After four to eight weeks under favourable conditions of rain and temperature, the basidiocarps are formed completing the life cycle of the pathogen (Purdy and Schmidt, 1996; Fig. 1).

Arthrospores were obtained in paired cultures originated from Trinidad and Tobago and Suriname. These spores quickly germinate and produce secondary mycelium with clamp connections. Both arthrospores and the originating secondary mycelium have failed to infect cocoa seedlings (Delgado and Cook, 1976).

4 Diversity, host range and symptomatology

4.1 Pathogen diversity and host range

There are reports of *M. perniciosa* causing witches' broom in plants of the families Malvaceae, Bixaceae, Solanaceae, Malpighiaceae and Bignoniaceae. The pathogen was initially classified on the basis of basidiocarp morphology into three varieties: var. *perniciosa*, var. *citriniceps* and var. *ecuatoriensis* (Pegler, 1978). Subsequent research employing cross-inoculations led to the classification of the species into four biotypes according to their host range: (1) biotype C infects the genera *Theobroma*, *Bixa* and *Herrania* of the family Malvaceae and Bixaceae; (2) biotype S infects species of the Solanaceae family, including the genera *Solanum* and *Capsicum*; (3) biotype H infects Malpighiaceae species in the genera *Mascagnia*, *Stigmaphyllon*, *Heteropterys* and *Banisteriopsis*; and (4) biotype L is an endophyte of the genus *Arrabidaea* from the Bignoniaceae family (Table 2; Fig. 2).

Isolates from biotype C do not infect solanaceous plants and biotype S isolates do not infect cocoa (Bastos and Evans, 1985; Bastos et al., 1981; Deganello et al., 2014). Interestingly, studies done in Brazil with isolates from *Mascagnia* cf. *cepium* from the

Table 2 Host range of the different biotypes of *Moniliophthora perniciosa*

Biotype	Host family	Host species	Reference
	Bixaceae	*Bixa orellana*	Bastos and Andebrhan, 1986
(C)	Malvaceae	*Herrania albiflora*	Evans, 1981
		H. nitida	Evans, 1981
		H. purpurea	Evans, 1981
		Theobroma bicolor	Stahel, 1915
		T. cacao	Silva et al., 1987
		T. grandiflorum	Gonçalves, 1965
		T. microcarpa	Baker and Holliday, 1957
		T. obovata	Baker and Holliday, 1957
		T. speciosum	Van Hall, 1910
		T. subincana	Baker and Holliday, 1957
		T. sylvestris	Luz et al., 1997
(S)	Solanaceae	*Athenaeum pogogena*	Bastos et al., 1991
		Capsicum annuum	Luz et al., 1997
		C. frutescens	Luz et al., 1997
		Solanum lycocarpum	Pereira et al., 1997
		S. cernuum	Evans and Barreto, 1996
		S. gilo	Luz et al., 1997
		S. grandiflorum	Lopes et al., 2001
		S. mauritianum	Bastos and Evans, 1985
		S. melongena	Luz et al., 1997
		S. lasiantherum	Bastos and Evans, 1985
		S. lycocarpum	Barreto and Evans, 1996
		S. paniculatum	Silva et al., 1992
		S. rugosum	Bastos and Evans, 1985
		S. stipulaceum	Luz et al., 1997
(H)	Malpighiaceae	*Heteropterys acutifolia*	Resende et al., 2000
		*H. nervosa**	Arruda et al., 2005
		Banisteriopsis caapi	Holmes et al., 2004
		Mascagnia cf. *sepium*	Bastos et al., 1998
		Stigmaphyllon cf. *blanche*	Bezerra et al., 1998
(L)	Bignoniaceae	*Arrabidaea verrucosa*	Griffith et al., 1994b

* The pathogen was described as *Crinipellis brasiliensis*.

Figure 2 Selected hosts of *Moniliophthora perniciosa*. (a) A green terminal broom on cupuassu (*Theobroma grandiflorum*) induced by biotype C. (b) Green broom on wolf's apple (*Solanum lycocarpum*) induced by biotype S. (c) Symptoms of witches' broom caused by biotype C on mocambo (*Theobroma bicolor*) characterized by a necrotic spot on the pod. (d) Green broom formation on *Herrania* sp. caused by biotype C.

Amazon, *Stigmaphyllon* cf. *blanchetii* collected in Bahia State and *Heteropterys acutifolia* in Minas Gerais State showed that these isolates were pathogenic to cocoa, although symptoms were not typical brooms (Bastos et al., 1998; Bezerra et al., 1998; Resende et al., 2000; Viana, 2001). Because of the striking molecular and morphological differences, an isolate from *H. acutifolia* was described as a new species, *Crinipellis brasiliensis* (Arruda et al., 2005), but it has not yet been transferred to *Moniliophthora*. Biotype L from *Arrabidaea verrucosa* found in Ecuador is non-pathogenic to cocoa and represents a benign endophyte (Griffith et al., 1994).

In general, the use of molecular techniques has shown that there is a significant level of genetic diversity in *M. perniciosa*, especially among isolates from different geographic regions and biotypes. Studies employing RAPD, ERIC-PCR, AFLP, karyotyping and telomere analysis detected variations in isolates of biotype C of *M. perniciosa* (Arruda et al., 2003b; Rincones et al., 2003; Ploetz et al., 2005). Such studies showed clusters of isolates in accordance to their geographical origin. The presence of two distinct genetic groups of *M. perniciosa* in southern Bahia indicates two deliberate and independent introduction events of isolates from the states of Rondônia and Pará (Andebrhan et al., 1999; Junior, 2006).

A phylogenetic analysis with three loci confirmed earlier findings (Griffith et al., 2003; Arruda et al., 2003a,b; Marelli et al., 2009) that biotypes C and S are indeed closely related, although separated into two distinct lineages, whereas biotypes H and L are not valid as currently defined (Tarnowski, 2009), probably owing to host misidentifications. This analysis, although limited by the small number of loci examined, indicates that hybridization events may have occurred among biotypes or lineages.

Biotypes C, S and H are homothallic, while biotype L is heterothallic. Griffith and Hedger (1994a,b) hypothesized that diversification of *M. perniciosa* in different biotypes or lineages was accompanied by a switch from a heterothallic to a homothallic mating strategy, which is advantageous for plant pathogens. While homothallism ensures reproduction without the need of compatible individuals, it also allows specialization to a given environment or lifestyle. It was posited that biotypes C and S evolved from an ancestor of biotype L in separate divergence events (Evans, 1981; Griffith et al., 1994). Biotype L has been shown to have higher levels of genetic diversity than biotypes C and S, which suggests that biotypes C and S diverged more recently as compared to biotypes L and H (Griffith et al., 1994; Hedger et al., 1994; Griffith, 2004). These authors speculated that other biotypes might be found that would represent older divergences from a heterothallic endophytic ancestor. Several host jumps occurred in *M. perniciosa* and the specialized genotypes are now in a process of speciation.

The mechanisms underlying the diversification of *M. perniciosa* are still poorly understood. Comparisons among different monosporic isolates from different basidiocarps revealed similarities among basidiocarps of the same broom, but differences were also found among monosporic cultures from the same basidiocarp (Gramacho et al., 2005). Genome sequencing data has shown the occurrence of a large number of transposons that may contribute to the great pathological and genetic diversity of *M. perniciosa* (Pereira et al., 2007).

4.2 Disease symptomatology

There are detailed descriptions of cocoa witches' broom symptoms on shoots, pods, flower cushions and seedlings, including less-frequently reported symptoms (Silva et al., 2003). In this section, the most common symptoms will be depicted (Fig. 3).

The fungus infects all meristematic tissues and infections of the apical buds of leaf flushes lead to the production of vegetative terminal brooms (Fig. 3b,g). The size of vegetative brooms is dependent on the level of resistance of the variety, tree vigour and fungal genotype. Besides terminal buds, petioles and pulvinii (Fig. 3f) may also suffer hyperplasia and hypertrophy. It is common to observe lateral budding of vegetative brooms as a result of loss of apical dominance. Terminal brooms may occur on orthotropic and plagiotropic branches due to the dimorphic growth habit of cocoa. Infected axillary buds produce lateral brooms and in some cases infected tissues respond with necrosis of the terminal bud. Dead terminal and lateral brooms are produced three to four months after infection and these remain attached to the trees (Fig. 3e). Dead brooms occurring in large quantities give the impression that the plant is burned (Fig. 3a).

Infected flower cushions can produce vegetative brooms; abnormal flowers; and parthenocarpic carrot-, strawberry-, custard apple-shaped pods (Fig. 3c,d), or remain asymptomatic, although endophytically colonized by *M. perniciosa* (Lana et al., 2011). Infected cushions suffer necrosis rapidly and become woody and mummified. When young pods or cherelles are directly infected, differently from a systemic flower infection, there is a longer period of time for the onset of necrosis that may be up to three months, rather than the few days for systemic flower infections. Infected pods suffer hypertrophy, remain

Figure 3 Witches' broom symptoms on cocoa. (a) Cocoa tree with a high number of terminal and lateral brooms showing a 'burned' aspect. (b) Green broom on a terminal cocoa branch. (c) Infected flower cushion with lateral green brooms. (d) Strawberry-shaped cocoa pods. (e) Dead broom with necrotic leaves and basidiocarps on an older dead broom. (f) Pulvinus and petiole swelling. (g) Green terminal broom on cocoa seedling. (h) Cocoa pod with witches' broom-induced necrosis. (i) Internal symptoms of witches' broom on pods. (j) Basidiocarps on dead broom and pod. (k) Close-up on basidiocarps.

relatively small and sometimes exhibit chlorosis (green islands surrounded by a yellow halo or vice versa), early ripening and necrotic lesions (Fig. 3h) that cause the pod to mummify or be colonized by saprophytic fungi. Internal damage vary from a dry to watery rot of the beans (Fig. 3i), which are unsuitable for harvest. Pods older than three months are resistant to the infection or suffer only a superficial husk infection that renders the beans still useful.

Cankers are formed both on dormant seedlings and on branches. Cankers with necrosis and cracking of the bark, exposing small woody portions, can lead to new brooms below or above the infected spot (Silva et al., 2002). All infected parts after dry and wet periods may produce fruiting bodies that release basidiospores (Fig. 3j,k).

5 Omics of *M. perniciosa*

The first draft genome of *M. perniciosa* was produced using Sanger sequencing (Mondego et al., 2008; Rincones et al., 2008), and had an estimated coverage of 1.5–2x and size of 30–40 Mbp (Rincones et al., 2003, 2006). The mitochondrial genome of *M. perniciosa* was also assembled and was shown to contain 109,103 bp, 31.9% GC, 14 protein-coding genes and a stably integrated linear plasmid similar to pKalilo. The role of this plasmid in *M. perniciosa* is unknown, but it is known to cause senescence in other host fungi (Formighieri et al., 2008).

Another assembly, this time produced with NGS and with a higher coverage, showed that the genome of *M. perniciosa* contained 17,008 putative genes was 44.6 Mbp in size compared to *M. roreri*, which was 52 Mbp (Meinhardt et al., 2014). The difference in size between the two species was mainly due to the presence of transposable elements, which corresponded to 1% and 7% of the genomes of *M. perniciosa* and *M. roreri*, respectively. The transposable elements were active in the genome of *M. perniciosa* (Pereira et al., 2013, 2015), possibly accounting for the considerable genetic diversity, resulting in biotypes with altered virulence levels, within this species. The genomes of these two pathogens were, however, very similar, supporting their taxonomic status as sister species (Meinhardt et al., 2014).

The genome annotation shows a significant capacity for synthesis of secondary metabolites, hormones, toxins and a complete reactive oxygen species (ROS) detoxification system, including a very high number of cytochrome P450 monooxygenases (188 gene models), which may be critical to detoxification and environmental adaptation as well as for disease development (Mondego at el., 2008). Putative fungal effectors and pathogenicity factors have also been annotated.

Genome analysis of the mating typing system of *M. perniciosa* revealed genes for homeodomain transcription factors and pheromone receptors, indicating that the fungus possesses genes orthologous to mating type genes (A- and B-mating-type-like) of heterothallic basidiomycetes (Mondego et al., 2008; Kues and Navarro-González, 2010), although it is homothallic. Based on these data, it was hypothesized that the mating type system is used to promote clamp connections and hyphae dikaryotization. Genome sequences have also revealed a high number of polymorphic microsatellites (SSR) that were successfully used to produce markers for diversity and population genetics (Gramacho et al., 2007).

Transcriptomic approaches have been used to study the differences between the biotrophic and necrotrophic phases of the pathogen through microarrays, ESTs and qPCR (Rincones et al., 2008). The interaction of *M. perniciosa* with susceptible and resistant cocoa varieties was studied through microarrays and cDNA library sequencing (Gesteira et al., 2007; Hora Junior et al., 2012). Recently, RNA-seq was used to study this plant–fungal interaction in green brooms (Teixeira et al., 2014).

6 Fungal pathogenicity and disease physiology

There were marked genetic, morphological and physiological differences between the biotrophic and the necrotrophic phases of *M. perniciosa*. The biotrophic mycelium grows in the apoplast of the host and absorbs nutrients without ever forming haustoria. During its unusual long biotrophic phase, it uses the numerous enzymes that encode ROS detoxification and other effectors to suppress plant defences (Mondego et al., 2008). The hormone imbalances promoted by the pathogen keep soluble sugars available in the apoplast (Scarpari et al., 2005), but when the supply decreases, the fungus promotes plant cell death and accelerates senescence producing the dry broom, where it survives and defends itself against competitors with toxins and secondary metabolites. Additionally, the lack of abscission layer between healthy and diseased tissues keeps the infected organ hanging on the tree, far from its enemies on the ground.

Moniliophthora perniciosa is known to produce hormones, such as indole acetic acid, abscisic acid, jasmonic acid and salicylic acid (Kilaru et al., 2007). A transcriptome analysis showed that the fungus causes hormonal imbalances by disturbing host genes involved in biosynthesis and/or perception of auxin, gibberellin, cytokinin and ethylene (Teixeira et al., 2014). Ethylene is produced in green brooms (Scarpari et al., 2005) and is responsible for stimulating tissue elongation, helping in symptom development and also cell death at the end of the cycle. Cytokinins are thought to induce loss of apical dominance (Teixeira et al., 2014).

Comparison of cDNA libraries of resistant and susceptible cocoa genotypes infected with *M. perniciosa* revealed qualitative differences (Gesteira et al., 2007). During the interaction between cocoa and the biotrophic mycelium of *M. perniciosa* the plant expresses a series of genes involved in defence responses, such as PR- and NB-LRR proteins involved in pathogen recognition and restrictive growth (Teixeira et al., 2014). Plant defence genes were activated earlier and at higher levels in resistant genotypes when compared to susceptible ones (Gesteira et al., 2007; Leal et al., 2007). Susceptible genotypes also produced defence molecules, but these could not successfully restrict the pathogen (Hora Junior et al., 2012).

Biochemical changes during the cocoa–*M. perniciosa* interaction include an increase in the concentration of caffeine, procyanidins, theobromine and phenolics, mainly tannins, during and immediately after the infection of cocoa (Aneja and Gianfagna, 2001; Scarpari et al., 2005; Chaves and Gianfagna, 2007), compounds possibly linked to plant defence responses to pathogen attack. Sugars such as sucrose and glucose accumulate during the formation of green brooms in comparison to healthy tissues. It is unclear whether the sugars originate from the degradation of the starch that also accumulates in green brooms or via translocation from other plant parts (Scarpari et al., 2005). Other biochemical changes observed were increases in the concentration of ethylene, glycerol and asparagine, changes in the proportion of fatty acids and decrease in pigment concentration (Scarpari et al., 2005). These biochemical changes can be coordinated by ethylene that ultimately leads to death of the infected part through programmed cell death (Scarpari et al., 2005). The pathogen appears to initiate and control nutrient accumulation in the brooms and induce subsequent tissue death to use it for reproduction. A recent study provided evidence that the decreased concentration of sugars in green brooms is a key factor in the switch from the biotrophic to necrotrophic phase (Barau et al., 2015).

The transition of the biotrophic to the necrotrophic phase involves programmed cell death triggered by ROS (Ceita et al., 2007) and nitric oxide (NO). The fact that calcium oxalate accumulated in green brooms and decreased in dead brooms (Ceita et al., 2007) may suggest that the production of calcium oxalate crystals by *M. perniciosa* may be involved in witches' broom pathogenesis (Rio et al., 2008). The production of ROS and NO by cocoa is an attempt to block the pathogen; however, *M. perniciosa* is able to protect itself by detoxifying these with enzymes such as catalases, superoxide dismutases and thioredoxin that are encoded by the genome (Mondego et al., 2008) and with an AOX that was found to be preferentially overexpressed in the biotrophic mycelium as compared to necrotrophic mycelium (Thomazella et al., 2012). The expression of AOX seems to decrease the proliferation of the biotrophic mycelium providing a larger amount of ATP for the growth of the necrotrophic mycelium (Thomazella et al., 2012). Thus, the AOX appears to be the switch from the biotrophic to the necrotrophic phase.

Following the transition to the necrotrophic phase, infected tissues undergo a process of senescence activated by necrosis and ethylene-inducing proteins (NEPs) (Barau et al., 2015). It is suggested that NEPs were probably acquired by *M. perniciosa* via horizontal transfer from oomycetes (Tiburcio et al., 2010). The necrotrophic mycelium of *M. perniciosa* has been shown to express the killer toxin protein, KP4, classified as polypeptides produced by viruses inhabiting fungal genomes (Rincones et al., 2008) to kill other competing fungi through the creation of pores in the target cell membranes (Clausen et al., 2000). Of the 13 thaumatin-like proteins that have antifungal properties, four were expressed in dry brooms and have antifungal properties and therefore may be involved in protection against competitors (Franco et al., 2015). Due to the lack of an abscission layer, infected parts remain attached to cocoa trees (Purdy and Schimidt, 1996), avoiding natural enemies such as *T. stromaticum*, that are more efficient competitors on the ground, where humidity is higher. The final stage is basidiocarp formation that has also been studied in detail (Pires et al., 2009).

7 Plant development and disease epidemiology

Developing tissues are the pathogen's target to initiate infection. In Bahia, leaf flushes occur from January to March and from September to October. However, it may also occur with less intensity in other times of the year. The cocoa tree has only a few weeks of dormancy, between July and August, the coldest period of the year, when flower cushions have no activity. Flowering and development of cherelles (young pods) occur during two seasons: main season, between March and May, and a secondary season, between November and January, although sporadic flowering and fruiting may occur during other months. Therefore, there is always susceptible tissue on cocoa trees, even during the dormant period, when growing pods may be present, which makes the disease cycle continuous (Luz et al., 2001).

The disease is considered monocyclic in one individual year and polycyclic over the years, because the inoculum formed in the plant re-infects shoots, flower cushions and developing pods of the same plant (Tovar, 1991). In Bahia, data from two experimental areas in the municipalities of Camacan and Ilhéus showed that production of basidiomata occurs twice a year – during the secondary season (first semester) and the main season (second semester). However, the disease is monocyclic because spores produced in one season have no opportunity to initiate new infections in the same season. Although basidiospores are released throughout the season, these come from basidiomata developing from infected tissues of the previous seasons. Under the conditions of the Brazilian Amazon, the disease is also monocyclic due to a well-defined dry season. In the Brazilian Amazon, the witches' broom is endemic in native populations of cocoa and is epidemic in commercial plantations where no control is applied (Bastos et al., 1990). In the cocoa-producing area of the Trans-Amazonian Highway, the disease is less severe when compared with other regions of the Amazon and to southern Bahia, where cocoa is extensively cultivated (Luz et al., 1997). Several factors may contribute to the lower disease severity in these areas: (1) farms with diversified cropping systems and small cocoa plantations (8–10ha) located far from each other and often separated by areas of forest; (2) genetic material cultivated with some level of resistance; and (3) weather conditions favourable to the disease control with a well-defined dry season, enabling the reduction of inoculum sources in the area (Andebrhan et al., 1993).

In Bahia, the peaks of higher incidence of vegetative brooms is recorded on average four weeks after the peak period of leaf flushing, although vegetative brooms may occur throughout the year (Luz et al., 1994, 1997). Lateral brooms usually occur during periods without leaf flushing. There is a sharp decline in the formation of new brooms during the months of July and August. Peak pod infection occurs predominantly in the main season, July to September, but in some years high incidence of infected pods are observed in the months of February and March. The highest production of fruiting bodies occur in June, July and August, while sporulation spans almost the whole year, with intervals never larger than nine weeks (Luz et al., 2002). Green brooms take an average of 7.5 weeks to dry and remain for 13 weeks on average in a dormant state (varying between 9 and 26 weeks). Dead brooms can produce basidiomata for 27 months and mummified pods for 21.5 months on average (Luz et al., 1994, 2002).

Cocoa is a native plant of the humid tropical American forests and the intensity and distribution of rainfall varies and this can exert a great influence on the incidence of brooms. In south-east Bahia rainfall is usually well distributed throughout the year with monthly totals exceeding 100 mm for every month. However, variations in rainfall may occur affecting the physiology of the cocoa tree, especially leaf flushing and fruiting (Alvim, 1984; Almeida, 2002). Witches' broom incidence is also affected by the distribution of rainfall because the pathogen requires alternating periods of rain and sunny days, nights with temperatures between 15 and 24°C and relative humidity above 90% for sporulation (Thorold, 1975; Bastos and Silva, 1980; Andebrhan et al., 1983; Almeida and Andebrhan, 1984; Bastos, 1990; Luz et al., 1997). Consistent rainy periods and relative humidity near saturation, as well as prolonged dry seasons, inhibit the production of basidiomata (Thorold, 1975). Experiments done in Pará State showed that monthly rainfall lower than 100 mm and above 300 mm decreased sporulation on dry brooms (Almeida and Andebrhan, 1984). Experiments done in broomeries (laboratory cabinets containing hanging brooms) showed that production of basidiomata on brooms or mummified pods is favoured by daily regimens with 16 h of dryness and 8 h of wetness, temperature around 25°C and 12 h photoperiod (Suárez, 1977; Rocha and Wheeler, 1982, 1985; Silva et al., 2000).

The spread of the pathogen results from basidiospores usually released early in the morning or at night, when relative humidity is high (>95%) and temperatures are between 20 and 25°C (Baker and Crowdy, 1943; Baker and Holliday, 1957; Bastos and Silva, 1980). Dispersal within the plant canopy and between neighbouring plants occurs by the action of rain and wind. Dissemination is more efficacious to plants closer (100–300 m) to the inoculum source (Baker et al., 1941), but basidiospores may reach 60 km and even longer distances (Frias et al., 1991; Griffith et al., 2003). Germination of basidiospores in moist conditions and ideal temperature occurs within 1–4 h (Bastos and Silva, 1980; Andebrhan et al., 1983).

Epidemiological parameters, such as incubation period, latency and infection, and the disease progression curves are difficult to dissect because they are highly variable from one region to another (Purdy and Schmidt, 1996). For this reason, it was not possible to establish through conventional statistics a correlation between dependent and independent variables with the typical effects of the epidemic in a series of international network experiments (Schmidt et al., 1993).

Moniliophthora perniciosa survives as a saprophyte in dry brooms, mummified pods, flower cushions and infected dormant buds. When basidiospores are deposited on dormant buds, although penetration occurs, symptoms do not develop and the infection becomes latent, assuming the appearance of small cankers or necrotic lesions that become active when the plant restarts flushing (Silva et al., 2002). Such infections are of epidemiological importance as they allow the survival of the fungus between successive periods of plant growth and fruiting.

8 Disease control: phytosanitation, and chemical and biological control

8.1 Phytosanitation

Phytosanitation, defined as the removal and destruction of diseased plant parts, may reduce pod losses and delay disease progression (Fig. 4). Studies done in Brazil showed that it is necessary to remove 95% of the brooms to achieve 50% reduction in pod losses (Rudgard and Butler, 1987). Pruning was recommended initially by Stahel (1915) and still stands as an effective practice to control the disease, despite the increased costs. The systematic removal of affected plant parts including brooms and pods and burial of the material to hasten decomposition prevents sporulation. The number and timing of these removals depends on the climatic conditions of each region. In the Amazon, for example, usually two removals, one in the dry season and the other before the rainy season, are

Figure 4 Phytosanitary broom removal in an old cocoa plantation heavily infected with *M. perniciosa*.

required, while in Bahia, at least four are necessary, usually in the months of February, May, August and November (Luz et al., 1997). Many farmers already incorporated this practice as part of other management activities to reduce the demand for labour and associated cost.

Fertilization is extremely important as pruning often leaves plants weaker, reducing productivity and resistance levels. Field experiments showed that phytosanitation without fertilization is detrimental to cocoa yields (Medeiros et al., 2010).

8.2 Chemical control

Research on chemical control of witches' broom started at the beginning of the twentieth century in Suriname (Van Hall and Drost, 1909), where Bordeaux mixture was used to control the disease. Since then chemical control has been studied in several countries. Copper-based fungicides have been the most investigated and have shown good results in some situations and inconsistencies in other cases, particularly when applied to new flushes and flower cushions (Laker and Rudgard, 1989; Laker and Ram, 1992). In Bahia, copper oxide at dosages of 3 or 6 g per plant of active ingredient, at monthly or every two month intervals, respectively, has been recommended to control witches' broom. These applications are preventive and must be done in association with pruning according to a timetable (Oliveira and Luz, 2005). However, protectants such as copper oxide have not always been economically viable, because of their low effectiveness under high rainfall conditions, difficulty in applying uniformly to trees and hence their activity is restricted to pods only, and the high labour costs.

Systemic fungicides have been intensively investigated and benzimidazoles, anilides, morpholines, trizoles and strobirulins have shown to be effective under *in vitro* conditions on basidiospore germination, mycelial growth and sporulation, and under greenhouse and field conditions decreased seedling, pod and flower cushion infections, and sporulation (Bastos and Medeiros, 1979; Silva et al., 1985; Lins, 1985; Bastos, 1987, 1989; Laker et al., 1988; McQuilken and Rudgard, 1988; Laker and Rudgard, 1989; Laker, 1991; Oliveira, 2004a,b; Albuquerque et al., 2005; Mota et al., 2010). Despite their relative efficacy, these fungicides have not been used frequently to control the disease in the field, mainly due to low cost effectiveness, negative impacts on health and environment and the risk of contaminating the cocoa beans (Meinhardt et al., 2008).

Synthetic or plant-derived inducers of systemic resistance have also been tested (Kessmann et al., 1994b; Costa et al., 2010; Carvalho et al., 2012; Gomes et al., 2014), but the cost and control levels attained do not make them promising candidates to be used in the field.

Chemical control, although intensively explored in the past, has never been technically and economically viable against witches' broom (Laker and Rudgard, 1989). Nevertheless, an interesting approach has been the application of fungicides at reduced dosages on varieties with susceptible pods, but with some level of genetic resistance in leaf flushes and flower cushions, as shown by Oliveira and Luz (2005), or in combination with biocontrol agents (Medeiros et al., 2010). In these instances chemical control has been effective.

8.3 Biological control

Diverse species have been studied for the biological control of witches' broom, such as *Botryosphaeria ribis*, *Clonostachys rosea*, *Cladobotryum amazonense*, *Gliocladium*

Figure 5 Biological control of the witches' broom disease. (a) *Trichoderma stromaticum* colonizing a dead broom. (b) Commercial product containing 40 g of *T. stromaticum* spores that are enough to apply in one hectare of cocoa. (c) *T. stromactium* producing asexual conidia (green) and sexual perithecia (orange) on the surface of a colonized dead broom. (d) *T. stromaticum* sporulating on autoclaved rice grains.

catenulatum, *Trichothecium roseum*, *Trichoderma atroviride*, *T. brevicompactum*, *T. harzianum*, *T. longibrachiatum*, *T. piluliferum*, *T. pseudokoningii*, *T. reesei*, *T. stromaticum*, *Candida* sp., *Streptomyces albovinaceus*, *S. caviscabies*, *S. griseus*, *S. setonii* and *S. virginiae* (Bastos et al., 1981; Samuels et al., 2000; Macagnan et al., 2008; Cabral et al., 2009; Oliveira et al., 2014). Among these, *T. stromaticum* (Fig. 5) has proven to be the most promising biocontrol agent of *M. perniciosa* (Bastos, 1988; de Souza et al., 2006).

The species *T. stromaticum* is only found in close association with cocoa and is thought to have co-evolved with the pathogen in the Amazon basin and is thus a specific biocontrol agent (Pomella et al., 2010). It acts through mycoparasitism, colonizing and producing spores on the surface of dead brooms and pods (de Souza et al., 2008; Pomella et al., 2010). The diversity and phylogeny of *T. stromaticum* have been studied in detail. Results showed the occurrence of two groups with distinct morphological, physiological and genetic characteristics (de Souza et al., 2006). However, these groups could not be unambiguously classified into different species (Samuels et al., 2012), indicating that speciation was incomplete. Studies on the effect of climatic conditions on *T. stromaticum* indicate that this agent needs temperatures of approximately 23°C and relative humidity higher than 90% to effectively colonize brooms in the field (Loguercio et al., 2009).

A product based on pure conidia of *T. stromaticum* known as Tricovab (Fig. 5) was developed by CEPLAC. The product is available in the market since 1999, although production has been interrupted for some time due to registration bureaucracy. It is produced on autoclaved rice grains and subsequently the spores are separated with a special machine known as mycoharvester. The new version of Tricovab is composed of 40 g of spores, which are adequate to apply to one hectare of cocoa (Pomella et al., 2010; Fig. 5).

Applications of *T. stromaticum* spores over brooms on the ground under cocoa trees reduced basidiocarp formation by 99% (Costa et al., 1996) and by 89% when the brooms were hanging on the trees (Hjorth et al., 2003). Field trials in Peru showed that *T. stromaticum* decreased the number of infected pods by 48% (Krauss and Soberanis, 2002). A three-year field experiment in Bahia showed that *T. stromaticum* alone or in combination with copper hydroxide increased healthy cocoa bean yield by 159% and 314%, respectively, when compared to the untreated control (Medeiros et al., 2010).

9 Disease control: genetic resistance and integrated management

9.1 Genetic resistance

The search for witches' broom resistance started in Ecuador in the 1920s, after the establishment of the disease in the coastal zone, through mass selection of seedlings derived from symptomless trees (Refractarios) (Zhang et al. 2008). This was followed by expeditions to the Amazon Valley, where F.J. Pound explored River Ucayali in Peru to collect disease-free plants in 1937–38 (Pound, 1938) and in 1942 (Pound, 1943). The resulting material was sent to Trinidad for evaluation, and among the collected accessions, one family of progenies identified with the name 'Scavina' received great attention. The accessions 'Scavina 6' and 'Scavina 12' were described as remarkably resistant in Trinidad in the 1950s (Bartley 1986, 2005). The resistance of these clones was hypothesized to be monogenic, with 'Scavina 6' supposedly homozygous dominant and 'Scavina 12' heterozygous (Bartley, 1986; Rudgard et al., 1993). Because of the poor agronomic characteristics, 'Scavina 6' and 'Scavina 12' were not directly used, but were crossed to develop hybrid varieties or clones, such as the 'Trinidad Selected Hybrids' (TSH) and 'Trinidad Selected Amazon' (TSA) series, successfully cultivated worldwide until today. TSH clones have been commercially used in Trinidad, where they appeared to have had a major role in the reduction of witches' broom incidence (Laker et al., 1988). Much later, in the beginning of the twentieth century, by genetic mapping, a major resistance locus was identified at linkage group nine of the cocoa consensus map (Brown et al., 2005; Faleiro et al., 2006). 'Scavina 6' has been widely used in cocoa breeding programmes as a source of resistance to witches' broom and other cocoa diseases. However, the resistance to *M. perniciosa* was overcome in Ecuador few years after its introduction (Bartley, 1986; Wheeler and Suarez, 1993), a trend also observed in the Brazilian Amazon (Fonseca and Albuquerque, 2000), Peru (Rios-Ruiz, 2001) and in the cocoa plantations of Bahia, Brazil (Paim et al., 2006; Albuquerque et al., 2010). This is likely because of the occurrence of more virulent isolates that were able to break down the resistance in 'Scavina 6' and 'Scavina 12'. This clearly reveals the risk of depending on single sources for long-term resistance deployment in this perennial crop.

The cocoa breeding programme conducted by CEPLAC in Ilhéus, Bahia and at Estação de Recursos Genéticos José Haroldo (ERJOH), in Marituba, Pará State, aims at developing witches' broom-resistant varieties with durable resistance and desirable agronomical traits. ERJOH's germplasm collection currently holds 1,817 accessions known as cocoa from the Brazilian Amazon (CAB), representing 36 out of the 186 Brazilian Amazon River basins (Almeida et al., 1995, 2005; Bartley 2005; Albuquerque et al., 2010). The collection in Ilhéus is composed of germplasm from countries such as Trinidad, Costa Rica, Ecuador and Peru and material from the Brazilian Amazon germplasm bank (Rudgard et al., 1993; Almeida et al., 1995; Bartley, 2005; Sereno et al., 2006; Monteiro and Ahnert, 2012). Crosses carefully designed to include different sources of resistance to witches' broom, other cocoa diseases and self-compatibility (Pires et al., 1996; Lopes et al., 2011) within three sources of germplasm viz. Upper Amazonian, Lower Amazonian and Trinitario materials (Cheesman, 1944) resulted in progenies (Pires et al., 1999), which were screened by using a large-scale seedlings inoculation method (Frias et al., 1995) under nursery conditions followed by field screening in both locations Ilhéus, Bahia and Marituba, Pará State (Luz et al., 1997, 2006; Paim et al., 2006; Albuquerque et al., 2010; Silva et al., 2010). The resulting selections were subjected to field evaluations over six years. Seven CAB accessions ('CAB 0064', 'CAB 0066', 'CAB 0156', 'CAB 0194', 'CAB 0195', 'CAB 0269' and 'CAB 0274') appeared to be significantly more resistant than 'Scavina 6' and 'Scavina 12' (Paim et al., 2006). It was hypothesized that cocoa accessions originally collected from river basins of the Upper Amazonian region would contain resistant individuals, particularly from Upper and Middle Solimões, Purus and Acre river basins, where *M. perniciosa* and *T. cacao* seem to have shared a common centre of diversity (Purdy and Schimdt, 1996). Three other CAB accessions from these regions evaluated also under nursery conditions and nine years of field evaluation were identified as important novel sources of resistance to witches' broom, namely 'CAB 0214', 'CAB 0208' and 'CAB 0270' (Albuquerque et al., 2010).

To date, 39 clones have been released to farmers by CEPLAC and more than 150000 hectares of susceptible varieties have been renewed with resistant clones in southern Bahia by using cuttings of resistant material (Fig. 6) or by grafting. More than 700 clones were tested in large-scale clonal trials and several base populations were established. Currently, large base populations are being formed using sources of resistance to *M. roreri* combined with witches' broom resistance as part of the preventive breeding strategy adopted by CEPLAC's breeding programme to mitigate the risk presented by possible introduction of *M. roreri* into the country.

The completion of the sequence of two cocoa genomes in 2010 opened new doors to improve the efficacy of identifying new segregants carrying genes of interest in breeding populations (Lopes et al., 2011).

Figure 6 Production of plantlets from plagiotropic cocoa branches by rooting cuttings of resistant genotypes. (a) Hormone treatment of the cuttings. (b) Containers filled with substrate for plantlet growth. (c) Rooting of cuttings in a screenhouse with heavy irrigation. (d) Acclimation of plantlets in the screenhouse before planting.

9.2 Integrated management

Problems in the control of plant diseases, including the witches' broom, have emerged as a result of the adoption of a single strategy of control. An integrated approach would be more effective than a single strategy and result in a more sustainable management of the disease. This approach takes into account the entire ecosystem where the crop is located. The adoption of an integrated strategy would ensure the effectiveness of fungicides and the durability of host resistance while reaping additional benefits such as reduced environmental impact and cost of control (Lucas, 1989).

The recommendation given by CEPLAC for integrated management of witches' broom includes the rational use of cupric and systemic fungicides, removal of infected tissue, replacing susceptible cocoa genotypes with more resistant ones and biological control with *Trichoderma stromaticum*. In fact, combining copper hydroxide with *T. stromaticum* consistently yielded a higher number of disease-free pods and increased the amount of healthy beans in a three-year field experiment (Medeiros et al., 2010). Although this experiment showed that broom removal was not considered cost effective under field conditions, this probably was because the cocoa stands were not fertilized. Phytosanitation debilitates plants and reduces pod production in the absence of proper fertilization.

In Bahia, despite the implementation of the recovery programme involving resistant varieties, it became clear that success in controlling losses depended on the adoption of an integrated approach by farmers, due to the peculiar characteristics and complexity of the witches' broom disease. Abandoned farms are sources of witches' broom inoculum to well-managed ones within a radius of at least 60 km (Frias et al., 1991; Griffith et al., 2003). Fertilization, weeding, thinning and pruning to reduce canopy overlap are essential components of cocoa health management.

10 Concluding remarks and outlook

Much is already known about this remarkable pathogen; however, more studies are still necessary to effectively lower its impact on cocoa production in affected areas.

The disease process may be further studied using the miniature tomato (*Solanum lycopersicum*) cultivar Micro-Tom, as a model organism to study its interaction with *M. perniciosa* (Marelli et al., 2009). The differential kinetics and intensity of gene expression between the compatible interactions with biotype S (pathogenic to solanaceous hosts) with an incompatible interaction with biotype C (non-host) was shown with this model (Deganello et al., 2014). The model may be used to study aspects of the interaction with *M. perniciosa* that are still obscure.

Teixeira et al. (2014) identified a set of 61 candidate-secreted effector proteins that may be involved with the pathogenicity of *M. perniciosa* in green brooms. Further studies should focus on the role of these proteins *in vivo*, which could be studied with the tomato model mentioned above, assuming that the pathogenicity genes are equivalent. If not, it is easy enough to obtain these pathogenicity genes from isolates of biotype S.

Techniques such as protoplast transformation (Lima et al., 2003; Barau et al., 2015) and RNAi-based gene silencing (Santos et al., 2009) may be applied to disrupt genes of *M. perniciosa* to identify key factors associated with infection that are still elusive.

Additionally, CRISPR-Cas9 (Wang et al., 2016) will certainly be helpful if implemented in *M. perniciosa*.

The diversity of *M. perniciosa* needs further investigation through more collections to understand host jumps, diversity of the different biotypes and their adaptation on cocoa genotypes with varying resistance levels. Since the diversity of what is named 'biotype H' is higher (Tarnowski, 2009), it probably constitutes new taxa, as shown for an isolate from *Heteropteris acutifolia* (Arruda et al., 2005). Most importantly, this diversity should be considered when designing future resistance pre-breeding and breeding programmes for cocoa and the corresponding movement of germplasm.

Several cocoa genotypes with higher levels of resistance as compared to the classical 'Scavinas' were identified (Paim et al., 2006). Additional research should concentrate on the characterization of the resistant genes in these accessions to plan their sustainable use in the field.

Integrated management will enable cocoa production in infected areas, but productivity levels will depend on the scale to which the measures are applied. The high cost of labour and other inputs such as agrochemicals need to be balanced with the productivity levels attained and the sustainability of the activity. The development of new, more specific and less toxic agrochemicals may lead to a more sustainable management of the disease. The identification of key fungal enzymes in the interaction with the plant, such as the AOX that is directly involved in the transition from the biotrophic to necrotrophic phase (Thomazella et al., 2012), is the first step in the development of such chemicals. Molecular modelling of the protein structure followed by docking with inhibitors may lead to the identification of potential candidates *in silico* (Santos Júnior et al., 2011, 2013).

11 Where to look for further information

During the elaboration of this chapter a new book on cocoa diseases was released and it is certainly an important source of information on witches' broom. It is edited by the group working on cocoa diseases at the USDA and the book is available from Springer: https://link.springer.com/book/10.1007%2F978-3-319-24789-2

Other sources of further information are:

1 CEPLAC, the Brazilian cocoa research institute was created in 1957 and since then conducts research on several aspects of cocoa, including diseases. A great number of freely available publications, unfortunately, many of them in Portuguese only, including scientific publications, technical bulletins and technical books on the witches' broom disease may be found at: htpp:// http://www.ceplac.gov.br/
2 Researchers at the LGE laboratory at Campinas University, Brazil, coordinated the *M. perniciosa* genome sequencing project and still conduct research on genomics and gene expression of *M. perniciosa* and *M. roreri*. A list of publications of the group may be found at: http://www.lge.ibi.unicamp.br/pagina/estrutura/corpo_publicacoes.php
3 The Sustainable Perennial Crops Laboratory at the USDA, Beltsville, Maryland, USA also conducts research on cocoa diseases. Further information on their activities may be found at: https://www.ars.usda.gov/northeast-area/beltsville-md/beltsville-agricultural-research-center/sustainable-perennial-crops-laboratory/

12 Acknowledgements

The authors acknowledge the financial support of CNPq and CAPES.

13 References

Aime, M. C. and Phillips-Mora, W. (2005). The causal agent of witches' broom and frosty pod rot of cacao (chocolate, *Theobroma cacao*) form a new lineage of Marasmiaceae. *Mycologia*, 97:1012–22.

Albuquerque, P. S. B., Bastos, C. N., Luz, E. D. M. N. and Silva, S. D. V. M. (2005). Doenças do cacaueiro. In: Kimati, H., Amorim, L., Resende, J. A. M., Bergamim Filho, A. and Camargo, L. E. A. (Eds), *Manual de Fitopatologia* (pp. 151–64). 4th ed., Volume 2, Piracicaba: Ceres.

Albuquerque, P. S. B., Silva, S. D. V. M., Luz, E. D. M. N., Pires, J. L., Vieira, A. M., Demetrio, C. G. B., Pascholatti, S. F. and Figueira, A. (2010). Novel sources of witches broom resistance (causal agent *Moniliophthora perniciosa*) from natural populations of *Theobroma cacao* from the Brazilian Amazon. *Euphytica*, 172:125–38.

Alger, K. and Caldas, M. (1994). The declining of the cocoa economy in the Atlantic forest of southern Bahia, Brazil: Conservation attitudes of cocoa planters. *Environmentalist*, 14:107–19.

Almeida, L. C. and Andebrhan, T. (1984). *Investigações Sobre Vassoura-de-bruxa Do Cacaueiro (Crinipellis perniciosa Stahel) Na Amazônia Brasileira*. Belém: Ceplac/Depea/Copes, p. 40.

Almeida, H. A. and Luz, E. D. M. N. (1995). Influência da chuva, temperatura e umidade relativa do ar na produção de basidiomas de *Crinipellis perniciosa*. *Fitopatol. Bras.*, 20:374.

Almeida, C.M.V.C., Machado, P.F.R., Barriga, J.P., Silva, F.C.O.D. (1995). Coleta de cacau (Theobroma cacao L.) da Amazônia Brasileira: uma abordagem histórica e analítica. Porto Velho: CEPLAC, SUPOCpp. 92.

Almeida, H. A. (2002). *Relação entre o fenômeno el niño-oscilação sul e a ocorrência de chuvas no sudeste da Bahia*. Ilhéus: Mapa/Ceplac, p. 19. (Boletim Técnico 183).

Almeida, C. M. C. V., Dias, L. A. S., Okabe, E. T., Raymundo, J. and Medeiros, P. (2005). Variability in genetic resources of cacao in Rondônia, Brazil. *Crop Breed. Appl. Biot.*, 5:318–24.

Alvim, P. T. (1984). Flowering of cacao. *Cacao Growers' Bull.*, 35:25–31.

Andebrhan, T., Almeida, L. C. and Fonseca, S. E. A. (1983). *Doenças do cacaueiro*. Belém: Ceplac/Depea/Copea, p. 91.

Andebrhan, T., Bonfim, J. C., Carvalho, A. L. P. and Albuquerque, P. S. B. (1993). Comparative epidemiology experiment: Brazil. In: Rudgard, S. A., Maddison, A. C. and Andebrhan, T. (Eds), *Disease Management in Cacao: Comparative Epidemiology of Witches' Broom of Cacao* (pp. 33–56). London: Chapman & Hall.

Andebrhan, T., Figueira, A., Yamada, M. M., Cascardo, J. and Furtek, D. B. (1999). Molecular fingerprinting suggests two primary outbreaks of witches' broom disease (*Crinipellis perniciosa*) of *Theobroma cacao* in Bahia, Brazil. *Eur. J. Plant Pathol.*, 105:167–75.

Andrade, B. S., Taranto, A. G., Góes-Neto, A. and Duarte, A. A. (2009). Comparative modeling of DNA and RNA polymerases from *Moniliophthora perniciosa* mitochondrial plasmid. *Theor. Biol. Med. Model.*, 6:1–6.

Aneja, M. and Gianfagna, T. (2001). Induction and accumulation of caffeine in young, actively growing leaves of cacao (*Theobroma cacao* L.) by wounding or infection with *Crinipellis perniciosa*. *Physiol. Mol. Plant Pathol.*, 59:13–16.

Arruda, M. C. C., Sepulveda, G. F., Miller, R. N. G., Resende, M. L. V., Dianese, J. C. and Felipe, M. S. S. (2005). *Crinipellis brasiliensis*, a new species based on morphological and molecular data. *Mycologia*, 97:1348–61.

Arruda, M. C. C., Miller, R. N. G., Ferreira, M. A. S. V. and Felipe, M. S. S. (2003a). Comparison of *Crinipellis perniciosa* isolates from Brazil by ERIC repetitive element sequence-based PCR genomic fingerprinting. *Plant Pathol.*, 52:236–44.

Arruda, M. C. C., Ferreira, M. A., Miller, R. N., Resende, M. L. V. and Felipe, M. S. S. (2003b). Nuclear and mitochondrial rDNA variability in *Crinipellis perniciosa* from different geographic origins and hosts. *Mycol. Res.*, 107:25–37.

Baker, R. E. D., Crowdy, S. H. and Thorold, C. A. (1941). *Witches' broom disease investigations.* I. Seasonal variations in the intensity of infections and their effect on control measures. *Trop. Agric.*, 18:107–16.

Baker, R. E. D. and Crowdy, S. H. (1943). Studies in the witches' broom disease of cacao caused by *Marasmius perniciosus* Stahel. Part I. Introduction, symptoms and etiology. *Memoirs of the Imperial College of Tropical Agriculture Trinidad*, 7:1–27.

Baker, R. E. D. and Holliday, P. (1957). *Witches' Broom Disease of Cacao (Marasmius perniciosus Stahel).* Richamond: Commonwealth Mycological Institute, p. 42. Phytopathological Papers, 2.

Barau, J., Grandis, A., Andrade Carvalho, V. M., Teixeira, G. S., Zaparoli, G. H. A., Rio, M. C. S., Rincones, J., Buckeridgem, M. S. and Pereira, G. A. G. (2015). Apoplastic and intracellular plant sugars regulate developmental transitions in witches' broom disease of cacao. *J. Exp. Bot.*, 66:1325–37.

Barreto, R. W. and Evans, H. C. (1996). Duas novas solanáceas selvagens hospedeiras de *Crinipellis perniciosa* em Minas Gerais: Uma explicação alternativa para a chegada da vassoura de bruxa do cacao na Bahia? *Fitopatol. Bras.*, 21:348.

Bartley, B. G. D. (1986). *Cacao, Theobroma Cacao.* FAO Plant Production and Protection Paper, 70, pp. 25–42.

Bartley, B. G. D. (2005). *The Genetic Diversity of Cacao and its Utilization.* London: CABI Publishing, p. 341.

Bastos, C. N. (1987). *Efeito Protetor e Curativo De Fungicidas Sobre C. Perniciosa Em Mudas De Cacaueiro.* Belém: Informe de Pesquisas, Ceplac/Depea, pp. 58–63.

Bastos, C. N. (1988). Resultados preliminares sobre eficacia de *Trichoderma viride* no controle da vassoura-de-bruxa (*Crinipellis* perniciosa) do cacaueiro. *Fitopatol. Bras.*, 13:340–342.

Bastos, C. N. (1989). Avaliação de fungicidas sistêmicos no controle da vassoura de bruxa do cacaueiro. *Agrotrópica*, 1:128–32.

Bastos, C. N. (1990). *Epifitiologia, Hospedeiros e Controle De Vassoura-de-bruxa (Crinipellis perniciosa) (Stahel) Singer.* Ilhéus: CEPLAC/CEPEC, p. 21. (Boletim Técnico, 168).

Bastos, C. N. and Almeida, T. (1988). Comparação morfológica de isolados de *Crinipellis perniciosa.* *Fitopatol. Bras.*, 13:202–5.

Bastos, C. N. and Silva, H. M. (1980). *Doenças Do Cacaueiro Na Amazônia Brasileira.* Belém: Ceplac/Depea/Copes, p. 42. (Boletim Técnico 2).

Bastos, C. N. and Medeiros, A. G. (1979). Seleção *in vitro* de fungicidas a *Crinipellis perniciosa* (Stahel) Singer causador da vassoura-de-bruxa do cacaueiro. *Revista Theobroma*, 9:129–35.

Bastos, C. N., Evans, H. C. and Samson, R. A. (1981). A new hyperparasitic fungus, *Cladobotryum amazonense*, with potential for control of fungal pathogens of cocoa. *Trans. Br. Mycol. Soc.*, 77:273–8.

Bastos, C. N. and Evans, H. C. (1985). A new pathotype of *Crinipellis perniciosa* (witches' broom disease) on solanaceous hosts. *Plant Pathol.*, 34:306–12.

Bastos, C. N. and Andebrhan, T. (1986). Urucu (*Bixa orellana*), a new host of witches' broom (*Crinipellis perniciosa*) of cacao. *Fitopatol. Bras.*, 11:963–5.

Bastos, C. N, Almeida, O. C. and Silva, S. D. V. M. (1998). Hospedeiros colaterais de *Crinipellis perniciosa* na Bahia, Brasil. *Fitopatol. Bras.*, 23(suppl.):228.

Bezerra, J. L., Almeida, O. C. and Silva, S. D. V. M. (1998). Hospedeiros colaterais de *Crinipellis perniciosa* na Bahia, Brasil. *Fitopatol. Bras.*, 23:228.

Bowers, J. H., Bailey, B. A., Hebbar, P. K., Sanogo, S. and Lumsden, R. D. (2001). The impact of plant diseases on world chocolate production. *Plant Health Progress*, 10, doi:10.1094/PHP-2001-0709-01-RV.

Brown, J. S., Schnell, R. J., Motamayor. J. C., Lopes. U., Kuhn, D. N. and Borrone, J. W. (2005). Resistance gene mapping for witches' broom disease in *Theobroma cacao* L. in an F2 population using SSR markers and candidate genes. *J. Am. Soc. Hortic. Sci.*, 130:366–73.

Cabral, A. D. S., Carvalho, P. M. B. D., Pinotti, T., Hagler, A. N., Mendonça-Hagler, L. C. S. and Macrae, A. (2009). Killer yeasts inhibit the growth of the phytopathogen *Moniliophthora perniciosa*, the causal agent of Witches' Broom disease. *Braz. J. Microbiol.*, 40:108–10.

Calle, H., Cook, A. A. and Fernando, S. Y. (1982). Histology of witches'-broom caused in cacao by *Crinipellis perniciosa*. *Phytopathology*, 72:1479–81.

Carvalho, L. E., Lima, M. P., Máximo, A. C., Pereira, E. C. S., Moreira, W. A. S., Ferreira, A. G., Véras, S. M. and Souza, M. G. (2012). Estudo em raiz e ráquis foliar de *Spathelia excelsa*: Fitoquímica e atividade frente ao fungo *Moniliophthora perniciosa* associado ao cupuaçuzeiro (*Theobroma grandiflorum*). *Quím. Nova*, 35:2237–40.

Ceita, G. O., Macêdo, J. N. A., Santos, T. B., Alemanno, L., Gesteira, A. S., Micheli, F., Mariano, A. C., Gramacho, K. P., Silva, D. C., Meinhardt, L., Mazzafera, P., Pereira, G. A. G. and Cascardo, J. C. M. (2007). Involvement of calcium oxalate degradation during programmed cell death in *Theobroma cacao* tissues triggered by the hemibiotrophic fungus *Moniliophthora perniciosa*. *Plant Sci.*, 173:106–17.

Charles, V. K. (1906). Occurrence of *Lasiodiplodia* on *Theobroma cacao* and *Mangifera indica*. *J. Mycol.*, 12:145–6.

Chaves, F. C. and Gianfagna, T. J. (2007). Cacao leaf procyanidins increase locally and systemically in response to infection by *Moniliophthora perniciosa* basidiospores. *Physiol. Mol. Plant Path.*, 70:174–9.

Cheesman, E. E. (1944). Notes on the nomenclature, classification and possible relationships of cacao populations. *Trop. Agric.*, 21:144–59.

Clausen, M., Kräuter, R., Schachermayr, G., Potrykus, I. and Sautter, C. (2000). Antifungal activity of a virally encoded gene in transgenic wheat. *Nat. Biotechnol.*, 18:446–9.

Costa, J. D. C. B., Resende, M. L. V. D., Ribeiro Júnior, P. M., Camilo, F. R., Monteiro, A. C. A. and Pereira, R. B. (2010). Induction of resistance in cacao seedlings against *Moniliophthora perniciosa* by a phosphorilated mannan oligosaccharide based product. *Trop. Plant Pathol.*, 35:285–94.

Costa, J. C. B., Bezerra, J. L. and Cazorla, I. M. (1996). Controle biológico da vassoura-de-bruxa do cacaueiro na Bahia com *Trichoderma polysporum*. *Fitopatol. Bras.*, 21:397.

Deganello, J., Leal, G. A., Rossi, M. L., Peres, L. E. P. and Figueira, A. (2014). Interaction of *Moniliophthora perniciosa* biotypes with Micro-Tom tomato: a model system to investigate the witches' broom disease of *Theobroma cacao*. *Plant Pathol.*, 63:1251–63.

De Souza, J. T., Pomella, A. W., Bowers, J. H., Pirovani, C. P., Loguercio, L. L. and Hebbar, K. P. (2006). Genetic and biological diversity of *Trichoderma stromaticum*, a mycoparasite of the cacao witches'-broom pathogen. *Phytopathology*, 96:61–7.

De Souza, J. T., Bailey, B. A., Pomella, A. W. V., Erbe, E. F., Murphy, C. A., Bae, H. and Hebbar, P. K. (2008). Colonization of cacao seedlings by *Trichoderma stromaticum*, a mycoparasite of the witches' broom pathogen, and its influence on plant growth and resistance. *Biol. Control*, 46:36–45.

Delgado, J. C. and Cook, A. A. (1976). Nuclear condition of the basidia, basidiospores and mycelium of *Marasmius perniciosus*. *Phytopathology*, 66: 717–718.

End, M. J., Daymond, A. J. and Hadley, P. (Eds). (2014). *Technical Guidelines for the Safe Movement of Cacao Germplasm*. Revised from the FAO/IPGRI Technical Guidelines No. 20 (Second Update, August 2014). Montpellier: Global Cacao Genetic Resources Network (CacaoNet), Bioversity International.

Espino, Z. D. (1989). Outbreaks and new records. Panama. Outbreak of cacao witches' broom disease, *Crinipellis perniciosa*. *FAO Plant Protect. Bull.*, 37:97.

Evans, H. C. (1978). Witches' broom disease of cacao (*Crinipellis perniciosa*) in Ecuador. *Ann. Appl. Biol.*, 89:185–92.

Evans, H. C. (1981). Witches' broom disease – a case study. *Cocoa Grower's Bull.*, 32:6–19.

Evans, H. C. (2002). Invasive neotropical pathogens of tree crops. In Watling, R., Frankland, J. C., Ainsworth, A. M., Isaac, S. and Robinson, C. H. (Eds.). *Tropical Mycology* (Micromycetes, Volume 2) (pp. 83–112). Wallingford: CABI Publishing.

Evans, H. C. and Prior, C. (1987). Cocoa pod diseases: Causal agents and control. *Outlook Agric.*, 16:35–41.

Evans, H. C. and Barreto R. W. (1996). *Crinipellis perniciosa*: A much investigated but little understood fungus. *Mycologist*, 10:58–61.

Evans, H. C., Holmes, K. A., Phillips, W. and Wilkinson, M. J. (2002). What's in a name: *Crinipellis*, the final resting place for the frosty pod rot pathogen of cacao? *Mycologist*, 16:148–52.

Faleiro, F. G., Queiroz, V. T., Lopes, U. V., Guimarães, C. T., Pires, J. L., Yamada, M. M., Araújo, I. S., Pereira, M. G., Schnell, R., Souza Filho, G. A., Ferreira, C. F., Barros, E. G. and Moreira, M. A. (2006). Mapping QTLs for witches' broom (*Crinipellis perniciosa*) resistance in cacao (*Theobroma cacao* L.). *Euphytica*, 149:227–35.

FAOSTAT Online Database. (2014). http://faostat.fao.org/

Formighieri, E. F., Tiburcio, R. A., Armas, E. D., Medrano, F. J., Shimo, H., Carels, N., Góes-Neto, A., Cotomacci, C., Carazzolle, M. F., Sardinha-Pinto, N., Rincones, J., Digiampietri, L., Carraro, D. M., Azeredo-Espin, A. M., Reis, S. F., Deckmann, A. C., Gramacho, K., Gonçalves, M. S., Moura Neto, J. P., Barbosa, L. V., Meinhardt, L. W., Cascardo, J. C. M. and Pereira, G. A. G. (2008). The mitochondrial genome of the phytopathogenic basidiomycete *Moniliophthora perniciosa* is 109 kb in size and contains a stable integrated plasmid. *Mycol. Res.*, 112:1136–52.

Fonseca, S. E.A. and Albuquerque, P. S. B. (2000). Avaliação de clones de cacau na Amazônia brasileira em relação a incidência de vassoura-de-bruxa. In: *International Cocoa Research Conference, 12, 1996. Salvador. Proceedings*. Lagos: Cocoa Producers' Aliiance, pp. 149–53.

Franco, S. F., Baroni, R. M., Carazzolle, M. F., Teixeira, P. J. P. L., Reis, O., Pereira, G. A. G. and Mondego, J. M. C. (2015). Genomic analyses and expression evaluation of thaumatin-like gene family in the cacao fungal pathogen *Moniliophthora perniciosa*. *Biochem. Biophys. Res. Commun.*, 466:629–36.

Frias, G. A., Purdy, L. H. and Schmidt, R. A. (1991). Infection biology of *Crinipellis perniciosa* on vegetative flushes of cacao. *Plant Dis.*, 75:552–6.

Frias, G. A., Purdy, L. H. and Schmidt, R. A. (1995). An inoculation method for evaluating resistance of cacao to *Crinipellis perniciosa*. *Plant Dis.*, 79:787–91.

Gesteira, A. S., Micheli, F., Carels, N., Silva, A. C., Gramacho, K. P., Schuster, I., Macedo, J. N., Pereira, G. A. and Cascardo, J. C. (2007). Comparative analysis of expressed genes from cacao meristems infected by *Moniliophthora perniciosa*. *Ann. Bot.*, 100:129–40.

Gomes, L. H., Duarte, K. M. R., Andrino, F. G., Leal Júnior, G. A., Garcia, L. M., Figueira, A. and Lira, S. P. (2014). Alpha-Tomatin against Witches' Broom Disease. *Am. J. Plant Sci.*, 5:596–604.

Gonçalves, J. R. C. (1965). *Theobroma grandiflorum* (Spreng.) Schum. as source of inoculum of witch-broom disease of *Theobroma cacao* L. *Trop. Agric.*, 42:261–3.

Gramacho, K. P., Lopes, U. V., Pires, J. L., Lopes, J. R. M., Bahia, R. C. S., Mahe, L. (2005). Phylogeography of the witches' broom pathogen in Bahia. *Agrotrópica*, 17:65–72.

Gramacho, K. P., Risterucci, A. M., Lanaud, C., Lima, L. S. and Lopes, U. V. (2007). Characterization of microsatellites in the fungal plant pathogen *Crinipellis perniciosa*. *Mol. Ecol. Notes*, 7:153–5.

Griffith, G. W. and Hedger, J. N. (1994a). The breeding biology of biotypes of the witches' broom pathogen of cocoa, *Crinipellis perniciosa*. *Heredity*, 72:278–89.

Griffith, G. W. and Hedger, J. N. (1994b). Spatial distribution of mycelia of the liana (L-) biotype of the agaric *Crinipellis perniciosa* (Stahel) Singer in tropical forest. *New Phytol.*, 127:243–59.

Griffith, G. W., Bravo-Velasquez, E., Wilson, F. J., Lewis, D. M. and Hedger, J. N. (1994). Autoecology and evolution of the witches' broom pathogen (Crinipellis perniciosa) of cocoa. In Blakeman, J. P. and Williamson, B. (Eds), *Ecology of Plant Pathogens* (pp. 245–67). Wallingford: CAB International.

Griffith, G. W., Nicholson, J., Nenninger, A., Birch, R. N. and Hedger, J. N. (2003). Witches' brooms and frosty pods: Two major pathogens of cacao. *New Zeal. J. Bot.*, 41:423–35.

Griffith, G. W. (2004). Witches' brooms and frosty pods: Threats to world cacao production. *Biologist*, 51:71–5.

Hanada, R. E., Pomella, A. W. V., Costa, H. S., Bezerra, J. L., Loguercio, L. L., Pereira, J. O. (2010). Endophytic fungal diversity in *Theobroma cacao* (cacao) and *T. grandiflorum* (cupuaçu) trees and their potential for growth promotion and biocontrol of black-pod disease. *Fungal biol.*, 114:901–10.

Hedger, J. N., Lewis, P. A., Gitay, H. (1994). Litter trapping by fungi in moist tropical forest.In: Isaac, S., Frankland, J. C., Watling, R., Whalley, A. J. S. (Eds), *Aspects of tropical mycology* (pp. 15–35). Cambridge: Cambridge University Press.

Hjorth, S., Pomella, A. W. V., Hockenhull, J. R. and Hebbar, P. K. (2003). Biological control of witches' broom disease, (*Crinipellis perniciosa*), with the co-evolved fungus, *Trichoderma stromaticum*: testing different delivery regimes. In: *XIV International Cocoa Research Conference*, Accra, Ghana, 2, pp. 691–7.

Holliday, P. (1952). *Witches' Broom Disease of Cacao (Marasmius perniciosus Stahel)*. Colonial Report, 286, pp. 1–8. London: HMSO.

Holliday, P. (1971). Some tropical plant pathogens of limited distribution. *Rev. Plant Pathol.*, 50:337–48.

Holmes, K. A., Schroers, H., Thomas, S. E., Evans, H. C. and Samuels, G. J. (2004). Taxonomy and biocontrol potential of a new species of *Trichoderma* from the Amazon basin of South America. *Mycol. Prog.*, 3:199–210.

Hora Júnior, B. T., Poloni, J.F, Lopes, M. A., Dias, C. V., Gramacho, K. P., Schuster, I., Schuster, I., Sabau, X., Cascardo, J. C. M., Mauro, S. M. Z., Gesteira, A. S., Bonatto, D. and Micheli, F. (2012). Transcriptomics and systems biology analysis in identification of specific pathways involved in cacao resistance and susceptibility to witches' broom disease. *Mol. Biosys.*, 8:1507–19.

Howard, A. (1901). Witch-broom of cacao. *W. Indian Bull.*, 2:205.

Junior, P. (2006). *Terrorismo Biológico*. Revista Veja: Junho, pp. 60–3.

Kelly, P. L., Reeder, R., Rhodes, S. and Edwards, N. (2009). First confirmed report of witches' broom caused by *Moniliophthora perniciosa* on cacao, *Theobroma cacao*, in Saint Lucia. *Plant Pathol.*, 58:798.

Kessmann, H., Staub, T., Ligon, J., Oostendorp, M. and Ryals, J. (1994b). Activation of systemic acquired disease resistance in plants. *Eur. J. Plant Pathol.*, 100:359–69.

Kilaru, A., Bailey, B. A. and Hasenstein, K. H. (2007). *Moniliophthora perniciosa* produces hormones and alters endogenous auxin and salicylic acid in infected cacao leaves. *FEMS Microbiol. Lett.*, 274:238–44.

Kilaru, A. and Hasenstein, K. H. (2005). Development and pathogenicity of the fungus *Crinipellis perniciosa* on interaction with cacao leaves. *Phytopathology*, 95:101–7.

Krauss, U. and Soberanis, W. (2002). Effect of fertilization and biocontrol application frequency on cocoa pod diseases. *Biol. control*, 24:82–9.

Kues, U., and Navarro-Gonzalez, M. (2010). Mating-type orthologous genes in the primarily homothallic *Moniliophthora perniciosa*, the causal agent of witches' broom disease in cacao. *J. Basic Microbiol.*, 50: 442–551.

Laker, H. A. (1991). Evaluation of systemic fungicides for control of witches' broom disease of cocoa in Trinidad. *Trop. Agric.*, 68:119–24.

Laker, H. A. and Ram, A. (1992). Investigation on integrated control of witches' broom disease of cocoa in Rondonia State, Brazil. *Trop. Pest Manage.*, 38:354–8.

Laker, H. A. and Rudgard, S. A. (1989). A review of the research on chemical control of witches' broom disease of cocoa. *Cocoa Growers' Bull.*, 42:12–24.

Laker, H. A., Sreenivasan, T. N. and Kumar, D. R. (1988). Present status of witches' broom disease of cocoa in Trinidad. *Int. J. Pest Manage.*, 34:318–23.

Lana, T. G., Azevedo, J. L., Pomella, A. W. V., Monteiro, R. T. R., Silva, C. B. and Araujo, W. L. (2011). Endophytic and pathogenic isolates of the cacao fungal pathogen *Moniliophthora perniciosa* (Tricholomataceae) are indistinguishable based on genetic and physiological traits. *Genet. Mol. Res.*, 10:326–34.

Leal, G. A., Albuquerque, P. S. B. and Figueira, A. (2007). Genes differentially expressed in *Theobroma cacao* associated with resistance to witches' broom disease caused by *Crinipellis perniciosa*. *Mol. Plant Pathol.*, 8:279–92.
Lemke, P. A. (1969). A reevaluation of homothallism, heterothallism and the species concept in *Sistotrema brinkmanii*. *Mycologia*, 61:57–76.
Lima, J. O., Santos, J. K., Pereira, J. F., Resende, M. L. V., Araújo, E. F. and Queiroz, M. V. (2003). Development of a transformation system for *Crinipellis perniciosa*, the causal agent of witches'broom in cacao plants. *Curr. Genet.*, 42:236–40.
Lins, A. C. R. (1985). *Witches' Broom Disease of Cocoa: A Search for New Methods of Control*. PhD Thesis, London: Imperial College of Science and Technology, p. 171.
Loguercio, L. L., Santos, L. S., Niella, G. R., Miranda, R. A. C., De Souza, J. T., Collins, R. T. and Pomella, A. W. V. (2009). Canopy-microclimate effects on the antagonism between *Trichoderma stromaticum* and *Moniliophthora perniciosa* in shaded cacao. *Plant Pathol.*, 58:1104–15.
Lopes, J. R. M., Luz, E. D. M. N. and Bezerra, J. L. (2001). Suscetibilidade do cupuaçuzeiro e outras espécies vegetais a isolados de *Crinipellis perniciosa* obtidos de quatro hospedeiros diferentes no sul da Bahia. *Fitopatol. Bras.*, 26:601–5.
Lopes, M. A., Gomes, D. S., Koblitz, M. G. B., Pirovani, C. P., de Mattos Cascardo, J. C., Góes-Neto, A. and Micheli, F. (2008). Use of response surface methodology to examine chitinase regulation in the basidiomycete *Moniliophthora perniciosa*. *Mycol. Res.*, 112:399–406.
Lopes, U. V., Monteiro, W. R., Pires, J. L., Clement, D., Yamada, M. M. and Gramacho, K. P. (2011). Cacao breeding in Bahia, Brazil: Strategies and results. *Crop Breed. Appl. Biotechnol.*, 11:73–81.
Lucas, J. A. (1989). *Plant Pathology and Plant Pathogens*. Oxford, UK: Black well Science, Ltd., p. 274.
Luz, E. D. M. N., Bezerra, J. L., Resende, M. L. V. and Oliveira, M. L. (1997). Cacau – controle de doenças. In: Vale, F. X. R. and Zambolim, L. (Eds), *Controle de Doenças de Plantas - Grandes Culturas*. Volume 2 (pp. 611–49). Viçosa, Minas Gerais: UFV.
Luz, E. D. M. N., Almeida, H. A. and Machado, R. C. R. (1994). Períodos de incubação, secamento de vassouras, produção de basidiomas e atividade de *Crinipellis perniciosa* em ramos e frutos de cacaueiro na Bahia. *Fitopatol. Bras.*, 19:341.
Luz, E. D. M. N., Almeida, H. A., Machado, R. C. R. and Santos Filho, L. P. (2001). Correlações entre dados climáticos e fenológicos no desenvolvimento de sintomas da vassoura-de-bruxa do cacaueiro na Bahia. *Fitopatol. Bras.*, 26:352.
Luz, E. D. M. N., Almeida, F. A. and Santos Filho, L. P. (2002). Contribuição da esporulação em frutos na epidemiologia da vassoura-de-bruxa do cacaueiro em Camacan-BA. *Fitopatol. Bras.*, 27:128.
Luz, E. D. M. N., Paim, M. C. A., Silva, S. D. V. M., Pires, J. L., Filho, L. S. and Monteiro, W. R. (2006). Correlations of cocoa clones' reactions to different species of *Phytophthora*. *Global Approaches to Cocoa Germplasm Utilization and Conservation*, 50:132.
Macagnan, D., Romeiro, R. D. S. and Pomella, A. W. (2008). Production of lytic enzymes and siderophores, and inhibition of germination of basidiospores of *Moniliophthora* (ex *Crinipellis*) *perniciosa* by phylloplane actinomycetes. *Biol. Control*, 47:309–14.
Marelli, J. P., Maximova, S. N., Gramacho, K. P., Kang, S. and Guiltinan, M. J. (2009). Infection biology of *Moniliophthora perniciosa* on *Theobroma cacao* and alternate solanaceous hosts. *Trop. Plant Biol.*, 2:149–60.
McQuilken, M. P. and Rudgard, S. A. (1988). Sensitivity of *Crinipellis perniciosa* to two triazole fungicides *in vitro* and their effect on development of the fungus in cocoa. *Plant Pathol.*, 37:499–506.
Medeiros, F. H. V., Pomella, A. W. V., de Souza, J. T., Niella, G. R., Valle, R., Bateman, R. P., Fravel, D., Vinyard, B. and Hebbar, P. K. (2010). A novel, integrated method for management of witches' broom disease in Cacao in Bahia, Brazil. *Crop Prot.*, 29:704–11.
Meinhardt, L. W., Costa, G. G. L., Thomazella, D. P. T., Teixeira, P. J. P. L., Carazzolle, M. F., Schuster, S. C., Carlson, J. E., Guiltinan, M. J., Mieczkowski, P., Farmer, A., Ramaraj, T., Crozier, J., Davis, R. E., Shao, J., Melnick, R. L., Pereira, G. A. G. and Bailey, B. A. (2014). Genome and secretome analysis of the hemibiotrophic fungal pathogen *Moniliophthora roreri*, which causes frosty pod

rot disease of cacao: Mechanisms of the biotrophic and necrotrophic phases. *BMC Genomics*, 15:1.
Meinhardt, L. W., Rincones, J., Bailey, B. A., Aime, M. C., Griffith, G. W., Zhang, D. and Pereira, G. A. (2008). *Moniliophthora perniciosa*, the causal agent of witches' broom disease of cacao: What's new from this old foe? *Mol. plant pathol.*, 9:577–88.
Meinhardt, L. W., Bellato, C. M., Rincones, J., Azevedo, R. A., Cascardo, J. C. M. and Pereira, G. A. G. (2006). *In vitro* production of biotrophic-like cultures of *Crinipellis perniciosa*, the causal agent of witches' broom disease of *Theobroma cacao*. *Curr. Microbiol.*, 52:191–6.
Mondego, J. M.C, Carazzolle, M. F., Costa, G. G. L., Formighieri, E. F., Parizzi, L. P., Rincones, J., Cotomacci, C., Carraro, D. M., Cunha, A. F., Carrer, H., Vidal, R. O., Estrela, R. C., García, O., Thomazella, D. P. T., Oliveira, B. V., Pires, A. B. L., Rio, M. C. S., Araújo, M. R. R., Moraes, M. H., Castro, L. A. B., Gramacho, K. P., Gonçalves, M. S., Moura Neto, J. P., Góes Neto, A., Barbosa, L. V., Guiltinan, M. J., Bailey, B. A., Meinhardt, L. W., Cascardo, J. C. M. and Pereira, G. A. G. (2008). A genome survey of *Moniliophthora perniciosa* gives new insights into witches' broom disease of cacao. *BMC Genomics*, 9: 548.
Monteiro, W. R., Ahnert, D. (2012). Melhoramento genético do cacaueiro. In: Valle, R. R. (Ed), *Ciência, tecnologia e manejo do cacaueiro* (pp. 1–16). Itabuna: Editora Vital Ltda.
Mota, S. G., Barros, T. F. and Castilho, M. S. (2010). *In vitro* screening and chemometrics analysis on a series of azole derivatives with fungicide activity against *Moniliophthora perniciosa*. *J. Braz. Chem. Soc.*, 21:510–19.
Niella, G. R., Resende, M. D., Castro, H. D., Silva, L. H. C. P. and Carvalho, J. A. (1999). Aperfeiçoamento da metodologia de produção artificial de basidiocarpos de *Crinipellis perniciosa*. *Fitopatol. Bras.*, 24:523–7.
Oliveira, M. M., Assis, F. G. D. V. and Leal, P. L. (2014). Efficiency of filamentous fungi in the biological control of witches' broom disease of cacao. *Sitientibus sér. Ci. Biol.*, 14:1–7.
Oliveira, M. L., Luz, E. D. M. N. (2005). *Identificação e Manejo Das Principais Doenças Do Cacaueiro No Brasil*. Ilhéus, BA: Cepec/Ceplac, p. 132.
Oliveira, M. L. (2004a). Eficácia de fungicidas triazóis no controle da vassoura-de-bruxa do cacaueiro causada por *Crinipellis perniciosa*. *Fitopatol. Bras.*, 29:S149.
Oliveira, M. L. (2004b). Estrobilurinas, novo grupo de fungicidas com eficácia contra a vassoura-de-bruxa do cacaueiro. *Fitopatol. Bras.*, 29:S285.
Padwick, G. W. (1956). *Losses Caused by Plant Diseases in the Colonies*. Phytopathological Papers, 1, pp. 1–60.
Paim, V. R. L. M., Luz, E. D. M. N., Pires, J. L., Silva, S. D. V. M., Souza, J. T., Albuquerque, P. S. B. and Santos Filho, L. P. (2006). Fontes de resistência a *Crinipellis perniciosa* em progênies de cacaueiros coletados na Amazônia brasileira. *Sci. Agr.*, 63:572–8.
Pegler, D. N. (1978). *Crinipellis perniciosa* (Agaricales). *Kew Bull.*, 32, 731–6.
Pereira, J. F., Ignacchiti, M. D. C., Araújo, E. F., Brommonshenkel, S. H., Cascardo, J. C. M., Pereira, G. A. G. and Queiroz, M. V. (2007). PCR amplification and sequence analyses of reverse transcriptase-like genes in *Crinipellis perniciosa* isolates. *Fitopatol. Bras.*, 32:373–81.
Pereira, J. F., Almeida, A. P. M. M., Cota, J., Pamphile, J. A., Silva, G. F., Araújo, E. F., Gramacho, K.P, Brommonschenkel, S. H., Pereira, G. A. G. and Queiroz, M. V. (2013). Boto, a class II transposon in *Moniliophthora perniciosa*, is the first representative of the PIF/Harbinger superfamily in a phytopathogenic fungus. *Microbiology*, 159:112–25.
Pereira, J. F., Araújo, E. F., Brommonschenkel, S. H., Queiroz, C. B., Costa, G. G. L., Carazzolle, M. F., Pereira, G. A. G. and Queiroz, M. V. (2015). MpSaci is a widespread gypsy-Ty3 retrotransposon highly represented by non-autonomous copies in the *Moniliophthora perniciosa* genome. *Curr. genet.*, 61:185–202.
Pereira, J. L., Barreto, R. W. and Bezerra, J. L. (1997). Distribuicão de *Crinipellis perniciosa* em solanáceas selvagens no Sudeste do Brasil. *Fitopatol. Bras.*, 22, 295.
Pereira, J. L., Ram, A., Figueredo, J. M. and Almeida, L. C. C. (1989). Primeira ocorrência da vassoura-de-bruxa na principal região produtora de cacau de Brasil. *Agrotrópica*, 1:79–81.

Pereira, J. L., Ram, A., Figueredo, J. M. and Almeida, L. C. C. (1990). First occurrence of witches' broom disease in the principal cocoa-growing region of Brazil. *Trop. Agric.*, 67:188–9.

Pires, A. B., Gramacho, K. P., Silva, D. C., Goes-Neto, A., Silva, M. M., Muniz-Sobrinho, J. S., Porto, R. F., Villela-Dias, C., Brendel, M., Cascardo, J. C. and Pereira, G. A. (2009). Early development of *Moniliophthora perniciosa* basidiomata and developmentally regulated genes. *BMC Microbiol.*, 9:158.

Pires, J. L., Monteiro, W. R., Luz, E. D. W. N., Silva, S. D. V. M., Pinto, L. R. M. and Brugnerotto, M. D. (1996). Cocoa Breeding for Witches' broom resistance at CEPEC in Brazil. In: *Proceedings International Workshop on the contribution of Disease Resistance*. Salvador, Bahia, Brazil (pp. 91–2).

Pires, J. L., Monteiro, W. R., Pinto, L. R. M. and Luz, E. D. M. N. (1999). Resistance to witches' broom evaluation of genotypes from different origins. In: *Proceedings 12th Int. Cocoa Res. Conf., Salvador, Bahia, Brazil*. Malaysia: Cocoa Producers' Alliance (COPAL) (pp. 389–97).

Ploetz, R. C., Schnell, R. J., Ying, Z., Zheng, Q., Olano, C. T., Motamayor, J. C. and Johnson, E. S. (2005). Analysis of molecular diversity in *Crinipellis perniciosa* with AFLP markers. *Eur. J. Plant Pathol.*, 111:317–26.

Pomella, A.W.V., De Souza, T.T., Niella, G.R., Bateman, R.P., Hebbar, P.K., Loguercio, L.L., and Lumsden, R.D. (2007). *Trichoderma stromaticum* for management of witches' broom of cacao in Brazil. In: Vincent, C., Goettel, M.S. and Lazarovits, G. (Eds), *Biological Control: a Global Perspective* (pp. 210–217). Wallingford: CABI Publishing.

Pound, F. J. (1938). *Cacao and witch-broom disease (Marasmius perniciosus) of South America*. With notes on other species of *Theobroma*. Report on a Visit to Ecuador, the Amazon Valley and Colombia. Trinidad and Tobago: Department of Agriculture, p. 58.

Pound, F. J. (1943). Cacao and witches' broom disease (Marasmius perniciosus). Report on a recent visit to the Amazon territory of Peru. Trinidad and Tobago: Department of Agriculture, p. 14.

Purdy, L. H. and Schmidt, R. A. (1996). Status of cacao witches' broom: Biology, epidemiology, and management. *Ann. Rev. phytopathol.*, 34:573–94.

Resende, M. L. V., Nojosa, G. B. A., Silva, L. H. C. P., Niella, G. R., Carvalho, G. A., Santiago, D. V. R. and Bezerra, J. L. (2000). *Crinipellis perniciosa* from a new host *Heteropterys acutifolia* is pathogenic to cacao. *Fitopatol. Bras.*, 25:88–91.

Rincones, J., Meinhardt, L. W., Vidal, B. C. and Pereira, G. A. (2003). Electrophoretic karyotype analysis of *Crinipellis perniciosa*, the causal agent of witches' broom disease of *Theobroma cacao*. *Mycol. Res.*, 107:452–8.

Rincones, J., Mazotti, G. D., Griffith, G. W., Pomela, A., Figueira, A., Leal Júnior, G. A., Queiroz, M. V., Pereira, J. F., Azevedo, R. A., Pereira, G. A., Meinhardt, L. W. (2006). Genetic variability and chromosome-length polymorphisms of the witches' broom pathogen *Crinipellis perniciosa* from various plant hosts in South America. *Mycol. Res.*, 110:821–32.

Rincones, J., Scarpari, L. M., Carazzolle, M. F., Mondego, J. M. C., Formighieri, E. F., Barau, J. G., Costa, G. G. L., Carraro, D. M., Brentani, H. P., Vilas-Boas, L. A., Oliveira, B. V., Sabha, M., Dias, R., Cascardo, J. M., Azevedo, R. A., Meinhardt, L. W. and Pereira, G. A. G. (2008). Differential gene expression between the biotrophic-like and saprotrophic mycelia of the Witches' broom pathogen *Moniliophthora perniciosa. Mol. Plant Microbe Interact.*, 21:891–908.

Rio, M. C. S., Oliveira, B. V., Thomazella, D. P., Silva, J. A. F. and Pereira, G. A. (2008). Production of calcium oxalate crystals by the basidiomycete *Moniliophthora perniciosa*, the causal agent of witches' broom disease of cacao. *Curr. Microbiol.*, 56:363–70.

Rios-Ruiz, R. A. (2001). Melhoramento para resistência a doenças. In: Dias, L. A. S. (Ed.), *Melhoramento Genético Do Cacaueiro*. Viçosa: FUNAPE, UFG, cap. 7, pp. 290–324.

Ritzema Bos, J. (1901). Over Krulloten en hexenbezems in de cacaoboomen in Suriname. *Z. Pflanzenkr.*, 11:26–30.

Rocha, H. M. and Wheeler, B. E. J. (1982). The water balance as an important factor in basidiocarp production by *Crinipellis perniciosa*, the causal fungus of cacao witches' broom. In: *8. International Cacao Research Conference 18–23 October 1981 Cartagena (Colombia)*. Lagos (Nigeria): Cacao Producers' Alliance.

Rocha, H. M. and Wheeler, B. E. J. (1985). Factors influencing the production of basidiocarps and the deposition and germination of basidiospores of *Crinipellis perniciosa*, the causal fungus of witches' broom on cacao (*Theobroma cacao*). *Plant Pathol.*, 34:319–28.

Rorer, J. B. (1913). The Suriname witch-broom disease of cacao. *Cir. Dept. Agric.*, 10:1–13.

Rorer, J. B. (1926). Ecuador cacao. *Trop. Agric.*, 3:46–7.

Rudgard, S. A., Andebrhan, T., Maddison, A. C. and Schmidt, R. A. (1993). Disease management: Recommendations. In: Rudgard, S. A., Maddison, A. C. and Andebrhan, T. (Eds) *Disease Management in Cocoa* (pp. 201–11). Netherlands: Springer. ISBN: 978-94-010-4943-6 (Print); 978-94-011-2126-2 (Online).

Rudgard, S. A. and Butler, D. R. (1987). Witches' broom disease in Rondonia, Brazil: pod infection in relation to pod susceptibility, wetness, inoculum, and phytosanitation. *Plant Pathol.*, 36:515–22.

Samuels, G. J., Ismaiel, A., De Souza, J. T., Chaverri, P. (2012). *Trichoderma stromaticum* and its overseas relatives. *Mycol. Progress*, 11:215–54.

Samuels, G. J., Pardo-Schultheiss, R., Hebbar, K. P., Lumsden, R. D., Bastos, C. N., Costa, J. C. and Bezerra, J. L. (2000). *Trichoderma stromaticum* sp. nov., a parasite of the cacao witches broom pathogen. *Mycol. Res.*, 104:760–4.

Santos Júnior, M. C., Assis, S. A., Góes-Neto, A., Duarte, Â. A., Alves, R. J., Junior, M. C. and Taranto, A. G. (2013). Structure-based drug design studies of UDP-N-acetylglucosamine pyrophosphosrylase, a key enzyme for the control of witches' broom disease. *Chem. Cent. J.*, 7:1–7.

Santos Junior, M. C., Gonçalves, P. A., Taranto, A. G., Koblitz, M.G.B, Góes-Neto, A., Pirovani, C.P., Cascardo, J. C. M., Cruz, S. H., Zingali, R. B., Pereira, G. A. G., Dias, C. V., Assis, S. A. (2011). Purification, characterization and structural determination of UDP-N-acetylglucosamine pyrophosphorylase produced by *Moniliophthora perniciosa*. *J. Braz. Chem. Soc.*, 22:1015–23.

Santos, A. C., Sena, J. A. L., Santos, S. C., Dias, C. V., Pirovani, C. P., Pungartnik, C., Valle, R. R., Cascardo, J. C. M. and Vicentz, M. (2009). dsRNA-induced gene silencing in *Moniliophthora perniciosa*, the causal agent of witches' broom disease of cacao. *Fungal Genet. Biol.*, 46:825–36.

Scarpari, L. M., Meinhardt, L. W., Mazzafera, P., Pomella, A. W., Schiavinato, M. A., Cascardo, J. C. and Pereira, G. A. (2005). Biochemical changes during the development of witches' broom: The most important disease of cocoa in Brazil caused by *Crinipellis perniciosa*. *J. Exp. Bot.*, 56:865–77.

Schmidt, R. A., Rudgard, S. A., Maddison, A. C., Andebrhan, T. (1993). Comparative epidemiology of the witches' broom pathogen. In: Rudgard, S. A., Maddison, A C., Andebrhan, T. (Eds), *Disease management in cocoa: Comparative epidemiology of witches' broom* (pp. 131–156). London: Chapman & Hall.

Sena, K., Alemanno, L. and Gramacho, K. P. (2014). The infection process of *Moniliophthora perniciosa* in cacao. *Plant Pathol.*, 63:1272–81.

Sereno, M. L., Albuquerque, P. S., Vencovsky, R. and Figueira, A. (2006). Genetic diversity and natural population structure of cacao (*Theobroma cacao* L.) from the Brazilian Amazon evaluated by microsatellite markers. *Conserv. Genet.*, 7:13–24.

Silva, J. A., Rodrigues, C. H., Almeida, L. C., Andebrhan, T. (1985). Efeito de fungicidas no controle da vassoura-de-bruxa do cacaueiro em condições de campo. In: CEPLAC (Ed), *Informe de pesquisas* (pp. 48–51). Belém: CEPLAC\DEPEA.

Silva, P. (1987). *Cacau e lagartão ou vassoura-de-bruxa: registros efetuados por Alexandre Rodrigues Ferreira nos anos 1785 a 1787 na Amazônia*. Ceplac/Cepec, 6 Boletim Técnico, n° 146.

Silva, S. D., and Matsuoka, K. (1999). Histologia da interação *Crinipellis perniciosa em* cacaueiros suscetível e resistente a vassoura-de-bruxa. *Fitopatol. Bras.*, 24:54–59.

Silva, S. D. V. M., Gramacho, K. P., Almeida, O. D. (1992). *Solanum paniculatum* hospedeiro de *Crinipellis perniciosa* na região sul da Bahia. *Agrotrópica*, 4:17–20.

Silva, S. D. V. M., Luz, E. D. M. N., Gramacho, K. P., Albuquerque, P. S. B. and Bezerra, J. L. (2000). Standardization of the belt system method to assess cacao genotypes to witches' broom resistance. In: *Proceedings of the 13th International Cocoa Research Conference: Towards the Effective and Optimum Promotion of Cocoa Through Research and Development* (pp. 9–14).

Silva, S. D. V. M., Luz, E. D. M. N., Almeida, O. D., Gramacho, K. and Bezerra, J. L. (2002). Redescrição da sintomatologia causada por *Crinipellis perniciosa* em cacaueiro. *Agrotrópica*, 1:1–23.

Silva, S. D. V. M., Luz, E. D. M. N., Pires, J. L., Yamada, M. M. and Santos Filho, L. P. D. (2010). Parent selection for cacao resistance to witches' broom. *Pesquisa Agropecuária Brasileira*, 45:680–5.

Singer, R. (1942). A monographic study of the genera *Crinipellis* and *Chaetocalathus*. *Lilloa*, 8:441–534.

Sreenivasan, T. N. and Dabydeen, S. (1989). Modes of penetration of young cocoa leaves by *Crinipellis perniciosa*. *Plant Disease*, 73:478–81.

Stahel, G. (1915). *Marasmius perniciosus* Nov. Spec. *Dept. Landbouw. Suriname Bull.*, 33:1–27.

Stahel, G. (1924). Verslag over het jaar 1923. *Review of Applied Mycology*, 4:723.

Stell, F. (1928). Witch broom disease of cacao. *Proc. Agric. Soc. Trinidad Tobago*, 28:375–7.

Suarez, C. (1977). Growth of Crinipellis perniciosa (Stahel) Singer in vivo and in vitro, p. 134. Tese (PhD Thesis in Plant Pathology). University of London.

Tarnowski, T. L. B. (2009). *Using molecular analysis to investigate phylogenetic relationships in two tropical pathosystems: Witches' broom of cacao caused by Moniliopohthora perniciosa and mango anthracnose caused by Colletotrichum spp*. PhD Thesis, Gainesville, FL: University of Florida.

Teixeira, P. J. P. L., Thomazella, D. P. T., Reis, O., Prado, P. F. V., Rio, M. C. S., Fiorin, G. L., José, J., Costa, G. G. L., Negri, V. A., Mondego, J. M. C., Mieczkowski, P. and Pereira, G. A. G. (2014). High-resolution transcript profiling of the atypical biotrophic interaction between *Theobroma cacao* and the fungal pathogen *Moniliophthora perniciosa*. *Plant Cell*, 26:4245–69.

Thomazella, D. P. T., Teixeira, P. J. P. L., Oliveira, H. C., Saviani, E. E., Rincones, J., Toni, I. M., Reis, O., Garcia, O., Meinhardt, L. W., Salgado, I. and Pereira, G. A. G. (2012). The hemibiotrophic cacao pathogen *Moniliophthora perniciosa* depends on a mitochondrial alternative oxidase for biotrophic development. *New Phytol.*, 194:1025–34.

Thorold, C. A. (1975). Witches' broom disease. In: Thorold, C. A. (Ed.), *Disease of Cacao*. Oxford: Claredon Press, pp. 11–31.

Tiburcio, R. A., Costa, G. G. L., Carazzolle, M. F., Mondego, J. M. C., Schuster, S. C., Carlson, J. E. and Pereira, G. A. G. (2010). Genes acquired by horizontal transfer are potentially involved in the evolution of phytopathogenicity in *Moniliophthora perniciosa* and *Moniliophthora roreri*, two of the major pathogens of cacao. *J. Mol. Evol.*, 70:85–97.

Tovar, G. (1991). La escoba de bruja del cacao [*Crinipellis perniciosa* (Stahel) Singer]: descripción de síntomas de la enfermedad. *Agron. Colomb.*, 8:227–39.

Trevizan, S. D. P. and Marques, M. (2002). Impactos socioeconomicos da crise do cacau: um estudo de comunidade-caso. *Agrotropica*, 14:127–36.

Van Hall, C. J. J. (1910). The "Krulloten" disease in a wild growing cacao species (*Theobroma speciosum* Spreng). *Agric. Soc. Trinidad Tobago*, 435:1–3.

Van Hall, C. J. J. and Drost, A. W. (1907). Les balais de sorciére du cacaoyer provoqués par *Colletotrichum luxificum* n. sp. Rec. Trav. Botan. Néerland., 4:243–321.

Van Hall, C. J. J. and Drost, A. W. (1909). The witch-broom disease in Suriname, its cause and treatment. *Bull. Dept. Landbouw*, 11:62–6.

Viana, C. A. C. (2001). *Compatibilidade somática e patogenicidade de Crinipellis perniciosa*. MSc Dissertation, MG, Brazil: Universidade Federal de Lavras.

Wang, H., La Russa, M., Qi, L. S. (2016). CRISPR/Cas9 in genome editing and beyond. Annu. Rev. Biochem., 85:227–64.

Wheeler, B. E. J. and Suárez, C. (1993). The pathosystem. In: *Disease Management in Cacao* (pp. 9–19). Netherlands: Springer.

Wheeler, B. E. J. and Mepsted, R. (1988). Pathogenic variability amongst isolates of *Crinipellis perniciosa* from cocoa (*Theobroma cacao*). *Plant Pathol.*, 37:475–488.

Zhang, D., Boccara, M., Motilal, L., Butler, D. R., Umaharan, P., Mischke, S. and Meinhardt, L. (2008). Microsatellite variation and population structure in the 'Refractario' cacao of Ecuador. *Conserv. Genet.*, 9:327–37.

Chapter 11

Frosty pod rot, caused by *Moniliophthora roreri*

Ulrike Krauss, Palm Integrated Services and Solutions (PISS) Ltd., Saint Lucia

1 Introduction: the pathogen

Frosty pod rot (FPR) of cacao (*Theobroma cacao*) is caused by the basidiomycete *Moniliophthora roreri*. Spores of *M. roreri* germinate and penetrate the pod surface soon after landing there and establish a systemic pod infection, which can remain latent for two months (Evans, 1981). During the essentially non-symptomatic phase, the seed chamber, where the actual cacao beans normally mature, is completely destroyed. Only then does the pathogen emerge to the pod surface, where it causes brown lesions, followed by rapid and prolific sporulation within one week (Leach et al., 2002). The fact that much of the pathogen development and damage takes place deep inside an intact pod poses significant challenges to FPR management by any agent that requires spatial proximity for effective control (ten Hoopen and Krauss, 2016 and references therein).

The pathogen is believed to have evolved in the forests of western Colombia and/or northwest Ecuador on *Theobroma gileri*. The morphologically distinct *M. roreri* var. *gileri* occurs on endemic *Theobroma* in the forests of north-west South America, is non-pathogenic to cacao, and possesses a unique molecular profile. Evans (2016) believes this to be the progenitor of the cacao pathogen – *M. roreri* var. *roreri*, which attacks all species of the genera *Theobroma* and *Herrania*; no other hosts have been reported for *M. roreri*.

M. roreri is a hemibiotroph with two distinct phases, which are distinguished morphologically, physiologically and genetically: the biotrophic monokaryophase and the necrotrophic dikaryophase, during which prolific sporulation occurs. Since reproduction in nature is not known on any substrate other than cacao pods, *M. roreri* is classified as an ecologically obligate parasite (CABI, undated). A similar pleomorphic life cycle of *Moniliophthora perniciosa* led the causal agent of cacao witches' broom to be transferred

http://dx.doi.org/10.19103/AS.2017.0021.17

to the genus *Moniliophthora*, as the only other member of this genus – after decades of taxonomic confusion (for more detail, see Evans, 2016).

Recent investigations employing a range of molecular analysis to compare sequences of rRNA, internal transcribed spacers (ITS) and five genes from nuclear DNA confirmed the hypothesized closer relationship between *M. roreri* and *M. perniciosa*. Furthermore, the observed extraordinary genetic variability between isolates of different geographic regions points towards genetic recombination in the form of meiosis taking place in *M. roreri*. Thus, the spores are not asexual conidia but multifunctional, sexual meiospores that are borne out of a radically modified basidium (Evans, 2016 and references therein).

2 Spread, invasion risk and impact of *Moniliophthora roreri*

2.1 Spread of *Moniliophthora roreri* and invasion risk

FPR was first described in 1817 in north-eastern Colombia (Ploetz, 2016 and references therein). The geographic spread throughout much of the Americas thus started relatively recently and was mediated by humans transporting cryptically infected cacao germplasm (Evans et al., 2003). The recalcitrant nature of cacao seeds (tendency to germinate as soon as they are removed from the pod) tempts traders and growers over and over again to carry healthy-looking cacao pods, often with disastrous outcome. The two main contributing risk factors are: (1) the long latency period of four to eight weeks (average seven) that lures smugglers into a false sense of security and (2) the exceedingly prolific sporulation of *M. roreri* – a mature pod can give rise to as many as 7 billion spores and go through multiple sporulation cycles (CABI, undated).

Over two centuries, the pathogen has expanded its range in South America: Colombia, Ecuador, Peru, Venezuela, and most recently (2012) Bolivia (Phillips-Mora et al., 2015). In 2009, *M. roreri* had overcome the natural barrier of the Andes Mountains and was recorded in the eastern Venezuelan states of Zulia, Táchira, Mérida, Trujillo, Barinas, Apure and Amazonas (Parra et al., 2009). This significantly increases the risk of introduction into Brazil, the Guyanas, and the island of Trinidad.

M. roreri first crossed into Central America, that is, Panama, in 1956; from there it progressed through Central America (Costa Rica, Nicaragua, Honduras) during the late 1900s. In recent years, FPR established itself in Guatemala (2002), Belize (2004), Mexico (2005) (Phillips-Mora et al., 2007), and El Salvador (Phillips-Mora et al., 2012).

A 1999 mention of FPR in Guyana (Corbin, 1999) is regarded erroneous (CABI, undated). Outbreaks in the Democratic Republic of São Tomé and Príncipe, Gulf of Guinea, off the western equatorial coast of Central Africa (Island Biodiversity Race, 2009), Cuba[1], Trinidad (CABI & EPPO, 2010), and Brazil (Anonymous, 2009) were feared, but later proved to be false alarms (Krauss, 2010). The status of some small Caribbean off-shore islands belonging to Central American countries is uncertain (CABI, undated). In May 2016, farmers in the Jamaica parish of Clarendon reported suspicious symptoms, which were subsequently

1. Based on a second-hand citation of Arnold, G.R.W., 1986. *Lista de Hongos Fitopatogenos de Cuba*. Ministerio de Cultura Editorial Cientifico-Tecnica, Habana, Cuba, pp. 207. This document actually refers to *Monilia* cf. *roreri* Cif. & Parodi, which is very unlikely to correspond to what we recognize today as *Moniliophthora roreri* (Cif.) H.C. Evans, Stalpers, Samson & Benny.

confirmed to correspond to *M. roreri*. Jamaica reported the detection of FPR to the International Plant Protection Convention (IPPC)[2] in September and triggered a campaign comprised of containment, eradication, public education, and surveillance. Within the following month, FPR had also been confirmed in the parishes of St. Catherine and St. Mary, suggesting that Jamaica is losing the battle against this invasive pathogen – with serious implications for the entire Caribbean.

In addition to Brazil, two Caribbean island nations are at particular risk of FPR introduction: the Dominican Republic and Trinidad & Tobago. The Dominican Republic is the world's largest producer of organic cocoa with a lot at stake. As a Spanish-speaking country, it maintains close cultural and economic ties with infested mainland neighbours. For example the Dominican Republic–Central America Free Trade Agreement (DR-CAFTA) encompasses the Central American countries of Costa Rica, El Salvador, Guatemala, Honduras and Nicaragua, as well as the United States and, since 2004, the Dominican Republic. The Dominican Republic and the United States are the only signatories that were still FPR-free in 2017. The United States is not a commercial cocoa producer, although some field cacao research is carried out in Hawaii, Florida and Puerto Rico. In the latter two locations cacao germplasm collections are also maintained.

The island of Trinidad is separated from Venezuela by the narrow Columbus Channel, which is traversed by both official and unofficial trade. Now that FPR is present in the Venezuelan provinces east of the Andes, there is an exceedingly high risk of *M. roreri* spreading across to the shores of Trinidad. This risk is exacerbated by a relatively high proportion of traffic that passes through unofficial entry points along Trinidad's south coast. Trinidad is a traditional producer of fine-flavoured cocoa and hosts an international cacao germplasm bank, with the concomitant regular exchange of experts and plant material, which requires utmost vigilance.

2.2 Impact of *Moniliophthora roreri*

Global losses due to FPR are still moderate, due to the limited geographic distribution of the pathogen. Nevertheless, FPR is potentially the most destructive of all cacao diseases, as it causes yield losses of around 70–80% in most areas within just a few years of pathogen establishment and can lead to complete crop failure (Krauss et al., 2003). FPR is also considered more difficult to manage than any other cacao disease (Ploetz, 2016). The combination of high losses and costly/difficult management frequently renders the production of cocoa unattractive, following the detection of FPR. This, in turn, leads to a range of negative secondary impacts.

Cacao is an environmentally sound understorey crop, typically produced by smallholders in highly diverse agroforestry systems. The abandonment of cacao, and the concomitant conversion of associated agroforestry systems to other land uses, results in deforestation, erosion of soil organic matter, and habitat loss and fragmentation of landscapes for wildlife. It also triggers unsustainable trends in livelihood strategies, such as conversion of forest into cattle pasture, moving towards monocultures of annual crops or urbanization with the concomitant real estate development. Thus, FPR constitutes the greatest potential threat to cocoa-producing regions, which are still regarded FPR-free in 2017. These include FPR-free Caribbean islands, the Guyanas and Brazil in the Americas, as well as countries in the

2. https://www.ippc.int/en/countries/jamaica/pestreports/2016/09/detection-of-frosty-pod-rot-in-jamaica/#.

continents of Africa and Asia, which produce the bulk of global cocoa supply. It is therefore imperative that the introduction of the pathogen into additional cocoa-producing states is prevented. However, few plant pathologists doubt that the spread of FPR throughout the tropical Americas is only a question of time. In fact, Evans (2016) expressed surprise that the lower Amazon region of Brazil, which is surrounded by countries with well-established infestations of the *M. roreri*, seems to have been spared from this highly invasive pathogen thus far.

3 Management options: management cascade for invasive alien species

Optimal management options for *M. roreri* are based on (1) the fact that this fungus almost certainly coevolved with other endophytic fungi in wild *T. gileri* and (2) its nature as an invasive alien species (IAS) in much of its current range. Thus, the full management cascade recommended for invasive plant pathogens is applicable at certain locations: prevention; EDRR; and impact mitigation, using various control approaches, with classical biocontrol being a particularly fascinating medium-term strategy.

As for any IAS, a hierarchical management cascade with three principal approaches is recognized as the recommended strategy for FPR management: (1) prevention, (2) EDRR, and (3) impact mitigation. These follow each other in terms of both the importance and the chronological order of deployment.

3.1 Prevention

Prevention of introduction is the most cost-effective measure for managing IAS. However, prioritizing exotic and still absent species to be kept out of a country, or a region within a nation, is a nearly insurmountable task, as it requires a rare combination of knowledge to assess risks in order to arrive at a list of priority species for preventative measures and capacity-building:

- Risk of arrival or introduction
- Local condition that may favour invasiveness
- Potential for invasiveness and impact severity under conducive conditions.

This information tends to be unavailable in most countries, particularly for pathogens that are still absent from the territory. External sourcing of information requires a good idea of where to start, that is, a pre-prioritization of organisms, leading to a circular task that can only partly be completed by a combination of good intuition, combined with careful and scientific iteration. In reality, the prioritization of mainstream preventative measures tends to focus on pathway management rather than particular species. The latter are targeted only when a known high risk of introduction exists and/or severe impact would be associated with their arrival, as is the case for *M. roreri* in cocoa-producing nations still free from this blight. Nevertheless, under-resourced phytosanitary inspection services often depend on poorly motivated personnel with limited awareness of the multitude of possible hiding places for tiny spores of a microscopic pathogen, such as hair, clothes or even 'chocolate pods'. Continuous capacity-building and institutional strengthening are prerequisites to make this system work: effective prevention of FPR entry into a country depends on efficient and uninterrupted enforcement of existing regulations.

The fact that *M. roreri* has an average latency period of seven weeks deceives many into believing that they are transporting healthy pods. By the time external symptoms appear, the cacao beans inside would have largely been destroyed. One week after the infected pods have been discarded, the fungus sporulates on the pod surface, releasing billions of microscopic, wind-dispersed spores into the air at the slightest agitation. On the other hand, *M. roreri* only infects the fruit and only of two closely related genera: *Theobroma* and *Herrania*. Thus prevention can focus on categorically prohibiting the transport of the pods, for example, for transporting the recalcitrant seeds, of these two species, whether they appear healthy or not, and enforcing this regulation effectively.

Both the Dominican Republic and Trinidad & Tobago were among the five participating countries in a regional project *Mitigating the Threats of Invasive Alien Species in the Insular Caribbean (MTIASIC)*, which is funded by the Global Environment Facility (GEF). One MTIASIC pilot project aimed to strengthen the preventative capacity of Trinidad & Tobago and used FPR as its model IAS. Interventions raised public awareness (Fig. 1a) and strengthened surveillance, quarantine and EDRR mechanisms. Lessons learnt are being exchanged with other project countries, which included Jamaica, as well as the Wider Caribbean Region. In response to the above-mentioned high risk of FPR introduction and other priority threats, the Dominican Republic embarked on the establishment of a nation-wide diagnostic network, which is linked also to the incipient Caribbean Pest Diagnostic

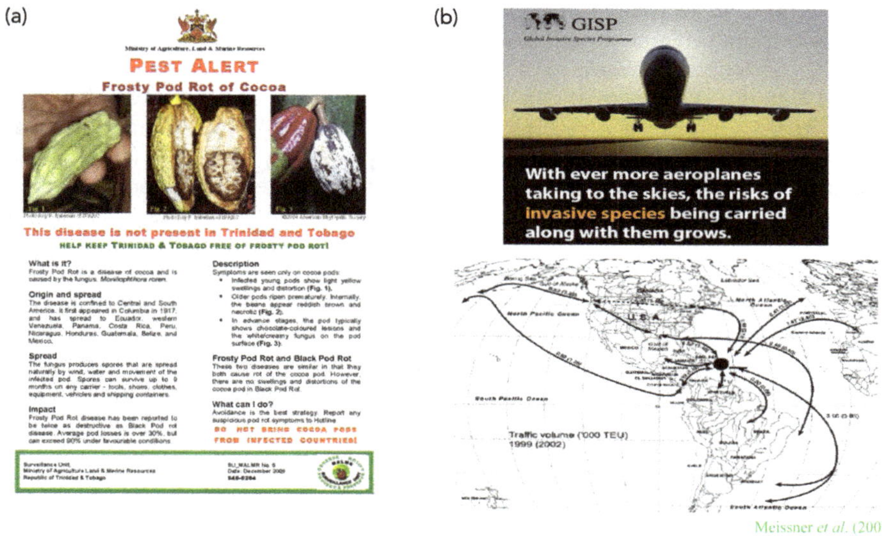

Pest Alert Poster in Trinidadian ports of entry [Source: V. Ferguson-Dewsbury, pers. comm.., 2013]

Inspections focus on high risk pathways of trade, tourism, transport, and travel (four Ts)

Figure 1 Frosty Pod Rot prevention in the insular Caribbean needs to focus on public awareness, alertness of inspectors along prioritized pathways of high risk. (a) Pest Alert Poster in Trinidadian ports of entry (Source: V. Ferguson-Dewsbury, pers. comm.., 2013). (b) Inspections focus on high risk pathways of trade, tourism, transport and travel (four Ts). Source: Meissner et al. (2009).

Network (Reyes Valentín et al., 2010). However, phytosanitary controls still leave significant room for improvement – even at the major airports.

Extreme alertness and regional cooperation are needed throughout the insular Caribbean, the Guyanas and Brazil if FPR introduction is to be prevented or at least delayed. On a global scale, strategic measures for intercontinental germplasm transfer, transport and trade are required to prevent the transatlantic spread of *M. roreri* to the cocoa-producing countries in West Africa. Since fungal dissemination in an eastward direction, that is, against the Trade Winds, is only conceivable if vectored by man, interventions should focus on the infamous 'four Ts': Trade, Tourism, Transport and Travel (Fig. 1b).

3.2 The international legislative framework for phytosanitary measures

Since IAS are international in character, prevention of IAS introductions at national borders requires an international legislative framework through global, regional or bilateral agreements. Trade is one major pathway (Waugh, 2009). The World Trade Organization's (WTO) Agreement on the Application of Sanitary and Phytosanitary Measures (SPS Agreement) is the major international instrument governing global trade and provides binding rules. The key objective of sanitary and phytosanitary (quarantine) measures is to protect humans, animals and plants (wild and cultivated) from damage due to pests and diseases. This is often achieved through the use of import and export control measures. The SPS Agreement also provides a set of basic rules as to how WTO members can apply measures to manage invasive alien pests and pathogens. Trade disputes are settled by a compulsory dispute settlement mechanism.

The IPPC, under the umbrella of the Food and Agriculture Organization (FAO) of the United Nations, is an international framework to prevent the spread of plant pests; it also promotes appropriate measures for their control. The IPPC Secretariat facilitates the development of International Standards for Phytosanitary Measures (ISPMs). These are designed to prevent the introduction or spread of plant pests of potential economic importance, while encouraging the international harmonization of phytosanitary measures to facilitate safe trade and avoid unjustified barriers. Parties are required to adopt standards and procedures to identify organisms that threaten plant health. Parties may prohibit the introduction of certain plants and restrict the import of plant products, execute inspections and detain consignments. Parties are also required to share information regarding plant pests as well as means of prevention and control. Each party is required to establish a National Plant Protection Organization (NPPO); some NPPOs are cooperating via Regional Plant Protection Organizations (RPPOs). One of the key components of many national phytosanitary systems is the three-stage pest risk analysis (PRA).

The RPPO of Brazil is the *Comité de Sanidad Vegetal* (COSAVE), where it joins other countries of the southern cone. The Dominican Republic is a member of the *Organismo Internacional Regional de Sanidad Agropecuaria* (OIRSA), which is dominated by Mesoamerican membership. The Dominican Republic and Trinidad & Tobago were members of the Caribbean Plant Protection Commission (CPPC), a RPPO, which was phased out in 2014 (FAO, 2014) and is gradually being replaced by the Caribbean Agricultural, Health and Food Safety Agency (CAHFSA). CAHFSA is mandated to coordinate the establishment and implement an effective regional (SPS) regime on behalf of Caribbean Community (CARICOM) member states, whenever a regional mechanism is the most efficient approach.

The Dominican Republic is not a member of CARICOM or CAHFSA. In recognition of the need to embrace the Dominican Republic into Caribbean programmes, several regional initiatives operate at a CARIFORUM level, that is, CARICOM plus the Dominican Republic. Examples are the European Commission-funded Agriculture Policy Programme[3] and moves towards a harmonized SPS legislation[4].

3.3 Early detection and rapid response (EDRR)

EDRR are the next steps in line whenever prevention lapsed and an incursion occurred. EDRR have a number of prerequisites for success. For early detection, quarantine and survey personnel need to know (i) what to look for, (ii) to whom they should report suspicious sightings for confirmation and (iii) if indicated, how to accomplish the destruction of the harmful invasive species. This requires training of field and border inspection/customs officers, which, for logistic reasons, cannot possibly address all threats, but instead tends to focus on interception along high risk pathways, that is, the 'four Ts'.

Early detection and rapid response have never been applied successfully against FPR! There are good reasons for that, and we need to learn from past failures. *M. roreri* has a latent infection phase of approximately seven weeks, followed by prolific sporulation within one week of diagnostic symptom development. This leaves an extremely narrow window of opportunity for early detection. Thus, farmers as well as field officers should be trained in the recognition of early symptoms of FPR. Practical tools for training and diagnostics have been developed (Krauss et al., 2006; Fig. 2).

An innovative approach has been devised in Brazil: classical spore trapping is combined with an enzyme-linked immunosorbent assay (ELISA; Pomella et al., 2005) to detect the inconspicuous spores of *M. roreri* and distinguish them from other fungal taxa. Sentinel

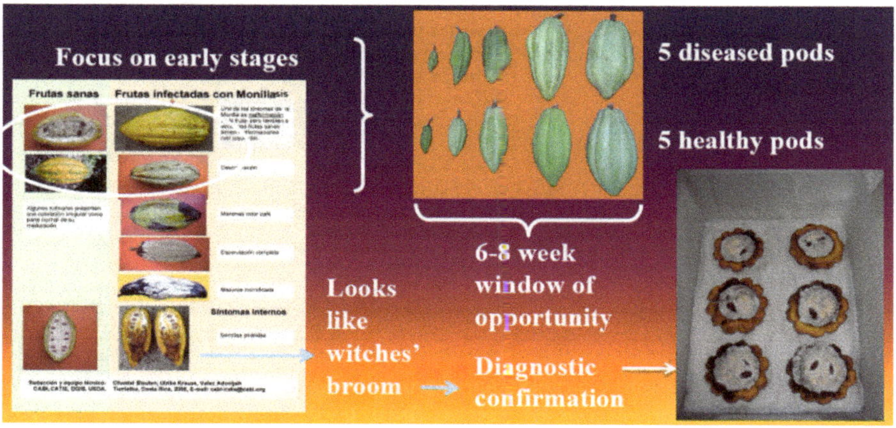

Figure 2 Visual aids help trainees to home in on the early phase of infections, during which symptoms are very subtle. A robust diagnostic test gives clarification within days.

3. http://capacity4dev.ec.europa.eu/intra-acp-app/blog/agriculture-policy-programme-app-launched%E2%80%A6euro-86-million-regional-agriculture.
4. http://caricom.org/SPS_Project_Document.pdf.

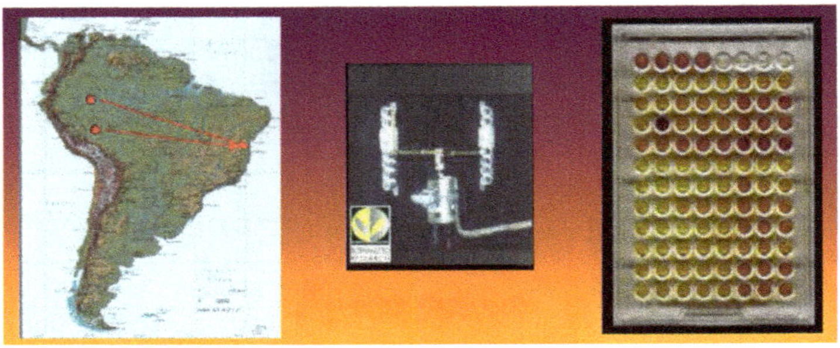

Figure 3 Brazil is attempting to detect border crossings of *Moniliophthora roreri* using a combination of spore trapping and immunological assays (Source: Pomella et al., 2005).

Table 1 Typical course of IRP action (first two columns), with example answers in the third column

S.no.	Action	Responsible/backup *[with Examples]*
1	Identify incursion	*Field inspector, extensionists, customs officer*
2	Confirm incursion	*Crop Protection Officer (with help of international expert?)*
3	Assess risk of the incursion	*Chief Crop Protection Officer*
4	Assess invasion severity and ability to respond	*Plant Quarantine:* • *High/intermediate in field* • *Intermediate/high at official entry points*
5	Based upon risk what action should occur	*Ministry of Agriculture:* *Rapid response - eradication*
6	Carries out the response action (in-house capacity or external contract?)	*Extension Department:* • *Complete cacao pod purging in area of X km² from detection site* • *Felling or radical pruning of excessively tall cocoa trees in radius of Y km² from detection site* • *Intensifies FPR surveillance* • *Assess farmers' losses* *Ministry of Agriculture, Cocoa Board:* • *Compensate for losses*
7	Reviews outcome of response	*Chief Crop Protection Officer*
8	Reporting to stakeholders	*Ministry of Agriculture, Cocoa Board*

stations near the borders with Colombia and Peru, where the disease is rife – but far away from the cacao-producing regions of Eastern Brazil (Fig. 3) – have been proposed, in the hope that distant rapid response could win enough time to save the Brazilian cocoa industry from an invasion by *M. roreri*. No update on experiences with this system could be obtained for this chapter. The intervention focuses on peoples crossing the Brazilian border in the Amazon basin, which is considered to be a high risk in the pathway analysis. Deliberate and illegal long-distance movement of cacao tissue – which is believed to have been responsible for the introduction and spread of witches' broom (*Moniliophthora perniciosa*) in Bahia (Andebrhan et al., 1999) – must be managed in parallel for EDRR measures to be effective.

For rapid response, it is essential that an incursion response plan (IRP) is in place prior to the detection of a specific IAS, so that specific action is triggered immediately upon detection of the IAS. Thus, the IRP needs to be prepared with anticipation, which is not a natural mentality characteristic of most cocoa-producing countries. In particular, Small Island Developing States (SIDS) suffer notorious understaffing of relevant positions. Thus, staff is forced to fire-fight and react to problems, rather than planning ahead, which is exactly what would be required to develop an IRP for FPR incursion. A generic template for key elements of an IRP is shown in Table 1.

A practical IRP requires an effective mechanism for prompt eradication of the newly detected IAS, that is, before it establishes, reproduces and spreads. This may entail host elimination, in which case a compensation scheme is also required. Thus, any realistic IRP must contemplate funding for the required interventions and should be accompanied by the establishment of a contingency fund that can be accessed without undue bureaucracy should certain circumstances (which are to be defined in the IRP) present themselves. Given the narrow host range of *M. roreri*, host destruction or stripping of developing pods are feasible options for rapid response, if thorough implementation can be achieved. This, however, will depend on the active collaboration of farmers, which can only be obtained via a fair loss compensation scheme. Thus, planning for rapid response must entail the establishment of a contingency fund, with a swift and unbureaucratic access to funds both for the mobilization of trained field technicians and to recompense growers.

4 Management options: impact mitigation

If the above-mentioned two approaches fail and an IAS becomes established, the only option left is damage limitation. Generally, integrated options that resort to several management tools are preferred. Their case-specific selection should be based on assessment of cost-effectiveness in the medium to long term. This is also true for FPR management, as detailed below. In this section, I focus on published research with a field phase of no less than four months in duration, given the long latency period of *M. roreri*.

4.1 Cultural control

Cultural control has proven to be the central pillar of any integrated control strategy for FPR in afflicted countries. Comprehensive phytosanitary pruning requires a manageable tree height. There is no shortcut to cultural measures, including formation and maintenance pruning.

All diseased pods must be removed from the tree at intervals before sporulation begins. The epidemiology of the pathogens dictates weekly phytosanitation, but this is not always cost-effective. Therefore Krauss et al. (2006) suggested more effort in training farmers to recognize earlier stages of infection. Economic analysis, for example, using modelling, needs to go hand in hand with training of field staff to find an optimum frequency for each locality (Krauss et al., 2003, 2006; Leach et al., 2002).

Severed, diseased pods can be left undisturbed on the ground, where they decompose quickly and spores lose their viability. Covering them with already decomposing leaf litter accelerates this process and involves only negligible extra labour. In contrast, mummified pods left in the canopy can go through 10–14 sporulation cycles for up to 80 days. Such spores can remain viable for up to nine months (CABI, undated and referenced therein) and thus contribute greatly to further pathogen dissemination, thereby counteracting any mitigating measures.

Cultural measures can be further enhanced by integrating cacao into a disease-resistant agroforestry system (AFS). A diversified AFS is not only a good risk avoidance strategy, but also a tool to optimize shade cover, temperature and aeration for a healthy cacao crop. Furthermore, non-hosts can intercept water splash and aerosols carrying the pathogen propagules, thereby reducing inoculum pressure (Krauss, 2004). The above tools have been available for considerable time and maintain their validity.

4.2 Chemical control

Fungicides are being employed as supplementary management options (Bateman et al., 2005a,b). Copper fungicides remain the most cost-effective to date. Copper hydroxide is the least hazardous compound and thus best suited for smallholders. Modern systemic agents could substitute copper during the early season, particularly if applied with a sticker (Krauss et al., 2010), but this increases the cost of chemical control.

The small size and narrow shape of cocoa pods as the target for fungicide delivery poses different challenges compared with foliar sprays to tree crops. Spray angles of <40° are particularly desirable. Nozzles that produced a very narrow angle (25–35°) and low flow rates at 200 kPa, such as the CD31-25 and D1.5-25, achieved substantially better deposits on cacao pods than the typical, adjustable D2-45 nozzle (Bateman, 2004). Research into improved cost-effectiveness of fungicide applications by better targeted delivery of the fungicide to cacao pod is in its infancy (Bateman et al., 2005b; Hidalgo et al., 2003).

4.3 Biological control

Biological control of *M. roreri* can be categorized into two approaches: classical and inundative biocontrol. Inundative biocontrol uses local antagonists with supposedly good adaptation to prevailing agro-ecological conditions. Nevertheless, these agents have to be applied regularly, which is costly. The classical approach is applicable in situations where a pest has been introduced into new regions. In the absence of natural enemies, the exotic pest escalates into a problem, frequently far more serious than in the country of origin. In such cases, the strategy is to introduce a coevolved, natural enemy of the pest into the new location to re-establish the natural equilibrium (Evans, 1999; Bateman et al., 2005b). Establishment can be the bottleneck with this approach, but, if successful, the introduced agent becomes self-perpetuating.

Mixtures of local mycoparasitic antagonists were highly successful in Peru, where inundatively applied agents controlled three cacao pod diseases (FPR, witches' broom and black pod) simultaneously, leading to yield increases of up to 16.7% (Krauss and Soberanis, 2001, 2002). Increasing application frequencies of a mixed biocontrol inoculum improved FPR and witches' broom control linearly; FPR exhibited the stronger response. Ten applications administered in two-week intervals increased yields by 15%. It was followed by three biocontrol applications adjusted to the production cycle (yield increase: 12%). The latter was the most economical biocontrol strategy under the conditions of eastern Peru: net returns were increased by 12% (Krauss and Soberanis, 2002).

However, this approach did not yield the same success in Costa Rica (Krauss et al., 2003). There, classical biocontrol may be more applicable, as both the crop and the pathogen are removed from their centre of origin. Evans et al. (2003) collected numerous candidates for FPR control from *T. gileri* in Western Ecuador, including several endophytes, that is, fungi that develop asymptomatic infections within healthy plants to form a mutualistic symbiosis. An endophytic and mycoparasitic isolate of *Trichoderma ovalisporum* (TK-1) yielded respectable results in both Ecuador and Costa Rica (Holmes et al., 2006; Krauss et al., 2010). Since there were indications that the formulations used had an effect on biocontrol efficacy, follow-up trials focusing on formulation improvement were prioritized. Crozier et al. (2015) showed that *T. ovalisporum* TK1, formulated in an invert 50% v/v corn oil, 2.5% v/v lecithin and 1% w/v potato dextrose broth, increased yield to 30.7% healthy pods compared with 9.7% in the water control.

Endophytic biocontrol agents have been implicated in induced systemic resistance (Bailey et al., 2006). They also create a much wider window of opportunity for antagonism through exclusion (competition), mycoparasitism and/or antibiosis, because they act where *M. roreri* is most active: within the cacao pod. Coevolved endophytic antagonists may thus be a particularly suitable medium-term solution for the classical biocontrol of *Moniliophthora* spp. in cacao planted outside their South American centre of origin (Krauss et al., 2010). For a more detailed treatise on FPR biocontrol, the reader is referred to ten Hoopen and Krauss (2016).

4.4 Disease resistance and breeding

With the exception of Ecuadorian '*Nacional*', there exists little genetic resistance to FPR in *T. cacao* (Evans, 2016; Sánchez-Mora et al., 2015), and the trait seems to be recessive in germplasm sources screened so far (Gutiérrez et al., 2016). In Colombia, during field trials with artificial inoculation, clone ICS-95 showed consistent resistance against several isolates that belong to four genetic groups of the pathogen (Phillips-Mora et al., 2005). A number of FPR-resistant cultivars have also been identified at the *Centro Agronómico Tropical de Investigación y Enseñanza* (CATIE) in Costa Rica, these are: ICS 43, UF 273, UF 712 and EET 75. The inheritance of resistance was found to be polygenic (Gutiérrez et al., 2016 and references therein). In Chiapas, Mexico, hybrids from the crosses UF 273 x PA 169 and PA 169 x UF 273 showed the highest field resistance and productivity (Solís et al., 2015).

Breeding for FPR resistance focuses on horizontal resistance, which is less complete but more durable, an important consideration in a perennial crop. At CATIE, resistant clones are being crossed progressively to obtain varieties with increased levels of resistance, thus exploiting the predominantly additive mode of inheritance of resistance to FPR (Gutiérrez et al., 2016). Schnell et al. (2007) have identified a quantitative trait marker(s) for each of FPR, witches' broom and black pod resistance, which are being used in accelerated breeding for resistance, as a longer-term perspective.

4.5 Integrating management options

Ten Hoopen et al. (2003) found no difference in the population dynamics of native mycoparasites on segregating F1 populations of cacao hybrids with different resistance reactions to *M. roreri* in Costa Rica. The authors concluded that genetic resistance to FPR and mycoparasite abundance are independent phenomena that should lead to additive effects if employed simultaneously in an integrated disease management programme.

The need for simultaneous control of FPR, witches' broom (*Moniliophthora perniciosa*) and black pod (*Phytophthora* spp.) prompted Krauss and Soberanis (2001) to work with mixtures of up to six mycoparasites in Peru. Several subsequent studies in Costa Rica looked deeper into the compatibility of mycoparasites with each other (Ten Hoopen et al., 2010, Krauss et al., 2013) and with entomopathogens for simultaneous biocontrol of fungal pathogens and insect pests (Krauss et al., 2004). Despite evidence of antagonism between certain biocontrol strains, co-application was more beneficial than would have been predicted from these compatibility studies, possibly because each strain has its own specific ecological niche, resulting in an overall broader spectrum of effectiveness of mixed biocontrol agents.

Cacao pods are most susceptible to infection during their first three months of development. Resistance reported in the Ecuadorian *Refractario* trees was found to be due to disease escape: main harvest develops during the extended dry season when conditions are unfavourable for FPR infection. As a result, crop manipulation through stimulation of flowering during the dry season has shown some promise (Evans, 2016 and references therein) and reiterates the central importance of cultural measures in any management strategy for FPR.

5 Conclusions

1 Prevention is the first choice management approach for IAS in general and *M. roreri* in particular. It should comprise:
   - Capacity-building and institutional strengthening
   - More efficient enforcement of existing regulations
   - Extreme alertness and regional cooperation in the FPR-free Caribbean islands, French Guiana, the Co-operative Republic of Guyana, Suriname, and Brazil
   - Strategic measures for intercontinental germplasm transfer, transport and trade

2 EDRR, the typical second line of defence against IAS, has never been applied successfully against FPR. A more rigorous approach, particularly the anticipatory development of an IRP, would be needed to improve this poor track record. Recommended actions are:
   - Training of quarantine and survey personnel in early detection of latent infections
   - Anticipatory IRP, based on pathway risk analysis, that clearly spells out the chronology of decision-making and responsibility for action
   - Effective mechanism for prompt host elimination and compensation scheme

3 Priority action points for the integrated management of *M. roreri* are:
   - Proactive operation of the management cascade
   - Awareness-raising among all stakeholders with respect to high risk pathways for FPR introduction, the cryptic nature of early infection and the devastating impact pathogen invasion would have on livelihoods and biodiversity. Phytosanitary inspectors and customs/quarantine officers need to be targeted as a priority
   - Training in the detection of early FPR symptoms for farmers, extensionists and field technicians
   - Comprehensive IRPs need to be developed prior to pathogen detection, including effective enforcement and funding mechanisms
   - Regional and international cooperation to prevent the long-distance dispersal of *M. roreri* by man

4 Impact mitigation for FPR must centre on sound cultural management. There are no shortcuts!
   - Design of disease-resilient agroforestry systems is an existing management strategy

- Chemical and biological agents can provide supplementary control. Copper hydroxide remains the most cost-effective and least hazardous fungicide. Both are short-term FPR management options
- Mixtures of inundatively applied local antagonists have the potential to simultaneously control the three predominant cacao pod diseases in South America: FPR, witches' broom and black pod
- Coevolved endophytic antagonists may be a particularly suitable medium-term solution for the classical biocontrol of *Moniliophthora* spp. in cacao planted outside their South American centre of origin
- Genetic and induced resistance are possible long-term options
- Regular review of the cost-effectiveness of measures is recommended, for example, modelling for decision-making.

6 Where to look for further information

The interested reader is referred to the monograph on *M. roreri* by Evans (1981). Recent findings, particularly on genetic and evolutionary studies, as well as control strategies, are covered in *Cacao Diseases: A History of Co-Evolution and Adaptation to Introductions* (Bailey, B. and Meinhardt, L. (Eds), 2016). Springer. Distribution data are up-dated regularly on CABI's Invasive Species Compendium: http://www.cabi.org/isc/datasheet/34779/aqb.

7 References

Andebrhan, T., Figueira, A., Yamada, M. M., Cascardo, J. and Furtek, D. B. 1999. Molecular fingerprinting suggests two primary outbreaks of witches' broom disease (*Crinipellis perniciosa*) of *Theobroma cacao* in Bahia, Brazil. *European Journal of Plant Pathology* 105: 167–75.

Anonymous. 2009. El gremio cacaotero dará asesoría técnica. Available in archives of www.eltiempo.com, 'Economía' section, 31 October 2009.

Bailey, B. A., Bae, H., Strem, M. D., Roberts, D. P., Thomas, S. E., Crozier, J., Samuels, G. J., Choi, I. Y. and Holmes, K. A., 2006. Fungal and plant gene expression during the colonization of cacao seedlings by endophytic isolates of four *Trichoderma* species. *Planta* 224: 1449–64.

Bateman, R. 2004. The use of narrow-angle cone nozzles to spray cocoa pods and other slender biological targets. *Crop Protection* 23(10): 989–99.

Bateman, R. P., Holmes, K. A., Krauss, U. and Padi, B. 2005a. Future tactics and tools for pest management. In *Cocoa Futures* (Flood, J. and Murphy, R. (Eds.)). CABI Commodities Press: Cali, Colombia, pp. 77–93.

Bateman, R. P., Hidalgo, E., García, J., Ten Hoopen, G. M., Krauss, U. and Adonijah, V. 2005b. Application of chemical and biological agents for the management of frosty pod rot (*Moniliophthora roreri*) in Costa Rican cocoa (*Theobroma cacao*). *Annals of Applied Biology* 147: 129–38.

CABI (undated). *Moniliophthora roreri* (frosty pod rot). Invasive Species Compendium, full data sheet, http://www.cabi.org/isc/datasheet/34779, retrieved 28 August 2016.

CABI & EPPO. 2010. Moniliophthora roreri. Distribution Maps of Plant Pathogens, No. 13, October 2010. CABI Head Office: Wallingford, UK.

Corbin, N. 1999. *Determination of the Suitability of Hosororo Hill for Organic Cocoa Production in Administrative Region 1, Guyana*. M.Sc. Thesis, University of the West Indies, Cave Hill Campus, Barbados, Faculty of Pure and Applied Sciences.

Crozier, J., Arroyo, C., Morales, H., Melnick, R. L., Strem, M. D., Vinyard, B. T. and et al. (2015). The influence of formulation on Trichoderma biological activity and frosty pod rot management in Theobroma cacao. *Plant Pathology*. 64(6), 1385–95, doi:10.1111/ppa.12383.

Evans, H. C. 1981. Pod rot of cocoa caused by *Moniliophthora (Monilia) roreri*. *Phytopathological Papers* 24. CAB Commonwealth Mycological Institute, UK: Kew, UK.

Evans, H. C. 1999. Classical biological control. In *Research Methodology in Biocontrol of Plant Diseases with Special Reference to Fungal Diseases of Cocoa* (Krauss, U. and Hebbar, P. (Eds)). CABI, CATIE, USDA, Turrialba, Costa Rica, pp. 29–37.

Evans, H. C. 2016. Frosty Pod Rot (*Moniliophthora roreri*). In *Cacao Diseases: A History of Co-Evolution and Adaptation to Introductions* (Bailey, B. and Meinhardt, L. (Eds)). Springer, Cham, Heidelberg, New York, Dordrecht and London, pp. 63–96.

Evans, H. C., Holmes, K. A. and Thomas, S. E. 2003. Endophytes and mycoparasites associated with an indigenous forest tree, *Theobroma gileri*, in Ecuador and a preliminary assessment of their potential as biocontrol agents of cocoa diseases. *Mycological Progress* 2, 149–60.

FAO. 2014. Abolition of the Caribbean Plant Protection Commission established in 1967 by Council Resolution 8/48. 99th session of the Committee on Constitutional and Legal Matters, Rome, Italy, 20–23 October 2014. http://www.fao.org/3/a-ml433e.pdf.

Gutiérrez, O. A., Campbell, A. S. and Phillips-Mora, W. 2016. Breeding for disease resistance in Cacao. In *Cacao Diseases: A History of Co-Evolution and Adaptation to Introductions* (Bailey, B. and Meinhardt, L. (Eds)). Springer, Cham, Heidelberg, New York, Dordrecht and London, pp. 567–609.

Hidalgo, E., Bateman, R., Krauss, U., Ten Hoopen, M. and Martínez, A. (2003). A field investigation into delivery systems for agents to control Moniliophthora roreri. *European Journal of Plant Pathology* 109, 953–61.

Holmes, K. A., Krauss, U., Samuels, G., Bateman, R. P., Thomas, S. E., Crozier, J., Hidalgo, E., García, J., Arroyo, C. and Evans, H. C. 2006. *Trichoderma ovalisporum*: A potential biocontrol agent of frosty pod rot (*Moniliophthora roreri*). Proc. 15th International Cocoa Research Conference, San José, Costa Rica, 9–14 October 2006, Vol. II, pp. 1001–6.

Island Biodiversity Race. 2009. The race: Bad news/good news. Wildlife Direct, 16 March and addendum December 2009, http://islandbiodiversityrace.wildlifedirect.org/category/chocolate/.

Krauss, U. 2004. Diseases in tropical agroforestry landscapes – the role of biodiversity. In *Agroforestry and Biological Conservation in Tropical Landscapes* (Schroth, G., Fonseca, G. A. B., Harvey, C. A., Gascon, C., Vasconcelos, H. F. and Izac, A. M. N. (Eds)). Island Press, Washington, Covelo and London, pp. 397–412.

Krauss, U. 2010. Integrated Management of the Invasive Cocoa Pathogen *Moniliophthora roreri*. Invited, oral paper presented at the 46th Annual Meeting of the Caribbean Food Crop Society, Boca Chica, Dominican Republic, 12–16 July, http://www.cedaf.org.do/eventos/cfcs_2010/presentaciones/03_martes/tstar11p.pdf and peer-reviewed proceedings paper, ISSN 95-07-0410, Vol 46(1), pp. 75–86, http://cfcs.eea.uprm.edu/article/t-star-invasive-species-symposium.

Krauss, U., Steuten, C., Adonijah, V. and Holmes, K. 2006. Early detection of frosty pod rot as key to cost-effective control. Oral and published paper in *Proceeding of 5th INCOPED Seminar*, Akrofi, A. Y. and Baah, F. (Eds), San José, Costa Rica, 17–18 October, pp. 8–13.

Krauss, U. and Soberanis, W. 2001. Biocontrol of cocoa pod diseases with mycoparasite mixtures. *Biological Control* 22: 149–58.

Krauss, U. and Soberanis, W. 2002. Effect of fertilization and biocontrol application frequency on cocoa pod diseases. *Biological Control* 24: 82–9.

Krauss, U., ten Hoopen, G. M., Hidalgo, E., Martínez, A., Arroyo, C., García, J., Portuguéz, A. and Sánchez, V. 2003. Manejo integrado de la moniliasis (*Moniliophthora roreri*) del cacao (*Theobroma cacao*) en Talamanca, Costa Rica. *Agroforestería en las Américas* 10: 52–8.

Krauss, U., Hidalgo, E., Arroyo, C. and Piper, S. R. 2004. Interaction between the entomopathogens *Beauveria bassiana*, *Metarhizium anisopliae* and *Paecilomyces fumosoroseus* and the mycoparasites *Clonostachys* spp., *Trichoderma harzianum* and *Lecanicillium lecanii*. *Biocontrol Science and Technology* 14, 331–46.

Krauss, U., Steuten, C., Adonijah, V. and Holmes, K. 2006. Early detection of frosty pod rot as key to cost-effective control. In A. Y. Akrofi and F. Baah [eds], *Proc. of 5th INCOPED Seminar*, San José, Costa Rica, 17–18 October, pp. 8–13.

Krauss, U., Hidalgo, E., Bateman, R., Adonijah, V., Arroyo, C., García, J., Crozier, J., Brown, N. A., ten Hoopen, M. and Holmes, K. A. 2010. Improving the formulation and timing of application of endophytic biocontrol and chemical agents against frosty pod rot (*Moniliophthora roreri*) in cocoa (*Theobroma cacao*). *Biological Control* 54: 230–40.

Krauss, U., ten Hoopen, M., Rees, R., Stirrup, T., Argyle, T., George, A., Arroyo, C., Corrales, E. and Fernando Casanoves, F. (2013). Mycoparasitism by *Clonostachys byssicola* and *Clonostachys rosea* on *Trichoderma* spp. from cocoa (*Theobroma cacao*) and implication for the design of mixed biocontrol agents. *Biological Control* 57, 317–27. DOI:10.1016/j.biocontrol.2013.09.011.

Leach, A. W., Mumford, J. D. and Krauss, U. 2002. Modelling *Moniliophthora roreri* in Costa Rica. *Crop Protection* 21: 317–26.

Meissner, H., Lemay, A., Bertone, C., Schwartzburg, K., Ferguson, L. and Newton, L. 2009. *Evaluation of Pathways for Exotic Plant Pest Movement into and within the Greater Caribbean Region*. Caribbean Invasive Species Working Group (CISWG), Plant Epidemiology and Risk Analysis Laboratory (PERAL), Center for Plant Health Science and Technology (CPHST), United States Department of Agriculture: North Carolina, USA.

Parra, D., Perez, S., Sosa, D., Rumbos, R., Gutierrez, B. and Moya, A. 2009. Avances en las investigaciones venezolanas sobre enfermedades del cacao. *Revista de Estudios Transdisciplinarios* 1(2).

Phillips-Mora, W., Castillo, J., Krauss, U., Rodríguez, E. and Wilkinson, M. J. 2005. Evaluation of cacao (*Theobroma cacao*) clones against seven Colombian isolates of *Moniliophthora roreri* from four pathogen genetic groups. *Plant Pathology* 54: 483–90.

Phillips-Mora, W., Ortíz, C. and Aime, C. 2007. Fifty years of frosty pod rot in Central America: Chronology of its spread and impact from Panama to Mexico. Proceedings 15th International Cocoa Research Conference, Vol. I, pp. 1039–47.

Phillips-Mora, W., Arciniegas-Leal A., Mata-Quiros A. and Motomajor-Arias J. C. 2012. *Catálogo de Clones de Cacao Seleccionados por el CATIE para la Siembra Comercial*. Centro Agronomico Tropical de Investigacion y Ensenanza (CATIE): Turrialba, Costa Rica, pp. 68.

Phillips-Mora, W., Baqueros, F., Melnick, R. L. and Bailey, B. A. 2015. First report of frosty pod rot caused by *Moniliophthora roreri* on cacao in Bolivia. *New Disease Reports* 31: 29. http://dx.doi.org/10.5197/j.2044-0588.2015.031.029.

Ploetz, R. 2016. The impact of diseases on cacao production: A global overview. In *Cacao Diseases: A History of Co-Evolution and Adaptation to Introductions* (Bailey, B. and Meinhardt, L. (Eds)). Springer, Cham, Heidelberg, New York, Dordrecht and London, pp. 33–59.

Pomella, A., Hebbar, P., Mitken, M, Roncato, K. and Leal, G. 2005. A potential technology to monitor FPR – Applied research in Brazil for *Crinipellis perniciosa*. In *Strategies to Control the Spread of Frosty Pod Rot in the Americas*. Turrialba, Costa Rica: CATIE, pp. 21–2. February 2005.

Reyes Valentín, M., Nolasco Alvarez, C. and Mercedes, L. 2009. The Caribbean Regional Diagnostic Network: Status and prospects for expansion towards coordinated Regional safeguarding. *Proceedings of Caribbean Food Crops Society* 45 (1): 63–70.

Sánchez-Mora, F. D., Medina-Jara, S. M., Díaz-Coronel, G. T., Ramos-Remache, R. A., F. Vera-Chang, J., Vásquez-Morán, V. F., Troya-Mera, F. A., Garcés-Fiallos, F. R. and Onofre-Nodari, R. 2015. Potencial Sanitario y Productivo de 12 clones de Cacao en Ecuador. *Revista Fitotecnia Mexicana* 38: 265–74.

Schnell, R. J., Kuhn, D. N., Brown, J. S., Olano, C. T., Phillips-Mora, W., Amores, F. M. and Motamajor, J. C. 2007. Development of a marker assisted selection program for cacao. *Phytopathology* 97: 1664–9.

Solís, B. J. L., Zamarripa, C. A., Quintero, V. P., Ramírez, E. G. and Gómez, E. H. 2015. Agronomic evaluation of cocoa hybrids (Theobroma cacao L.) for high-yield selection and field resistance to moniliasis. *Revista Mexicana de Ciencias Agrícolas* 6 (1): 71–82.

Ten Hoopen, G. M. and Krauss, U. 2016. Biological control of cocoa diseases. In *Cacao Diseases: A History of Co-Evolution and Adaptation to Introductions* (Bailey, B. and Meinhardt, L. (Eds)). Springer, Cham, Heidelberg, New York, Dordrecht and London, pp. 511–66.

Ten Hoopen, G. M., Rees, R., Aisa, P., Stirrup, T., Krauss, U. (2003) Population dynamics of epiphytic mycoparasites of the genera Clonostachys and Fusarium for the biocontrol of black pod (Phytophthora palmivora) and moniliasis (Moniliophthora roreri) on cocoa (Theobroma cacao). *Mycological Research* 107, 587–96.

Ten Hoopen, G. M., George, A., Martínez, A., Stirrup, T., Flood, J. and Krauss, U. 2010. Compatibility between *Clonostachys* isolates with a view to mixed inocula for biocontrol. *Mycologia* 102, 1204–15.

Waugh, J. D. 2009. *Trade and Invasive Species in the Caribbean: A Universe of Risk*. IUCN: Gland, Switzerland, p. 52.

Chapter 12

Cacao diseases: vascular-streak dieback

David I. Guest, University of Sydney, Australia; and Philip J. Keane, LaTrobe University, Australia

1 Introduction

A devastating dieback disease of cacao was first reported in Papua New Guinea (PNG) in the early 1960s (Bridgland et al. 1966a,b, 1967; Shaw 1962). The distinctive pattern of symptom development associated with this new disease differed from the previously known stress and insect related dieback syndromes of cacao (Turner 1967). Characteristic green spotted chlorosis (Fig. 1a) first appeared on a single leaf on the second or third growth flush behind the tip of vigorously growing branches (Keane et al. 1972; Prior 1980). These leaves fell and the chlorosis appeared on adjacent leaves, which also fell, until the entire branch was defoliated or the tip was killed. The symptoms were also distinct from sudden death symptoms characterised by the rapid wilting and death of the entire canopy caused by stem canker (*Phytophthora palmivora*), *Ceratocystis* wilt (*Ceratocystis cacaofunesta*) or root rot fungi (including *Armillaria* sp., *Verticillium dahliae*, *Rigidoporus microporus*, *Rosellinia* spp. and *Phellinus noxius*). In the susceptible genotypes of cacao first exposed to the disease in the initial epidemics, the fungus commonly spread to the main trunk and killed mature trees. Infection of the main shoot of seedlings was invariably lethal. The disease was named vascular-streak dieback (VSD) after the characteristic brown streaking in the xylem of infected shoots visible after scraping off the dry surface of the leaf scars left by the abscission of infected leaves. The epidemic caused heavy losses of the mature Trinitario trees that were the common type in PNG at the time and of seedlings planted near older cacao. After the first report of the disease in PNG, it was realised that a devastating dieback disease that had prevented establishment of Amelonado cacao in Malaysia in the late 1950s was this

http://dx.doi.org/10.19103/AS.2017.0021.18

(a)

(b)

Figure 1 Leaf symptoms of vascular-streak dieback in Vietnam: (a) typical 'green island' symptom seen commonly until 2004; (b) symptom of marginal leaf necrosis observed commonly in Malaysia and Indonesia (Java, Bali, Sulawesi and Papua) after 2004.

same disease (Turner and Shepherd 1980) and the disease was found to be common in the new cacao plantings in Malaysia in the 1960s and 1970s. During the initial epidemics in PNG, white basidiocarps were observed commonly on leaf scars following the abscission of diseased leaves during periods of wet weather. The fungus was shown to be a new genus and species of tullasneloid basidiomycete, *Oncobasidium theobromae* (Talbot and Keane 1971), and was found growing only in the xylem vessels of infected leaves and stems. The pathogen is a near-obligate parasite; unlike the witches' broom pathogen (*Moniliophthora perniciosa*) it does not survive in dead branches, and has proven difficult to isolate into pure culture – it emerges readily from infected stem or leaf tissue onto agar media, but is unable to be subcultured and usually does not sporulate in culture. This fungus was shown to be the cause of the disease by inoculation using basidiospores shed from fresh sporocarps collected from cacao plantings (Keane et al. 1972; Keane and Prior 1991). In PNG and Malaysia in the 1960s (and later in Indonesia in the 1980s), the epidemic destroyed many new plantings of cacao that were generally highly susceptible to the disease. Planting seedlings and clonal plants derived from surviving, less susceptible trees saved the industry (Keane and Prior 1991; Efron et al. 2002). Marked differences in susceptibility between Trinitario clones were evident in clonal trials during the initial epidemic in PNG (J.B. O'Donohue, pers. comm.).

VSD has since been found in most cacao-growing areas in South and Southeast Asia and Melanesia, from Kerala (India) in the west, through Myanmar, Thailand, Hainan Island (China) and Vietnam, Malaysia, Indonesia and the southern Philippines, to New Britain (PNG) in the east. The disease is estimated to directly cause yield losses of 30 000 tonnes, comparable to other geographically restricted diseases like frosty pod rot in Central America (Ploetz et al. 2007). VSD and cocoa pod borer (*Conopomorpha cramerella*) were key factors in the decline of the Malaysian cacao industry as rising costs of management led farmers to replace cacao with alternative crops such as oil palm. VSD is widespread in Indonesia, including in the fine-flavour cacao plantations in East and West Java, and in the main cacao plantings in Sulawesi. VSD poses a major threat to the cacao industries in Africa and Latin America, where much of the Amelonado planting material is known to be susceptible – West African Amelonado cacao planted in Malaysia in the 1950s was destroyed and Amelonado types planted in PNG have been very susceptible.

The taxonomy of *C. theobromae* has been revised since Talbot and Keane's first description of the pathogen as a unique aerially dispersed, leaf-infecting, xylem-colonising, binucleate, near-obligate *Rhizoctonia*-like fungus in 1971, and continues to be debated. Talbot and Keane (1971) considered the cacao pathogen to be distinct from *Ceratobasidium* based on its unique pathogenic biology, which contrasted with ceratobasidiums and rhizoctonias that are mainly saprotrophic orchid mycorrhizal fungi or pathogens. Roberts (1999) renamed the fungus as *Thanatephorus theobromae* and then Samuels et al. (2012) using DNA sequence data placed the species within *Ceratobasidium*, where it currently stands. However, Oberwinkler et al. (2013) later supported Roberts' arguments, although they proposed the more widely used genus name *Rhizoctonia*.

The only known host of *C. theobromae* other than cacao is avocado (*Persea americana*: Lauraceae), also an exotic plant in Southeast Asia and the Pacific (Anderson 1989). The fungus has not been reported from outside Southeast Asia and parts of PNG, and it is believed that it evolved on an as-yet-unidentified host indigenous to Southeast Asia and the islands of New Guinea and New Britain in PNG and has transferred to introduced cacao. The occurrence of the disease in widely separated localities throughout the region, coupled with the evidence of highly restricted dispersal ability of the fungus, suggests that the fungus has transferred repeatedly from its indigenous host to cacao in different locations. A testable hypothesis is that this indigenous host does not occur on the islands of New Ireland and Bougainville (to the east of New Britain in PNG), where VSD has never been reported to occur naturally despite the extensive cacao plantings there. Thus, VSD is another example of a new encounter disease in cacao.

2 VSD symptoms and pathogen biology

The most characteristic initial symptom of VSD is the chlorosis of a single leaf, usually on the second or third flush behind the shoot apex, with scattered islets of green tissue 2–5 mm in diameter (Keane et al. 1972; Keane and Prior 1991; Fig. 1a). Affected leaves are shed within a few days of becoming fully chlorotic, and symptoms progressively develop in adjacent leaves up and down the stem. Lenticels on infected trees usually become enlarged, causing roughening of the bark (Fig. 2). Three discoloured vascular traces are visible when the dry surface is scraped off scars resulting from the fall of diseased leaves

Figure 2 Swollen lenticels on infected stem and basidiocarps of *Ceratobasidium theobromae* formed on the leaf abscission scars on a cacao branch affected by vascular-streak dieback in Sumatra, Indonesia (photo P.D. Turner).

or when newly infected leaves are pulled off the branches (Fig. 3). When stems are split longitudinally, infected xylem is visible as dark streaks within the vascular tissue (Fig. 4). Eventually leaf fall occurs right to the growing tip, which then dies, usually several months after the initial infection. Lateral buds may proliferate on the infected stem, but then die, as they are infected from the main stem, causing 'broomstick' symptoms to develop (Fig. 5). The fungus may spread internally to other branches and usually kills the tree if it spreads to the trunk, as was common in the early epidemics in PNG and Malaysia on very susceptible genotypes.

When an infected leaf falls during wet weather, hyphae may emerge from the leaf scar and develop into a tullasneloid basidiocarp. If leaf fall occurs on a dry day, the scar heals over and the fungus fails to emerge. Continued development of the basidiocarp occurs only during wet weather and results in the formation of a white, flat, velvety coating over the leaf scar and adjacent bark (Fig. 2). Basidia develop on the basidiocarp as they dry out somewhat in the evening after being wetted by late afternoon rainfall. Formation and forcible discharge of basidiospores occur mainly about midnight into the early morning. Basidiospores remain viable for only a few hours into the morning and require free water for germ tube growth (Prior 1979). Basidiocarps remain fertile for about a week on attached branches, but only for a day or two on cut branches. Extended periods of wetness, rather than intense rainfall, are required for basidiocarp formation, basidiospore release, dissemination of spores and infection of leaves. Thus, there is a critical link between wetness, as influenced by rainfall and humidity, and infection periods (Keane 1981; Dennis and Keane 1992).

Figure 3 Discolouration of the vascular traces on leaf abscission scars on a cacao branch affected by vascular-streak dieback.

Figure 4 Vascular streaking on a cacao branch affected by vascular-streak dieback.

The distinctive leaf symptoms of VSD as first described (chlorosis of infected leaves with remnant green islands; leaves abscising cleanly as soon as they became fully chlorotic) remained relatively uniform across its distribution for many years. However, around 2004 an apparently sudden change in symptoms was reported in several widely separated locations, including in Malaysia (Peninsular Malaysia and Sabah) and Indonesia (East Java, Bali, Sulawesi, Papua and West Papua) (Keane and Guest 2007; Purwantara et al. 2009).

Figure 5 'Broomsticking' caused by successive leaf drop and axillary bud development on a cacao branch affected by vascular-streak dieback.

The new symptoms featured the development of necrotic patches on green leaves often on the margins, rather than the familiar green islands on yellowing leaves (Fig. 1b). Infected leaves tended to remain firmly attached.

In many reported cases, the new symptoms occurred alongside the more traditional symptoms, occasionally on the same tree or even the same branch and leaf, although often the traditional symptoms were less distinctive (e.g. green islands were less sharply defined). The internal symptom of discolouration of vascular tissue in the stems and vascular traces of infected leaves remained, except that it was much blacker. Infection was still confined to xylem vessels and first symptoms still appeared on the second or third flush behind the shoot apex. Ultimately defoliation and shoot dieback occurred.

Cracks commonly develop on the main veins of infected leaves, especially near the petiole, probably as a consequence of delayed leaf fall and the necrotic reaction of tissues. Basidiocarps (that formed basidia and basidiospores typical of *Ceratobasidium theobromae*) frequently developed on these cracks, as never previously observed, while the formation of basidiocarps on leaf scars was much rarer because leaves tended to remain attached to branches. Further, the basidiocarps emerging from cracks on the leaf veins were often observed outside periods of wet weather, and occurred much more commonly than basidiocarps on leaf scars. It is likely that the condensation of water on leaf laminae that commonly occurs at night in the wet tropics is sufficient to allow the emergence of hyphae through the cracks, whereas heavy rain is required to wet leaf scars on stems sufficiently for hyphal emergence and formation of basidiocarps. The more frequent occurrence of basidiocarps may explain why an increase in disease incidence has generally been observed in association with the necrotic symptoms.

The reason for this widespread change in symptoms is puzzling, and few clues exist. The incidence of VSD was observed to decline with increasing altitude, with the incidence of necrotic symptoms declining more rapidly than the traditional symptoms, possibly indicating a link to environmental factors (S. Bryceson, BSc(Hons) Thesis, LaTrobe University 2014). The aetiology of this symptom is unknown, and while it is similar to marginal necrosis induced by potassium deficiency, there is no clear link to soil nutrition or changes in cacao genotype (McMahon and Purwantara 2016). Until more is known about the population structure of *C. theobromae*, which at present can be investigated only by molecular means, it is impossible to determine if the change follows a change in the pathogen genotype.

Attempts to observe infection of unhardened seedling leaves with basidiospores falling directly from sporulating basidiocarps revealed very occasional germ tubes penetrating directly through the cuticle above the leaf veins (Prior 1979; Keane and Prior 1991). Penetration of hyphae into xylem vessels has not been observed, and the details of the initial penetration of xylem in leaves remain unknown. Infected leaves do not show symptoms for three to five months, by which time the pathogen has ramified through the xylem and into the adjacent stem. On rare occasions, it was shown by microscopic observation of veins that the fungus was still confined to the first infected leaf by the time chlorosis was evident in the leaf, providing evidence that the fungus first infected the leaf lamina. The observed incubation period explains why first symptoms usually appear on the second or third flush behind the growing tip (the shoot has added three to five months growth, i.e. two to three growth flushes, during the incubation period), and why the peak in the appearance of disease symptoms often occurs three to five months after seasonal rainfall peaks.

Attempts to axenically culture *C. theobromae* have been only partially successful. The fungus grows readily from surface-sterilised petioles, leaf pieces and stem pieces onto water agar or cacao callus, but cannot be subcultured or induced to sporulate (Prior 1977). Lam et al. (1988) and Vadamalai (1999) reported limited sporulation by growing hyphae on nutrient-poor medium under a stream of moist air, but numerous other attempts to induce sporulation in culture have failed. While these techniques do not produce enough basidiospores for artificial inoculation, the mycelium emerging from plated tissues yields enough pure DNA for recent attempts at genome sequencing (Bryan Bailey, pers. comm.). This will become an invaluable tool for future research.

3 VSD epidemiology

Basidiocarps develop only when leaf fall exposes abscission scars during wet weather. Basidiocarps have a short lifespan and only release spores at night, and then only as the basidiocarps dry out slowly after being wetted by afternoon and evening rains. Basidiospores are dispersed by wind, although effective spore dispersal is probably limited to a few hours on the early morning of their discharge when humidity is high and leaves are wet (Keane 1981). As a result, very few infections occur beyond 100 m from diseased cacao, or, presumably, from the putative indigenous host. The rate of disease spread is also limited by the relatively low sporulation rate of the fungus, linked to the very precise environmental conditions required for sporulation. There has been no evidence of wind-borne dispersal of the pathogen from East New Britain, where the disease has been very common, to widespread cacao plantings in New Ireland only 60 km to the east, nor to the extensive cacao plantings on Bougainville, 250 km to the southeast.

The rate of symptom development is closely correlated with rainfall incidence during a few prior months, and, in PNG and West Malaysia, the disease is most common in wetter regions where annual rainfall exceeds 2500 mm (Dennis and Keane 1992). Heavy shade and unpruned cacao increases canopy humidity and leaf wetness, thus increasing sporulation and infection (Dennis and Keane 1992).

While wetness is required for sporulation, as with other vascular wilt diseases, dieback symptoms become most severe during periods of water stress. The disease is most damaging in seedlings less than ten months old, which are killed by infection of the single stem. Only the most susceptible genotypes are killed by infections beginning in outer branches on older trees (Keane 1981).

4 VSD management

4.1 Quarantine

The appearance of VSD in newly planted areas of cacao could result from its introduction on infected planting material or from infection from a locally present indigenous host. This has significant implications for quarantine interventions. Despite intensive studies (Keane and Prior 1991), dispersal of the fungus has not been demonstrated to occur through pods or seed (the main means of dispersal of cacao in the early days of plantation development in Southeast Asia and PNG). Dispersal in infected budwood sticks is highly unlikely because the fungus rapidly kills infected buds or grafts made from infected material. The most likely method of spread by human agency is through transport of living plants in pots or polybags (or bare-rooted plants).

Unintentional movement of infected planting material presents a significant biosecurity risk. However, appropriate quarantine precautions based on our current understanding of the biology and transmission of the pathogen can allow dissemination of cacao planting material while reducing the risk. For example, strict quarantine protocols implemented in PNG have enabled transfer of superior clones from New Britain to the disease-free islands of Bougainville and New Ireland (Prior 1985).

Nursery plants can be protected by any structure that prevents foliage becoming wet for more than an hour or so, as this prevents infection and sporulation. It was demonstrated in PNG that plants raised under plastic shelters rarely became infected, even when surrounded by infected trees (J.B. O'Donohue, pers. comm.).

Given the success of quarantine protocols in preventing the movement of VSD to nearby islands, it is interesting to note that the Cocoa Pod Borer moth spreads so rapidly and widely following the first reported outbreak in East New Britain in 2006. Human movement of infested pods is thought to be largely responsible for the rapid dispersal of this pest, suggesting that pods are transported frequently between islands outside these quarantine protocols. Also, the small highly active moth is likely to be spread on the wind much more widely than the VSD pathogen. It appears that the distribution of VSD depends more on the distribution of the indigenous host than on anthropogenic movement of cacao. Therefore, the quarantine focus should be on identifying this host and avoiding transport of plants that are asymptomatic carriers.

4.2 Disease resistance

Detailed studies of resistance have been hampered by the difficulty in obtaining regular supplies of inoculum for bioassays, as the fungus cannot be maintained in culture. Field-collected spores are rarely available and sporulation from field-collected basidiocarps occurs for only about a day after collection, and even then requires that the basidiocarps be exposed to very precise moisture conditions to stimulate basidiospore formation and release.

The VSD epidemic in PNG during the 1960s devastated plantings established during the previous decade, killing a high proportion of mature trees and young seedlings. The epidemic slowed by the 1970s as the most susceptible genotypes succumbed to VSD and growers had no choice but to plant seed from surviving trees that were less susceptible. Selection for resistance to the disease occurred among seedlings in the heterogeneous, partly out-breeding Trinitario population of cacao in PNG (Prior 1979). Replicated clonal trials of Trinitario material planted for agronomic testing at the Lowlands Agricultural Experiment Station (LAES) in the early 1960s were exposed to natural infection during the initial epidemic. Some clones sustained many infections that progressed rapidly through the branches; all trees of these clones were killed and the clones became extinct. Other clones sustained relatively few infections, and these infections progressed more slowly in the plants and never killed them unless they occurred in the main stem of a young plant. Later, Upper Amazon planting material introduced into PNG in the 1960s was screened by exposure to natural epidemics and was shown to include types that showed similar partial resistance. Clones derived from this material were subsequently used as male parents together with resistant Trinitario female parents to produce hybrids having a useful degree of resistance. These are now widely grown throughout PNG. Similar screening for resistance using natural infection in the field was carried out in Malaysia (Cheng-Tuck Ho, pers. comm.) and Indonesia, resulting in very useful levels of resistance.

Less susceptible genotypes still become infected, but less commonly, and symptoms are less severe and the pathogen grows more slowly and only rarely sporulates; invasion into larger branches and the main trunk also appears to be restricted in these genotypes. In PNG, breeding with surviving genotypes has been very successful, and has relegated VSD to minor importance in most years. Cacao in Sulawesi also appears to have a degree of resistance to VSD, probably because of natural selection operating in the genetically diverse cacao germplasm there. Under conducive environmental conditions the disease occurs very commonly in the outer branches but rarely kills the trees as it did in the first wave of epidemics in PNG and Malaysia.

VSD resistance has been shown to be highly heritable and additively inherited (Tan and Tan 1987). It has proved to be highly durable and has not broken down in the 50 years since its first identification and widespread deployment. Quantitative trait linkages (QTLs) for VSD resistance among crosses between the Trinitario types K82 (susceptible to VSD) and KA2-101 (resistant to VSD) have been identified on all chromosomes of cacao, clustering with QTLs for resistance to other diseases (Peter Epaina, PhD Thesis, The University of Sydney 2012). These QTLs thus have great potential for marker-assisted selection of breeding progeny. Other research conducted in East Java by the Indonesian Coffee and Cocoa Research Institute suggests one to two genes might be responsible for resistance (Susilo and Anita-Sari 2011).

4.3 Cultural practices

Raising clean nursery stock well away from diseased cacao ensures that stock transplanted into the field is initially disease free. This is important as the disease generally kills young plants infected in their main stem. Nursery stock can be further protected by growing it in a shade house or under a shelter that keeps leaves dry for all but a few hours following watering. Very high levels of infection and subsequent death of young plants have been observed in unprotected plants raised near or under older, diseased cacao (which used to be the standard smallholder practice).

Pruning diseased material about 30 cm below discoloured xylem prevents further extension of infections and reduces inoculum levels by removing potential sites of sporulation. In a mature planting in Java, where highly trained teams detected and pruned out infected branches every two weeks for nearly two years, the incidence of infected trees was kept below 1%; in a similar unpruned planting, disease incidence increased from about 30% to 90% in a ten-month period (Parwirosoemardjo and Purwantara 1992). During regular pruning of cacao to maintain smaller, more manageable trees and open canopies, VSD infections can be targeted, allowing effective disease control with minimal extra effort. It is therefore crucial that growers recognise the earliest symptoms of the disease. Thorough pruning out of infected branches before the onset of the wet season will be the most effective in reducing disease by reducing the incidence of sporulation.

Shade and canopy management to increase aeration and sunlight on the foliage is of critical importance as sporulation and infection require moist conditions (Dennis and Keane 1992). These practices are best combined with integrated management practices designed to control *Phytophthora palmivora* and other pests and diseases.

Providing appropriate fertiliser applications that improve vigour reduces the stress-related dieback symptoms. However, increased flushing produces more young leaves that are vulnerable to infection.

4.4 Biological control

C. theobromae is a vascular pathogen that infects immature leaves. It is possible that endophytic fungi and bacteria may protect against vascular colonisation. Biological control agents could also be targeted at the basidiocarp stage, although basidiocarps are so ephemeral that this is unlikely to be effective.

4.5 Fungicides

Systemic ergosterol biosynthesis-inhibiting fungicides, including flutriafol, hexaconazole, propiconazole, tebuconazole and triadimenol, have been successfully used to control VSD under experimental and nursery conditions in Malaysia. However, none of these fungicides have been shown to be commercially viable in established cacao plantations (Holderness 1990, Guest, unpublished). Protection of planting material is best achieved by growing the plants in nurseries enclosed in plastic sheeting and hand watering in the morning so that the foliage dries out rapidly.

5 Case study: the importance of field studies

Shaw (1962) and Bridgland (1966a,b, 1967) described a destructive lethal dieback disease of the new, post–Second World War cacao plantings in PNG in the 1960s and distinguished it from stress and insect-induced dieback. They named it 'Botryodiplodia dieback' because *Botryodiplodia theobromae* was commonly cultured from the dead branches. Distinguishing symptoms included typical chlorosis with green islets of infected leaves prior to their abscission, brown streaks in the wood, enlarged lenticels and the proliferation of lateral buds associated with tip dieback.

As a young PhD student supported by cacao growers, Philip Keane was recruited to work at the LAES, Keravat, in the centre of the epidemic of the disease in East New Britain

Province. His research built on the first descriptions of the disease, and was guided by Leon Bridgland, an agronomist at LAES; John O'Donohue; and growers. Steep gradients of infection in young cacao planted adjacent to older stands had been observed, and by counting infections on a range of cacao clones teams of labourers working for O'Donohue were able to identify that symptoms first appeared on leaves that were invariably on the second or third flush behind the actively growing shoot apex (whether on main stems of seedlings or outer branches on mature trees). It was evident that much early symptom development occurred over many weeks before the eventual death of branches that were then invaded by *Botryodiplodia* and other saprotrophs.

Keane found that xylem vessels in symptomatic stems were colonised by *Rhizoctonia*-like hyphae, and that when chlorotic leaves abscissed during wet weather, these hyphae emerged from the three vascular traces in the leaf scars and formed tullasneloid basidiocarps on and around the scars. During persistent wet weather these basidiocarps produced basidia and basidiospores of a previously undescribed fungus, which was named *Oncobasidium theobromae* by Pat Talbot at the Waite Agricultural Research Institute, Adelaide, Australia, a specialist in the Tulasnellales (mainly isolated from orchids). Keane noticed that basidiocarps formed basidia only at night after late afternoon rains, forming and releasing basidiospores by about midnight. Spores germinated only on wet leaves and lost viability within a few hours on the following day. These studies were confirmed and expanded by Chris Prior, who came to PNG as a postdoctoral fellow after completing his PhD at Cambridge University.

While the fungus emerged prolifically from cut stem pieces onto agar media, it never sporulated in culture and could not be subcultured. By placing vigorously growing seedlings at night under infected branches collected from the field and wetted to simulate afternoon rains and stimulate formation and release of basidiospores, he found that typical VSD symptoms developed on most of the seedlings after about three months, but not on similar batches of seedlings not exposed to spores. The three-month incubation period for the disease explained why the first symptoms invariably occurred on the second or third flush behind the apex – if infection of leaves occurred at the apex and it took three months for symptoms to develop, the apex would have added another three months growth (i.e. two or three flushes) by the time they appeared. Also, it explained why the disease occurred most commonly in the field three to five months after prolonged periods of wet weather. Although Koch's postulates for this obligate biotroph could not be fulfilled, constant association of the fungus with the disease and other biological evidence was overwhelming and has only strengthened over time.

While the identification of the pathogen as a unique, airborne, xylem-infecting, basidiomycete causing tip dieback in cacao was biologically significant, it also showed that reducing sporulation and protecting planting material from exposure to spores could be useful management strategies. O'Donohue had shown during the height of the initial epidemics that seedlings grown in plastic-covered enclosures in the field did not get infected, and with the demonstration that the disease was caused by airborne spores this led to the recommendation that plants should be raised in plastic-covered nurseries. Also, sanitary pruning of symptomatic branches reduced the inoculum sources, and opening up the cacao canopy by pruning of cacao and shade trees reduced the canopy wetness required for sporulation and infection (later clearly shown experimentally by another PhD student at LAES, Jason Dennis).

Prior to any demonstration of causation, the most susceptible cacao genotypes were killed by the first waves of the VSD epidemic and growers had no choice but to propagate from the less susceptible survivors. Replicated clonal trials established by

O'Donohue for agronomic studies prior to the epidemic clearly showed the range of susceptibility in the genetically diverse Trinitario cacao genotypes that dominated the PNG plantings – some very susceptible clones that sustained many infections and were a prolific source of basidiocarps for experimental studies were killed and became extinct, while others sustained far fewer infections, and these did not progress to the main trunk and kill the trees. Further selection trials identified less susceptible Trinitario and Upper Amazonian genotypes that came to dominate cacao production in PNG, and much of Southeast Asia.

Similar partial resistance was selected in Malaysia – for example, the famous Prang Besar Clone (PBC) 123 selected by Cheng-Tuck Ho on Prang Besar Estate. Breeding programmes in Asia and Melanesia now routinely include VSD screening, and together with good nursery management and regular pruning, the disease can be controlled. Nevertheless, VSD causes significant losses in poorly managed farms. If the pathogen is ever transported to cacao-growing regions in Africa and the Americas it is likely to be very destructive in cacao genotypes that have not been selected for resistance to the disease.

6 Conclusion

VSD results from a new encounter interaction between an indigenous fungus, *Ceratobasidium (Oncobasidium) theobromae*, and cacao introduced into Southeast Asia and parts of Melanesia. Infection causes a progressive dieback starting in outer branches, resulting in xylem necrosis and defoliation. The disease is invariably fatal if infection occurs in the main stem of seedlings or clonal plants, but if the infection reaches the main stem of mature plants the disease is usually fatal only in susceptible genotypes. While much of the infection biology remains a mystery, integrated management, including the production of disease-free plants in covered nurseries, canopy management and regular pruning of infected branches, maintaining only low levels of shade, and use of partially resistant genotypes of cacao, provides adequate control of the disease.

7 Future trends

In the eastern provinces of Indonesia where cocoa production has expanded rapidly, it is important that resistance to VSD be incorporated into the planting material. Following a rapid expansion of cacao planting, the disease has not been as destructive in Sulawesi as it was in PNG and Malaysia in the 1960s. Sulawesi benefited from the fact that the cacao plantings included types already selected for resistance in Malaysia, the Indonesian Coffee and Cocoa Research Institute in East Java, and the planting of seed from hybrid cacao pods that produced wide genetic diversity.

Cacao breeding for VSD resistance would be greatly facilitated if markers for resistance could be developed to rapidly screen progeny rather than waiting for symptoms to develop in field plantings. Recent research indicates that useful genetic markers for both VSD and *Phytophthora* resistance exist, but these need to be screened more widely to evaluate their potential in large collections. Screening of breeding populations in Africa and Latin America for these markers would help establish the potential risk VSD poses to the cacao industries globally.

There is a lot still to be discovered about the biology of the pathogen and impact of local environment before an accurate evaluation of the global threat posed by VSD is possible. The devastating epidemic in PNG and Malaysia in the 1960s highlights the potential impact of the disease on susceptible cacao genotypes under favourable environmental conditions. While the probability of the pathogen spreading in infected cacao seed or budwood is small, the potential for the pathogen to spread in unidentified indigenous hosts is unknown. What, for example, if the pathogen was an orchid endophyte, and what if the orchid host was shipped around the world as an ornamental? Until the true distribution and host range of the pathogen is known, we cannot assess this risk.

There are still gaps in our understanding of the life cycle and genetic diversity of the pathogen. We do not know why new disease symptoms on cacao appeared simultaneously in a number of places in the mid-2000s and what this means. We still have little understanding of the initial stages of infection of *C. theobromae* on cacao. Early studies by Prior, Keane and Dennis point to basidiospore infection of unhardened flush leaves, but the evidence remains scarce because of the dependence on natural inoculum for infection studies and the fact that sporulation of the fungus in the field is ephemeral and limited. This means research has to be done in field stations where advanced facilities are not always available. Discovering the indigenous host may provide a more reliable source of inoculum until we learn how to culture the pathogen and produce spores axenically. The lack of a reliable infection assay limits our understanding of the infection and early colonisation process.

Further, if basidiospores are the inoculum that initiates infection, what does this mean for genetic diversity? Infection is initiated by uninucleate basidiospores formed after meiosis; however, hyphae colonising the leaf are binucleate (Keane et al. 1972). When does anastomosis and plasmogamy occur? Is the VSD pathogen a distinct lineage of *C. theobromae*? Further studies will be greatly facilitated by the use of tools such as single nucleotide polymorphisms (SNPs) or microsatellite markers to reveal the population structure.

The browning and necrosis symptoms of VSD seen in a more accentuated form since 2004 are essentially a host response indicating incompatibility between the host and pathogen. These symptoms may not appear in the compatible interaction between the co-evolved indigenous host and *C. theobromae*. Are there other asymptomatic hosts? Are there genotypes of cacao that result in asymptomatic infections, and if so, what are the biosecurity implications of quarantine protocols based on identifying symptoms?

VSD currently impacts cacao grown only in Asia and parts of Melanesia. The precautionary principle, however, suggests that all possible steps should be taken to prevent this disease from spreading. Any cacao planting material transported between regions has to go through an intermediate quarantine facility such as International Cocoa Quarantine Centre at the University of Reading or that at USDA-ARS, Miami, to minimise the risk of unintentional introductions.

8 Where to look for further information

Most research on VSD has been centred at the PNG Cocoa Coconut Institute in East New Britain, PNG; the Indonesian Coffee and Cocoa Research Institute in Jember, Indonesia; and the Malaysian Cocoa Board. A collaboration based at the USDA-ARS laboratory in Beltsville, Maryland, is attempting to sequence the genome of *C. theobromae*, which may generate tools that help to answer some of the questions raised in this review (Bryan Bailey, pers. comm. 19 April 2016).

9 References

Anderson, R. D. 1989. Avocado, an alternate host for *Oncobasidium theobromae*. *Australas. Plant Pathol.* 18, 96–7.

Bridgland, L. A., Richardson, I. M. and Edward, I. L. 1966a. Dieback diseases of cacao (Part 1). *S. Pac. Planter* 1(6), 13–20, 28.

Bridgland, L. A., Richardson, J. M. and Edward, I. L. 1966b. Dieback diseases of cacao (Part 2). *S. Pac. Planter* 1(7), 3–6, 22.

Bridgland, L. A., Richardson, J. M. and Edward, I. L. 1967. Dieback disease of cacao (Part 3). *S. Pac. Planter* 1(8), 9–11, 16–17.

Dennis, J. J. C. and Keane, P. J. 1992. Management strategies for control of vascular-streak dieback of cocoa. *Cocoa Growers' Bull.* 45, 31–44.

Efron, Y., Marfu, J., Faure, M. and Epaina, P. 2002. Screening of segregating cocoa genotypes for resistance to vascular-streak dieback under natural conditions in Papua New Guinea. *Australas. Plant Pathol.* 31, 315–19.

Holderness, M. 1990. Control of vascular-streak dieback of cocoa with triazole fungicides and the problem of phytotoxicity. *Plant Pathol.* 39, 286–93.

Keane, P. J. 1981. Epidemiology of vascular-streak dieback of cocoa. *Ann. Appl. Biol.* 98, 227–41.

Keane, P. J. and Guest, D. I. 2007. Vascular-streak dieback: A new encounter disease of cacao in Papua New Guinea and Southeast Asia caused by the obligate Basidiomycete *Oncobasidium theobromae*. *Phytopathology* 97, 1654–7.

Keane, P. J. and Prior, C. 1991. *Vascular-streak Dieback of Cocoa*. Phytopathological Papers No. 33. CAB International: Wallingford, UK, p. 39.

Keane, P. J., Flentje, N. T. and Lamb, K. P. 1972. Investigation of vascular-streak dieback of cocoa in Papua New Guinea. *Aust. J. Biol. Sci.* 25, 553–64.

Lam, C. H., Varghese, G. and Zainal Abidan, M. A. 1988. In-vitro production of *Oncobasidium theobromae* basidiospores. *Trans. Br. Mycol. Soc.* 90, 505–8.

McMahon, P. and Purwantara, A. 2016. Vascular-streak dieback (*Ceratobasidium theobromae*): History and biology. In Bailey, B. A. and Meinhardt, L. W. (Eds), *Cacao Diseases*. Springer International Publishing: Switzerland, pp. 307–35.

Oberwinkler, F., Riess, K., Bauer, R., Kirschner, R. and Garnica, S. 2013. Taxonomic re-evaluation of the *Ceratobasidium-Rhizoctonia* complex and *Rhizoctonia butinii*, a new species attacking spruce. *Mycol. Prog.* 12, 763–76.

Pawirosoemardjo, S. and Purwantara, A. 1992. Occurrence and control of VSD in Java and South-East Sulawesi. In Keane, P. J. and Putter, C. A. J. (Eds), *Cocoa Pest and Disease Management in South-East Asia and Australasia*. FAO Plant Production and Protection Paper No. 112. FAO: Rome, Italy, pp. 209–13.

Ploetz, R. C. 2007. Cacao diseases: Important threats to chocolate production worldwide. *Phytopathology* 97, 1634–9.

Prior, C. 1977. Growth of *Oncobasidium theobromae* in dual culture with callus tissue of *Theobroma cacao*. *J. Gen. Microbiol.* 99, 219–22.

Prior, C. 1979. Resistance of cocoa to vascular-streak dieback disease. *Ann. Appl. Biol.* 92, 369–76.

Prior, C. 1980. Vascular-streak dieback. *Cocoa Growers' Bull.* 29, 21–6.

Prior, C. 1985. Cocoa quarantine: measures to prevent the spread of vascular-streak dieback in planting material. *Plant Pathol.* 34, 603–8.

Purwantara, A., Iswanto, A., Sukamto, S., McMahon, P. J., bin Purung, H., Lambert, S., Guest, D. I. and Keane, P. J. 2009. New symptoms of vascular-streak dieback of cocoa in Southeast Asia

– Possible causes, studies required and control. In *Proceedings of the 16th International Cocoa Research Conference*, Bali, Indonesia, 16–21 November 2009. Cocoa Producers Alliance.

Roberts, P. M. 1999. *Rhizoctonia-forming Fungi: A Taxonomic Guide*. Royal Botanic Gardens: Kew, UK.

Shaw, D. E. 1962. Diseases of cacao in Papua New Guinea. *Papua New Guinea Agr. J.* 15, 79–90.

Susilo, A. W. and Anita-Sari, I. 2011. Respons ketahanan beberapa hibrida kakao (*Theobroma cacao* L.) terhadap serangan penyakit pembulah kauyu (vascular-streak dieback). *Pelita Perkebunan* 27, 77–87.

Talbot, P. H. B. and Keane, P. J. 1971. *Oncobasidium*, a new genus of tulasnelloid fungi. *Austral. J. Bot.* 19, 203–6.

Tan, G. Y. and Tan, W. K. 1987. Genetic variation in resistance to vascular-streak dieback in cocoa. *Theor. Appl. Genet.* 75, 761–6.

Turner, P. D. 1967. Cocoa dieback – a review of present knowledge. *FAO Plant Prot. Bull.* 15, 81–101.

Turner, P. D. and Shepherd, R. 1980. Cocoa diseases in Malaysia and Indonesia: their present and potential importance, pp. 308–21. In: *Proceedings International Conference on Cocoa and Coconuts Kuala Lumpur 1978*. Incorporated Society of Planters, Kuala Lumpur.

Vadamalai, G. 1999. In vitro production and infectivity of *Oncobasidium theobromae* Talbot and Keane basidiospores. Masters Thesis, Universiti Putra Malaysia.

Chapter 13

Insect pests affecting cacao

Leïla Bagny Beilhe, Régis Babin and Martijn ten Hoopen, CIRAD, France

1 Introduction

Cacao is affected by a large number of pests and diseases, which have variable impact on cacao production. It is estimated that at least 30–40% of potential global production is lost due to pests and diseases (Flood, 2004; Ploetz, 2016). The damage caused by pests and diseases depends on a number of factors including environmental conditions, the plant species itself, the socio-economic conditions of farmers and the level of technology used (Oliveira et al., 2014). For organisms that directly affect fruit development, estimates of potential losses are relatively easy to make. However, accurate estimates of indirect losses caused by pests that affect tree production capacity are often much more difficult to obtain. Cacao in Africa and Asia is more under threat of pests than the Americas, the continent of origin of cacao, where diseases are of much greater importance. Mirid bugs and cacao pod borer (CPB) are the major pests of cacao worldwide with mirids being especially harmful in Africa and CPB in Asia. These major pests cause significant yield reductions if no control strategies are put into place. Next to these major pests, numerous 'minor' pests exist which generally do not significantly affect yield, except in cases of localized outbreak

http://dx.doi.org/10.19103/AS.2017.0021.19

events which can occur under particular environmental or crop management conditions. Nonetheless, taken together, these minor pests probably do constitute an important constraint and farmers faced with outbreaks will certainly not consider their damage minor.

Among all these pests, some lead to direct losses to ongoing production, for example, CPB attacks pods and reduces bean quantity and quality. Others, like mirids, cause damage to vegetative parts of the cacao tree, thereby affecting cacao production capacity not only in the current crop season but over several years of production. Cumulative damage over consecutive years can severely hamper cacao production. It is this cumulative and 'indirect' impact which is most difficult to quantify and is generally underestimated. Uncertainty regarding the actual impact that pest damage has on cacao productivity makes it difficult to formulate rational decisions regarding intervention strategies.

In addition to crop losses caused by insect pests, their control has an economic cost, related to the purchase and application of insecticides. These direct costs are often offset by the gains in productivity. However, the negative impacts on farmer health, product quality and the environment are often neglected. Overall, pest and disease management is usually problematic for cacao farmers, especially in Africa and Asia, because of limited farmer knowledge and access to resistant varieties, agrochemicals and training (ICCO, 2013).

This chapter will first provide a general overview, by geographic region, of pests in cacao agrosystems. Subsequently, current advances in pest control based on plant breeding and agro-ecological approaches to pest control in cacao agroforestry systems will be presented. Finally, knowledge gaps will be identified and recommendations for future research provided.

2 Cacao pests in the Americas and the Caribbean

Cacao has its origin in the upper Amazon basin of South America (Ji et al., 2013). Here, two pathogens co-evolved with cacao which are now considered to be the two most dangerous cacao diseases: *Moniliophthora roreri*, the causal organism of frosty pod rot, and *M. perniciosa* (previously known as *Crinipellis perniciosa*), the causal organism of witches' broom (Ploetz, 2016). Curiously though, no major cacao pests co-evolved or originated in the Americas with cacao. The major pests are found in Africa and Asia, which are outside the native range of cacao. It is possible that local beneficial fauna in the centre of origin limits outbreaks of insect pests on cacao plantations. Nonetheless, a large number of minor insect pests of cacao do exist in the Americas and Caribbean.

In general, insect pests in the Americas and Caribbean are most damaging when attacking young plants in nurseries or those that have been recently transplanted in the field. Many pests that cause insignificant damage in established plantations can be important pests of young plants, for example, mites, aphids, coleopteran stem borers, cutworms and lepidopteran defoliators (Saunders, 1981).

Although the diversity of pest species in established plantations is quite impressive, the ones that occasionally cause substantial damage, most often on a localized scale, are but a few. The most important ones are species of *Monalonion* (Hemiptera: Miridae), lepidopteran defoliators, defoliating ants and thrips. Some insects have been implicated in disease transmission, most notably Ambrosia beetles of the genus *Xyleborus* suspected to aid the dissemination of *Ceratocystis cacaofunesta* (Engelbrecht et al., 2007).

2.1 Major cacao pests

Cacao mirids

Cacao mirids are ubiquitous wherever cacao is cultivated but with variable impact on the crop. The cacao mirids belong to the insect family Miridae, also known as 'plant bugs', and to the subfamily Bryocorinae. The subfamily Bryocorinae includes several taxonomic subgroups and until recently, cacao mirids were split between two different tribes, the Monaloniini and the Odoniellini, according to their general morphology and life history (Lavabre, 1977). A recent study on the Bryocorinae taxonomy considered the two tribes as synonymous, bringing all cacao mirids together into the tribe Monaloniini (Namyatova et al., 2016). The most well-known cacao insect pests in the Americas and the Caribbean are mirids of the genus *Monalonion*, which gave their name to the tribe Monaloniini. By their morphology and biology, *Monalonion* spp. are more closely related to the mirid bugs of genus *Helopeltis*, which causes serious damage in Asia, yet which are also present in Africa as *Afropeltis* spp., as compared to *Sahlbergella singularis* and *Distantiella theobroma*, the two main cacao pests in West and Central Africa (Babin, 2009; Babin et al., 2010) (see also Section 3). Seven species of the genus *Monalonion,* distributed in Latin America from Mexico to Bolivia, are found on cacao (de Abreu, 1977; Vargas et al., 2005). Nymphs and adults of these species preferentially feed on pods, but also feed on leaf petioles and young shoots. Intensive feeding on pods can lead to young fruit abortion and lowering of bean quality. Intensive stem feeding can cause necrosis and defoliation. Although in comparison with their relatives from Africa, these insects are generally considered minor pests of cacao, the species *M. dissimulatum* (Fig. 1), which is the most widely distributed on cacao, was the cause of considerable production losses in Venezuela and Peru in the first half of the twentieth century (Entwistle, 1972). In several countries of Latin America, such as Bolivia and Ecuador, *M. dissimulatum* is considered to be the most important cacao pest, yet generally causing little damage (Ferrari et al., 2014; Vargas et al., 2005). Since damage is generally higher on full sun-grown cacao compared with shaded cacao, as demonstrated in Costa Rica (Villacorta, 1977), shade management

Figure 1 *Monalonion dissimulatum* on cacao pod in Central America.

should aim at providing a homogeneous shade canopy in order to reduce the impact of these pests.

Thrips

The red-banded thrips, *Selenothrips rubrocinctus* (Thysanoptera: Thripidae), was first described from Guadeloupe, West Indies, where it was causing considerable damage to cacao (Denmark and Wolfenbarger, 1999). As a result, it was referred to as the cacao or cacao thrips. It is now one of the world's major insect pests, damaging a range of tropical and subtropical tree crops including cacao (Guzman, 1985; McCallan, 1975; Peng and Christian, 2004). It is indigenous to tropical America and is an important pest here as well as in the West Indies. Gregarious feeding of immature stages produces necrotic spots and can kill leaves and cause severe defoliation. Heavy attacks are often associated with high temperatures and cacao under water stress (Guzman, 1985; Saunders, 1981). In the Waslala region in Nicaragua, many organic cacao farmers mentioned thrips as a serious pest of cacao, although according to these farmers, thrips can relatively easily be controlled using neem oil applications (derived from the neem tree, *Azadirachta indica*) (G. Martijn ten Hoopen, personal observation, 2004).

Control

Chemical control is most widely used and is effective for both mirid and thrips, as well as many other cacao insect pests. Shade management (uniform shade) and removal of chupons in order to reduce feeding sites, especially when pods are absent, are also useful. Some mirids of the family *Termatophylidae* are known predators of thrips (McCallan, 1975), and ants control thrips as well as mirids (Delabie et al., 2007; Philpott and Armbrecht, 2006).

2.2 Minor cacao pests

Although ants (Hymenoptera: Formicidae) can be useful in controlling certain insect pests, ants by themselves can also be pests. There are many ant species that either feed on or are in other ways associated with cacao production systems. Several species of leaf-cutting ants (Myrmicinae: Attini) such as *Atta cephalotes*, *A. sexdens* and *A. laevigata*, as well as *Acromyrmex balzani*, *A. rugosus* and *A. brunneus subterraneus* occur in cacao plantations. All of these ant species are considered pests because they cut cacao leaves, sometimes destroying considerable areas of foliage, bark, flowers and even small pods, especially on young plants (de Abreu and Soria, 1981; Delabie et al., 2007; Rudolph and Loudon, 1986). Moreover, *Atta* spp. can cause considerable damage to some of the other plants of (economic) interest present in cacao-based production systems.

Some insects do not directly attack cacao but can perpetuate or even aggravate losses due to, for example, diseases. Such is the case for Ambrosia beetles of the genus *Xyleborus* (Coleoptera: Scolytidae), especially *X. ferrugineus* and possibly to a lesser extent *X. posticus* (Saunders, 1981). These Ambrosia beetles often attack trees affected by *Ceratocystis cacaofunesta* (previously known as *C. fimbriata*), liberating frass as they excavate their galleries. The frass contains aleurioconidia (chlamydospores) and conidia of the pathogen and these may be carried by wind or rain splash to wounds on other trees. Fungal-feeding nitidulid beetles are vectors of ascospores of other members of the *C. fimbriata* species complex but have not been associated with cacao wilt (Engelbrecht et al., 2007; Mbenoun et al., 2016). The dragonfly *Gynacantha bifida* has been identified as an important predator of *Xyleborus* beetles (Soria and Machado, 1982), but to our knowledge no studies have looked at their role in controlling *Xyleborus* populations. More recently, bottle and funnel traps filled to a quarter with 70% ethanol proved useful in capturing bark and Ambrosia beetles, with the funnel traps being most efficient (Mazón et al., 2013).

Several mealybug species (Hemiptera: Pseudococcidae) are widely distributed in the Americas and the Caribbean which could potentially damage cacao. However, to our knowledge there are no clear reports from the Americas on serious crop losses in cacao due to mealybugs. This is probably due, in part, to the absence of major virus diseases for which these pests are often transmitters (see also Section 3 on mealybugs and cacao swollen shoot virus (CSSV)). In Trinidad, however, virus-like disease symptoms have been reported on cacao trees as early as 1943 by Posnette (1944). Symptoms consisted of two distinct phenotypes, 'red mottling' and 'vein clearing' and were referred to as strain A and strain B of the suspect plant virus. Transmission studies showed that the mealybugs *Planococcus citri*, *P. brevipes*, *Pseudococcus comstocki* and *Ferrisia virgata* were vectors of both virus strains (Baker and Dale, 1947; Kirkpatrick, 1950). Recent research identified these two strains as two new Badnaviruses (Chingandu et al., 2017). *Planococcus citri* occurs occasionally on cacao fruits and *P. brevipes* forms colonies on fruit peduncles. Other mealybugs such as the passionvine mealybug *P. minor* (Francis et al., 2012), and the invasive mealybug *Maconellicoccus hirsutus* (Green) (Culik et al., 2013), native to Southern Asia or Australia, are present in several Neotropical locations including the island of Trinidad in the Southern Caribbean and have been recorded on cacao.

Termites (Isoptera: Termitidae) are ubiquitous in many cacao plantations in the Americas and Caribbean and can cause damage to cacao trees as well as associated trees. In the Neotropics, the Nasutitermitinae subfamily, especially species of the *Nasutitermes* genus, represents the taxon with the greatest number of species that construct conspicuous nests. They can tunnel through trees and destroy its cortex, thus weakening trees.

The most common aphid (Homoptera: Aphidae) pest of cacao in Central America is probably *Toxoptera aurantii*, the black citrus aphid. It is mostly considered a nursery pest, where it may severely affect young plant growth. Heavily attacked leaves may become cupped or rolled (Flores, 1976; Saunders, 1981). Cortez-Madrigal et al. (2003) reported the entomopathogenic fungus *Lecanicillium lecanii* (Deuteromycotina: Hyphomycetes) infecting the aphid *T. aurantii* in cacao crops in Tabasco, Mexico, which could possibly be used as a biocontrol agent of this pest. In Brazil, the aphid *Hoplophorion pertusum* can occasionally become a problem (de Abreu and Milanez, 1982). Other pests of cacao in the Neotropics are, for example, longicorn beetles of the genus *Steirastoma* (Coleoptera: Cerambycidae) widely distributed on various host plants. *Steirastoma breve* damages young cacao trees: adult females feed on the bark and nymphs bore into the cacao stems making galleries, which opens the door for pathogenic microorganisms to colonize the plants. The beetle is considered an important pest of cacao in Venezuela, Suriname and some parts of Brazil, as well as in the Caribbean islands of Trinidad, Grenada, Martinique, Puerto Rico and Jamaica (de Abreu and Soria, 1981; Liendo-Barandiaran et al., 2010; Liendo et al., 2005). Other Coleopteran species, mainly attacking cacao leaves, include those belonging to the family Chrysomelidae, which consists of species of the genera *Colapsis* and *Nodonata* that make small feeding holes in leaves, and those belonging to the families of the Curculionidae and Scarabaeidae, for example the adults of *Phyllophaga* and *Anomala* spp. are nocturnal foliar feeders. However, most of these pests cause insignificant damage (Saunders, 1981).

The cacao fruit borer, *Carmenta theobromae* (Busck) (Lepidoptera: Sesiidae), is a small moth recently considered to be an important pest of cacao in some areas of Venezuela and has been reported in Colombia and Panama (Morillo et al., 2009). Another species, *C. foraseminis* increasingly affects cacao production in Colombia (Figueroa Medina et al., 2013). Figueroa Medina et al. (2013) have found several entomopathogens that could be used for control of this cacao fruit borer. The larvae of the cacao plume moth, *Michaelophorus nubilus* belonging to the family Pterophoridae, seem to feed on young cacao leaves in Latin America (Matthews and Miller, 2010). Other Lepidopteran insect pests belong to family Paloniidae, including *Catephoides zuelana*; family Noctuidae, including species of *Agrotis* and *Spodoptera*; family Pyralidae, including species such as *Lygropia paraxalis*; family Stenomidae, including larvae of *Ceranota* spp. which are skeletonizers of mature leaves; and family Gracillaridae, including *Marmara* spp., which are common pod miners (Saunders, 1981).

3 Cacao pests in West Africa

West Africa has both the largest cacao-producing countries and the most destructive insect pests of cacao. Crop losses due to these insects not only threaten the livelihood of millions of cacao smallholders, but also impact adversely on the economies of the cacao-producing countries and consequently the world cacao market. West Africa is where better management of cacao pests can have a significant positive impact on the livelihoods of cacao farmers, and on the cacao market.

In West Africa, major pests of cacao belong to two insect groups, mirids and mealybugs. Mirids feed on cacao fruits and young vegetative parts, whereas mealybugs transmit a virus, cacao swollen shoot virus (CSSV) which causes CSSV disease. This disease weakens cacao trees and hampers productivity. Both pests can cause a rapid degeneration of the cacao tree leading to its eventual death if not properly managed.

3.1 Major cacao pests

Cacao mirids

In West Africa, two distinct groups of cacao mirids can be distinguished. The genus *Afropeltis* (*Helopeltis*) includes a dozen closely related species of coloured and gracile insects that are sometimes called 'mosquito bugs'. Adults fly during the day and are easily detected feeding on cacao pods along with nymphs. Damaged pods show black spots (feeding lesions) evenly distributed on the pod surface. Pod growth can be disturbed especially with early lesions, producing deformed pods. However, cacao mirids of the *Afropeltis* group are usually not considered to be major pests, although control measures might be necessary on localized scale in case of outbreaks. The second group, formerly known as Odoniellini, includes a few species of stout-bodied and usually brown coloured insects of which *Sahlbergella singularis* Hagl. (Fig. 2) and *Distantiella theobroma* Distant are considered to be the most important. These two closely related species have similar distributions in West Africa, and are found on cacao along the forest belt from Sierra Leone to Cameroon for *D. theobroma* and into the Democratic Republic of Congo for *S. singularis*.

Figure 2 *Sahlbergella singularis* adult on cacao pod in Cameroon.

The cacao mirids *Sahlbergella singularis* and *Distantiella theobroma* are cryptic insects which hide in the cacao bark crevices during the sunniest hours of the day. *Sahlbergella singularis* females insert eggs into the bark of shoots, chupons (suckers growing from the trunk) and pod peduncles. Nymphs and adults feed on pods (Fig. 3A), suckers and green shoots, the latter being most detrimental to the trees (Fig. 3B). However, young pods seem to be the most favourable food source (Babin et al., 2011). Cacao mirids multiply slowly (Babin et al., 2008) and densities in cacao plantations rarely exceed two or three insects per tree. Contrary to expectations, recent studies have shown that cacao mirids are more numerous and their damage more pronounced in plantations exposed to full sunlight and in exposed areas of shaded plantations (Babin et al., 2010). On well-developed pods, mirid feeding leads to areas of dead black tissues usually limited to the cortex. On green shoots, feeding lesions can cut sap supply to the terminal part of branches causing the drying up of leaves and buds and a reduction of photosynthetic activity of the trees. Moreover, with time, feeding lesions develop into cankers that weaken the trees and increase the risk of dieback due to cryptogamic diseases (Williams, 1953). Damage to the vegetative parts of the cacao tree hampers its productive capacity, although the actual extent of production loss due to vegetative damage is difficult to assess. However, most authors agree that it cannot be less than 20% and is most probably between 25% and 30% for most of the cacao-producing countries of West Africa (Lavabre, 1977).

Mealybugs as vectors of CSSV

Mealybugs (Hemiptera: Pseudococcidae) are widely distributed sucking insects that can be pests of many crops. In West Africa, cacao hosts some twenty mealybug species that are not considered major pests due to their limited direct impact on yield. However, three-quarters of them have been recorded as potential vectors of a devastating badnavirus infection, commonly known as CSSV disease (Wetten et al., 2016). The main mealybug species involved in the transmission of the virus are the pseudococcids *Planococcoides njalensis* (Laing), *Planococcus citri* (Rossi) and *Ferrisia virgata* (Cockerell). These species are widely distributed in West Africa and CSSV has been reported in nearly all of the cacao-producing countries of the area, from Nigeria to Sierra Leone (Ameyaw et al., 2014), with Cameroon being a notable exception (for CSSV presence) among the big cacao-producing countries in Africa.

Figure 3A *Sahlbergella singularis* nymphs and damage on cocoa pods in Cameroon.

Figure 3B Old damage of *Sahlbergella singularis* on cacao tree in Cameroon.

Cacao mealybugs usually live in large colonies developing on various parts of the cacao tree, from fruits at different development and maturation stages, to shoots, buds and leaves. Mealybug nymphs, commonly called crawlers (Strickland, 1951), can quickly move over relatively long distances, whereas adults are hardly mobile, and consequently nymphs are the main stage for CSSV transmission. Adult males are winged and fly to reach reproductive females, but have no mouthparts and are unable to feed and transmit the disease (Ameyaw et al., 2014). Ecological preferences vary between species: *P. njalensis* thrives in mature and shaded plantations, whereas young and unshaded plantations are more favourable to *P. citri* (Dufour, 1991). Relationships with attendant ants are a crucial component of CSSV-mealybug ecology. For some species such as the dominant *P. njalensis*, association with ants is strictly needed for colony development.

Typical symptoms of the CSSV disease are a swelling of shoots, nodes and internodes of branches and roots. Severe infections usually cause defoliation and pod deformation, leading to low yield and rapid death of trees (Ameyaw et al., 2014; Domfeh et al., 2011).

Control

Pesticide spraying has always been and still is the most commonly used control method for cacao mirids, which usually show a good sensitivity to insecticides. Current spraying recommendations are based on a good knowledge of mirid seasonal variations and include more specific active ingredients like imidacloprid and thiamethoxam, both neonicotinoids with strong systemic action. Considering that chemical insecticide use is controversial now, agricultural research recommends a range of agronomic management options to control mirid outbreaks, including best practices for plant associations, shade management (see Section 6.1) and cacao pruning. When cacao does not bear fruits, chupons are the alternative food source and oviposition sites that enable mirid populations to grow. Cacao pruning recommendations include the removal of chupons for mirid control. Some studies presented the use of neem extracts as promising to control mirid populations (Ayenor et al., 2007; N'Guessan et al., 2006; Padi et al., 2003). In Ghana (Ayenor et al., 2007; Sarfo et al., 2010) and Cameroon (Mahob et al., 2011) the use of pheromone traps has been explored for mirid control with promising results. However, to date no commercial traps are available on the market.

Mealybugs are naturally protected by a white wax that makes contact insecticides ineffective. Spraying of potassium-based soap is recommended to kill them. Removal of infected trees is the only recommended practice to stop the spread of the disease, although it does not appear to be effective enough to halt the progression of the disease (Domfeh et al., 2011).

3.2 Minor cacao pests

In West Africa, cacao production can be hampered at localized levels by several minor pests. Among cacao mirids, *Bryocoropsis laticollis* Schum. which is similar to *S. singularis* Hagl. and *D. theobroma* in shape and biology is of secondary importance. The species can be found mixed with other mirids in cacao plantations but is found in limited numbers (around 1% of the total mirid population in Côte d'Ivoire (Kouamé et al., 2014). Although *Bryocoropsis laticollis* can feed on pods it only has a limited impact on cacao trees. The species has been recorded on other host plants of economic importance, like the sugar-apple tree, *Annona squamosa* L. (Annonaceae) (Entwistle, 1972). The cacao shield bug *Bathycoelia thalassina* (Herrich-Schaeffer) (Hemiptera: Pentatomidae) is a stink bug that feeds directly on developing cacao beans, causing premature pod ripening. The damage is commonly known as immature ripening. Green pods fed on by *B. thalassina* stop growing and turn yellow. The adults have long feeding stylets (about 2.2 cm) which can penetrate the pod cortex and suck liquid from the developing beans resulting in malformed or atrophied beans (Gerald, 1965). Cacao with a hard (stony) mesocarp shows some degree of tolerance to *B. thalassina* (Owusu Manu et al., 1990). Under natural conditions, *B. thalassina* is kept under control by its natural enemies (Owusu-Manu, 1976). The pest also causes severe damage to cacao pods in some regions of Cameroon and Côte d'Ivoire but there is no estimation of crop losses. The use of insecticides for the control of mirids has apparently caused an increase in the population of the pest in Nigeria and in Ghana (for more information see http://www.plantwise.org/KnowledgeBank/Datasheet.aspx?dsid=8560). This example illustrates the fact that the balance between major and minor pests populations can be disturbed by both environmental conditions and practices. Finally, the larvae of the cacao stem borer *Eulophonotus myrmeleon* Fldr. (Lepidoptera: Cossidae) are capable of boring galleries into cacao trunks and branches, weakening trees and leading to low yield.

4 Cacao pests in Asia

In Asia, as in the Americas and Africa, a wide variety of insect species feed on and can cause damage to cacao. Unfortunately, and similar to the situation in Africa, a few indigenous insects have taken such a liking to this exotic crop that they now cause substantial economic losses.

The most important cacao pest in Asia is the cacao pod borer (CPB), *Conopomorpha cramerella* (Lepidoptera: Gracillariidae). In Indonesia, CPB is present in all cacao-producing provinces and in 2000 had infested 60 000 hectares causing losses of up to US$40 million a year[1]. The second most important cacao pest in Asia is mirid bug, closely related to the mirid bugs attacking cacao in Africa and the Americas. Other insects attacking cacao are of less importance, although occasionally they can cause considerable losses on a localized scale.

4.1 Major cacao pests

Cacao pod borer

The CPB *Conopomorpha cramerella* (formerly *Acrocercops cramerella*), also known as the cacao moth, is a small moth endemic to Southeast Asia affecting different native fruit trees, such as rambutan (*Nephelium lappaceum*), pulasan (*Nephelium mutabile*), nam-nam (*Cynometra cauliflora*), kasai (*Pometia pinnata*) and different *Cola* species (Lim, 1992; Mumford, 1986; Mumford and Ho, 1988). Following its appearance at Manado in the 1840s it was quickly recognized to be a threat to the cacao industry (Day, 1989; Toxopeus and Giesberger, 1983). The Dutch studied the pest in the early part of the century and made detailed investigations into some aspects of the pest (Van Hall, 1914). However, relatively little else was done until the 1980s (Day, 1989), when CPB was discovered in Sabah, East Malaysia, and shortly after, in 1986 in Melaka, West Malaysia (Mumford and Ho, 1988). Subsequently CPB was recognized as a major pest of cacao in Southeast Asia. As of 2006, CPB has also been declared a serious pest in Papua New Guinea, following the discovery of an outbreak in the Kerevat area. CPB is now considered the main threat for cacao production in most Asian producing countries (Posada et al., 2011; Yen et al., 2010). Since *C. cramerella* borers are known to travel long distances, rambutans exported from Thailand to Saudi Arabia containing live pupae were found in a Riyadh supermarket (Mumford and Ho, 1988), this pest also poses a serious threat to cacao production elsewhere. Adult *Conopomorpha cramerella* females mostly lay eggs on the surface of

1. http://www.icco.org/about-cacao/pest-a-diseases.html (consulted 25 October 2016).

immature cacao pods. Newly hatched (2–3 days) nymphs bore into the pod epidermis to reach the placenta (pulp) on which they feed for about two weeks, disturbing the development of beans (Beevor et al., 1993; Lim, 1992). Cacao yield can be reduced by 60–84% in case of severe infestation, and dry bean quality is also affected. However, Day (1989) reported that pod infestation by CPB at rates of 60% or lower resulted in little loss, while losses increased markedly above this threshold. Lim and Phua (1986) identified a similar relation between total incidence and pod losses, but found losses increased dramatically when total incidence exceeded 70%.

Cacao mirids

In Southeast Asia, Monaloniini mirids from the genus *Helopeltis* (Hemiptera: Miridae), are considered serious pests by farmers and agriculturalists. Several *Helopeltis* species are considered important pests of cash crops in Asia, such as black pepper (*Piper nigrum*), cashew (*Anacardium occidentale*), cinchona (*Cinchona* spp.), tea (*Camellia sinensis*) and, of course, cacao (Stonedahl, 1991). For example, the mirid *Helopeltis sulawesi* feeds on pods of all ages and young shoots of cacao (Giesberger, 1983). *Helopeltis antonii* and *H. theivora* (with *H. theobromae* as a subspecies) are nowadays widely distributed on cacao, causing damage similar to that of African cacao mirids, although economic loss is mainly due to damage on pods. They make characteristic necrotic lesions that can kill young pods and shoots (Fig. 4). The surface of damaged pods is covered with scars and a thick sclerotic layer, and this may promote the abscission of young fruits (cherelles) (Muhamad and Way, 1995). Fungi such as *Lasiodiplodia theobromae* sometimes enter the wounds and cause pods to rot. Lesions on shoots may coalesce to ring the shoot, resulting in dieback (Tan, 1974). In Malaysia, maximum damage to pods was estimated as 85% during the fruiting months and yield losses of around 50% have been reported (Tong-Kwee et al., 1989). Although short periods of heavy *H. theivora* damage on medium-sized pods can make them wilt and die, even greater damage yet spread over a longer period may not. This suggests that pods can cope with non-lethal injury. Moreover, yield seems unaffected when full sized but not yet ripe pods were heavily damaged (Muhamad and Way, 1995). A study by Bos et al. (2007), investigating the impact of shade on fruit abortion, insect pests and pathogens of cacao, found that nearly all pods attacked by insects were attacked by the mirid bug *H. sulawesi*, although the majority of these pods (62%) still reached maturity. Overall, mirids are considered a minor pest when compared with CPB (Wielgoss et al., 2012).

Figure 4 *Helopeltis* spp. damage on pods. (a and b) old damage on cherelles involving pod distortion; (c) old damage on an adult pod; (d) new damage caused by mirid at different development stages.

Control

As in the Americas and Africa, the most common method to try to control cacao pests in Asia is the use of insecticides (e.g. Mumford and Ho, 1988). However, in the case of CPB, since the nymph lives inside the pod, it is mostly out of reach of insecticide spraying (Day, 1989; Shapiro et al., 2008). Nonetheless, a modelling exercise to assess the effects of CPB management, through pesticide spraying, on the productive outputs and the ecological functioning of a cacao agro-ecosystem, showed that pesticide spraying can be economical, yet yield strongly varies according to both pesticide type and timing of pesticide application (Sabatier et al., 2013).

More ecologically friendly methods have also been explored, for example, pod sleeving for CPB (Rosmana et al., 2010; Saripah and Azhar, 2007; Saripah et al., 2005), which is, however, very laborious. Farmers use plastic bags to sleeve pods and prevent the moth from laying eggs on them. A recent study conducted in Indonesia showed that 50% of the pods are protected this way from CPB and discussed the importance of timing of pod sleeving, which seems to depend on cacao variety and season (Rosmana et al., 2010). Sleeving pods at an earlier developmental stage can reduce CPB infestation by 85–100% but also increases the risk of production losses due to physiological death of pods and *Phytophthora* pod rot. The issue of plastic bag pollution has led researchers to test the method using biodegradable plastic bags.

Zehnthner (1903) proposed complete stripping of immature and mature pods for a short period. In theory, this would cause a disruption in the pod borers' life cycle. Yet this method has not proven to be very effective (Mumford and Ho, 1988). Another method, pheromone-based mass trapping, has shown potential (Beevor et al., 1993; Mumford and Ho, 1988) but not yet sufficient to be used for CPB control (Zhang et al., 2008).

Very promising and interesting, however, is the use of ants to control both CPB and mirids. However, ant species that are effective predators of *Helopeltis* might indirectly increase the probability of *C. cramerella* infection. Wielgoss et al. (2012) showed that ovipositing *C. cramerella* females avoid pods damaged by the mirid *H. sulawesi*. Moreover, given that the CPB and mirids have contrasting spatial and temporal activity patterns, they are unlikely to be equally exposed to particular predator species. Wielgoss et al. (2012) found that the differential yield impacts by the two pests and the avoidance by *C. cramerella* of pods damaged by *H. sulawesi* lead to a yield optimum at a *H. sulawesi* incidence of 51%.

4.2 Minor cacao pests

Although there is mention of other serious pests of cacao in Asia, relatively little information is available on these pests. The bee bug *Platyngomiriodes apiformis* Ghauri is considered a pest of cacao in Sabah, Malaysia (Lim, 1992). *Pseudodoniella laensis* Miller, *P. pacifica* China & Carvalho and *P. typica* (China & Carvalho) are known as pests of cacao in New Guinea (Entwistle, 1972).

Weevils from the genus *Pantorhytes* (Coleoptera: Curculionidae) are important pests of cacao in New Guinea. Of 11 species injurious to cacao, six have been found to be of economic importance. Larvae tunnel into the cacao stem between the bark and the wood leading to weakened trees, infection by microbes and often sudden tree death. Totally destroyed plantations have been reported in some areas of Papua New Guinea and the weevil has contributed to the collapse of cacao production in some regions of the country (Moxon, 1992). The crazy ant, *Anoplolepsis longipes*, has shown promise for the control of *Pantorhytes* spp. in Papua New Guinea and methods have been developed for establishing colonies in cacao plantations (Moxon, 1992). Yet the use of crazy ants as a biological control agent is questioned due to the difficulty of maintaining large colonies in plantations over time.

Leaf-feeding moths and beetles are numerous on cacao, although rarely associated with damage of economic importance. The cacao armyworm *Tiracola plagiata*, a noctuid moth, is a pest of cacao in Asia and has been shown to be more abundant in plantations shaded with *Leucaena leucocephala* (Room and Smith, 1975). Certain pod-boring beetles (Coleoptera: Scolytidae and Nitidulidae) feed on pods and could be involved in disease dissemination of, for example, black pod disease (caused by *Phytophthora palmivora*) (Konam and Guest, 2004). Some leaf-feeding beetles are reported as pests of cacao, among which is the Scarabaeidae *Adoretus versutus*, an Asian polyphagous chafer beetle, outbreaks of which caused serious defoliations in cacao plantations in Vanuatu in the 1980s (Beaudoin et al., 1995). Chafer beetles from the genera *Apogonia*, *Anomala* and *Chaetadoretus* include leaf-feeding pests of cacao as well (Entwistle, 1985). Long-horned beetles (Cerambycidae) include several species of cacao stem borers damaging cacao branches and stems worldwide.

From the previous sections (2, 3 and 4) it becomes clear that although there are many pests attacking cacao, on a global scale and considering their economic impact, only mirid bugs and CPB are considered major pests. It could be argued that mealybugs should be added to this group, even though by themselves they cause little damage. Yet in their role as CSSV vector they have a devastating effect on the cacao industry in West Africa. Most of the work to control pests in cacao has thus focused on control of mirids and CPB. Direct intervention strategies based on the use of chemicals are still the most effective to control pests. However, to limit their adverse effects, recommendations advocate a more intelligent and careful use. Chemicals with the least harmful effect on non-target beneficial insects (like natural enemies and pollinators) are preferred. At the same time, biopesticides are experiencing a huge upswing. In cacao agroforestry systems, ecological processes are exploited to develop sustainable crop protection. In the following section, some advances with regard to non-chemical means of pest control in cacao are discussed.

5 Advances in control of cacao pests: breeding for pest resistance

Cacao breeding for resistance towards diseases has been the focus of a great deal of research worldwide, especially over the last three decades. Varietal selection for insect control, however, has globally received much less attention and been less successful than for diseases. Here, we will focus on the little progress that has been made for the two most significant cacao pests worldwide, cacao mirids in West Africa and CPB in Asia.

Cacao resistance and tolerance to the two major mirid pests, *Sahlbergella singularis* and *Distantiella theobroma*, have been measured in various ways in Ghana, Côte d'Ivoire and Cameroon. Ranking of a range of international cacao clones for their susceptibility to mirids was performed in Côte d'Ivoire by recording cumulative mirid damage in a progeny trial (Sounigo et al., 2003). Results of the study showed that the discrimination of clones for their resistance to mirids based on cumulative damage on their progeny was possible.

Upper Amazon Forastero varieties were less susceptible to mirids than Trinitarios and lower Amazon Forasteros. Later studies detailed methodologies developed in more controlled conditions to study mechanisms involved in resistance. Antixenosis was tested in laboratory behavioural bioassays, where starved mirids were allowed to make a choice between twigs of different cacao varieties (Anikwe et al., 2009; Dibog et al., 2008; N'Guessan et al., 2008). Additional experiments assessed the tolerance of cacao genotypes by enclosing mirid nymphs on shoots with sleeves and monitoring the level of damage and the ability of the plant to recover from it (Babin et al., 2004; N'Guessan et al., 2008). These studies allowed the ranking of cacao genotypes for resistance and tolerance to mirids which was in general concordance with field observations of tolerance, yet sometimes the results were inconsistent, demonstrating the complexity of the mechanisms involved in resistance. Recently, a more participatory approach has been adopted for cacao selection that incorporates farmer perception of the susceptibility of their own cacao to pests and diseases (Eskes, 2011).

A similar participatory approach has been adopted in Sulawesi, Indonesia, for the selection of genotypes against CPB (McMahon et al., 2009). Studies focusing on the mechanisms governing resistance showed that pods with smooth and harder surfaces were less favourable for the development of the pest (Teh et al., 2006). Ranking of clones based on related morphological characteristics (husk and sclerotic layer hardness and thickness) has resulted in identification of some promising clones for breeding in Malaysia (Azhar et al., 1995). In a participatory on-farm study, McMahon et al. (2009) emphasized the role of farmer perception and the importance of high level of genetic diversity in smallholder farms in improving resistance to insect pests. They also hypothesized that, compared to single genotype stands, farms with mixed genotypes may delay pod borer infestation due to the time spent by the moth in choosing optimal oviposition sites. The search for cacao varieties resistant to CPB is complicated as cacao trees face not only CPB but also diseases such as black pod disease and vascular streak dieback, and hence breeders should be able to combine tolerances to these problems simultaneously. For example, clones resistant to CPB were susceptible to black pod disease (McMahon et al., 2015) or other pests/diseases, and vice versa. Although progress has been made in the selection of cacao varieties with resistance to insect pests, much more remains to be done. For example, until now, no research programme has looked at mealybug resistance in cacao, breeding programmes have focused mainly on resistance to the virus. Maybe it would be interesting to look at resistance to mealybugs as a means to avoid infection by CSSV. Other challenges remain, such as elucidating the way to combine resistance to several pests and diseases that can jointly threaten cacao production. Another challenge is the development of sustainable programmes for improved seed material production and dissemination that are still weak in many areas of the world.

6 Advances in control of cacao pests: agro-ecological control of pests in cacao-based agroforestry systems

6.1 Shade management and pest control

Traditional cacao plantations often have a highly diverse shade tree stratum either retained from the original forest (de Rouw, 1986; Johns, 1999; Schroth et al., 2000) or planted by farmers. Some of the practices recommended for cacao/tree association are specific to pest control. Generally, researchers recommend to eliminate tree species that serve as alternative hosts for pests and diseases from the cacao farms. In West Africa, cacao mirids have adapted to cacao from native forest trees of the Malvaceae family (e.g. *Ceiba pentandra, Cola acuminata, Cola lateritia, Sterculia rhinopetala* and *Triplochiton scleroxylon*) (Entwistle, 1972). These alternative hosts contribute to maintaining mirid populations in the plantation even when the physiological conditions of the cacao tree are unsuitable for their development. However, due to their economic and traditional values, it is not uncommon to find stands of these species on cacao farms even though there is a campaign in West African countries against them (Asare, 2006). In fact, farmers use biodiversity in the cacao farm to meet their household demands and needs and to improve their revenues. It is important then to evaluate when the benefits from the associated tree (soil fertility, carbone storage, biodiversity conservation) are offset by the cacao losses due to increased mirid damage. The tree associations within cacao plantations are often the result of a trade-off between the different services a farmer wants to obtain from his production system and it is not easy for managers and farmers to choose which services to favour.

Optimization of overhead shade in agroforestry systems can be an efficient control strategy for many pests and diseases given the effects of shade on the micro-climatic conditions under which pest and disease organisms and their natural enemies develop. Shaded cacao agroforestry suffers less from insect pest problems (Rice and Greenberg, 2000), due to suckers such as thrips and mirid bugs, or due to leaf herbivory (Schroth et al., 2000; Tscharntke et al., 2011). Shade has proven to be a key factor in determining cacao mirids, *Distantiella theobroma* and *Sahlbergella singularis*, infestation and damage in West Africa. Heavy shade reduces mirid density, probably by limiting cacao vegetative growth, that is, the availability of young shoots, which are a primary food resource for mirids. In shaded plantations, mirid populations aggregate in groups of 20–30 adjacent cacao trees exposed to direct sunlight through gaps in the shade tree canopy layer, which are known as 'mirid pockets' (Babin et al., 2010). Thus, it is recommended to maintain homogeneous shade levels to control mirids in cacao plantations in West Africa (Babin et al., 2010; Gidoin et al., 2014; Padi et al., 2002). In order to provide homogeneous shading it is recommended to avoid shade tree aggregation since aggregation introduces heterogeneity in light distribution. Large forest trees, unlike fruit trees, tend to homogenize the distribution of the light resource more and consequently are more suitable to reduce mirid populations (Gidoin et al., 2014). Whereas a certain equilibrium develops between cacao and its mirid pests under shaded conditions, unshaded plots have to be sprayed at frequent intervals.

Some natural enemies like entomopathogenic fungi can be also more efficient under shade (Schroth et al., 2000) helping to reduce pest populations. Optimization of shade in cacao plantations depends on local environmental conditions (e.g. temperature and rainfall) and the principal pests and pathogens at the respective site. In some cases, shade management for pest control should be given priority over shade management for the control of diseases (e.g. cacao in Africa and Asia). An important principle is also to manage shade in order to minimize physiological stress of the crops themselves (e.g. to avoid excessive irradiation, extreme temperatures and drought stress), thereby increasing their resistance to pest and disease attack. Since many shade trees can also provide additional products and services, shading is thus an 'effective insurance strategy' against insect pests (Cassano et al., 2009; Johns, 1999) even if it reduces short-term yield (Clough et al., 2009; Johns, 1999; Steffan-Dewenter et al., 2007).

6.2 Ants for pest control in cacao

Natural enemies are organisms that have a negative effect on others. When natural enemies are capable of limiting pest populations they provide an important biological control service and this service can be exploited and become an important component of integrated pest management (IPM) programmes. Important natural enemies of insect pests include predators, parasites and pathogens. Among predators, some are specific to certain groups of species like parasitoids and others like ants are more generalists. Both can be efficient in controlling crop pests.

Ants are a key component of cacao agrosystems (Philpott and Armbrecht, 2006). In Bahia, Brazil, two ant species (*Azteca chartifex spiriti* and *Ectatomma tuberculatum*) are recognized for their role as biocontrol agents of cacao thrips and mirids (Delabie et al., 2007). Farmers in Bahia have promoted this species of *Azteca* in their plantations for its biological control activity since the beginning of the twentieth century (Delabie et al., 2007).

In Asia, the use of ants to control cacao pests is a strategy implemented with success in most producing countries for several centuries now (Way and Khoo, 1992). The Asian weaver ant *Oecophylla smaragdina* (Fabr.) is known to be a beneficial predator of many pests such as the mirids *Helopeltis theobroma* (Miller) and *Pseudodoniella laensis*, or the cacao weevils (*Pantorhytes* spp.) in Papua New Guinea (Anon, 1984; Way and Khoo, 1991; Way and Khoo, 1992). Farmers regularly introduce nests of *O. smaragdina* to control populations of *Helopeltis* spp. in cacao farms. They practice weaver ant husbandry and they provide shelter, food and construct ropes between trees populated with weaver ants in order to maintain the ant colonies' presence (Offenberg, 2015). However, due to its agressiveness, this species is not always welcome in plantations. The black ant *Dolichoderus thoracicus* (Smith) (Fig. 5) is relatively less efficient than *O. smaragdina* but since it is not aggressive it is also valued by farmers in Indonesia and Malaysia as biological control agent of mirids (Saripah and Azhar, 2012; Way and Khoo, 1992). The establishment of *Dolichoderus thoracicus* and *Oecophylla smaragdina* in cacao fields can also reduce damage caused by CPB (Gassa et al., 2015; Gassa et al., 2003). Currently, recommendations exist to favour the establishment of ant colonies in plantations. These include among others, the destruction of antagonist ant species, the improvement of nesting conditions by planting coconut trees (Way and Khoo, 1992), introduction of artificial nests (Gassa et al., 2015) and the artificial infestation of cacao trees with mealybugs which are tended by black ants for honeydew (Fig. 5) (Tuck, 1994; Way and Khoo, 1992). Considering the importance of CSSV disease in West Africa (Ameyaw et al., 2014), it would, however, probably be better to avoid recommendations concerning the enhancement of ants that develop mutualistic relationships with mealybug species.

The crazy ant, *Anoplolepsis longipes*, has shown promise for the control of *Pantorhytes* spp. in Papua New Guinea, and methods have been developed for establishing colonies in cacao plantations (Moxon, 1992). Nonetheless, the use of crazy ants as a biological control agent is questioned due to the difficulty of maintaining large colonies in plantations over time. In Africa, a lot is still to be learned about mirid/ant interactions before appropriate recommendations can be formulated, especially in complex agroforestry systems. Only two recent studies investigated the role of the predatory weaver ant, *O. longinoda*, on mirid populations in cacao plantations in Africa and especially in Ghana (Ayenor et al., 2007; van Wijngaarden et al., 2007). Van Wijngaarden et al. (2007) showed that an increase of *O. longinoda* nests caused a decrease in the percentage of pods damaged by mirids and that the trees harbouring high abundance of these ants presented thicker canopies. Biological control with *Oecophylla* spp. can be as

Figure 5 The black ant *Dolichoderus thoracicus* tending mealybugs (*Cataenococcus* spp.) in Vietnam. Pods, pod peduncles, bark and green shoots are feeding sites for the mealybugs (white).

efficient as insecticide use and even less expensive (Offenberg, 2015). Van Wijngaarden et al. (2007) suggested that the protection of pods observed in cacao systems results more from exclusion of mirids via interference competition for space when ant population is very high, than from direct predation interactions. In some cases, ants have been introduced in order to help with pest control in cacao. A good example is that of the introduction of the little fire ant (*Wasmannia auropunctata*) into Cameroon for mirid control (Bruneau de Miré, 1969). This invasive ant, native to the Americas (Central America), was first reported in Cameroon in 1969 and it actively spread in many parts of the country displacing local species. Finally, along with its negative impact on local fauna and its aggressiveness towards the farmers, the relevance and effectiveness of this species introduction are questionable.

Although ants are one of the most promising agents for sustainable management of cacao as they can control pests over the long term, it is important, however, to optimize their use as biological control agents. It is best to enhance the presence and abundance of already existing species instead of introducing exotic ones that can disturb ecological equilibrium. In agroforestry systems, it is important to have diversity, yet in order to control pests it is better to have controlled diversity with a functional role to control pests.

7 Conclusion

Fortunately, only a few pests of cacao have a measurable impact on global cacao production. However, it is really challenging to study (and estimate) their actual damage, especially since some pests like mirid bugs in Africa have an indirect effect on cacao productivity. A better knowledge regarding the interactions between cacao and its pests will help to develop tools that could more accurately estimate losses. Crop loss estimates due to insect damage will be an important tool for the improvement of IPM.

IPM is necessary for sustained control of pests in perennial crops under tropical environments, especially since global changes (climate change, land use changes) will have a significant impact on the emergence of insect pests. Some major pests may lose relative importance, and others, previously considered secondary may become major challenges. However, there is a lack of knowledge on the impact these global changes will have on pests of cacao.

8 Future trends

- There is much information and many statistics on breeding for pest resistance, which is still in its infancy
- Research should focus on the impact of global changes on importance of insect pests (accompanied by programmes to monitor and evaluate losses)
- Improved control methods (non-chemical) based on agro-ecological principles have to be developed

9 Where to look for further information

There are many information and statistics on cacao pest and disease available on the FAO website (http://agris.fao.org/agris-search/index.do) and on the CABI website (http://www.plantwise.org, http://www.cabi.org/cpc/ and http://www.cabi.org/isc/).

The PestNet network can help in pest identifications from digital images, pest outbreak alerts, pest management (biological, cultural and chemical) and quarantine interceptions (http://www.pestnet.org).

10 References

Ameyaw, G. A., Dzahini-Obiatey, H. K. and Domfeh, O. (2014) Perspectives on cocoa swollen shoot virus disease (CSSVD) management in Ghana. *Crop Protection* 65:64–70. DOI:http://dx.doi.org/10.1016/j.cropro.2014.07.001.

Anikwe, J. C., Omoloye, A. A., Aikpokpodion, P. O., Okelana, F. A. and Eskes, A. B. (2009) Evaluation of resistance in selected cocoa genotypes to the brown cocoa mirid, Sahlbergella singularis Haglund in Nigeria. *Crop Protection* 28:350–5. DOI:10.1016/j.cropro.2008.11.014.

Anon. (1984) Ants encouraged in Papua New Guinea cocoa farms. *Tropical Pest Management* 30:455.

Asare, R. (2006) A review on cocoa agroforestry as a means for biodiversity conservation, World Cocoa Foundation Partnership Conference, Brussels, Citeseer, p. 15.

Ayenor, G. K., Huis, A.V., Obeng, O. D., Padi, B. and Roling, N. G. (2007) Facilitating the use of alternative capsid control methods towards sustainable production of organic cocoa in Ghana. *International Journal of Tropical Insect Science* 27:85–94.

Azhar, I., Long, G. E. and Musa, M. J. (1995) Qualitative and multivariate analyses of clonal cocoa resistance to cocoa pod borer (Lepidoptera: Gracillariidae). *Planter* 832:307–21.

Babin, R. (2009) Contribution à l'amélioration de la lutte contre le miride du cacaoyer *Sahlbergella singularis* Hagl. (Hemiptera: Miridae). Influence des facteurs agro-écologiques sur la dynamique des populations du ravageur., Ecole doctorale 60 'Territoires, Temps, Sociétés et

Développement' Département Biologie Ecologie Environnement, Université de Montpellier III, Montpellier, p. 201.

Babin, R., Anikwe, J. C., Dibog, L. and Lumaret, J. P. (2011) Effects of cocoa tree phenology and canopy microclimate on the performance of the mirid bug Sahlbergella singularis. *Entomologia Experimentalis et Applicata* 141:25–34.

Babin, R., Bisseleua, D. H. B., Dibog, L. and Lumaret, J. P. (2008) Rearing method and life-table data for the cocoa mirid bug Sahlbergella singularis Haglund (Hemiptera: Miridae). *Journal of Applied Entomology* 132:366–74. DOI:10.1111/j.1439-0418.2008.01273.x.

Babin, R., Hoopen, G. M. T., Cilas, C., Enjalric, F., Yédé, Gendre, P. and Lumaret J. P. (2010) Impact of shade on the spatial distribution of *Sahlbergella singularis* in traditional cocoa agroforests. *Agricultural and Forest Entomology* 12:69–79. DOI:10.1111/j.1461-9563.2009.00453.x.

Babin, R., Sounigo, O., Dibog, L. and Nyassé S. (2004) Field tests for antixenosis and tolerance of cocoa towards mirids. *Ingenic newsletter* (9):45–50.

Baker, R. E. D. and Dale, W. T. (1947) Virus disease of cacao in Trinidad II. *Tropical Agriculture* 24:127–30.

Beaudoin, L., Morin, J. P., Nguyen, C. and Decazy B. (1995) Study of underground Adoretus versutus Har. (Col., Scarabaeidae) populations in Vanuatu: Detection of cohabitation with other white grubs. *Journal of Applied Entomology* 119:391–7. DOI:10.1111/j.1439-0418.1995.tb01306.x.

Beevor, P. S., Mumford, J. D., Shah, S., Day, R. K. and Hall, D. R. (1993) Observations on pheromone-baited mass trapping for control of cocoa pod borer, Conopomorpha cramerella, in Sabah, East Malaysia. *Crop Protection* 12:134–40. DOI:http://dx.doi.org/10.1016/0261-2194(93)90140-E.

Bruneau de Miré, P. (1969) Une fourmi utilisée au Cameroun dans la lutte contre les Mirides du cacaoyer: *Wasmannia auropunctata* Roger. *Café Cacao Thé* 13:209–12.

Cassano, C. R., Schroth, G., Faria, D., Delabie, J. H. C. and Bede, L. (2009) Landscape and farm scale management to enhance biodiversity conservation in the cocoa producing region of southern Bahia, Brazil. *Biodiversity and Conservation* 18:577–603. DOI:10.1007/s10531-008-9526-x.

Clough, Y., Faust, H. and Tscharntke, T. (2009) Cacao boom and bust: Sustainability of agroforests and opportunities for biodiversity conservation. *Conservation Letter* 2:197–205.

Cortez-Madrigal, H., Alatorre-Rosas, R., Mora-Aguilera, G., Bravo-Mojica, H., Ortiz-García, C. F. and Aceves-Navarro, L. A. (2003) Characterization of multisporic and monosporic isolates of Lecanicillium (= Verticillium) lecanii for the management of Toxoptera aurantii in cocoa. *BioControl* 48:321–34. DOI:10.1023/a:1023663629826.

Culik, M. P., Fornazier, M. J., dos Santos Martins, D., Zanuncio, J. S., Ventura, J. A., Peronti, A. L. B. G. and Zanuncio, J. C. (2013) The invasive mealybug Maconellicoccus hirsutus: Lessons for its current range expansion in South America and invasive pest management in general. *Journal of Pest Science* 86:387–98. DOI:10.1007/s10340-013-0512-z.

Day, R. K. (1989) Effect of cocoa pod borer, Conopomorpha cramerella, on cocoa yield and quality in Sabah, Malaysia. *Crop Protection* 8:332–9. DOI:http://dx.doi.org/10.1016/0261-2194(89)90052-5.

de Abreu, J. M. (1977) Mirideos neotropicais associados ao cacaueiro. In E. M. Lavabre (Ed.), *Les Mirides du Cacaoyer*. G-P. Maisonneuve et Larose: Paris, pp. 85–106.

de Abreu, J. M. and Milanez J. M. (1982) Aviliação de piretroides sintéticos no controle de Hoplophorion pertusum (Gemar) (Homoptera: Membracidae) Praga do cacueiro na Bahia, 8th International Cocoa Research Conference, Cocoa Producers Alliance (COPAL), Cartagena, Colombia, pp. 291–3.

de Abreu, J. M. and Soria, S. J. (1981) Controle de pragas do cacaueiro na América do sul, 7th International Cocoa Research Conference, Cocoa Producers Alliance (COPAL), Douala, Cameroon, pp. 433–41.

de Rouw, H. J. (1986) Tree management as part of two farming systems in the wet forest zone (Ivory Coast). *Acta Oecologica. Oecologia Applicata* 8:39–51.

Delabie, J. H. C., Jahyny, B., do Nascimento, I. C., Mariano, C. S. F., Lacau, S., Campiolo, S., Philpott, S. M. and Leponce, M. (2007) Contribution of cocoa plantations to the conservation of native ants (Insecta: Hymenoptera: Formicidae) with a special emphasis on the Atlantic Forest fauna of southern Bahia, Brazil. *Biodiversity and Conservation* 16:2359–84. DOI:10.1007/s10531-007-9190-6.

Dibog, L., Babin, R., Mbang, J. A. A., Decazy, B., Nyasse, S., Cilas, C. and Eskes, A. B. (2008) Effect of genotype of cocoa (Theobroma cacao) on attractiveness to the mirid Sahlbergella singularis (Hemiptera: Miridae) in the laboratory. *Pest Management Science* 64:977–80. DOI:10.1002/Ps.1589.

Domfeh, O., Dzahini-Obiatey, H., Ameyaw, G. A., Abaka-Ewusie, K. and Opoku, G. (2011) Cocoa swollen shoot virus disease situation in Ghana: A review of current trends. *African Journal of Agricultural Research* 6:5033–9.

Dufour, B. (1991) The role and importance of different insect species in the ecology of CSSV (cacao swollen shoot virus) in Togo. *Café, Cacao, Thé* 35:197–204.

Engelbrecht, C. J., Harrington, T. C. and Alfenas A. (2007) Ceratocystis wilt of cacao-a disease of increasing importance. *Phytopathology* 97:1648–9. DOI:10.1094/phyto-97-12-1648.

Entwistle, P. F. (1972) *Pests of Cocoa*. Longman Group Limited: London.

Entwistle, P. F. (1985) Insects and cacao. In G. A. R. Wood and Laas. (Eds), *Cacao*. Longman: London.

Eskes A. (Ed.) (2011) Collaborative and participatory approaches to cocoa variety improvement: Final report of the CFC/ICCO/Biodiversity international project on 'Cocoa productivity and quality improvement: A participatory approach' (2004–2010). CFC, Amsterdam, Pays-Bas.

Ferrari, L., Flores, A., Velásquez, F., Schneider, M., Andres, C., Milz, J., Trujillo, G., Alcon, F. and Studer, C. (2014) Evaluation of organic pest management strategies to control the cocoa mirid (*Monalonion dissimulatum* Dist.), Alto Beni, Bolivia, International Agriculture in a Changing World: Good News from the Field, 19 June 2014, HAFL, Zollikofen, Switzerland.

Figueroa Medina, W., Ramírez Sulvarán, J. A. and Sigarroa Rieche, A. K. (2013) Efecto de las cepas nativas Paecilomyces sp. (Bainier) y Lecanicillium sp. (Zimm) en el control de Carmenta foraseminis Eichlin (Lepidoptera: Sesiidae) en cultivos de cacao (Theobroma cacao L.). *Acta Agronómica* 62:279–86.

Flores, F. J. D. (1976) Insectos asociados con el cultivo del cacaotero, fluctuación de las principales especies fitófagas y su combate químico en el estado de Tabasco, México Colegio Superior de Agricultura Tropical, Tesis de Maestría, Cárdenas, p. 137.

Francis, A., Kairo, M., Roda, A., Liburd, O. and Polar, P. (2012) The passionvine mealybug, Planococcus minor (Maskell) (Hemiptera: Pseudococcidae), and its natural enemies in the cocoa agroecosystem in Trinidad. *Biological Control* 60:290–6.

Gassa, A., Abdullah, T., Fatahuddin and Junaid, M. (2015) The use of several types of artificial diet to increase population and aggressive behavior of weaver ants (Oecophylla Smaragdina F.) in reducing cocoa pod borer infestation (Conopomorpha Cramerella Sn.). *Academic Research International* 6:63–72.

Gassa, A., Daha, L. and Rosmana, A. (2003) The impact of Crematogaster spp. on cocoa pod borer, Conopomorpha cramerella. In A. Rosmana, P. van Grinsven, La Daha and G. Sarbini (Eds), Summary and Highlights of Technical Brain-Storming Meeting on Bio-control Technologies for Integrated Pest Management (IPM) of Cocoa, Effem Prima Project, 2Acdi-Voca, USAID, and Hasanuddin University, pp. 39–40.

Gerald, B. M. (1965) Bathycoelia thalassina (Herrich-Schaeffer), (Hemiptera: Pentatomidae); a Pest of Theobroma cacao L. *Nature* 207:881. https://doi.org/10.1038/207881a0

Gidoin, C., Babin, R., Bagny Beilhe, L., Cilas, C., ten Hoopen, G. M. and Bieng, M. A. (2014) Tree spatial structure, host composition and resource availability influence mirid density or black pod prevalence in cacao agroforests in Cameroon. *PLoS ONE* 9:e109405. DOI:10.1371/journal.pone.0109405 PONE-D-14-17577 [pii].

Giesberger, G. (1983) Biological control of the *Helopeltis* pest of cacao in Java. *Archives of Cacao Research* 2:91–180.

Guzman, V.d.C. (1985) Fluctuaciones de poblaciones del *Selenothrips rubrocinctus* (Giard) en el cultivo del cacao (Theobroma cacao L.) en zona 17–20 latitud Norte, 9th International Cocoa Research Conference, 12–18 February 1984, Cocoa Producers Alliance (COPAL), Lomé, Togo, pp. 529–33.

Ji, K., Zhang, D., Motilal, L. A., Boccara, M., Lachenaud, P. and Meinhardt, L. W. (2013) Genetic diversity and parentage in farmer varieties of cacao (Theobroma cacao L.) from Honduras and

Nicaragua as revealed by single nucleotide polymorphism (SNP) markers. *Genetic Resources and Crop Evolution* 60:441–53. DOI: 10.1007/s10722-012-9847-1.
Johns, N. D. (1999) Conservation in Brazil's chocolate forest: The unlikely persistence of the traditional cocoa agroecosystem. *Environmental Management* 23:31–47.
Kirkpatrick, T. W. (1950) Insect transmission of cacao virus disease in Trinidad. *Bulletin of Entomological Research* 41:99–118.
Konam, J. K. and Guest, D. I. (2004) Role of beetles (Coleoptera: Scolytidae and Nitidulidae) in the spread of Phytophthora palmivora pod rot of cocoa in Papua New Guinea. *Australasian Plant Pathology* 33:55–9. DOI: 10.1071/ap03082.
Lavabre, E. M. (1977) Importance économique des Mirides dans la cacaoculture mondiale. In E. M. Lavabre (Ed.), *Les Mirides du Cacaoyer.* G-P. Maisonneuve et Larose: Paris, pp. 139–53.
Liendo-Barandiaran, C. V., Herrera-Malaver, B., Morillo, F., Sanchez, P. and Hernandez, J. V. (2010) Behavioral responses of *Steirastoma breve* (Sulzer) (Coleoptera: Cerambycidae) to host plant *Theobroma cacao* L., brushwood piles, under field conditions. *Applied Entomology and Zoology* 45:489–96. DOI: 10.1303/aez.2010.489.
Liendo, C. V., Morillo, F., Sanchez, P., Muñoz, W., Guerra, J., Cabrera, A. and Hernandez, J. V. (2005) Olfactory behavior and electroantennographic responses of the cocoa beetle *Steiratoma breve* (Coleoptera: Cerambycidae). *Florida Entomologist* 88:117–22.
Lim, G. T. (1992) Biology, ecology and control of cacao podborer *Conopomorpha cramerella* (Snellen). In P. J. Keane and C. A. J. Putter (Eds), *Cacao Pest and Disease Management in Southeast Asia and Australasia* (FAO Plant Production and Protection Paper No 112), Rome. pp. 85–100.
Lim, G. T. and Phua, P. K. (1986) Effect of cacao pod borer, Acrocercops cramerella (Snellen) on yield and bean size. In E. Pushparaja and P. S. Chew (Eds), *Cacao and Coconuts: Progress and Outlook.* Incorporated Society of Planters: Kuala Lumpur, pp. 325–36.
Mahob, R. J., Babin, R., ten Hoopen, G. M. and Dibog, L., Yede, Hall, D. R. and Bilong, C. F. B. (2011) Field evaluation of synthetic sex pheromone traps for the cocoa mirid Sahlbergella singularis (Hemiptera: Miridae). *Pest Management Science* 67:672–6. DOI: 10.1002/Ps.2107.
Matthews, D. L. and Miller, J. Y. (2010) Notes on the cacao plume moth in Honduras and description of the larvae and pupae (Lepidoptera: Pterophoridae). *Tropical Lepidoptera Research* 20:28–34.
Mazón, M., Díaz, F. and Gaviria, J. C. (2013) Effectiveness of different trap types for control of bark and ambrosia beetles (Scolytinae) in Criollo cacao farms of Mérida, Venezuela. *International Journal of Pest Management* 59:189–96. DOI: 10.1080/09670874.2013.810794.
Mbenoun, M., Wingfield, M. J., Begoude Boyogueno, A. D., Nsouga Amougou, F., Petchayo Tigang, S., ten Hoopen, G. M., Mfegue, C. V., Dibog, L., Nyassé, S., Wingfield, B. D. and Roux, J. (2016) Diversity and pathogenicity of the Ceratocystidaceae associated with cacao agroforests in Cameroon. *Plant Pathology* 65:64–78. DOI: 10.1111/ppa.12400.
McCallan, E. (1975) Miridae of the genusTermatophylidea [Hemiptera] as predators of cacao thrips. *Entomophaga* 20:389–91. DOI: 10.1007/bf02371593.
McMahon, P., Iswanto, A., Susilo, A. W., Sulistyowati, E., Wahab, A., Imron, M., Purwantara, A., Mufrihati, E., Dewi, V. S., Lambert, S., Guest, D. and Keane, P. (2009) On-farm selection for quality and resistance to pest/diseases of cocoa in Sulawesi: (i) performance of selections against cocoa pod borer, *Conopomorpha cramerella*. *International Journal of Pest Management* 55:325–37. https://doi.org/10.1080/09670870902923438
McMahon, P., Purung, H. B., Lambert, S., Mulia, S., Nurlaila, Susilo, A. W., Sulistyowati, E., Sukamto, S., Israel, M., Saftar, A., Amir, A., Purwantara, A., Iswanto, A., Guest, D. and Keane, P. (2015) Testing local cocoa selections in three provinces in Sulawesi: (i) Productivity and resistance to cocoa pod borer and Phytophthora pod rot (black pod). *Crop Protection* 70:28–39. DOI: http://dx.doi.org/10.1016/j.cropro.2015.01.001.
Morillo, F., Sánchez, P., Herrera, B., Liendo-Barandiaran, C., Muñoz, W. and Vicente Hernández, J. (2009) Pupal development, longevity and behavior of carmenta theobromae (Lepidoptera: Sesiidae). *Florida Entomologist* 92:355–61. DOI: 10.1653/024.092.0222.

Moxon, J. E. (1992) Insect pests of cacao in Papua New Guinea, importance and control. In P. J. Keane and C. A. J. Putter (Ed.), *Cacao Pest and Disease Management in Southeast Asia and Australasia* (FAO Plant Production and Protection Paper No 112). Rome, pp. 129–44.
Muhamad, R. and Way, M. J. (1995) Damage and crop loss relationships of Helopeltis theivora, Hemiptera, Miridae and cocoa in Malaysia. *Crop Protection* 14:117–21. DOI: http://dx.doi.org/10.1016/0261-2194(95)92865-K.
Mumford, J. (1986) Control of the cocoa pod borer (Acrocercops cramerella): A critical review. In E. Pushparajah and Chew, P. S. (Eds), *Cocoa and Coconut: Progress and Outlook*. Incorporated Soc. of Planters: Kuala Lumpur, Malaysia, pp. 277–86.
Mumford, J. D. and Ho, S. H. (1988) Control of the cacao pod borer (Conopomorpha cramerella). *Cocoa Grower's Bulletin* 40:19–29.
N'Guessan, F. K., Kouassi, A. F. and Atindehou, K. (2006) Study of the effect of the neem, Azadiractha indica Juss (Meliaceae) on Sahlbergella singularis (Hemiptera: Miridae), an important pest of cocoa. 15th International Cocoa Research Conference.
N'Guessan, K. F., N'Goran, J. A. K. and Eskes, A. B. (2008) Resistance of cacao (*Theobroma cacao* L.) to *Sahlbergella singularis* (Hemiptera: Miridae): Investigation of antixenosis, antibiosis and tolerance. *International Journal of Tropical Insect Science* 28:201–10.
Namyatova, A. A., Konstantinov, F. V. and Cassis, G. (2016) Phylogeny and systematics of the subfamily Bryocorinae based on morphology with emphasis on the tribe Dicyphini sensu Schuh. *Systematic Entomology* 41:3–40. DOI: 10.1111/syen.12140.
Offenberg, J. (2015) Ants as tools in sustainable agriculture. *Journal of Applied Ecology* 52:1197–205. DOI: 10.1111/1365-2664.12496.
Oliveira, C., Auad, A., Mendes, S. and Frizzas, M. (2014) Crop losses and the economic impact of insect pests on Brazilian agriculture. *Crop Protection* 56:50–4.
Owusu-Manu, E. (1976) Natural enemies of Bathycoelia thalassina (Herrich-Schaeffer) (Hemiptera: Pentatomidae), a pest of cocoa in Ghana. *Biological Journal of the Linnean Society* 8:217–44. DOI: 10.1111/j.1095-8312.1976.tb00247.x.
Owusu Manu, E., Manu, O., Mensah R. E., Dakwa, E., Mensah, R. J. T., Asante, S., Mordjifa, D. E. and Manteaw, F. E. (1990) *Feeding behaviour and the damage caused by Bathycoelia thalassina (Herrich-Schaeffer)(Hemiptera: Pentatomidae)*. Cocoa Research Institute: Tafo (Ghana).
Padi, B., Ackonor, J., Opoku, I., Vos, J. and Neuenschwander, P. (2002) Cocoa IPM research and implementation in Ghana, West Africa Regional Cocoa IPM Workshop, Cotonou, Benin, 13–15 November 2001. Proceedings., CPL Press.
Padi, B., Adu-Acheampong, R., Asamoah, M. and Aneani, F. (2003) Screening of Neem extracts for the control of cocoa capsids (Heteroptera: miridae). In A. Y. Akrofi, J. B. Ackonor and L. A. A. Ollennu (Eds), *Incoped 4th International Seminar*, 19–21 October 2003, Accra, Ghana, pp. 44–52.
Peng, R. K. and Christian, K. (2004) The weaver ant, Oecophylla smaragdina (Hymenoptera: Formicidae), an effective biological control agent of the red-banded thrips, Selenothrips rubrocinctus (Thysanoptera: Thripidae) in mango crops in the Northern Territory of Australia. *International Journal of Pest Management* 50:107–14. DOI: 10.1080/09670870410001658125.
Philpott, S. M. and Armbrecht, I. (2006) Biodiversity in tropical agroforests and the ecological role of ants and ant diversity in predatory function. *Ecological Entomology* 31:369–77.
Ploetz, R. (2016) The impact of diseases on cacao production: A global overview. In B. A. Bailey and L. W. Meinhardt (Eds), *Cacao Diseases: A History of Old Enemies and New Encounters*. Springer International Publishing: Cham, pp. 33–59.
Posada, F. J., Virdiana, I., Navies, M., Pava-Ripoll, M. and Hebbar, P. (2011) Sexual dimorphism of pupae and adults of the cocoa pod borer, *Conopomorpha cramerella*. *Journal of Insect Science* 11:52.
Posnette, A. F. (1944) Virus diseases of cacao in Trinidad. *Tropical Agriculture* 21:105–6.
Rice, R. A. and Greenberg, R. (2000) Cacao cultivation and the conservation of biological diversity. *Ambio* 29:167–73.

Room, P. M. and Smith, E. S. C. (1975) Relative abundance and distribution of insect pests, ants and other components of the cocoa ecosystem in Papua New Guinea. *Journal of Applied Ecology* 12:31–46. DOI: 10.2307/2401716.

Rosmana, A., Shepard, M., Hebbar P. and Mustari, A. (2010) Control of cacao pod borer and *Phytophthora* pod rot using degradable plastic pod sleeves and a nematode, *Steinernema carpocapsae*. *Indonesian Journal of Agricultural Science* 11:41–7.

Rudolph, S. G. and Loudon, C. (1986) Load size selection by foraging leaf-cutter ants (Atta cephalotes). *Ecological Entomology* 11:401–10. DOI: 10.1111/j.1365-2311.1986.tb00319.x.

Sabatier, R., Meyer, K., Wiegand, K. and Clough, Y. (2013) Non-linear effects of pesticide application on biodiversity-driven ecosystem services and disservices in a cacao agroecosystem: A modeling study. *Basic and Applied Ecology* 14:115–25. DOI: http://dx.doi.org/10.1016/j.baae.2012.12.006.

Sarfo, J. E., Hall, D. H., Campbell, C. A. M., Downham, M. C. A. and Cudjoe, A. R. (2010) Improving the management of cocoa mirids, Distantiella theobroma (Dist.) and Sahlbergella singularis Hagl. (Heteroptera: Miridae) through pheromone trapping. 16th International Cocoa Research Conference.

Saripah, B. and Azhar, I. (2007) Implementation of cocoa pod sleeving in controlling cocoa pod borer infestation, 2007 Conference on Plantation Commodities, Putra World Trade Centre, Kuala Lumpur (Malaysia), 3–4 Jul 2007, Malaysian Cocoa Board.

Saripah, B. and Azhar, I. (2012) Five years of cocoa black ants, *Dolichoderus thoracicus* (Smith) at farmer's plot – An epilogue. *Malaysian Cocoa Journal* 7:8–15.

Saripah B., Azhar I., Roslan S. and Abdul Shukor, M. (2005) Efficacy of Pod Sleeving in Cacao Pod Borer Management, 18–19 July 2005, 4th Malaysian International Cacao Conference (MICC), Malaysian Cocoa Board, Kuala Lumpur, Malaysia, pp. 169–75.

Saunders, J. L. (1981) Cacao Pests in Central America, 7th International Cocoa Research Conference, 4–12 November 1979, Cocoa Producers Alliance (COPAL), Douala, Cameroun, pp. 429–32.

Schroth, G., Krauss, U., Gasparotto, L., Aguilar, J. A. D. and Vohland, K. (2000) Pests and diseases in agroforestry systems of the humid tropics. *Agroforestry Systems* 50:199–241. DOI: 10.1023/a:1006468103914.

Shapiro, L. H., Scheffer, S. J., Maisin, N., Lambert, S., Purung, H. B., Sulistyowati, E., Vega, F. E., Gende, P., Laup, S., Rosmana, A., Djam, S. and Hebbar, P. K. (2008) Conopomorpha cramerella (Lepidoptera: Gracillariidae) in the Malay Archipelago: Genetic Signature of a Bottlenecked Population? *Annals of the Entomological Society of America* 101:930–8. DOI: 10.1603/0013-8746(2008)101[930:cclgit]2.0.co;2.

Soria, S. J. and Machado, A. B. M. (1982) Contribuicao ao conhecimento dos inimigos naturais das coleobrocas (Xyleborus spp.) pragas do cacaueiro na Bahia, Brasil, 8th International Cocoa Research Conference, 18–23 October 1981, Cocoa Producers Alliance (COPAL), Cartagena, Colombia, pp. 289–90.

Sounigo, O., Coulibaly, N., Brun, L., N'Goran, J., Cilas, C. and Eskes A. B. (2003) Evaluation of resistance of *Theobroma cacao*L. to mirids in Côte d'Ivoire: Results of comparative progeny trials. *Crop Protection* 22:615–21.

Steffan-Dewenter, I., Kessler, M., Barkmann, J., Bos, M. M., Buchori, D., Erasmi, S., Faust, H., Gerold, G., Glenk, K., Gradstein, S. R., Guhardja, E., Harteveld, M., Hertel, D., Höhn, P., Kappas, M., Köhler, S., Leuschner, C., Maertens, M., Marggraf, R., Migge-Kleian, S., Mogea, J., Pitopang, R., Schaefer, M., Schwarze, S., Sporn, S. G., Steingrebe, A., Tjitrosoedirdjo, S. S., Tjitrosoemito, S., Twele, A., Weber, R., Woltmann, L., Zeller, M. and Tscharntke, T. (2007) Tradeoffs between income, biodiversity, and ecosystem functioning during tropical rainforest conversion and agroforestry intensification. *Proceedings of the National Academy of Sciences* 104:4973–8. DOI: 10.1073/pnas.0608409104.

Stonedahl, G. M. (1991) The Oriental species of *Helopeltis* (Heteroptera: Miridae): A review of economic literature and guide to identification. *Bulletin of Entomological Research* 81:465–90.

Strickland, A. H. (1951) The entomology of swollen shoot of cacao. II. – The bionomics and ecology of the species involved. *Bulletin of Entomological Research* 42:65–103. DOI: 10.1017/S0007485300025177.

Tan, G. S. (1974) Helopeltis theivora theobromae on cocoa in Malaysia. II. Damage and control. *Malaysian Agricultural Research* 3:204–12.

Teh, C.-L., Pang, J. T.-Y. and Ho, C.-T. (2006) Variation of the response of clonal cocoa to attack by cocoa pod borer Conopomorpha cramerella (Lepidoptera: Gracillariidae) in Sabah. *Crop Protection* 25:712–17. DOI: http://dx.doi.org/10.1016/j.cropro.2005.10.009.

Tong-Kwee, L., Muhamad, R., Fee, C. G. and Lan, C. C. (1989) Studies on Beauveria bassiana isolated from the cocoa mirid, Helopeltis theobromae. *Crop Protection* 8:358–62. DOI: http://dx.doi.org/10.1016/0261-2194(89)90055-0.

Toxopeus, H. and Giesberger, G. (1983) History of cacao and cacao research in Indonesia. *Archives of Cacao Research* 2:7–34.

Tscharntke, T., Clough, Y., Bhagwat, S. A., Buchori, D., Faust, H., Hertel, D., Hölscher, D., Juhrbandt, J., Kessler, M., Perfecto, I., Scherber, C., Schroth, G., Veldkamp, E. and Wanger, T. C. (2011) Multifunctional shade-tree management in tropical agroforestry landscapes - A review. *Journal of Applied Ecology* 48:619–29. DOI: 10.1111/j.1365-2664.2010.01939.x.

Tuck, H. C. (1994) Methods towards efficient establishment of introduced black cocoa ant, Dolichoderus thoracicus for natural control of Helopeltis theivora damage in cocoa. *Planter* 70:487–95.

Van Hall, C. J. J. (1914) *Cacao*. Macmillan and Co.: London, p. 514.

van Wijngaarden, P. M., van Kessel, M. and van Huis A. (2007) *Oecophylla longinoda* (Hymenoptera: Formicidae) as a biological control agent for cocoa capsids (Hemiptera: Miridae) Netherlands Entomological Society Meeting 18, Netherland, pp. 21–30.

Vargas, A., Somarriba, E. and Carballo M. (2005) Dinámica poblacional del chinche (Monalonion dissimulatum Dist.) y daño de mazorcas en plantaciones orgánicas de cacao del Alto Beni, Bolivia. [Population dynamics of the thrip (Monalonion dissimulatum Dist.) and pod damage in organic cacao plantations in Alto Beni, Bolivia]. *Agroforestería en las Américas (CATIE)* 43–44:72–6.

Villacorta, A. (1977) Annual fluctuation of the Monalonion annulipes population and its relationship with the dieback of Theobroma cacao in Costa Rica. *Anais da Sociedade Entomologica do Brasil* 6:215–23.

Way M. J. and Khoo K. C. (1991) Colony dispersion and nesting habits of the ants, Dolichoderus thoracicus and Oecophylla smaragdina (Hymenoptera: Formicidae), in relation to their success as biological control agents on cocoa. *Bulletin of Entomological Research* 81:341–50. DOI:10.1017/s0007485300033629.

Way M. J. and Khoo K. C. (1992) Role of ants in pest management. *Annual Review of Entomology* 37:479–503.

Wetten, A., Campbell, C. and Allainguillaume, J. (2016) High-resolution melt and morphological analyses of mealybugs (Hemiptera: Pseudococcidae) from cacao: Tools for the control of Cacao swollen shoot virus spread. *Pest Management Science* 72:527–33. DOI: 10.1002/ps.4017.

Wielgoss, A., Clough, Y., Fiala, B., Rumede, A. and Tscharntke, T. (2012) A minor pest reduces yield losses by a major pest: Plant-mediated herbivore interactions in Indonesian cacao. *Journal of Applied Ecology* 49:465–73.

Williams, G. (1953) Field Observations on the Cacao Mirids, *Sahlbergella singularis* Hagl. and *Distantiella theobroma* (Dist.), in the Gold Coast, Part I. Mirid Damage. *Bulletin of Entomological Research* 44:101–19.

Yen, J. D., Waters, E. K. and Hamilton, A. J. (2010) Cocoa Pod Borer (Conopomorpha cramerella Snellen) in Papua New Guinea: Biosecurity models for New Ireland and the autonomous region of Bougainville. *Risk Anal* 30:293–309. DOI: 10.1111/j.1539-6924.2009.01297.x.

Zhang, A., Kuang, L. F., Maisin, N., Karumuru, B., Hall, D. R., Virdiana, I., Lambert, S., Bin Purung, H., Wang, S. and Hebbar, P. (2008) Activity evaluation of cocoa pod borer sex pheromone in cacao fields. *Environmental Entomology* 37:719–24.

Chapter 14

Nematode pests of cocoa

Samuel Orisajo, Cocoa Research Institute of Nigeria, Nigeria

1 Introduction

Cacao (*Theobroma cacao* L.) plays an important role in the growing nations' economies. It is native to the upper Amazon region of South America (Bartley, 2005). The fruit of cacao, termed cocoa, is a part of the tree that is most valued and, together with seeds, is most closely connected with the crop commercial outcome (Bartley, 2005). Cacao is cultivated in many countries of South and Central America, Africa, Asia and Oceania, located mostly between 10° North and South of the Equator. West Africa has dominated the world cocoa market. The main cacao-growing countries of West Africa, namely Côte d'Ivoire, Ghana, Nigeria, Cameroon and Togo make up above 70% of the world's annual cocoa production (ICCO, 2017). It may be possible to produce cocoa beans in large volumes also in other regions, but replicating the unique flavour, chemical and physical characteristics of the West African cocoa, may prove to be a task very difficult to achieve (Taylor and Taylor, 2006).

It is noteworthy that cacao suffers from severe losses due to pests and diseases in any growing area (Bartley, 2005). Although the most important plant health constraints of cacao worldwide are caused by *Phytophthora* spp., *Crinipellis perniciosa*, *Moniliophthora roreri* and Cocoa swollen shoot virus (CSSV), also plant-parasitic nematodes, are known to have caused cocoa yield losses, sudden death of trees and retardation of seedling growth in nurseries (Campos and Villain, 2005; Orisajo, 2009).

http://dx.doi.org/10.19103/AS.2017.0021.20

2 Hidden nature of plant-parasitic nematodes

One difficulty with assessing the nematode impact is that the damage resulting from nematode attacks is often less evident than that caused by many other pests or diseases (Nicol et al., 2011). Because of the ubiquitous interaction of nematodes with other pathogens, the concept of economic threshold, well established for many crops and pathogens in temperate environments, is difficult to establish under field conditions in tropical regions. The presence of multiple pathogens makes the diagnosis often confusing, frequently resulting in underestimates of the damage caused by nematodes.

Losses due to the nematode attack may not necessarily be a consequence of direct cell death, necrosis or 'diseased' tissue, but may derive from other more insidious aspects, such as the interference of the histopathological changes induced with the root system physiology, reducing roots efficiency in terms of access and uptake of nutrients and water. Nematode-affected plants present typical drought and nutrient stress symptoms, which are easily and often misdiagnosed (Nicol et al., 2011). The masking effects of the damage caused by nematodes can also lead to the adoption of erroneous control measures. For example, root-knot nematodes (*Meloidogyne* spp.) attack cacao seedlings leading to dieback (Fig. 1), but farmers/growers erroneously consider the damage as due to a fungal infection (Orisajo, 2012).

3 Plant-parasitic nematodes associated with cacao

Many genera and species of root-feeding nematodes have been found in association with cacao (Table 1), although the pathogenic relationship, for most of them, has not been experimentally proved. Root-knot nematodes are the most important group found on cacao, due to their pathogenicity and wide distribution in cocoa-producing regions of the world (Sharma and Sher, 1974a; Campos and Villain, 2005). *Meloidogyne* spp. have been found in cacao since 1900 (Ritzema Bos in Sosamma et al. (1980a)), and have been reported from Zaire,

Figure 1 Dieback of cacao caused by root-knot nematodes (left) compared to a healthy plant.

Table 1 List of endoparasitic and ectoparasitic nematodes associated with cocoa roots*

Allotrichodorus brasiliensis	*Meloidogyne javanica*
Allotrichodorus campanulatus	*Meloidogyne thamesi*
Allotrichodorus sharmae	*Neodolichodorus*
Allotrichodorus westindicus	*Ogma decalineatum*
Aorolaimus banoae	*Paralongidorus* sp.
Aorolaimus holdemani	*Paratrichodorus minor*
Aorolaimus levicaudatus	*Paratylenchus* sp.
Aorolaimus nigeriensis	*Peltamigratus christiei*
Aorolaimus vigiae	*Peltamigratus holdemani*
Criconema braziliense	*Peltamigratus macbethi*
Criconema demani	*Pratylenchus brachyurus*
Criconemoides ferniae	*Pratylenchus coffeae*
Criconemoides onoensis	*Pratylenchus zeae*
Criconemoides paradenoudeni	*Radopholus similis*
Criconemoides paralineolata	*Rotylenchulus microstriatus*
Discocriconemella degrissei	*Rotylenchulus reniformis*
Discocriconemella limitanea	*Scutellonema brachyurus*
Dolichodorus minor	*Scutellonema clathricaudatum*
Eutylenchus africanus	*Trichodorus monohystera*
Helicotylenchus dihystera	*Trophorus imperialis*
Helicothylenchus erythrinae	*Tylenchorhynchus annulatus*
Helicotylenchus multicinctus	*Tylenchorhynchus queirozi*
Hemicriconemoides cocophillus	*Tylenchus* sp.
Hemicriconemoides mangiferae	*Xiphinema abeokutae*
Hemicycliophora chilensis	*Xiphinema americanum*
Hemicycliophora loofi	*Xiphinema bergeri*
Hemicycliophora thienemanni Heterodera sp.	*Xiphinema brasiliense*
Hoplolaimus sp.	*Xiphinema brevicollum*
Longidoroides sp.	*Xiphinema ifacolum*
Longidorus sp.	*Xiphinema krugi*
Meloidogyne arenaria	*Xiphidorus minor*
Meloidogyne exigua	*Xiphinema vulgare*
Meloidogyne incognita	

* Sources = Afolami and Caveness (1983); Badaru et al. (1999); Bridge et al. (1982); Campos and Villain, 2005; Crozzoli et al. (2001); Freire and Monteiro (1978); Loof and Sharma (1980); López (1994); López et al. (1980); Luc and Coomans (1993); Mança, 1969; Orisajo, 2009; Pinochet and Raski (1976); Sharma (1982); Sharma and Loof (1974); Sharma and Sher (1973, 1974a); Sosamma et al. (1980a,b); Sudha and Sundararaju (2002); Thorold (1975); Whitehead (1969).

São Tomé, Java (Fluitter and Mulholland, 1941), Ghana, Malawi, Côte d'Ivoire (Martin, 1961), Nigeria (Afolami and Caveness, 1983), Venezuela, Brazil and India (Sosamma et al., 1980b).

Root-knot nematodes are most frequently found in cacao (Sharma and Sher, 1974a) and are a common pest in West Africa (Whitehead, 1969), including Nigeria where members of the genus appear as the most economically important (Afolami and Caveness, 1983; Badaru et al., 1999; Fademi et al., 2006). *Meloidogyne incognita* is also common in India (Sosamma et al., 1980b; Sudha and Sundararaju, 2002), Malaysia (Razak, 1981) and Venezuela (Crozzoli et al., 2001), and is widespread in cacao -growing regions of Brazil (Sharma and Sher, 1974a,b). However, other species of *Meloidogyne* have also been found on cacao including *Meloidogyne exigua* in Bolivia (Bridge et al., 1982), *Meloidogyne javanica* in Malawi, Venezuela (Crozzoli et al., 2001) and in Central Africa (Martin, 1961), and *Meloidogyne arenaria* and *Meloidogyne thamesi* in Brazil (Sharma, 1982).

In addition, the root-lesion nematode, *Pratylenchus brachyurus*, has been widely reported from cacao in Bahia, Brazil (Sharma and Sher, 1973), and also occurs in Western Africa. In Java, *Pratylenchus coffeae* was reported from roots of cacao. Sudha and Sundararaju (2002) found *Pratylenchus coffeae* on this crop in Kerala State, India, and Kumar et al. (1971) reported the multiplication of this nematode on cacao in glasshouse experiments. *Pratylenchus coffeae* is also reported on cacao in Indonesia. A nother species, *Pratylenchus zeae*, occurs on cacao in Venezuela (Crozzoli et al., 2001).

Other root-feeding nematodes of cacao include *Hemicycliophora* spp., *Hoplolaimus* spp., *Radopholus similis*, *Rotylenchulus reniformis*, *Scutellonema brachyurus*, *Trichodorus* spp. and *Xiphinema* spp. (Sharma, 1982). *Helicotylenchus* spp. are widespread on cacao in South America and Asia, and *Helicotylenchus pseudorobustus* was found to reproduce on cacao in Liberia (Lamberti et al., 1992). *Helicotylenchus dihystera* was reported in the Bahia State (Brazil) as the most widespread species on cacao, occurring in 70% of the samples (Sharma, 1982). In Nigeria, Afolami and Caveness (1983) reported the frequency and geographic distribution of plant-parasitic nematodes associated with cacao. The common genera found were *Helicotylenchus*, *Hemicycliophora*, *Meloidogyne*, *Radopholus*, *Scutellonema*, *Trichodorus* and *Xiphinema*.

3.1 Symptoms of root-knot nematode damage on cacao

In artificially infected seedlings, *Meloidogyne incognita* caused dieback, stunting, wilting, yellowing and size reduction of leaves. Tiny galls, juveniles and females of *Meloidogyne incognita* with egg masses can be observed on roots (Campos and Villain, 2005; Orisajo and Fademi, 2005; Meyer et al., 2011, Fig. 2). In Nigeria, seedlings of cv. Amelonado grown in soil inoculated with *Meloidogyne incognita* showed visible symptoms at the 16th week, leading to wilting by the 24th week (Afolami, 1981a; Afolami and Caveness, 1983; Afolami and Ojo, 1984). Amazon cultivars also showed symptoms, but only after the 24th week, with no signs of wilting (Afolami, 1981b; Orisajo, 2009; Orisajo and Dongo, 2005; Orisajo and Fademi, 2005; Orisajo et al., 2012).

Meloidogyne incognita produces galls with exposed egg masses on roots, dieback and sudden death of the infested plants in the field (Orisajo et al., 2007). According to Sharma and Sher (1973), when the dieback conditions occur, the trees die down to their roots. The roots, however, remain alive and send up shoots in the next growing season and also when the dead terminals are pruned off (Fig. 1). The syndrome of sudden death disease

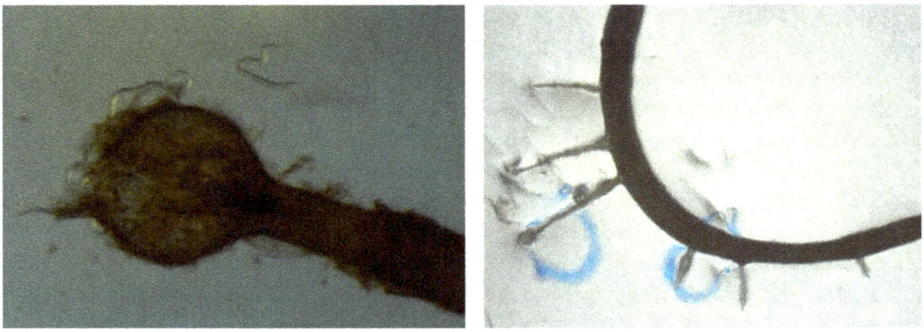

Figure 2 Tiny galls and *M. incognita* juveniles on roots of cacao.

Figure 3 Sudden death of cacao seedlings by parasitic nematodes (left) compared to a healthy one.

Figure 4 An enlarged aboveground hypocotyl swelling of young cacao two years after transplanting.

is a permanent wilting, with green leaves suddenly turning yellow and brown, drying up to remain hanging (Fig. 3). Sharma and Sher (1973) associated the occurrence of sudden death with root-knot nematodes. In Congo, Ghesquière (1921) showed that *Meloidogyne* sp. caused a sudden death of cacao as a consequence of intensive attacks on the roots. In Brazil, a combination of attacks by root-knot and root-lesion nematodes was considered possible (Saenz, 1969).

Reports from Ghana and Nigeria revealed that *Meloidogyne incognita* infection of cacao seedlings resulted in above-ground hypocotyl swellings (Fig. 4) in addition to large galls on tap roots and smaller ones on feeding roots (Asare-Nyako and Owusu, 1979; Orisajo et al., 2012). The head end of the swollen female nematode was found in the cambial region in both hypocotyl and root, with the typical hyperplasia and hypertrophy of parenchyma cells associated with root-knot attacks. No secondary thickening of cells was found in the infected root areas. There were also significant reductions in root weight and seedling height. Asare-Nyako and Owusu (1979) suggested that the reduction in roots, and not the extent of gall formation, accounts for the height reduction observed. It was also suggested that galling of roots was not a sufficiently clear criterion to assess *Meloidogyne* parasitism, as on some crops infection can be overlooked (Whitehead, 1969).

In Central African environmental conditions, cacao roots infested with *Meloidogyne javanica* form galls (Martin, 1961). In Malawi, the slow growth rate of cacao and the failure of young seedlings to get established in the field have been associated with soils heavily infested with *Meloidogyne javanica* (Corbett, 1961). Damage symptoms were also observed on cacao roots infested by *Meloidogyne exigua* in Bolivia (Bridge et al., 1982). General galling of secondary roots and atrophy of primary roots were caused by *Meloidogyne arenaria* in Brazil (Manço, 1969).

3.2 Other plant-parasitic nematode damage on cacao

Endoparasitic species include the root-lesion nematode, *Pratylenchus brachyurus*, found around cacao roots in Côte d'Ivoire (Luc and de Guiran, 1960), Nigeria (Caveness, 1967) and Brazil (Sharma and Sher, 1973), and *Pratylenchus coffeae* reported in cacao roots in Java (Fluitter and Mulholland, 1941). In India, *Pratylenchus coffeae* and *Rotylenchulus reniformis*, also reported on cacao in Jamaica (Dixon, 1961), were found to actively multiply on local cacao crops (Kumar et al., 1971).

Ectoparasitic species reports include the pathogenic *Helicotylenchus dihystera* on cacao in Brazil (Campelo and Galli, 1980). A significant reduction was observed in dry root weight and leaf number of inoculated seedlings. Dieback and death of cacao seedlings in nurseries in Brazil have also been attributed to attacks by *Dolichodorus minor* (Sharma, 1971). The entire root system was reduced, blackened and showed disintegrated cortex and bead-like gall formations, with the gall portions turning reddish-brown and hard.

4 Disease complexes

The fungus *Mycoleptodiscus terrestris* was found to be associated with the sudden death of cacao in Bahia, Brazil (Ram, 1973). It is interesting to speculate on the possibility of an interaction with root-knot nematodes. Nematodes have been found in an extract with zoospores of *Phytophthora megakarya* from black pod of cocoa in Nigeria (unpublished).

On the other hand, disease complexes involving nematodes and fungi have been documented, not much can be said of disease complexes involving nematodes and bacteria or viruses on tree crops. Nevertheless, and in spite of the fact that this aspect has not received attention as it relates to cacao, the possibility of a real threat cannot be excluded (Orisajo, 2009).

By the late 1980s, 22 longidorid and 14 trichodorid nematodes had been reported as vectors of viruses for a wide range of plants (Lamberti and Roca, 1987). In Longidoridae, 11 species belonged to the genus *Xiphinema*, 10 to the genus *Longidorus* and one to *Paralongidorus*. Concerning Trichodoridae, five species belonged to the genus *Trichodorus* and nine to *Paratrichodorus*. Nematodes belonging to the above-mentioned genera were among those associated with cacao in Brazil, Côte d'Ivoire, Ghana and Nigeria. Although few concerted efforts have been made on nematode–virus associations in cacao, apart from the much known CSSV, a nematode-transmitted viral disease – Cacao necrosis nepovirus – was first reported in Ghana (Kenten, 1972) and then documented (Brunt et al., 1996).

5 Management of plant-parasitic nematodes of cacao

Management and control of nematodes are more difficult in perennial than in annual crops, as the long-time nature of the former makes rotation schemes impractical. In perennial crops such as cacao, nematodes that survive the control practices have time to recover and build up again to destructive levels (Orisajo, 2009). Nematode control, therefore, becomes essential in order to reduce crop losses and to ensure self-sufficiency in food and industrial raw materials. Several nematode management options are available. These can be broadly categorized into chemical and non-chemical control methods. The latter includes good farm hygiene, use of resistant or tolerant genotypes, biological control and organic amendments, while the former essentially involves the use of nematicides.

5.1 Chemical control

Historically, nematicides have been highly effective and reliable in controlling a wide range of nematode species in certain crop situations. The advantages to be derived are the fast effects on plant-parasitic nematodes leading to high increases in yields, especially of high-valued crops. Sosamma et al. (1980a) have reported an increase in the number of cocoa pods by the application of fensulfothion (Dasanit™) and fenamifos (Nemacur™). Nemagon™ (1,2–dibromo-B-chloropropane), injected into the soil around infected trees, can control nematode damage temporarily and increase crop yields (Entwistle and Caveness, 1963; Ichinohe, 1967) but it is not always effective, toxic to humans and is not available any more.

Elimination of nematodes by the sterilization of soil or application of nematicides has improved the health of the cacao crop (Afolami, 1981a; Afolami and Ojo, 1984). Afolami (1993) reported some degree of effectiveness on the use of dazomet (Basamid™) in controlling plant-parasitic nematodes in bagged nursery soils. In the report, 2 g of granules in a 2-l nursery bag placed at 0.25–0.5 inch depth eliminated plant-parasitic nematodes, but preserved free-living *Rhabditis* and nematophagus *Mononchids*. A higher rate of 4 g per 2-l bag, though equally effective, was highly phytotoxic.

5.1.1 Limitations of chemical control

The economic threshold level for chemical control of plant-parasitic nematodes in cacao is not yet determined. While chemical control can reduce the impact of nematode infestations (Afolami, 1993), available and effective nematicides are too expensive for resource-poor farmers and often result in a noticeable decrease in many soil biological processes, notably by suppressing predators and parasites of nematodes. Further negative outcomes include the contamination of aquifers, health hazards, environmental pollution and potential atmospheric ozone depletion. Moreover, soils, especially those with a low microbial population, are more vulnerable to the reinvasion of pathogens even after fumigation (Gamliel et al., 2000), and nematode populations might become sensitized or resistant to repeated nematicide applications.

The incidence of pesticide poisoning and mortality in some countries serves as a grim warning of the risks that arise when these products are widely used under poor management. Increased social and legislative pressure to restrict the use of methyl bromide, an effective soil fumigant and biocide, used extensively to control a broad spectrum of pests, promoted the search for alternative management approaches. Since the first use of carbon disulphide (a fumigant) as a soil nematicide, several nematicides, fumigant and non-fumigant, have been developed. Developed products include organophosphates or carbamates, all with dual action as nematicides and insecticides. Many of these chemicals ended up unregistered because they were phytotoxic, while some, though initially registered, had to be withdrawn (Bateman, 2008; Orisajo, 2009). Hence, there is an urgent need to develop new management tools that are environmentally and toxicologically safe. Genetic resistance and organic soil amendment are some of the new and alternative technologies proposed.

5.2 Resistant or tolerant cultivars

Resistant cultivars hold out most promises for the effective and economic control of nematodes. A resistant cultivar without nematode treatment may often yield as much as a high-yielding susceptible one, treated with nematicides (Epps et al., 1981).

Conveniently, the use of resistant planting materials is the most economic approach to nematode control in plants. However, this method is unfortunately not very applicable to cacao. Although this crop was not listed among the eight nut/fruit trees which have locally available nematode-resistant or nematode-tolerant seeds or rootstocks (Sasser and Frekman, 1987), the assessment of materials in the Nigerian cacao germ plasm has shown some degree of resistance (Afolami and Ojo, 1984; Okeniyi, et al., 2009). These, unfortunately, do not fall within the categories selected by breeders for broad-based multiplication and distribution to farmers. Nematodes, thus far, still pose a threat to cacao production that cannot be addressed through resistance or tolerance based on the genetic resources of the available planting materials. There is still a possibility of unlocking the lack of tolerance, by applying modern biotechnology tools.

5.3 Biological control

Biological control using antagonistic microorganisms, alone or in combination with other control methods in integrated pest management programmes, may offer a possible solution to root-knot nematodes. The most studied group among the nematode-antagonistic organisms is that of the nematophagous fungi (Orisajo, 2009). Vesicular-arbuscular mycorrhiza fungi have been reported to increase plant tolerance and offset the growth reductions caused by *Meloidogyne arenaria* in groundnut (Carling et al., 1996). Inoculation of tomato plants with the mycorrhizal fungus *Glomus mosseae* suppressed gall index and the average number of galls/root system by 52% and 66%, respectively, compared with seedlings inoculated with *Meloidogyne javanica* alone (Al Raddad, 1995).

The association of mycorrhiza with cacao roots has been reported in Malaysia (Nadarajah, 1980), Brazil (Ezeta and Santos, 1981), Ecuador and Indonesia (Kramadibrata and Hedger, 1987). It has been reported that mycorrhiza likely do play a role in early cacao seedling growth, when the proportion of infected roots to total tree biomass is much greater than that in established trees (Kramadibrata and Hedger, 1987).

5.4 Organic soil amendments

The need for alternatives to nematicides has stimulated research work focusing on sustainable tactics for the management of plant-parasitic nematodes. The addition of organic amendments that stimulate the growth of antagonistic microorganisms or release toxins during decomposition has been advocated as an alternative method of nematode control. Some research works have focused on the utilization of animal and industrial waste products for the management of nematodes. The addition of nitrogenous soil amendments results in the formation of ammonia that has nematicidal properties. A range of organic amendments, including poultry litter or other animal wastes, have provided considerable protection for plants against plant-parasitic nematodes (Koenning et al., 2003).

Nutritive and nematicidal effects of poultry litter and compost on cacao seedlings have been reported (Orisajo and Afolami, 2009; Meyer et al., 2011; Orisajo et al., 2012). The nutrient content (%) of poultry litter that was used to amend soil by the latter researchers was as follows: N 3.24, C 27.9, P 1.86, K 0.05, Ca 1.40, Mg 0.64, Na 0.02, Mn 0.08, Fe 0.10, Cu 0.09 and Zn 0.04 (Orisajo et al., 2012). The C:N ratio was 9:1. In general, amending the soil with a low C:N ratio (less than 20:1) substrate resulted in an abundance of enrichment–opportunist antagonistic microbes (Wang et al., 2004, 2006) and rapid mineralization of N in the form of NH_4^+ or NO_3^- for the absorption and uptake by plant roots (Powers and McSorley, 2000).

Bacteria and fungi identified from the litter-amended plots were species commonly found in soil. These included *Arthrobacter globiformis*, *Arthrobacter ilicis*, *Arthrobacter pascens*, *Bacillus macerans*, *Bacillus thuringiensis*, *Cellulomonas fimi*, *Pasteuria penetrans*, *Pseudomonas fluorescens*, *Pseudomonas corrugata* and *Pseudomonas vesicularis*. *Arthrobacter*, *Bacillus* and *Pseudomonas* were among the most frequently identified species. Fungi identified included *Drechslerella* (*Arthrobotrys*) *dactyloides*, *Aspergillus niger*, *Aspergillus flavus*, *Trichoderma harzianum*, *Trichoderma viride*, *Paecilomyces lilacinus* and *Pochonia chlamydosporia*. Bacteria are quite diverse in nutritional requirements, and several species identified in the genera *Arthrobacter*, *Bacillus* and *Cellulomonas* have the ability to use ammonium salts or nitrates as their sole source of nitrogen. Poultry litter has high levels of nitrogen in the organic and inorganic forms that would be readily available to these organisms. Bacteria, such as *B. macerans*, use ammonia to fix N_2 under anaerobic conditions. In addition, the genus *Cellulomonas* has been isolated from substrates which are high in cellulose and has the ability to reduce nitrate to nitrite (Edwards and Daniel, 1992).

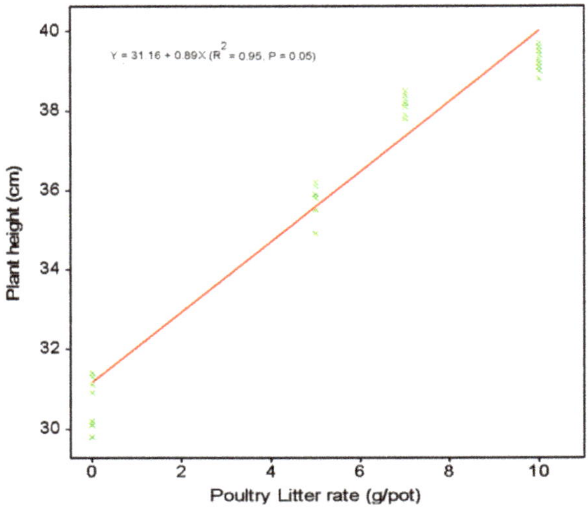

Figure 5 Influence of poultry litter applications on plant height of nematode-infested cacao seedlings.

In the nursery, *Meloidogyne incognita* population densities decreased in response to increasing rates of poultry litter in amended soil (Fig. 5). This result was consistent with the report by Riegel and Noe (2000) showing that the application of litter 14 days before planting was optimal for effects on nematode population densities. Plant height, leaf area, dry shoot and root weights of cacao seedlings were all stimulated by the addition of litter in *Meloidogyne incognita*-infested soil. Agronomic studies on the plant growth-enhancing effects of poultry manure have shown an increase in the stem and leaf biomass of plants (Mohanty et al., 2006). Phosphorus uptake is enhanced by the application of poultry manure, and this could be attributed to the greater leaf biomass yield in poultry-manure-treated soil (Mohanty et al., 2006).

The addition of poultry litter to soil leads to a better environment for root growth. This enhances the utilization of soil nutrients, as a consequence of which the nematode damage might have been markedly reduced (Van Den Boogert et al., 1994). A significant increase in the root mass in soil amended with poultry litter and subsequent decrease in the number of nematodes may be observed (Fig. 6), responsible for the increased growth of the cacao seedlings (Orisajo and Dongo, 2007, Orisajo et al., 2012). This is in agreement with previous reports (Van Den Boogert et al., 1994) since any decrease allows fewer seedlings disturbances, resulting in an unhindered growth.

Similarly, poultry litter compost extracts reportedly inhibited hatch of *Meloidogyne incognita* eggs and immobilized infective second-stage juveniles (J2) on cacao seedlings, indicating that chemical compounds might account for the compost action against nematodes (Meyer et al., 2011). Plant vigour was also enhanced, confirming the nutritive effects of the compost.

Results with the poultry litter compost indicate that there is a potential for use of this amendment for suppressing root-knot nematode populations on cacao seedlings (Meyer et al., 2011). Local soils can be tested for optimal amendment rates. In addition, systems have been developed for large trees to be grown in buried containers (Hoitink and Ramos, 2004), with a potential for use of compost addition to soil, that is not infested

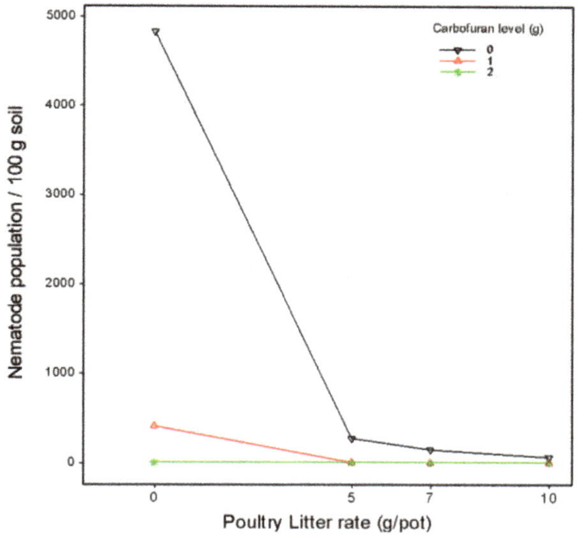

Figure 6 Effects of poultry litter on the soil populations of *M. incognita* on infested cacao seedlings.

with nematodes, thereby decreasing their damage while allowing for beneficial compost effects on trees (Meyer et al., 2011). Compost can be prepared to eradicate any plant-parasitic nematodes that might already be present in the composting materials. For examples, species of *Meloidogyne* and most other tested nematodes were generally killed by compost with a temperature (constant or average) of 60°C for ca. two days (Noble and Roberts, 2004). For compost to be routinely beneficial, composting methods need to be established and followed so that the finished material becomes a reliable product.

6 Future trends and conclusion

Feeding the ever-increasing world population on a finite land and balancing the ecosystem health pressure are among the future challenges the world has to face. One way of meeting the food demand is by developing high-yielding crops (Good et al., 2004). These may be, however, limited by nutrient management in depleted soils and by changes affecting the food production systems (Baligar et al., 2001). Soil nutrient management has been a fundamental factor sustaining food production. A major concern with nutrient management for breeders and soil scientists has been the efficient use of fertilizers and nutrients, about which many advances have been made (Baligar et al., 2001; Good et al., 2004). For example, approximately 90 million metric tons of N is applied annually to crops worldwide (Good et al., 2004). Around 60% of N applied to wheat in developing countries is lost (Heng et al., 2006). Among other factors, such losses of N have negative economic, agronomic and environmental implications. In fact, over 170 million metric tonnes of fertilizers per year are applied globally to produce food and fibre, and to improve the nutritionally and structurally depleted capacity of soils (Heffer, 2013). However, there are few soil nutrient management recommendations that incorporate nematode-driven yield-limiting conditions.

Organic soil amendment is a nematode management option with nutritional implications, and numerous aspects of research would have practical applications in commercial agriculture for the amelioration of pest problems. Organic amendments alone and/or in combination with other tactics like resistance management may meet some of the food production demands. For example, with large livestock production entities, countries like Brazil, the USA and even Nigeria will have plenty of organic sources of fertilizer supply (Anon., 2003; 2004; Zasada and Tenuta, 2004).

Relationships among parasitic nematodes, soil environment, organic amendments and plant health are obviously complex and make it difficult to forecast the activities occurring in soil. Research efforts should be directed towards revealing these relationships. The observed increase in other soil-borne microbial populations using organic amendments may have an essential role in nematode-suppressive mechanisms. More detailed studies on microbial population dynamics in amended soils are needed to identify the specific roles that fungi and bacteria may have in nematode regulation.

The widespread distribution of nematode pests in many cacao-producing areas makes in-depth studies urgent if farmers are to be assisted. They currently need to overcome the frustration they face due to the failure of *Theobroma cacao* seedlings to establish in new and rehabilitated plantations. The pathogenic relationship of nematodes with several arable and tree crops has been investigated and reported in depth, but it is only recently that studies on pathogenic relationship with cacao are being carried out. Even then, the studies appear to be restricted to biology and effects of root-knot nematodes on cacao and exclude their interactions with other soil pathogens (Orisajo, 2009).

The economic importance of cocoa justifies that a wide range of approaches should be adapted to breeding and plant pathology research works. Therefore, it is pertinent to acknowledge the fact that breeding programmes (especially for disease resistance) in cacao can no longer neglect the effect of plant-parasitic nematodes, particularly *Meloidogyne* spp. (Orisajo, 2009). It is also very important to note that research and development agencies have not paid due attention to other biotic agents, concentrating efforts on only black pod disease, cocoa swollen shoot viral disease and further problems. It will, therefore, be useful to adapt a network approach for a regenerated research on the role of nematodes in cacao, especially with regard to the estimation of yield losses, distribution, pathogenicity and interaction with other biotic constraints.

7 Where to look for further information

1. ATTRA Sustainable Agriculture Nematodes: Alternative Controls https://attra.ncat.org/attra-pub/viewhtml.php?id=149
2. Cocoa Research Institute of Nigeria www.crin-ng.org
3. World Cocoa Foundation www.worldcocoafoundation.org
4. CropLife International http://www.croplife.org/
5. CAB International http://www.cabi.org/index.asp
6. International Cocoa Organisation (ICCO) http://www.icco.org/
7. Organic production - IFOAM http://www.ifoam.org/
8. United States Department of Agriculture (USDA) Nematology laboratory https://iapreview.ars.usda.gov/main/site_main.htm?modecode=80-42-05-65

8 References

Afolami, S. O. (1981a), 'Changes in nematode types and population levels in the first two years of cocoa establishment on a field inter cropped with yam, maize and cassava', In: *Proceedings, 8th International. Cocoa Research Conference*, Cartagena, Columbia, South America. 18–24 October 1981, pp. 301–6.

Afolami, S. O. (1981b), 'Symptoms of root –knot nematode infection on *Theobroma cacao* seedlings in Nigeria – a preliminary investigation', In: *Proceedings, 3rd Regional Conference of the International Meloidogyne Project*, IITA, Ibadan, Nigeria, pp. 148–56.

Afolami, S. O. (1993), 'The Effect of Basamid Granular (Dazomet) on nematodes parasitic on cacao seedlings in the nursery', In: *Proceedings, 11th International Cocoa Research Conference*, Bahia, Brazil, pp. 237–40.

Afolami, S. O. and Caveness, F. E. (1983), 'The frequency of occurrence and geographical distribution of plant-parasitic nematodes associated with *Theobroma cacao* in Nigeria', *Turrialba*, 33(1), 97–100.

Afolami, S. O. and Ojo, A. A. (1984), 'Screening of *Theobroma cacao* germplasm for resistance against a root- knot nematode – *Meloidogyne incognita* in Nigeria', In: *Proceedings, 9th International Cocoa Research Conference*, Lome, Togo. 12–18 February 1984, pp. 237–42.

Al Raddad, A. M. (1995), 'Interaction of *Glomus mosseae* and *Paecilomyces lilacinus* on *Meloidogyne javanica* of tomato', *Mycorrhiza*, 5(3), 233–6.

Anon. (2003), '*Managing Manure Nutrients for Crop Production*', Extension Bulletin PM1811, Iowa State University.

Anon. (2004), '*Fertilizer Use by Crop in Brazil*', FAO, Rome.

Asare-Nyako, A. and Owusu, G. K. (1979), '*Meloidogyne incognita* infection of cocoa seedlings', In: *Proceedings, 7th International Cocoa Research Conference*, Douala, Cameroun. 4–12 November 1979, pp. 457–61.

Badaru, K., Eran, E. B., Williams, J. A. and Eskes, A. (1999), 'A review of breeding work on disease resistance of cocoa in Nigeria', In: F. Bekele and M. End (Eds) *Proceedings of the International Workshop on the Contribution of Disease Resistance to Cocoa Variety Improvement*. International Group for Genetic Improvement of Cocoa, University of Reading, Reading, UK, pp. 167–71.

Baligar, V. C., Fageria, N. K. and He, Z. L. (2001), 'Nutrient use efficiency in plants', *Communication in Soil Science and Plant Analysis*, 32, 921–50.

Bartley, B. G. D. (2005), *The Genetic Diversity of Cacao and Its Utilization*, CABI Publishing, Wallingford, UK.

Bateman, R. (2008), *Pesticide Use in Cocoa: A Guide for Training Administrative and Research Staff*. Ist Edition. ICCO/IPARC, London, UK, 57p.

Bridge, J., Page, S. L. J. and Waller, J. M. (1982), *Plant Parasitic Nematodes and Diseases of Crops in the Santa Cruz Department of Bolivia*, UK Overseas Development Administration Report.

Brunt, A. A., Crabtree, K., Dallwitz, M. J., Gibbs, A. J. and Watson, L. (1996), *Viruses of Plants. Descriptions and Lists from the Wide Database*, CAB International, Oxon.

Campelo, A. M. F. and Galli, F. (1980), 'Patogenicidade de *Helicotylenchus dihystera* (Cobb) Sher em *Theobroma cacao* L.' *Revista theobroma*, 10(1), 7–14.

Campos, V. P. and Villain, L. (2005), 'Nematode parasites of coffee and cocoa', In: M. Luc, R. A. Sikora and J. Bridge (Eds) *Plant Parasitic Nematodes in Subtropical and Tropical Agriculture*. CAB International, London, pp. 529–79.

Carling, D. E., Roncadori, R. W. and Hussey, R. S. (1996), 'Interactions of arbuscular mycorrhizae, *Meloidogyne arenaria*, and phosphorus fertilization on peanut', *Mycorrhiza*, 6 (1), 9–13.

Caveness, F. E. (1967), *End of Tour Progress Report on the Nematology Project*, USAID/Nigeria project 620–11-110–050, Min. of Agric. And Natural Resources, Ibadan, 135p.

Corbett, D. C. M. (1961), *Report of Department of Agriculture 1959/1960, Part II*. Nyasaland, pp. 157–8.

Crozzoli, R., Lamberti, F., Greco, N. and Rivas, D. (2001), 'Plant parasitic nematodes associated with cocoa in Choroni, Cumboto and Cuyagua, Aragua State', *Fitopatologia Venezolana*, 14 (1), 5–12.

Dixon, W. B. (1961), 'Nematological investigations 1958–61', *Bulletin Ministry of Agricultural Lds Jamaica*, 59, 1–35.

Edwards, D. R. and Daniel, T. C. (1992), 'Environmental impact of on-farm poultry waste disposal - a review', *Bioresource Technology*, 41, 9–33.

Entwistle, P. F. and Caveness, F. E. (1963), *Nematology*, Annual Report, West Africa Cocoa Research Institute, 1961–62, pp. 113–14.

Epps, J. M., Young, L. D. and Hartwig, E. E. (1981), 'Evaluation of nematicides and resistant cultivar for control of soybean cyst nematode race 4', *Plant Disease*, 65, 665–6.

Ezeta, F. N. and Santos, O. M. (1981), Importancia da endomicorriza na nutricao mineral do cacauiero. *Revista Brasileira de Ciência do Solo*, 5, 22–7.

Fademi, O. A., Orisajo, S. B. and Afolami, S. O. (2006), 'Impact of plant parasitic nematodes on cocoa production (in Nigeria) and outlook for future containment of the problem', In: *Proceedings 15th International Cocoa Research Conference*, San Jose, Costa Rica.

Fluitter, H. J. and Mulholland, J. J. (1941), 'Gegevens, Verkregen biji het onderzoek naar de waardplanten van *Tylenchus coffeae'*, *Bergcultures*, 15, 1588–93.

Freire, F. C. O. and Monteiro, A. R. (1978), 'Nematoides da região Amazonica II: Ematóides parasitos e de vida livre associados a pimenta do reino (*Piper nigrum* L.) e ao cacaueiro (*Theobroma cacao* L.)', *Acta Amazonica*, 8, 561–4.

Gamliel, A., Austerweil, M. and Kritzman, G. (2000). 'Non-chemical approach to soilborne pest management-organic amendments', *Crop Protection*, 19, 847–53.

Ghesquière, J. (1921), 'Nouveaux parasites du cacoyer. Maladie vermiculire du cacaoyer *Tylenchus* (*Heterodera*) *radicicola* [Greef.] et relation avec la maladie du Diplodia', *Bullentin de Agriculture*, 12, 709–18.

Good, A. G., Skrawat, A. K. and Muench, D. G. (2004), 'Can less yield more? Is reducing nutrient input into the environment compatible with maintaining crop production?', *Trends in Plant Science*, 9, 697–05.

Heffer, P. (2013), *Assessment of Fertilizer Use by Crop at the Global Level 2010–10/11*, International Fertilizer Industry Association (IFA), Paris – France.

Heng, L. K., Moutonnet, P. and Baethgen, W. A. (2006), *Nuclear Techniques and DSSAT for Optimizing Nitrogen Fertilizer Application under Irrigated Wheat*, FAO, Rome.

Hoitink, H. A. J and Ramos, L. Z. (2004), 'Disease suppression with compost: History, principles and future', In: *Proceedings, SoilACE 1st International Conference on Soil and Compost Eco-Biology*, León, Spain. 15–17 September 2004, pp. 185–97.

ICCO (International Cocoa Organization). (2017), *ICCO Quarterly Bulletin of Cocoa Statistics*, Vol. XLIII, No. 1, Cocoa year 2016/17, https://www.icco.org/about-us/international-cocoa-agreements/cat_view/30-related-documents/46-statistics-production.html (accessed 15 April 2017).

Ichinohe, M. (1967), 'Plant protection in Japan. VII. Control of plant nematodes', *Agriculture Asia*, 5, 93–8.

Kerten, K. H. (1972), 'The purification and some properties of cocoa necrosis virus, a serotype of tomato black ring virus', *Annals of Applied Biology*, 71, 119–26.

Koenning, S. R., Edmisten, K. L., Barker, K. R., Bowman, D. T. and Morrison, D. E. (2003), 'Effects of rate and time of application of poultry litter on *Hoplolaimus columbus* on cotton', *Plant Disease*, 87(10), 1244–9.

Kramadibrata, K. and Hedger, J. N. (1987), 'Comparative studies on the mycorrhizal symbionts of cocoa in Ecuador and Indonesia', In: *Proceedings 10th International Cocoa Research Conference*, Santo Domingo, Dominican Republic.

Kumar, A. C., Viswanathan, P. R. K. and D'Souza, G. I. (1971), 'A study on plant parasitic nematodes of certain commercial crops in coffee tracts of south India', *Indian Coffee*, 35(6), 222–4.

Lamberti, F., Boiboi, J. B., Ciancio, A., Tuopay, D. K., Jimehez, E. A. and Elia, F. (1992), 'Plant parasitic nematodes associated with tree crops in Liberia', *Nematologia Mediterranea*, 20, 79–85.

Lamberti, F. and Roca, F. (1987), 'Present status of nematodes as vectors of plant viruses', In: J. A. Veech and D. W. Dickson (Eds) *Vistas on Nematology*, Society of Nematologists Inc., Hyattsville, Maryland, pp. 321–8.

Loof, P. A. A. and Sharma, R. D. (1980), '*Discocriconemella* species from Bahia state, Brazil (Nematoda:Criconematidae). *Mededelingen Fakulteit Landbouwwetenschappen Rijksunivérsiteit', Gent* 45, 795–805.

López, R. (1994), 'Morfología de *Dolichodorus minor* (Nemata: Dolichodoridae) asociado al cacao en el sureste de Costa Rica', *Agronomia Costarricense*, 18 (1), 87–92.

López, R., Salazar, F. L. and Asofeifa, C. J. (1980), 'Observaciones sobre la distribucion espacial de nematodos associados al cacao en Costa Rica', *Nematropica*, 10(1), 3–4.

Luc, M. and Coomans, A. V. (1993), 'Description of *Xiphinema abeokutae* n.sp. and redescription of *X. paulistanum* Carvalho, 1965, two rare species (Nemata: Longidoridae)', *Fundamental and Applied Nematology*, 16, 393–9.

Luc, M. and de Guiran, G. (1960), 'Les nematodes associes aux plantes de l'ouest Africain. Liste preliminaire', *L'Agronomie Tropicale Nogeut*, 15, 434–49.

Manço, G. R. (1969), 'Nematóides em plântulas de cacau', In: *Informe Technico 1968–69*, Itabuna, Bahia, Brazil, pp. 99–101.

Martin, G. C. (1961), 'Plant species attacked by root-knot nematodes (*Meloidogyne* spp.) in the Federation of Rhodesia and Nyasaland', *Nematologica*, 6, 130–4.

Meyer, S. L. F., Orisajo, S. B., Chitwood, D. J., Vinyard, B. T. and Millner, P. D. (2011), 'Poultry litter compost for suppression of root-knot nematode on cacao plants', *International Journal of Nematology*, 21(2), 153–62.

Mohanty, S., Paikaray, N. K. and Rajan, A. R. (2006), 'Availability and uptake of phosphorus from organic manures in groundnut (*Arachis hypogea* L.)-corn (*Zea mays* L.) sequence using radio tracer technique', *Geoderma*, 133, 225–30.

Nadarajah, P. (1980), 'Species of Endoganaceae and mycorrhizal association of Elaesis guineensis and Theobroma cacao', In: P. Nikola (Ed.), *Tropical Mycorrhiza Research*, Clarendon Press, Oxford, pp. 232–7.

Nicol, J. M., Turner, S. J., Coyne, D. L., den Nijs, L., Hockland, S. and Tahna Maafi, Z. (2011), 'Current nematode threats to world agriculture', In: J. Jones, G. Gheysen and C. Fenoll (Eds), *Genomics and Molecular Genetics of Plant-Nematode Interactions*, Springer, pp. 21–43pp., http://www.springer.com/978–94-007–0433-6 (accessed 12 April 2017).

Noble, R. and Roberts, S. J. (2004), 'Eradication of plant pathogens and nematodes during composting: A review', *Plant Pathology*, 53, 548–68.

Okeniyi, M. O., Afolami, S. O., Fademi, A. O. and Aikpokpodion, P. (2009), 'Evaluation of cacao (*Theobroma cacao* L.) clones for resistance to root-knot nematode *Meloidogyne incognita* (Kofoid & White) Chitwood', *Journal of Applied Biosciences*, 17, 913–21.

Orisajo, S. B. (2009), 'Nematodes of cacao and their integrated management', In: A. Ciancio and K. G. Mukerji (Eds), *Integrated Management of Fruit Crops and Forest Nematodes*, Springer, NL, pp. 119–34.

Orisajo, S. B. (2012), 'Integrated pest management', *Manufacturing Confectioner*, 92(5), 52– 63.

Orisajo, S. B. and Afolami, S. O. (2009), 'Amelioration of nematode parasitism on cocoa (*Theobroma cacao* L.) seedlings with poultry litter as soil amendment in the nursery and during field establishment', In: *Proceedings, 9th Conference of the African Crop Science Society*, Cape Town, South Africa, 9, pp. 687–90.

Orisajo, S. B., Afolami, S. O., Fademi, O. A., Okelana, M. A. O. and Atungwu, J. J. (2012), 'Effects of poultry litter on establishment of cocoa seedlings and plantain infected with parasitic nematodes', *International Journal of Research in Chemistry and Environment*, 2(4), 278–89.

Orisajo, S. B. and Dongo, L. N. (2005), 'Nematicidal potential of some indigenous plant extracts against root-knot nematode on cacao', *African Scientist*, 6(4), 129–34.

Orisajo, S. B. and Dongo, L N. (2007), 'Comparative effects of organic soil amendments and carbofuran on *Meloidogyne incognita* and cacao growth', *International Journal of Agriculture and Food Systems*, 1(1), 13–17 .

Orisajo, S. B. and Fademi, O. A. (2005), 'Influence of neem-fortified cocoa pod husks soil amendment on *Meloidogyne incognita* in cocoa', *African Scientist*, 6(3), 125–8.

Orisajo, S. B., Okeniyi, M. O., Fademi, O. A. and Dongo, L. N. (2007), 'Nematicidal effects of water leaf extracts of *Acalypha ciliata, Jatropha gossypifolia, Azadirachta indica* and *Allium ascalonicum* on *Meloidogyne incognita* infection on cacao seedlings', *Journal of Research in Bioscience*, 3(3), 49–53.

Pinochet, J. and Raski, D. J. (1976), '*Discocriconemella repleta* n.sp., and the male of *Criconemoides inusitatus* Hoffmann, 1974 (Criconematidae; Nematoda)', *Journal of Nematology*, 8, 327–35.

Powers, L. E. and McSorley, R. (2000), '*Ecological Principles of Agriculture*', Delmar Thomson Learning, Albany, NY.

Ram, A. (1973), '*Relatorio Annual de Fitopatologia*', Centro de Pesquisao do Cacau, Ilheus, Bahia, Brazil, pp. 24–6.

Razak, A. R. (1981), 'The economic importance and identification of root-knot nematode isolates of Malaysia', In: *Proceedings of the 3rd Research Planning Conference on Root-knot Nematodes, Meloidogyne spp.*, Region VI, 20–4 July 1981, Jakarta, Indonesia, pp. 31–9.

Riegel, C. and Noe, J. P. (2000), 'Chicken litter soil amendment effects on soilborne microbes and *Meloidogyne incognita* on cotton', *Plant Disease*, 84, 1275–81.

Saenz, E. J. (1969), 'Relacion entre el ataque de nematodes y la muerte subita del cacau (Theobroma cacao L.) en Bahia, Brazil', *Turrialba*, 19, 255–60.

Sasser, J. N. and Freckman, D. W. (1987), 'A world perspective on nematology: The role of society', In: J. A. Veech & D. W. Dukson (Eds), *Vistas on Nematology*, Society of Nematologists Inc, Hyattsville, Maryland, pp. 7–14.

Sharma, R. D. (1971), 'Nematode associated with cacao and rubber in Bahia, Brazil', *Revista Theobroma*, 1, 43–5.

Sharma, R. D. (1982), 'Nematodes associated with cocoa hybrids and clones in Bahia, Brazil', *Nematologia Brasileira*, 6, 85–91.

Sharma, R. D. and Loof, P. A. A. (1974), 'Nematódes of the cocoa region of Bahia, Brazil. III – Plant parasitic and free-living nematodes in the rhizospheres of six different plant species', *Revista Theobroma*, 4, 39–43.

Sharma, R. D. and Sher, S. A. (1973), 'Nematodes associated with coffee in Bahia, Brazil', *Archivos do Instituto Biologico de Sao Paulo*, 40, 131–5.

Sharma, R. D. and Sher, S. A. (1974a), 'Nematodes of the cocoa region of Bahia, Brazil. II. Occurrence and distribution of plant parasitic nematodes associated with cocoa (*Theobroma cacao* L.)', *Revista Theobroma*, 3, 17–24.

Sharma, R. D. and Sher, S. A. (1974b), 'Nematoides da regiao cacaueira do Espirito Santos, Brasil. I. Nematoides associates ao cacaueiro (*Theobroma cacao* L.)', *Revista Theobroma*, 4, 26–31.

Sosamma V. K., Koshy, P. K. and Sundararaju, P. (1980a), *Nematodes Associated with Cacao – A Review*, Technical Document No. 123, Plant Protection Committee for the South East Asia and Pacific region, 14p.

Sosamma V. K., Koshy, P. K. and Sundararaju, P. (1980b), 'Plant parasitic nematodes associated with cocoa', *Cocoa Growers Bulletin*, 29, 27–30.

Sudha, S. and Sundararaju, P. (2002), 'Occurrence and distribution of *Radopholus similis* (Cobb, 1893) Thorne, 1949 and other plant parasitic nematodes in arecanut-based cropping system in Kerala', *Indian Journal of Nematology*, 32, 219.

Taylor, C. and Taylor, L. (2006), 'Future trends in cocoa industry - a perspective', *Coffee & Cocoa International*, 33(3), 39–41.

Thorold, C. A. (1975), *Disease of Cocoa*, Clarendon Press, Oxford.

Van Den Boogert, P. H., Velvis, H., Hettema, C. H. and Bouwman, L. A. (1994), 'The role of organic matter in population dynamics of the endoparasitic nematophagous fungus *Drechmeria coniospora* in microcosms', *Nematologica*, 40, 249–57.

Wang, K. H., McSorley, R., Marshall, A. J. and Gallaher, R. N. (2004), 'Nematode community changes following decomposition of *Crotalaria juncea* amendment in litterbags', *Applied Soil Ecology*, 27, 31–45.

Wang, K. H., McSorley, R., Marshall, A. J. and Gallaher, R.N. (2006), 'Influence of organic *Crotalaria juncea* hay and ammonium nitrate fertilizers on soil nematode communities', *Applied Soil Ecology*, 31, 186–198.
Whitehead, A. G. (1969), 'The distribution of root-knot nematodes *Meloidogyne* spp. in Tropical Africa', *Nematologica*, 15, 315–33.
Zasada, I. A. and Tenuta, M. (2004), 'Chemical-mediated toxicity of N-Viro Soil to *Heterodera glycines* and *Meloidogyne incognita*', *Journal of Nematology*, 36, 297–302.

Chapter 15

Advances in pest- and disease-resistant cocoa varieties

Christian Cilas and Olivier Sounigo, CIRAD, France; Bruno Efombagn and Salomon Nyassé, Institute of Agricultural Research for Development (IRAD), Cameroon; Mathias Tahi, CNRA, Côte d'Ivoire; and Sarah M. Bharath, Meridian Cacao, USA

1 Introduction

The deforestation seen in most cacao-producing countries has now reached limits that are unsustainable. Securing cacao supplies will require better management of current cultivation. Such sustainable management will require rational control of the main threats to cocoa health: pests and diseases. Currently, agroforestry systems are developed and could help to prevent certain pest and disease constraints, but these systems need improvement, particularly the use of adapted plant material. Improving resistance in planting material is the main way of improving crop protection and reducing pesticide use.

The cacao plant (*Theobroma cacao* L.) is a neotropical, small, evergreen tree native to South America (Motamayor et al., 2002). This plant, formerly included in the Sterculiaceae family, has been recently placed in the Malvaceae family (Alverson et al., 1999) and is grown for its fruits, known as cacao pods, although botanically these fruits are actually berries (von Balthazar et al., 2004). Since its discovery by Europeans after the conquest of the New World, and subsequent introduction into Africa in the nineteenth century, its cultivation has increased significantly. The first attempt to cultivate cacao outside the American continent was made in 1670 when a plant from Mexico was introduced into the Philippines (Blanco, 1837). In West Africa the first material was introduced in 1822 through São Tomé and Príncipe and was of 'lower' Amazonian origin (Bartley, 2005). Today, Africa is the leading producer of cocoa, accounting for around 74% of world

http://dx.doi.org/10.19103/AS.2017.0021.22

production, with Côte d'Ivoire alone accounting for 40% of world produce. The pods contain seeds, whose surrounding mucilage (pulp) is fermented and then the seeds are dried to give fermented dried cocoa, the raw material used to make chocolate. The cacao tree is a small tree whose fruits grow on both the trunk and branches. The fruits arise from the pollination of flowers grouped in flower cushions (a condition known as cauliflory). A distinction is traditionally made between three major groups of cultivated cacao trees: 'Criollo', 'Forastero' and 'Trinitario' (Cheesman, 1944). More recently, other classifications have been proposed based on molecular markers. As opposed to the traditional groups recognized with *T. cacao*, a new classification of cacao germplasm into ten major genetic clusters has been proposed (Motamayor et al., 2008). These clusters or groups include: Marañon, Curaray, Criollo, Iquitos, Nanay, Contamana, Amelonado, Purus, Nacional and Guiana. This new classification reflects more accurately the genetic diversity now available to breeders.

Cacao production is significantly affected by a diverse array of pests and diseases. It is estimated that over 40% of all cocoa produced is lost annually to just five diseases, with additional losses caused by insect and vertebrate pests (Flood et al., 2004). Mirids and Phytophthora pod rot (also called black pod, caused by several *Phytophthora* spp.) are global threats causing great losses, especially in West Africa. Cacao Swollen Shoot Virus (CSSV, of the genus *Badnavirus*) is a 'new encounter' disease in Côte d'Ivoire, the highest producing country (Kouakou et al., 2012). Cocoa pod borer (CPB) and Vascular Streak Dieback (VSD, causal agent: *Oncobasidium theobromae* Talbot & Keane) cause localized but significant problems in Southeast Asia and the Pacific (Flood et al., 2004). Witches' broom, caused by *Moniliophthora perniciosa* (Stahel), and frosty pod rot, caused by *Moniliophthora roreri* (Cif. & Par.), lead to significant losses in South and Central America (Evans et al., 1998; Krauss et al., 2003). Currently many of these pests and diseases are geographically restricted but international trade and the movement of germplasm constitute a serious threat with regard to the spread of pests and diseases. Should witches' broom or frosty pod (South and Central America), *Phytophthora megakarya* Brasier & Griffin or CSSV (Africa), or VSD or CPB (Southeast Asia) be introduced into other cocoa-producing areas, this could lead to serious problems for many smallholder cocoa farmers (Flood et al., 2004).

One of the best ways to deal with pest and disease attacks is to increase the resistance level of the cultivated material. However, breeding for resistance is often complicated due to several difficulties: (1) the assessment of resistance is, most of the time, difficult; (2) the breeding process is long, particularly when pest and disease attack fruits and (3) resistances are often overcome after several years.

2 History of diseases and pests of cocoa

The cacao tree is native to the American continent. Some of the diseases that affect cocoa are from the same area and therefore have co-evolved with it, as is the case for witches' broom caused by *Moniliophthora perniciosa* and frosty pod rot, also known as Moniliasis, caused by *Moniliophthora roreri*. These diseases pose real threats to worldwide cocoa production; they explain in part why the leading producers of cocoa are not found in the American continent but in the African continent. Diseases of perennial crops are often responsible for migration of these crop cultures. The example of the rubber tree is particularly illustrative: the South American leaf blight caused by *Pseudocercospora ulei* (formerly *Microcyclus ulei*) prevents the development of this crop on the American continent, and the main producing continent of natural rubber is nowadays Asia, with Thailand being the biggest producer in the world. Similarly, coffee, originally from Africa, is mainly cultivated in the Americas. Two serious diseases attacking cocoa in Africa are the coffee wilt disease caused by *Fusarium xylarioides* and coffee berry disease caused by *Colletotrichum kahawae*. Disease prevalence does not always fully explain the migration of certain crops, but diseases are often key criteria involved in understanding the current distribution of crops, particularly perennial crops. Plant populations are often genetically polymorphic for resistance to pathogens. The effectiveness of this resistance is limited because the pathogens are, in turn, polymorphic for virulence genes that can evade plant resistance, particularly when plant and pathogen populations co-evolve (Frank, 1992).

However, some of the major diseases affecting cocoa are from other continents, implying that 'host jumps' occurred; that is to say, a disease from certain native plant species moved to new species cultivated nearby, like cocoa in Africa or in Asia. This is the case of the CSSV which exists only in Africa; the African plants hosting the virus are probably *Commelina* spp. In Asia, VSD caused by *Oncobasidium theobromae* is one of the main constraints on cocoa farming, especially in Indonesia. VSD, which is found in other plants like forest trees, is also adapted to cocoa. There was no co-evolution between the diseases and the new host plant, cocoa tree (Table 1).

Cocoa is attacked more frequently by insect species that did not originally exist on the American continent, as seen with mirids on cocoa in Africa (*Sahlbergella singularis* and *Distantiella theobroma*) or the CPB (*Conopomorpha cramerella*) in Southeast Asia. In fact, there is no particular insect damaging cocoa trees in the Americas despite the number and diversity of insects in the Amazonian forest (Table 1).

Table 1 Most important cacao pests and diseases

Region	Disease/pest
West Africa	Black pod (mainly *Phytophthora megakarya*) Cocoa Swollen Shoot Virus (*Badnavirus*) Mirids (*Sahlbergella singularis* and *Distantiella theobroma*)
South America	Witches' broom (until Panama) (*M. perniciosa*) Moniliophthora pod rot (except Brazil) (*M. roreri*) Black pod (mainly *P. palmivora*) Mirids
Mesoamerica	Moniliophthora pod rot (*M. roreri*) Black pod (*P. palmivora*) Mirids
Southeast Asia	Vascular streak dieback (*Oncobasidium theobromae*) Cocoa pod borer (*Conopomorpha cramerella*) Mirids (*Helopeltis* spp.)

Source: ICCO (2015), http://www.icco.org/about-cocoa/pest-a-diseases.html.

3 Assessment of resistance

One of the most difficult challenges associated with breeding for resistance involves the development of an effective assessment system that is relatively easy to execute and highly replicable. How should we assess the level of plant resistance for each disease? The assessment varies as a function of the affected organs, and of the expression of the disease.

In natural epidemics, it is generally not possible to predict the resistance value of plants because of the other factors influencing the behaviour of the plants, for example, the environmental factor(s) having an impact on disease level and the phenological cycle which is not the same during experimental observation in the field.

For instance, certain authors propose the use of the area under the disease-progress curve as a quantitative measure of disease resistance (Jeger and Viljanen-Rollinson, 2001). In the case of virus diseases like CSSV, quantification is difficult and the time taken for symptoms to manifest after inoculation is a possible measure.

The rest of this chapter will aim to focus on the advances made in developing resistant varieties against *Phytophthora* spp. This disease was selected for research because of its widespread nature, the fact that it is highly studied across every infected region and a range of genotypes. The successes and failures of the research findings can therefore be used in one way or another to guide approaches for the other pest and disease scenarios.

4 Breeding for *Phytophthora* disease resistance

Selecting cacao trees less susceptible to black pod rot for agroforestry systems is a priority objective for reducing disease impact in the various producing countries, especially in Africa where the species *Phytophthora megakarya* exists. The search for cacao trees displaying total resistance to this disease has so far been unsuccessful, and numerous authors suggest that differences in reaction to *Phytophthora* spp. arise from partial, probably, polygenic resistance (Tan and Tan, 1990). In addition, it has been demonstrated for *Phytophthora megakarya* (Berry and Cilas, 1994) and for *Phytophthora palmivora* (Tan and Tan, 1990; Cilas et al., 1999) that transmission of the field resistance, under natural infection conditions, is mainly additive. Different ways of assessing resistance of plant material have been studied (Blaha, 1974), with observation of the field performance of trees under natural infection conditions, and artificial inoculation tests on pods or leaves being the most common. In this section, we shall be concentrating particularly on the genetic parameters of resistance assessed in the field.

4.1 Field resistance heritability

These results were obtained from mating designs set up in three countries:

- a 6 × 6 complete diallel, planted at the Barombi-Kang station (IRAD) in Cameroon in 1974.
- a 12 × 12 triangular diallel, planted at the Tové station (CRA/F, ex IRCC) in Togo in 1987.
- a 16 ♀ × 4 ♂ factorial mating design (North Carolina 2 Design), set up at the Bingerville station (CNRA) in Côte d'Ivoire in 1978.

The cacao trees observed in Cameroon were derived from a 6 × 6 complete diallel design (without the selfs). This trial, which was planted at the Barombi-Kang station in 1974, comprised six blocks. Each block contained 12 trees per family set out totally at random, at a density of 1330 plants/ha. The six parents used were: SNK 10, UPA 134, IMC 67, ICS 95, SNK 413 and ICS 84. SNK 10 and SNK 413 were local Trinitarios and ICS 95 and ICS 84 were Trinitarios from Trinidad selected by Imperial College, while UPA 134 was a clone derived from Upper Amazon Forastero (UAF) progeny from Ghana and IMC 67 was an Upper Amazon collected at Iquitos (Peru).

The trees observed in Togo were obtained from a 12 × 12 triangular diallel design (without the selfs). The plot, which was planted in 1987 in the Litimé region, comprised two blocks. Each block contained six trees per family set out totally at random, at a density of 1330 plants/ha.

The 12 parents used were:

SCA 6	(UAF of wild origin)
IMC 67	(UAF of wild origin)
NA 32	(UAF of wild origin)
T60/887	(UA derived from cross Pa 7 × Na 32)
T85/799	(UA derived from cross IMC 60 × Na 34)
T86/45	(UA derived from cross Pa 35 × Pa 7)
UPA 134	(Progeny of UA selected in Ghana)
IFC 5	(L.A. – Lower Amazon Forastero – selected in Côte d'Ivoire)
SNK 64	(L.A. selected in Cameroon)
UF 676	(Trinitario selected in Costa Rica)
ICS 40	(Trinitario from Trinidad selected by ICTA)
ICS 100	(Trinitario from Trinidad selected by ICTA)

The cacao trees in Côte d'Ivoire were obtained from a factorial mating design between 16 Upper Amazon female parents and 4 Lower Amazon male parents. The trees of the different crosses were planted in a total random design of plots comprising a single individual.

The 16 female parents used were:

P 7	(U.A. of wild origin)
NA 79	(U.A. of wild origin)
SCA 6	(U.A. of wild origin)
IMC 67	(U.A. of wild origin)
PA 150	(U.A. of wild origin)
NA 32	(U.A. of wild origin)
PA 35	(Poorly identified Trinitario)

PA 7	(U.A. of wild origin)
IMC 78	(U.A. of wild origin)
T60/887	(U.A. derived from cross Pa 7 × Na 32)
T79/501	(U.A. derived from cross Na 32 × Pa 7)
T79/416	(U.A. derived from cross Na 32 × Pa 7)
T79/467	(U.A. derived from cross Na 32 × Pa 7)
T63/971	(U.A. derived from cross Pa 35 × Na 32)
T63/967	(U.A. derived from cross Pa 35 × Na 32)
T85/799	(U.A. derived from cross IMC 60 × Na 34)

The four male parents used were:

IFC 1	(L.A. selected locally)
IFC 2	(L.A. selected locally)
IFC 5	(L.A. selected locally)
IFC 15	(L.A. selected locally)

Each trial tree was observed during the fruiting period (May–November), for seven years in Cameroon (1986, 1987, 1988, 1989, 1990, 1995 and 1996) and in 1991 in Togo.

Rotten pods (affected by black pod), wilted pods (physiological desiccation), rodent-damaged pods and healthy ripe pods were counted each week. A sanitary harvest was carried out during the counting operations. The topographical situation of the different trees was also recorded in Togo, so that pod rot distribution in the plot could be visualized.

Losses due to *Phytophthora* spp. were estimated in relation to potential production by the formula:

Pod rot rate, Prr = (number of rotten pods)/(number of rotten pods + number of ripe pods + number of healthy pods on last count).

In Côte d'Ivoire, the number of rotten pods and the number of healthy ripe pods were recorded on each harvesting round during the fruiting period. For each of the trial trees, the total number of pods, production in kg of fresh beans and the pod rot rate were estimated from nine years of observation.

After adjusting for any block effects, diallel analyses were carried out using Keuls and Garretsen's general method (Keuls and Garretsen, 1977) adapted to unbalanced mating designs, possibly with missing crosses:

$$Pijk = \mu + gi + gj + mi - mj + sij + rij + Eijk$$

where:

gi: general combining ability of parent i
mi: general reciprocal effect (or maternal effect) of parent i
sij: specific combining ability of pair ij
rij: specific reciprocal effect of pair ij.

The different characteristics studied were:

- potential production: total number of pods formed, log-transformed to satisfy conditions for application of analysis of variance models: Log(yp);
- actual production: number of ripe pods harvested, log-transformed: Log(y);
- pod rot rate: prr.

Analysis of the factorial mating design in Côte d'Ivoire was based on the 'North Carolina 2' model:

$$Pijk = \mu + fi + mj + sij + Eijk$$

where:
fi: effect of female parent i
mj: effect of male parent j
sij: interaction or specific combining ability of pair ij.

The different traits studied were:

- potential production (total number of pods formed, log-transformed, log(yp);
- actual production (fresh bean weight), yb;
- pod rot rate, prr.

Narrow- and broad-sense heritabilities were estimated for each of the mating designs studied (Cilas, 1995). Lastly, phenotypic, genetic and environmental correlations were calculated between the pod rot rates and the production variables.

The trial analyses were mostly carried out with Diogene (formerly OPEP) genetics software (Baradat et al., 1995).

4.2 Trial analyses

4.2.1 Cameroon

There were substantial pod losses due to rodents, with average rates of 20–50%.

The block effects were highly significant for all the characteristics and years, indicating that the blocks controlled some major environmental effects. The complete diallel analyses are shown in Table 2. The genetic effects were limited for potential production: Log(yp); this trait therefore appeared to have fairly low heritability. On the other hand, the GCA effect was substantial for the actual production characteristics: Log(y), and especially for the pod rot rate: prr.

The GCAs of the different parents for the pod rot rate in the field could be estimated and compared by a multiple comparison of means test (Table 5).

4.2.2 Togo

The diallel analyses were conducted after adjusting the data to account for block effects due to the design of the trial (Table 3). These analyses involved actual and potential production (Log-transformed) and the pod rot rates per tree.

As found previously, the GCAs were preponderant, meaning that these characteristics were primarily transmitted additively. However, the effects relative to the reaction to black pod were less marked than in the Cameroon trial, probably because the observations were only carried out over one year.

Table 2 Analysis of the complete diallel for the different variables considered in Cameroon

	Variables		
Effects	*Log(y)*	*Log(yp)*	*prr*
GCA F	4.126	1.465	11.976
α (%)	(0.114)	(19.779)	(<0.001)
SCA F	2.644	2.847	1.604
α (%)	(0.516)	(0.271)	(10.894)
GRE F	5.231	4.136	3.818
α (%)	(0.012)	(0.112)	(0. 213)
SRE F	0.448	0.434	1.245
α (%)	(92.250)	(93.014)	(25.741)

GCA: General combining ability, SCA: Specific combining ability, GRE: General reciprocal effect (maternal), SRE: Specific reciprocal effect.

Table 3 Analysis of the triangular diallel in Togo

	Variables		
Effects	*Log(yp)*	*Log(y)*	*prr*
GCA F	4.27	5.23	2.73
α (%)	(<0.001)	(0.001)	(0.20)
SCA F	1.71	1.73	1.01
α (%)	(0.176)	(0.145)	(45.17)

The GCAs of the different parents could be estimated and compared by a multiple comparison of means test (Table 4).

4.2.3 Côte d'Ivoire

The analysis of the factorial mating design is shown in Table 4. Production (in kg of fresh beans) and the pod rot rate per tree were analysed in line with the factorial model.

The female parent effects were the main variation factors, confirming primarily additive transmission of the characteristics considered.

Table 4 Analysis of the 16 ♀ × 4 ♂ factorial mating design (Côte d'Ivoire)

	Variables		
Effects	*Log(yp)*	*yb*	*prr*
Female F	5.86	8.15	13.57
α (%)	(<0.001)	(<0.001)	(<0.001)
Male F	1.37	1.24	3.37
α (%)	(28.40)	(29.35)	(1.82)
SCA F	1.27	1.63	0.912
α (%)	(13.59)	(1.18)	(62.13)

4.3 Classification of the parents in the different countries

Based on the above analyses, a parent classification was established, based on the pod rot rate observed in their progeny. For the diallels, it involved a multiple comparison of the estimated GCAs per parent. For the factorial design in Côte d'Ivoire, independent classifications for the female and male parents are presented (Table 5).

The lower susceptibility to black pod rot due to parent UPA 134 was confirmed in Cameroon (Berry and Cilas, 1994). It should be noted that this classification, obtained from the GCAs, only involved a single inversion compared to that for the specific values of clones estimated in the clonal trial at the same experimental station (Berry and Cilas, 1994).

On the other hand, these classifications differed from those observed in pod inoculation tests (Blaha and Lotodé, 1976). The relative classifications of parents UPA 134 and IMC 67 were confirmed in the diallel trial in Togo. However, the Togo results suggest the need to use a more resistant parent than UPA 134 in Cameroon. The better performance of the Lower Amazons in Togo needs to be confirmed after several years of observations. The classifications obtained in Côte d'Ivoire tallied with those obtained in Togo, though the pathogen species was different. In fact, the superiority of parents SCA 6, T85/799 and T60/887 over parents NA 32 and IMC 67 was confirmed.

Narrow-sense and broad-sense heritabilities were estimated for the pod rot rate (Table 6).

Table 5 Classification of the different parents for their susceptibility to *Phytophthora* spp. using the Newman and Keuls test (5%)

Susceptible	Cameroon (3 years) *P. megakarya*	Togo (1 year) *P. megakarya*	Côte d'Ivoire (9 years) *P. palmivora*
↓		IFC 5 a	PA 150 a
		SNK 64 ab	SCA 6 a
		T86/45 ab	P 7 a
		T85/799 ab	T85/799 b
		T60/887 ab	T60/887 bc
		SCA 6 ab	T79/416 bc
		ICS 100 ab	T79/501 bc
	UPA 134 a	UPA 134 abc	PA 7 bcd
		ICS 40 abc	T63/967 bcde
		NA 32 abc	NA32 bcde
		UF 676 bc	
	IMC 67 b	IMC 67 c	IMC 67 bcde
	SNK 413 b		T63/971 bcde
	ICS 84 b		T79/467 bcde
	ICS 95 c		NA 79 cde
	SNK 10 c		IMC 78 de
			PA 35 e

a–e: means followed by different letters are significantly different (Newman–Keuls at 5 %).

Table 6 Individual heritability values for pod rot rate (prr)

	Pathogen species	Number of years	h_n^2	h_b^2
Cameroon	*P. megakarya*	7	0.155	0.195
Togo	*P. megakarya*	1	0.061	0.061
Côte d'Ivoire	*P. palmivora*	9	0.681	0.681

h_n^2 : Narrow-sense heritability; h_b^2 : Broad-sense heritability.

The broad-sense heritabilities were identical to the narrow-sense heritabilities in Togo and Côte d'Ivoire, and of the same magnitude in Cameroon. Primarily, additive-type heritability therefore seemed to govern transmission of these traits, which confirmed results from earlier studies (Tan and Tan, 1990; Berry and Cilas, 1994). The heritability values increased with the number of years taken into account. The accuracy of the individual values, hence the family values, was indeed better when observations covered a greater number of years. In that respect, the data for the design in Côte d'Ivoire should be considered the most reliable, whilst further analyses need to be carried out in Togo.

Phenotypic, genetic and environmental correlations were calculated for the potential production and pod rot rate variables. It was mainly a matter of assessing how the quantity of fruits produced on the trees affected disease expression in the field. These different correlations are shown in Table 7. It should be remembered that the phenotypic correlations corresponded to tree-by-tree correlations, and did not take into account the genetic structure of the study population. In the three designs, this population consisted of full-sib families connected to each other by half-sib relations. The genetic correlations corresponded to links between the trees that were explained by the family structure of the study populations. If the parents were homozygous, these genetic correlations would correspond to the correlations between the family means, and the environmental correlations would correspond to the correlations between the residuals, once the family effects were removed. The environmental correlations therefore corresponded to the correlations between trees, which could not be explained by kinship relations.

The phenotypic correlations between the production and pod rot rate variables were not stable according to site. This correlation was maximum for Cameroon, where the pod rot rates were highest: in that country, high-yielding trees were in fact more severely attacked irrespective of the family to which they belonged.

The genetic correlations were negative in Côte d'Ivoire and Togo, meaning that parental material that showed high production also demonstrated good disease resistance. It may be that the severely attacked families bore fewer pods due to attacks that might have occurred on young fruits or flowers. This correlation was positive in Cameroon, indicating

Table 7 Phenotypic, genetic and environmental correlations between pod load and pod rot rate (tolog and prr)

	Phenotypic	Genetic	Environmental
Cameroon	0.397	0.339	0.407
Togo	0.041	−0.278	0.296
Côte d'Ivoire	0.090	−0.236	0.449

that under severe attack conditions, it will be more difficult to select planting material that is both high-yielding and resistant.

The environmental correlations between pod rot rate and potential production were systematically positive. This means that the highest yielding trees in a given family also tended to be the most severely attacked. This correlation was undoubtedly due to secondary infections, from pod to pod, which increased in line with fruit density on the trees.

4.4 Index-based selection of individuals

The main aim of genetic improvement of cacao is to increase production per unit area planted. In addition, sustainable development of cacao cultivation involves, among other things, the provision of planting material that is less susceptible to diseases, so as to reduce phytosanitary inputs. With this dual objective in mind, it is necessary for *T. cacao* L. breeding programmes to take into account adversities that might affect production. Among them, pod rot caused by various species of the genus *Phytophthora* causes the greatest losses. With a view to disseminating high-yielding clonal material that is less susceptible to the disease, selection based on an index combining productivity and resistance to *Phytophthora* spp. has been undertaken in the experimental designs previously studied in Côte d'Ivoire and Cameroon (NDoumbé et al., 2001).

4.4.1 Côte d'Ivoire

The three selection criteria adopted for this study were:

- production, measured as the weight of pods produced per tree over nine years (yb);
- resistance to rot, assessed by the pod rot rate (prr);
- bulk, estimated by the canopy diameter at ten years (cd).

Under plantation conditions, cocoa trees are generally subjected to competition from the age of 5 to 6 years. It is therefore necessary to limit the bulkiness of the trees so as to lessen competition at the current planting densities of 1330 trees/ha. This is why we included a bulkiness aspect in the selection index.

This study therefore involved combined individual-family selection, for subsequent dissemination of clonal material. It was a multi-trait selection based on an index combining the three target traits already mentioned. Predictor traits were added to these criteria to improve index precision; the predictor traits adopted were girth diameter increase between 1 and 2 years (dg) and trunk circumference at 11 years (tc). This study used the (16 females × 4 males) factorial mating design used to evaluate the genetic parameters of resistance.

In the immature period, girth diameter increase between 1 and 2 years was therefore calculated for each tree in the trial (dg). In 1989, that is, 11 years after planting, tree vigour and bulk, that is, trunk circumference (tc) and canopy diameter in a horizontal plane (cd), were measured. Genetic prediction coefficients were used to quantify the efficacy of indirect selection, for example, selection efficacy for girth diameter increase (dg) against production (yb) (Table 8).

The genetic parameters previously estimated were used to construct a selection index with a view to selecting trees that perform well, for their dissemination in budding or cutting

Table 8 Genetic prediction coefficients and broad-sense heritability on the diagonal

	dg	*tc*	*cd*	*yb*	*prr*
dg	0.186				
tc	0.196	0.292			
cd	0.159	0.255	0.416		
yb	0.312	0.261	0.230	0.524	
prr	−0.138	−0.067	−0.035	−0.335	0.681

form. Total genetic values were used to identify high-yielding trees with low susceptibility to black pod rot, and with limited vegetative development comparable to the plot mean.

It was possible to test several ways of weighting the target traits. The choice of weighting system depended on the relative genetic gains desired for each of the traits and the quality of the index, estimated by r(h,i) (correlation between merit and the index). This correlation was a measurement of the stability of the calculated indexes.

First of all, we estimated the genetic gains obtained by performing the maximum genetic gain for each of the main two traits (yb and prr) (Table 9).

Selection with maximum genetic gain for prr involved substantial genetic gain, not only for the trait (−105.89% for a 1% selection rate), but also for production, due to the previously mentioned genetic correlation. This genetic progress corresponded to a reduction of more than half of the pod rot rates per tree.

Selection with maximum genetic gain for yb acted in the same way, obviously with a higher genetic gain for that trait. However, selection for production alone induced a

Table 9 Relative genetic gains for selection based on rot rate and production, with different selection rates

Trait	Weighting	10% rate	5% rate	1% rate
dg	0	15.69	18.45	23.83
tc	0	4.00	4.70	6.08
cd	0	6.62	8.01	10.35
yb	0	44.51	52.33	67.60
prr	−1	−69.73	−81.97	−105.89
				r(h,i) = 0.861
dg	0	25.60	32.44	41.91
tc	0	10.94	12.86	16.62
cd	0	23.61	27.76	35.86
yb	1	63.66	74.83	96.66
prr	0	−48.76	−57.32	−74.05
				r(h,i) = 0.824

r(h,i): correlation between the calculated index and the 'merit' (theoretical index, if the genetic values were known).

modest genetic gain for the vigour traits, notably the cd variable, which represented the ground area projection of the canopy of each tree. To select trees to be planted at the same planting density, it was therefore necessary to limit genetic gain for that trait. In addition, it is also advisable to balance the genetic gains for the two main traits: yb and prr. After several simulations, the choice of weighting coefficients was fixed in accordance with those objectives (Table 10).

This selection index was therefore chosen to best meet the selection objectives previously mentioned. The normality of index distribution was checked. The list of trees selected with a 1% selection rate is given in Table 11.

A selection index was therefore established from a factorial mating design in a process of combined individual-family selection based on total genetic values. It amounted to the selection of individuals for vegetative multiplication, which need, first of all, to be

Table 10 Relative genetic gains depending on the chosen weighting

Traits	Weighting	10% rate	5% rate	1% rate
dg	0	22.67	26.65	34.42
tc	0	6.59	7.75	10.00
cd	−1	5.99	7.04	9.09
yb	1	57.41	67.48	87.18
prr	−20	−57.88	−68.03	−87.89
				r(h,i) = 0.835

Table 11 List of index-selected trees at a 1% selection rate

Female parent	Male parent	Row no.	Tree no.	Index
SCA 6	IFC 1	18	33	2.387
SCA 6	IFC 15	7	3	2.398
P 7	IFC 5	6	24	2.491
SCA 6	IFC 15	7	25	2.505
SCA 6	IFC 2	16	21	2.520
T79/501	IFC 5	18	6	2.588
P 7	IFC 5	3	32	2.593
SCA 6	IFC 2	13	4	2.598
SCA 6	IFC 2	15	39	2.598
SCA 6	IFC 5	8	31	2.767
SCA 6	IFC 2	13	2	2.803
Na 32	IFC 5	4	21	2.947
SCA 6	IFC 2	10	37	3.081
SCA 6	IFC 5	10	41	3.392
PA 150	IFC 1	7	35	3.728

confirmed in a clonal trial. Substantial genetic progress can be expected for both the production trait and for less susceptibility to pod rot.

4.4.2 Cameroon

The purpose of this selection operation was to choose high-yielding planting material with resistance to pod rot, for propagation in cutting or budding form. It thus consisted of combined individual-family selection based on an index integrating total genetic effects (additive + dominance).

The two traits to be improved were therefore y (increase) and ppr (reduction), with yp being a trait that could be used as an associated predictor.

It was possible to test several ways of weighting the two target traits. The choice of weighting system depended on the relative genetic gains desired for each of the target traits and the quality of the index [estimated by r(h,i)]. In fact, h, called 'merit', corresponded to the theoretical value of the index (i), if the genetic values were known and not estimated. The correlation r(h,i) was therefore a measurement of the stability of the calculated indexes.

First of all, the genetic gains obtained by performing maximum genetic gain for each of the main two traits (y and prr) were estimated without using any predictor (Table 12).

Introducing the potential production trait as a predictor increased genetic gains and gave better correlations between merit and index (Table 13).

Table 12 Relative genetic gains for selection based on production and pod rot rate, respectively

Trait	Weighting	10% rate	5% rate	1% rate
y	1	27.42	32.11	41.48
prr	0	−3.67	−4.31	−5.57
				r(h,i) = 0.541
y	0	3.77	4.43	5.72
prr	−1	−24.92	−29.29	−37.84
				r(h,i) = 0.692

Table 13 Relative genetic gains for selection based on production and pod rot rate, respectively, with the introduction of *yp* as a predictor

Traits	Weighting	10% rate	5% rate	1% rate
y	1	27.43	32.24	41.65
yp	0	25.46	29.93	38.66
prr	0	−4.46	−5.25	−6.78
				r(h,i) = 0.543
y	0	4.21	4.95	6.39
yp	0	−9.46	−11.12	−14.36
prr	−1	−27.32	−32.11	−41.48
				r(h,i) = 0.758

Table 14 Relative genetic gains depending on the chosen weighting

Traits	Weighting	10% rate	5% rate	1% rate
y	1	17.63	20.72	26.77
yp	0	5.39	6.33	8.18
prr	−3	−23.21	−27.28	−35.24
				r(h,i) = 0.697

The coefficients were then weighted to best meet the selection objectives fixed by the breeders (Table 14).

This selection index gave worthwhile genetic gains for each of the target traits. The normality of index distribution was checked. The list of trees selected for a 1% selection rate is given in Table 15.

Index quality, which was measured by the correlation between merit and the index, increased with the use and the number of predictors. This situation was not a generality, as introducing many traits often led to problems with the inversion of variance-covariance matrices between phenotypic predictors. Associated predictors need to be added gradually (Vincourt and Gallais, 1981; Baradat et al., 1995), until the best prediction of merit (H) by the index (I) is obtained.

The index adopted, using an associated predictor trait, suggests that a genetic gain of 26.77% can be expected for the total number of pods and a genetic gain of −35.24% for the pod rot rate, with a 1% selection rate.

A selection index was therefore established from a diallel mating design in a process of combined individual-family selection based on total genetic values. It involved selecting individuals for vegetative propagation, which need, first of all, to be confirmed in a clonal trial. Substantial genetic progress can be expected for both the production trait and for less susceptibility to *Phytophthora megakarya*.

The selected trees were multiplied by budding and a confirmation trial was set up at the IRAD Barombi-Kang station in Cameroon. These trees will also have to be multiplied by the somatic embryogenesis technique, with a view to setting up multi-site trials in several countries.

Table 15 List of index-selected trees with a 1% selection rate

Female parent	Male parent	Block no.	Tree no.	Index
SNK 413	UPA 134	1	140	2.338
UPA 134	SNK 413	6	1606	2.371
UPA 134	IMC 67	4	943	2.380
UPA 134	SNK 413	3	377	2.397
IMC 67	UPA 134	2	726	2.467
UPA 134	SNK 413	3	693	2.541
SNK 413	UPA 134	4	887	2.594
UPA 134	IMC 67	3	347	2.667

4.5 Leaf inoculation test to assess resistance to *Phytophthora* at early stage

A leaf inoculation method was developed (Nyassé et al., 1995) to assess the resistance of cacao to black pod disease caused by *Phytophthora* spp. The leaf test is an artificial inoculation method that can be used to assess the resistance of genotypes, on a scale of 0–5. The test was first carried out on whole leaves, and later on leaf discs 15 mm in diameter, with a purpose of reducing the space required to compare genotypes. Young cacao leaves being more susceptible were selected to discriminate cocoa clones. The results obtained using this early screening test on clones strongly correlated to the results obtained using field observations. An advantage of this leaf test is that it can be applied as soon as leaves are available, for example, 8–10 weeks after sowing, compared to field observation on resistance that cannot be started before four years after planting.

The validity of this leaf inoculation test was confirmed when assessing the plant's resistance to *Phytophthora megakarya* in a cacao diallel mating design (Nyassé et al., 2002) and in local and introduced cacao varieties in Cameroon (Nyassé et al., 2003a). The results obtained confirmed that the transmission of resistance to pod rot caused by this fungus was mostly additive. The leaf disc test is also convenient to assess resistance to *Phytophthora palmivora* (Tahi et al., 2006).

Research on resistance to black pod disease carried out in Cameroon for 40 years has yielded a significant database. This database is currently exploited for the modelling of disease resistance levels of the major clones selected as candidate parents for the creation of new resistant and consequently more productive cacao varieties (Nyassé et al., 2003b). The basic knowledge developed from this local database has enabled ranking of the clones by resistance levels. Subsequent prediction of the yield gains expected from the use of different cacao clones or possible combinations (progenies) available on-station and in farmers' stands in different agro-ecological growing conditions is also possible.

The leaf disc test was improved in Côte d'Ivoire (Tahi et al., 2007). Particularly, the experimental conditions were optimized in order to standardize the test. As a standard method, it is recommended to collect 50- to 60-day old leaves growing under medium shade conditions in the early morning (07:00–09:00 am); it is also recommended to prepare leaf discs on the same day for inoculation at the end of the afternoon. However, when large experiments need to be undertaken, with a more extended time needed for leaf collection, the leaves should be kept overnight in the dark at 100% humidity with leaf discs being prepared and inoculated the following day. When these conditions cannot be fulfilled (for example, no leaves of the best age are available for all clones), the use of more replications in time will be needed to obtain reliable results.

4.6 The use of molecular markers

4.6.1 Genetic mapping of QTLs (quantitative trait loci) for black pod resistance in cocoa

Genome mapping studies were developed in several projects in order to precisely determine the genetic bases of resistance to *Phytophthora*. Using this approach it is possible to identify several genetic components, or quantitative trait loci (QTL), affecting a complex quantitative trait. The QTL approach could allow identification of the number of genetic factors involved in the resistance trait and localizing them in the cocoa genome (Lanaud et al., 2009).

From trials in Côte d'Ivoire, several significant QTLs of resistance evaluated by the % of rotten pods in the field were identified in the different parents: one QTL was located on chromosome 1 of UF676, another QTL was located on chromosome 10 of T60/887 and another QTL was located on chromosome 4 of IMC 78. In Trinidad, using a progeny derived from the cross between IMC 57 and CATONGO, QTLs for resistance to *Phytophthora palmivora* were detected in IMC 57. In Cameroon a QTL was detected for leaf resistance traits (*Phytophthora megakarya*) in UPA134 on chromosome 9 by interval mapping at an LOD of 4 and confirmed by Kruskal and Wallis at P = 0.0005 (Flament et al., 2001). Moreover, this QTL was also detected by variance analysis when the number of individuals was increased to 104. Another QTL was detected for fruit resistance traits in UPA123 on chromosome 2. This QTL was detected in two different sets of experiments, confirming its existence. The different regions identified for resistance to *Phytophthora* showed that a pyramiding of different resistance genes from various parents is possible between parents to improve their level of resistance. Marker-assisted selection (MAS) would facilitate this accumulation of resistance genes. These methods could allow an identification of candidate genes involved in disease resistances.

4.6.2 Genomic selection for black pod resistance in cocoa

Molecular breeding and, particularly, genomic selection (GS) are important strategies in developing varieties that can resist pests and diseases without the need for pesticides. The last ten years were particularly rich in scientific advances in the field of GS, which has been detailed in the seminal paper by Meuwissen et al (2001).

GS is a statistical approach that utilizes all available genome-wide markers simultaneously to estimate breeding values or total genetic values. GS consists of two steps: (1) estimation of the effects of all markers in a 'training data set', where the individuals are phenotyped and genotyped and (2) prediction of the breeding values of a 'candidate data set' by combining their marker genotypes with the estimates obtained in step (1).

All markers are included in the model to estimate their effects. The challenge is that there are usually more parameters to estimate (genetic marker values, knowing that hundreds or thousands of markers are used) than phenotypic data available for their estimation. The Bayesian approach is used for estimating the marker effects and its specificity is to model the data at two levels: at the level of data, like the best linear unbiased prediction regression, and at the level of the variances of the markers. Two main methods have been developed by Meuwissen et al. (2001): BayesA and BayesB. In these two methods, hierarchical models are applied with different prior distributions assuming unequal variances across the markers. In BayesA, the probability that the variance is equal to zero is infinitesimal. Whereas in BayesB another prior distribution is used, allowing a variance equal to zero with a high density. The results of Bayesian approaches have given more accurate predictions of breeding values, in some contexts, than the methods assuming an equal variance at each marker, which is an unrealistic assumption.

GS is now applied to select for *Phytophthora megakarya* resistance in Cameroon. The data will be analysed with several statistical models in order to choose the best one.

5 Conclusion

Breeding for resistance will never be over because of the perpetual evolution of pathogens. The aim of breeding is also to accumulate, in same or similar individuals, favourable traits like high yield, resistance for several diseases, adaptation to climate change and good quality of beans. Consequently, selection on index remains an indispensable route. Traditional breeding continues on cacao as for other crops. The challenge for the future will be to know how to integrate the new tools of molecular biology in the breeding processes. Two options could be explored: (1) MAS with an integration of the markers in a global index and (2) GS, if the cost of sequencing continues to decline.

6 References

Alverson, W. S., Whitlock, B. A., Nyffeler, R., Bayer, C. and Baum, D. A. 1999. Phylogeny of the core Malvales: evidence from ndhF sequence data. *American Journal of Botany*, Vol. 86, pp. 1474–86.

Baradat, P., Labbé, T. and Bouvet, J. M. 1995. Conception d'index pour la sélection réciproque récurrente. Aspects génétiques, statistiques et informatiques. In: *Traitements statistiques des essais de sélection: stratégies d'amélioration des plantes pérennes. Actes du Séminaire de Biométrie et Génétique Quantitative*. Montpellier, pp. 101–50.

Bartley, B. G. D. 2005. *The Genetic Diversity of Cacao and Its Utilization*. CABI Publishing, Wallingford, UK. 341pp.

Berry, D. and Cilas, C. 1994. Etude génétique de la réaction à la pourriture brune des cabosses de cacaoyers (*Theobroma cacao* L.) issus d'un plan de croisements diallèle. *Agronomie*, Vol. 14, pp. 599–609.

Blaha, G. 1974. Methods of testing for resistance. In: P. H. Gregory (Ed.), *Phytophthora Disease of Cocoa*. Longman, London, UK, pp. 259–68.

Blaha, G. and Lotodé, R. 1976. Un caractère primordial de sélection du cacaoyer au Cameroun: la résistance à la pourriture des cabosses. *Café Cacao Thé*, Vol. 20, pp. 97–116.

Blanco, M. 1837. *Flora de Filipinas*. Imprenta de Santo Thome, Manila, Philippines, 887pp.

Cheesman, E. E. 1944. Notes on the nomenclature, classification and possible relationships of cacao populations. *Tropical Agriculture*, Vol. 22, pp. 144–59.

Cilas, C., Verschave, P. and Berry, D. 1995. Recherche d'un index de sélection pour deux caractères (production et résistance à la pourriture brune des cabosses) chez le cacaoyer. In: *Traitements statistiques des essais de sélection, actes du séminaire de biométrie et de génétique quantitative*, Montpellier (France), 12–14 September 1994, pp. 333–41.

Cilas, C., Paulin, D., Clément, D. and Baradat, Ph. 1999. Sélection multi-caractères dans un plan factoriel de croisements en Côte d'Ivoire. Définition d'un index de sélection. In: *Proceedings 12th COPAL*, 20–25 October 1996, Salvador de Bahia, Brazil, pp. 411–16.

Evans, H. C., Krauss, U., Rios, R., Zecevich, T., and Arévalo, E. 1998. Cocoa in Peru. *Cocoa Growers' Bulletin*, Vol. 51, pp. 7–22.

Flament, M. H., Kébé, I. S., Clément, D., Pieretti, I., Risterucci, A. M., N'Goran, J. K. A., Cilas, C., Despréaux, D. and Lanaud, C. 2001. Genetic mapping of resistance factors to *Phytophthora palmivora* in cocoa. *Genome*, Vol. 44(1), pp. 79–85.

Flood, J., Guest, D., Holmes, K. A., Keane, P., Padi, B. and Sulistyowati, E. 2004. Cocoa under attack. In: J. Flood and R. Murphy (Eds), *Cocoa Futures*. CABI-FEDERACAFÉ, Chinciná, Colombia, p. 164.

Frank, S. A. (1992). Models of plant-pathogen coevolution. *Trends in Genetics*, 8(6), 213–19.

Jeger, M. J., and Viljanen-Rollinson, S. L. H. (2001). The use of the area under the disease-progress curve (AUDPC) to assess quantitative disease resistance in crop cultivars. *Theoretical and Applied Genetics*, 102(1), 32–40.

Keuls, M. and Garretsen, F. 1977. A general method for the analysis of genetic variation in complete and incomplete diallels and North Carolina II designs. Part 1: procedures and general formulas for the random model. *Euphytica*, Vol. 26, pp. 537–51.

Kouakou, K., Kébé, B. I., Kouassi, N., Ake, S., Cilas, C. and Muller, E. 2012. Geographical distribution of Cacao swollen shoot virus Molecular Variability in Côte d'Ivoire. *Plant Disease*, Vol. 96, pp. 1445–50.

Krauss, U., ten Hoopen, G. M., Hidalgo, E., Martínez, A., Arroyo, C., García, J., Portugués, A. and Sánchez, V. 2003. Manejo integrado de la moniliasis (*Moniliophthora roreri*) del cacao (*Theobroma cacao*) en Talamanca, Costa Rica. *Revista Agroforestería en las Américas*, Vol. 10, pp. 52–8.

Lanaud, C., Fouet, O., Clément, D., Boccara, M., Risterucci, A. M., Surujdeo-Maharaj, S., Legavre, T. and Argout, X. 2009. A meta-QTL analysis of disease resistance traits of *Theobroma cacao* L. *Molecular Breeding*, Vol. 24, pp. 361–74.

Meuwissen T. H. E., Hayes, B. J., and M. E. Goddard. 2001. Prediction of total genetic value using genome-wide dense marker maps. *Genetics*, 157(4): 1819–29

Motamayor, J. C., Risterucci, A. M., Lopez, P. A., Ortiz, C. F., Moreno, A. and Lanaud, C. 2002. Cacao domestication I: the origin of the cacao cultivated by the Mayas. *Heredity*, Vol. 89, pp. 308–86.

Motamayor, J. C., Lachenaud, P., Da Silva, E, Mota, J. W., Loor, R., Kuhn, D. N., Brown, J. S. and Schnell, R. J. 2008. Geographic and genetic population differentiation of the Amazonian chocolate tree (*Theobroma cacao* L). *PLoS One*, 3(10): e3311 (8p).

NDoumbé, M., Bieysse, D. and Cilas, C. 2001. Multi-trait selection in a diallel crossing scheme of cocoa. *Plant Breeding*, Vol. 120, pp. 365–7.

Nyassé, S., Cilas, C., Hérail, C. and Blaha, G. 1995. Leaf inoculation as an early screening test for cocoa (*Theobroma cacao* L.) resistance to Phytophthora black pod disease. *Crop Protection*, Vol. 14(8), pp. 657–63.

Nyassé, S., Despréaux, D. and Cilas, C. 2002. Validity of a leaf inoculation test to assess the resistance to *Phytophthora megakarya* in a cocoa (*Theobroma cacao* L.) diallel mating design. *Euphytica*, Vol. 123, pp. 395–9.

Nyassé, S., Efombagn M. I. B., Bouambi, E., Ndoumbè-Keng M Eskes, A. B. 2003a. Early selection for resistance to *Phytophthora megakarya* in local and introduced varieties in Cameroon. *Tropical Science*, Vol. 43(1), pp. 96–102.

Nyassé, S., Efombagn, M. I. B. and Eskes, A. B. 2003b. Selection for resistance to black pod disease and yield gains prediction by use of selected cocoa varieties in Cameroon. *Plant Genetic Resources*, Vol. 1(2–3), pp. 157–60.

Tahi, G. M., Kébé, B. I., N'Goran, J. K. A., Sangaré, A., Mondeil, F., Cilas, C. and Eskes, A. 2006. Expected selection efficiency for resistance to cacao pod rot (*Phytophthora palmivora*) comparing leaf disc inoculations with field observations. *Euphytica*, Vol. 149(1), pp. 35–44.

Tahi G. M., Kébé I. S., Sangaré A., Cilas C., Eskes A. 2007. Foliar resistance of cacao (*Theobroma cacao*) to *Phytophthora palmivora* as an indicator of pod resistance in the field: the effect of light intensity and time of day of leaf collection. *Plant Pathology*, 56(2), pp. 219–26.

Tan, G. Y. and Tan, W. K. 1990. Additive inheritance of resistance to pod rot caused by *Phytophthora palmivora* in cocoa. *Theoretical and Applied Genetics*, Vol. 80, pp. 258–64.

Vincourt, P. and Gallais, A., 1981. Application of index selection in maize breeding for silage. In: A. Gallais (Ed.), *Proceedings of 4th Meeting of the Section Eucarpia, 'Biometrics in Plant Breeding'*, Lusignan, pp. 285–94.

von Balthazar, M., Alverson, W. S., Schönenberger, J. and Baum, D. A. 2004. Comparative floral development and androecium structure in Malvoideae (Malvaceae sl). *International Journal of Plant Sciences*, Vol. 165, 445–73.

Part 4

Safety and sensory quality

Chapter 16

Improving best practice with regard to pesticide use in cocoa

M. A. Rutherford, J. Crozier and J. Flood, CABI, UK; and S. Sastroutomo, CABI-SEA, Malaysia

1 Introduction

Pesticides are undoubtedly beneficial to agricultural production. Lomberg (2001) stated that 'if pesticides were abolished, the lives saved would be outnumbered by a factor of around 1000 by lives lost due to poorer diets'. Primary benefits of pesticides include increases in yield due to decreased pest losses, while their secondary benefits include reducing the emission of greenhouse gases, for example, herbicides are less energy intensive then mechanical cultivation (Cooper and Dobson, 2007). Nevertheless, public perception of pesticides and pesticide residues is poor, with many groups surveyed overestimating the risk posed by pesticides (Hibbitt, 1990). This perception has continued into the twenty-first century as consumers have become increasingly aware of food safety issues following a number of high-profile 'food scares' related to traceability, bacterial contamination and pesticide usage.

This consumer pressure has been translated into policy changes for crop production, and cocoa production is no exception. For instance in the the European Union (EU), from 1 September 2008, European Commission (EC) regulation EC/396/2005 and its annexes concerning maximum residue levels (MRLs) of pesticides in products of plant and animal origin (EC/149/2008 and EC/839/2008) came into force in order to protect European

http://dx.doi.org/10.19103/AS.2017.0021.24

consumers (Rutherford, 2008). Under this regulation, all foodstuffs intended for human or animal consumption in the EU are subject to an MRL, defined as 'the maximum concentration of pesticide residue likely to occur in or on a specific food commodity after the pesticide has been used under GAP (good agricultural practice) as described on the label'. The regulation included raw tropical commodities such as cocoa beans. Similar regulations exist in other lucrative cocoa markets, such as Japan and the USA. However, one key challenge for producers and other stakeholders in the cocoa supply chain is the fact that different markets impose different regulations with regard to the same active ingredients and also use different approaches for residue analysis. In the EU and the USA, residue analysis is conducted on cocoa beans once the seed coat (testa) has been removed, as is the case when cocoa beans are processed to make chocolate. In Japan, however, analysis is conducted on whole beans with the seed coat intact. The latter is more likely to increase detection and rejection rates as contaminants can be present on the surface of the seed coat which would normally be removed before processing (Bateman, 2015).

Lack of harmonization in importing countries makes it much more difficult for smallholders in cocoa-producing countries to comply with these regulations. This is often further compounded by national regulatory bodies and trade organizations possessing insufficient knowledge of, and information on, a number of key areas namely: pesticide regulations; pesticide products and active ingredients; potential pesticide adulteration ('counterfeit' or 'fake' pesticides); and the actual supply and use of pesticides on the ground. The dynamic nature of regulatory changes as well as the changing situation through the supply chain emphasizes the need for discussion between all interested parties. There is a need for proactive dissemination of relevant and trusted information to all supply chain stakeholders, so that they can act quickly to avoid problems with compliance.

2 Pesticide use in cocoa-producing countries

Over 40% of all cocoa production is estimated to be lost annually to just five diseases (Flood and Murphy, 2004; Flood et al., 2004). When additional losses from insects and vertebrates are considered, it is clear that pesticide application, when conducted appropriately, is an important element of the 'management toolbox' against cocoa pests and diseases. However, when policymakers in importing countries first began to consider setting MRLs on imported tropical commodities such as cocoa, it was evident that there was very little information available about trade and actual use of pesticide products and active substances in producing countries or the extent of their usage along the supply chain of production, processing, transportation and storage. There was similarly very little information on counterfeit materials or the use of chemicals that were not registered for use in cocoa. Many chemicals are available specifically for the control of pests and diseases in cocoa. In producing countries, where the crop is largely cultivated by smallholder farmers whose resources are limited, chemical suppliers range from large registered companies in cities to individual input suppliers passing through villages on a moped or bicycle. The products being provided and the level of information communicated to and understood by the buyer can therefore vary markedly. Furthermore, monitoring and management of product supply and use by national regulatory authorities present a major challenge, with supporting information being scarce.

One key group of stakeholders within the supply chain are the providers of pesticides (input suppliers), from whom many producers seek advice as well as purchase products. This group should, in theory, be able to provide up-to-date information and advice on best practice to achieve compliance. Yet many have insufficient or inappropriate knowledge to provide effective advice to producers on appropriate pesticide selection, pesticide use and indeed alternative management options. Further, in some cases, input suppliers may actively promote inappropriate products due to market pressures; so pesticide end users do not receive the appropriate information to enable them to handle, apply and dispose of products in the correct manner.

The few studies on pesticide use in cocoa production would suggest relatively limited use by many smallholders. A study in Southern Cameroon concluded that over 50% of cocoa farmers used pesticides but that their use was limited by cost and availability (Sonwa et al., 2008). A study of pesticide use in Ghana found low adoption rates, for example, for mirid management with insecticides (10%) and black pod disease management with fungicides (7%) (Aneani et al., 2012). Nevertheless, Okoffo et al. (2016) analysed the levels of organochlorine pesticide (OCP) residues in fermented dried cocoa beans in Ghana as compared to the EU regulations; pesticide residues did not exceed EU MRLs. These authors suggested that there was a need for more regular monitoring of pesticide residues in cocoa and to enforce regulations. Oyekunle et al. (2017) similarly reported levels of OCP residues in dried cocoa beans in south western Nigeria; detected levels were within the MRLs set by the WHO/FAO. The authors also emphasized the need for greater monitoring of pesticides used in the crop.

3 Improving pesticide regulation

There is international support from trade organizations, national producing and importing governments to hold discussions and facilitate the harmonization of individual country standards. This includes both harmonizing standards on specific MRLs for active ingredient and the methods for analysing MRLs. Although this is taking place, it is a slow process and is currently only carried out on a substance-by-substance basis (Bateman, 2015). As cocoa-producing countries rarely export their cocoa to one single market, harmonization of standards internationally would make it much easier for cocoa farmers to comply with pesticide residue regulations.

An additional challenge is that, by virtue of improved detection methodologies, policymakers are continuously seeking to reduce permitted levels of pesticides in food, so regulations are frequently reviewed and amended with associated impacts on the producers and other stakeholders in the supply chain. For example, in 2016, it was proposed by the EU (ED(MEM) 1009 and ED(ME) 1015) to reduce the MRLs of allowable metalaxyl and metalaxyl M in cocoa from 0.1 to 0.01 mg/kg. However, these chemicals represent the most effective management of *Phytophthora megakarya* which is a major constraint to cocoa production in West Africa, and there are currently few alternatives for effective management. The European Cocoa Association (ECA) discussed its critical usage with the EU policymakers and as a result of those discussions, a WTO notification (G/SPS/N/EU/174) was published on 12 December 2016, with the European Commission deciding that the current MRL (0.1 m/kg) would be maintained for the present time. This illustrates the benefits of consultation and collaboration between the various stakeholders.

Several initiatives have been carried out in recent years in collaboration with producer countries (discussed in the case studies below) to support and empower national authorities with respect to strengthening pesticide regulations for use in cocoa and to encourage the authorities to enforce regulations more effectively. Many products are now pretested in country on cocoa and the production and dissemination of lists of pesticides recommended for use in cocoa will go some way in reducing the use of banned or inappropriate pesticides in cocoa. However, dissemination of this information to remote smallholder farmers is again a challenge that needs to be more fully addressed.

4 Developing best practice for pesticide use

In conjunction with the development of recommended pesticide lists, information products on best practices for cocoa production, post-harvest processing and transport have been developed and continue to be updated. Such resources provide quality information on best practice for pesticide use for producers, processors and others through the supply chain, including pesticide suppliers.

Best practice guidelines have been commissioned by the cocoa private sector through trade associations such as ECA/Association of Chocolate, Biscuits & Confectionery of Europe (CAOBISCO)/The Federation of Cocoa Commerce (FCC); Cocoa Beans: Chocolate & Cocoa Industry Quality Requirements; CAOBISCO/ECA/FCC (2015) and by other cocoa stakeholders such as the International Cocoa Organization (ICCO) and Pesticide Use in Cocoa: A Guide for Training Administrative and Research Staff (Bateman, 2015). A recent example of best pesticide practice being promoted in West Africa was the Sanitary and Phytosanitary (SPS) capacity building project (ICCO, 2014) and in South East Asia and the Pacific was the 'CocoaSafe' project (www.cocoasafe.org).

These initiatives highlight the need not only to provide up-to-date information and training to cocoa stakeholders but, crucially, to provide it in a format that stakeholders can understand and apply. Literacy and linguistic barriers are often encountered in developing countries; so translation of validated information is needed for all stakeholders. Hence the need for radio or TV broadcasts, as well as posters and flyers, in local languages arises. Discovery learning approaches such as farmer field schools can be useful tools, where the participants learn by doing, and have been widely recommended and implemented for commodities including cocoa for a number of years (Vos et al., 2003; David, 2005).

Another challenge for trainers in developing curricula and generally keeping information accurately updated, is the rate at which the content and nature of pesticide regulatory information can change. Thus, information on which pesticides are permitted to be used for any crop in any given market, which pesticides are available in any given producing country, which pesticides are registered for a given crop in each producing country and how these pesticides are actually being used at any given time or location – all of this can change quickly and training curricula will need to adapt to reflect these changes.

Once best practices are established, the role of training and dissemination of information often falls to extension service providers. Unfortunately, in many cocoa-producing countries, such services are under great pressure, largely due to the very large number of smallholder producers – many of whom are located in remote and inaccessible areas. Many major commodity producing countries have specific extension services that specialize in the individual commodity crop, which adds to the difficulty of reaching large numbers of remote producers. Alternative approaches are needed to reach producers and other supply chain participants. The increasing use of ICTs to enable overstretched extension staff and producers themselves to receive information directly is another option that is being explored in producing countries. One such information service 'CocoaLink' was an industry-sponsored service in Ghana, but as is seen in many countries, when projects come to a close or sponsorship is withdrawn, sustainability of the service is a challenge. Embedding such initiatives within national systems is likely to enhance longer-term sustainability. Handover and governments' buy-in need to be assured and incorporated when initiatives are in the planning stage.

An example of how cocoa extension services can be supported by general agricultural extension can be seen in Ghana through the Plantwise Programme. This CABI-led programme is a multi-donor plant health initiative currently running in 34 countries worldwide. Its key aims are to strengthen plant health networks in country and to provide farmers with practical advice on plant health issues (www.plantwise.org). The initiative is usually coordinated in country by the National Plant Protection Organisation or the Agricultural Extension Service Department. It builds on services which already exist in country and trains extension staff to offer more comprehensive advice to farmers through Plant Clinics, supported by a knowledge hub called the Knowledge Bank. In Ghana, Plantwise has helped to link general extension services with cocoa extension services and research to allow them reach a greater number of farmers with current validated information on management of cocoa pests and diseases and also agronomic practices.

Another major challenge is that although information on best practice is available including in local languages and even when training is available, adoption of best practice is often limited and the reasons may be a complicated mixture of sociological and financial factors. Sonwa et al. (2008) reported constraints in acquisition and application of pesticides in the southern Cameroon. Over 50% of the cocoa farmers used pesticides, but the high cost and lack of availability in rural areas are cited, respectively, by 65% and 55% of cocoa farmers as major constraints to using pesticides. Only 21% of cocoa farmers bought their pesticides in the village and even when pesticides were bought, many farmers failed to apply the pesticides correctly, so yields remained low. This suggests lack of sound advice and training support for these producers. If producers buy expensive pesticides and yields do not improve, they are unlikely to purchase these expensive inputs again.

Similarly, Aneani et al. (2012) studied adoption of the cocoa production technologies in Ghana as recommended to cocoa farmers by the Cocoa Research Institute of Ghana. Adoption rates had been low, for example: mirid management with insecticides (10%), black pod disease management with fungicides (7%), weed control manually or with herbicides (4%), planting hybrid cocoa varieties (44%) and fertilizer application (33%), leading to yield and production levels below those expected. The study revealed that the reasons for lack of adoption were complex and interrelated, with access to credit being a major factor particularly with female farmers. If a farmer owned a number of farms, he/she was more likely to adopt, while age and educational status were also important factors. Cocoa yields and prices were also important considerations for adoption decisions. Female farmers were more inclined to conduct weeding than male farmers, and women usually hired labourers for extra work, but that cost impacted on the total production costs. Labour availability was also cited as a factor in adoption as many younger people are seeking work outside cocoa production often in urban centres.

Undoubtedly, adoption of best practice including that of best practice with pesticides is an extremely complex problem and much is linked to the profitability of cocoa, access to labour, credit and knowledge. Yet as consumers become more and more concerned about pesticide residues in their foods, policymakers in consuming countries will seek to minimize pesticide residues in food with associated challenges of compliance in production countries. The pressure on producing countries will increase to have necessary regulation and in some cases legislation in place, to monitor the production and use of pesticides and to facilitate the implementation of best practice. Ultimately, countries will need to introduce and adopt legislation and practices to ensure issues relating to counterfeit pesticides and pesticide residues are adequately tackled and enforced. Failure to do so, and to ensure regulatory compliance, will otherwise lead to restricted access to lucrative markets.

5 Case studies: establishing baselines, raising awareness and promoting best practice

5.1 Case study 1: establishing a baseline

In 2006, a study entitled 'Safe cocoa, sustainable production: A concerted programme by the cocoa and chocolate industries to understand and address market changes regarding pesticide acceptability' was undertaken in West Africa. The project was funded by the Netherlands Ministry of Agriculture, managed by ECA/CAOBISCO and implemented by CABI together with the National Cocoa Research Organisations of Cameroon, Nigeria, Ghana and Ivory Coast. Each of these producing countries is a major supplier of cocoa for European and other markets. The aim of the study was to provide essential baseline data on the supply and use of pesticides in cocoa to inform and enable policymakers in both producer and consumer countries to set realistic recommendations and regulations.

Through structured questionnaire surveys, completed in direct consultation with key stakeholder groups within the in-country supply chains – producers, intermediary traders (local buying companies), exporters and input suppliers – it was possible to ascertain which pesticide products and related active ingredients were being provided and used for cocoa production and trade. Furthermore, data were gathered on who was using and who was providing each individual product, where they were used, for what purpose and in what manner with respect to product labelling, regulations and the use of appropriate equipment. The findings revealed that, across the four countries, more than 120 different pesticide products comprising a total of 35 active ingredients were used by farmers alone (3014 consulted) solely for cocoa production. Crucially, this work also identified areas where the provision and use of particular chemicals was deemed inappropriate. In many instances, products were officially registered and marketed legitimately, but were not intended for use on cocoa. In other cases, products had not been registered by the national authorities.

Based on the findings of this study, measures have subsequently been introduced to raise in-country awareness of areas of risk presented by chemical use, to modify or eliminate the use of certain products and to provide information and training to improve many aspects of pesticide practice as a component of good agricultural practice (GAP). Where a need exists to withdraw a pesticide, recommendations were provided for effective and affordable alternatives available to all supply chain stakeholders. Importantly, national authorities and members of the cocoa trade community have become better empowered to effectively negotiate with national and international regulatory authorities to ensure that new or amended regulations are appropriate and will not impact negatively on the sustainable supply of an important commodity (Rutherford, 2009, 2011; Rutherford et al., 2010).

5.2 Case study 2: raising awareness and promoting best practice

Following on from the initial baseline study described in Case Study 1, a project was launched in 2011 to build Sanitary and Phytosanitary (SPS) capacity and promote best pesticide practice in West Africa. The project to 'Mitigate the harmful effects of pesticide residues in order to maintain market access' was led by the International Cocoa Organization (ICCO) in partnership with the NARS (national agricultural research systems) from the West African cocoa-producing countries of Ghana, Ivory Coast, Nigeria, Cameroon and Togo, Crop Life, EDES/COLEACP, UNIDO and funded by the Standards and Trade Development Facility (STDF) and partners (ICCO, 2014).

The project helped create awareness among producers and other stakeholders in country about SPS standards as well as promoting best pesticide practice and helped build capacity in country to monitor and enforce adherence to SPS standards. The project also disseminated information on the nature of food safety concerns in cocoa and their implication for the cocoa trade in the wider sense. This much-needed raising awareness of food safety concerns was conducted through flyers, posters and radio and TV programmes. An information sharing website (https://www.icco.org/sites/sps/) was initiated to encourage regional interaction. There was an emphasis on enhancing capacity of participating countries with a self-assessment guide system developed to assist the countries to assess SPS risks at every stage of production, harvesting, transport, packaging, processing and distribution. This increased the capacity of stakeholders to comply with SPS regulatory requirements as well as with product quality requirements.

As a result, training modules on GAP and good warehouse practice were developed and trainers were trained as resource people to cascade knowledge to supply chain stakeholders. Training on anti-counterfeit measures and how to recognize fake, obsolete and banned substances was provided for a wide range of stakeholders including input suppliers, customs and immigration services as well as extension staff and agronomists. In addition, the project improved the capacity of the countries to carry out pesticide residues testing as well as testing for other harmful substances and contaminants in cocoa through acquisition of laboratory equipment and training of laboratory personnel. Thus, the project helped build capacity in country and helped producing countries to understand the responsibility that is being placed on them when it comes to their continued access to lucrative markets.

5.3 Case study 3: expanding the scope

Complementing the African SPS project, a similar initiative was launched in South East Asia and the Pacific in 2013. 'CocoaSafe' Capacity Building and Knowledge Sharing in SPS in Cocoa in South East Asia was led by CABI in partnership with in-country cocoa responsible organizations in Indonesia, Malaysia and Papua New Guinea, ICCO, Crop Life Asia, Mars, funded by STDF with co-financing being provided by partners. The main objective of the project was to improve the safety and quality of cocoa by strengthening SPS capacity along in-country supply chains and to reduce harmful residues and contaminants through implementation of best practices.

A training syllabus was developed and adapted to each country's needs. The scope of the training syllabus was expanded to include all aspects of cocoa agronomy, processing and storage to provide a comprehensive country-based cocoa manual that could be adopted for all subsequent cocoa training programmes and ensure awareness is raised on food safety issues including reducing contamination caused by pesticide residues, heavy metals, polycyclic aromatic hydrocarbons and mycotoxins.

A Training of Trainers approach was employed to train Master Facilitators who, in turn, trained staff from extension services and lead farmers groups, post-harvest processors and input suppliers. These facilitators were then able to pass this knowledge to their peers along the supply chain. Other awareness-raising materials such as best practice posters and videos were also produced in local languages by the country partners to make it easier for farmers and processors to understand the information provided. The training syllabus has been adopted by the project partners and is being scaled up across the countries.

In addition to the activities in the individual countries, a project website was established where all of the training materials and resources can be accessed (www.cocoasafe.org) which is linked to the ICCO Cocoa SPS Africa website. Indonesia and Malaysia have developed their own webpages on their institution websites in local language which will allow the information to be accessed by a wider audience and improve sustainability in going forward.

6 Conclusion

As outlined above, a number of major challenges exist that constrain or prevent the application of good pesticide practice as a component of GAP. Regardless of whether these are related to lack of information and resources, selection of inappropriate products, technical difficulties or failure to adhere to recommendations and regulations, they create unnecessary and unacceptable risks for those who may become exposed and to the environment. They also constitute a waste of valuable time and resources, including financial. In many instances, it may be possible to swiftly introduce and apply measures that will help to address these challenges, or to develop new measures. However, and as highlighted by Case Study 1 in particular, it is essential in the first instance to gain a clear understanding of existing practice and how this relates to, and perhaps conflicts with, what is required. Only then can specific problem areas be identified and measures be introduced that are appropriate, effective, available, practical, affordable and safe. The processes should not be undertaken in isolation but with consideration of the broader supply chain, as the needs and actions of a particular stakeholder group will have consequences for others. Changes in market regulations, for example, can and should influence the thoughts and actions of cocoa producers, traders, processor and retailers and will consequently make their life more difficult or perhaps more straightforward. Equally, the actions of producers may raise concerns and necessitate modification in the practice of immediate traders and processors.

As shown by the research in West Africa, the process of gathering baseline information on the actions and interactions of a multitude of diverse players is not straightforward and can be time-consuming. Once obtained, interpretation of data is also critical in terms of deciding what measures are necessary, how and when these should be introduced, for whom and by whom. Numerous possibilities exist for modifying and improving pesticide practice including, specifically, reducing or eliminating residues of active ingredients on the cultivated crop and crop produce. Examples of differing approaches, which were adopted in West Africa, include: modification of government policy; regulation, prohibition or replacement of pesticide products and substances; awareness-raising campaigns, through mass media and local stakeholder gatherings; increased and more targeted training; and provision of relevant information and advisory materials, such as pest identification aids and manuals on best practice. Each measure varies in its level of outreach to the intended beneficiaries, but once adopted a noticeable improvement in practice, associated decline in pest prevalence and crop damage and consequential increase in yield (and perhaps quality) should become observable. This should ultimately lead to other socioeconomic and environmental benefits, such as increased income and available cash, reduced labour, food security, safer food, improved health and education and an increase in natural diversity as components of an overall improvement in quality of life. These benefits may be felt well beyond those directly involved with pesticides and perhaps extend nationally and regionally. Inclusion of measures referred to here with non-chemical pest management options may reap further rewards, although the resources required for the latter and the management package in its entirety must also be given careful consideration.

7 Future trends

Research on cocoa management in recent years has focused on a more holistic integrated approach to improve productivity and quality. Pest and disease management is considered a component of this broader approach. The development of Integrated Pest Management (IPM) strategies is a methodology centred around the economic reduction of pest levels to an acceptable threshold with minimum risk to human health and the environment. This can be achieved by combining several different management options including preventative cultural methods, developing monitoring schedules, mechanical control, biological control and promoting the responsible use of pesticides (RUP). A combination of these methods is advocated as part of an integrated approach for many crops and can be appropriate for smallholder production (Morse and Buhler, 1997). Many authors have advocated integrated approaches for cocoa smallholders including several authors in Bailey and Meinhardt (2016) and this will be discussed in greater detail in another chapter.

Many smallholder farmers in developing countries view pesticides as an easy 'one stop' solution for their pest and disease problems but are often unaware of country regulations on which pesticides are suitable for use on cocoa, how to apply them appropriately and the use of protective clothing. RUP aims to improve selectivity of products and increase the precision and timing of the applications – improving overall safety and efficiency as well as reducing costs (Bateman, 2015). RUP is now being more widely considered by research organizations in cocoa-producing countries as a component of IPM strategies. New, less toxic active ingredients are being routinely screened and recommended against pest and diseases in many countries. Further research is being conducted to determine more efficient application systems such as low volume spraying equipment as well as to determine application timings for optimum pest and disease control. Combined, these aspects can improve human and environmental safety and reduce the amount of labour and pesticide required to manage pests and diseases. Much more emphasis is also being placed on research to define and incorporate recommendations for application regimes which take into account the mode of action of different pesticides, to reduce the likelihood of pest and diseases developing resistance.

Research into cocoa pest and disease management and development of best practices has advanced reasonably well over recent years, but an area which has not been advanced sufficiently is research into adoption of new technologies by farmers and other stakeholders in the cocoa supply chain. The cocoa community is now slowly recognizing the importance of further investigation into the social and economic factors affecting the rate of technology uptake. The authors hope that this is one area where efforts will be concentrated in the near future. Low cocoa prices have often been attributed to the lack of technology adoption by farmers, when the price of cocoa falls below a certain threshold it becomes uneconomic to invest funds in purchasing inputs and additional labour. The cocoa industry is supporting the move to a more mechanized approach to cocoa production systems, to improve production efficiency and to make cocoa farming a more economic proposition for rural smallholders. Currently, there are few technologies available to achieve this. In the meantime, global demand for cocoa is rising and there is increasing pressure to intensify production on land that is currently farmed. Reducing the current losses to pests and diseases through more efficient and safer management strategies could go a long way to bridge the predicted production gap in the short to medium term.

8 Where to look for further information

A wide range of resources are available, on various platforms and in differing formats, that provide further information, advice and guidance on chemical pesticide (active substances), pesticide residues, pesticide products and good pesticide practice. Some useful examples are provided below:

Bateman, R. (2015). *Pesticide Use in Cocoa: A Guide for Training Administrative and Research Staff.* 3rd ed. London, International Cocoa Organization (ICCO). https://www.icco.org/about-us/icco-news/292-new-pesticide-use-in-cocoa-manual-now-available.html

Bateman, R. P., Holmes, K. A., Krauss, U. and Padi, B. (2004). Future tactics and tools for pest management, pp. 33–53. In: J. Flood and R. Murphy (Eds), *Cocoa Futures.* Chinciná, Colombia, CABI-FEDERACAFE. ISBN: 9589744117.

CAOBISCO/ECA/FCC. (2015). Cocoa Beans: Chocolate and Cocoa Industry Quality Requirements. September 2015 (End, M. J. and Dand, R. (Eds)). ISBN: 978-2-9601817-0-8. http://www.cocoaquality.eu/data/Cocoa%20Beans%20Industry%20Quality%20Requirements%20Apr%202016_En.pdf

Cocoa Crop Protection. http://www.dropdata.org/cocoa/

Cocoa SPS Africa Project Website: Mitigating the harmful effects of pesticide residues in Cocoa in Africa. https://www.icco.org/sites/sps/

CocoaSafe Project Website: Capacity Building and Knowledge Sharing in SPS for Cocoa in Southeast Asia. http://www.cocoasafe.org/

Codex: pesticide maximum residue levels (MRLs) in cocoa. http://www.fao.org/fao-who-codexalimentarius/standards/pestres/commodities-detail/en/?c_id=239

Cocoa Pests and Diseases. International Cocoa Organization (ICCO) website. https://www.icco.org/about-cocoa/pest-a-diseases.html

Compendium of Pesticide Common Names. http://www.alanwood.net/pesticides/index.html
Dohmen, M. M. and Quinn, D. (2010). *Counterfeit and Illegal Pesticides: A Training Manual for Master Trainers*. CropLife Africa Middle East. https://croplife.org/wp-content/uploads/2016/04/Training-Manual-on-Counterfeit-and-illegal-products.pdf
Environmental Handbook – Documentation on Monitoring and Evaluating Environmental Impacts. German Federal Ministry for Economic Cooperation and Development (Bundesministerium für wirtschaftliche Zusammenarbeit und Entwicklung [BMZ]).
European Commission. http://ec.europa.eu/agriculture/envir/pesticides/index_en.htm
European and Mediterranean Plant Protection Organization (EPPO). http://www.eppo.org/
European Food Safety Agency (EFSA). http://www.efsa.europa.eu/en/panels/pesticides.htm
European Union (EU) Pesticides Database. Directorate General for Health and Consumer Protection (DG SANCO), European Commission 2008. http://ec.europa.eu/food/plant/pesticides/eu-pesticides-database/public/?event=homepage&language=EN
Extension Toxicology Network, Pesticide Information Profiles. http://extoxnet.orst.edu/pips/ghindex.html
Food and Agriculture Organisation (FAO). http://www.fao.org/
Food and Agriculture Organisation (FAO) statistics database (FAOSTAT). Food and Agriculture Organization of the United Nations. Accessed 15 December 2009. http://faostat.fao.org/site/567/DesktopDefault.aspx?PageID=567#ancor
International Federation of Organic Agriculture Movements (IFOAM). http://www.ifoam.org/
Information on chemicals subject to the PIC procedure. European Chemicals Agency. https://echa.europa.eu/information-on-chemicals/pic/chemicals
International Cocoa Organization (ICCO) Website. https://www.icco.org/
International Code of Conduct on the Distribution and Use of Pesticides (2003). Food and Agriculture Organization of the United Nations. Rome, Italy. ISBN: 92-5-104914-9. http://www.fao.org/docrep/005/Y4544E/y4544e00.htm
International Union of Pure and Applied Chemistry Pesticide Properties Database. Hatfield, University of Hertfordshire. http://sitem.herts.ac.uk/aeru/iupac/
Pesticide Action Network International. http://pan-international.org/
Pesticide Action Network Pesticide Database. http://www.pesticideinfo.org/
Pesticide Residues in Food? http://www.who.int/features/qa/87/en/(last reviewed 2016). World Health Organisation
Plantwise Knowledge Bank. http://www.plantwise.org/KnowledgeBank/home.aspx. CABI, United Kingdom.
Recognition and Management of Pesticide Poisonings (6th edition). Roberts, J. R. and Reigart, J. R. (Eds) (2013). United States Environmental Protection Agency, Office of Pesticide Programs. https://www.epa.gov/pesticide-worker-safety/recognition-and-management-pesticide-poisonings
Rotterdam Convention on the Prior Informed Consent Procedure for Certain Hazardous Chemicals and Pesticides in International Trade. Rotterdam Convention Secretariat, Rome, Italy. http://www.pic.int/home.php?type=s&id=77
Roundtable for Sustainable Cocoa (RSCE-3). http://www.roundtablecocoa.org
The International Code of Conduct on Pesticide Management (2014). World Health Organization. Food and Agriculture Organization of the United Nations, Rome.

http://www.fao.org/fileadmin/templates/agphome/documents/Pests_Pesticides/Code/CODE_2014Sep_ENG.pdf

The List of Lists: a catalogue of lists of pesticides identifying those associated with particularly harmful health or environmental impacts. Briefing Paper, 3rd edition 2009. Pesticide Action Network, London, UK. http://www.pan-uk.org/List%20of%20Lists.html

The Pesticide Manual: A World Compendium. 17th edition (2015). J. A. Turner (Ed.). British Crop Production Council, Alton, UK, 1440pp. ISBN: 978-1-901396-88-1. http://www.bcpc.org/page_Pesticide-Manual_100.html?name=pesticide-manual&type=page

The UK Pesticide Guide (2016). M. A. Lainsbury (Ed.). CAB International and British Crop Protection Council. ISBN: 978-1-780647-93-7. http://www.ukpesticideguide.co.uk/

The WHO Recommended Classification of Pesticides by Hazard and Guidelines to Classification, 2009. Geneva, World Health Organization Press. ISBN: 978-92-4-154796 3. http://www.who.int/ipcs/publications/pesticides_hazard/en/

United States of America Environmental Protection Agency (EPA) Regulation of Pesticide Residues on Food. https://www.epa.gov/pesticide-tolerances

World Health Organisation (WHO). http://www.who.int/en/

9 References

Aneani, F., Anchirinah, V. M., Owusu-Ansah, F. and Asamoah, M. (2012). Adoption of some cocoa production technologies by cocoa farmers in Ghana. *Sustainable Agriculture Research*, 1(1), 103–17.

Bailey, B. A. and Meinhardt, L. W. (Eds). (2016). *Cacao Diseases: A History of Old Enemies and New Encounters.* Springer, Cham, Switzerland, pp. 307–35. ISBN: 978-3-319-24787-8.

Bateman, R. (2015). *Pesticide Use in Cocoa: A Guide for Training Administrative and Research Staff.* 3rd ed. ICCO, London, UK. Retrieved from ICCO SPS: http://www.icco.org/sites/sps/manual.html

CAOBISCO/ECA/FCC. (2015). *Cocoa Beans: Chocolate and Cocoa Industry Quality Requirements.* M. J. End and Dand, R. (Eds). September 2015. ISBN: 978-2-9601817-0-8. http://www.cocoaquality.eu/data/Cocoa%20Beans%20Industry%20Quality%20Requirements%20Apr%202016_En.pdf

Cooper, J. and Dobson, H. (2007). The benefits of pesticides to mankind and the environment. *Crop Protection*, 26(9), 1337–48.

David, S. (2005). *Learning about Sustainable Cocoa Production: A Guide for Participatory Farmer Training. 1. Integrated Crop and Pest Management.* Sustainable Tree Crops Program, International Institute of Tropical Agriculture, Yaoundé, Cameroon.

Flood, J. and Murphy, R. (2004). *Cocoa Futures: A Source Book of Critical Issues Facing the Cocoa Industry.* Colombia/CAB International, UK: The Commodities Press, Cenicafe. ISBN: 958-97441-1-7.

Flood, J., Guest, D., Holmes, K.A., Keane, P., Padi, B. and Sulistyowati, E. (2004). Cocoa under attack. In: J. Flood and R. Murphy (Eds), *Cocoa Futures.* CABI-FEDERACAFE, Chinciná, Colombia, pp. 33–53. ISBN: 9589744117.

Hibbitt, C. (1990). Putting pesticides into perspective. *The Agricultural Engineer*, 45(2), 61.

ICCO. (2014). Final project report: Mitigating the harmful effects of pesticide residues in cocoa in Africa. Web: http://www.standardsfacility.org/PG-298

Lomberg, B. (2001). *The Skeptical Environmentalist.* Cambridge University Press, New York, NY, 496p. Reprint edition 28 August 2001. ISBN: 0521010683.

Morse, S. and Buhler, W. (1997). *Integrated Pest Management: Ideals and Realities in Developing Countries.* Lynne Rienner Publishers Inc., Boulder, CO, 171p. ISBN: 1555876854.

Okoffo, E., Fosu-Mensah, B. and Gordon, C. (2016). Persistent organochlorine pesticide residues in cocoa beans from Ghana: A concern for public health. *International Journal of Food Contamination*, 3(5), 1–5.
Oyekunle, J., Akindolani, O., Sosan M. and Adekunle, A. (2017). Organochlorine pesticide residues in dried cocoa beans obtained from cocoa stores in Ondo and Ile-Ife, Southwestern Nigeria. *Toxicology Reports*, 4, 151–9.
Rutherford, M. (2008). EU pesticide residues legislation: The cocoa industry response. *International Pest Control*, 50(4), 215–18. Web: http://www.researchinformation.co.uk/.
Rutherford, M. A. (2009). Helping cocoa farmers adjust to new EU pesticide laws. *Pesticide News*, 85, 16–18.
Rutherford, M. A. (2011). Assessing and addressing pesticide practice in cocoa producing countries to meet regulatory standards. *Proceedings International Workshop on Safe Use of Pesticides in Cocoa and Harmonized Legislation for Food Safety*. Kuala Lumpur, Malaysia, 25–27 January 2011.
Rutherford, M. A., Nyassé, S., N'guessan, K. F., Agbeniyi, S. O. and Baah, F. (2010). Cocoa production and trade in West Africa – meeting the needs of new EU pesticide legislation. *Proceedings 16th International Cocoa Research Conference*. Denpasar, Bali, Indonesia, 16–21 November 2009, pp. 903–908. ISBN: 978-065-959-5.
Sonwa, D. J., Coulibaly, O., Weise, S. and Adesina, A. A. (2008). Management of cocoa: Constraints during acquisition and application of pesticides in the humid forest zones of southern Cameroon. *Crop Protection*, 27, 1159–64.
Vos, J. Ritchie, B. and Flood, J. (2003). *Discovery Learning about Cocoa. An Inspirational Guide for Training Facilitators*. CABI Bioscience, UK, 112p.

Chapter 17

Mycotoxins in cocoa: causes, detection and control

Mary A. Egbuta, Southern Cross University, Australia

1 Introduction

Mycotoxins are structurally diverse chemical compounds produced as secondary metabolites by different species of filamentous fungi (Yu, 2012). Most mycotoxins are generally lipophilic (Wild and Gong, 2010) and are categorized according to the fungi which produce them, their structure and mode of action (Zain, 2011; Ismaiel and Papenbrock, 2015). There is an association between fungal infestation and mycotoxin contamination of many agricultural crops and products (Yin et al., 2008; Egbuta, 2012; Copetti et al., 2014; Chilaka et al., 2017), although the presence of fungi on a crop is not always tantamount to mycotoxin production, except where conditions are favourable.

The occurrence of mycotoxins in cocoa has been reported in studies from different parts of the world with aflatoxins and ochratoxin A (OTA) commonly mentioned (Mounjouenpou et al., 2008; Copetti et al., 2010, 2012c, 2013a; Mounjouenpou et al., 2011). In most instances, aflatoxins and OTA co-occur in cocoa seeds and their production is influenced by a range of factors (Magan et al., 1984; Mitchell et al., 2004; Ribeiro et al., 2006, Copetti et al., 2012c). The degree of mycotoxin contamination in cocoa can be detected through several methods ranging from simple solvent extraction (Egbuta et al., 2013) to complex molecular methods (Turcotte et al., 2013). When isolated, these mycotoxins can be quantified using some analytical techniques (Amézqueta et al., 2005; Bandeira et al., 2013) to determine their concentration in the crop, thereby evaluating health risks associated with their consumption.

As a result of the potential health risks associated with mycotoxins, particularly aflatoxins and OTA, different strategies have been developed to control mycotoxin contamination of

http://dx.doi.org/10.19103/AS.2017.0021.25

agricultural products (Niderkorn et al., 2007; Alberti, 2011; Lanubile et al., 2014; Alberts et al., 2016). This chapter discusses the different mycotoxins which have been reported in cocoa contamination, the different factors that influence their production, as well as methods of detection and ways they can be controlled in the crop.

2 Mycotoxins contaminating cocoa

The occurrence of mycotoxin-producing filamentous fungi in cocoa is a major reason for mycotoxin contamination of cocoa. Cocoa is susceptible to a wide range of filamentous fungi species (Copetti et al., 2011a; Egbuta et al., 2015), some of which have the ability to produce mycotoxins. *Aspergillus*, *Penicillium* and *Fusarium* are the main genera of filamentous fungi found to contaminate cocoa and its by-products (Sánchez-Hervás et al., 2008; Copetti et al., 2011a).

There are over 300 secondary metabolites, including mycotoxins, produced by filamentous fungi, but some of these mycotoxins are found to occur more frequently in food products than others (Hussein and Brasel, 2001). These common mycotoxins are referred to as major mycotoxins, and include aflatoxins, ochratoxins, fumonisins, zearalenone (ZEA), trichothecenes, tremorgenic toxins and ergot alkaloids. Most of these common mycotoxins are produced by species of the filamentous fungi genera *Aspergillus*, *Fusarium* and *Penicillium*. In cocoa, mycotoxins reported to occur in cocoa are predominantly aflatoxins and OTA (Raters and Matissek, 2005; Copetti et al., 2011a,b; Egbuta et al., 2013). Other mycotoxins have also been found contaminating the food crop and they are produced by some *Fusarium* species (Egbuta et al., 2013).

2.1 Aflatoxins

Aflatoxins are produced mainly by produced mainly by some species of the *Aspergillus* genus. Apart from other species such as *A. nomius* and *A. pseudotamarii*, *A. flavus* and *A. parasiticus* are the main producers of aflatoxins (Frisvad et al., 2005), with *A. flavus* regarded as the most serious due to its ability to grow at high temperatures (up to 48°C), as well as to its survival on a wide range of substrates, including plant debris, crops, animal fodder and compost (Klich, 1998; Yu, 2012).

Aflatoxins are basically difuranocoumarin compounds such as aflatoxin B_1, B_2, B_{2a}, B_3, G_1, GM_1, G_2, G_{2a}, M_1, M_2, M_{2a}, GM_2, P_1, Q_1, R_0, RB_1, RB_2, AFL, AFLH, AFLM and methoxy, ethoxy and aceto derivatives (Sinha and Bhatnagar, 1998; Egbuta, 2012). However, only four of these compounds occur frequently (Sinha and Bhatnagar, 1998), and they include AFB_1 and AFB_2 produced by *A. flavus*, whereas, AFB_1, AFB_2, AFG_1 and AFG_2 are produced by *A. parasiticus* (Fig. 1). Recently, AFM_1 and AFM_2, which are hydroxylated derivatives of AFB_1 and AFB_2, have been reported to be secreted in the milk of ruminants (also shown in Fig. 1) (Mwanza et al., 2015).

Aflatoxins are the most studied of all the mycotoxins due to their high toxicity and wide distribution in foods (Copetti et al., 2012c), with aflatoxin B_1 regarded as the most toxic of all (Khlangwiset et al., 2011) because of its association with a wide of range of health issues affecting both humans and animals (Shuaib et al., 2010; Zain, 2011; Valencia-Quintana et al., 2012; Torres et al., 2015). The International Agency for Research on Cancer (IARC) has recognized the carcinogenicity of aflatoxins as a Group 1 carcinogen. Given the proven knowledge of aflatoxin B_1 as a primary carcinogen for humans, the presence of these

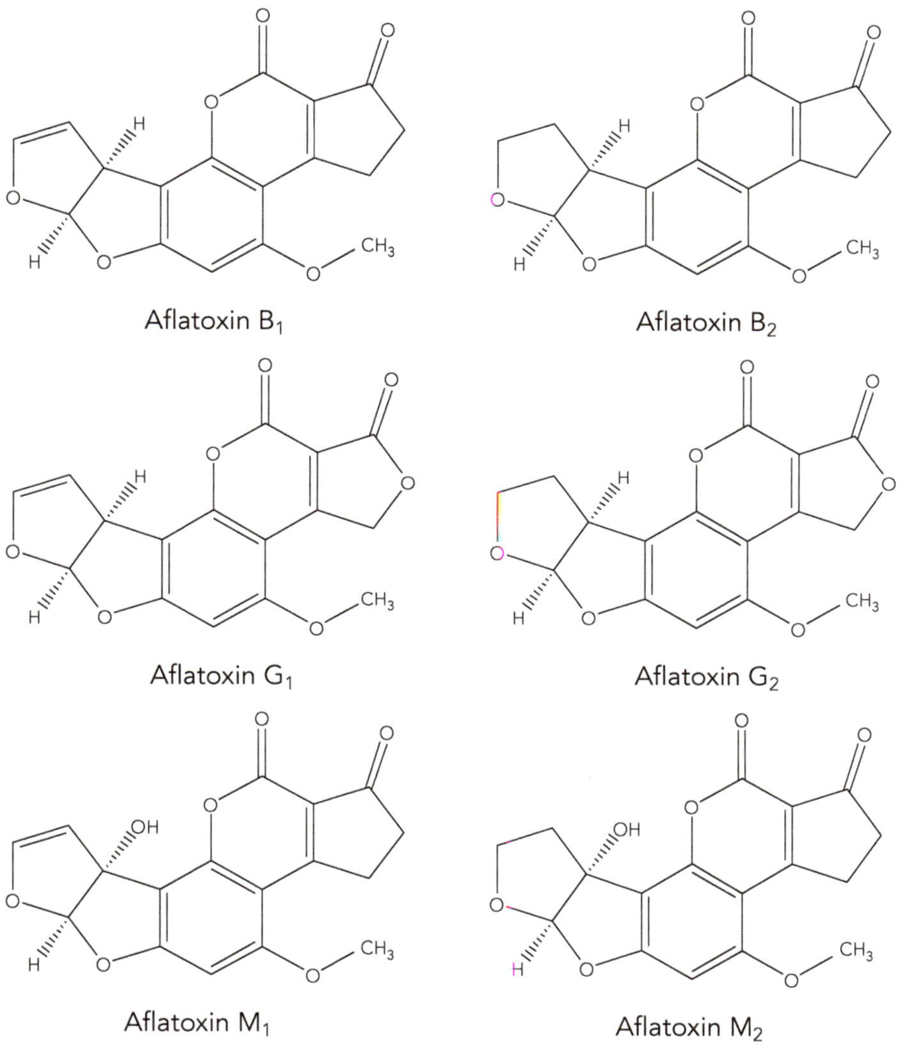

Figure 1 Chemical structure of common types of aflatoxins.

contaminants in the diet should be as low as possible. The European Food Safety Agency has established a tolerable weekly intake (TWI) of 120 ng/kg bodyweight (EFSA, 2006, 2010). The Joint FAO/WHO Expert Committee on Food Additives (JECFA) has established a TWI of 100 ng/kg bodyweight (JECFA, 2007).

AFB_1 and AFB_2 are common types of aflatoxins contaminating cocoa crops from field to storage (Copetti et al., 2011a, 2014). There are also reports of aflatoxin contamination of cocoa and cocoa-based products, including chocolate (Copetti et al., 2012c; Egbuta, 2012). The degree of aflatoxin contamination in cocoa varies across the different stages of cocoa cultivation and production, but the levels found are generally low. A study by

Copetti et al. (2011b) showed that aflatoxin contamination of cocoa seeds began mainly at the drying stage after harvest, when the seeds began to lose water. This increase in aflatoxin contamination at the drying stage was due to reduction of other competitors such as bacteria, which are sensitive to low water activity. Their study showed that aflatoxin contamination of cocoa was higher at the sun drying stage and lower at the storage stage.

2.2 Ochratoxins

The ochratoxins are a group of mycotoxins produced by some *Aspergillus* and *Penicillium* species including, *A. ochraceus*, *A. niger*, *A. carbonarius* and *P. verrucosum* (Wang et al., 2016). They comprise a polyketide-derived dihydroisocoumarin moiety linked via its 12-carbon group by an amide bond to L-β-phenylalanine. The ochratoxins include OTA, ochratoxin B (OTB) and ochratoxin C (OTC), with OTA being the most toxic of the three and occurring most frequently. As a result of its toxic nature, OTA has significant health implications for both human and animals such that it is classified as a Group 2B carcinogen by IARC (Fig. 2).

A large number of studies have shown OTA to contaminate cocoa, its products and by-products (Serra, 2004; Copetti et al., 2010, 2012c, 2013a; Turcotte et al., 2013), although there are no reports of OTB or OTC in cocoa (Mounjouenpou et al., 2008, 2011; Copetti et al., 2014). These studies confirm OTA as the main mycotoxin contaminating cocoa, with some studies showing significantly higher concentrations of OTA in comparison to aflatoxins (Egbuta et al., 2013; Turcotte et al., 2013). The occurrence of OTA in cocoa corresponds to contamination of the crop mainly by black aspergilli and *A. ochraceus* (Mounjouenpou et al., 2008; Sánchez-Hervás et al., 2008; Copetti et al., 2011a), with the latter being a more common potential producer of the toxin.

2.3 Other mycotoxins

Fumonisins, ZEA and deoxynivalenol (DON) are seldom reported to contaminate cocoa or its products. Nevertheless, some studies have shown that, despite the predominance of aflatoxins and ochratoxins in cocoa, other major mycotoxins occur in the crop, mostly at the storage stage (Egbuta, 2012). The fumonisins are reportedly produced by two species of the *Fusarium* genera, *F. verticillioides* and *F. proliferatum* (Nelson et al., 1992; Dilkin et al., 2002), with a recent discovery of *A. niger* producing a type of fumonisin (Frisvad et al., 2011). Although there are different types of fumonisins, FB_1, FB_2 and FB_3 (Fig. 3) have received the most attention. The occurrence of fumonisins in cocoa is lower than that of aflatoxins and OTA (Egbuta et al., 2013).

Also known as vomitoxin, DON is one of the most common *Fusarium* mycotoxins (Sobrova et al., 2010). It is produced predominantly by *F. graminearum* (Kushiro, 2008) and occasionally by *F. oxysporum* (Molto et al., 1997). DON is a common contaminant of many agricultural products including grains and seeds (Sobrova et al., 2010). DON in cocoa has been reported by different studies. Raters and Matissek (2008) reported contamination of cocoa beans by DON at low levels across Africa, America and Asia. The presence of DON has been confirmed in a more recent study by Egbuta et al. (2013) of cocoa from Nigeria in West Africa. The toxin has the ability to withstand very high temperatures between 170°C and 350°C and is therefore able to survive cocoa processing (Hughes et al., 1999). The toxin is associated with vomiting and nausea, although it is not seen as a serious threat to public health (Perkowski et al., 1990; Sobrova et al., 2010) (Fig. 4).

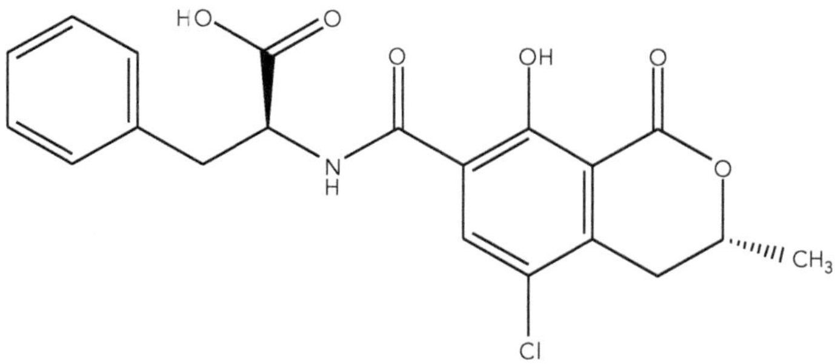

Ochratoxin A

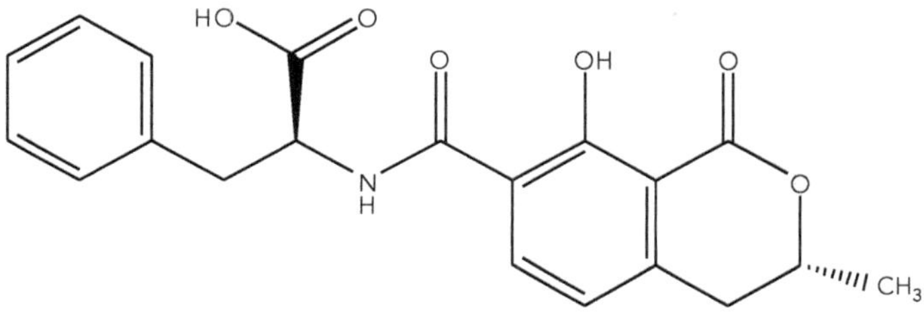

Ochratoxin B

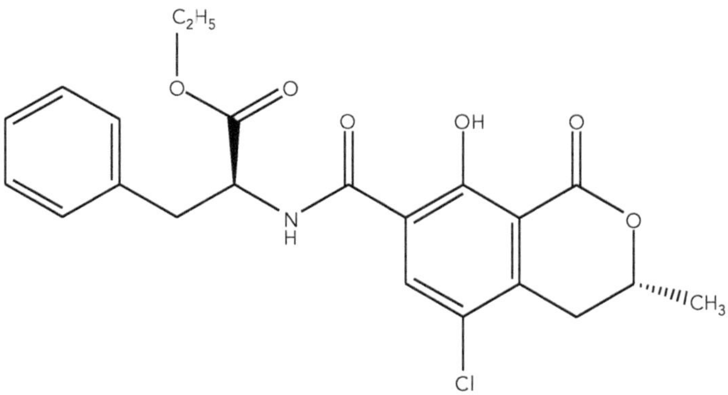

Ochratoxin C

Figure 2 Chemical structure of common types of ochratoxins.

ZEA, which is also known as F-2 toxin, is a potent estrogenic metabolite produced by some species of the *Fusarium* and *Gibberella* genera. The toxin (shown in Fig. 5) is most associated with cereals and grains (Campbell and Armstrong, 2007; Zain, 2011) and less associated with cocoa contamination. A study by Egbuta et al. (2013) reported the occurrence of this mycotoxin in cocoa and its by-products, with a high occurrence of up to 83.6 μg/kg of the toxin in cocoa seeds and up to 531.9 μg/kg in processed cocoa samples.

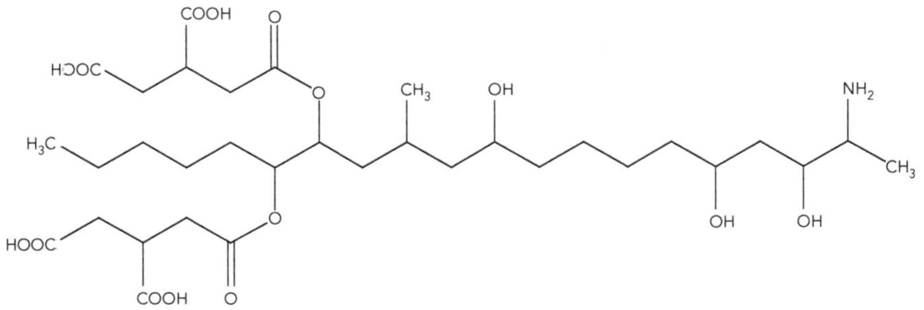

Fumonisin B_1

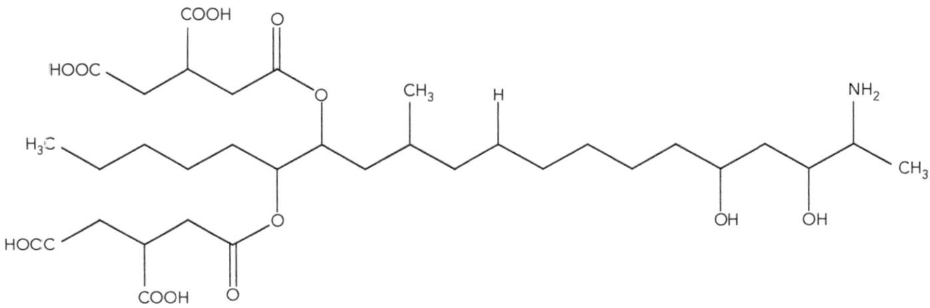

Fumonisin B_2

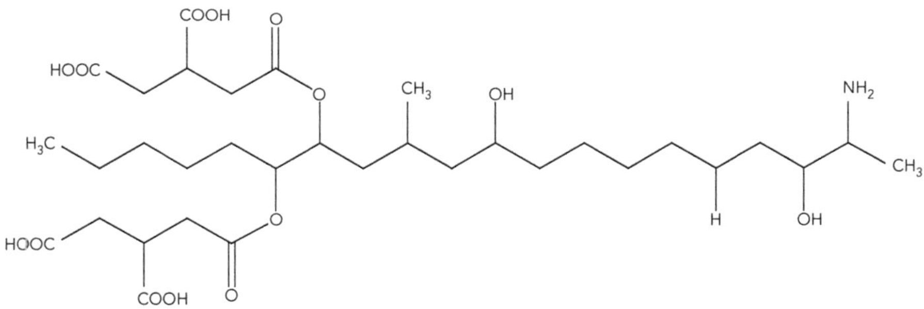

Fumonisin B_3

Figure 3 Chemical structure of common types of fumonisins.

3 Detection of mycotoxins in cocoa

Mycotoxin detection in cocoa is similar to that in other agricultural products affected by these metabolites of filamentous fungi. Screening of mycotoxins generally involves procedures for extraction and clean-up of the compounds, followed by analytical techniques for identification and quantification of toxins.

Methods of extraction include multi-mycotoxin extraction using different organic and inorganic solvents (Egbuta et al., 2013) and immunoassays (Copetti et al., 2013a;

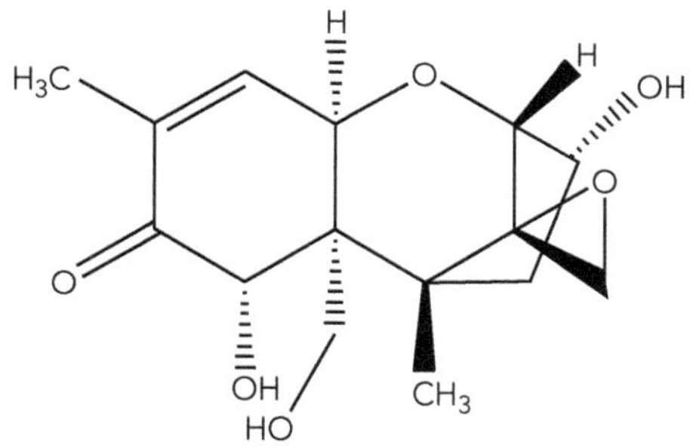

Figure 4 Chemical structure of deoxynivalenol.

Turcotte et al., 2013). Immunoassays involve use of immune-affinity columns specific for target compounds and have been used to extract mycotoxins from cocoa as well as other agricultural products (Amézqueta et al., 2005; Turcotte et al., 2013; Copetti et al., 2014). Most columns are specific for one or in some cases two mycotoxins, including aflatoxin and OTA. Immunoassays are regarded as very effective methods of extraction because of specificity of the compounds in the column which act as antigens for their target mycotoxins.

3.1 Enzyme-linked immunoassay (ELISA)

The ELISA is an example of immunochemical methods used in mycotoxin detection. The assay allows rapid, quantitative mycotoxin analysis of large numbers of samples. The assay is a very sensitive procedure that ensures highly accurate quantification of mycotoxins down to parts per trillion (ppt) at a relatively low cost. This method of mycotoxin analysis is effective across different food matrices (Leszczyńska et al., 2001). Application of ELISA for mycotoxin detection in cocoa and its products has not been reported, although there could be possible future application of the method to mycotoxin detection in the crop and its by-products.

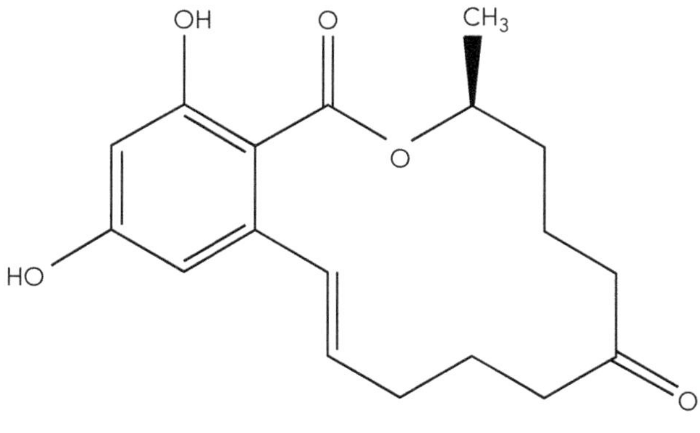

Figure 5 Chemical structure of zearalenone.

3.2 High-performance liquid chromatography (HPLC)

The detection of mycotoxins by HPLC involves a process whereby samples are passed through C-18 columns. Mycotoxins are eluted in these columns according to their solubility and detected by either fluorescence detectors or UV detectors, depending on the nature of the compounds. Eluted compounds are represented as peaks which are matched to peaks of known standard concentrations. Mycotoxins such as OTA and the aflatoxins in cocoa and cocoa products have been detected by HPLC (Brera et al., 2003, 2005; Amézqueta et al., 2005; Copetti et al., 2012c). Other mycotoxins including DON and ZEA have also been detected in cocoa via HPLC (Raters and Matissek, 2008; Egbuta et al., 2013).

3.3 Liquid chromatography–tandem mass spectrometry

Liquid chromatography–tandem mass spectrometry is a powerful and highly sensitive technique, which combines the separation capabilities of liquid chromatography with the mass analysis capabilities of mass spectrometry to enable separation, detection and identification of chemical compounds (Arpino, 1992; Pitt, 2009). This method has been used for the detection of mycotoxins in cocoa, particularly aflatoxins and OTA (Şenyuva et al., 2008; Bandeira et al., 2012, 2013; Turcotte et al., 2013).

4 Controlling mycotoxins in cocoa production

It is widely accepted that beans inside undamaged pods are sterile (CAC, 2012, 2013). Research suggests that OTA contamination is primarily associated with mechanical damage to pods when they are harvested (Bastide et al., 2006; Gilmour and Lindblom, 2008; Dembele et al., 2009). This exposes beans to environmental sources of mycotoxin contamination which become mainly located in the shell. Microbial growth is determined by factors such as pH, temperature, water activity and competition from organic acids produced during fermentation (Magan et al., 1984; Ribeiro et al., 2006; Copetti et al., 2012a, 2014). There have been various studies establishing thresholds for water activity and temperature which promote OTA and aflatoxin production (Mitchell et al., 2004; Pardo et al., 2004; Lahouar et al., 2016).

Although there have been attempts to breed cereals with resistance to mycotoxin-producing fungi, this approach has yet to be adopted in the case of cocoa (Williams, 2006; Lanubile et al., 2014; Lemmens et al., 2016). Good agricultural practices (GAPs) can help prevent mycotoxin contamination. Plants damaged by insect pests or disease are particularly vulnerable to mycotoxin contamination, highlighting the importance of integrated pest and disease management (Bush et al., 2006; Medeiros et al., 2012; Alberts et al., 2016). It is advisable to ensure that there is even distribution of water during irrigation, although irrigation during anthesis and ripening of crops should be avoided (Alberti, 2011). Developing and maintaining a crop rotation schedule is also essential to controlling pre-harvest mycotoxin contamination. A practice of rotating cocoa with other crops that are less susceptible to aflatoxin and ochratoxin producing fungi can help to reduce mycotoxin production in the crop at this stage. Removal of residues such as old seed heads, stalks, leaves and other debris which could act as substrates for growth of mycotoxin-producing fungi from the seed beds, can also help to reduce mycotoxin contamination of cocoa and other crops at pre-harvest stage of production (Alberti, 2011).

Once pods are harvested and opened to remove the beans, the beans are fermented and then dried before being quality-checked and graded, bagged and stored for shipping. Whilst fermentation and drying are not direct causes of OTA contamination, poorly controlled fermentation and drying can exacerbate contamination, which first becomes detectable during the fermentation process. Poor control by smallholders and conditions during the dry season in particular have been linked to inadequate fermentation (Bastide et al., 2006). Other research has highlighted the role of poor drying practices in contamination (Copetti et al., 2010, 2012b). High humidity during storage can also encourage mould growth (Copetti et al., 2014). Whilst subsequent processes such as roasting do not eradicate contamination entirely, processing steps such as de-shelling can significantly reduce contamination levels (Gilmour and Lindblom, 2008; Manda et al., 2009; CAC, 2012; Copetti et al., 2012b, 2013).

There have been attempts to develop techniques to decontaminate crops infected with mycotoxins (Bhat et al., 2010). Methods used to achieve mycotoxin control include thermal treatment (Betina, 1989; La Pera et al., 2008); alkaline treatment (Hendrich et al., 1993); radiation processing (Ogbadu, 1980; Paster et al., 1985); use of chemical agents such as acids, bases oxidants, reducing agents and chlorinated agents (Scott, 1998); and use of natural organic binders such as glucomannans isolated from cell wall of yeast and fermentation bacteria, for example lactic acid and propionic acid bacteria (Niderkorn et al., 2007).

A draft code of practice for preventing and reducing OTA contamination of cocoa has been put together by the Codex Alimentarius Commission (CAC) Committee on Contaminants in Food (CAC, 2013). Recommendations include the following:

- Pre-harvest practices should include using GAP to maintain healthy plants, as well as to avoid overhead irrigation which might promote spore dispersal and proper management of organic waste to prevent spreading OTA-producing fungi.
- Harvest practices should include prompt removal of diseased pods and avoiding mechanical damage to pods during harvesting. Damaged pods should not be stored for more than 1 day before opening and fermentation.
- Pod breaking should be undertaken promptly to avoid fungal proliferation and should avoid damaging seeds. Any damaged or diseased beans should be separated out and disposed of.
- Fermentation should take place in dry conditions and last between 4 and 7 days with frequent turning of fermentation heaps to ensure uniform heating.
- Drying should be undertaken immediately after fermentation. Drying needs to be uniform (e.g. by using thin layers of beans for drying with frequent turning) with a moisture content of 6–8%.
- Sorting needs to remove damaged or diseased beans.
- Storage needs to ensure clean and dry conditions as well as protection from pests.
- Transport and shipping must ensure clean and dry conditions, with cleaning between cargoes to prevent cross-contamination.

There are no specific sampling procedures for analysis of OTA in cocoa and its products (CAC, 2012). General sampling procedures for mycotoxins in food have been outlined in the EU Commission Regulation 401/2006 (EC, 2006).

5 Conclusion

It is clear that more attention has been given to mycotoxin contamination of other agricultural crops such as cereals rather than cocoa. There is therefore a need for more research into the various mycotoxins that can infect cocoa as well as into ways of reducing the presence of these chemical metabolites of filamentous fungi in cocoa.

6 Where to look for further information

Different sources for more information on mycotoxins in cocoa include publications which have been cited in this chapter. Two studies in particular are those of Raters and Matissek (2005) which investigated mycotoxins in cocoa samples and Copetti et al. (2014), who published a review on fungi and mycotoxins contaminating cocoa through the processing chain of the food crop. There are also published book chapters which provide more information on mycotoxin contamination of cocoa. An example of such book chapters is the one by Copetti et al. (2013b).

There are organizations involved in mycotoxin research on cocoa as well as other food crops such as CSIRO Food and Nutritional Sciences (https://csiropedia.csiro.au/food-and-nutritional-sciences-2010-2012) and the International Institute of Tropical Agriculture (http://www.iita.org/search_gcse/?q=cocoa%20crop) from which more information on mycotoxin contamination in cocoa can also be sourced.

7 References

Alberti, I. (2011). Factors influencing mycotoxin production pre and post harvest. *Workshop on Scientifically-based Control of Mycotoxins in Food and Management of Contaminated Lands*, 21st–22nd February 2011, Baku, Azerbaijan.

Alberts, J. F., van Zyl, W. H. and Gelderblom, W. C. A. (2016). Biologically based methods for control of fumonisin-producing *Fusarium* species and reduction of the fumonisins. *Frontiers in Microbiology* 7, 548. doi:10.3389/fmicb.2016.00548.

Amézqueta, S., González-Peñas, E., Murillo, M. and López de Cerain, A. (2005). Occurrence of ochratoxin A in cocoa beans: Effect of shelling. *Food Additives & Contaminants* 22(6), 590–6. doi:10.1080/02652030500130160.

Arpino, P. (1992). Combined liquid chromatography mass spectrometry. Part III. Applications of thermospray. *Mass Spectrometry Reviews* 11(1), 3–40. doi:10.1002/mas.1280110103.

Bandeira, R. D. d. C. C., Uekane, T. M., Cunha, C. P. d., Cunha, V. S. d., Rodrigues, J. M., Godoy, R. L. d. O. and de la Cruz, M. H. C. (2012). Development and validation of a method for detection and quantification of ochratoxin A in green coffee using liquid chromatography coupled to mass spectrometry. *Food Science and Technology (Campinas)* 32, 775–82.

Bandeira, R. D. d. C. C., Uekane, T. M., Cunha, C. P. d., Rodrigues, J. M., la Cruz, M. H. C. d., Godoy, R. L d. O. and Fioravante, A. d. L. (2013). Comparison of high performance liquid chromatography with fluorescence detector and with tandem mass spectrometry methods for detection and quantification of Ochratoxin A in green and roasted coffee beans. *Brazilian Archives of Biology and Technology* 56, 911–20.

Bastide, P., Fourny, G., Durand, N., Petithuguenin, P, Guyot, P., Gilmour, M. and Lindblom, M. (2006). Identification of Ochratoxin A sources during cocoa post-harvest processing: Influence of harvest quality and climatic factors. *15th International Cocoa Research Conference*, 9th–17th October 2006, San José, Costa Rica.

Betina, V. (1989). Mycotoxins. Chemical, biological and environmental aspects. In: *Bioactive Molecules*, edited by V. Betina. London: Elsevier Applied Science, pp. 114–50.

Bhat, R., Rai, R. V. and Karim, A. A. (2010). Mycotoxins in food and feed: Present status and future concerns. *Comprehensive Reviews in Food Science and Food Safety* 9(1), 57–81. doi:10.1111/j.1541-4337.2009.00094.x.

Brera, C., Grossi, S., De Santis, B. and Miraglia M. (2003). High performance liquid chromatographic method for the determination of Ochratoxin A in cocoa powder. *Journal of Liquid Chromatography & Related Technologies* 26, 585–98.

Brera, C., Grossi, S. and Miraglia, M. (2005). Interlaboratory study for Ochratoxin A determination in cocoa powder samples. *Journal of Liquid Chromatography & Related Technologies* 28, 35–61.

Bush, J., Jander, G. and Ausubel, F. M. (2006). Prevention and control of pests and diseases. In: *Arabidopsis Protocols*, edited by J. Salinas and J. J. Sanchez-Serrano. Totowa, NJ: Humana Press, pp. 13–25.

CAC. (2012). Discussion paper on Ochratoxin A in cocoa. *Codex Alimentarius Commission (CAC) Committee on Contaminants in Food*, Sixth Session, 26th–30th March 2012, Maastricht, the Netherlands.

CAC. (2013). Proposed draft code of practice for the prevention and reduction of Ochratoxin A in contamination of cocoa. *Codex Alimentarius Commission (CAC) Committee on Contaminants in Food*, Seventh Session, 8th–12th April 2013, Moscow, the Russian Federation.

Campbell, H. M. and Armstrong, J. F. (2007). Determination of zearalenone in cereal grains, animal feed, and feed ingredients using immunoaffinity column chromatography and liquid chromatography: Interlaboratory study. *Journal of AOAC International* 90(6), 1610–22.

Chilaka, C. A., De Boevre, M., Atanda, O. O. and De Saeger, S. (2017). The status of *Fusarium* mycotoxins in sub-Saharan Africa: A review of emerging trends and post-harvest mitigation strategies towards food control. *Toxins* 9(1), 19. doi:10.3390/toxins9010019.

Copetti, M. V., Pereira, J. L., Iamanaka, B. T., Pitt, J. I. and Taniwaki, M. H. (2010). Ochratoxigenic fungi and Ochratoxin A in cocoa during farm processing. *International Journal of Food Microbiology* 143(1–2), 67–70. doi:http://dx.doi.org/10.1016/j.ijfoodmicro.2010.07.031.

Copetti, M. V., Iamanaka, B. T., Frisvad, J. C., Pereira, J. L. and Taniwaki, M. H. (2011a). Mycobiota of cocoa: From farm to chocolate. *Food Microbiology* 28(8), 1499–504. doi:http://dx.doi.org/10.1016/j.fm.2011.08.005.

Copetti, M. V., Iamanaka, B. T., Pereira, J. L., Fungaro, M. H. and Taniwaki, M. H. (2011b). Aflatoxigenic fungi and aflatoxin in cocoa. *International Journal of Food Microbiology* 148(2), 141–4. doi:http://dx.doi.org/10.1016/j.ijfoodmicro.2011.05.020.

Copetti, M., Iamanaka, B., Mororó, R., Pereira, J., Frisvad, J. and Taniwaki, M. (2012a). The effect of cocoa fermentation and weak organic acids on growth and Ochratoxin A production by *Aspergillus* species. *International Journal of Food Microbiology* 155, 158–64.

Copetti, M., Iamanaka, B., Pereira, J., Lemes, D., Nakano, F. and Taniwaki, M. (2012b). Co-occurrence of Ochratoxin A and aflatoxins in chocolate marketed in Brazil. *Food Control* 26, 36–41

Copetti, M. V., Iamanaka, B. T., Pereira, J. L., Lemes, D. P., Nakano, F. and Taniwaki, M. H. (2012c). Determination of aflatoxins in by-products of industrial processing of cocoa beans. *Food Additives & Contaminants: Part A* 29(6), 972–8. doi:10.1080/19440049.2012.660657.

Copetti, M. V., Iamanaka, B. T., Nester, M. A., Efraim, P. and Taniwaki, M. H. (2013a). Occurrence of ochratoxin A in cocoa by-products and determination of its reduction during chocolate manufacture. *Food Chemistry* 136(1), 100–4. doi:http://dx.doi.org/10.1016/j.foodchem.2012.07.093.

Copetti, M. V., Iamanaka, B. T. and Taniwaki, M. H. (2013b). Fungi and mycotoxin occurrence in cocoa. In: *Chocolate in Health and Nutrition*, edited by R. Watson, V. Preedy and S. Zibadi. Nutrition and Health, volume 7. Totowa, NJ: Humana Press, pp. 61–71.

Copetti, M. V., Iamanaka, B. T., Pitt, J. I. and Taniwaki, M. H. (2014). Fungi and mycotoxins in cocoa: From farm to chocolate. *International Journal of Food Microbiology* 178, 13–20. doi:http://dx.doi.org/10.1016/j.ijfoodmicro.2014.02.023.

Dembele, A., Coulibaly, A., Traoré, S., Mamadou, K., Silue, N. and Abba Touré, A. (2009). Détermination du niveau de contamination de l'ochratoxine A (OTA) dans les fèves de cacao à l'exportation. *Tropicultura* 27(1), 26–30.

Dilkin, P., Mallmann, C. A., Almeida, C. A. A.d., Stefanon, E. B., Fontana, F. Z. and Milbradt, E. L. (2002). Production of fumonisins by strains of *Fusarium moniliforme* according to temperature, moisture and growth period. *Brazilian Journal of Microbiology* 33, 111–18.

EC. (2006). Commission Regulation (EC) No. 401/2006 of 23rd February, 2006 laying down the methods of sampling and analysis for the official control of the levels of mycotoxins in foodstuffs. *Official Journal of the European Union* OJL70, 12–34. http://eur-lex.europa.eu/legal-content/EN/TXT/?uri=CELEX%3A32006R0401.

EFSA. (2006). Opinion of the scientific panel on contaminants in the food chain on a request from the Commission related to Ochratoxin A in food. *EFSA Journal* 365, 1–56.

EFSA. (2010). Statement on recent scientific information on the toxicity of Ochratoxin A. *EFSA Journal* 8(6), 1626–33.

Egbuta, M. A. (2012). Occurrence of mycotoxins in some Nigerian food commodities and their health risk assessment. M.Tech. thesis, University of Johannesburg, Johannesburg, South Africa.

Egbuta, M. A., Chilaka, C. A., Phoku, J. Z., Mwanza, M. and Dutton, M. F. (2013). Co-contamination of Nigerian cocoa and cocoa-based powder beverages destined for human consumption by mycotoxins. *Ethno Medicine* 7(3), 187–94.

Egbuta, M. A., Mwanza, M., Njobeh, P. B., Phoku, J. Z., Chilaka, C. A. and Dutton, M. F. (2015). Isolation of filamentous fungi species contaminating some Nigerian food commodities. *Journal of Food Research* 4(1). doi:10.5539/jfr.v4n1p38.

Frisvad, J. C., Skouboe, P. and Samson, R. A. (2005). Taxonomic comparison of three different groups of aflatoxin producers and a new efficient producer of aflatoxin B_1, sterigmatocystin and 3-O-methylsterigmatocystin, *Aspergillus rambellii* sp. nov. *Systematic and Applied Microbiology* 28, 442–53.

Frisvad, J. C., Larsen, T. O., Thrane, U., Meijer, M., Varga, J., Samson, R. A. and Nielsen, K. F. (2011). Fumonisin and ochratoxin production in industrial Aspergillus niger strains. *PLoS ONE* 6(8), e23496. doi:10.1371/journal.pone.0023496.

Gilmour, M. and Lindblom, M. (2008). Management of ochratoxin A in the cocoa supply chain: A summary of work by the CAOBISCO/ECA/FCC working group on ochratoxin A. In: *Mycotoxins: Detection Methods, Management, Public Health and Agricultural Trade*, edited by J. Leshe, R. Bandyopadhyay and A. Visconti. Wallingford – Oxford, UK: CABI, pp. 231–43.

Hendrich, S., Miller, K., Wilson, T. and Murphy, P. (1993). Toxicity of *Fusarium proliferatum* fermented nixtamalized corn-based diets fed to rats: effect of nutritional status. *Journal of Agriculture and Food Chemistry* 41, 1649–54.

Hughes, D. M., Gahl, M. J., Graham, C. H. and Grieb, S. L. (1999). Overt signs of toxicity to dogs and cats of dietary deoxynivalenol. *Journal of Animal Science* 77(3), 693–700. doi:10.2527/1999.773693x.

Hussein, H. S. and Brasel, J. M. (2001). Toxicity, metabolism, and impact of mycotoxins on humans and animals. *Toxicology* 167(2), 101–34. doi:http://dx.doi.org/10.1016/S0300-483X(01)00471-1.

Ismaiel, A. and Papenbrock, J. (2015). Mycotoxins: Producing fungi and mechanisms of phytotoxicity. *Agriculture* 5(3), 492.

JECFA. (2007). *Joint FAO/WHO Expert Committee on Food Additives 68th Meeting (19th-28th June 2007): Summary and Conclusions (JECFA/68/SC)*, Geneva, Switzerland.

Khlangwiset, P., Shephard, G. S. and Wu, F. (2011). Aflatoxins and growth impairment: A review. *Critical Reviews in Toxicology* 41(9), 740–55. doi:10.3109/10408444.2011.575766.

Klich, M. A. (1998). Soil fungi of some low-altitude desert cotton fields and ability of their extracts to inhibit *Aspergillus flavus*. *Mycopathologia* 142, 97–100.

Kushiro, M. (2008). Effects of milling and cooking processes on the deoxynivalenol content in wheat. *International Journal of Molecular Sciences* 9(11), 2127–45. doi:10.3390/ijms9112127.

La Pera, L., Avellone, G., Lo Turco, V., Di Bella, G., Agozzino, P. and Dugo, G. (2008). Influence of roasting and different brewing processes on the ochratoxin A content in coffee determined by high-performance liquid chromatography-fluorescence detection (HPLC-FLD). *Food Additives & Contaminants: Part A* 25(10), 1257–63. doi:10.1080/02652030802112627.

Lahouar, A., Marin, S., Crespo-Sempere, A., Saïd, S. and Sanchis, V. (2016). Effects of temperature, water activity and incubation time on fungal growth and aflatoxin B_1 production by toxinogenic Aspergillus flavus isolates on sorghum seeds. *Revista Argentina de Microbiología* 48(1), 78–85. doi:http://dx.doi.org/10.1016/j.ram.2015.10.001.

Lanubile, A., Maschietto, V. and Marocco, A. (2014). Breeding maize for resistance to mycotoxins. In: *Mycotoxin Reduction in Grain Chains*, edited by J. F. Leslie and A. F. Logrieco. Chichester, UK: John Wiley & Sons, Ltd, pp. 37–58.

Lemmens, M., Steiner, B., Sulyok, M., Nicholson, P., Mesterhazy, A. and Buerstmayr, H. (2016). Masked mycotoxins: Does breeding for enhanced *Fusarium* head blight resistance result in more deoxynivalenol-3-glucoside in new wheat varieties? *World Mycotoxin Journal* 9(5), 741–54. doi:10.3920/wmj2015.2029.

Leszczyńska, J., Masłowska, J., Owczarek, A. and Kucharska, U. (2001). Determination of aflatoxins in food products by the ELISA method. *Czech Journal of Food Science* 19(1), 8–12.

Magan, N., Cayley, G. R. and Lacey, J. (1984). Effect of water activity and temperature on mycotoxin production by Alternaria alternata in culture and on wheat grain. *Applied and Environmental Microbiology* 47(5), 1113–17.

Manda, P., Dano, D., Kouadio, J., Diakité, A., Sangaré-Tigori, B., Ezoulin, M. J., Soumahoro, A., Dembele, A. and Fourny, G. (2009). Impact of industrial treatments on Ochratoxin A content in artificially-contaminated chocolate. *Food Additives and Contaminants* 26(7), 1081–8.

Medeiros, F. H. V. d., Martins, S. J., Zucchi, T. D., Melo, I. S. d., Batista, L. R. and Machado, J. d. C. (2012). Biological control of mycotoxin-producing molds. *Ciência e Agrotecnologia* 36, 483–97.

Mitchell, D., Parra, R., Aldred, D. and Magan, N. (2004). Water and temperature relations of growth and ochratoxin A production by Aspergillus carbonarius strains from grapes in Europe and Israel. *Journal of Applied Microbiology* 97(2), 439–45. doi:10.1111/j.1365-2672.2004.02321.x.

Molto, G. A., Gonzalez, H. H. L., Resnik, S. L. and Gonzalez, A. P. (1997). Production of trichothecenes and zearalenone by isolates of Fusarium spp. from Argentinian maize. *Food Additives & Contaminants* 14(3), 263–8. doi:10.1080/02652039709374523.

Mounjouenpou, P., Gueule, D., Fontana-Tachon, A., Guyot, B., Tondje, P. R. and Guiraud, J.-P. (2008). Filamentous fungi producing ochratoxin a during cocoa processing in Cameroon. *International Journal of Food Microbiology* 121(2), 234–41. doi:http://dx.doi.org/10.1016/j.ijfoodmicro.2007.11.017.

Mounjouenpou, P., Gueule, D., Ntoupka, M., Durand, N., Fontana-Tachon, A., Guyot, B. and Guiraud, J. (2011). Influence of post-harvest processing on ochratoxin A content in cocoa and on consumer exposure in Cameroon. *World Mycotoxin Journal* 4(2), 141–6. doi:10.3920/wmj2010.1255.

Mwanza, M., Abdel-Hadi, A., Ali, A. M. and Egbuta, M. (2015). Evaluation of analytical assays efficiency to detect aflatoxin M_1 in milk from selected areas in Egypt and South Africa. *Journal of Dairy Science* 98(10), 6660–7. doi:http://dx.doi.org/10.3168/jds.2014-9220.

Nelson, P. E., Plattner, R. D., Shackelford, D. D. and Desjardins, A. E. (1992). Fumonisin B_1 production by *Fusarium* species other than F. moniliforme in section Liseola and by some related species. *Applied and Environmental Microbiology* 58(3), 984–9.

Niderkorn, V., Morgavi, D. P., Pujos, E., Tissandier, A. and Boudra, H. (2007). Screening of fermentative bacteria for their ability to bind and biotransform deoxynivalenol, zearalenone and fumonisins in an in vitro simulated corn silage model. *Food Additives & Contaminants* 24(4), 406–15. doi:10.1080/02652030601101110.

Ogbadu, G. (1980). Influence of gamma-irradiation of aflatoxin B_1 production by *Aspergillus flavus* growing on some Nigerian foodstuffs. *Microbios* 27, 19–26.

Pardo, E., Marín, S., Sanchis, V. and Ramos, A. J. (2004). Prediction of fungal growth and ochratoxin A production by Aspergillus ochraceus on irradiated barley grain as influenced by temperature and water activity. *International Journal of Food Microbiology* 95(1), 79–88. doi:http://dx.doi.org/10.1016/j.ijfoodmicro.2004.02.003.

Paster, N., Barkai-Golan, R. and Padova, R. (1985). Effect of gamma radiation on ochratoxin production by the fungus Aspergillus ochraceus. *Journal of the Science of Food and Agriculture* 36(6), 445–9. doi:10.1002/jsfa.2740360604.

Perkowski, J., Chelkowski, J. and Wakulinski, W. (1990). Deoxynivalenol and 3-acetyl-deoxynivalenol in wheat kernels and chaff with head fusariosis symptoms. *Nahr Food* 34, 325–8.

Pitt, J. J. (2009). Principles and applications of liquid chromatography-mass spectrometry in clinical biochemistry. *The Clinical Biochemist Reviews* 30(1), 19–34.

Raters, M. and Matissek, R. (2005). Study on distribution of mycotoxins in cocoa beans. *Mycotoxin Research* 21(3), 182–6. doi:10.1007/bf02959259.

Raters, M. and Matissek, R. (2008). Analysis and occurrence of deoxynivalenol (DON) in cocoa. *European Food Research and Technology* 226(5), 1107–12. doi:10.1007/s00217-007-0638-4.

Ribeiro, J. M. M., Cavaglieri, L. R., Fraga, M. E., Direito, G. M., Dalcero, A. M. and Rosa, C. A. R. (2006). Influence of water activity, temperature and time on mycotoxins production on barley rootlets. *Letters in Applied Microbiology* 42(2), 179–84. doi:10.1111/j.1472-765X.2005.01830.x.

Sánchez-Hervás, M., Gil, J. V., Bisbal, F., Ramón, D. and Martínez-Culebras, P. V. (2008). Mycobiota and mycotoxin producing fungi from cocoa beans. *International Journal of Food Microbiology* 125(3), 336–40. doi:http://dx.doi.org/10.1016/j.ijfoodmicro.2008.04.021.

Scott, P. (1998). Industrial and farm detoxification processes for mycotoxins. *Veterinary Medicine Review* 149, 543–8.

Şenyuva, H., Gilbert, J., Özcan, S. and Gürel, N. (2008). Rapid LC and LC/MS for routine analysis of mycotoxins in foods. *World Mycotoxin Journal* 1(3), 229–35. doi:10.3920/WMJ2008.x033.

Serra, B. J. (2004). Occurrence of ochratoxin A in cocoa products and chocolate. *Journal of Agricultural and Food Chemistry* 52(20), 6347–52. doi:10.1021/jf040153w.

Shuaib, F. M. B., Ehiri, J., Abdullahi, A., Williams, J. H. and Jolly, P. E. (2010). Reproductive health effects of aflatoxins: A review of the literature. *Reproductive Toxicology* 29(3), 262–70. doi:http://dx.doi.org/10.1016/j.reprotox.2009.12.005.

Sinha, K. K. and Bhatnagar, D. (1998). *Mycotoxins in Agriculture and Food Safety*. New York: Marcel and Delkker Inc.

Sobrova, P., Adam, V., Vasatkova, A., Beklova, M., Zeman, L. and Kizek, R. (2010). Deoxynivalenol and its toxicity. *Interdisciplinary Toxicology* 3(3), 94–9. doi:10.2478/v10102-010-0019-x.

Torres, O., Matute, J., Gelineau-van Waes, J., Maddox, J. R., Gregory, S. G., Ashley-Koch, A. E., Showker, J. L., Voss, K. A. and Riley, R. T. (2015). Human health implications from co-exposure to aflatoxins and fumonisins in maize-based foods in Latin America: Guatemala as a case study. *World Mycotoxin Journal* 8(2), 143–59. doi:10.3920/wmj2014.1736.

Turcotte, A.-M., Scott, P. M. and Tague, B. (2013). Analysis of cocoa products for ochratoxin A and aflatoxins. *Mycotoxin Research* 29(3), 193–201. doi:10.1007/s12550-013-0167-x.

Valencia-Quintana, R., Sánchez-Alarcón, J., Tenorio, M. G., Deng, Y., Waliszewski, S. M. and Valera, M. Á. (2012). Preventive strategies aimed at reducing the health risks of Aflatoxin B_1. *Toxicology and Environmental Health Sciences* 4(2), 71–9. doi:10.1007/s13530-012-0119-4.

Wang, Y., Wang, L., Liu, F., Wang, Q., Selvaraj, J., Xing, F., Zhao, Y. and Liu, Y. (2016). Ochratoxin A producing fungi, biosynthetic pathway and regulatory mechanisms. *Toxins* 8(3), 83.

Wild, C. P. and Gong, Y. Y. (2010). Mycotoxins and human disease: A largely ignored global health issue. *Carcinogenesis* 31(1), 71–82. doi:10.1093/carcin/bgp264.

Williams, W. P. (2006). Breeding for resistance to aflatoxin accumulation in maize. *Mycotoxin Research* 22(1), 27–32.

Yin, Y.-N., Yan, L.-Y., Jiang, J.-H. and Ma, Z.-H. (2008). Biological control of aflatoxin contamination of crops. *Journal of Zhejiang University. Science. B* 9(10), 787–92. doi:10.1631/jzus.B0860003.

Yu, J. (2012). Current understanding on aflatoxin biosynthesis and future perspective in reducing aflatoxin contamination. *Toxins* 4(11), 1024–57. doi:10.3390/toxins4111024.

Zain, M. (2011). Impact of mycotoxins on humans and animals. *Journal of Saudi Chemical Society* 15, 129–44.

Chapter 18

Analysing sensory and processing quality of cocoa

Darin A. Sukha and Naailah A. Ali, The University of the West Indies, Trinidad and Tobago

1 Introduction

Cacao belongs to the family Malvaceae (Bayer et al., 1999) and originated as an understorey tree species in the tropical rainforests of the upper Amazonian region of South America and its current distribution extends across a belt 20°N and 15°S latitude (Butler and Umaharan, 2004; Mossu, 1992). This tropical equatorial zone is characterised by heavy, well-distributed rainfall (1500–2000 mm) with a distinct dry season, consistent high humidity and uniformly high temperatures throughout the year (26–36°C) (78.7–96.8°F) (Mossu, 1992; Wood and Lass, 1985). The genus *Theobroma* contains some 22 species and cacao is the only species that is cultivated on a large scale (Wood and Lass, 1985). Traditionally three different morpho-geographical groups (Criollo, Forastero and Trinitario) were recognised within cacao, based on genetic origin, pod morphology and size, as well as, colour and flavour of beans (Engels, 1981; Laurent et al., 1994; Toxopeus, 1985).

'Cacao' usually refers to the tree, the pods and the unfermented beans from the pods, whilst 'cocoa' generally refers to the manufactured product sold for drinking or food-manufacturing purposes, but recently it has also been frequently used to describe the fermented cocoa beans in bulk (Badrie et al., 2015). This format will be followed in the chapter.

Cocoa has always been appreciated for its sensory properties and positive effects on the body and is now classed into that category of foods whose characteristic flavour is

http://dx.doi.org/10.19103/AS.2017.0021.27

developed through a fermentation process (Lopez and Dimick, 1995; Schwan and Wheals, 2004; Badrie et al., 2015). The first intimations of cocoa consumption date back to 1600 BC and its consumption was linked to divinity and longevity (Corti et al., 2009). Both the Maya and Aztec civilisations believed the cacao tree to be of divine origin (Lippi, 2009). When Spanish conquistador Hernán Cortés met the Aztecs, he observed that cocoa beans were used in the preparation of a drink – 'chocolatl' by roasting the beans, grinding and mixing them with maize meal, vanilla and chilli (Wood and Lass, 1985). No doubt the divine origin of cacao was considered when Swedish botanist Carl Linnaeus published his first edition of the *Species Plantarum* in 1753 where the scientific name of cacao was taken from the Greek words *theo* (god) and *broma* (drink) (Linnaeus, 1753).

Recent figures from the International Cocoa Organization (ICCO) indicate that world production of cocoa beans in 2015/2016 is now estimated to be higher than previously envisaged, at 4.031 million tons. However, this figure represents a decline of 5% compared to the previous season according to ICCO-revised crop estimates (ICCO, 2016a). In terms of its share of total world production, Africa remains by far the largest cocoa-producing region, accounting for 73% of world cocoa output in 2015/2016. The shares of the Americas and of the Asia and Oceania regions are 16% and 11% respectively (ICCO, 2016a). Côte d'Ivoire is currently the world's top cocoa-producing country, accounting for more than 40% of world production, followed by Ghana, Indonesia, Nigeria, Brazil, Cameroon and Ecuador in descending order of production volume. Collectively, these countries produce more than 90% of the world's cocoa, and the global chocolate market was forecasted to be worth US$98.3 billion dollars in 2016 according to Markets and Markets (2011).

According to the ICCO (2016b), the world cocoa market distinguishes between two broad categories of cocoa beans that are traded: 'bulk' or 'ordinary' and 'fine or flavour' cocoa beans. As a broad generalisation, bulk or ordinary cocoa beans, which account for approximately 95% of the world cocoa production, are produced mainly in West Africa and come from Forastero trees, whilst fine or flavour cocoa beans are produced from Criollo or Trinitario cacao tree varieties. There are, however, known exceptions to this generalisation. 'Nacional' trees in Ecuador, considered to be Forastero-type trees, produce fine or flavour cocoa. On the other hand, cocoa beans from Cameroon, produced by Trinitario-type trees and whose cocoa powder has a distinct and much sought-after red colour, have so far been classified as bulk cocoa beans (ICCO, 2016b).

The share of fine or flavour cocoa in the total world production of cocoa beans is relatively small and has been falling over the years, from a 40–50% share at the beginning of the twentieth century, when Ecuador and Trinidad and Tobago were the major producers of fine or flavour cocoa (ICCO, 2016c). Currently, approximately 5% of world cocoa production is classified as being fine or flavour, with 23 countries included in Annex 'C' of the International Cocoa Agreement 2010 as partial or exclusive producers and exporters of fine or flavour cocoa (see Table 1) (ICCO, 2016d).

Apart from being small compared to the international market for bulk cocoa, the fine or flavour cocoa market is regarded as being highly specialised and separate, with its own supply and demand characteristics, but lucrative, with an annual value of over US$4 billion. Fig. 1, adapted from the ICCO (ICCO, 2016c), provides a summary of the price ranges (in US$ per metric ton) and tiered marketing structure for bulk or ordinary and fine or flavour cocoas with estimated quantities available at current production levels showing the very small premium specialty cocoa market that now exists for cocoas under the current system of genetic and market definitions. Typically, fine or flavour cocoa commands between US$3700 and 5000 per metric ton with some origins fetching between US$6000 and

Table 1 Countries to be included in Annex 'C' of the International Cocoa Agreement 2010 as partial or exclusive producers and exporters of fine and flavour cocoa

Countries	Share of total exports of the country classified as fine and flavour cocoa
Belize	50%
Bolivia	100%
Colombia	95%
Costa Rica	100%
Dominica	100%
Dominican Republic	40%
Ecuador	75%
Grenada	100%
Guatemala	50%
Honduras	50%
Indonesia	1%
Jamaica	95%
Madagascar	100%
Mexico	100%
Nicaragua	100%
Panama	50%
Papua New Guinea	90%
Peru	75%
Saint Lucia	100%
São Tomé and Principé	35%
Trinidad and Tobago	100%
Venezuela, Bolivarian Rep. of	100%
Vietnam	40%

Source: (ICCO, 2016d).

10 000 per metric ton compared to US$3000–3500 per metric ton paid for bulk or ordinary cocoa and US$3100–3700 for certified bulk cocoa (ICCO, 2016c). Presently, Latin America and the Caribbean produce 80% of the world's fine or flavour cocoa. Ecuador is the world's largest exporter, accounting for 65% of the world production (ICCO, 2016c).

From these illustrious beginnings, the cocoa bean continues to enrich the lives of us all. Indeed, high-quality cocoa beans offer chocolate and confectionery manufacturers a cornucopia of delight in the products that can be made. Overall growth of the chocolate industry over the last decade as well as demand for fine or flavour cocoa has been growing rapidly over the past five years. This is as a result of increasing awareness of the health benefits of certain types of chocolate, growing popularity for chocolate, especially in Asian

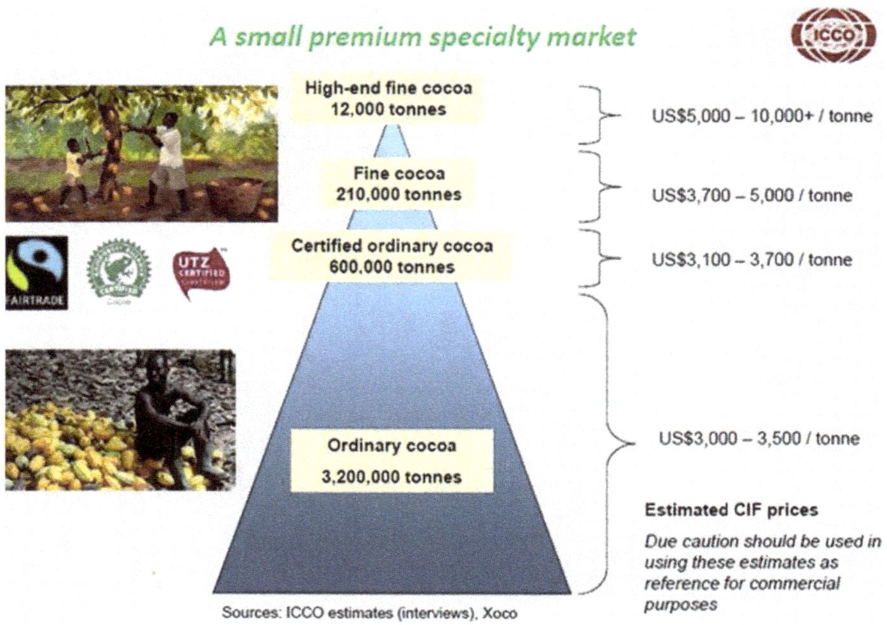

Figure 1 A summary of the price ranges (in US$ per metric ton) and tiered marketing structure for bulk or ordinary and fine or flavour cocoas, with estimated quantities available at current production levels showing the very small premium speciality cocoa market share. (Adapted from ICCO, 2016c).

and Pacific countries, with particular attention to origin-specific dark chocolates becoming increasingly popular, particularly in the more mature markets of Western Europe, North America and Japan (Franchise Help 2016; Hustler, 2014).

1.1 Standards and protocols in cocoa sensory and processing quality

At first glance, the practical reality faced by millions of cacao farmers all over the world is that there seems not much they can do to affect the end result in terms of cocoa quality. They have very little influence over the cultural factors of cacao growing, as parameters such as the chemical characteristics of the soil available to them, the genetic make-up of the planting material used and the climatic environment are imposed on them by nature and science. The farmer may exercise some choice in the cacao varieties planted but this is constrained by the diversity and characteristics of the cacao varieties available from research and extension services in that country or region. The farmer may also be able to exercise some choice regarding the most suitable location on the farm where cocoa is planted.

In many instances, the only thing a farmer can realistically exercise some control over affecting bean quality is by employing 'Good Agricultural Practices' (GAP). These are best achieved by concentrating on good cultivation and post-harvest practices through good husbandry of the cocoa farm, pest and disease control and harvest and post-harvest handling whilst considering social and environmental dimensions of these activities into production practices.

Addressing the complex challenge of achieving a sustainable balance between cocoa farming and improving farmer livelihoods whilst producing cocoa in the quality and quantity needed by the cocoa and chocolate industry represents an elusive concept or a 'zero sum game'[1] between GAP and rigid industry-quality certification criteria and specifications. In light of the limitations farmers face, their concerted attempts to meet often poorly defined or communicated industry-quality certification requirements offer little sustainable and holistic economic benefit to them.

This was a harsh reality that the industry faced in the mid-1990s, as consumer awareness, access to information and the industry as a whole started to become more 'quality-centric'. Deep concerns were expressed over the deterioration in the physical quality of cocoa beans supplied to the industry and lack of economic sustainability at the critical initial supply point in the cocoa value chain. It appeared that farmers were not consistently harvesting, fermenting and drying their cocoa in line with recommended practices. There was also a lack of knowledge or application of best agronomic practices and inefficient supply chains in producing countries where cocoas of widely different fermentation qualities and bean size were often blended down to the minimum limits which contracts allowed to produce Fair Average Quality (FAQ)-traded cocoa.

These problems were partly attributed to a lack of clearly established quality criteria and unclear communication coupled with lapses or difficulty in enforcement and measurement of these criteria in the origin countries. A more integrated approach to cocoa production was urgently needed. This reality and need identified in the mid-1990s spawned the many long-term cocoa sustainability initiatives we see operating today in the sector that are supported by various industry interests.

Arising from all this good intention around the holistic appreciation of both sensory and processing quality along the cocoa value chain and direct interventions at origin by many chocolate companies and industry groups (large and small) were a plethora of different quality criteria for cocoa being developed and communicated to cocoa farmers with varying levels of depth, clarity and consistency. Indeed, Fowler (1996) surmised at the start of all this attention that cacao growers or researchers should not be surprised if different companies have different quality certification criteria for accepting or rejecting cocoa beans. He added that even different factories of the same company may have different requirements (Fowler, 1996).

Rewind to the last 8–10 years and we have a situation where there is some confusion caused by the barrage of information and good intentions all focused on quality and targeted at the key stakeholders – farmers, cooperatives, producer and marketing groups. Among stakeholders, it appears that we are all taking different roads to reach a similar destination – 'QUALITY – with no real coordination between ourselves. There have been different versions of 'cocoa and chocolate quality' definitions used, as well as, different approaches to arrive at and assess for cocoa and chocolate quality at origin with different bean to bar and larger-scale chocolate producers working at origin.

As a corollary to this, there are number of international initiatives recognising cocoa quality that have been launched within the last 8–10 years such as the Cocoa of Excellence Programme and International Cocoa Awards, Heirloom Cocoa Preservation Fund and the

1. Zero-sum is a situation in economic or game theory in which one person's gain is equivalent to another's loss, so the net change in wealth or benefit is zero. A zero-sum game may have as few as two players, or millions of participants.

International Chocolate Awards, to mention a few. These initiatives build on the trend of heightened consumer awareness for recognising bean origins and genetics used in various chocolates and essentially take the form of receiving bean samples, processing them according to some defined set of protocols followed by qualitative and/or quantitative assessment by a trained or experienced tasting panel. The modalities and execution between each initiative differ but all try to link flavour quality to genetics, farm or region with the aim of recognising desirable quality attributes and/or facilitating niche marketing.

This situational analysis summarises the need for and genesis of standards and protocols in cocoa sensory and processing quality. It also highlights the inextricable link between GAP and Quality Certification. Both are critical considerations in this chapter and highlight the urgent need to create closer links between cocoa producers and consumers to improve the recognition, preservation and valuation of cacao diversity by promoting awareness of different qualities of cocoa in the different cocoa markets and how quality is assessed.

1.2 The knowledge divide

Amidst the positive growth in demand forecasted for the dark and specialty chocolate sector, there seems to be a disconnect or lag between the output of relatively recent research and the ability of the cocoa industry to adopt and capitalise on this research to expand existing marketing opportunities. This is especially the case in the areas of processing quality and sensory evaluation which raises critical questions on how quality in cocoa should be defined and assessed with its link to more effective marketing and improved incomes across the value chain.

The continued use of the traditional definition of cacao into three genetic groups, that is, Criollo, Forastero and Trinitario, by the industry and the narrow classification of the industry into 'bulk' or 'ordinary' and 'fine or flavour' cocoa seem counter-intuitive in the light of new information and current research.

Motomayor et al. (2008) in a study of cacao germplasm from Central and South America delineated 10 distinct genetic clusters using microsatellite markers as opposed to the two genetic groups and one hybrid group traditionally recognised (Criollo, Forastero and Trinitario). The results of this new classification of cacao germplasm with 10 genetic clusters support the palaeoarches hypothesis of species diversification and further research is still being done, with more groups, or sub-groups, being defined as more information on the genetic structure of wild populations becomes available. A better understanding of population differentiation within the species facilitates enhanced management and utilisation (Motomayor et al., 2008). However, according to a market study of 11 selected fine or flavour cocoa-producing countries conducted by the Royal Tropical Institute (van der Kooij, 2013), the elucidation of the 10 new genetic groups raises important questions regarding their definition as either 'bulk' or 'ordinary' and 'fine or flavour' in the context of the current ICCO (ICCO, 2016b) definitions. The Royal Tropical Institute (van der Kooij, 2013) also stated in their market study that the word 'Criollo' is not only used to describe the variety of cacao, but also used to mean 'local' or 'native' cacao, which adds to the confusion around the use of the word Criollo in defining varieties.

Five years of work carried out in Ecuador, Papua New Guinea, Trinidad and Tobago and Venezuela coordinated by the ICCO between 2001 and 2006 in a project to determine physical, chemical and organoleptic parameters to differentiate between fine or flavour and bulk cocoas (ICCO, 2007) provided strong evidence that, in general, fine or flavour cocoas from different origins have distinct flavour profiles, thus reducing market competition among them because they occupy distinct flavour niches in the market. This project laid the foundation for standardisation of quality evaluation protocols. The findings also provided a strong incentive for future collaborative research among fine or flavour cocoa-producing countries to advance the development of this market segment in the context of analysing the sensory and processing quality of cocoas from different origins. Yet even after this project, the findings were not implemented and the industry still grapples with the fundamental issue of defining and understanding quality in cocoa and having universally accepted criteria to differentiate between fine or flavour and bulk cocoas.

Building on the project to determine physical, chemical and organoleptic parameters to differentiate between fine or flavour and bulk cocoas, work by Sukha et al. (2008, 2014) optimised flavour assessment protocols and has linked flavour and intrinsic quality attributes to a particular growing origin (country, region or farm), cacao varieties and set of processing conditions, thereby applying the concept of 'terroir' – which has already been well established for wines – to cocoa (Sukha et al., 2014). Aspects of this work are now being considered in producing a harmonised approach towards cocoa quality assessment and in providing a scientific basis for cocoa quality standards and certification programmes that legitimises true 'origin-specific' marketing claims made with fine or flavour cocoa origins.

Although the general trend shows an expansion of chocolate consumption, there is still a considerable knowledge gap among consumers about quality, flavours and origins. Many consumers of chocolate products are unaware of the difference in qualities of cocoa beans and the efforts undertaken by smallholder cocoa farmers to achieve better quality. This, in turn, discourages farmers from producing high-quality cocoa beans, as there is no financial recognition for these efforts.

Closing this divide involves raising awareness to tackle the knowledge gap of consumers and to also raise awareness among cocoa farmers about consumer demand for quality cocoa beans and the quality requirements linked to that demand.

2 The evolution of sensory evaluation as a science

The broad history of sensory evaluation is inextricably bound to the history of mankind as the quest for food has fuelled our evolution as a civilisation. Even the most primitive animals have well-developed preferences for certain types of foods. Although Palaeolithic man may have had little choice in his never-ending quest for food as a hunter-gatherer, he would have obviously rejected certain foods by trial and error, such as foods that made him ill, and favoured certain types of foods that were more pleasurable or in relatively abundant supply across the seasons. The inherent need for survival superseded considerations of flavour which most likely did not occur until man accidently discovered fire and roasting. The art of early roasting of food allowed for some level of preservation and facilitated the change from hunter-gatherer to one of a more settled existence and evolution to Neolithic man.

Neolithic man planted cereals, domesticated animals, irrigated fertile land and settled into villages. This heralded the birth of early civilisations where, as food became abundant, society differentiated itself into rulers, producers, artisans, priests and warriors, etc. Leisure time allowed for relatively rapid flavour distinctions and food preferences to evolve especially along the lines of class and status (Amerine et al., 2013).

The evolution of sensory evaluation continued as mankind engaged in trading, where buyers sampled food in the hope that it would be representative of the whole before purchase or barter. Likewise, sellers adjusted their prices or exchange demand based on the responses and preferences displayed. This led to ritualistic or more formalised schemes for grading early commodities such as wine, coffee, tea, etc., which continues to this day albeit in a more systematic and structured way (Meilgaard et al., 2006). Sensory evaluation also evolved as a more formal discipline in response to warfare and the need to feed soldiers going into battle. The maxim that *'an army marches on its stomach'* has great historical importance in deciding the fates of many conquests that have shaped our modern world.

Indeed, modern sensory evaluation as we know it today, is a discipline born after the Second World War in response to post-war urbanisation and the more mainstream entry of women into the workplace. According to Skinner (1989) as cited in Lawless and Heymann (2010), *'modern sensory evaluation is a child of industry spawned in the late 1940's by the rapid growth of processed food and consumer product companies'*. This was the time when structured sensory tests were developed and techniques for accurate measurements of human responses to foods with minimisation of bias and the development of brand identity based on consumer perception were codified. The application of instrumentation and more quantitative techniques towards sensory evaluation with codification of methodology only relatively recently have evolved to meet the challenges faced with space travel, habitation in space and keeping astronauts fed whilst in orbit. Towards this end, Stone and Sidel (1993) defined modern sensory evaluation as 'a scientific method used to evoke, measure, analyse and interpret those responses to products as perceived through the senses of sight, smell, touch, taste and hearing'. This definition has been accepted and endorsed by the Institute of Food Technologists as well as the American Society for Testing and Materials. Other organisations use similar definitions for sensory but all consider these core principles as set out by Stone and Sidel (1993).

Taste is the sensation perceived on the tongue, of which five are recognised, that is, salt, sweet, bitter, acid/sour and umami (glutamate/savoury). A sixth taste for fat 'oleogustus' is being debated (Laugerette et al., 2007) and is linked to rancidity. It is important to note that the work by Collings (1974) debunked the taste map concept proposed in work done by Hänig (1901), which stated that different sections of the tongue were exclusively responsible for different basic tastes. It is illustrated with a schematic map of the tongue, with certain parts of the tongue labelled for each taste. Collings revisited this early work and found that some facts were misinterpreted during translation from the original German text to English and proved that there is sensitivity to the basic tastes on the entire surface of the tongue and no segmentation of perception exists where only certain areas on the tongue perceive certain tastes.

Flavour, on the other hand, is the sensation produced by a material taken in the mouth. Flavour perception is a more complex sensation involving our senses of both smell and taste that become responsive to pressure, touch, temperature, common chemical sensitivity, vision and hearing as mediated by the trigeminal nerves in what are termed 'Chemesthetic' sensations (Lawless and Heymann, 2010). This adds to our mental library of sensations for education and comparison each time we experience a flavour event.

3 Cocoa quality and the link to flavour

The quality of cocoa beans is a key factor that will highly influence the various end products made from cocoa beans (chocolates, cocoa liquor, cocoa butter, nutraceuticals, etc.). In the broadest sense of the word, 'quality' may be considered as a specification or set of specifications which are to be met within given tolerances or limits (Kramer and Twigg, 1970). The *Concise Oxford Dictionary* goes on to further define 'quality' as '...possessing a degree of excellence...and to be concerned with the maintenance of high quality' (Alle, 1990). Quality is therefore concerned with meeting the requirements or specifications that define high quality for that particular item or product.

Maximising the potential of cocoa is linked to quality and in this context according to a definition from the Association of Chocolate, Biscuits and Confectionery of Europe (CAOBISCO)/European Cocoa Association (ECA)/Federation of Cocoa Commerce (FCC) (2015) in their revised 'Cocoa Beans: Chocolate and Cocoa Industry Quality Requirements', cocoa quality includes *'the all-important aspects of flavour and purity, and physical characteristics that have a direct bearing on manufacturing performance and aspects such as traceability, geographical indicators and certification to indicate the sustainability of the production methods'*. According to CAOBISCO/ECA/FCC (2015), aspects or specifications of quality that affect the value and price paid for a parcel of beans in cocoa therefore include:

- Flavour
- Food safety and wholesomeness
- Physical characteristics
- Consistency
- Yield of edible material
- Cocoa butter characteristics
- Colour potential – 'colourability'
- Traceability, geographical indications and certification

A key challenge faced by cocoa industry continues to reside in the inconsistency and subjectivity of some of the protocols employed to assess the quality of all types of cocoa that are traded either bulk or ordinary and fine or flavour. According to ICCO (2016c), usually, a combination of criteria and protocols are used and include:

- Identifying the genetic origin of planting material
- Morphological characteristics of the plant
- Assessing the flavour characteristics of the cocoa beans produced
- Analysing chemical characteristics of the cocoa beans
- Colour of the cocoa beans and nibs
- Degree of fermentation and drying
- Presence of acidity and off-flavours
- Percentage of internal mould and insect infestation
- Percentage of impurities

3.1 Unlocking genetic flavour potential in cacao

Each cacao bean (regardless of its traded end use as either 'bulk' or 'fine or flavour') has a genetic flavour potential that is either expressed or depressed as it moves along the cocoa-processing chain during the primary processing activities that include fermentation, drying, as well as, storage and other handling operations. Secondary processing follows, with roasting and conching being the two main operations that have a substantial impact on final flavour and aroma quality in chocolate (Kamphuis, 2011). Roasting is an integral part of the manufacturing process to produce the final flavour that we associate with chocolate. It is also described as a dry heat treatment that produces certain flavour and aroma compounds mainly via Maillard reactions and Strecker degradation (non-enzymatic browning reactions). Maillard-type reactions involve the reaction of free amino groups of amino acids and reducing sugars to produce heterocyclic compounds with distinct aromas and flavours even at low thresholds. Roasting also inhibits the activity of polyphenoloxidase enzymes and decreases free acetic acidity which eliminates some undesirable volatile compounds of cocoa (Rohsius et ai., 2006a; Bonvehí, 2005; Jeanjean et al., 1994; Cros, 1996, 1998). Flavour modifications during conching arise from the elimination of flavour volatiles during the initial hours of processing (Manière and Dimick, 1979 cited in Cros, 1996).

The first consideration in the definition of cocoa quality from CAOBISCO/ECA/FCC (2015) refers to flavour, which is a key criterion of quality for manufacturers of cocoa products. The flavour criterion includes both the intensity of the cocoa or chocolate flavour (typical in many traded 'bulk' cocoas) together with ancillary flavour notes (dominant in many traded 'fine or flavour' cocoas) and the absence of flavour defects regardless of traded cocoa type. It should be noted, according to the ICCO (2016b), that the difference between bulk or ordinary cocoa and fine or flavour cocoa is primarily in the flavour but can include other quality factors such as colour and origin. Fine flavours include fruit (fresh and browned, mature fruits), floral, herbal, and wood notes, nut and caramel type notes, as well as rich and balanced chocolate bases (ICCO, 2016b).

3.2 Primary processing of cocoa

Best practices during primary processing have the greatest impact on expressing this genetic potential, whilst poor practices depress or reduce the expression of full genetic potential. Having a clear understanding of what this potential is, being able to communicate this information and most importantly being able to assess for these attributes are fundamental and have a direct impact on final cocoa bean quality.

3.2.1 Cacao fermentation

Beans inside the unopened cacao pod are microbiologically sterile with fresh cocoa seeds being characterised by an astringent and bitter taste due to the high content of phenolics, especially anthocyanins (Jinap et al., 2005) and because they do not contain the necessary aroma precursors (Ziegleder and Biehl, 1988). Once the pod is opened, the pectinaceous pulp surrounding the beans is exposed to many sources of inoculation and fermentation begins. The process of fermentation has been widely researched for many years, with early work being carried out by Knapp (1937) and De Witt (1955).

During fermentation, there is a microbial succession of a wide range of yeasts, lactic-acid, and acetic-acid bacteria during which high temperatures of up to 50–52°C (122–126°F) and microbial products, such as ethanol, lactic acid, and acetic acid, permeate through the differentially permeable seed coat into the seed tissue, causing seed death and the production of flavour precursors by degradation of the polyphenols, storage proteins and carbohydrates (Ostovar and Keeney, 1973; Schwan 1998; Gálvez et al., 2007; Nielsen et al., 2007; Camu et al., 2008; Kostinek et al., 2008; De Bruyne et al., 2008; Badrie et al., 2015). The microbial succession, ingress of acidity, proteolysis and oxidative reactions yield aroma precursors including peptides, free amino acids and reducing sugars that unlock the genetic flavour potential that resides within each cacao bean (Afoakwa et al., 2008; Andersson et al., 2006; Rohsius et al., 2006b; Kadow et al., 2013; Schwan and Wheals, 2004; Ziegleder and Biehl, 1988; Badrie et al., 2015).

Over-fermentation leads to a rise in bacilli and filamentous fungi that can cause off-flavours. The physiological roles of the predominant microorganisms are now reasonably well understood and the crucial importance of a well-ordered microbial succession in cocoa aroma development has been established and reported by authors such as Schwan and Wheals (2004), Afoakwa et al. (2008) and Badrie et al. (2015).

In addition to the endogenous formation of flavour and aroma precursors from fermentation, there is also mounting evidence for the exogenous influence of the pulp directly on aroma and flavour development. Kadow et al. (2013) highlighted quantitative and qualitative differences in the aromatic profile of fruit pulp from different cocoa cultivars and that aroma components present in the fruit pulp migrate into the cotyledon tissue during ripening in addition to those derived from fermentation. Whilst Eskes et al. (2007, 2012) and Ali et al. (2013) discussed the direct permeation of aroma components from the pulp into the seed tissue during fermentation and the addition of pulp from other fruits as an indicator that affects final flavour quality which may be retained during the drying process.

Recent work by Sukha et al. (2017) failed to detect direct Xenia effects for most of the important flavour attributes, that is, cocoa flavour, acidity, fruitiness and floral flavour which agree with the preliminary work of Clapperton et al. (1994a,b) and Lockwood and Eskes (1996), suggesting a general lack of pollen donor (Xenia[2]) effect on flavour in cacao and implying that the flavour quality of cocoa beans is determined principally by the genotype of the female parent.

Current work by Ali et al. (2016) is looking at the fermentation requirements, flavour and other quality attributes of six of the ten genetic clusters identified by Motomayor et al. (2008) to determine the optimal fermentation end point allowing for the exploitation of the inherent diversity in pulp quality and bean chemical characteristics towards genetic branding.

3.2.2 Cocoa drying

Following fermentation, cocoa beans are dried to reduce the moisture content from about 60% to 6–7% for safe storage and transportation. Drying also continues oxidative chemical reactions that reduce the astringency and bitterness of the bean (Wood, 1961; Hii et al., 2008; Fagunwa et al., 2009). Sun-drying of cocoa beans is the most widely practised method of drying due to the simplicity and use of natural energy, but this method is labour intensive, weather dependent and requiring large amounts of space to locate drying floors (Bravo, 1974; McDonald and Lopez, 1981; Bonaparte et al., 1998; Hii et al., 2006). In areas where rainy weather is prevalent during peak cocoa crop times and/or when production capacity outstrips available drying capacity or space, artificial drying systems are used to shorten drying time and increase the rate of moisture removal (Wood and Lass, 1985; Faborode et al., 1995; Ndukwu, 2009).

Regardless of the drying method used, the final moisture content of the bean and the drying rate to arrive at this moisture content have direct implications on physical, chemical, sensory, and food safe quality of the beans (Faborode et al., 1995; Schwan and Wheals,

2. The Xenia effect in plants is the effect of pollen on seeds and fruit of the fertilised plant. The effect is separate to the contribution of the pollen towards the next generation.

2004; Rodriguez-Campos et al., 2012; Jinap, 1994; Jinap et al., 1994). The rate of drying depends on two factors: transfer of heat into the bean and the movement of water vapour from the bean to the surrounding air (Wood and Lass, 1985; Wan Daud et al., 1996; Hii et al., 2008, 2009; Chinenye et al., 2010).

In artificial drying extra care should be taken to have a relatively slow drying rate with the air temperature not exceeding 40°C until the moisture content of the beans is below 20% (Clapperton et al., 1994a). A final moisture content of below 5% can result in brittle beans, whilst moisture contents of more than 8% can increase the risk of mould growth (Wood and Lass, 1985; Schwan and Wheals, 2004). Slow drying can sometimes result in over-fermentation of the beans and mould growth outside and inside the bean, which can contribute to strong putrid, rancid and hammy off-flavours (Schwan and Wheals, 2004; Hii et al., 2006). Too rapid drying can result in acidic beans with shrivelling and/or case hardening of the shells (Augber et al., 1999; Jinap, 1994; Jinap et al., 1994). Smoke contamination on the other hand arises from directly fired artificial dryers without heat exchangers that use either fuel oil (diesel) or wood (Lowor et al., 2012).

To offset the drawbacks associated with sun-drying and artificial drying of cocoa beans using fuelled heating, artificial solar dryers are being used (McDonald and Lopez, 1981; Sankat and Ramlochan, 1997; Hollywood et al., 1996; Bonaparte et al., 1998; Fagunwa et al., 2009; Lasisi, 2014) with varying degrees of success on final flavour, physical bean characteristics and drying efficiency.

3.2.3 Cleaning, bagging and storage of cocoa beans

Freshly fermented and dried beans usually have either very strong volatile fruit notes, lower cocoa flavour notes, acid notes (especially acetic acid), and a range of possible off notes. It is therefore recommended that beans are stored for a period of time under appropriate conditions to 'age' before they are transformed into powder and/or liquor. Cleaning, bagging and physical assessments are usually carried out during this time which can vary between 6 and 12 weeks.

Cleaning of fermented and dried beans has an impact on factors associated with consistency and purity that are discussed later in the text. Bags used to store cocoa beans should be food safe, clean (ideally new) and strong enough to discourage pest contamination, as well as, properly sewn or sealed to withstand transport and storage. They should also be well labelled and where there is an overlap of cocoa production with other allergenic crops, only new or cocoa dedicated bags should be used to avoid cross-contamination.

Storage in the tropics poses a unique set of challenges related to preventing re-humidification, insect infestation and contamination of the beans, as well as providing appropriate ventilation. In general cocoa beans should be stored in an area that is clean with appropriate temperature and humidity control, free from insect and rodent pests or either at 20°C (68°F) for immediate and intermediate storage or at −20°C (−4°F) for longer-term storage (CAOBISCO/ECA/FCC, 2015). Storage periods in the tropics should not exceed three months unless special precautions are taken. Secondary mould growth from re-humidification caused by moisture contents above 8% leads to fat degradation, with a rise in free fatty acids (FFAs) leading to negative flavour and other quality effects later on during the chocolate-making process (Wood and Lass, 1985; CAOBISCO/ECA/FCC, 2015).

Apart from bulk storage, bean samples (sampling is discussed later in this chapter) taken for quality assessments should be placed in suitable food-safe containers (plastic container, high-quality snapseal plastic bags, drawstring cotton jute or burlap bag or vacuum sealable bag). The latter is a specific consideration if aroma volatiles or other chemical assessments are planned for the sample (CAOBISCO/ECA/FCC, 2015).

3.3 Processing quality in cocoa

In addition to flavour, processing quality is another critical criterion that considers aspects of cocoa beans that are more physical in nature and according to CAOBISCO/ECA/FCC (2015) are linked to:

- Consistency
- Purity
- Yield of edible material
- Cocoa butter characteristics

The Model Ordinance of the International Cocoa Standards (Wood and Lass, 1985) indicates that cocoa of merchantable quality is defined as being:

- Fermented, thoroughly dry, free from smoky beans, free from abnormal or foreign odours and free from any evidence of adulteration;
- Must be reasonably free from living insects; and
- Reasonably uniform in size, reasonably free from broken beans, fragments, and pieces of shell, and be virtually free from foreign matter.

3.3.1 Consistency

The first part of the Model Ordinance of the International Cocoa Standards definition (Wood and Lass, 1985) considers flavour quality, but consistency of flavour as well as other attributes are also important considerations impacting processing quality. As stated by CAOBISCO/ECA/FCC (2015), no matter what the genetic origin, quality, both from bag to bag within a particular consignment or lot of beans, and between different consignments or lots of the same mark, needs to be consistent. A consistent quality and a mark or origin which can be relied upon to supply beans of consistent quality will be valued more highly than one of heterogeneous quality as chocolate manufacturers of all sizes (bean to bar, intermediate and industrial scale) aim to produce chocolate of consistent quality (CAOBISCO/ECA/FCC, 2015).

Other aspects of consistency refer to the amount of defective beans, bean size distribution, average bean size and degree of fermentation. The weight of a cocoa bean should be at least 1.0 g and bean weight is easily measured. There is a negative correlation between bean weight and shell content, that is, smaller beans have proportionately more shell which is considered waste. Shell content (with residues of mucilage and testa) of finished cocoa beans varies from 11 to 17% depending on genotype (Lockwood and Eskes, 1996), and although smaller beans can be used in chocolate making, their higher shell contents (and proportionately more waste compared to nib[3]) make their use uneconomical.

Whole bean roasting produces more delicate flavours due to the preservation of volatile cocoa flavour notes in the shell during roasting; therefore too much variability in individual

3. Cocoa nibs are cocoa beans that have been roasted, separated from their husks, and broken into smaller pieces.

bean weight in a roasting batch is undesirable if beans are roasted whole as it prevents uniform roasting. Smaller beans tend to get over-roasted and larger beans tend to have under-roasted centres (Kamphuis, 2011).

Grading according to bean size is therefore an important activity and for this reason many traded lots of beans are sold on the basis of size classification as the number of beans per 100 g, with lower bean counts indicating larger beans and higher bean counts (>110 beans in 100 g) indicating smaller beans. Blending or mixing beans from different lots or sizes creates a challenge to optimal processing quality (Fowler, 1996), consistency is achieved only if the blended beans are of the same size and quality. However, mixing lots of different qualities creates FAQ where the intrinsic value of higher quality cocoa is diminished but some value is added to poor-quality cocoa (CAOBISCO/ECA/FCC, 2015). The ideal situation is to have a slightly skewed, normal but tight bean size distribution (Fowler, 1996).

Industry has tried to deal with the issue of inconsistency in bean size and its impact on whole bean roasting by developing 'nib roasting' techniques that pre-treat cocoa beans to separate the shell from the broken cotyledon and then roast the broken beans. The pre-treatment step is called 'Micronising' and uses infrared energy to quickly heat the whole bean (115°C in 45 s). The radiant heat causes the shell to dry out very quickly, therefore loosening it. The infrared energy absorbed by the bean causes a rapid rise in water vapour pressure at the surface, and the shell splits away from the cotyledon. The hot broken cocoa beans and shells are then winnowed and the nibs roasted afterwards. This process has many advantages such as ensuring cleanliness, optimum shell separation (with less than 0.3% loss of nib in shell), larger nib %, fewer fines, no fat migration or measurable loss of cocoa butter, improved roaster performance, major reductions in bacterial plate counts, insect fragments and rodent hair contamination, etc., but are not suitable for fine or flavour cocoas where delicate flavours need to be developed and retained during roasting (Kamphuis, 2011; Micronising Company, 2017).

As a guide, for uniformity in size, not more than 12% of beans should be outside the range of plus or minus one-third of the average weight. Shrivelled beans tend to be more difficult to process in terms of breaking and winnowing operations after roasting (Kamphuis, 2011) and flat beans are not considered as a whole bean. In addition to homogeneity of bean size, moisture content and cocoa butter fat content, as well as quality (including FFA level), are important considerations under processing quality of cocoa (CAOBISCO/ECA/FCC, 2015).

3.3.2 Purity

Processing quality is also linked to physically removing materials that pose a risk to human health and could damage the chocolate-processing machinery. Cleaning is therefore done to remove foreign matter from beans together with grading as the first activities undertaken when making chocolate. Foreign matter is often divided into two types, one that has no commercial value to the manufacturer and the other that has reduced value, also known as 'residue' or 'cocoa-related matter' (CAOBISCO/ECA/FCC, 2015). Manufacturers set up the bean cleaning part of their operations to carry out mechanical sieving, air aspiration, stone separation on a fluidised bed and removal of pieces of metal via a powerful magnet (Fowler, 1996). It can be alarming at times to see the vast array of foreign material that comes out from these operations which can include very large stones and pieces of concrete, pieces of plastic, bits of string, an assortment of vegetative matter including sticks and pieces of wood, pod husk and placenta, and various pieces of metal and metallic objects including nails and screws. Small beans, flat beans, very large beans, double and clumped beans will also get removed along with the waste which affects yield and final value of the lot of cocoa used.

3.3.3 Cocoa butter

Cocoa butter is still generally considered to be the most valuable part of the bean with ranges of 50% to just over 60% (dry nib basis) according to cacao variety (Wood and Lass, 1985). Cocoa butter is unique among vegetable fats due its composition and crystallisation behaviour, which is linked to hardness. Temperature during pod maturation affects cocoa butter hardness, which is influenced by triglycerides that represent 97–98% of cocoa butter, with three main fatty acids – palmitic (C16), stearic (C18) and oleic fatty acid (C18:1) – being present. Higher temperatures during the latter stages of pod development lead to harder cocoa butter, which is considered more valuable in the context of processing quality for cocoa and is linked to the amount of FFA present (Wood and Lass, 1985).

Higher FFA contents give a softening effect to cocoa butter and are often linked to poor post-harvest practices where lipase enzymes release FFAs from the triglycerides present. The use of high FFA butter makes poor-quality chocolate and affects both tempering ability and appearance of fat bloom, as well as having a negative effect on flavour (CAOBISCO/ECA/FCC, 2015; Fowler, 1996; Lockwood and Eskes, 1996; Lehrian et al., 1980; Saldaña et al., 2002).

4 Analysing sensory and processing quality of cocoa

This section considers and presents a detailed step-by-step approach for analysing physical and sensory quality of cocoa that includes (as much as possible) existing recognised international standards, experience and recommendations by the authors and recent research output.

4.1 Sampling

According to the International Organization for Standardization (ISO) Standard 2292 (1973), correct sampling is a difficult operation which requires most careful attention and the importance of obtaining a properly representative sample[4] of the lot of cocoa beans for examination cannot be over-emphasised. Also, sampling during fermentation must consider not only the top 30 cm of a fermentation mass but also the middle and bottom. Samples from a cocoa drying floor must be drawn from various locations for the same batch of beans. Similarly, beans selected for processing in a flavour lab, whether at origin or in an industry-quality control lab, should be representative of the lot.

4. A representative sample is a subset of a statistical population that accurately reflects the members of the entire population. A representative sample should be an unbiased indication of what the population is like.

4.1.1 Sampling procedures

Suggested sampling procedures vary according to the lot sizes (number of bags). For lots of beans over 30 bags it is suggested that every other bag be sampled, with individual bean samples being drawn from the top middle and bottom of the bag using an appropriate stab sampler. For lots of beans less than 30 bags, it is suggested that every bag be sampled, drawing beans from the top, middle and bottom of the bag (CAOBISCO/ECA/FCC, 2015).

The amount of beans drawn using these sampling frequencies might exceed the amount required for analysis. The coning and quartering process is then used to reduce this amount of sample to the desired quantity. This is achieved by forming all the bean samples drawn into a cone on a suitable food safe surface. The cone is then flattened and divided into four equal parts (in a cross) and only two opposing quarters are retained. The other half is returned to the lot that was sampled. This process is repeated until the desired quantity of beans for quality assessment is obtained. It should be noted that enough sample should be retained to either conduct quality assessment in triplicate or at least have three samples (including one for retention in case more beans are needed).

4.1.2 Sample passport data

The sample should have some passport data associated with it which could include:

1 Country
2 Farmer or farm location
3 Processor
4 Date fermented
5 Dates of drying and method used
6 Turning regime used, etc.

Some additional passport data that should also be collected from the sample and logged before it is handled further could include:

1 Date arrived
2 Sample identity and origin
3 Bag or container the samples were received in
4 General comments about the sample condition (visual, aroma, presence of waste, insects, mould, broken beans, etc.)
5 Internal assessment code (if used)
6 Total weight received
7 Sample storage conditions prior to receipt and before assessment

These passport data can be logged in a dedicated hard cover notebook or virtually on a suitable spreadsheet or database software application allowing for easy reference. Each sample should also be assigned a unique identifier code in this database separate and apart from the submitted name.

4.2 Physical assessments

Physical assessments of cocoa bean quality are well documented and have been developed to ensure that commercial cocoa consignments can be assessed and classified using agreed terminology and methodology. This is the most organised/formalised, well understood and communicated aspect of cocoa quality assessment. There are existing ISO standards governing activities carried out under physical assessments (ISO Standards 2451 (2014); 1114 (1977); 2291 (1980) and 2292 (1973)) as well as older standards from the United States Food and Drug Administration (FDA, 1968). The ISO standards are currently under review to adapt new rules and processes to ensure that the correlation between standards and commerce is properly maintained.

Physical assessments are usually carried out during the storage period before the samples are processed into powder, cocoa liquor or chocolate. Assessments include:

1 Average moisture content
2 Bean count
3 Individual bean weight
4 Cut tests

4.2.1 Moisture content

The exact range of acceptable moisture contents varies but almost all agree that the minimum value should be around 6.5% and should not exceed 8.5%. The desired range of limits is 6.5–7.5%. There is an existing ISO standard for determination of moisture content using gravimetric methods ISO Standard 2291 (1980) but most people use moisture meters that are either handheld or benchtop. A protocol specific to any one brand or make of moisture meter will not be presented as there are various types available to assess moisture content. These vary from probe samplers that are inserted into the large 70 kg (Ca 150 lb) jute bags used to store cocoa to small meters that have a chamber to manually crush beans to assess moisture. The most commonly used meters use changes in capacitance to measure moisture. Whichever moisture meter used, the correct weight of beans must be added as per operating instructions and it must be calibrated for cocoa beans in the range of moisture contents being assessed. It is important that the beans going into the moisture meter are at ambient temperature and not hot or cold. They also must be free from foreign material such as stones, leaves, sticks, etc., to prevent sticking and damage.

4.2.2 Bean count and individual bean weight

Bean counts are simply the quantity of beans required to make up 100 g. The ISO standard 2541 (2014) for cocoa bean specification provides a methodology for bean counts. Individual bean weight is derived from the bean count by the simple formula: 100 g/ Bean Count. Usually individual bean weight above 1 g is considered desirable and bean standard sizes considered important in industry are:

- Standard beans – bean count ≤100
- Medium beans – bean count 101–110
- Small beans – bean count 111–120
- Very small beans – bean count >120

There are various bean count allowances aligned to time in the cocoa crop year as well as allowances for cocoa-related matter and foreign matter that are available from CAOBISCO/ECA/FCC (2015).

4.2.3 Cut tests

Cut tests are done on fermented and dried beans and are one of the most commonly used methods to visually assess bean quality. The ISO standard 1114 (1977) documents a procedure for this but actual practices may vary from this and there is no cut test chart associated with this standard. For the cut test, a representative sub-sample of beans (300 according to ISO (1977) rules or 100 according to FDA (1968), sometimes fewer beans depending on local standards and lot size) are cut centrally along the longitudinal axis to expose to equal halves that are quickly examined in good light before surface oxidation has occurred.

Cutting is achieved in different ways starting on individual beans with a simple sharp knife whose blade could be serrated or not, to anvil pruning shares or via the Magra Bean cutter that allows one to cut 50 beans at once with a guillotine action. Hand cutting gives all the information you need for a defect count as well as colour and fissuring evaluation; it is more time-consuming, and the delicate aromas released from a cut bean can quickly fade. The Magra Bean Cutter has the advantage of allowing you to immediately smell the aroma from the beans by cutting 50 beans at once.

Beans selected for cutting with the Magra Bean Cutter must be reasonably plump to avoid slightly flat beans from being 'missed' by the cutter. Beans should be carefully oriented with the longitudinal axis of the beans lined up with the longitudinal axis of the cavity in the cutting bed. One bean must be placed per chamber in the cutter. After filling, the cutter is closed and latch engaged by finger tightening to hold the cutter closed when the guillotine is used. The cutter is placed on its two feet on the floor or a low table and the guillotine blade inserted to begin cutting. There is only one way that the blade can enter the cutting chamber since one of the brass rails is larger than the other. Applying careful pressure (not in a jerking fashion) ensures that the blade goes down the rails vertically until it comes to a stop with the handle on the top of the cutting block. Applying even pressure whilst cutting allows for a smooth, even cut. Applying uneven pressure often results in bean breakage and a rough cut which makes examination less easy. The guillotine is opened before the blade is removed by laying the cutter flat on a table and undoing the latch opening and top part of the cutter and removing the blade out (Seguine, 2014).

4.2.4 Aroma

Immediately after cutting, aroma is best assessed by putting the beans about 2 cm (0.8 inches) away from your nose moving up and down the rows of cut beans on the Magra cutting bed or the individual cut beans if using a knife or anvil pruning shears. This gives a good initial indication of what dominant flavours might be present in the finished liquor or chocolate. Defects such as smoke and oily taints, and even putrid or hammy notes, from over-fermentation can be detected at this stage and it may be a good reason to reject or discontinue further evaluations after these defects are identified (Seguine, 2014).

4.2.5 Cut test appearance and colour

Examining the cut halves of the beans in full sunlight or equivalent light is desirable (Fig. 2). The cocoa cut test chart that is presented in Fig. 3 (Adapted from Sukha and

Rohsius, 2004) is useful in examining the cut surface of the beans based on the colour and appearance. Typical categories for the cut test include:

1. Well fermented (for light brown, brown, and dark brown bean categories)
2. Unfermented/slaty beans
3. Under-fermented beans (violet to partly purple/partly brown)
4. Over-fermented beans
5. Mouldy beans

Figure 2 Magra bean cutter laid open after beans have been cut and close-up of longitudinal section cuts of various degrees of fermentation progression and bean qualities (Adapted from, Cacao & Chocolate Institute, 2016).

Figure 3 A cut test chart showing longitudinal section cuts on fermented and dried cocoa beans of various degrees of fermentation progression and bean qualities (Adapted from Sukha and Rohsius, 2004).

6 Germinated beans
7 Flat beans
8 Insect-damaged/infested beans
9 Smoky beans are assessed by aroma.

There are different grades and categories identified by ISO, FCC and FDA based on the extent of defects and % fully fermented. A maximum 3% mouldy – maximum 3% infested are permitted by ISO standard 1114 (1977). More details on these are presented in CAOBISCO/ECA/FCC (2015). The time window after cutting to do the colour evaluation before beans start to fade and change is about 30 min. If photographs are taken of the cut tests, they must be taken within this time and should include a MacBeth ColorChecker (Pascale, 2006) or equivalent to allow standardisation of the colours due to lighting differences and camera exposure settings.

4.3 Coarse powder, liquor and chocolate preparation

While the visual and odour cues obtained from the cut test give some inference of flavour, for obvious microbiological and food safety concerns, cocoa beans should be roasted before tasting either coarse powder, liquor or chocolate. Appendix B of CAOBISCO/ECA/FCC (2015) provides protocols for the preparation and flavour assessment of samples and small-scale fermentation techniques.

4.3.1 Roasting

Roasting technique is one of the ways that chocolate makers can bring their own unique artistic vision to their chocolate making to best express the genetic favour potential of the cocoa they are working with. Variables such as temperature, the temperature curve (how long the temperature remains at a particular point), the degree of roast, the type of roaster used, and a whole host of other variables all make a significant impact on the flavour of the final chocolate. There is no single *'right'* way to roast cocoa beans though there are many *'wrong'* ways to execute this activity with a detrimental effect on genetic flavour potential expression. Appendix B of CAOBISCO/ECA/FCC (2015) describes in detail the options for roasters, roasting trays and roasting conditions.

The type of roaster to use is governed firstly by the objective of the roast and secondly by location of the roast as well as budget and facilities available. These can range from rapid quality screening to detailed flavour × roast mapping. An overarching consideration is to avoid over-roasting or burning of the bean samples.

The main purpose of rapid quality screening is to identify flavour defects that were not apparent from visual and odour cues in the bean sample. This type of roasting does not require simultaneous temperature and time control but should be light to moderate. Converted small-scale rotary-type coffee roasters and rotisserie ovens are relatively simple and affordable options to consider. These units typically do not have either temperature control or time control so some degree of manipulation of the non-fixed variable in the system is needed to arrive at a suitable protocol for this.

Following successfully screening for the presence of defects in rapid quality screening, tracking flavour profile development of particular beans with different roast temperatures and times may be necessary. However, this may not be feasible in the country of cocoa bean origin due to limitations of facilities and budget. More often, precise roast mapping is done in the consuming country or end-use facility at better-equipped quality control labs with mechanical convection ovens that offer the best opportunity to achieve this objective at the quality control level with a relatively small budget. However, in recent years, a few cocoa bean origin countries have set up well-equipped labs. The mechanical convection ovens used in these quality control labs should have the desired capacity, variable control capabilities, appropriate inner chamber airflow and thermal uniformity, and a recovery time of less than 10 min from when the door is closed to 2°C (35.6°F) below the set point.

Roasting tray options are also discussed in Appendix B of CAOBISCO/ECA/FCC (2015) and optimally should be made from a wide (0.6 cm or 0.25 inch) stainless steel mesh with enough stiffness in the wire so that it can be formed into a tray that is mounted onto the rack to promote air flow across the beans and loaded at a fixed quantity at a single bean depth.

At the small oven capacities suggested earlier, for specific varietal or origin investigations, typically only one tray should be roasted at a time loaded with 300–400 g of sorted beans of relatively uniform size. For roasting not linked to different cocoa origins or varieties two trays could be roasted. As long as the bean size is in the range of 70–130 beans/100 g, the roasting conditions do not need adjustment for bean size. Where there is insufficient sample to fill the tray, *'filler beans'* of good quality should be used so that all roasts are carried out with the same bean loading. Dividers should be mounted in the trays when filler beans are used to keep them separate from the sample of interest.

Roasting conditions should be chosen to maximise the flavour potential for each type of cocoa bean and would need to be mapped according to temperature, time and loading capacity for each variety and for the specific type of roaster used.

Typically, according to CAOBISCO/ECA/FCC (2015), the following is used and suggested as a starting guide for individual roast mapping using a convection oven with trays:

1 Trinitario types – 120°C for 25 min
2 Varieties considered in formerly classified 'Forastero' types – 130°C for 25 min
3 Ancient Criollo types – 112°C for 25 min
4 Modern Criollo types – 120°C for 25 min

The roast time should be measured starting from 2°C below the set point. These times are based on an oven recovery time of 5–7 min from the time that the door is closed to 2°C below the set point. Generally, most modern Criollo types can be roasted following the Trinitario conditions but the selection of specific roasting conditions for this type of cocoa should seek to express intrinsic fruity and floral ancillary flavours of Trinitarios whilst preserving delicate nutty/caramel notes of Criollo types. Given the wide variety of formerly classified 'Forastero' types (Motomayor et al., 2008), roasting conditions should be refined to bring out any ancillary flavours as well as the maximum cocoa flavour inherent in these types.

Bean size and initial moisture content of the beans prior to roasting are important considerations, and samples with very low moisture contents (<6.5%) or very high moisture contents (>8.5%) may require adjustments to be made to the roasting conditions to ensure a standardised roast for flavour evaluation.

After roasting, the beans need to be cooled to room temperature by placing them on an elevated rack and cooled at ambient conditions or with a small fan on them to accelerate the cooling if necessary. Care must be taken to avoid exposure to strong odours when cooling and for appropriate food safety control; from this point onward, there should be no contact between this sample and unroasted beans. Unroasted beans are considered to be a raw agricultural product that may contain pathogens prior to roasting. Ideally, handling of raw beans should be in a separate room from roasted beans for sanitation purposes.

A batch system should be used when handling roasted cocoa and unroasted cocoa, if for logistical and budget constraints they must be handled in the same space. There should also be appropriate cleaning of all counter space and contact surfaces with approved cleaning agents and sanitisers between handling unroasted and roasted bean batches or separating activities by time by performing them on separate days.

4.3.2 Breaking and winnowing

Breaking and winnowing must not be conducted in close proximity to the unroasted beans to avoid microbial contamination of roasted beans – and winnowing should occur immediately following cooling of the beans after roasting (ideally between 20 and 60 min and not more than 3–4 h) for efficient breaking – and to ensure that no off-flavours are picked up from the environment and that the beans are not too cold. Shells come off poorly when the beans are still warm and when they have been out too long.

Breaking and winnowing can be done most simply by placing cooled beans in a high-quality snapseal plastic bag, removing as much of the air as possible and using a rolling pin to lightly break the beans. Afterwards a home use handheld hairdryer can be used to blow off the free shell from the nibs in a flat tray in a well-ventilated area. This approach is however only feasible for a few samples. Care must be taken not to make too many small pieces of nibs with shell (fines) through too aggressive breaking; this will reduce yield, since fine nibs do not separate well from shell in most of the small-scale winnowers and are lost with the shell stream.

Higher throughput sample preparation for flavour evaluation will require at least a mechanised winnowing system. There are a variety of options of small- and medium-scale winnowers and breakers that can be hand crank operated or mechanised. Separation with these systems could be less efficient with nib yields frequently as low as 62–78% of the starting raw beans. This is an important consideration in determining the sample size of beans needed for roasting to meet the volume needs for flavour and other evaluations. Residual shell in the winnowed nibs, including both loose shell fragments and pieces of shell adhering to a piece of nib, can be manually removed with tweezers. This process can take the shell content effectively to zero, increasing nib yields from small and/or scarce samples.

Nibs should be converted into either powder or liquor for evaluation within 48 h of winnowing. Storage of winnowed nibs is important to avoid re-humidification and picking up odours and taints. Immediate storage should be done in clean food grade containers or high-quality plastic snapseal bags, preferably a multi-layer, barrier film vacuum seal type to provide barrier film protection without vacuuming.

For precise flavour profiling, nibs should not be stored longer than a week (even in a sealed bag) prior to liquor milling. Storage temperature should be 10–24°C (50–75°F). If nibs are stored at temperatures less than 18°C (64°F), they must be allowed to warm to room temperature prior to opening the bag.

4.3.3 Preparation of coarse powders

Rapid flavour screening can be carried out on coarse powders evaluated with or without the addition of sugar. Coarse powders between 500 μm can be prepared by using a 100 g

(Ca 3 oz) stainless steel blade coffee grinder. About 50 g (1.5 oz) of roasted and winnowed nibs can be added to the grinder one teaspoon at a time whilst engaging the grinder in 10 × 1 second pulses and shaking it up and down to assist in mixing. This coarse powder can then be stored in four oz sterile specimen containers at 20°C (68°F) for a maximum of 2–3 h before tasting.

4.3.4 Preparation of cocoa liquors

There are a number of options available for milling nibs into liquor:

- Table top liquidisers for coarse grinding and coarse liquor milling (up to 100 g of nibs)
- Table top and free-standing mortar and pestle mills of varying capacities (100–500 g of nibs)
- Laboratory-scale melangeurs capable of handling from 200 g up to 2.5 kg of nibs.

It is important that the nibs should be gently warmed (not more than 40°C (104°F) before milling and equipment such as bowls, pestles and the stones from melangeurs should be pre-warmed to ensure that the cocoa butter in the sample melts and facilitates liquor grinding. The temperature of the milling mass can be measured using an infrared thermometer and should remain below 55°C (130°F). Should the temperature of the mass rise above this value, the room can be ventilated (cooled) or the mill can be turned off to allow the sample to cool off.

Exact milling times cannot be specified as this is dependent on a number of factors such as fat content of the nibs, degree of fermentation of the beans, specific mill used, condition of the stones in the mill, etc. But milling should be accomplished gently without the addition of significant external mechanical pressure. The objective is to produce liquor that will have no discernible grit. Therefore, regardless of the milling equipment used, particle size, as determined by a micrometer, is a critical parameter in determining milling end point. A particle size ranging between 14 and 25 μm is optimal for effective flavour evaluation without sample grit distracting significantly from the evaluation.

After milling, liquors should be transferred to labelled 4 oz sterile specimen containers, covered, allowed to cool to room temperature and then stored at −20°C (−4°F) if the processing stops here and only liquors are assessed.

4.3.5 Chocolate making

Sometimes based on the skill set of the tasting panel it is more desirable to prepare chocolate for tasting. It is also sometimes desirable to project how liquor flavour potential translates into chocolate as sugar and other ingredients in the matrix become important factors affecting or influencing the expression of the genetic flavour potential of the bean sample. Cocoa liquor tasting is also more technically challenging to perform and does require some more specialised training compared to chocolate evaluation due to more dominant acidity, bitterness and astringency. Also, the liquor alone does not often display the full flavour potential that will be present in the chocolate and sometimes flavours that are present in the liquor are not present in the chocolate and vice versa.

The recipe used in chocolate making for the purpose of creating test samples for flavour evaluation of beans from different varieties or origins is very important and standard formulations range between 65% and 70% cocoa mass with 2–10% added deodorised cocoa butter used. Lecithin is added to give a more desirable mouthfeel when tasting.

Table 2 Recipes that have been successfully used for semi-sweet chocolate evaluations

Ingredients	Recipe 1 (%) ingredients	Recipe 2 (%) ingredients	Recipe 3 (%) ingredients
Chocolate liquor	65.10	61.00	55.00
Deodorised cocoa butter	3.00	5.00	10.00
Sugar	31.55	33.65	35.00
Soya lecithin (double bleached)	0.35	0.35	0.35

Source: Appendix B of CAOBISCO/ECA/FCC (2015).

There are established protocols for making small-scale chocolate samples, such as from the International Office of Cocoa, Chocolate and Sugar Confectionery (IOCCC, now CAOBISCO) (1996a) or more recent modern recipes and methods as presented in Appendix B of CAOBISCO/ECA/FCC (2015). These recipes that have been successfully used for semi-sweet chocolate evaluations internationally in both the Heirloom Cacao Preservation (Recipe 1) and Cocoa of Excellence (CoEx) (Recipe 2) initiatives as well as an average recipe from SeguineCacao (Recipe 3) are presented in Table 2.

The cocoa butter, sugar and soya lecithin used must be neutral tasting to avoid influencing the flavour inherent in the liquor. Sugar can be evaluated by placing a 50–120 g sample in a jar twice that size, securely capping the jar and warming it to 50°C (122°F). It should be held at this temperature for at least 1 h, and then when uncapped and immediately smelled, there should be no inherent odour to the sugar (CAOBISCO/ECA/FCC, 2015).

Just like liquor preparation, an exact milling time for chocolate cannot be accurately pre-determined, the fineness of grind determines the end point. The chocolate should be refined to less than 20 μm and tested using a micrometer.

Conching should be kept at a minimum for making these semi-sweet chocolates for flavour evaluation and should also be at a low temperature (not more than 40°C (104°F)). The actual conching time should be long enough to remove 'peaky' volatile acid flavours and short enough to retain the intrinsic flavour potential of the bean as much as possible to best express its flavour potential and gauge its performance as a chocolate (Seguine, 2014).

Chocolate samples for flavour evaluation can be assessed as either un-tempered or tempered pieces. There are three main ways to achieve well-tempered chocolate by careful manipulation of temperature during the cocoa butter crystallisation process. These are as follows:

1. Manually using a marble slab,
2. Using a double boiler and
3. Using a small table top tempering machine.

Regardless of the equipment or method used, the chocolate must first be heated to 45°C (113°F) to melt all six forms of crystals. Next, it should be cooled to about 27°C (80.6°F) to allow crystal types IV and V to form. At this temperature, the chocolate is agitated to create many small crystal 'seeds', which will serve as nuclei to create small crystals in the chocolate. The chocolate is then heated to about 31°C (87.8°F) to eliminate any type IV crystals, leaving just type V. Any excessive heating of the chocolate after this point will destroy the temper, and this process will have to be repeated. Moulding and cooling into small bars or pieces immediately follow tempering (CAOBISCO/ECA/FCC, 2015).

Refrigerators are not ideal for cooling moulded chocolate but are often used and their temperature setting should not be very cold but closer to 12°C (53°F)–15°C (59°F), air-conditioned rooms set to 19–21°C (66–69°F) or wine chillers set to 16°C (60°F) are often used to cool filled chocolate moulds but both need to be checked prior to use to ensure they are neutral smelling and do not contain any off-odours.

The time frame before flavour evaluation is important to note because the flavour profile of chocolate (particularly semi-sweet chocolate) changes (becomes more mellow) for about 2–4 months after being made (CAOBISCO/ECA/FCC, 2015).

5 Flavour assessment of cocoa

Flavour assessment is also covered in some detail in the CAOBISCO/ECA/FCC (2015) document and the only standard that exists for flavour assessment in cocoa liquors is the detection of specific off-flavours in cocoa beans IOCCC (1996); however, flavour assessment of powders, liquors and chocolate can take the following formats:

- Evaluation by a panel of skilled tasters for presence or absence of defects
- Evaluation by a panel of unskilled tasters using hedonic preference indicators
- Evaluation by a panel of skilled tasters providing both quantitative and qualitative assessments (including presence or absence of defects), as well as an overall global quality or preference score
- Evaluation by a single or few highly skilled tasters providing both quantitative and qualitative assessments (including presence or absence of defects), as well as an overall global quality or preference score

Each flavour assessment format identified above has a direct implication on the type of sample assessed, amount of sample needed, the size of the panel and number of repetitions of tasting required for a robust dataset based on the inherent purpose and need of the evaluation exercise. Critical elements in this process therefore include:

- Tasting area and layout
- Panellist training or experience
- Tasting design and/or sample randomisation
- Sample storage and thawing
- Sample presentation
- The evaluation process
- Flavour descriptors
- Data collection and collation
- Interpreting the results via appropriate statistical analyses and presentation in tables and/or graphs where appropriate

5.1 Tasting area and layout

Ideally, flavour assessments should be carried out in white tasting booths with appropriate light (red light) and temperature control, warm water and a sink, etc. However, this setup could be prohibitively expensive and impractical. A simple air-conditioned room that is clean, free from distractions (noise and visual) and strong odours with a large enough table and comfortable chairs for the panellists to sit is sufficient. There should, however, be easy access to a sink. Movable partitions can also be constructed out of non-resinous wood and painted a neutral colour with low odour water-based paint to separate panellists.

Panellists should not be distracted when tasting, so the layout of the room (location of samples, water and expectorant cups, scoring sheets and pencils, etc.) should be the same each time. As presented in CAOBISCO/ECA/FCC (2015), panellists should also consider the following guidelines carefully and try to put them into practice whenever they participate in a tasting session:

- Excessively strenuous exercise, smoking, drinking alcohol or coffee or eating food that is spicy or hot that could burn or desensitise the taste buds or alter a panellists' sense of taste should be avoided at least 1 h or preferably longer before a tasting session.
- The use of strong scents, perfumes and aftershaves should be avoided by panellists and anyone else involved in the setup of the tasting area or sample handling. Hands should be washed prior to tasting using perfume-free soap.
- Persons suffering from colds should not attend or participate in tasting sessions or setup.
- Instructions handed out at tasting sessions should be read carefully and understood before commencing. Panellists should feel free to ask any questions if they are unsure about the instructions.
- Panellists should avoid talking until everyone has finished tasting.
- Panellists should strive to follow their first instinct about a particular flavour attribute and trust in their ability.

5.2 Panellist training

Some level of intensive training is required for cocoa liquor assessments, whilst training for chocolate tasting is not required but very desirable. Only hedonic (preference tasting) can be done with an untrained panel (Stone and Sidel, 1993).

An intensive panellist training programme in cocoa liquor tasting that spans two weeks is recommended at the start of the process, but it takes some time to gain expertise in flavour recognition and anchoring of associations to flavour descriptors over a range of attributes and developing confidence in cocoa flavour evaluation of liquors. Ideally between eight and ten persons should be selected for training to make up a core of six panellists. A critical element of panel training is the mental association or anchoring of flavours to the full range of flavour descriptors. This is achieved by presenting calibrated solutions, odour references, reference liquors and chocolate samples to panellists for both attribute identification and scaling over a range of intensities (Sukha et al., 2008; CAOBISCO/ECA/FCC, 2015).

Levels of expertise can be categorised into the following based on the level of training or taste experience: Beginner, Intermediate, Advanced and Master, with only Advanced and Master levels being allowed to train other panellists. Sukha et al. (2008) used an optimised sensory assessment protocol to describe and quantify different flavour attributes of cocoa liquors made from Ghana and Trinitario beans, and elements of this training protocol are presented below.

Prior to selecting sensory panellists for further training, the panellists' attitude towards tasting, their time and availability, health and allergies, smoking and tobacco usage should be considered. Further screening involves basic taste identification and threshold-level sensitivity identification using the following aqueous solutions presented in Table 3.

This involves making solutions up to the concentrations as listed in Table 3 and having trainee panellists taste and identify these at normal and threshold levels of concentration.

Following the development of basic taste recognition skills, training is extended to odour recognition with vocabulary generation for association anchoring. Both positive and negative odour references applicable to cocoa liquor assessment can be found in commercially available odour kits such as Le Nez du Vin (Lenoir, 2016) and Projet Chocolat (Rea, 2015). Vocabulary generation is a relatively simple process where trainee panellists smell or taste coded samples and write down a few words on a blank sheet of paper to describe what they smelled or tasted. The panel trainer then asks each person to state what they have written down and this is either written on a white board or projected onto a screen. Whilst there is usually a consensus about the attribute in question, if there are differences in descriptors, the panel trainer should facilitate the associations of differing terms into the common vocabulary or glossary of terms used by the panel.

The next step in the panel training process is vocabulary generation and mental association to flavour descriptors. After the initial taste identification part of training and odour recognition, panellists can now progress to associate specific flavour descriptions for cocoa liquors on suitably prepared cocoa liquor reference samples and mixtures of reference liquors together with previous taste experiences to gain agreement on the same sensory language. This can be achieved firstly as vocabulary generation exercises and then ranking and scoring exercises using the cocoa liquor reference samples that dominantly demonstrate attributes across the 10 core flavour descriptors, namely cocoa, acid, astringent, bitter, fruity, floral, nutty, woody, sweet/caramel/malt and spicy flavours. In addition, identifiable off-flavours such as smoky, hammy, mouldy and unfermented can be included. Panellists should also be encouraged to identify any other ancillary flavours or defects that are apparent in the cocoa liquors (recorded under 'other' flavours) (Sukha et al., 2008; CAOBISCO/ECA/FCC, 2015).

Table 3 Flavour attributes, solutions used and concentrations for panellist training in basic taste identification and threshold sensitivity identification

Flavour attribute	Solution	Concentration
Basic taste identification		
Sweet	Sucrose	5.0 g/500 mL
Bitter	Quinine chloride	0.072 g/500 mL
Salt	Sodium chloride	0.8 g/500 mL
Acid	Citric acid	0.25 g/500 mL
Astringent	Maleic acid	0.25 g/500 mL
Fruity	Kola flavour	2 mL/500 mL
Floral	Orange blossom water	2 mL/500 mL
Identification at threshold concentration		
Acid	Citric acid	0.1 g/500 mL
Bitter	Quinine chloride	0.009 g/500 mL
Astringent	Maleic acid	0.15 g/500 mL

Source: Sukha et al. (2008).

5.3 Flavour profiling of cocoa liquors

More specific and focused flavour profiling of cocoa liquors follows next with scaling and intensity training based on reference liquors that are coded with three-digit numbers and presented to trainee panellists in order of increasing intensity (least intense first). Prior to panelling, the cocoa liquor samples should be removed from storage in the sterile specimen containers, brought to room temperature and about 1.5–2.0 g of liquor sample placed into small labelled plastic soufflé cups with random 3–4 digit numbers to ensure that all flavour evaluations are blind. Random numbers should avoid codes starting with double digit zeroes (like 001–009) (Sukha et al., 2008; CAOBISCO/ECA/FCC, 2015). Cups can be labelled directly with no odour, thin point permanent markers or by writing the codes on small paper labels that are stuck onto the cups. The plastic soufflé cups should be covered with matching tight-fitting lids. The cocoa liquor samples should be melted at no more than 50°C (122°F) for 15–20 min prior to flavour evaluation and a maximum of six liquors should be tasted in any one session to prevent panellist fatigue (Sukha et al., 2008; CAOBISCO/ECA/FCC, 2015).

Flavour evaluation of cocoa liquors starts with stirring and smelling the heated sample and recording observations about positive and negative odours and intensity. One gram of melted cocoa liquor is placed on a small spatula or tasting spoon and spread directly on the tongue, keeping it there for about 20–30 s and periodically rubbing the tongue on the roof of the mouth. During this time the different attributes making up the flavour profile become apparent at three contiguous time intervals, namely initial front flavour notes, middle flavour notes and residual end flavour notes. Panellists should note that some flavours either appear or disappear very quickly or are easily masked, whilst other flavours could linger for a longer time with distinct aftertaste. During flavour evaluation, panellists should identify and score any 'other' ancillary flavours (positive notes) or defects (such as smoky, hammy, mouldy, over-ripe, dirty and unfermented) that are apparent in the cocoa liquors in a section for 'other' flavour as well as any general comments about the smell and taste of the sample (Sukha et al., 2008; CAOBISCO/ECA/FCC, 2015).

After liquor evaluation, the following clearing procedure is used:

- Expectorate the sample
- Rinse with warm water, expectorate rinse water
- Chew 1/8–1/6th of a Table Water Cracker/Wafer (non-yeast based) with the front incisors, and not the molars, and swallow
- Rinse with warm water, expectorate rinse water
- Rinse again with warm water, this time swallowing the rinse water

The performance of the sensory panel can be optimised during evaluations by including a hidden reference sample to check panellist consistency between repetitions during training and evaluation sessions (CAOBISCO/ECA/FCC, 2015)

5.4 Coarse powder assessment

Instead of flavour assessment of cocoa liquors, training could also continue if desired, with coarse powder produced from roasted bean samples prior to conversion into milled cocoa liquors. Using a protocol formulated by the Fine Cacao and Chocolate Institute (FCCI)

(FCCI, 2016) this begins firstly by removing the coarse powders from storage in the sterile specimen containers and placing the ground material in a stemless wine glass labelled with a random three-digit identifier, taking care to avoid codes starting with double digit zeroes (like 001–009).

Each sample is evaluated without sugar first by agitating the material whilst holding your nose over the cup via tumbling or gentle stirring with a flavour and odour neutral implement. Then one teaspoon of the material is placed into your mouth and chewed by moving the material around your mouth, periodically opening your mouth while chewing, detecting aromas for 30 s. Expectorate the sample and record impressions on taste and aroma. Palate cleansing follows the same procedure for liquor. Samples are evaluated with sugar by repeating the procedure with the addition of 1/8 tsp of sugar or xylitol and note any differences in aroma detection results (FCCI, 2016).

5.5 Chocolate assessment

Chocolates can be evaluated either as solid moulded blocks or as pieces melted at 45°C (113°F). Evaluation using solid blocks is recommended unless an assessment of the melting performance on the palate is not important. Training for chocolate tasting should include an exposure to a wide variety of different origin chocolates to build a mental library of associations linked to key chocolate flavour descriptors (CAOBISCO/ECA/FCC, 2015). The degree and quality of tempering in the chocolate are initially assessed from its surface sheen/gloss and 'snap' when broken. The latter is a function of the serving temperature of the chocolate. There should be a nice sheen on the surface of the chocolate without white spots or swirl marks. Assessment of physical attributes such as 'snap' and texture can be perceived by breaking and holding a piece of chocolate between your thumb and index finger, letting it melt, and gently rubbing your fingers together to feel the texture once it begins to melt. The texture should be smooth, and you should not be able to feel the particles with your fingers.

Aroma is assessed by bringing the chocolate close to your nose to inhale in order to determine if there is a distinct or subtle smell. Determine if there are any other scents – such as vanilla, any other spices, floral or fruit flavours or even off-odours.

Finally, a small piece of chocolate is placed in your mouth, then let it to sit on your tongue to melt. Inhaling through your mouth and out through your nose allows for the flavours and aromas to become fully apparent. The piece is now chewed three to four times whilst concentrating on the taste and texture and paying attention to the front, middle, end and aftertaste/finish of the chocolate. Impressions can be recorded on a suitably prepared form with categories for physical appearance, aroma, texture, flavour (front, middle and end) and aftertaste.

5.6 Intensity scales

There are many scaling systems used in flavour descriptor assessment, 0–5, 0–10 and 0–15. The most commonly used are 0–10 and 0–5. For many, scoring on a 0–10-point scale is easier because we often think in base 10 or in a denary fashion. In such a system, the attribute intensity scores as seen in Table 4 will have the following meanings where the higher numbers denote stronger flavour intensities. The full use of the intensity scale is a matter of preference, training and experience (CAOBISCO/ECA/FCC, 2015).

Table 4 Flavour attribute intensity meaning on a 0–10-point scale

Attribute intensity	Meaning
0	None present
1	Just a trace and may not be found if tasted again
2	Present in the sample
3–5	Clearly characterising the sample
6–8	Dominant
9–10	Maximum that you have experienced in a cocoa sample

Source: Seguine and Sukha, Cocoa of Excellence Programme 2015, cited in CAOBISCO/ECA/FCC (2015).

5.7 Flavour descriptors

According to CAOBISCO/ECA/FCC (2015) one of the most difficult parts of flavour evaluation of both liquors and chocolates is finding the right words to describe the perceived flavours since this relies heavily on a mental association with the flavour descriptors. Glossaries of flavour descriptors with comments and flavour wheels have been developed to group terms used to describe the flavours and can be used by panellists to ensure they use a common language when describing their perceptions and to aid in the interpretation of results.

The glossary of flavour descriptors used for the Cocoa of Excellence Programme is presented in Table 5, which shows descriptors for flavour attributes with examples of some origins/reference notes for calibration (CAOBISCO/ECA/FCC, 2015). Apart from tabulated glossaries, flavour wheels are also convenient ways to display flavour descriptors where main categories and subcategories for descriptors are clearly defined. Fig. 4 shows the Seguine and Sukha flavour wheel with main and subcategories for both cocoa liquor and chocolates (CAOBISCO/ECA/FCC, 2015).

Having a global quality indicator in the flavour testing both liquor and chocolate is very useful as it goes beyond simple attributes and reflects an overall attribute standing. It should not be a score derived using a formula or calculation from the attributes but stands on its own for each evaluator to indicate their impression of overall quality (CAOBISCO/ECA/FCC, 2015). There are different approaches that can be used in scoring global quality: the first is a simple 0 (worst) to 0 (best) rating, which is easiest to do or a more complicated composite calculated score this is derived from physical to flavour assessments. In this system, scores are assigned to each level of assessment which are then weighted in a mathematical transformation to give a composite value to which an overall quality designation is assigned. Adequate panellist experience and exposure to a wide range of samples are critical considerations before having persons assign global quality scores.

5.8 Calibration of persons

Calibration of persons as part of initial and then ongoing flavour evaluation training is a continuous process that is best achieved only through experience. Panellist drift is not

Table 5 Glossary of terms for flavour evaluations with matching descriptors and examples of some origins/reference notes for calibration used in the Cocoa of Excellence Programme

Descriptor	Description	Examples of origins/references (for calibration)
Cocoa	• Describes the typical flavour of cocoa beans that are well fermented, roasted and free of defects – chocolate bars, fermented/roasted cocoa	• Low intensity = 1–2 (unfermented cocoa) • Medium intensity = 4–6 (fully fermented Indonesia, PNG and Arriba) Strong intensity = 8 (West Africa)
Acidity	• Citric acid – fruit • Acetic acid – vinegar (you can smell it in the sample) • Lactic acid – vomit like, like in sour milk or molasses • Mineral acid – metallic tasting	• Low intensity = 0–2 (West Africa) • Medium = 3 (Arriba) • Strong = 4–8 (PNG and Malaysia)
Bitterness	• Usually due to a lack of fermentation; perceived on the rear of the tongue/ throat – caffeine (Coffee), beer, grapefruit	• Low intensity = 1–2 (ancient Criollo) • Low/moderate (normal) intensity = 3–4 (West Africa) • Strong intensity = 7–8 (unfermented cocoa)
Astringency	• Usually due to a lack of fermentation; mouth drying and/or puckering effect which boosts the production of saliva; perceived between tongue and palate or at the back of the front teeth – raw nut skins, banana skins, some wines	• Low intensity = 2–3 (ancient Criollo) • Moderate (normal) intensity = 3–4 (some West Africa) • Medium intensity = 5–6 (Arriba) • Strong intensity = 7–8 (unfermented cocoa)
Sweet	• Describes liquors with a characteristic flavour of unrefined caramelised cane juice (Panela) – caramel, brown sugar, fudge	• Low intensity = 0 (West Africa – Ghana) • Strong intensity = 5–8, ancient Criollo (Venezuela)
Fresh fruit	• Broad range of fresh fruits • Fruit berry – currants, not fully ripe raspberry • Fruit citrus – essence of citrus • Fruit tropical – banana, passion fruit, orange, almost always some citrus note involved	• Low intensity = 1–2 (West Africa) • Medium intensity = 3–5 (Fully Fermented Indonesia) • Strong intensity = 6–7 (PNG, some Trinidad (TSH))
Browned fruit	• Fruit dark tree – plum, dark cherry • Fruit dried – dried apricot, banana etc., caramelisation of fruit sugar, essence of a fruit that has undergone the drying process, sulphur and nutty notes also • Fruit over-ripe – beginning of over-fermentation, over-ripe fruit as a step to over-fermentation • Fruit brown – prunes or dates	• Low intensity = 2, (West Africa) • Medium intensity = 3–5 (fully fermented Indonesia) • Strong intensity = 6–7, (PNG, some Caribbean origins)

(Continued)

Table 5 (*Continued*)

Descriptor	Description	Examples of origins/references (for calibration)
Nutty	• Nutty – nut meat • Nut skins – associated with some astringent sensation like skins of almond and peanuts, etc.	• Frequently just as a note = 2–3 but can be pronounced 5–8 (ancient Criollo types) • Too low roast or under fermentation of most origins
Floral	• Broad range from green grassy vegetative to flowers and perfumed types notes • Floral – coming from natural environment you can get this by taking a walk in your garden, green earthy, herbal and woodsy • Floral grassy – green on fresh cut grass, very fresh grass, young leaf (green floral) • Floral green vegetative (dark green) – green vegetative, old cocoa leaf crushed, dark green note. Green beans, cooked bell peppers (dark green vegetables) • Floral woodsy (generic) – was grown now dried essential oil, structural bases, going for walk in forest before winter, dried flowers • Floral mushroom – mushroom, meaty, savoury, MSG • Floral earthy – forest after the rain, smell of dampness coming up from the cocoa estate soil • Floral herbal – aged dried spices. Commonality of all the dried herbs and linked with astringency at times • Floral perfumey – a persistence that lingers like when fixatives (e.g. Vanilla) added to perfume to kick the smell into a persistent mode • Floral flowers - breathe it in and it's gone. Difference between most roses and a Lincoln rose • Floral orange blossom – is essentially floral-flowers but with orange blossom flavor specifically	• Absent – low intensity = 0–2, (West Africa) • Medium – strong intensity = 3–7, (Arriba, Scavina, some Trinidad (TSH)) • Floral orange blossom – Peru, Ecuador • Ecuador origin beans tend to be floral, herbal, floral orange blossom, earthy • Floral mushroom and earthy are positive attributes (not earthy associated with off-flavours musty, mouldy, etc.)
Woody	• Woody light wood – ash, beach, maple, white pine, cut cocoa tree • Woody dark wood – oak, walnut, teak • Woody resin – pitch pine, balsam of dark or light tree resins	• Woody light wood and Woody resin are often (but not always) associated with under-fermented and acidic beans • Woody dark wood is often (but not always) associated with well fermented beans

(*Continued*)

Table 5 (*Continued*)

Descriptor	Description	Examples of origins/references (for calibration)
Spicy	• Spice tobacco – Tobacco spice is the smell outside a tobacco shop, not ashy and dirty but rather like pipe tobacco, sweet • Spice peppery – spicy, peppery, savoury	• As a note = 1, or clearly present in the sample = 2–3 • Spice tobacco from West Africa, particularly Ivorian beans • Spice peppery – mainly Colombian and Peruvian origins
Global quality	• Goes beyond simple attributes but is intended to reflect an overall attribute standing. It is NOT a score derived from any formula or calculation from the attributes but stands on its own for each evaluator to indicate • It gives an impression of overall quality	• No off flavour must be present in giving a high scoring (>5) for Global quality • We expect the beans to be good and not have those defects. If they do, the overall quality would be not very good = 1–4 • Zero for Global quality means a serious flaw is present. This is *not* a 'veto' but is a very clear opinion of the quality or in this case lack thereof
Off-flavours	• Hammy – carved meats, ham, and improper fermentation • Smoky – happens when burning vegetative matter (wood, grass, cocoa hulls, etc.). Other off-flavours – cocoa contaminated with diesel fumes • Leather – not freshly tanned in a leather store, but rather more like leather with sweat and urine, like horse saddles • Over-fermented manure – farm yard, manure • Over-fermented putrid – Faeces • Dirty – unpleasant dirty character, like dirty utensils, often associated with quality of astringency, increased astringency = increased dirty flavour etc. Function of dusty • Bark wood – not good, typically unpleasant, dry, dusty, smelly, not a clean smell. Under-fermented, astringency, raw, leather, dirty tend to be associated with bark wood as well	

Source: Seguine and Sukha, Cocoa of Excellence Programme 2015, cited in CAOBISCO/ECA/FCC (2015).

uncommon and having factorial statistical sensory design that incorporates hidden reference liquors is a critical element to check for and manage this process (Sukha et al., 2008; CAOBISCO/ECA/FCC, 2015). Tasting sighted calibration control samples with pointed or a range of familiar flavour attributes at the start of each flavour profiling session (such as a Ghana, Madagascar origin (for example) or some other reference liquor) is also very important.

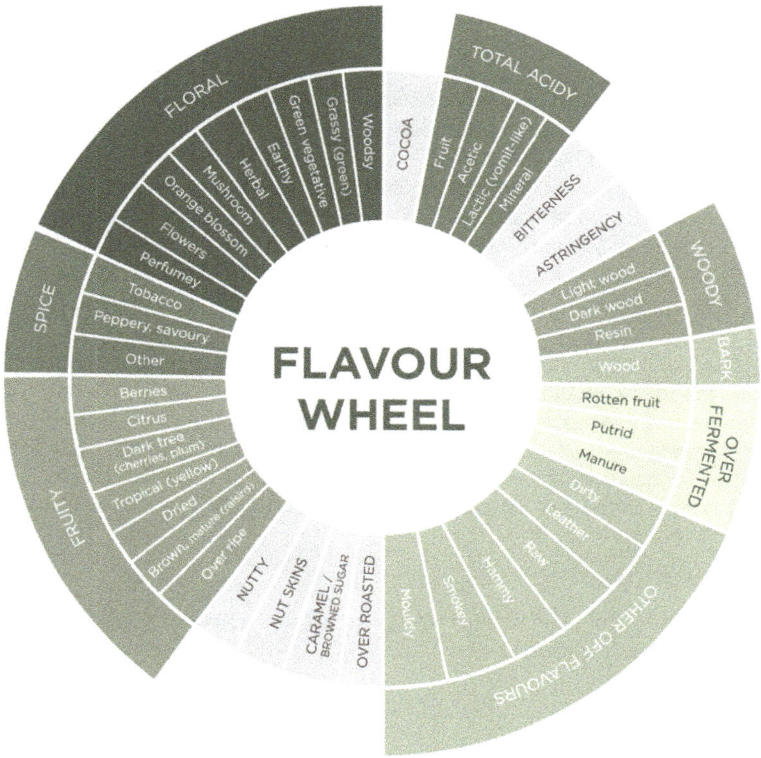

Figure 4 Seguine and Sukha flavour wheel with main and subcategories for both cocoa liquor and chocolates (Adapted from CAOBISCO/ECA/FCC, 2015).

Additionally, tasting at least three replications of the same samples helps in dealing with positional bias of sample ordering and managing carry-over effects after tasting samples. The need for appropriate reference liquor and chocolate samples and following a rigorous palate cleaning protocol cannot be over-emphasised in both panel training and maintaining panellist calibration. Group tasting of blinded reference samples followed by a discussion session is also important in achieving and maintaining panellist calibration.

5.9 Sensory statistics

The data generated from sensory evaluations can be captured directly either in some version of a spreadsheet software programme or on paper-based forms, where in the case of the latter, data will have to be transcribed into a spreadsheet for statistical treatment. Sensory statistics in panel training either test panellist consistency or identify inconsistent data, but they can also show how panellists group together based on the samples assessed.

There is a wide variety of general-purpose statistical software available to choose from and some come with extra capability to handle sensory data. Some relatively simple and effective statistical analyses that effectively treat sensory panel training and evaluation data include:

- Calculating means for individual panellists
- Calculating Standard Error of the Mean (SEM)
- Calculating Standard Deviation (SD)
- Analysis of Variance (ANOVA) looking at interactions between factors and panellists
- Correlation and Regression Analysis of individual panellists against the panel mean or against each other to pinpoint problem panellists
- Principal Component Analysis or Linear Discriminant Analysis on larger panellist datasets to examine groupings or clustering of panellists and/or samples

Inconsistencies can be corrected by weighing the score of the panellist in question so that it is closer to the mode of the panel. Other statistical techniques of data smoothing can be used; however, these approaches are only feasible if the panellist is 'consistent' in the error that is being made. Random scoring errors would indicate that the panellist is unsure of their flavour associations or anchoring to descriptors. In most instances, it is best to remove data from inconsistent panellists from the global panel dataset and retrain them.

5.10 Visualisation of results

Visualisation of results from sensory training or sample assessment must be fit for purpose and can take the following forms:

- Simple tables of mean data from individual panellists or the entire panel for samples with:
  - SEM, SD or ANOVA significance interactions.
- Simple graphs such as:
  - Line graphs
  - X-Y scatter plots with best fit lines and R values
  - Bar graphs with SEM or ANOVA significance interactions represented
  - Radar plots or spider graphs

Each visualisation mentioned above has particular advantages: they may be static and stable, which make them fit for purpose and for print and online use. They however all lack the ability to be dynamic or be animated to see very quickly how trends change over time.

Recently, different and more dynamic approaches are being taken to visualise sensory data that consider animation of how values are presented to look at changes and trends over time. This was discussed during an interview with Clay Gordon (creator and moderator of TheChocolateLife.com) on 27 April 2016. The graphs resemble a pie chart but the pie slices do not get wider or smaller to show changes in value. Instead, this new representation technique fills the individual pie slices to show increase in intensity. The width does not change. The pie is divided into 11 slices covering 11 flavour attributes that are grouped intuitively, with groups separated by a heavier shaded line. The name and numerical value of the attribute intensity are presented on the outer rim of the slice for easy reference.

This novel approach will allow one to animate how the slices are filled and one can use this to track the change of flavour from powder, to liquor, to chocolate. Additionally, one can focus on a particular flavour attribute to see how this changes with time as a sub-chart

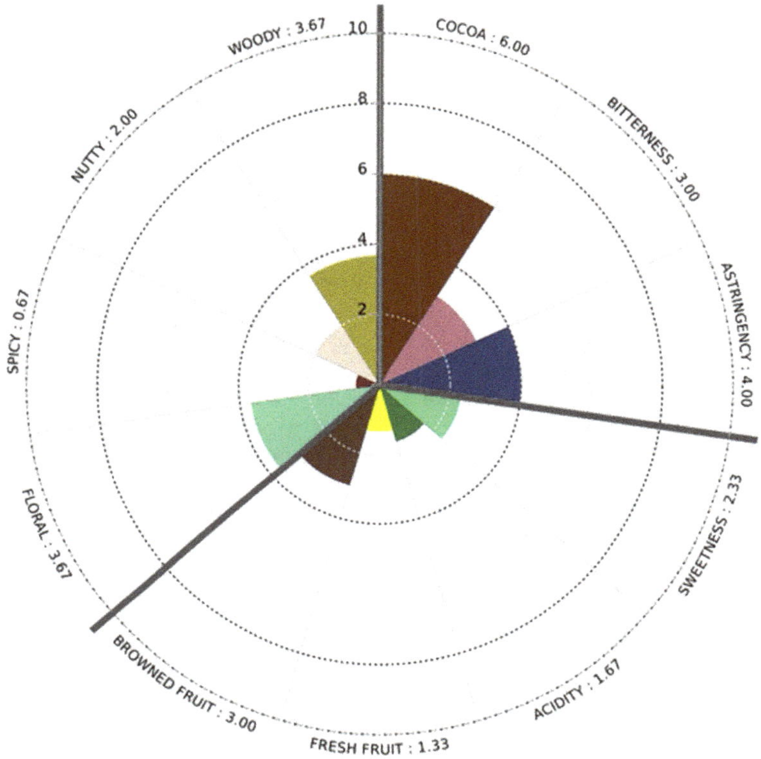

Figure 5 Filled pie slice plot of a sensory results from cocoa liquor sample on a 0–10-point scale over 11 flavour attributes (Adapted from Cocoa of Excellence Programme, 2015. Produced by Gastronomy 2016).

or use it for panellist calibration to show how taste scores change or align with experience or drift with time. An example of one of these filled 'pie slice' plots is shown in Fig. 5.

Even black-and-white versus colour versions of this representation show that with greyscale printing the graphs are still effective at showing differences between attributes. The intention is to have the option to create these graphs as part of an online service with a database that can also be used to share sensory descriptors and compare an expert panel opinion with that of others. Also, this visualisation can be used to identify people who, though not formally trained, know what they are tasting and compare their scores to a trained panel.

6 Conclusion

This chapter provided an overview of growth in the international cocoa sector in the context of the global chocolate market- and origin-specific dark and speciality chocolates to which sensory and processing quality are key drivers of growth in this segment. A timeline with highlights of some of the major research carried out in sensory and cocoa-processing quality was presented, showing the gaps and disconnect that exists in the application of this research to addressing needs specific to sensory and processing quality and the chocolate market segment that can benefit most from it.

A background on the evolution of sensory evaluation as a science, its use in defining quality for cocoa and identifying critical cocoa quality criteria leading to a recently revised definition of cocoa quality was provided and pertinent elements in this definition related to flavour and processing quality were expanded upon. The harsh reality and challenge faced by cacao farmers in their dual role as farmer and processer were highlighted in the context of needing to create closer and more harmonised links between producers and consumers. Raising awareness among cocoa farmers about consumer demand for quality cocoa beans was also highlighted in the context of improving recognition, preservation and valuation of cacao diversity, both for the bulk or ordinary and fine or flavour cocoa markets and how quality is assessed.

Finally, the chapter went on to provide a detailed step-by-step approach for analysing physical and sensory quality of cocoa including existing recognised international standards, experience and recommendations by the authors and recent research output whilst linking quality with the genetic expression of flavour potential.

7 Future trends

Cocoa has long been considered an orphaned commodity in which there has sometimes been a significant lag in the application of cutting-edge technology and even relatively old technology to address needs and challenges faced. This gap between new research and appropriate application is now closing and there are a number of exciting future trends to consider in the thematic areas of sensory evaluation and processing quality of cocoa. This section will briefly explore some of these trends, giving some brief background on their operational principle and application to cacao and cocoa.

7.1 Near-Infrared Reflectance Spectroscopy

Near-infrared reflectance spectroscopy (NIRS) is an accurate, precise and rapid alternative to wet chemistry procedures for determining concentrations of major classes of chemical compounds in organic materials. NIRS is the measurement of the wavelength and intensity of the absorption of near-infrared radiation by a sample. Near-infrared radiation spans the 800–2500 nm (wave numbers: 12 500–4000 cm^{-1}) range and is energetic enough to excite overtones and combinations of molecular vibrations to higher energy levels. The method utilises reflectance signals resulting from bending and stretching vibrations in molecular bonds between carbon, nitrogen, hydrogen and oxygen to quantitatively measure these organic functional groups. Calibration is required to correlate the spectral response of each sample at individual wavelengths to known chemical concentrations from laboratory analyses. Detection limits are typically 0.1% and applications include pharmaceutical, agricultural, polymer and biological analysis (Forest Ecosystems Group, 2006; Linde Gas, 2006).

NIRS has been applied to several agricultural activities such as in the production of organic fertiliser and how fertiliser is transformed in the soil (Thuriès et al., 2007a,b). It is also applied to a number of crops such as rice (*Oryza sativa* L.) where it is used to distinguish between aromatic and non-aromatic varieties (Davrieux et al., 2007a) and develop equations for grain quality traits in sorghum (*Sorghum bicolor* (L.) Moench) (De Alencar Figueiredo et al., 2006). NIRS has been used to develop prediction models to discriminate between coffee (*Coffea arabica*) samples from the same plantation, which underwent the same post-harvest treatment, independently of the harvest date and degree of ripeness based on both quantitative and qualitative coffee attributes (Davrieux et al., 2004). NIRS has also recently been applied to cocoa beans (Davrieux et al., 2005, 2007b) on a limited scale and there are many opportunities for its expanded application to cocoa with respect to quantitative and qualitative cocoa attributes. The work by Sukha et al. (2014) demonstrated processing location and growing environment effects on the flavour attributes of selected cacao genotypes with strongly supporting NIRS results.

7.2 Solid-Phase Micro-Extraction and head space sampling with Coupled Gas Chromatographic Mass Spectrometry

Solid-Phase Micro-Extraction (SPME) was developed in the early 1990s. It is not a new technology, but its application to better understanding the aromatic volatile compounds linked to flavour attributes in cocoa and chocolate is relatively recent (Pini et al., 2004; Ducki et al., 2008; Ariza, 2011). It uses a fused silica fibre that looks a modified syringe coated with a thin film of several stationary phase polymers to act like a sponge to concentrate organic analytes on the surface of the fibre via adsorption and absorption. After a suitable extraction time the fibre is withdrawn and injected directly into a gas chromatograph (GC) to elucidate and identify the organic analytes (Kataoka et al., 2000).

Headspace sampling involves taking a sample from the gas space in a chromatography vial above the sample without the use of solvent extraction. Headspace analysis is therefore the analysis of the components present in that gas. In static headspace sampling, the sample is sealed into a vessel, warmed, and then a sample of the atmosphere surrounding the sample is withdrawn and injected into the injection port of the GC. Dynamic headspace sampling, on the other hand, is a technique which uses a flow of carrier gas through the sample vessel to increase the headspace sample size, and thus the sensitivity of the technique. Instead of allowing the sample to come to equilibrium in a sealed container, the sample is warmed and the headspace atmosphere is constantly purged out of the sample vessel through a trap. The analytes are collected in this trap while the carrier gas is vented. After the collection step, the trap is heated and back flushed to transfer the adsorbed compounds to the GC for analysis. Since no solvents are used to prepare the sample for the GC, there is no solvent peak, no dilution of the compounds of interest, and no waste solvent to discard. Headspace sampling allows one to introduce volatile compounds automatically, from virtually any sample matrix, directly into a GC or GC/MS (CDS Analytical, 2016)

SPME or head space sampling can be used together with Gas Chromatography Mass Spectrometry (GC–MS) to measure significant differences in volatile compound concentrations of roasted and unroasted cocoa bean samples. The odours associated with these compounds can then be compared to the sensory attributes of the samples to search for links that may exist between these two quality criteria (Pini et al., 2004; Ducki et al., 2008; Ariza, 2011).

7.3 Artificial Neural Network

An artificial neural network (ANN) is an information-processing paradigm that is based on the cognitive process of the human brain. The key element of this paradigm is the novel structure of the information-processing system. It is composed of a large number of highly interconnected processing elements (neurones) which mimics the human nervous system (Persaud and Pelosi, 1992) working in unison to solve specific problems. An ANN, like human brain, learns by example. An ANN is configured for a specific application, such as pattern recognition or data classification, through a learning process. Learning in biological systems involves adjustments to the synaptic connections that exist between the neurones. In an ANN, a weight is randomly assigned to each neuron and then adjusted by means of an iterative or 'learning' process such as error back-propagation, until the desired outputs are obtained. The resultant set of weights and functions is then saved as a 'neural network'.

Neural networks have the ability to derive meaning from complicated or imprecise data and can be used to extract patterns and detect trends that are too complex to be noticed by either humans or other computer techniques. A trained neural network can be thought of as an 'expert' in the category of information it has been given to analyse. This can then be used to provide projections given new situations of interest and answer 'what if' questions.

Other advantages of ANN include:

- Adaptive Learning: An ability to learn how to do tasks based on the data given for training or initial experience.
- Self-Organisation: An ANN can create its own organisation or representation of the information it receives during learning time.
- Real-Time Operation: ANN computations may be carried out in parallel, and special hardware devices which take advantage of this capability are designed and manufactured.
- Fault Tolerance via Redundant Information Coding: Partial destruction of a network leads to the corresponding degradation of performance. However, some network capabilities may be retained even with major network damage (Stergiou and Siganos, 2010).

Electronic nose or 'e-nose' is an instrument consisting of an array of weakly specific or broad-spectrum chemical sensors that are intended to mimic the human olfactory system, by converting sensor signals to digital data and then analysing the data by appropriate software (Gardner and Bartlett, 1994; Keller, 1995).

An ANN-based electronic nose system with coupled gas chromatography, which uses a combination of e-nose system with ANN, is currently being researched at the University of Georgia to determine the quality of cocoa beans (over-fermented/putrid or not), to grade of the beans, and to check fermentation quality and roasting degree (Personal communication Juzhong Tan 28 March 2017). Traditional evaluation methods, such as GC–MS with SPME extraction, need considerable time and effort in training and calibration without which it can be arbitrary. The e-nose-based ANN can be a fast and cheap way to monitoring the processing of chocolate in bean grading, fermentation, roasting and conching by mapping the volatile compound profiles.

In this ongoing study by Tan and Kerr at the University of Georgia, Athens, Georgia, USA, a self-made e-nose system consisting of gas sensors was used to detect the gas generated by five types of graded cocoa beans and the signals collected by gas sensors were used to train a three-layer ANN. Chemical analysis of the graded beans was operated by a GC and the contents of volatile chemical compounds were used to train another ANN as a reference to electronic nose signals trained ANN. Both trained ANNs (e-nose and GC) were used to grade the graded beans over 96.7% and 83.8% accuracy respectively.

7.4 Olfactory genomics

Olfactory genomics is the newest approach to sensory perception of aroma and flavour that is currently being marketed by the Aromyx Corporation in California, USA, as the Aromyx EssenceChip™. The approach puts human olfactory receptors on a disposable bioassay format in a convenient, stable and cost-effective way that customers can use on their own premises.

Unlike chemical sensors or mass spectrometry, the Aromyx EssenceChip™ system shows the biochemical signals that a cup of coffee, a sniff of perfume or a glass of wine send to the brain, as mediated by the human nose and tongue. The bioassay reveals the same biochemical signals that the nose and tongue send to the brain in response to a flavour or fragrance. It records flavours and fragrances, with reproducible results. The output of the bioassay has a direct correlation to what a person smells and what the brain experiences. The Aromyx Corporation states that their approach maintains both sensitivity and specificity – the ability to discriminate – of the human nose and tongue, and the collection of biochemical signals is represented by a digital aromagraph (Aromxy, 2016).

8 Where to look for further information

- Review article or text books:

Afoakwa, E. O., 2016. Chocolate science and technology. John Wiley & Sons. ISBN: 978-1-4051-9906-3.
Beckett, S. T. ed., 2011. *Industrial chocolate manufacture and use*. John Wiley & Sons. ISBN: 978-1-4051-3949-6.
CAOBISCO/ECA/FCC *Cocoa Beans: Chocolate and Cocoa Industry Quality Requirements*. September 2015 (End, M. J. and Dand, R., Editors) Published by ECA-Caobisco-FCC Cocoa Research Fund www.cocoaquality.eu ISBN: 978-2-9601817-0-8.
Wood, G. A. R. and Lass, R. A., 1995. Cocoa 4th Edition. John Wiley & Sons. ISBN:0-632-06398-X.

- Seminal articles or books

Badrie, N., Bekele, F., Sikora, E. and Sikora, M., 2015. Cocoa agronomy, quality, nutritional, and health aspects. *Critical Reviews In Food Science And Nutrition*, 55(5), pp. 620–659.
Schwan R. F. and Alan E. Wheals. 2004. The Microbiology of Cocoa Fermentation and its Role in Chocolate Quality. *Critical Reviews In Food Science And Nutrition*, 44(4), 2004.

- Key societies, organisations or websites

The International Cocoa Organisation: https://www.icco.org/
The World Cocoa Foundation: http://www.worldcocoafoundation.org/

The Association of Chocolate, Biscuit and Confectionery Industries of Europe: http://caobisco.eu/
The Federation of Cocoa Commerce: http://www.cocoafederation.com/
The European Cocoa Association: http://www.eurococoa.com/
The Cocoa Merchants' Association of America, Inc.: http://www.cocoamerchants.com/
Bioversity International: https://www.bioversityinternational.org/
The Fine Chocolate Industry Association: http://www.finechocolateindustry.org/
The Cocoa of Excellence Programme: http://www.cocoaofexcellence.org/
The Heirloom Cacao Preservation Fund: http://hcpcacao.org/
The International Chocolate Awards: http://www.internationalchocolateawards.com/

- Top five research centres

The Cocoa Research Centre: https://sta.uwi.edu/cru/
French Agricultural Research Centre for International Development: http://www.cirad.fr/en
The Cocoa Research Institute of Ghana: http://www.crig.org.gh/
The Cocoa Research Institute of Nigeria: http://www.crin-ng.org/
Malaysian Cocoa Board: http://www.koko.gov.my/

9 References

Afoakwa, E. O., Paterson, A., Fowler, M. and Ryan, A. (2008). 'Flavor formation and character in cocoa and chocolate: A critical review', *Critical Reviews in Food Science and Nutrition*, 48(9), 840–57.

Ali, N. A., Baccus-Taylor, G. S. H., Sukha, D. A. and Umaharan, P. (2013). A preliminary investigation of the effect of cacao (*Theobroma cacao* L.) pulp on final flavour. In: *The Malaysian International Cocoa Conference*. Kuala Lumpur, Malaysia: MICC, pp. 188–93.

Ali, N. A., Baccus-Taylor, G. S.H; Sukha, D. A. and Umaharan, P. (2016). Fermentation dynamics and quality in various genetic groups of *Theobroma cacao* L. In: *Third International Congress of Biotechnology and Biodiversity (III CIBB)*. Guyaquil, Ecuador: Ecuadorian Center for Biotechnology Research (CIBE) of the Escuela Superior Politécnica del Litoral (EPSOL).

Alle, R. E. (1990). Quality. In: *Concise Oxford English Dictionary of Current English*, 8th ed. New York: Oxford University Press, pp. 909.

Amerine, M. A., Pangborn, R. M. and Roessler, E. B. (2013). *Principles of Sensory Evaluation of Food*. Elsevier, pp. 1–22.

Andersson, M., Koch, G. and Lieberei, R. (2006). 'Structure and function of the seed coat of *Theobroma cacao* L. and its possible impact on flavour precursor development during fermentation', *Journal of Applied Botany and Food Quality*, 80(1), 48–62.

Ariza, B. T. S. (2011). 'Use of gas chromatography–olfactometry in combination with solid phase micro extraction for cocoa liquor aroma analysis', *International Food Research Journal*, 18(2), 829–35.

Aromyx. (2016). Aromyx Technology. [online] Available at http://www.aromyx.com/technology.html [Accessed on 13 December 2016].

Augber, F., Nganhou, J., Benet, J. C., Berthomieu, G. and Barel, M. (1999). 'Experimental study of matter transfer in cocoa beans during fermentation and drying', *Drying Technology*, 17(6), 1027–42.

Badrie, N., Bekele, F., Sikora, E. and Sikora, M. (2015). 'Cocoa agronomy, quality, nutritional, and health aspects', *Critical Reviews in Food Science and Nutrition*, 55(5), 620–59.

Bayer, C., Fay, M. F., Bruijn, A. Y., Savolainen, V., Morton, C. M., Kubitzki, K., Alverson, W. S. and Chase, M. W. (1999). 'Support for an expanded family concept of Malvaceae within a recircumscribed order Malvales: A combined analysis of plastid atpB and rbcL DNA sequences', *Botanical Journal of the Linnean society*, 129(4), 267–303.

Bonaparte, A., Alikhani, Z., Madramootoo, C. A. and Raghavan, V. (1998). 'Some quality characteristics of solar-dried cocoa beans in St Lucia', *Journal of the Science of Food and Agriculture*, 76(4), 553–8.

Bonvehí, J. S. (2005). 'Investigation of aromatic compounds in roasted cocoa powder', *European Food Research and Technology*, 221(1–2), 19–29.

Bravo, A. M. (1974). 'Fundamental artificial drying characteristics of cocoa beans', *Tropical Agriculture* (Trinidad and Tobago), 51(3), 395–406.

Butler, D. R. and Umaharan, P. (2004). Working with cocoa germplasm. In: J. Flood and R. Murphy (Eds), *Cocoa Futures – A Source Book of Some Important Issues Confronting the Cocoa Industry*. Chinchiná, Colombia: Commodities Press, pp. 54–64.

Cacao & Chocolate Institute. (2016). Caoao Bean Cutting. [online] Available at http://cacaochocolate.org/?act=board&bbs_code=webzine&bbs_mode=view&bbs_seq=7 [Accessed on 9 June 2017].

Camu, N., De Winter, T., Addo, S. K., Takrama, J. S., Bernaert, H. and De Vuyst, L. (2008). 'Fermentation of cocoa beans: Influence of microbial activities and polyphenol concentrations on the flavour of chocolate', *Journal of the Science of Food and Agriculture*, 88(13), 2288–97.

CDS Analytical. (2016). CDS Analytical Headspace. [online] Available at: http://www.cdsanalytical.com/prod_headspace.html [Accessed on 18 November 2016].

Chinenye, N. M., Ogunlowo, A. S. and Olukunle, O. J. (2010). 'Cocoa bean (*Theobroma cacao* L.) drying kinetics', *Chilean Journal of Agricultural Research*, 70(4), 633–9.

Clapperton, J. F., Lockwood, G., Yow, S. T. K. and Lim, D. H. K. (1994a). 'Effects of planting materials on flavour', *Cocoa Growers' Bulletin 48*, Birmingham, England: Cadbury Limited, pp. 47–57.

Clapperton, J. F., Yow, S. T. K., Lim, D. H. K. and Lockwood, G. (1994b). Genetic variation in cocoa flavor. In: *11th International Cocoa Research Conference*. Yamoussoukro, Cote d'Ivoire: Cocoa Producers' Alliance COPAL, pp. 749–58.

Collings, V. B. (1974). 'Human taste response as a function of locus of stimulation on the tongue and soft palate', *Perception & Psychophysics*, 16(1), 169–74.

Corti, R., Flammer, A. J., Hollenberg, N. K. and Lüscher. T. F. (2009). 'Cocoa and cardiovascular health', *Circulation*, 119(10), 1433–41.

Cros, E. (1996). Cocoa aroma formation. In: *Proc. Cocoa Meetings – The Various Aspects of Quality*. Montpellier, France: Centre de coopération internationale en recherche agronomique pour le développement, pp. 169–79.

Cros, E. (1998). Aspects of the quality of fine cocoa: Differences compared to bulk cocoa. In: *9th Meeting of Advisory Group on the World Cocoa Economy*. Seville, Spain: ICCO.

Davrieux, F., Bertrand, B. Bastianelli, D. and Guyot, B. (2004). Authentication of three green coffee cultivars from Costa Rica by near infrared spectroscopy: Preliminary study. In Near infrared spectroscopy. In: *11th International Conference*. Cordoba, Spain: NIR Publications.

Davrieux, F., Assemat, S., Boulanger, R. and Cros, E. (2005). Determination of cocoa purine content by near infrared spectroscopy. In: *14th International Cocoa Research Conference*. Accra, Ghana: COPAL, pp. 861–7.

Davrieux, F., El Ouadrhiri, Y., Pons, B. and Bastianelli, D. (2007a). Discrimination between aromatic and non-aromatic rice by near infrared spectroscopy: A preliminary study. In Near infrared spectroscopy. In: *12th International Conference, Auckland, New Zealand*. Chichester: IM Publications, pp. 394–6.

Davrieux, F., Assemat, S., Sukha, D. A., Bastianelli, D., Boulanger, R. and Cros, E., (2007b). Genotype characterisation of cocoa into genetic groups through caffeine and theobromine content predicted by NIRS. In Near infrared spectroscopy. In: *12th International Conference, Auckland, New Zealand, 2005*. Chichester: IM Publications, pp. 382–6.

de Alencar Figueiredo, L. F., Davrieux, F., Fliedel, G., Rami, J. F., Chantereau, J., Deu, M., Courtois, B. and Mestres, C. (2006). 'Development of NIRS equations for food grain quality traits through

exploitation of a core collection of cultivated sorghum', *Journal of Agricultural and Food Chemistry*, 54(22), 8501–9.
De Bruyne, K., Camu, N., Lefebvre, K., De Vuyst, L. and Vandamme, P. (2008). 'Weissella ghanensis sp. nov., isolated from a Ghanaian cocoa fermentation', *International Journal of Systematic and Evolutionary Microbiology*, 58(12), 2721–5.
De Witt, K. W. (1955). The flavour assessment of cacao. *A Report on Cacao Research 1954*. The Imperial College of Tropical Agriculture, Trinidad and Tobago, pp. 77–81.
Ducki, S., Miralles-Garcia, J., Zumbé, A., Tornero, A. and Storey, D. M. (2008). 'Evaluation of solid-phase micro-extraction coupled to gas chromatography – mass spectrometry for the headspace analysis of volatile compounds in cocoa products', *Talanta*, 74(5), 1166-74.
End, M. J. and Dand, R. (2015). *Cocoa Beans: Chocolate and Cocoa Industry Quality Requirements*. London: CAOBISCO/ECA/FCC.
Engels, J. M. M. (1981). 'Genetic resources of cacao: A catalogue of the Centro Agronómico Tropical de Investigación y Enseñanza collection', *Technical Bulletin 7*, Turrialba, Costa Rica: CATIE.
Eskes, A. B., Guarda, D., Garcia, L. and Garcia, P. (2007). 'Is genetic variation for sensory traits of cocoa pulp related to fine flavor cocoa traits?', *INGENIC Newsletter*, 11, 22–8.
Eskes, A. B., Ahnert, D., Garcia, L. C., Seguine, E., Assemat, S., Guarda, D. and Garcia, P. R. (2012). *Evidence on the Effect of the Cocoa Pulp Flavour Environment During Fermentation on the Flavour Profile of Chocolates*. [online] Available at: http://agritrop.cirad.fr/568108/1/document_568108.pdf [Accessed on 21 July 2016].
Faborode, M. O., Favier, J. F. and Ajayi, O. A. (1995). 'On the effects of forced air drying on cocoa quality', *Journal of Food Engineering*, 25(4), 455–72.
Fagunwa, A. O., Koya, O. A. and Faborode, M. O. (2009). 'Development of an intermittent solar dryer for cocoa beans', *Agricultural Engineering International: CIGR Journal*, XI, 14.
Fine Cacao and Chocolate Institute (FCCI). (2016). *FCCI Cacao Sampling Protocol*. Massachusetts: Fine Cacao and Chocolate Institute, pp. 1–2.
Forest Ecosystems Group. (2006). *Near Infrared Reflectance Spectroscopy*. [online] Available at: http://www.forest.sr.unh.edu/nir/nir.html [Accessed on 4 June 2006].
Fowler, M. (1996). Quality of cocoa beans for chocolate manufacturers. In: *Proc. Cocoa Meetings – The Various Aspects of Quality*. Montpellier, France: Centre de coopération internationale en recherche agronomique pour le développement, pp. 147–53.
Franchise Help. (2016). *Chocolate Industry Analysis 2016 – Cost & Trends*. FranchiseHelp Holdings LLC. [online] Available at: https://www.franchisehelp.com/industry-reports/chocolate-industry-report/ [Accessed on 26 November 2016].
Gálvez, S. L., Loiseau, G., Paredes, J. L., Barel, M. and Guiraud, J. P. (2007). 'Study on the microflora and biochemistry of cocoa fermentation in the Dominican Republic', *International Journal of Food Microbiology*, 114(1), 124–30.
Gardner, J. W. and Bartlett, P. N. (1994). 'A brief history of electronic noses', *Sensors and Actuators B: Chemical*, 18(1), 210–11.
Hänig, D. (1901). 'Zur Psychophysik des Geschmackssinnes', *Philosophische Studien*, 17, 576–623.
Hii, C. L., Abdul Rahman, R., Jinap, S. and Che Man, Y. B. (2006). 'Quality of cocoa beans dried using a direct solar dryer at different loadings', *Journal of the Science of Food and Agriculture*, 86(8), 1237–43.
Hii, C. L., Law, C. L. and Cloke, M. (2008). 'Modelling of thin layer drying kinetics of cocoa beans during artificial and natural drying', *Journal of Engineering Science and Technology*, 3(1), 1–10.
Hii, C. L., Law, C. L., Cloke, M. and Suzannah, S. (2009). 'Thin layer drying kinetics of cocoa and dried product quality', *Biosystems Engineering*, 102(2), 153–61.
Hollywood, N., Brown, S. and Toreu, B. (1996). 'A design for improved efficiency in solar drying of cocoa', *Cocoa Growers' Bulletin 50*, Cadbury Ltd, England, pp. 38–45.
Hustler, M. (2014). The Cocoa Industry in CARIFORUM Background Paper. In: *3rd CARIFORUM-EU Business Forum*. Montego Bay, Jamaica: Enabling Environments Ltd., 34p.
International cocoa organization ICCO. (2007). *Project to Determine the Physical, Chemical and Organoleptic Parameters to Differentiate Between Fine and Bulk Cocoa – Project Completion*

Report. [online] Available at: http://www.icco.org/about-us/international-cocoa-agreements/doc_download/198-the-differencebetween-fine-and-bulk-cocoa.html [Accessed on 26 November 2016].

International cocoa organization ICCO. (2016a). *Quarterly Bulletin of Cocoa Statistics*, Vol. XLII – No. 4 – Cocoa year 2015/2016. [online] Available at: https://www.icco.org/about-us/icco-news/333-quarterly-bulletin-of-cocoa-statistics-november-2016.html [Accessed on 26 November 2016].

International cocoa organization ICCO. (2016b). What is Fine or Flavour Cocoa? [online]. Available at: https://www.icco.org/about-cocoa/fine-or-flavour-cocoa.html [Accessed on 26 November 2016].

International cocoa organization ICCO. (2016c). Characteristics of the Fine or Flavour Cocoa Market. [online] Available at: https://www.icco.org/about-cocoa/fine-or-flavour-cocoa.html [Accessed on 26 November 2016].

International cocoa organization ICCO. (2016d). *ICCO Panel Recognizes 23 Countries as Fine and Flavour Cocoa Exporters*. [online] Available at: https://www.icco.org/about-us/icco-news/319-icco-panel-recognizes-23-countries-as-fine-and-flavour-cocoa-exporters.html [Accessed on 26 November 2016].

International office of cocoa, chocolate and sugar confectionery (IOCCC). (1996a). *Manufacturing Process of Small Scale Chocolate Samples*. Analytical Method 12-1971. Brussels: International Office of Cocoa, Chocolate and Sugar Confectionery.

International office of cocoa, chocolate and sugar confectionery (IOCCC). (1996b). *The Detection of Specific Off-flavours in Cocoa Beans*. Analytical Method 44-1996. Brussels: International Office of Cocoa, Chocolate and Sugar Confectionery.

International Organization for Standardization ISO. (1973). *Standard 2292: Cocoa Beans – Sampling*. Geneva: International Organization for Standardization.

International Organization for Standardization ISO. (1977). *Standard 1114: Cocoa Beans – Cut Test*. Geneva: International Organization for Standardization.

International Organization for Standardization ISO. (1980). *Standard 2291: Cocoa Beans – Determination of Moisture Content (Routine method)*. Geneva: International Organization for Standardization.

International Organization for Standardization ISO. (2014). *Standard 2541: Cocoa Beans – Specification and Quality Requirements*. Geneva: International Organization for Standardization.

Jeanjean, N., Cros, E. and Flori, A. (1994). Cocoa Flavour: Correlation between non-volatile and volatile compounds. In: *International Cocoa Conference: Challenges in the 1990's*. Kota Kinabalu, Malaysia: Malaysian Cocoa Board, pp. 296–308.

Jinap, S. (1994). 'Organic acids in cocoa beans-a review', *ASEAN Food Journal (Malaysia)*, 9, 3–12.

Jinap, S., Thien, J. and Yap, T. N. (1994). 'Effect of drying on acidity and volatile fatty acids content of cocoa beans', *Journal of the Science of Food and Agriculture*, 65(1), 67–75.

Jinap, S., Jamilah, B. and Nazamid, S. (2005). 'Changes in polyphenol ability to produce astringency during roasting of cocoa liquor', *Journal of the Science of Food and Agriculture*, 85(6), 917–24.

Kadow, D., Bohlmann, J., Phillips, W. and Lieberei, R. (2013). 'Identification of main fine flavour components in two genotypes of the cocoa tree (*Theobroma cacao* L.)', *Journal of Applied Botany and Food Quality*, 86(1), 90–8.

Kamphuis, H. J. (2011). Production and quality standards of cocoa mass, cocoa butter and cocoa powder. In: S. T. Beckett (Ed.), *Industrial Chocolate Manufacture and Use*, 4th ed. Chichester: John Wiley & Sons, pp. 121–41

Kataoka, H., Lord, H. L. and Pawliszyn, J. (2000). 'Applications of solid-phase microextraction in food analysis', *Journal of Chromatography A*, 880(1), 35–62.

Keller, P. E. (1995). Electronic noses and their applications. In: *Northcon 95. IEEE Technical Applications Conference and Workshops Northcon95*. Boston: IEEE, p. 116.

Knapp, A. W. (1937). *Cacao Fermentation*. London: John Bale, Sons and Curnow.

Kostinek, M., Ban-Koffi, L., Ottah-Atikpo, M., Teniola, D., Schillinger, U., Holzapfel, W. H. and Franz, C. M. (2008). 'Diversity of predominant lactic acid bacteria associated with cocoa fermentation in Nigeria', *Current Microbiology*, 56(4), 306–14.

Kramer, A., and Twigg B. A. (1970). *Quality Control for the Food Industry, Vol. 1. Fundamentals*, 3rd ed., Westport, Connecticut: The AVI Publishing Company, Inc., p. 197.

Lasisi, D. (2014). 'A comparative study of effects of drying methods on quality of cocoa beans', *International Journal of Engineering*, 3(1), 991–6.

Laugerette, F., Gaillard, D., Passilly-Degrace, P., Niot, I. and Besnard, P. (2007). 'Do we taste fat?', *Biochimie*, 89(2), 265–9.

Laurent, V., Risterucci, A. M. and Lanaud, C. (1994). 'Genetic diversity in cocoa revealed by cDNA probes', *Theoretical and Applied Genetics*, 88(2), 193–8.

Lawless, H. T. and Heymann, H. (2010). *Sensory Evaluation of Food: Principles and Practices*. NewYork: Springer Science & Business Media.

Lehrian, D. W., Keeney, P. G. and Butler, D. R. (1980). 'Triglyceride characteristics of cocoa butter from cacao fruit matured in a microclimate of elevated temperature1', *Journal of the American Oil Chemists' Society*, 57(2), 66–9.

Lenoir, J. (2016). *The Masterkit 54 Aromas*. [online] Available at https://www.lenez.com/en/kits/wine/masterkit_54 [Accessed on28 November 2016].

Linde Gas. (2006). *Near Infrared Spectroscopy*. [online] Available at http://hiq.linde-gas.com/international/web/lg/spg/likelgspg.nsf/docbyalias/anal_ninfra [Accessed on 4 June 2006].

Linnaeus, C. (1753). *Species Plantarum*. Stockholm: Laurentii Salvii. [online] Available at: http://taxonomicon.taxonomy.nl/Reference.aspx?id=4673 [Accessed on 26 November 2016].

Lippi, D. (2009). 'Chocolate and medicine: Dangerous liaisons?', *Nutrition*, 25(11), 1100–3.

Lockwood, G. and A. B. Eskes. (1996). Relationship between cocoa variety and quality. In: *Proc. Cocoa Meetings – The various Aspects of Quality*. Montpellier, France: Centre de coopération internationale en recherche agronomique pour le développement, pp. 159–67.

Lopez, A. S. and Dimick, P. S., (1995). Cocoa fermentation. In: H. J. Rehm and G. Reed (Eds), *Biotechnology Set*, 2nd ed. Weinheim, Germany: Wiley-VCH Verlag GmbH, pp. 561–77.

Lowor, S. T., Jacquet, M., Vrielink, T., Aculey, P., Cros, E. and Takrama, J. (2012). 'Post-harvest sources of polycyclic aromatic hydrocarbon contamination of cocoa beans: A simulation', *International Journal of AgriScience*, 2(11), 1043–52.

Manière, F. Y., and Dimick P. S. (1979). 'Effects of conching on the flavour and volatile components of dark semi-sweet chocolat', *Lebensmittel-Wiss.-Technology*, 12, 102–7.

Markets and Markets. (2011). *Report Code: CG 1111Global Chocolate, Cocoa Beans, Lecithin, Sugar and Vanilla Market By Market Share, Trade, Prices, Geography Trend and Forecast (2011–2016)*. [online] Available at: http://www.marketsandmarkets.com/Market-Reports/global-chocolate-market-164.html [Accessed on26 November 2016].

McDonald, C. R. and RA López, S. A. F. (1981). 'Cocoa drying-a review', *Cocoa Growers' Bulletin 31*, Birmingham, England: Cadbury Limited, pp. 5–41.

Meilgaard, M. C., Carr, B. T. and Civille, G. V. (2006). *Sensory Evaluation Techniques*. 4th ed. London: CRC Press, Taylor and Francis Group.

Micronising Company. (2017). 'Cocoa Beans', Micronizing infrared food processing systems [online] Available at: http://www.micronizing.com/?page_id=93 [Accessed on 31 May 2017].

Mossu, G. (1992). Cocoa. In: René Coste (ed.), *The Tropical Agriculturalist*. London, UK: MacMillan, pp. 10–22.

Motamayor, J. C., Lachenaud, P., e Mota, J. W. D. S., Loor, R., Kuhn, D. N., Brown, J. S. and Schnell, R. J. (2008). 'Geographic and genetic population differentiation of the Amazonian chocolate tree (*Theobroma cacao* L.)', *PLoS One*, 3(10), e3311.

Ndukwu, M. C. (2009). 'Effect of drying temperature and drying air velocity on the drying rate and drying constant of cocoa bean', *Agricultural Engineering International: CIGR Journal Manuscript 1091*. Vol. XI, 7.

Nielsen, D. S., Teniola, O. D., Ban-Koffi, L., Owusu, M., Andersson, T. S. and Holzapfel, W. H. (2007). 'The microbiology of Ghanaian cocoa fermentations analysed using culture-dependent and culture-independent methods', *International journal of food microbiology*, 114(2), 168–86.

Ostovar, K. and Keeney, P. G. (1973). 'Isolation and characterization of microorganisms involved in the fermentation of Trinidad's cacao beans', *Journal of Food Science*, 38(4), 611–17.

Pascale, D. (2006). *RGB coordinates of the Macbeth ColorChecker*. Quebec, Canada: The BabelColor Company, pp. 1–16.

Persaud, K. C. and Pelosi, P. (1992). Sensor arrays using conducting polymers for an artificial nose. In: J. W. Gardener and P. N. Bartlett (Eds), *Sensors and Sensory Systems for an Electronic Nose*. the Netherlands: Springer, pp. 237–56.

Pini, G. F., Brito, E. S. D., García, N. H., Valente, A. L. and Augusto, F. (2004). 'A Headspace Solid Phase Microextraction (HS-SPME) method for the chromatographic determination of alkylpyrazines in cocoa samples', *Journal of the Brazilian Chemical Society*, 15(2), 267–71.

Rea, S. (2015). Projet Chocolat. [online] Available at http://projetchocolat.com/ [Accessed on 9 June 2017].

Rodriguez-Campos, J., Escalona-Buendía, H. B., Contreras-Ramos, S. M., Orozco-Avila, I., Jaramillo-Flores, E. and Lugo-Cervantes, E. (2012). 'Effect of fermentation time and drying temperature on volatile compounds in cocoa', *Food chemistry*, 132(1), 277–88.

Rohsius, C., Matissek, R. and Lieberei, R. (2006a). 'Free amino acid amounts in raw cocoas from different origins', *European Food Research & Technology*, 222, 432, https://doi.org/10.1007/s00217-005-0130-y

Rohsius, C. R., Andersson, M., Niemenak, N., Sukha, D. and Lieberei, R. (2006b). Fermentation quality and its dependence on the testa structure and transport processes. In: *15th International Cocoa Research Conference*. San José, Costa Rica: COPAL, p. 81.

Saldaña M. D., Mohamed, R. S. and Mazzafera, P. (2002). 'Extraction of cocoa butter from Brazilian cocoa beans using supercritical CO 2 and ethane', *Fluid phase equilibria*, 194, 885–94.

Sankat, C. K. and Ramlochan, M. (1997). The performance of a cocoa bean solar fermenting/drying system for small farmers. In: *9. Seminar on Agricultural Research and Development: Advances and Prospects for Sustainable Agriculture for Export and National Food Security in the New Millennium*. Port of Spain, Trinidad and Tobago: NIHERST, pp. 62–72.

Schwan, R. F. (1998). 'Cocoa fermentations conducted with a defined microbial cocktail inoculum', *Applied and Environmental Microbiology*, 64(4), 1477–83.

Schwan R. F. and Wheals, A. E. (2004). 'The microbiology of cocoa fermentation and its role in chocolate quality', *Critical Reviews in Food Science and Nutrition*, 44(4), 205–21.

Seguine, E. (2014). *Operating Procedures and Recommendations for Equipment Operation - Laboratory Evaluation of Cocoa Beans*, Version 1.0, 23 November 2014. Pennsylvania: Seguine Cacao Cocoa & Chocolate Advisors.

Stergiou, C. and Siganos, D. (2010). *Neural Networks. 1996*. [online] Available at: https://www.doc.ic.ac.uk/~nd/surprise_96/journal/vol4/cs11/report.html [Accessed on 10 December 2016].

Stone, H. and Sidel, J. L. (1993). *Sensory Evaluation Practices*. 2nd ed. San Diego: Academic Press.

Sukha, D. A., Umaharan, P. and Butler, D. R. (2017). 'The Impact of Pollen Donor on Flavor in Cocoa', *Journal of American Society for Horticultural Science*, 142(1), 13–19.

Sukha, D. A., Butler, D. R., Umaharan, P. and Boult, E. (2008). 'The use of an optimized organoleptic assessment protocol to describe and quantify different flavour attributes of cocoa liquors made from Ghana and Trinitario beans', *European Food Research and Technology*, 226(3), 405–13.

Sukha, D. A., Butler, D.R., Comissiong, E. A. and Umaharan, P. (2014). The impact of Processing Location and Growing Environment on flavor in cocoa (*Theobroma cacao* L.) – implications for 'Terroir' and Certification – Processing Location study. In: *III International Conference on Postharvest and Quality Management of Horticultural Products of Interest for Tropical Regions*. Port of Spain, Trinidad and Tobago: Acta Horticulturae. (International Society for Horticultural Science) 1047, pp. 255–62.

Sukha, D. A. and Rohsius, C. (2004). *Cocoa Cut Test Chart*. St. Augustine, Trinidad and Tobago: Cocoa Research Centre.

Thuriès, L., Bastianelli, D., Davrieux, F., Bonnal, L. and Oliver, R. (2007a). Prediction by near infrared spectroscopy of the biochemical composition of various raw materials used in the organic fertiliser industry. In Near infrared spectroscopy. In: *12th International Conference*. Auckland, New Zealand, Chichester: IM Publications, pp. 781–5.

Thuriès, L., Bastianelli, D., Davrieux, F., Bonnal, L. and Oliver, R. (2007b). Near infrared spectroscopy for predicting quality indices in the organic fertiliser industry. In Near infrared spectroscopy. In: *12th International Conference*. Auckland, New Zealand, Chichester: IM Publications, pp. 789–92.

Toxopeus, H. (1985). Botany, types and populations. In: G. A. R. Wood and R. A. Lass (Eds), *Cocoa*. 4th ed. London: Longman, pp. 11–37.

United States Food and Drug Administration (FDA). (1968). *Chocolate, Sugars, and Related Products – Method for Cocoa Beans (v-18)*. (Macroanalytical Procedures Manual MPM: V-4). Silver Spring: Food and Drug Administration.

van der Kooij, S. (2013). Market study of fine flavour cocoa in 11 selected countries – revised version. *Royal Tropical Institute KIT Development Policy and Practice*. Amsterdam: Royal Tropical Institute Inforum.

Wan Daud, W. R., Meor Talib, M. Z. and Hakimi Ibrahim, M. (1996). 'Characteristic drying curves of cocoa beans', *Drying Technology*, 14(10), 2387–96.

Wood, G. A. R. (1961). Experiments on cocoa drying in the Cameroon. *Tropical Agriculture (Trinidad)*, 44, 125–32.

Wood, G. A. R. and Lass, R. A. (1985). From harvest to store and Quality and inspection. In: G. A. R Wood and R. A. Lass (Eds), *Cocoa*. 4th ed. London: Longman, pp. 444–527.

Ziegleder, V. G. and Biehl B. (1988). Analysis of cocoa flavor components and flavor precursors In: H. J. Likens and J. F. Jackson (Eds), *Modern Methods of Plant Analysis*. Berlin: Springer-Verlag, pp. 321–90.

Part 5

Sustainability

Chapter 19

Climate change and cocoa cultivation

Christian Bunn, Fabio Castro and Mark Lundy, International Center for Tropical Agriculture (CIAT), Colombia; and Peter Läderach, International Center for Tropical Agriculture (CIAT), Vietnam

1 Introduction

It is virtually certain that global temperatures have increased by about 0.85°C during the period 1880–2012 (Stocker et al., 2013). Anthropogenic greenhouse gas (GHG) emissions were identified as the most likely cause of the observed radiative forcing. Global circulation models (GCMs) have projected the effects of continued emissions on the global climate, and an additional increase of 1.4–3.1°C by the end of this century was found to be likely (Stocker et al., 2013). The implications of increased atmospheric energy on the distribution of precipitation were less certain. However, there is increasing consensus that the risk of drought has increased over the past decades. GCMs were found to reproduce these trends and projected additional drought risk for the coming 30–90 yrs, either through decreased precipitation or increased evaporation (Dai, 2013).

Global cocoa production occupies about ten million ha in tropical countries close to the equator (Fig. 1). Most of the 4.5 million tons of cocoa produced are exported to developed countries, mainly North America and Europe. Over the past decades roughly 70% of production originated in Africa, with just two countries, Ivory Coast and Ghana, accounting for about 50% of the global production over the last five decades (FAO, 2016). All of the tropical Americas and Southern Asia account for about 15% each.

http://dx.doi.org/10.19103/AS.2017.0021.28

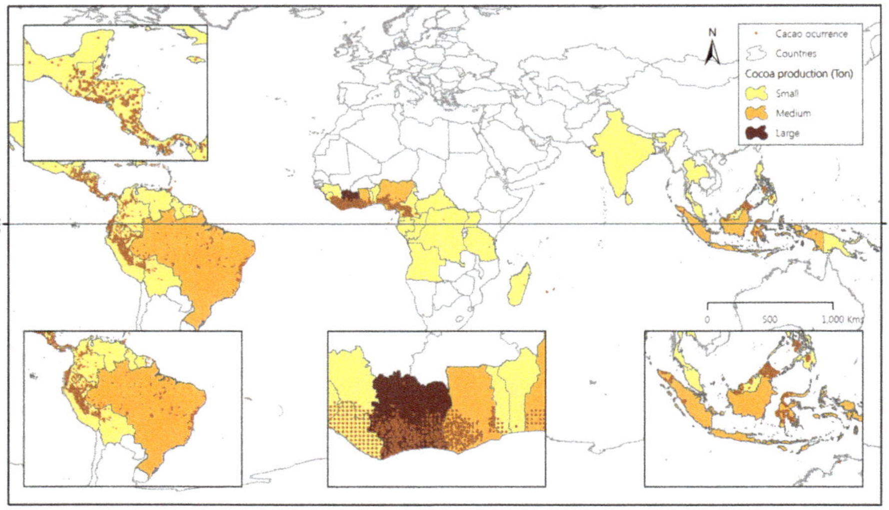

Figure 1 Distribution of cocoa occurrence locations of *T. cacao* and produced quantities (1998–2002) (FAO, 2016).

Thus, production is highly concentrated in a relatively small region where it contributes significant shares of GDP and export revenues of countries concerned. Negative climate change impacts would therefore not only affect producers in West Africa but also have repercussions for the economies of the region and the entire cocoa sector.

The cocoa tree, *Theobroma cacao L.*, originated from the wet forests of South America, very close to the equator. In its natural habitat, rainfall is heavy and the temperature is relatively uniform. High precipitation of about 1500–2000 mm well distributed throughout the entire year is considered optimal for the growth of cocoa crop. However, in the major production regions in West Africa, production systems have been adapted to a weather pattern with one or two dry seasons per year. Months with less than 100 mm are considered dry months (Lass and Wood, 1985), as little as 60 mm is tolerated. While a dry season induces uniform flowering, overall yields are higher without such a period (Zuidema et al., 2005). This may be associated with low assimilation rates caused by vapour pressure deficit in the dry season (Acheampong et al., 2013). In general, for optimal conditions, temperatures should be within 18°C and 32°C for high assimilation rates (Almeida and Valle, 2007). Temperatures below 10°C can be lethal (FAO, 2007). For maladapted genotypes, fruit size and bean quality may be affected by excessive growing season temperatures (Daymond and Hadley, 2008). Heat has been shown to induce unfavourable root development (Sena Gomes and Kozlowski, 1987).

Some authors have specifically researched putative impacts of climate change and concluded that negative impacts may be related to dry season precipitation (Schroth et al., 2016b) and temperatures (Anim-Kwapong and Frimpong, 2005; Schroth et al., 2016b), or growing season evapotranspiration (Läderach et al., 2013). The extent of suitable climates for cocoa production (Läderach et al., 2013; Schroth et al., 2016b) and yields (Anim-Kwapong and Frimpong, 2005) were found to be reduced as a result of projected future conditions in these studies. It is thus likely that climate change will shape the future production of cocoa. Adaptation to novel climatic hazards will therefore be a concern for farmers.

Climate-smart agriculture (CSA) is an approach to confront the new challenge of climate change (Lipper et al., 2014). It expands the concept of sustainability of agricultural production by adding adaptation to climate change and mitigation of GHG emissions from production as a new dimension. In CSA the sustainable intensification of production by efficiently using available resources is the priority. The second pillar is adaptation of production systems to climate changes. Third, the mitigation element emphasizes explicitly the reduction of GHG emissions from the production of cocoa by using alternative practices. The CSA concept accounts for synergies and trade-offs between the three objectives. For example, the sustainable improvement of cocoa productivity on existing land will be an important contributor to the reduced conversion of forest to cocoa production.

Climate-smart cocoa (CSC) will thus be produced using sustainable practices that aim to increase farm income and production on existing farmland while becoming adapted to the hazards of climate change. In some countries cocoa production has been closely linked to deforestation (Ruf et al., 2015). Its role as a frontier crop is undeniable and as such, mitigation through avoidance of deforestation should be a major CSA consideration for the sector. However, it has been argued that in the context of developing countries, sustainable intensification and adaptation ought to be prioritized because food security and economic growth are the primary objectives (Lipper et al., 2014). Sustainable intensification is the topic of this book which is most appropriate in this context. Increased output while focusing on efficient resource use will likely result in reduced GHG emissions per produced bag as a climate-smart co-benefit. Therefore, here we will present the implications of climatic changes for a sector-wide adaptation strategy.

Adaptation to climate change in agriculture refers to a change of cultivation practices as a result of observed or projected changes in the climatic environment of the production system. This change may either be spontaneous (e.g. migration during harsh drought) or planned (e.g. adopting a drought-tolerant variety) (Füssel, 2007). Planned adaptation always bears a cost. In its most direct form, adaptation requires monetary investments in physical infrastructure. The effort required to plan and implement the change of practices also comes at the opportunity cost of not being able to invest the devoted time to alternative action. Thus, adaptation to climatic changes is often perceived as costly, while inaction is assumed to be cost-free.

GCMs that project future climatic changes are unable to provide highly certain predictions of the future climate. As a result stakeholders shirk investing in adaptive action because the returns to adaptation bear a high risk. But rating non-investment as preferential over adaptive action implicitly assumes that climate change will not have economic consequences. This conflicts with the consensus that climate change is increasingly becoming a reality, and that projected changes can reasonably be assumed to have significant negative impacts on cocoa production. Thus, cost-effective adaptation constitutes a vital challenge for the sector.

The stronger the climate change signal is, the stronger the benefit from adaptation. CSC production and effective adaptive action therefore need to take into account the severity of climatic changes. A gradient of climate impacts is defined by the ratio of current climatic variability and the degree of climatic changes (Vermeulen et al., 2013). Associated with the climatic variability is the coping range, which is the economically viable tolerance of suboptimal growing conditions. Usually the climatic variability remains within the coping range and yields are sufficiently high because the production system is well adapted to such events. However, due to climatic variation, low-frequency events may occur that cause severe damage, for example, an unexpectedly long drought. Expanding the coping range of the system would require investments, in this case in irrigation equipment, that in the past were not beneficial because of the low frequency of such events. With climate change the question arises whether such an event is isolated or a signal of a trend, which would increase the benefits from adaptation.

The degree of impact versus the feasible coping range is therefore an important determinant of the adaptation strategy. At locations where changed climatic variability is minor compared to historic climatic variability, the available coping range may be sufficient. Incremental adaptations may be implemented by transferring adequate existing technological solutions (Vermeulen et al., 2013), such as changed pest and disease control. With increasing degree of climate impacts, systemic adaptation measures will have to be developed in joint public–private efforts that expand the climatic range in which cocoa can be produced economically (e.g. novel varieties). Finally, when climate change causes frequent events that make adaptation uneconomical when comparing with the benefits from alternative crops, a transformation out of the system has to be considered.

Thus, a forward-looking adaptation approach to sustainable cocoa production will enable stakeholders to avoid catastrophic impacts and to develop hazard-specific responses. The effective design of response strategies will first evaluate the degree of the climate change impact and then identify hazards to the system from analysis of practitioner's knowledge, data analysis and climate simulations.

In the following two case studies we first present global climate change projections for important cocoa-growing regions and then present a study specific for Ivory Coast. An understanding of changes at the global scale will support the prioritization of interventions in a sector-wide adaptation strategy. However, climate change impacts are spatially and temporally differentiated, and climatic limits to cocoa production equally vary between regions. Therefore we additionally demonstrate region-specific impacts for the world's largest cocoa producer, Ivory Coast.

2 Case study on global climate projections at cocoa-producing regions: overview, methods and data

2.1 Overview

The objective of this first case study is to demonstrate projected climatic changes at current cocoa-growing locations as a first step towards a sector-wide adaptation approach. The distribution of cocoa production is determined by the sufficient availability of precipitation throughout the year (Lass and Wood, 1985). Wet months are productive, while arid months are not (Zuidema et al., 2005). Consequently, drought and high temperatures were previously suggested to affect the distribution of suitable climates (Carr and Lockwood, 2011): negative impacts may be related to low dry season precipitation (Lass and Wood, 1985; Schroth et al., 2016b) and temperatures (Schroth et al., 2016b), high growing season evapotranspiration (Läderach et al., 2013), low total water availability and high average temperatures (Anim-Kwapong and Frimpong, 2005), high growing season temperatures (Daymond and Hadley, 2008) or vapour pressure deficit in the dry season (Acheampong et al., 2013).

For our review of projected climate change at cocoa occurrence locations, we used indicators that were suggested in the literature. Crop water availability in dry months has been discussed as a determinant of cocoa productivity (Acheampong et al., 2013; Asare et al., 2016). High growing season temperatures might affect quality (Daymond and Hadley, 2008) and it was suggested that excessive dry season temperatures may be a future limitation for the future geographic distribution of cocoa production (Schroth et al., 2016b). Therefore, we here used four indicator variables to investigate the magnitude and direction of climatic changes at current cocoa occurrence locations: the number of consecutive arid months, the total water deficit during this period, growing season average temperatures and mean dry season maximum temperatures.

2.2 Methods and data

We used data of the current distribution of *T. cacao* to characterize the climates that are suitable for cultivation. We assembled a global data set of cocoa occurrence locations from a range of sources with the objective to include all major climatic regions in which cocoa is produced (Fig. 1). The data were obtained from several sources:

1. We reached out to stakeholders in West Africa and South and Central America to request data contributions. Several occurrences thus represent existing databases, maps, expert knowledge and field missions where coordinates were gathered using a global positioning system (GPS). Stakeholders included private sector actors, research institutions, certification programmes (e.g. Rainforest Alliance) and farmer organizations.
2. Previous publications on the climate change impacts in West Africa (Läderach et al., 2013; Schroth et al., 2016b).
3. GBIF (Global Biodiversity Information Facility) (Global Biodiversity Information Facility, 2015).
4. Centroids of census units of the 2012 Censo Nacional Agropecuario, in which relevant cocoa quantities were reported (INEI, 2012).
5. For important producer countries (Ecuador, Mexico and Honduras) where available GPS data did not cover all production regions, we generated cocoa occurrence locations by drawing random samples from administrative areas with cocoa production (FAO, 2014) and agricultural land use (European Commission Joint Research Centre, 2003) with a minimum distance of 1 ArcMin.

The resulting raw databases included a total of 67304 cocoa occurrence locations. We reduced this initial data set to 3489 unique occurrence pixels at 10 Arcmin (about 20 km at the equator) to eliminate bias from highly clustered occurrences. We further stratified the sample by eliminating locations with mean monthly minimum temperatures (in the coldest month) below 10°C and total annual precipitation below 900 mm/yr (putatively irrigated sites).

The final database included 2981 cocoa occurrence locations, of which 830 were in Western Africa, 1319 in South America, 337 in Southeastern Asia, 321 in Central America and 162 in other regions such as Oceania and East Africa. The database was thus representative of most production regions across the globe but was not bias-corrected to represent produced quantities.

For the current climate (1950–2000), we used the WorldClim data sets (Version 1.4) at 10 Arcmin resolution (Hijmans et al., 2005). WorldClim provides data of monthly precipitation and mean monthly minimum and maximum temperatures. Version 2beta in addition provides maps of solar radiation. For future conditions no simulated solar radiation data were available so we assumed that they remain constant.

Future climate data layers were generated by downscaling the native outputs of 10 GCMs (BCC-CSM1-1, CCSM4, GISS-E2-R, HadGEM2-AO, IPSL-CM5A-LR, MIROC-ESM-CHEM, MIROC-ESM, MIROC5, MRI-CGCM3 and NorESM1-M) from the 5th assessment report for the representative concentration pathways (RCP) 2.6 (very low GHG emissions that peak before mid-century and then decline towards the end of the century (van Vuuren et al., 2007)), RCP 6.0 (intermediate scenario in which radiative forcing continues to increase until the end of the century (Fujino et al., 2006)) and RCP 8.5 (increasing GHG emissions over time resulting in high GHG concentrations (Riahi et al., 2007)) using the delta method (Ramirez and Jarvis, 2010): The difference between model outputs for current conditions and the average for the periods 2040–69 (2050ies) and 2059–89 (2070ies) was computed. The resulting layers were smoothed to a 10-ArcMin resolution and applied to the WorldClim layers for current climate.

We here defined arid months as months with excess potential evapotranspiration over precipitation. Potential evapotranspiration is considered a good proxy of vapour pressure (Allen et al., 1998) which may have direct effects on the cocoa tree (Mielke et al., 2005). We estimated potential evapotranspiration following Hargreaves and Samani (1985) as recommended by the Food and Agriculture Organization of the United Nations (FAO) (Allen et al., 1998). The method was found to produce useful results for West Africa (Kra and Ofosu-Anim, 2010), cocoa's most important region. In general evapotranspiration methods have an uncertain fate in climate change, but the Hargreaves was found to be within the likely range around equator (Kingston et al., 2009).

$$ETP = 0.0023 * Ra * (T - t)^{0.5} * (tm + 17.8) mm/day$$

ETP – evapotranspiration in mm/day
Ra – extraterrestrial solar radiation expressed in water equivalent (mm/day)
T–t – difference between monthly maximum and minimum mean temperature (°C)
tm – mean air temperature (°C)

We derived the indicator variables for current conditions using the WorldClim data and for future conditions using the CMIP5 data.

3 Case study on global climate projections at cocoa-producing regions: results and discussion

3.1 Results

Globally, most cocoa production has a dry season of up to 3 months with a total potential evapotranspiration deficit of 130 mm. Mean growing season temperatures were found to range from 20°C to 28°C at current cocoa occurrence locations, though approximately half are within a range of 24–26.5°C (Fig. 2). 90% of cocoa occurrence locations had mean maximum dry season temperatures of 26–33°C, and half of the locations were within 29.2°C and 32°C.

In Central America, growing areas were found to be relatively dry with a comparatively long dry season and a significant dry season water deficit (Fig. 2). Under current conditions

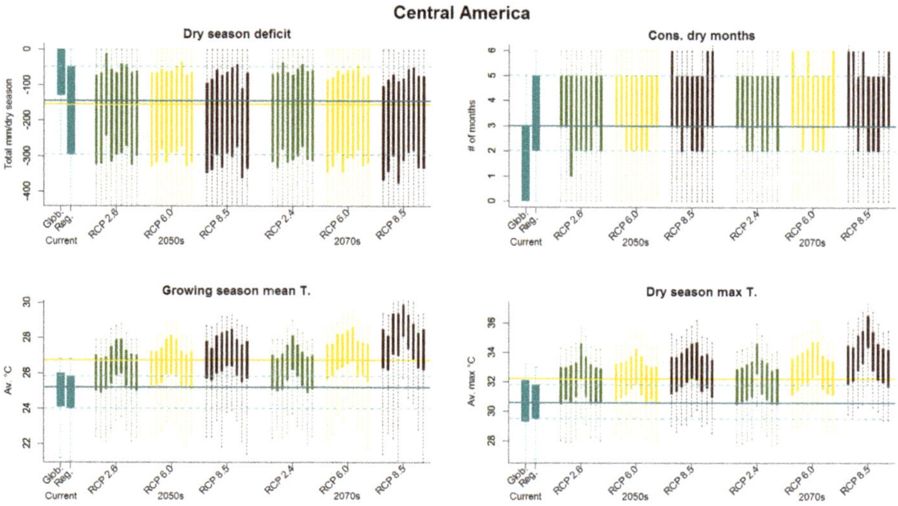

Figure 2 Comparison of observed and simulated climate at cocoa occurrence locations in Central America; teal coloured bars show current climate conditions at all cocoa occurrence locations and only the locations in Central America; green, yellow and brown bars show the projections in the RCP's 2.6, 6.0 and 8.5, respectively, for the 2050s and 2070s. Bars represent the 1st and 3rd interquartile ranges, whiskers 5% and 95% percentiles. Teal horizontal lines represent the current median (solid) and the 1st and 3rd quartiles at the region's cocoa occurrence locations. The yellow horizontal line represents median conditions at cocoa occurrence locations in the 2050s in the RCP 6.0 emission scenario.

temperatures were near the global average. In the climate change simulations, the climate was projected to become drier and warmer. Few GCMs projected a reduced water deficit or a shorter dry season. In the RCP 6.0 and 8.5 emission scenarios, slightly more GCMs projected increasingly severe dry seasons. For temperature variables an unambiguous trend towards increased heat across both time slices and all emission scenarios could be observed. In the intermediate emission scenario RCP 6.0, the median temperature in the 2050s would be hotter than those at 95% of locations under current conditions both during the growing and the dry seasons.

In South America, growing areas were found to be relatively humid with a comparatively short dry season and a low dry season water deficit (Fig. 3). Under current conditions, temperatures were near or below the global average. In the climate change simulations, the climate was projected to become drier and warmer. A reduced water deficit or a shorter dry season will be unlikely. In the RCP 6.0 and 8.5 emission scenarios, slightly more GCMs projected increasingly severe dry seasons. For temperature an unambiguous trend towards increased heat could be observed across both time slices and all emission scenarios. In the intermediate emission scenario RCP 6.0, the median temperatures in the 2050s would be hotter than those at 95% of South American locations under current conditions. However, projected median conditions were found to remain within the interquartile range of global conditions.

In Southeastern Asia, growing areas were found to be humid with no dry season (Fig. 4). Under current conditions, growing season temperatures were near or above the global average. At the few locations where a dry season occurred, dry season temperatures were

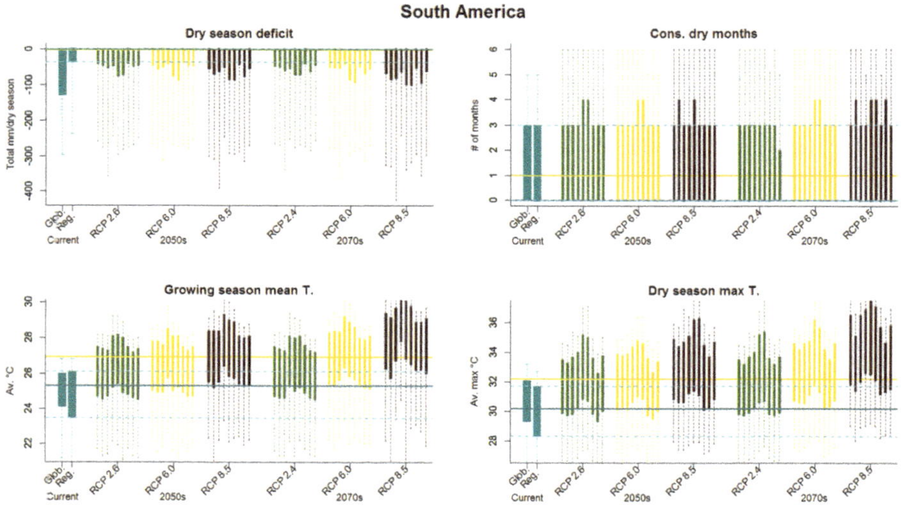

Figure 3 Comparison of observed and simulated climate at cocoa occurrence locations in South America; teal coloured bars show current climate conditions at all cocoa occurrence locations and only the locations in South America; green, yellow and brown bars show the projections in the RCP's 2.6, 6.0 and 8.5, respectively, for the 2050s and 2070s. Bars represent the 1st and 3rd interquartile ranges, whiskers 5% and 95% percentiles. Teal horizontal lines represent the current median (solid) and the 1st and 3rd quartiles at the region's cocoa occurrence locations. The yellow horizontal line represents median conditions at cocoa occurrence locations in the 2050s in the RCP 6.0 emission scenario.

below average. Precipitation projections in the climate change simulations were uncertain. In the RCP 6.0 and 8.5 emission scenarios, slightly more GCMs projected increasingly severe dry seasons, but overall the humid conditions were projected to prevail. For temperature an unambiguous trend towards increased growing season heat could be observed across both time slices and all emission scenarios. In the intermediate emission scenario RCP 6.0, the median growing season temperature in the 2050s would be hotter than those at 95% of global locations under current conditions.

In West Africa growing areas were found to have a pronounced drought both in length and in deficit compared to other global locations (Fig. 5). Under current conditions growing and dry season temperatures were above most other global locations. Precipitation projections in the climate change simulations were uncertain. In the RCP 6.0 and 8.5 emission scenarios, slightly more GCMs projected increasingly severe dry seasons but overall the humid conditions were projected to prevail. Temperature projections show temperature increases but not as uniform as in the other regions. In the intermediate emission scenario RCP 6.0, the median temperatures in the 2050s would be hotter than those at 98% of global locations under current conditions.

3.2 Discussion

Projected climatic changes of four indicator variables at current cocoa occurrence locations were found to be uncertain for precipitation variables but exhibited a clear upward trend for growing and dry season temperatures. Currently cocoa is grown within a rather narrow

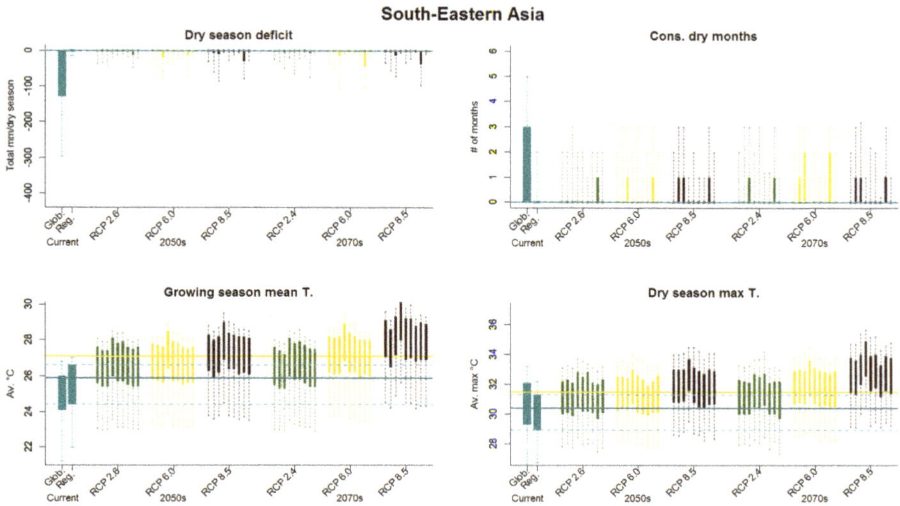

Figure 4 Comparison of observed and simulated climate at cocoa occurrence locations in South Eastern Asia; tealcoloured bars show current climate conditions at all cocoa occurrence locations and only the locations in South Eastern Asia; green, yellow and brown bars show the projections in the RCP's 2.6, 6.0 and 8.5, respectively, for the 2050s and 2070s. Bars represent the 1st and 3rd interquartile ranges, whiskers 5% and 95% percentiles. Teal horizontal lines represent the current median (solid) and the 1st and 3rd quartiles at the region's cocoa occurrence locations. The yellow horizontal line represents median conditions at cocoa occurrence locations in the 2050s in the RCP 6.0 emission scenario.

temperature niche with a defined upper temperature limit. Climatic change will raise temperatures beyond this limit with uncertain effects on the crop. Despite the uncertainty in precipitation projections, dry season severity is likely to become a threat in some regions. The following case study will demonstrate that effects are spatially differentiated, that is, precipitation increases in one region may outweigh decreases elsewhere in an aggregated analysis.

The increase in both growing season and dry season temperatures was found to exceed currently experienced temperature with high certainty. The current global range of average growing season temperature is rather narrow at about 2.5°C. Therefore, even in a best case scenario in which GHG emissions are curbed to limit global warming to 2°C, temperatures will exceed this range. Previous studies that have researched the effect of growing season temperatures focused on varietal differences and their suitability for historic climatic conditions. Among others, the variety Amelonado which is popular in West Africa was found to exhibit reduced lipid content in hot environments (Daymond and Hadley, 2008). In addition, high temperatures may reduce root development (Sena Gomes and Kozlowski, 1987) in seedlings. However, these studies limited environmental conditions to average growing season temperatures of about 26°C as they were typical for historic conditions. The effect of a 2°C increase therefore remains unclear, but altered crop behaviour appears likely. The choice of cocoa varieties for cultivation should take into account adaptability to future climate scenarios.

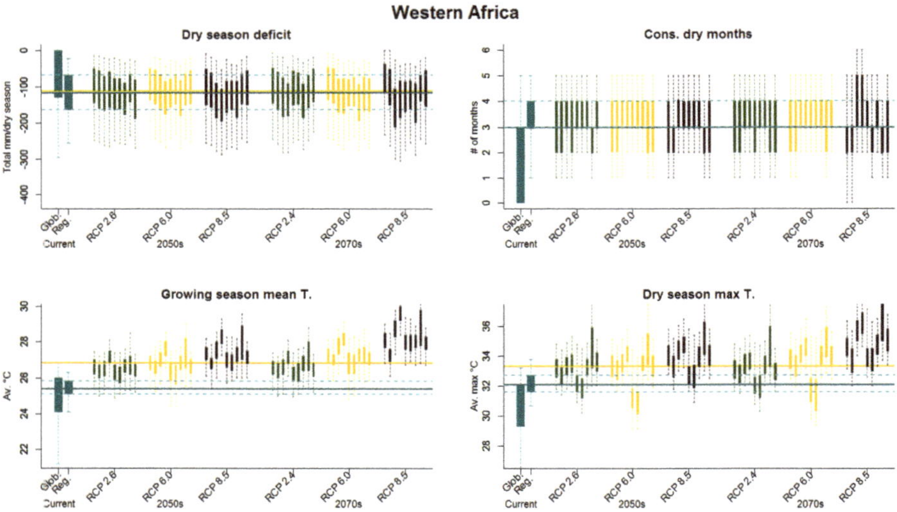

Figure 5 Comparison of observed and simulated climate at cocoa occurrence locations in West Africa; teal coloured bars show current climate conditions at all cocoa occurrence locations and only the locations in West Africa; green, yellow and brown bars show the projections in the RCP's 2.6, 6.0 and 8.5, respectively, for the 2050s and 2070s. Bars represent the 1st and 3rd interquartile ranges, whiskers 5% and 95% percentiles. Teal horizontal lines represent the current median (solid) and the 1st and 3rd quartiles at the region's cocoa occurrence locations. The yellow horizontal line represents median conditions at cocoa occurrence locations in the 2050s in the RCP 6.0 emission scenario.

Dry season temperatures were previously hypothesized to negatively affect changes of crop suitability near the Savanna transition zone in Western Africa (Schroth et al., 2016b). Maximum temperatures from WorldClim at current cocoa occurrences were similar to the values reported in the literature. Temperatures near the upper limit or above were shown to negatively impact the physiology of cocoa. High temperatures induce faster leaf growth, ageing and abscission possibly caused by high maintenance respiration rates (Almeida and Valle, 2007). In addition high temperatures may also further reduce transpiration rates (Amorim and Valle, 1992) when in the dry season net carbon assimilation is already reduced (Carr and Lockwood, 2011). The increased temperatures could thus aggravate the effects even of an otherwise unchanged dry season.

Cocoa is commonly grown below a more or less dense shade canopy. We here used interpolated climate data that are based on historic temperature records. Such data have been criticized as biased because records may have been taken from populated locations and therefore may overestimate temperatures in rural areas with dense forest cover or high moisture (Deblauwe et al., 2016). However, recent work has shown that there was little effect of canopy cover on air temperatures (Asare et al., 2016). Differences between shaded and unshaded plots were limited to a maximum 1°C. We therefore conclude that a possible bias in temperature data does not compromise our finding that future temperatures will exceed current limits.

We found a slightly average tendency towards drier conditions across global cocoa occurrence locations with marked disagreement between global climate models. In general, GCMs have been criticized to show high uncertainty for precipitation projections (Stocker et al., 2013). There is some agreement that differences between wet and dry locations and wet and dry seasons will increase (Stocker et al., 2013). Additionally, inter-annual variability has also been suggested to increase. Cocoa is generally grown in wet regions with high precipitation. However, important West African production locations have a pronounced dry season and are located at the Savanna transition zone. Therefore when averaging across geographically distant cocoa occurrence locations, differentiated effects may have been masked. Furthermore, we made the assumption that solar radiation remains constant across time because adequate data were not available to us. It appears plausible that less precipitation would imply less cloud cover and thus increased solar radiation and evapotranspiration. The findings presented here could therefore be exacerbated with better data. Nevertheless, our finding that dry conditions will not increase uniformly across cocoa occurrence locations is in line with general descriptions of future climate change for the equatorial region.

We here found regional differences that were largely aligned with previous literature. African and Asian locations were found to have comparatively higher growing season temperatures than American locations. Central American and West African locations had a more pronounced dry period than South American and Southeast Asian locations. Thus, South American origins appear to be better situated to confront climate change as precipitation is high and along the Andes high temperatures may be avoided by altitudinal migration. West Africa on the other hand may experience unprecedented heat events and has a higher risk of increasing drought severity. However, arguably pest and diseases play an important part in determining economic benefits from cocoa production (Bowers et al., 2001). The risk from pests and diseases may be heightened or curtailed by the same climatic changes that threaten cocoa production (Almeida and Valle, 2007). Therefore, from this analysis it may not be concluded that one region may be at more risk than another.

A limitation of our analysis results from our use of climate data aggregated to monthly means over multiple decades. This implicitly assumed that long-term climate conditions determine the viability of cocoa production. By doing so, we disregarded the importance of catastrophic extreme events for the survival of cocoa. For example, extreme leaf temperatures have been shown to potentially lead to crop failure for coffee (Marias et al., 2017). Such leaf temperature stress was exacerbated in tomato during water limited conditions by the inability of some cultivars to regulate leaf temperatures (Nankishore and Farrell, 2016). In a dry season with elevated temperatures and thus reduced cloud cover, similar effects would likely affect cocoa trees in full sun systems. Thus, even short but extreme temperature events are likely to have consequences even when multi-decadal mean conditions are favourable despite the possible positive effects of increased CO_2 levels (Lahive et al., 2017).

Therefore, additional research is required to understand the effect of growing and dry season temperatures on cocoa productivity and quality. Currently available literature is limited to historic climate patterns and does not account for temperatures that were historically uncommon but may become more dominant in coming decades. Furthermore, the implications of increased uncertainty from seasonality of precipitation deserve attention especially with regard to intraregional differences as will be demonstrated in the following case study. Finally, little is understood about the impact of climate and climatic changes on the distribution of pests and diseases of cocoa.

4 Case study on regionally differentiated climate change impacts in Ivory Coast: overview and methodology

4.1 Overview

Ivory Coast is the world's largest cocoa producer and has accounted for roughly half of global cocoa production over the last decades (FAO, 2016). Negative climate change impacts in this country would likely have negative repercussions across the entire sector. As discussed in the previous section, West Africa has a challenging climate for cocoa production with a strong dry season and high growing season temperatures. However, from the aggregated analysis of the previous section it was not possible to investigate spatially differentiated impacts. It therefore was unclear whether changed precipitation and increased temperature would have a homogenous effect across the country. Previously it was projected that some areas could be negatively affected in the region, especially towards the Savanna transition zone (Läderach et al., 2013; Schroth et al., 2016a,b).

A regional impact gradient for adaptation planning is a requirement for effective adaptation. It may support decision makers in their assessment of where to focus public and private adaptation interventions, and where adaptation appears unviable (Schroth et al., 2016a). In this case study we sought to guide adaptation strategies by using a multiclass suitability distribution model.

A gradient of impacts was developed to differentiate fundamental types of adaptation strategies: expansion of production, incremental changes, systemic changes, increased resilience or transformational change (Vermeulen et al., 2013). The first three categories will become or remain climatically suitable for cocoa production in the future. We differentiated between production locations that will not fundamentally change their climatic characteristics, that is, they remain in the same agroecological zone (AEZ). These sites will be able to cope and only incremental changes to agronomic practices will be required. On the other hand, sites that will experience changes that fundamentally change climate characteristics from one AEZ to a different AEZ between current and 2050s' climate conditions were described as adjustment sites that will need systemic changes. At these locations agronomic practices will have to change to adapt to the new cocoa AEZ. For some regions future climate projections were uncertain. We suggest adaptation by augmenting the resilience through systemic changes. Finally, some areas were found to become unsuitable for cocoa production in the future. These sites will need to *transform*. Production in these zones will likely become unviable and other crops should be considered. Finally, locations that are currently unsuitable but show a more favourable climate in the future are described as *opportunity* sites. At these sites cocoa production could be established in coming decades if soil conditions and protection status do not prohibit this.

4.2 Methodology

AEZs for cocoa production in Ivory Coast were defined using a combination of spatial climate data from WorldClim (Hijmans et al., 2005) and a reduced database of occurrences of cocoa production. The database of occurrences was assembled from data used in

Läderach et al. (2013) and Schroth et al. (2016b) and substantial data sets received from private sector partners. We stratified the data set from 19504 total occurrences in Ivory Coast to 796 unique occurrence pixels on a 5-Arcmin grid.

To determine distinct suitability zones for cocoa, we used hierarchical clustering with the Ward algorithm on Euclidean distances calculated from six uncorrelated climate variables after outlier removal. Suitability-type descriptions were based on the differences of the group means from the grand mean for a subset of variables and their confidence intervals as calculated by R's multcomp package (Hothorn et al., 2008). We then used the Random Forest (RF) algorithm (Breiman, 2001) to extrapolate their spatial distribution under current and future climate. RFs are machine learning classifiers that are formed by ensembles of classification trees. RFs are very popular because of their efficiency on large data sets without overfitting. We used the randomForest package (Liaw and Wiener, 2002) in the statistical software R (R Core Team, 2014) that implements the RF approach. We here used the RF classifier in two distinct applications. (1) We used the RF classifier to classify climate data of current and future conditions into the resulting suitability types. (2) We then used RF to evaluate individual or sets of climate variables to distinguish insignificant changes in climate from changes that pose hazards to cocoa production.

Input variables were 20 bioclimatic variables at 2.5′ resolution from WorldClim (Hijmans et al., 2005) for current conditions and the respective data for 2050s in the intermediate RCP 6.0 emission scenario (Fujino et al., 2006) from the CCAFS climate portal (Ramirez and Jarvis, 2010). In addition we constructed a soil suitability variable that indicated water holding capacity for cocoa. We did this by first asking regional cocoa experts to prioritize soil variables that improve the soil water holding capacity for cocoa from AfSys data (Leenaars et al., 2015) and then used the RF classifier in regression mode to calculate a probability of similarity between cocoa soils and other soils.

Regional experts evaluated and verified cocoa occurrence data for model input, prioritized climate and soil variables for modelling use and confirmed the validity of the resulting model.

5 Case study on regionally differentiated climate change impacts in Ivory Coast: results and discussion

5.1 Results

Analysis of a dendrogram of distances between occurrence pixels suggested four distinct suitability clusters. Clusters were named according to their features. The confidence intervals for each cluster resulting from RF classification were compared to the grand mean (Fig. 6). Differences to the grand mean of all cocoa occurrence locations were used to describe the suitability types as follows:

- Cluster 1 had high average and dry season temperatures. Precipitation was high throughout the year and the dry season relatively short. This suitability cluster was therefore described as 'hot/wet'.
- Cluster 2 had similarly high temperatures on average and during the dry season. However, total precipitation was found to be less than that at other locations. We therefore described this cluster as 'hot/dry'.

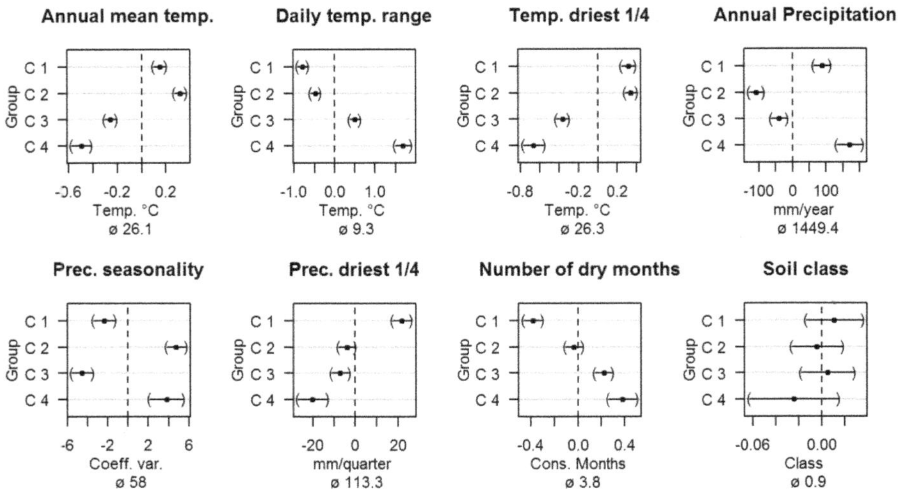

Figure 6 Contrast plots of group confidence intervals to grand mean by cluster; confidence intervals for each cluster are shown relative to the grand mean (dotted line), the average value is given below each plot and the x-axis indicates the deviation.

- Cluster 3 was found to have lower than average temperatures and average precipitation conditions with low seasonality. It was described as 'cool/moist'.
- Cluster 4 had comparatively low temperatures and high precipitation, but the dry season is quite long and pronounced. We therefore named it 'cool/long dry season'.

Extrapolation of the RF classifiers that were trained on the clustered occurrence pixels resulted in maps of the current distribution of these clusters (Fig. 7). The southern part of the current cocoa belt was thus characterized by high average temperatures. Precipitation was found to be high towards the eastern border with Ghana and towards the west. In the 'hot/dry' suitability zone, north of Abidjan precipitation was lower than elsewhere in the cocoa belt. This is also a zone with low cocoa prevalence. The northwestern patch had cooler temperatures than the south. Total precipitation was found to be highest towards the border with Liberia but with a relatively long dry season. Between suitability zones the algorithm showed a lower classification confidence as is expected with spatially continuous climate variables. Towards the north and the south of the cocoa belt, areas were classified as marginal. Such areas share some climate characteristics with suitable climate zones but are most likely unsuitable.

The most likely climate change impact scenarios were calculated as the modal classification across 19 individual GCM projections. The map described the most likely suitability type with future climate change but not the overall aptitude for cocoa production. Most suitability types were no longer unequivocally identified by the algorithm with the exception of the 'hot/dry' climate region with low cocoa prevalence (Fig. 8). Most of the current cocoa belt was projected to remain more similar to suitable conditions than to unsuitable conditions. But the uncertainty to which suitability type the future regions will belong was high and most area therefore indicated as 'mixed'. The total area with 'marginal' conditions was projected to expand towards the north.

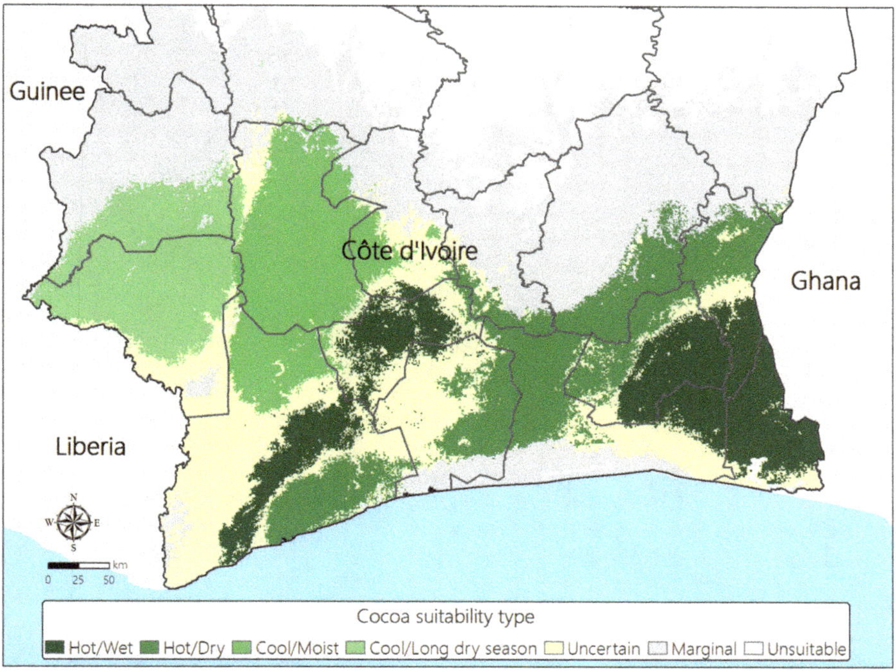

Figure 7 Distribution of suitability types for cocoa production in Ivory Coast under current climate conditions. Green colours indicate suitable climates; classification confidence of crème coloured areas was uncertain; grey areas indicate marginally suitable zones; white area was not classified (outside of Ivory Coast) or unsuitable for cocoa production (within Ivory Coast).

Uncertainty for future extrapolations had two sources: a) independent repeats with subsets of training data and differences between future GCM projections for Ivory Coast. The importance of the latter effect is here exemplified in Fig. 9. In the GISS E2 H GCM projection average temperatures at cocoa occurrence locations would increase by 1.1°C and precipitation by 357 mm/yr. This would result in an expansion of suitable climates and only the 'hot–wet' climate would remain unaltered (Fig. 9-Left). The intermediate GCM would result in a similar distribution as the mode across all GCM projections. In a scenario with a high temperature increase (+2.1°C) and a reduction in total precipitation (–154 mm/yr) large parts of the cocoa belt would become unsuitable (Fig. 9-Right).

Comparison of current distribution of suitability zones and the future projection yielded a gradient of likely climate change impacts. Climate change will threaten the viability of cocoa production towards the southern and northern margins of the cocoa belt in Ivory Coast (Fig. 10). The central cocoa zone will be least affected with only incremental adaptation requirements. Towards the west and the border of Ghana, systemic changes will be necessary to keep cocoa production viable. In most of the cocoa zone projected climate change impacts were highly uncertain because of disagreement between global climate models. These regions will require an adaptation strategy that aims at increasing the resilience of the system.

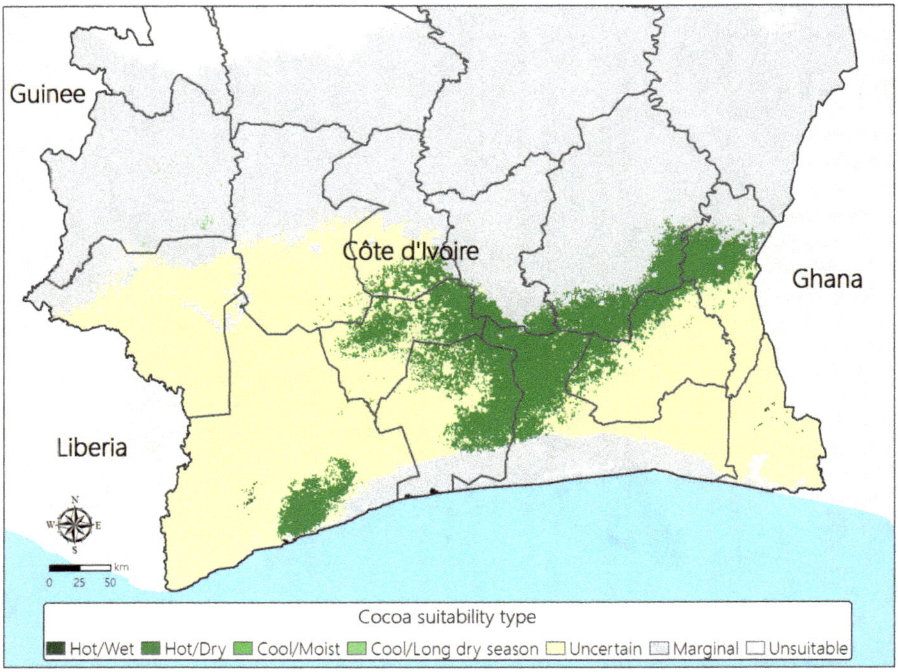

Figure 8 Distribution of suitability types for cocoa production in Ivory Coast under climate conditions (RCP 6.0) in the 2050s. Green colours indicate suitable climates; classification confidence of crèmecoloured areas was uncertain; grey areas indicate marginally suitable zones; white area was not classified (outside of Ivory Coast) or unsuitable for cocoa production (within Ivory Coast).

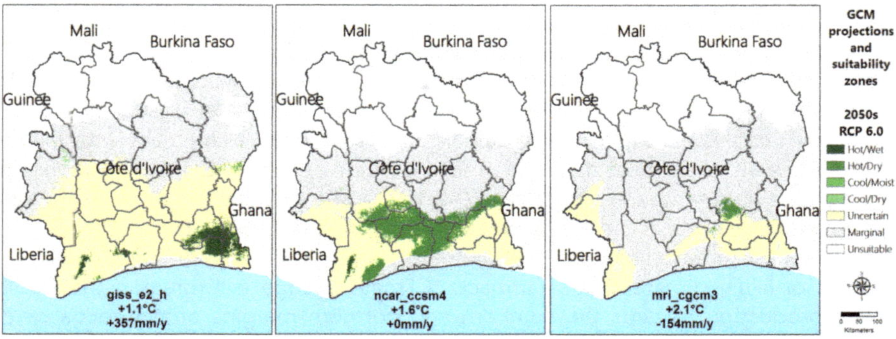

Figure 9 Uncertainty in the distribution of suitability types for cocoa production in Ivory Coast under climate conditions (RCP 6.0) in the 2050s. Inserts show distributions under different GCM projections (Left – low impact – GISS E2 H; Middle – intermediate impact – Ncar_CCSM4; Right – high-impact – MRI CGCM3). Green colours indicate suitable climates; classification confidence of crème coloured areas was uncertain; grey areas indicate marginally suitable zones; white area was not classified (outside of Ivory Coast) or unsuitable for cocoa production (within Ivory Coast).

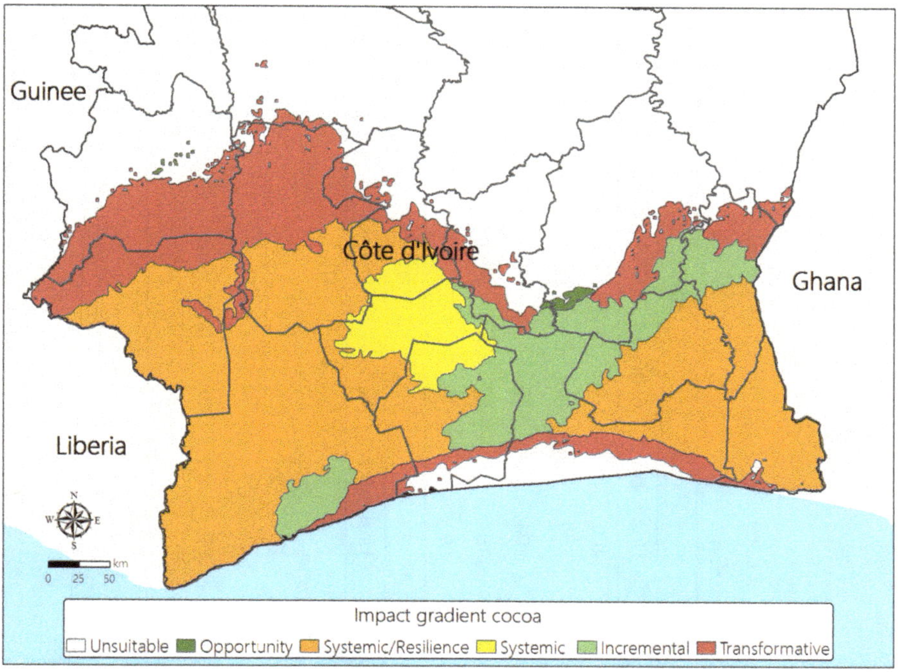

Figure 10 Distribution of climate change impact zones (2050s-RCP6.0) and dominant adaptation typology in Ivory Coast: dark green areas are opportunity areas; light green areas are adjustment or incremental adaptation areas; yellow areas are designated systemic adaptation zones; for orange areas, climate projections are not unambiguous and should increase resilience and red areas will transition to other crops without substantial changes in production systems.

For each adaptation zone, the suitability of a reduced set of indicator variables was evaluated for current and future conditions. We then calculated the percentage of GCM projections that will result in worse, unchanged or improved suitability for each adaptation zone. Additionally we evaluated in which share of the scenarios the future state of the indicator variable becomes unsuitable, that is, lower than currently observed at cocoa occurrence locations (Table 1). Precipitation projections were found to show uncertainty with unclear direction of impact. Furthermore, even if reductions in suitability were projected it remained unclear whether these would negatively impact cocoa production. Heat was found to change with negative implications for cocoa suitability in nearly all zones and GCMs. In most scenarios heat will show values beyond the suitable range in the 2050s.

The transformation zone will likely be challenged by reduced suitability based on drought, heat and dry season length, while total precipitation may increase (Table 1). In Ivory Coast, most areas will require heightened resilience because of uncertain GCM projections. Suitability of drought, heat, dry season length and total precipitation is more likely to decrease than to increase, but impacts are unlikely to cause a loss of suitability, with the exception of heat. Projections for the opportunity zone in which expansion may be possible were diverging for most indicators. Suitability improvements were more likely than reductions, but in several scenarios the zone would remain unsuitable.

Table 1 Percentage of GCM projections in which the suitability of indicators becomes unsuitable, reduced, unchanged or improved

Adaptation zone	GCM projection							
	Drought				Heat			
	Unsuitable	Reduced	Unchanged	Improved	Unsuitable	Reduced	Unchanged	Improved
Transform	26	58	5	37	100	100	0	0
Opportunity	21	42	0	58	100	89	0	11
Incremental	0	95	5	0	100	100	0	0
Systemic	0	11	89	0	16	42	0	58
Resilience	0	100	0	0	89	100	0	0
	Dry season length				Total precipitation			
	Unsuitable	Reduced	Unchanged	Improved	Unsuitable	Reduced	Unchanged	Improved
Transform	5	58	0	42	5	37	0	63
Opportunity	5	47	11	42	58	32	0	68
Incremental	0	74	11	16	21	53	0	47
Systemic	0	11	89	0	16	42	0	58
Resilience	0	89	11	0	0	47	42	11

5.2 Discussion

We here presented impact gradient mapping as a decision support tool for climate change adaptation at cocoa's most important origin, the Ivory Coast. Our research suggested that drought, caused by reduced rainfall in the dry season, and increased temperatures threaten the viability of current cocoa-growing areas. GCM projections showed that total precipitation will increase, but an increase in high temperatures outweighs this positive effect, and as a result conditions for cocoa production may worsen. Adaptation needs in Ivory Coast were therefore high across the current cocoa belt.

These findings are in line with those of previous research for the region that identified similar regions of high impacts (Läderach et al., 2013) and also suggested increased dry season temperatures as the most likely cause (Schroth et al., 2016b). Previous research, however, limited itself to presenting relative changes of suitability which made it difficult to discern significant from insignificant impacts. Furthermore, GCM model disagreement masked the possible impact of changed precipitation in the previously chosen 'most likely' impact scenarios. Here we made this uncertainty explicit and showed for which regions this uncertainty is highest and suggested that production systems will require measures to increase general resilience.

The most affected region in the northwest was characterized by mean temperatures that are below the national average at cocoa occurrence locations under current climate conditions. However, dry season precipitation was lower than elsewhere and the number of consecutive dry months is higher than further south. The less affected regions along the cocoa belt had high mean temperatures under current conditions and higher precipitation. Findings of our model are thus in line with those of previous research that pointed out dry season conditions as critical for the viability of cocoa production.

We evaluated sets of bioclimatic variables as indicators for projected hazards. Instead of looking at changes of individual variables, we evaluated the resulting change in suitability for cocoa production. Suitability of heat indicators showed unambiguous negative impacts. Severity of drought is highly likely to increase, while the likelihood for negative changes in the length of the dry season is only 60%. The projected trend for total annual precipitation is uncertain with some models suggesting increases and other decreases. These projections were in line with trends reported in other studies of the region (Ruf et al., 2015).

6 Conclusions

CSC broadens the scope of sustainable intensification of production with explicit objectives to improve adaptation and mitigation to confront the challenge of climate change. We focused on effective adaptation because reduced emissions per produced bag and avoidance of deforestation will likely be co-benefits of sustainable and adapted cocoa production on existing land. To reduce economic risk from investments in climate-smart practices, adaptive action must address site-specific hazards in a forward-looking manner. Because of the high uncertainties attached to each method, an approach to guide adaptation strategies was suggested that builds on analysis of perceptions, trends and simulations of climate change.

On a global scale current cocoa occurrence locations will likely experience temperatures that have not been previously observed. Historically, regions with average temperatures above 28°C existed but nearly none of these had adequate precipitation for cocoa production. Therefore it will be difficult to prepare for future climate change based on past experience. Controlled experiments in the literature focused on regional differences and implications for varietal choice (Daymond and Hadley, 2008) but stopped short of researching the effects of future climate trends. Exceedingly high temperatures therefore pose an unknown risk to cocoa production.

Projections of changes to precipitation were highly uncertain at cocoa occurrence locations. The seasonal distribution and total availability of precipitation are the dominant yield determining factors (Zuidema et al., 2005). Increasing differences between dry and wet regions, years and seasons may therefore increase economic risks for cocoa producers and the sector (Carr and Lockwood, 2011; Schroth et al., 2016a). However, at a global scale, no clear trend was found in projections. Increased drought risk in West Africa appeared to be more likely than favourable changes. Global climate projections are unlikely to improve in time to accurately guide adaptation measures with long lead time. Therefore, for adaptation a focus should be on measures that increase the overall resilience of the production system to precipitation risk.

No-regret adaptation measures will have a net positive impact for the producer under all future scenarios (Heltberg et al., 2009). In cocoa production such measures would be expected to generate substantial benefits beyond reduced climate risk. For example, improved soil health with high organic matter may improve productivity through improved nutrient availability and disease resistance. Improved water holding capacity in case of drought would then be an additional benefit. In contrast, for example, irrigation equipment causes costs which may not be recovered in some climate change scenarios. Research that seeks to reduce precipitation-related risk should therefore account for synergies with other challenges to production.

Losses to cocoa production caused by pests and diseases have threatened cocoa production in some regions in the past (Bowers et al., 2001). Changing temperatures and uncertain dry–wet cycles may be beneficial for one disease but harmful for another. Thus, changing climate patterns may alter the relative importance of biotic threats to cocoa production. A detailed understanding of the relationship between cocoa pests and diseases, their management and climate may support the sector in avoiding additional losses caused by unexpected or novel threats in the coming decades.

Management practices are often optimized to site-specific conditions through farmer-led innovation processes. Climate change progresses at a rate that will challenge the effectiveness of such innovation. Research will have to develop site-specific agronomic practices that account for changing climate. For example, the optimal type of shade trees and level of shade required may change over time at a given location depending on how climate change progresses. In order to develop forward-looking recommendations, novel research approaches will be needed.

One such approach may be to benefit from the use of modern information technology. The speed of climate change demands that research findings are disseminated faster than before. Modern communication tools could serve to bypass long decision-making processes and ease communication between research and producers. Science could regard each cocoa plot as an experimental location and collect and integrate information in big data approaches. Findings could then be communicated widely through the same tools.

However, before producers turn information into action, the right incentives will have to be in place. Beyond the physiology and agronomy of cocoa production, the economic environment is challenging. Often farmers deliberately do not optimize their cocoa productivity because the required financial means and trade-offs with other household activities are prohibitively high. Adaptation to climate change will not least be an economic problem, especially when the focus has to be on no-regret measures as is the case for cocoa production. Climate change threatens to become a poverty trap in which the most exposed producers lack the means to invest in adaptation resulting in a vicious circle that further reduces their endowments to confront challenges after failed harvests.

One of the challenges for CSC production is therefore scale. Implementation of such practices in lighthouse projects through sustainability initiatives constitutes the first step towards dissemination to the hundreds of thousands of smallholder farmers in remote regions. Widespread adoption will require the inclusion of actors along the entire value chain including policy makers, processors, traders and local organizations to provide producers with the right incentives to reduce the overall vulnerability of the cocoa sector to climate change. This approach means moving beyond individual responses and identifying leverage points for more systemic change. Clearly this challenge cannot be successfully met by one actor in one geography. Through collective efforts the necessary technical solution may be developed, and novel finance mechanisms such as weather insurance or climate finance will require coordination. Not least producers and private actors will need to lobby for a policy environment that is conducive to CSC. Without successful collective action, the widespread establishment of resilient cocoa systems that are able to thrive under future climates remains doubtful.

7 Acknowledgements

This work was implemented by the International Center for Tropical Agriculture (CIAT) as part of the CGIAR Research Program on Climate Change, Agriculture and Food Security (CCAFS), which is carried out with support from CGIAR Fund Donors and through bilateral funding agreements. For details, please visit https://ccafs.cgiar.org/donors. The views expressed in this document cannot be taken to reflect the official opinions of these organisations.

The International Center for Tropical Agriculture (CIAT) – a CGIAR Research Center – develops technologies, tools and new knowledge that better enable farmers, especially smallholders, to make agriculture eco-efficient – that is, competitive and profitable as well as sustainable and resilient. Headquartered near Cali, Colombia, CIAT conducts research for development in tropical regions of Latin America, Africa and Asia. www.ciat.cgiar.org

8 References

Acheampong, K., Hadley, P. and Daymond, A. J., 2013. Photosynthetic activity and early growth of four cacao genotypes as influenced by different shade regimes under West African dry and wet season conditions. *Exp. Agric.* 49, 31–42.

Allen, R. G., Pereira, L. S., Raes, D. and Smith, M., 1998. Crop evapotranspiration-Guidelines for computing crop water requirements-FAO Irrigation and drainage paper 56. FAO, Rome 300, D05109.

Almeida, A.-A. F. de and Valle, R. R., 2007. Ecophysiology of the cacao tree. *Braz. J. Plant Physiol.* 19, 425–48.
Amorim, S. M. C. de and Valle, R. R., 1992. Efeito da temperatura radicular na resistência ao movimento de água no cacaueiro (Theobroma cacao L.). *Acta Bot. Bras.* 6, 55–64.
Anim-Kwapong, G. J. and Frimpong, E. B., 2005. Vulnerability of agriculture to climate change-impact of climate change on cocoa production. Final Rep. Submitt. Neth. Clim. Change Stud. Assist. Programme NCAP 2.
Asare, R., Asare, R. A., Asante, W. A., Markussen, B. and RÆbild, A., 2016. Influences of shading and fertilization on on-farm yields of cocoa in Ghana. *Exp. Agric.* 53(3), 416–31.
Bowers, J. H., Bailey, B. A., Hebbar, P. K., Sanogo, S. and Lumsden, R. D., 2001. The impact of plant diseases on world chocolate production. *Plant Health Prog.* doi:10.1094/PHP-2001-0709-01-RV.
Breiman, L., 2001. Random forests. *Mach. Learn.* 45, 5–32.
Carr, M. K. V. and Lockwood, G., 2011. The water relations and irrigation requirements of cocoa (theobroma cacao l.): A review. *Exp. Agric.* 47, 653–76.
Dai, A., 2013. Increasing drought under global warming in observations and models. *Nat. Clim. Change* 3, 52–8.
Daymond, A. J. and Hadley, P., 2008. Differential effects of temperature on fruit development and bean quality of contrasting genotypes of cacao (Theobroma cacao). *Ann. Appl. Biol.* 153, 175–85.
Deblauwe, V., Droissart, V., Bose, R., Sonké, B., Blach-Overgaard, A., Svenning, J.-C., Wieringa, J. J., Ramesh, B. R., Stévart, T. and Couvreur, T. L. P., 2016. Remotely sensed temperature and precipitation data improve species distribution modelling in the tropics. *Glob. Ecol. Biogeogr.* 25, 443–54.
European Commission Joint Research Centre, 2003. The Global Land Cover Map for the Year 2000 [WWW Document]. *GLC2000 Database*. URL: http://www-gem.jrc.it/glc2000 (accessed 20 January 2014).
FAO, 2007. ECOCROP [WWW Document]. URL: http://ecocrop.fao.org/ecocrop/srv/en/home (accessed11 June 2014).
FAO, 2014. Agro Maps [WWW Document]. URL: http://kids.fao.org/agromaps/ (accessed 19 May 2014).
FAO, 2016. FAOSTAT [WWW Document]. URL: http://www.fao.org/faostat/en/#home (accessed 7 February 2017).
Fujino, J., Nair, R., Kainuma, M., Masui, T. and Matsuoka, Y., 2006. Multi-gas mitigation analysis on stabilization scenarios using AIM global model. *Energy J.* 27, 343–54.
Füssel, H., 2007. Adaptation planning for climate change: concepts, assessment approaches, and key lessons. *Sustain. Sci.* 2, 265–75.
Global Biodiversity Information Facility, 2015. GBIF [WWW Document]. Free Open Access Biodivers. Data GBIForg. URL http://www.gbif.org/(accessed 13 October 15).
Hargreaves, G. H. and Samani, Z. A., 1985. Reference crop evapotranspiration from temperature. *Appl. Eng. Agric.* 1, 96–9.
Heltberg, R., Siegel, P. B. and Jorgensen, S. L., 2009. Addressing human vulnerability to climate change: Toward a 'no-regrets' approach. *Glob. Environ. Change* 19, 89–99.
Hijmans, R. J., Cameron, S. E., Parra, J. L., Jones, P. G. and Jarvis, A., 2005. Very high resolution interpolated climate surfaces for global land areas. *Int. J. Climatol.* 25, 1965–78.
Hothorn, T., Bretz, F. and Westfall, P., 2008. Simultaneous inference in general parametric models. *Biom. J.* 50, 346–63.
INEI, 2012. *IV Censo Nacional Agropecuario*. Instituto Nacional De Estadistica E Informatica, Lima, Peru.
Kingston, D. G., Todd, M. C., Taylor, R. G., Thompson, J. R. and Arnell, N. W., 2009. Uncertainty in the estimation of potential evapotranspiration under climate change. *Geophys. Res. Lett.* 36, L20403.
Kra, E. Y. and Ofosu-Anim, J., 2010. Modeling maize planting date to minimize irrigation water requirements. *Aust. J. Agric. Eng.* 1, 66.
Läderach, P., Martinez-Valle, A., Schroth, G. and Castro, N., 2013. Predicting the future climatic suitability for cocoa farming of the world's leading producer countries, Ghana and Côte d'Ivoire. *Clim. Change* 119, 841–54.

Lahive, F., Hadley, P. and Daymond, A. J., 2017. The impact of elevated CO2 and water deficit stress on growth and photosynthesis of juvenile cacao (Theobroma cacao L.). *Photosynthetica* 1–10. DOI: 10.1007/s11099-017-0743-y.

Lass, R. A. and Wood, G. A. R., 1985. *Cocoa Production*, World Bank Technical Paper, Washington, DC, USA.

Leenaars, J., Hengl, T., González, M. R., de Jesus, J. M., Heuvelink, G., Wolf, J., van Bussel, L., Claessens, L., Yang, H. and Cassman, K., 2015. *Root Zone Plant-Available Water Holding Capacity of the Sub-Saharan Africa Soil, Version 1.0*. Gridded Funct. Soil Inf. Dataset RZ-PAWHC SSA V 10.

Liaw, A. and Wiener, M., 2002. Classification and regression by random forest. *R News* 2, 18–22.

Lipper, L., Thornton, P., Campbell, B. M., Baedeker, T., Braimoh, A., Bwalya, M., Caron, P., Cattaneo, A., Garrity, D. and Henry, K., 2014. Climate-smart agriculture for food security. *Nat. Clim. Change* 4, 1068–72.

Marias, D. E., Meinzer, F. C. and Still, C., 2017. Impacts of leaf age and heat stress duration on photosynthetic gas exchange and foliar nonstructural carbohydrates in *Coffea arabica*. *Ecol. Evol.* 7, 1297–310.

Mielke, M. S., Almeida, A.-A. F. de and Gomes, F. P., 2005. Photosynthetic traits of five neotropical rainforest tree species: Interactions between light response curves and leaf-to-air vapour pressure deficit. *Braz. Arch. Biol. Technol.* 48, 815–24.

Nankishore, A. and Farrell, A. D., 2016. The response of contrasting tomato genotypes to combined heat and drought stress. *J. Plant Physiol.* 202, 75–82.

R Core Team, 2014. *R: A Language and Environment for Statistical Computing*. R Foundation for Statistical Computing, Vienna, Austria.

Ramirez, J. and Jarvis, A., 2010. Downscaling global circulation model outputs: the delta method decision and policy analysis Working Paper No. 1. *Policy Analysis*, 1:1–18.

Riahi, K., Grübler, A. and Nakicenovic, N., 2007. Scenarios of long-term socio-economic and environmental development under climate stabilization. *Technol. Forecast. Soc. Change* 74, 887–935.

Ruf, F., Schroth, G. and Doffangui, K., 2015. Climate change, cocoa migrations and deforestation in West Africa: What does the past tell us about the future? *Sustain. Sci.* 10, 101–11.

Schroth, G., Läderach, P., Martinez-Valle, A. I. and Bunn, C., 2016a. From site-level to regional adaptation planning for tropical commodities: Cocoa in West Africa. *Mitig. Adapt. Strateg. Glob. Change* 22(6), 903–27.

Schroth, G., Läderach, P., Martinez-Valle, A. I., Bunn, C. and Jassogne, L., 2016b. Vulnerability to climate change of cocoa in West Africa: patterns, opportunities and limits to adaptation. *Sci. Total Environ.* 556, 231–41.

Sena Gomes, A. R. and Kozlowski, T. T., 1987. Effects of temperature on growth and water relations of cacao (Theobroma cacao var.Comum) seedlings. *Plant Soil* 103, 3–11.

Stocker, T. F., Qin, D., Plattner, G. K., Tignor, M., Allen, S. K., Boschung, J., Nauels, A., Xia, Y., Bex, B. and Midgley, B. M., 2013. IPCC, 2013: climate change 2013: the physical science basis. Contribution of working group I to the fifth assessment report of the intergovernmental panel on climate change.

van Vuuren, D. P., Den Elzen, M. G., Lucas, P. L., Eickhout, B., Strengers, B. J., van Ruijven, B., Wonink, S. and van Houdt, R., 2007. Stabilizing greenhouse gas concentrations at low levels: An assessment of reduction strategies and costs. *Clim. Change* 81, 119–59.

Vermeulen, S. J., Challinor, A. J., Thornton, P. K., Campbell, B. M., Eriyagama, N., Vervoort, J. M., Kinyangi, J., Jarvis, A., Läderach, P., Ramirez-Villegas, J., Nicklin, K. J., Hawkins, E. and Smith, D. R., 2013. Addressing uncertainty in adaptation planning for agriculture. *Proc. Natl. Acad. Sci.* 110, 8357–62.

Zuidema, P. A., Leffelaar, P. A., Gerritsma, W., Mommer, L. and Anten, N. P., 2005. A physiological production model for cocoa (Theobroma cacao): model presentation, validation and application. *Agric. Syst.* 84, 195–225.

Chapter 20

Analysis and design of the shade canopy of cocoa-based agroforestry systems

Eduardo Somarriba, CATIE, Costa Rica; Luis Orozco-Aguilar, University of Melbourne, Australia; Rolando Cerda, CATIE, Costa Rica; and Arlene López-Sampson, James Cook University, Australia

1 Introduction

Agroforestry (AF), the art and science of farming with trees (Van Noordwijk et al., 2016; Torquebiau, 2000) covers 800–1000 million ha worldwide (Nair et al., 2009; Zomer et al., 2009). Shaded tree-crop systems [involving crops such as coffee (Coffea spp.) and cacao (*Theobroma cacao* L.)] are prominent examples of AF systems. Three million smallholders cultivate 7–8 million ha of cacao worldwide (www.worldcocoafoundation.org), with at least 80% of the cacao cultivated under a shade tree canopy, that is, in AF systems (Somarriba et al., 2012). Cocoa-based AF systems have been top-ranked as land use alternatives to cope with climate change because of their high levels of species diversity, year-round soil cover, high levels of stored carbon in both the soil and wood (above and below ground biomass) and other desirable attributes (Cerda et al., 2014; Harvey et al., 2014; Vaast and Somarriba, 2014; Tscharntke et al., 2011).

Farmers retain and plant a large number of tree species to provide shade and shelter to cocoa trees and help sustain high cacao yields (Koko et al., 2013; Anglaaere et al., 2011; Somarriba and Beer, 2011; Bos et al., 2007). Trees in the cacao shade canopy also produce timber, fruits and other goods for either family consumption or sale (Deheuvels et al., 2012; Somarriba, 2007); generate income (Cerda et al., 2014; Oke and Odebiyi, 2007); conserve biodiversity at the plot and landscape levels (de Beenhouwer et al., 2013;

http://dx.doi.org/10.19103/AS.2017.0021.29

Clough et al., 2011; Stenchly et al., 2012); increase landscape connectivity (Harvey et al., 2014; Asare et al., 2014); sequester carbon (Saj et al., 2017; Abou Rajab et al., 2016; Schroth et al., 2016; Jadan et al., 2015; Jacobi et al., 2014; Norgrove and Hauser, 2013; Leuschner et al., 2013; Somarriba et al., 2013; Oke and Olatiilu, 2011; Wade et al., 2010; Smiley and Kroschel, 2008); enhance the pollination of cocoa (Toledo-Hernández et al., 2017; Frimpong et al., 2011); sustain soil fertility (Mortimer et al., 2017; Rousseau et al., 2012; Moço et al., 2010); and reduce farmers' vulnerability to the effect of extreme climatic events (Schwendenmann et al., 2010; Läderach et al., 2013).

Farmers have ample knowledge on the products and services they obtain from a diverse, tree-shade canopy in the cocoa plantation (Silva et al., 2013; Ortiz-González, 2006). Shade trees: 1) modify the light regime (in both quality and quantity), air temperature, humidity and wind movement within the plantation, directly affecting photosynthesis, growth and yield of cocoa (de Almeida and Valle, 2007; Zuidema et al., 2005); 2) favour or hamper the population dynamics and incidence of pests and diseases (and of their natural enemies) that reduce yields of both cocoa and its companion species (Mortimer et al., 2017; Schroth and da Mota, 2014); and 3) produce significant quantities of organic matter, recycle nutrients and help to maintain the natural fertility of the site (Hartemink, 2005), a service that is of utmost importance since most cocoa plantations are not fertilized (Tscharntke et al., 2011).

The relationships between shade cover and cocoa yields are inconsistent between sites (Asare, 2006; Asare et al., 2017). Blaser et al. (2017) while studying the effect of shade trees on soil fertility and soil carbon stock in cocoa AF systems at the plot level found no positive effect of shade trees on soil fertility under increasing shade cover. However, individual shade trees had a positive effect on soil carbon, nitrogen fertility and soil aggregation at these localized sites. A comprehensive review of the effects of shade trees on the maintenance of soil fertility was provided by Mortimer et al. (2017). A recent study showed that ants and birds caused variation in yields between 100 and 800 kg^{-1} ha^{-1} $year^{-1}$ depending on the level of shade tree cover. For example, yield was higher under 30–40% shade cover and with an undisturbed ant community (Gras et al., 2016). On the other hand, Schroth et al. (2014) found that highest yields of cocoa agroforests (585 kg ha^{-1} or twice the current regional average) were compatible with a tree cover up to 55% shade and above-ground C stock in the large shade trees (>30 cm DBH) of up to 65 Mg ha^{-1}.

In this chapter we: 1) revisit the different cocoa system typologies described in the literature, placing special emphasis on cocoa–timber systems; 2) present a simple guide for the analysis of the cocoa shade canopy of shaded cocoa systems; 3) explore the trade-offs and synergies implicit in the optimal design of cocoa shade canopies, with emphasis on the analysis of carbon storage and cocoa yields; and 4) provide a brief look at the critical issues to be taken into consideration to optimally use AF science/knowledge in cocoa cultivation programmes.

2 Cocoa shade typologies

The diversity, botanical composition and structural complexity of cocoa shade canopies vary widely between geographical regions, between farms within a region and even between sections within a plantation. Some cocoa shade canopies are species-rich and structurally complex, with several vertical strata and diverse spatial and temporal configurations (Asare, 2006; Bentley et al., 2004). However, most cocoa shade canopies have low species richness and a simple vertical and horizontal structure to facilitate crop management (Deheuvels et al., 2014; Gockowski and Sonwa, 2011). Irrespective of their complexity, most cocoa shade canopies are suboptimal in design and management due to: 1) excessive or deficient shade levels, 2) uneven spatial distribution of shade cover, 3) suboptimal botanical composition in relation to fulfilling farmers' objectives and 4) poor agronomic and silvicultural management, for example, inadequate pruning and thinning regimes of timber trees (Somarriba et al., 2001).

Six shade canopy typologies have been identified (Table 1, Fig. 1) (Rice and Greenberg, 2000), namely: 1) cocoa without shade; 2) cocoa with a mono-specific 'service' shade, that is, with a legume tree species of the genera *Erythrina, Gliricidia, Inga* or *Albizia* (Smiley and Kroschel, 2009); 3) cocoa with one or a few productive shade tree species that is, cocoa–fruit, cocoa–timber, cocoa–rubber (*Hevea brasiliensis*) or cocoa–bananas (Sonwa et al., 2014; Utomo et al., 2016; Wu et al., 2016); 4) cocoa with mixed shade (Jagoret et al., 2012; Deheuvels et al., 2012); 5) rustic cocoa, also known as 'cabrucas' in Brazil (Sambuichi et al., 2012); and 6) successional agroforests (Somarriba and Lachenaud, 2013).

Unshaded cacao can be found mostly in West Africa, Indonesia, and Brazil and Ecuador in South America (Tondoh et al., 2015; Jadan et al., 2015; Siebert, 2002). In unshaded systems, cacao planting density and yield are up to five times greater than some shaded cacao typologies (Table 1). Monospecific, 'service' shade and productive cacao shade canopies are found in various countries in West and Central Africa and in South America; in these typologies cacao density, cacao productivity and shade tree diversity are moderate (Tondoh et al., 2015; Asare et al., 2009). Mixed shaded cacao is the dominant typology across Mesoamerica; shade canopy in this typology has high tree diversity, high levels of stored carbon, low cacao planting density and low cacao yields (Table 1). Rustic cacao is common in the Atlantic forest region of southern Brazil; 'cabrucas' have high tree species richness, complex rainforest structure and low cacao productivity. Finally, successional cacao agroforests are found in the Amazon riverine system and have a forest-like physiognomy, high species diversity, low cacao planting density and very low cacao yields.

2.1 Cocoa–timber systems

Timber trees are a frequent component in cocoa shade canopies, and various authors have recommended cocoa–timber systems to increase the sustainability and financial efficiency of cocoa farming (Jaimez et al., 2013; Gockowski and Sonwa, 2011; Ofori-Bah and Asafu-Adjaye, 2011; Ruf, 2011; Somarriba and Beer, 2011), and as a viable strategy for the intensification of cocoa cultivation (Somarriba et al., 2014; Vaast and Somarriba, 2014). Two basic models for timber production in cocoa plantations can be identified; 1) Trees are planted for shade and lumber at the onset of the cocoa plantation and 2) Trees are recruited from the natural regeneration at the site. In the first model, trees are harvested (clear cut) at rotation age, hopefully coinciding with renovation of the cocoa stand. Because all timber trees are planted at the same time, the tree stand is even-aged. In the second model of timber production, tree recruitment is a continuous process and mortality/harvest occurs during the entire cycle of the cocoa plantation; the tree stand is uneven-aged.

Table 1 General characteristics of the six major typologies of cacao agroforestry systems worldwide

Typology	Species richness	Density (trees ha^{-1})	Shade trees basal area ($m^2\ ha^{-1}$)	Cacao density (plants ha^{-1})	Cacao yield (kg ha^{-1})	Location	Reference
Unshaded	1–5	2–3	0.2–0.4	≥1500	1200–1500	Sulawesi, Indonesia	Siebert (2002)
	4–13	3–16	N/A	≥1500	N/A	Ivory Coast, Africa	Tondoh et al. (2015)
	1–7	20–25	0.5–1	1600–2250	1500–2000	Napo province, Ecuador	Jadan et al. (2015)
Mono-specific service shade	7–12	12–37	2.7–4.4	700–1250	700–1000	Ivory Coast and Ghana	Tondoh et al. (2015)
	8–16	30–97	1.7–2.4	900–1111	500–800	Ghana and Nigeria, Africa	Asare et al. (2009)
	15–26	25–57	N/A	1000–1250	600–900	Southern Cameroon	Sonwa et al. (2007)
	2–5	68–100	N/A	1111–1250	N/A	Mérida, Venezuela	Jaimez et al. (2013)
	2–3	150–200	N/A	1111–1250	1000–1500	Central Sulawesi, Indonesia	Moser et al. (2010)
	3–6	3–23	2.3–6.4	≥ 1250	N/A	Ondo Stage, Nigeria	Oke and Odebiyi (2007)
Productive shade	15–26	23–39	8.1–18.9	500–625	500–700	Alto Beni, Bolivia	Jacobi et al. (2014)
	43–96	120–275	N/A	450–600	300–450	Choco Region, Ecuador	Waldron et al. (2012)
	35–95	245–350	1.15–1.97	540–690	50–450	Talamanca, Costa Rica	Deheuvels et al. (2012)
	21–47	240–620	N/A	1200–1800	150–850	Central Cameroon, Africa	Saj et al. (2013)

Mixed shade	22–27	10–25	9–22	N/A	250–350	Soconusco, Chiapas, Mexico	Salgado-Mora et al. (2007)
	40–80	190–300	15–23	550–850	350–450	Central America	Somarriba et al. (2013)
	16–33	90–200	5–17	625–833	400–700	San Alejandro, Peru	Vebrova et al. (2013)
	10–50	120–170	1–10	500–650	450–550	Napo province, Ecuador	Jadan et al. (2015)
	37–63	42–227	N/A	450–700	550–700	Alto Beni, Bolivia	Schneider et al. (2017)
	45–89	67–107	6.6–10.7	1200–1900	350–750	Central Cameroon, Africa	Jagoret et al. (2017)
Rustic cacao	46–180	45–355	11–28	200–300	200–300	Bahia, Brazil	Sambuichi and Haridasan (2007), Rolim and Chiarello (2004)
Successional agroforests	≥100	600–700	N/A	≤200	≤100	Ecuador, Colombia, Peru, Brazil and French, Guyana	Somarriba and Lachenaud (2013), de Almeida and Valle (2007)

Figure 1 Horizontal and vertical structure in a mixed shade canopy, cocoa-based agroforestry system.

Planted, even-aged, timber stands used as shade in cocoa plantations typically follow the canopy development pattern depicted in Fig. 2. As trees grow and develop their crowns (as a proxy for tree size, standing timber volume, biomass or carbon), the shade level ncreases until a maximum threshold level is reached; cocoa yields are depressed

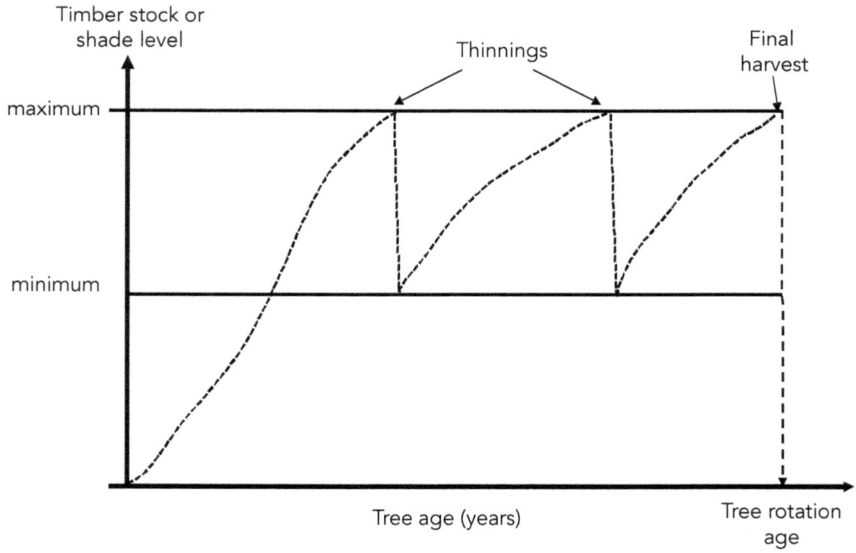

Figure 2 Typical temporal trend in the development of the shade canopy in a planted, even-aged timber tree stand used as shade over cocoa.

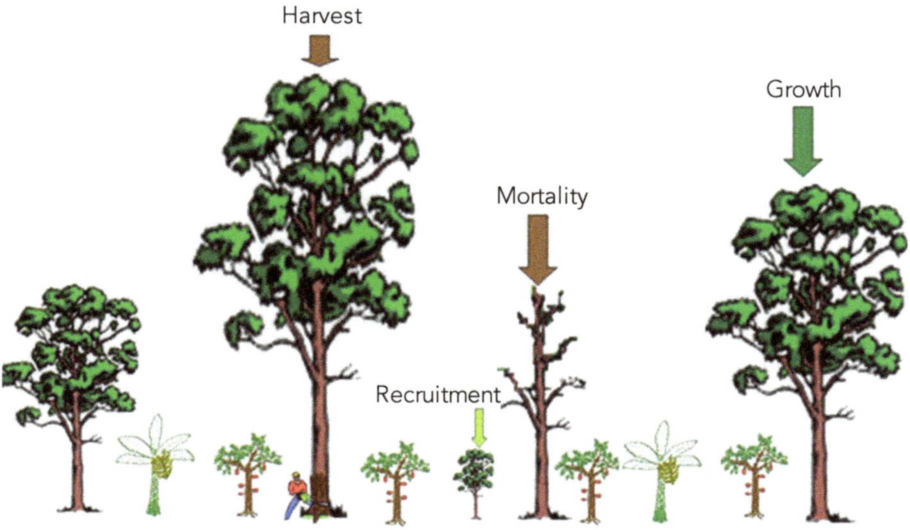

Figure 3 Tree population dynamics parameters.

when tree stand density passes this threshold. Tree thinning and pruning keep stand density above a minimum threshold level below which timber per hectare is insignificant. The rate of thinning (trees ha^{-1} removed) over the life cycle of the cocoa plantation is determined by the initial planting density, crown growth rate, crown closure and leaf fall patterns of the timber trees (Somarriba and Beer, 2011).

Naturally regenerated timber species in cocoa shade canopies belong to a small group of successfully reproducing, native species representative of the local flora. Notable examples include *Cordia alliodora* in Latin America (Somarriba et al., 2014), and *Terminalia ivorensis, Terminalia superba* in West Africa (Gockowski and Sonwa, 2008; Sonwa et al., 2007; Anglaaere, 2005; Asare, 2005; Nomo, 2005). Uneven-aged timber tree stands used as shade cover require the analysis of the population dynamics of a structured (age and size) tree population (Fig. 3). Parameters such as stem diameter frequency distribution (Fig. 4), recruitment of new individuals and growth, mortality and harvest rates are evaluated to project population size and structure over time; all population dynamics parameters act simultaneously and their relationships are treated analytically using probability transition matrices (Somarriba et al., 2014).

2.2 Timber harvest and damage to the cacao crop

The damage inflicted on cocoa during the harvest of timber trees (a concern for farmers and extension workers alike) is not as severe as usually expected. Income from timber sales easily offset the costs of both renewing the cocoa plants and bringing the plantation back to pre-damage condition. Damage to cocoa due to the harvest of timber should not be an argument to limit the promotion and use of timber trees in cocoa systems.

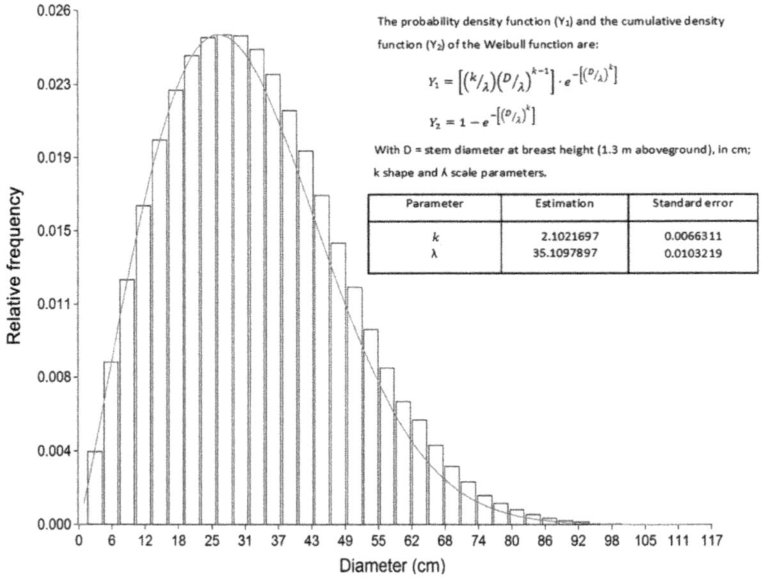

Figure 4 Stem diameter frequency distribution of naturally regenerated *Cordia alliodora* trees in cocoa plantations, Changuinola, Panama.

To assess the feasibility and financial outcome of harvesting timber trees in cocoa plantations, a simple, three-step guide is proposed (Ryan et al., 2009):

- First, damaged cocoa plants resulting from the harvest of one timber tree are inventoried, and then each damaged plant is categorized according to the six severity levels (Table 2). Damage classes are defined based on the activities required to bring back the damaged cocoa plant to the pre-damage condition.
- Second, damage to cocoa must be translated into yield losses and their corresponding economic values, using a cocoa yield-age model which can be easily developed

Table 2 Damage classes used to depict the impact of harvesting timber trees used as shade in cocoa plantations

Damage class	Description	Corrective activity
1	Uprooted tree, 100% crown loss	Replant a new cocoa plant
2	Snapped off or all primary branches damaged, 100% crown loss	Stump and coppice selection to reconstruct tree crown
3a	Primary branches damaged, 75–99% crown loss	Very heavy pruning required
3b	Primary branches damaged, 50–74% crown loss	Heavy pruning required
4a	Secondary branches damaged, 25–49% crown loss	Moderate pruning required
4b	Secondary branches damaged, 0–24% crown loss	Light pruning required

using expert knowledge. The reduction in cocoa yield and economic loss, and the cost of renewing the damaged cocoa plant must be determined per year over the full duration of the recovery period.

- Third, a sensitivity analysis should be conducted to evaluate the effect of varying cocoa planting densities, cocoa yield levels and prices for both cocoa and timber, on the feasibility of harvesting the timber.

3 A four-step guide for analysis of the shade canopy

A proper design and management of the shade canopy is critical for achieving optimal, sustainable yields (Jaimez et al., 2013; Somarriba and Beer, 2011; Beer et al., 1998), and healthy landscapes (Harvey et al., 2014). Farmers have a significant amount of knowledge on the roles, uses, advantages and disadvantages of shade trees in cocoa farming (Silva et al., 2013; Anglaaere et al., 2011; Asare, 2005), but 'flunk' in complex issues such as the management of the interactions between shade–soil fertility–cocoa yields, and in understanding how changes in exposure, latitude and the daily and yearly movement of the sun affects shading (Somarriba, 2007).

Shade canopy design and management is a complex issue, involving a range of factors related to plantation management, site conditions and selection and management of shade species. There are no recipes, each cocoa plantation is unique and requires tailored design according to all the intervening factors. In what follows we present a guide describing the main steps for the analysis of cocoa shade canopies. The design of improved shade canopies is treated in a subsequent section (Fig. 5).

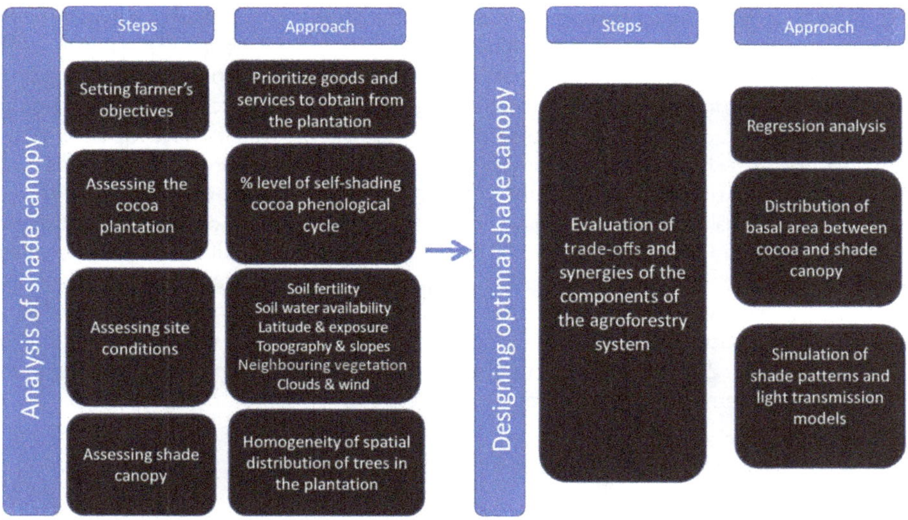

Figure 5 Framework for the analysis and design of the shade of cocoa-based agroforestry system.

3.1 Setting farmers' objectives

The analysis of the shade canopy and the design of improvements (shade canopy botanical composition, spatial and temporal patterns, and management, as well as other crop husbandry practices) are geared towards achieving farmers' objectives. Setting clear goals is the first necessary step for a successful shade canopy analysis. The objectives of a farmer are the benefits (ecosystem services) that (s)he wants to obtain from the shade canopy. Ecosystem services are grouped in four categories: provisioning services (cocoa, fruits, timber, construction materials, medicine, fresh water); regulating services (pests and diseases regulation, pollination, carbon sequestration, extreme events protection, reduction of soil erosion, maintenance of soil fertility); cultural services (recreation, tourism) and habitat services (species, genetic diversity; Müller et al., 2015). Each farmer has a prioritized list of ecosystem services (and dis-services such as competition for soil, water and nutrients, pollution, etc.) to optimize (Dale and Polasky, 2007; Zhang et al., 2007).

For a given set of objectives, the analysis of the shade canopy requires the systematic evaluation of the three major sources of shading: the cocoa plantation itself, site conditions and the shade canopy itself. Each source of shading requires the assessment of several contributing factors; each factor has a proper, expected response in the assessment (Somarriba et al., 2011).

3.2 Assessing the cocoa plantation

Two key subfactors should be assessed: level of self-shading and phenological cycle of the cacao plant. Self-shading denotes the shade projected by upper leaves and branches on lower leaves in the same crown plus the shade projected by the crowns of neighbouring cocoa trees. When self-shading is high, less shade from the tree canopy is needed, thus reducing the potential of the cocoa plantation to produce goods (timber, fruit, etc.) and services (cultural, environmental). Self-shading is the result of 'crowding' in the plantation, which in turn is determined by a combination of the shape and size of the crown of the cocoa tree, and the planting configuration and spacing of the cocoa. This includes, for instance, plant age (young cocoa plants have small crowns, hence low self-shading); pruning regimes (infrequent and light pruning of cocoa lead to more self-shading); genetics (grafted cacao trees from plagiotropic buds tend to be more open-crowned than trees from seeds; less self-shading is expected in open-crowned cocoa trees); and spacing and planting configuration (high planting density of cocoa will result in high self-shading; triangular planting arrangements can 'pack' more densely cocoa trees in the plot than square or rectangular configurations, thus resulting in more self-shading). Every year, cocoa trees transit over a sequence of phenological stages such as leaf flushing, flowering, fruit development and filling, fruit maturation, radial growth of stem, branches and roots and a quiescent stage. More solar radiation is needed at flowering and pod filling (de Almeida and Valle, 2007). Crop phenology determines the monthly pattern of shading (and crown management) needed to optimize the performance of the cacao crop.

3.3 Assessing site conditions

Cocoa photosynthetic rate depends on both the intensity and quality of solar radiation reaching the site and the supply of nutrients: more solar radiation will result in a higher demand for soil nutrients (sourced either from the natural fertility stock of the soil or from applied fertilizers, either organic or inorganic) and water. Fertile soils can sustain a cocoa plantation with low shade levels; more shade will be needed on infertile soils. The relationship between optimal shade levels and water availability in the soil is a complex, non-linear one: 1) water consumption by trees may advise the reduction (or even total removal) of shade trees in very dry places (trees consume water that would otherwise be available for cocoa); 2) at intermediate levels of water availability, the amount of water 'saved' due to reduced evapotranspiration (lower air temperature, year-round soil cover, etc.) is larger than the amount of water used up by the tree (Niether et al., 2017); and 3) in very humid sites the use of water by the tree has no effect on cocoa and consequently the use of shade trees is optional. Threshold values of soil water availability separating these three zones have yet to be determined for each site and shade canopy botanical composition, tree density and management.

High cloudiness leads to a reduced need of shade either from self-shading or from the shade canopy. Winds force farmers to plant trees in lines (ideally, perpendicular to wind direction) instead of evenly spacing them in the plot. Tree rows project 'lateral shade' over the cocoa, which is of 'lower quality' than the shade projected by evenly spaced trees over the entire plantation (that ensures that all cocoa plants experience similar growing conditions). When tree lines are used as shade, the amount of shade projected onto the cocoa trees is determined by both the proximity of the cocoa tree to the tree line, and the orientation of the tree line with respect to the hourly and daily movement of the sun.

Hills and tall vegetation at the border of the cocoa plantation also cast lateral shade over the plantation. A cocoa plantation positioned at the base of a steep, long hill facing west will cast lateral shade over the plantation during morning hours, and will be fully exposed to the sun in the afternoon. The degree and length of the slope will determine the number of hours of lateral shade per day. Steep, long slopes will cast more hours of lateral shade than short, relatively flat slopes. The latitude of the site will also influence the amount of solar radiation received by a cocoa plantation. North-facing slopes will receive more solar radiation in southern latitudes; conversely, exposure to solar radiation will be higher for south-facing slopes in the northern hemisphere.

The position of the Sun, as seen by an observer at a given latitude, day of the year, and hour of the day is described by two angles: azimuth and solar elevation (Fig. 6). The following equations estimate these two angles (Quesada et al., 1987):

$$hour = 15(hour - 12)$$

$$declination = 23{,}45\ \sin\left(\left(\frac{360}{365}\right)(day + 284)\right)$$

$$elevation = \sin^{-1}\cos(latitude)\cos(hour) + \sin(latitude)\sin(declination)$$

$$azimuth = \frac{\sin^{-1}\left[\cos(latitude)\cos(declination) - (\sin(latitude)\cos(declination)\cos(hour)\right]}{\cos(elevation)}$$

where hour = hour of the day; declination = solar declination; day = day of the year, day one is 1 January; elevation = solar elevation angle; latitude = geographical latitude, ranging from 0° to 90° with a positive sign in the northern hemisphere and a negative sign in the southern hemisphere; azimuth = azimuth angle, ranging from 0° to 360°, measured clockwise on the ground.

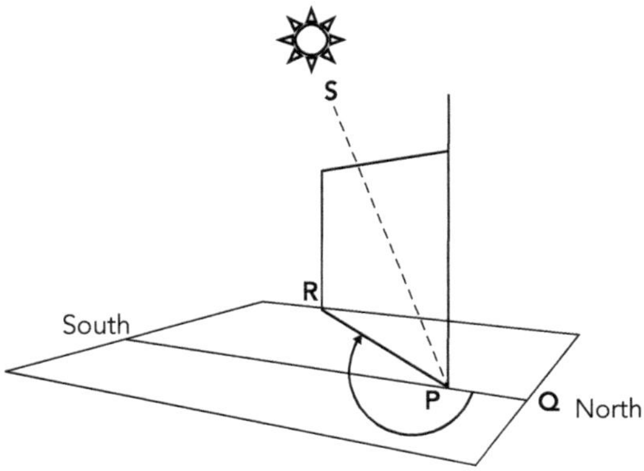

Figure 6 Two angles (elevation and azimuth) describe the position of the Sun to the eyes of an observer on Earth. Elevation angle is SPR and azimuth QPR.

3.4 Assessing the shade canopy

Farmers manipulate the botanical composition, population density, spatial and temporal patterns and management of tree cover in the plantation. Ideally, tree cover should be as spatially homogeneous as possible to ensure that all cocoa trees experience similar shading conditions. Most cocoa plantations, however, are suboptimal in terms of spatial homogeneity of canopy cover, with gaps (little or no shade) or high tree density (excessive shade) spots in the same plantation. Tree species will differ in height, crown shape, crown size and crown density (porosity or its inverse, opacity), and monthly defoliation patterns. These variables allow us to estimate tree cover and spacing (Somarriba, 2002).

The area covered by tree crowns (*C*) is proportional to the population density of canopy trees (N, in trees ha^{-1}); tree crown diameter (*D*, in m); and crown opacity (*P*, expressed in the range 0–1; a 0.5 opacity indicates that only 50% of crown projection area on the ground is actually covered by the crown's leaves and branches). In analytical form:

$$C = N * A * P$$

Assuming that crowns are circular, tree crown area (*A*):

$$A = \frac{\pi}{4} * D^2 = 0.7854 * D^2$$

$$N = \frac{C}{(A * P)}$$

The average spacing between trees (S, in m), assuming a regular, square, planting arrangement of trees is:

$$S = \sqrt{\frac{10000}{N}}$$

For example, for a given cocoa plantation with a maximum of 30% of canopy cover (equivalent to C =3000 m^2 ha^{-1} of effective canopy cover); crown diameter 8 m; and crown opacity 50% (0.5 in a 0–1 scale):

$$N = \frac{3000}{\left(\frac{\pi}{4} * 8^2 * 0.5\right)} \rightarrow \frac{3000}{25.13} \rightarrow 119 \text{ trees ha}^{-1}$$

and

$$S = \sqrt{(10000/119)} = 9.17 \text{ m}.$$

Tree height also influences the shade levels experienced by cacao plants. Farmers know that tall trees project a lighter shade than short trees. However, a better explanation is that tall trees project a faster shade than short ones. For instance (Fig. 7), in one day (say between 10 am and 2 pm) the length of the path of the shade projected on the ground by the crown of a tall tree (L2) will be longer than the path of the shade projected by a short tree (L1). This can be demonstrated by considering that in Fig. 7:

$$L1 = 2 * \left[h1 * \cot(\propto)\right]$$

and

$$L2 = 2 * \left[(h1 + \beta) * \cot(\propto)\right]$$

where α = solar elevation angle.

Consequently, L2 is larger than L1 by the amount: $2 * \beta * \cot(\propto)$. The shade projected by a taller tree travels a longer distance (in the same amount of time) than the shade projected by a shorter tree. Therefore, the shade projected by the crown of a tall tree moves faster along the path than the shade projected by a short tree. A cocoa tree in the path of the shade of a tall tree will experience fewer hours of shade per day than under a short tree. This has practical applications, for instance, you can use tall trees with dense crowns and yet project acceptable levels of shade on the cocoa canopy. When a short tree species is used as companion, it better have a sparse crown to avoid excessive shading of the cocoa.

Crown size, tree spacing and planting spatial configuration (triangle, square and rectangle) have profound effects on the temporal dynamics of canopy closure and shading of the cocoa (Fig. 8). Three critical moments in the closure of the canopy need to be considered: 1) crown diameter is smaller than tree spacing, and there is open canopy space between neighbour trees, 2) crown diameter equals tree spacing and neighbour tree crowns 'touch' each other without overlapping and 3) crown diameter is larger than tree spacing and crown overlapping occurs (although some canopy gaps may still exist). The interactions between crown diameter and tree spacing results in three shading conditions: shade without overlap, no shade and shade with overlap. The mathematical conditions determining the magnitude of these three shading

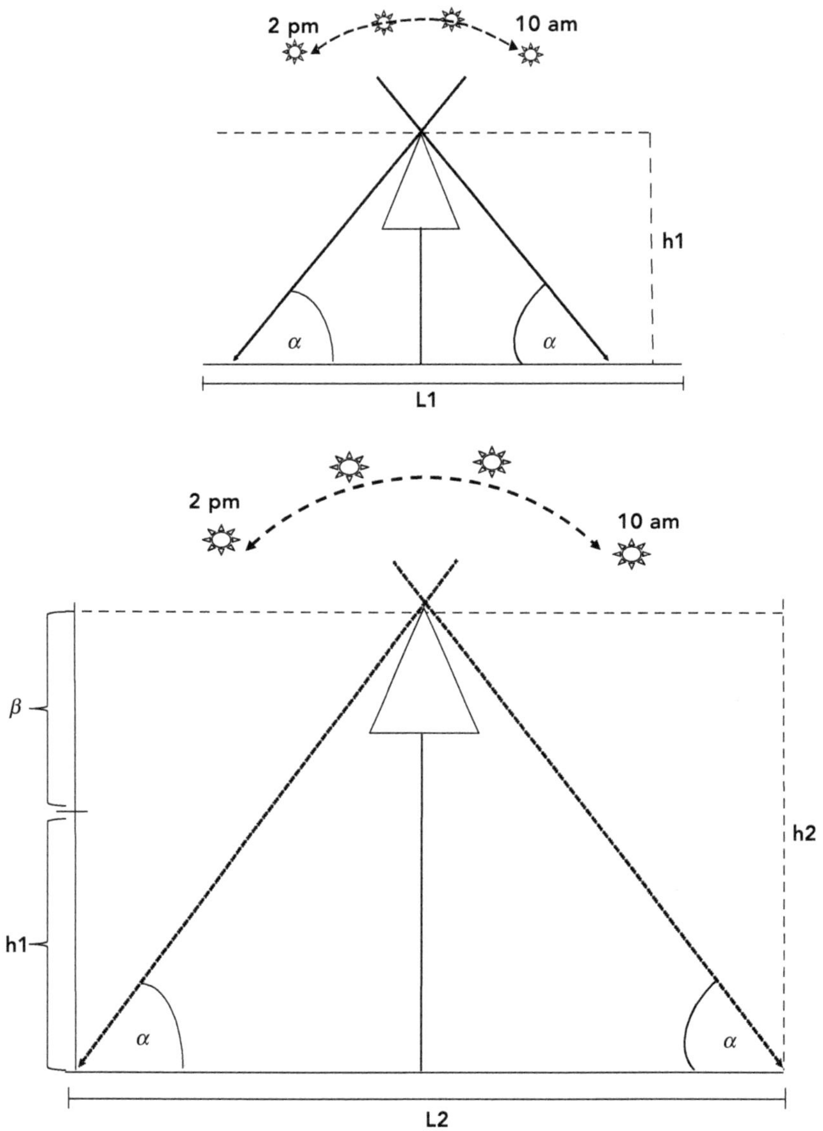

Figure 7 Tree height and the length of the path transited by the shade projected on the ground by a tall and a short tree.

conditions are presented in Somarriba (2002). These equations show that with trees planted in square configurations, when crown diameter equals tree spacing (i.e. crowns of neighbour trees 'touch' each other) still a significant 21.46% of the canopy is open space. The amount of open space is reduced to 9% if trees are planted in a triangular spacing pattern (Somarriba, 2002).

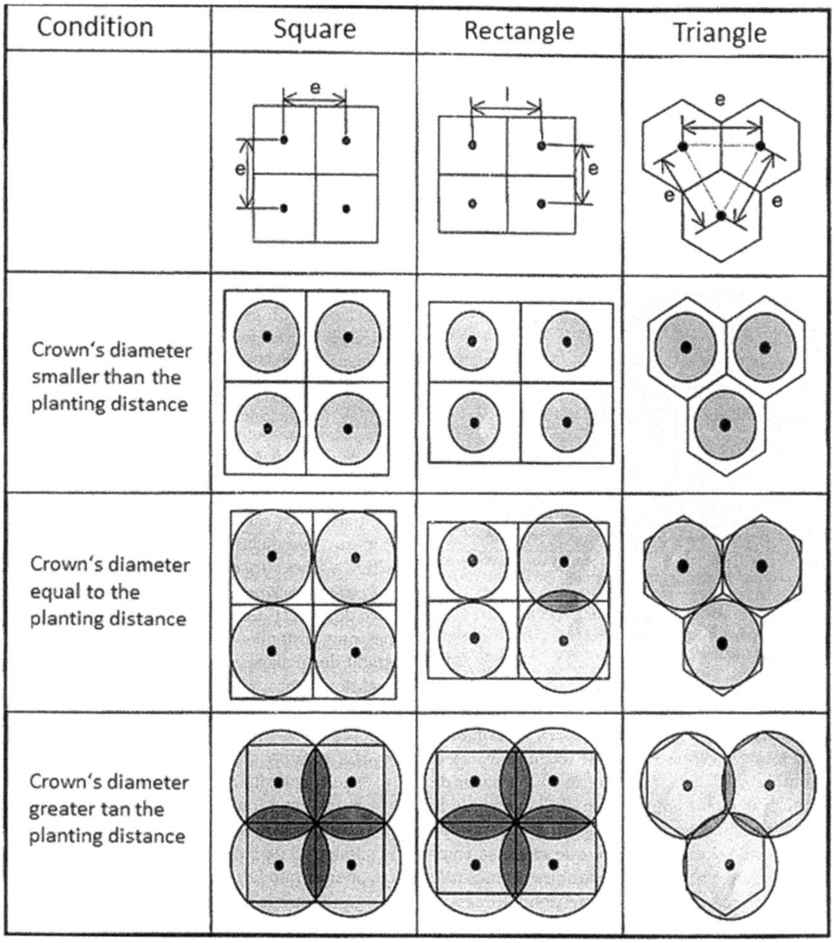

Figure 8 Canopy closure as a function of spacing, crown diameter and planting configuration. Source: Somarriba (2002).

4 Designing improved shade canopies

Designing shade canopies in cocoa requires a careful consideration of the four basic diagnostic steps presented in previous sections (setting farmers' objectives, assessing the effects of the site, the crop, and the shade canopy). However, one question remains to be answered. Is the current shade canopy design optimal? To address this question, different approaches can be used to assess the synergies and trade-offs between crop husbandry, shade canopy design and management, and the ecosystem services the system can provide. Trade-offs and synergies can be assessed by regression analyses, system optimization, simulation of shade patterns and light transmission models, and other techniques. In what follows we provide a brief description of three approaches: 1) regression analysis, 2) competitive allocation of stand basal area (biomass or carbon) between cocoa and the shade canopy plants and 3) simulation of shading patterns.

4.1 Analysing trade-offs and synergies by regression analysis

Trade-offs between plant biodiversity and indicators of ecosystem services have been the subject of recent research showing that high biodiversity levels are compatible with high cocoa yields (Cerda et al., 2014; de Beenhouwer et al., 2013; Clough et al., 2011) and with high carbon sequestration (Somarriba et al., 2013; Kessler et al., 2012; Wade et al., 2010). Other studies have highlighted the impacts of management intensification and agricultural practices on the conservation of biodiversity and provision of ecosystem services (Vaast and Somarriba, 2014; Gockowski and Sonwa, 2011; Bisseleua et al., 2009; Steffan-Dewenter et al., 2007). For instance, the intensification of crop husbandry (pests and diseases control) can reduce both shade canopy biodiversity and the provision of other ecosystem services (Cheatham et al., 2009).

It seems possible to reduce trade-offs, or even achieve win-win situations (synergies), by adjusting the spatial/temporal arrangement, the botanical composition and the management of both cocoa and the shade canopy plants. Rapidel et al. (2015) demonstrated the importance of the analysis of relationships between indicators of ecosystem services as a key step for the design of optimal agroecosystems with perennial crops. Inspired by that work, we propose a four-step guide to use trade-off analysis in the design of optimal cocoa AF systems:

1. Identify the ecosystem services provided by the AF system and prioritize (rank) them according to the objectives of the farmer (this comes from the 'Setting farmers' objectives' section).
2. Estimate the level of each expected service.
3. Estimate regression functions between ecosystem services and biodiversity. For example, if a farmer prioritizes, in decreasing order of importance, three ecosystem services (ES1 > ES2 > ES3) and also wants to retain high levels of plant biodiversity in the shade canopy, then ES1 should always be the most important dependent variable. Trade-offs, synergies or independence between factors can be determined by looking at the slopes (sign and magnitude) of the regression functions (Fig. 9).
4. Iteratively, adjust the botanical composition, relative abundances, spatial/temporal patterns and management of both the crop and the shade canopy plants to enhance synergies and minimize the most problematic trade-offs, until an ideal cocoa AF system is achieved (in terms of reaching farmer's goals).

In the hypothetical case presented in Fig. 9, the most important ecosystem services for the farmer (ES1 and ES2) have trade-offs with the plant biodiversity desired in the system. ES1 and ES2 have synergy between each other; and the ES3 has no relationship with ES1 and ES2 (independence). Given that ES1 and ES2 are priorities for the farmer, the botanical composition, planting density and spatial arrangement, and the management of plant biodiversity in the improved system should be adjusted to minimize trade-offs (or set them to zero) and maximize synergies. In this example, since ES3 is not a priority for the farmer,

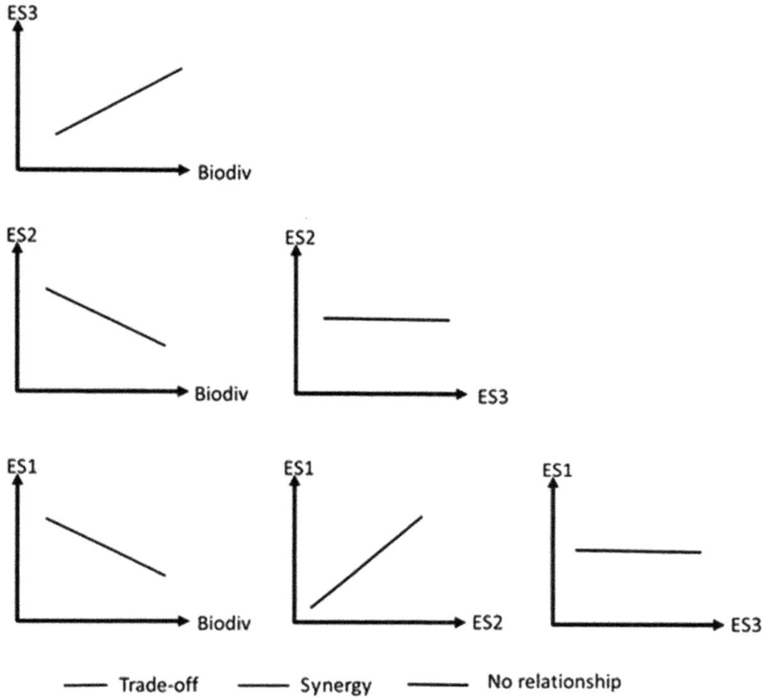

Figure 9 Analysis of relationships (trade-offs, synergies, independence = no relationships) between indicators of ecosystem services (ES1–ES3) and biodiversity, and between different ecosystem services.

the improved design may involve a loss of synergy between ES3 and plant biodiversity, and this would be acceptable as long as ES1 and ES2 are not reduced.

4.2 The competitive allocation of basal area between cocoa and the shade canopy

A model based on the competitive allocation of the basal area of the cocoa plantation to either the cocoa or the shade canopy plants can help design optimal cocoa plantations. A six-step analysis is required for the optimal design of the system.

1. First, we consider the limits in stocking imposed by the tree growth factors in any given site. For instance, maximum attainable basal area is likely to vary along large-scale gradients of temperature (latitudinal and altitudinal gradients), water (from deserts to very humid forest zones) and soil.
2. Second, at any given location along this gradient, the stocking of cocoa plantations will vary depending on the typology, with low stocking in mono-cropped cocoa and maximum in rustic and successional cocoa agroforests (Somarriba et al., 2013).

3 Third, at a given site and typology, optimization of the total productivity of the cocoa plantation requires the competitive allocation of plantation basal area between: 1) cocoa and the shade canopy as a whole; and 2) between groups of species producing particular goods and services within the shade canopy (Fig. 10). For instance, high stocking of cocoa, as is commonly the case in West Africa with planting densities typically between 1500 and 1700 cocoa trees ha^{-1} (Sonwa et al., 2007, Zapfack et al., 2002, Vernon and Morris, 1964), will result in higher levels of self-shading, and consequently, a reduced need for trees in the shade canopy. And vice-versa: high stocking in the shade canopy will be conducive to lower stocking levels in the cocoa, as is commonly the case in Central America with cocoa planting densities typically between 500 and 1000 trees ha^{-1} and highly stocked shade canopies (Cerda et al., 2014). Rustic systems, with high basal area in the shade canopy, also show the competitive, two-way interaction, between stocking in the cocoa and the shade canopy (Sambuichi, 2002; Sambuichi and Haridasan, 2007).

4 Fourth, simultaneous to the competitive assignment of stand density (basal area) between cocoa and the shade canopy, farmers must competitively allocate total

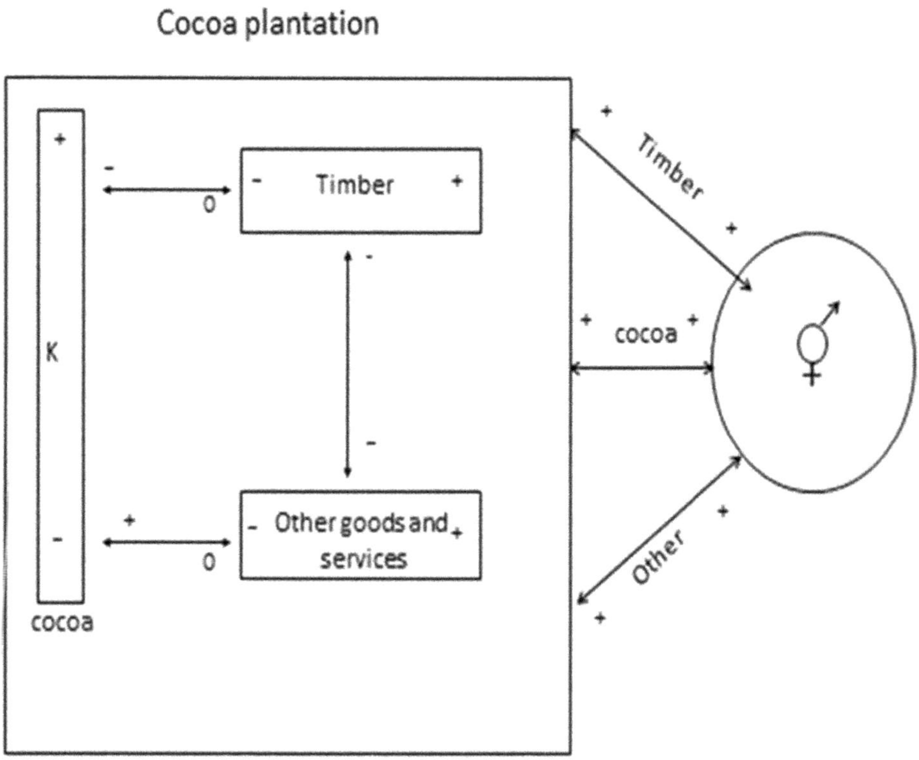

Figure 10 Analysis of trade-offs based on the competitive allocation of basal area between cocoa and the shade canopy.

basal area in the shade canopy plants between groups of useful species (timber, fruit and other goods and services expected from the shade canopy). For instance, if a farmer is interested in increasing timber production, s(he) must allocate more basal area to timber species in the shade canopy, but with a concomitant reduction in basal area allocated to species valued for their production of fruit, amelioration of soil fertility or any other desirable product/service. Farmers may also need to adjust the basal area of cacao (or prune it more heavily) to reduce self-shading and to be able to absorb the increase in shade from the canopy.

5 Fifth, we need to assess (qualitatively or quantitatively) the relationships between the components of the system. For instance, in Fig. 10, all arrows may be represented by a boundary condition, a functional relationship expressed as an equation or no effect (zero). To give this a formal treatment, let *P* stand for the entire cocoa plantation (cocoa + shade canopy), *K* stand for cacao, *D* for canopy, *T* for timber, *Q* for 'other goods and services' and *G* for basal area (biomass, carbon or shade). Then, the following conditions may hold:

$$P = K + D$$

There is a maximum *G* for each *P* depending upon local conditions (for instance, cloudiness), type of cocoa production system and AF management. Expert knowledge suggests that in most cases:

$$P = (K + D) \leq 40\ m^2\ ha^{-1}$$

G is competitively allocated between *K* and *D* (for instance, if *K* increases, self-shading increases, leading to reductions in the need for shading from the canopy trees, that is, *D* has to decrease)

$$D = T + Q$$

$$Yield\,of\,cacao(YK) = f1(K) + f2(T) + f3(Q)$$

$$Yield\,of\,timber(YT) = f4(T) + f5(Q)$$

$$Yield\ of\,other\,goods\,and\,services(YQ) = 1 - f6(T)$$

6 With *f1–f6* expressing yet undetermined functional relationships between components, the optimal design is determined using a set of differential equations and numerical optimization procedures. This has not yet been attempted in cocoa AF, and is a promising research field.

Various management options can modify the competitive effects between cacao and shade canopy plants, and between the production of timber and other goods and services in the shade canopy. For instance, pruning lower branches increases the length of the bole of the tree with clean wood (free of knots fetching higher prices), and improves the shading conditions in the plantation; the crown of a tall tree casts a faster and lighter shade than the crown of a short tree as shown before. Some morphological and functional traits can be used also to reduce competitive relationships between cocoa and shade trees. For instance, trees with small leaves will favour the transmission of light in the plantation; trees with reversed phenology could reduce shading during rainy, cloudy periods; trees with deep root systems would reduce competition for nutrients and water with cocoa; tall trees with clean boles can accumulate a lot of carbon while casting light shade on the cacao, and so forth.

Some crop and tree management practices must also be adjusted to fulfil the needs of cocoa, timber and other goods. For instance, timber trees shall not be pruned or thinned considering only the light regime needed by the cocoa, but also to ensure that: 1) high-quality timber is produced, 2) only the best formed and fastest growing trees are kept for future harvest (a systematic thinning, that would be appropriate to distribute canopy cover homogeneously over the cocoa plot, may not be a good thinning strategy from a timber production standpoint since it may eliminate good trees and leave out poorly formed, slow-growing ones); and 3) a significant number of timber or fruit trees are retained in the plot to yield significant returns, which may result in some over shading and suboptimal cocoa yields.

In the analysis of shade canopy design using basal area allocation as proxy for tree biomass, carbon stock or shade level, basal area is considered a good predictor of yield, which may not always be the case. For instance, two cocoa plantations may have the same canopy basal area or carbon level, but in Plantation 1 basal area is concentrated in a few, big, tall trees, and in Plantation 2 distributed in many, small, short trees. Plantations 1 and 2 will have significant differences in the pattern of transmission of solar radiation inside the cocoa plantation and this may result in significant differences in cocoa yield. In yet another example, two plantations with the same canopy carbon level may have radically different botanical compositions, resulting again in very different patterns of transmission of solar radiation into the plantation and possibly in cocoa yields. Canopy species have an array of morphological (e.g. leaf size and morphology, root systems, etc.) and functional attributes (i.e. leaf fall patterns) that drastically influence the transmission of solar radiation into the cocoa plantation, thereby affecting the growth and yield of cocoa, and altering the microclimate that is a major determinant of the life cycles of pest and diseases that reduce yields.

The manipulation of the morphological and functional traits of canopy species has been used to optimize shade canopy design in coffee (Linkimer et al., 2002; Bellow and Muschler, 1999) and could be used in cocoa to maintain high yields and carbon stocks. To stock high levels of carbon in the plot without adversely affecting cocoa yields, one could select tree species with:

1 Tall, cylindrical and thick stems such as *Terminalia ivorensis* A. Chev. in West African cocoa plantations. Large trees store most of forest above-ground biomass (Lindner, 2010)
2 Small crowns and light foliage (i.e. *Albizia* spp.)
3 Large, deep, thick roots (Nair et al., 2009)
4 Rapid growth and high-density timber; and
5 Inverted phenology which would be of special interest (i.e. *Faidherbia albida* (Delile) A. Chev.) in Africa, which loses foliage during the rainy season and keeps it during the dry season. This inverted phenology behaviour has been observed in *Dalbergia glomerata* Hemsley (a highly valuable timber) used as shade over cacao in Honduras (A. Dubón, FHIA, Honduras, pers. comm., 2011).

Research has shown that it is possible to design cocoa-based AF systems that provide both good yields (from both cocoa and valuable products from the shade canopy) and high carbon stocks (i.e. Clough et al., 2011; Wade et al., 2010; Somarriba et al., 2013).

4.3 Simulation of shade patterns and light transmission models

The presence of a shade canopy in a cocoa-based AF system places cocoa in a 'disadvantageous' position regarding the capture of light (and nutrients) for photosynthesis. The competitive relationship for light between shade canopy plants and the cocoa in the understorey is usually analysed in terms of: 1) the fractioning of the light captured by all plants (to establish a functional relationship between light intensity and quality, photosynthesis, plant growth, and yield), or 2) the estimation of the amount of shade shed by the canopy plants on the cocoa canopy. Light transmission models usually involve repeated measurements of radiation fluxes, specialized instrumentation, detailed plant growth models and physiological measurements of plant function (Lamanda et al., 2007).

A simpler approach to assess the interaction between shade plants and cocoa trees is to focus on the characteristics of the shade canopy plants (crown form, size, spatial pattern, leaf fall patterns, etc.) that block the transmission of light. The assessment of crown canopy cover (Korhonen et al., 2006) can be done with hemispherical photography (Bellow and Nair, 2003), handheld (spherical or point) optical densitometers (Smith et al., 2008), visual estimation (Somarriba, 2002) or by simulation (www.shademotion.net). The software ShadeMotion facilitates the exploration of the shade pattern resulting from different shade canopy designs (species composition, plant size and growth, crown form, leaf phenology, planting configurations, population density of tree species, etc.) by estimating the number of hours of shade cast on each point in the cocoa plantation. The amount of shade is estimated for any cocoa-based AF system, at any location on earth, in horizontal or tilted planes, with any number of trees, belonging to any number of species, and over variable temporal periods. The software provides both 3-D layouts (profiles) of the trees on the plot and shade contour maps of the plot (Fig. 11).

Finally, it should be noted that shade regulation and canopy design and management is only one of many good husbandry practices required to sustainably produce cocoa and other goods and services from cocoa-based AF systems. In many instances, shade regulation will be one of the first innovations applied when optimizing a cocoa production system (Fig. 12). Shade regulation practices and other crop husbandry practices such as plant protection measures, fertilization and the use of improved genetics will have simultaneous effects on the production of both cacao and tree products from the shade canopy. A hypothetical example from Central American cocoa plantations is depicted in Fig. 12.

5 Future needs: a call for action

AF science/knowledge has much to contribute to sustainable cocoa production worldwide. Farmers and technicians need to acquire and apply the knowledge and skills to manipulate cocoa shade canopies to optimally satisfy their needs. To consolidate cocoa AF science and technology, we need developments in (at least) the following areas:

Figure 11 Architectural 3D layouts (profiles) of shade canopy trees in a cocoa plantation. Shade contour map on the floor (www.shademotion.net).

1 Understanding of key ecological interactions in the cocoa-shade canopy system. For instance: (a) shade design and management and nutrient cycles; b) light–fertility–yield interactions; c) micro-environment and pathogens; water–shade–yields across all shade canopy typologies.
2 Understanding the role of shade canopy trees in farm financial diversification, income stability and financial risks.
3 Understanding the effects of the household life cycle on the adoption of innovations, given the aversions and preferences usually present regarding the use of trees on the farm.
4 Modelling approaches to assess trade-offs and synergies between cocoa and shade canopy design to fulfil farmers' multiple objectives.
5 Understanding the relationships between shade trees, cocoa and water balance in the soil. Climate change will alter the availability of water (reduced and more erratic rainfall) in the most important cocoa-producing regions of the world.
6 Understanding the legal and policy framework that encourages or discourages households from fully realizing the benefits derived from cocoa AF. For instance, we should evaluate (a) the forestry and environmental legislation that deters tree planting and utilization on farms; (b) sanitary barriers – for fruits, especially; (c)

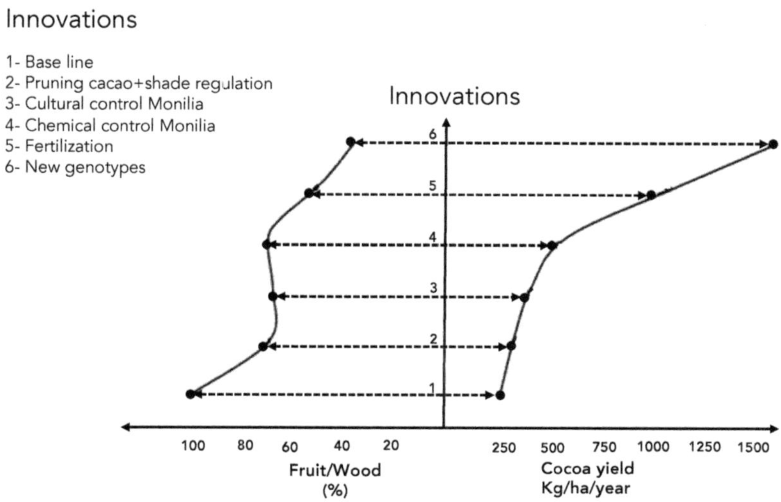

Figure 12 Technical innovations and their potential impacts on cacao productivity and fruit/timber yield in Central America. Proposed innovations are as follows: 1) Baseline scenario (BL); 2) BL + yearly pruning of cocoa (P) + and shade regulation (SR); 3) BL + P + SR + Cultural control of Monilia (CM); 4) BL + SR + CM + Chemical control of Monilia (QM); 5) BL + SR + CM + QM + Application of Fertilizer (AF); and 6) BL + SR + CM + QM + AF + Improved planting material (IM).

limitations on germplasm rights for useful native fruit and timber species (and facilitate exchange between cocoa-growing regions); and (d) the impact of global trends, conventions, trade agreements, technology bottlenecks, climate change, and so forth.

7 Understanding the role of cocoa plantations in landscape biological connectivity, conservation of biodiversity and provision of ecosystem services at the landscape level.

6 Where to look for further information

Several sources of information are available on cocoa-agroforestry. A non-exhaustive list of key sources of information is presented in Table 3.

7 Acknowledgements

This research was funded by the CGIAR research program on Forests, Trees and Agroforestry (FTA), CATIE, The University of Melbourne and James Cook University, Australia.

Table 3 Selected sources of information on cocoa-based agroforestry

Country	Institution	Information, web links
Venezuela	National Institute of Agricultural Research (INIA)	Technical bulletins http://www.inia.gov.ve
Colombia	National Federation of Cocoa Farmers (FEDECACAO)	Technical bulletins and farmers training materials http://www.fedecacao.com.co
	CIAT (International Center for Tropical Agriculture).	Technical bulletins, research papers and policy brief. All the publications, tools, and data-sets are also available on CGspace, CGIAR's repository of Agricultural Research Outputs. https://cgspace.cgiar.org
	Colombian Corporation of Agricultural Research (CORPOICA).	Technical bulletins and farmers training materials. http://www.corpoica.org.co
Ecuador	National Association of Cacao Farmers and Exporters (ANECACAO)	Genetic improvement program, technical bulletins, training materials and Anecacao technical magazine http://www.anecacao.com
	National Institute of Agricultural Research (INIAP)	Genetic improvement program, technical bulletins and training materials. Digital repository http://www.iniap.gob.ec
Peru	National Institute of Agricultural Innovation (INIA)	Genetic improvement program, technical bulletins and training materials. Digital repository. http://www.inia.gob.pe
Bolivia	Research Institute of Organic Agriculture (FiBL)	Peer-reviewed articles, technical bulletins, training materials, book chapters, policy briefs http://www.systems-comparison.fibl.org
Brazil	The Brazilian Agricultural Research Corporation (EMBRAPA)	Peer-reviewed articles, technical bulletins, training materials, book chapters, policy briefs. https://www.embrapa.br
	Executive Commission for the Development of Cacao Cultivation (CEPLAC)	Specialized *Journal do Cacao*, *Agrotropica*, technical bulletins, peer-reviewed papers and farmers training materials. http://www.ceplac.gov.br
Honduras	The Honduran Agricultural Research Foundation (FHIA)	Articles, technical bulletins, training materials, book chapters, policy briefs. http://www.fhia.org.hn/htdocs/cacao_y_agroforesteria.html

Costa Rica	National Chambers Association of fine cacao of Costa Rica (CANACACAO)	Training materials, governance, policy briefs. juanpablo@canacacao.org
Mexico	Thematic Network in agroforestry system of Mexico (RED SAM)	Peer-reviewed articles, technical bulletins, training materials, book chapters, policy brief. http://www.red-sam.enesmorelia.unam.mx/
Dominican Republic	Centre for Agroforestry Development of the Dominican Republic (CEDAF)	Technical bulletins, training materials, policy briefs. http://cedaf.org.do
Cuba	Institute for Agroforestry Research of Cuba (INAF)	Articles, technical bulletins, training materials, book chapters, policy briefs. http://www.inaf.co.cu/
Trinidad and Tobago	Cocoa Research Centre	Peer-reviewed articles, technical bulletins, training materials, book chapters, policy briefs. http://sta.uwi.edu/cru/index.asp
Regional	CATIE (Tropical Agricultural Research and Higher Education Center)	Peer-reviewed articles, technical bulletins, training materials, book chapters, policy briefs. www.catie.ac.cr
Regional	Mesoamerican Scientific Partnership Platform for Agroforestry Systems with Perennial Crops (PCP)	Peer-reviewed articles, technical bulletins, training materials, book chapters, policy briefs. jacques.avelino@cirad.fr ; rcerda@catie.ac.cr
Regional	CIRAD, UMR System	Peer-reviewed articles, technical bulletins, training materials, book chapters, policy briefs. https://umr-system.cirad.fr/
Global	INAFORESTA	Peer-reviewed articles, technical bulletins, training materials, book chapters, policy briefs. http://orton.catie.ac.cr/inaforesta
Global	The Alliance of Cocoa Producing Countries (COPAL)	Peer-reviewed articles, technical bulletins, training materials, book chapters, policy briefs. http://www.copal-cpa.org/
Global	World Cocoa Foundation	Peer-reviewed articles, technical bulletins, training materials, book chapters, policy briefs. www.worldcocoafoundation.org

8 References

Abou Rajab, Y., Leuschner, C., Barus, H., Tjoa, A. and Hertel, D. 2016. Cacao cultivation under diverse shade tree cover allows high carbon storage and sequestration without yield losses. *PLoS ONE*, 11(2): e0149949. doi:10.1371/journal.pone.0149949

Anglaaere, L. C. N. 2005. Improving the sustainability of cocoa farms in Ghana through utilization of native forest trees in agroforestry systems. PhD Dissertation, University of Wales, Bangor.

Anglaaere, L. C. N., Cobbina, J., Sinclair, F. L. and McDonald, M. A. 2011. The effect of land use systems on tree diversity: Farmer preference and species composition of cocoa-based agroecosystems in Ghana. *Agroforest. Syst.*, 81(3): 249–265. doi:10.1007/s10457-010-9366-z

Asare, R. 2005. Cocoa agroforest in West Africa. A look at activities on preferred trees in the farming systems. Forest and Landscape Working paper No. 6. Danish Centre for Forest, Landscape and Planning, Royal Agricultural and Veterinary University (KVL), Denmark, 89p.

Asare, R. 2006. A review on cocoa agroforestry as a means for biodiversity conservation. In: Centre for Forest, L.a.P. (Ed.), *Forest and Landscape*. University of Life Science, Brussels, Denmark, p. 21.

Asare, R., Ofori-Frimpong, K. and Ekpe, P. K. 2009. Impact of cocoa farming on vegetation in an agricultural landscape in Ghana. *Afr. J. Ecol.*, 58, 338–46.

Asare, R., Afari, V., Osei, Y. and Pabi, O. 2014. Cocoa agroforestry for increasing forest connectivity in a fragmented landscape in Ghana. *Agroforest. Syst.*, 88(6), 1143–56. doi:10.1007/s10457-014-9688-3.

Asare, R., Asare, R. A., Asante, W. A., Markussen, B. and Raebild, A. 2017. Influences of shading and fertilization on on-farm yields of cocoa in Ghana. *Exp. Agr.*, 53(3): 416–31. doi:10.1017/S0014479716000466.

Beer, J., Muschler, R., Somarriba, E. and Kass, D. 1998. Shade management in coffee and cocoa plantations. *Agroforest. Syst.*, 38: 139–4.

Bellow, J. G. and Muschler, R. G. 1999. Screening for promising tree associates for coffee in Central America. In: Jiménez, F. and Beer, J. (Compilers), *Multi-strata Agroforestry Systems with Perennial Crops. Proceedings, IUFRO Symposium*. CATIE, Turrialba, Costa Rica, pp. 171–4.

Bellow, J. G. and Nair, P. K. R. 2003. Comparing common methods for assessing understory light availability in shaded-perennial agroforestry systems. *Agr. For. Meteorol.*, 114(3–4): 197–211.

Bentley, J. W., Boa, E. and Stonehouse, J. 2004. Neighbor trees: Shade, intercropping and cacao in Ecuador. *Hum. Ecol.*, 32(2): 241–70.

Bisseleua, D. H., Missoup, A. D. and Vidal, S. 2009. Biodiversity conservation, ecosystem functioning, and economic incentives under cocoa agroforestry intensification. *Conserv. Biol.*, 23: 1176–84.

Blaser, W. J., Oppong, J., Yeboah, E. and Six, J. 2017. Shade trees have limited benefits for soil fertility in cocoa agroforests. *Agr. Ecosyst. Environ.*, 243, 83–91.

Bos, M. M., Steffan-Dewenter, I. and Tscharntke, T. 2007. Shade tree management affects fruit abortion, insect pests and pathogens of cacao. *Agr. Ecosyst. Environ.*, 120(2–4): 201–5. doi:10.1016/j.agee.2006.09.004.

Cerda, R., Deheuvels, O., Calvache, D., Niehaus, L., Saenz, Y., Kent, J., Vilchez, S., Villota, A., Martinez, C. and Somarriba, E. 2014. Contribution of cocoa agroforestry systems to family income and domestic consumption: Looking toward intensification. *Agroforest. Syst.*, 88(6): 957–81. doi:10.1007/s10457-014-9691-8.

Cheatham, M. R., Rouse, M. N., Esker, P. D., Ignacio, S., Pradel, W., Raymundo, R., Sparks, A. H., Forbes, G. A., Gordon, T. R. and Garrett, K. A. 2009. Beyond yield: Plant disease in the context of ecosystem services. *Phytopathology*, 99, 1228–36.

Clough, Y., Barkmann, J., Juhrbandt, J., Kessler, M., Wanger, T. C., Anshary, A., Buchori, D., Cicuzza, D., Darras, K., Putra, D. D., Erasmi, S., Pitopang, R., Schmidt, C., Schulze, C. H., Seidel, D., Steffan-Dewenter, I., Stenchly, K., Vidak, S., Weist, M., Wielgoss, A. C. and Tscharntke, T. 2011. Combining high biodiversity with high yields in tropical agroforests. *Proc. Natl. Acad. Sci. (PNAS)*, 108(20): 8311–16. doi:10.1073/pnas.1016799108.

Dale, V. H. and Polasky, S. 2007. Measures of the effects of agricultural practices on ecosystem services. *Ecol. Econ.*, 64, 286–96.

de Almeida, A. A. and Valle, R. R. 2007. Ecophysiology of the cacao tree. *Braz. J. Plant Physiol.*, 19(6): 425–48.

de Beenhouwer, M., Aerts, R., Honnay, O. 2013. A global meta-analysis of the biodiversity and ecosystem service benefits of coffee and cacao agroforestry. *Agr. Ecosyst. Environ.*, 175: 1–7. doi:10.1016/j.agee.2013.05.003

Deheuvels, O., Avelino, J., Somarriba, E. and Malezieux, E. 2012. Vegetation structure and productivity in cocoa-based agroforestry systems in Talamanca, Costa Rica. *Agr. Ecosyst. Environ.*, 149: 181–8.

Deheuvels, O., Rousseau, G. X., Soto Quiroga, G., Decker, F. M., Cerda, R., Vilchez Mendoza, S. J. and Somarriba, E. 2014. Biodiversity is affected by changes in management intensity of cocoa-based agroforests. *Agroforest. Syst.*, 88: 1081–99. Doi: 10.1007/s10457-014-9710-9.

Frimpong, E. A., Gemmill-Herren, B., Gordon, I. and Kwapong, P. K. 2011. Dynamics of insect pollinators as influenced by cocoa production systems in Ghana. *J. Pollin. Ecol.*, 5(10): 74–80.

Gockowski, J. and Sonwa, D. 2008. Biodiversity and smallholder cocoa production systems in West Africa. Sustainable tree crop program, Working paper series 6 (Version January 2008). International Institute of Tropical Agriculture (IITA), Accra, p. 21.

Gockowski, J. and Sonwa, D. 2011. Cocoa intensification scenarios and their predicted impact on CO_2 emissions, biodiversity conservation, and rural livelihoods in the Guinea rain forest of West Africa. *Environ. Manage.*, 48(2): 307–21.

Gras, P., Tscharntke, T., Maas, B., Tjoa, A., Hafsah, A., Clough, Y. and Manning, P. 2016. How ants, birds and bats affect crop yield along shade gradients in tropical cacao agroforestry. *J. Appl. Ecol.*, 53: 953–63.

Hartemink, A. E. 2005. Nutrient stocks, nutrient cycling, and soil changes in cocoa ecosystems: A review. *Adv. Agron.*, 86: 227–53. doi:10.1016/s0065-2113(05)86005-5.

Harvey, C. A., Chacón, M., Donatti, C. I., Garen, E., Hannah, L., Andrade, A., Bede, L., Brown, D., Calle, A., Chará, J., Clement, C., Gray, E., Hoang, M. H., Minang, P., Rodríguez, A. M., Seeberg-Elverfeldt, C., Semroc, B., Shames, S., Smukler, S., Somarriba, E., Torquebiau, E., van Etten, J. and Wollenberg E. 2014. Climate-smart landscapes: Opportunities and challenges for integrating adaptation and mitigation in tropical agriculture. *Conserv. Lett.*, 7(2): 77–90.

Jacobi, J., Andres, C., Schneider, M. and Pillco, M. 2014. Carbon stocks, tree diversity, and the role of organic certification in different cocoa production systems in Alto Beni, Bolivia. *Agroforest. Syst.*, 88: 1117–32. DOI 10.1007/s10457-013-9643-8.

Jadán, O., Cifuentes, M., Torres, B., Selesi, D., Veintimilla, D. and Günter, S. 2015. Influence of tree cover on diversity, carbon sequestration and productivity of cocoa systems in the Ecuadorian Amazon. *Bois et forêts des tropiques* 325(3): 35–47.

Jagoret, P., Michel-Dounias, I., Snoeck, D., Ngnogue, H. T. and Malezieux, E. 2012. Afforestation of savannah with cocoa agroforestry systems: A small-farmer innovation in central Cameroon *Agroforest. Syst.*, 86 (3): 493–504.

Jagoret, P., Snoeck, D., Bouambi, E., Ngongue, H. T., Nyasse, S. and Saj, S. 2017. Rehabilitation practices that shape cocoa agroforestry systems in Central Cameroon: Key management strategies for long-term exploitation. *Agroforest. Syst.* doi: 10.1007/s10457-016-0055-4.

Jaimez, R. E., Araque, O., Guzman, D., Mora, A., Espinoza, W. and Tezara, W. 2013. Agroforestry systems of timber species and cacao: Survival and growth during the early stages. *J. Agr. Rural Dev. Trop.*, 114(1): 1–11.

Kessler, M., Hertel, D., Jungkunst, H. F., Kluge, J., Abrahamczyk, S., Bos, M., Buchori, D., Gerold, G., Gradstein, S. R., Kohler, S., Leuschner, C., Moser, G., Pitopang, R., Saleh, S., Schulze, C. H., Sporn, S. G., Steffan-Dewenter, I., Tjitrosoedirdjo, S. S. and Tscharntke, T. 2012. Can joint carbon and biodiversity management in tropical agroforestry landscapes be optimized? *PLoS One*, 7: e47192.

Koko, L. K., Snoeck, D., Lekadou, T. T. and Assiri, A. A. 2013. Cacao-fruit tree intercropping effects on cocoa yield, plant vigour and light interception in Côte d'Ivoire. *Agroforest. Syst.*, 87(5): 1043–52. doi:10.1007/s10457-013-9619-8

Korhonen, L., Korhonen, K. T., Rautiainen, M. and Stenberg, P. 2006. Estimation of forest canopy cover: A comparison of field measurement techniques. *Silva Fenn.*, 40(4): 577–88.

Läderach, P., Martinez-Valle, A., Schroth, G. and Castro, N. 2013. Predicting the future climatic suitability for cocoa farming of the world's leading producer countries, Ghana and Côte d'Ivoire. *Climat. Change*, 119(3–4): 841–54. doi:10.1007/s10584-013-0774-8

Lamanda, N., Dauzat, J., Jourdan, C., Martin, P. and Malézieux, E. 2007. Using 3D architectural models to assess light availability and root bulkiness in coconut agroforestry systems. *Agroforest. Syst.*, 72(1): 63–74. doi:10.1007/s10457-007-9068-3.

Leuschner, C., Moser, G., Hertel, D., Erasmi, S., Leitner, D., Culmsee, H., Schuldt, B. and Schwendenmann, L. 2013. Conversion of tropical moist forest into cacao agroforest: Consequences for carbon pools and annual C sequestration. *Agroforest. Syst.*, 87: 1173–87. doi 10.1007/s10457–013-9628-7.

Lindner, A. 2010. Biomass storage and stand structure in a conservation unit in the Atlantic Rainforest–The role of big trees. *Ecol. Eng.*, 36: 1769–73.

Linkimer, M., Muschler, R., Benjamin, T. and Harvey, C. 2002. Árboles nativos para diversificar cafetales en la zona Atlántica de Costa Rica. *Agroforest. Ame.*, 9(35–36): 37–43.

Moço, M. K. S., Gama-Rodrigues, E. F., Gama-Rodrigues, A. C., Machado, R. C. R. and Baligar, V. C. 2010. Relationships between invertebrate communities, litter quality and soil attributes under different cacao agroforestry systems in the south of Bahia, Brazil. *App. Soil Ecol.*, 46(3): 347–54. doi:10.1016/j.apsoil.2010.10.006

Mortimer, R., Saj, S. and David, C. 2017. Supporting and regulating ecosystem services in cacao agroforestry systems. *Agroforest. Syst.* doi:doi.org/10.1007/s10457-017-0113-6.

Moser, G., Leuschner, C., Hertel, D., Hölscher, D., Köhler, M., Leitner, D., Michalzik, B., Prihastanti, E., Tjitrosemito, S. and Schwendenmann, L. 2010. Response of cocoa trees (*Theobroma cacao*) to a 13-month desiccation period in Sulawesi, Indonesia. *Agroforest. Syst.*, 79: 171–87.

Müller, A., Sukhdev, P., Miller, D., Sharma, K. and Hussain, S. 2015. *TEEB for Agriculture & Food. Towards a Global Study on the Economics of Eco-Agri-Food Systems.* UNEP, Geneva, 19p.

Nair, P. K. R., Kumar, B. M. and Nair, V. D. 2009. Agroforestry as a strategy for carbon sequestration. *J. Plant Nutr. Soil Sc.*, 172: 10–23.

Niether, W., Schneidewind, U., Armengot, L., Adamtey, N., Schneider, M. and Gerold, G. 2017. Spatial-temporal soil moisture dynamics under different cocoa production systems. *Catena*, 158: 340–9.

Nomo, B. 2005. Farmers' ecological and agronomic knowledge about the management of multistrata cocoa systems in Southern Cameroon. Doctoral dissertation, PhD thesis, School of Agricultural and Forest Sciences, University of Wales, Bangor, Wales, UK.

Norgrove, L. and Hauser, S. 2013. Carbon stocks in shaded *Theobroma cacao* farms and adjacent secondary forests of similar age in Cameroon. *Trop. Ecol.*, 54(1): 15–22.

Ofori-Bah, A. and Asafu-Adjaye, J. 2011. Scope economies and technical efficiency of cocoa agroforestry systems in Ghana. *Ecol. Econ.*, 70: 1508–18.

Oke, D. O. and Odebiyi, K. A., 2007. Traditional cocoa-based agroforestry and forest species conservation in Ondo State, Nigeria. *Agr. Ecosyst. Environ.*, 122(3): 305–11.

Oke, D. and Olatiilu, A. 2011. Carbon storage in agroecosystems: A case study of the cocoa based agroforestry in Ogbese Forest Reserve, Ekiti State, Nigeria. *J. Environ. Prot.*, 2: 1069–75. doi:10.4236/jep.2011.28123.

Ortíz-González, M. 2006. Conocimiento local y decisiones de los productores de Alto Beni, Bolivia, sobre el diseño y manejo de la sombra en sus cacaotales. Tesis MSc., CATIE, Turrialba, Costa Rica, 76p.

Quesada, F., Somarriba, E. and Vargas, E. 1987. Modelo para la simulación de patrones de sombra de árboles. Serie Técnica, Informe Técnico 118, CATIE, Turrialba, Costa Rica, 91p.

Rapidel, B., Ripoche, A., Allinne, C., Metay, A., Deheuvels, O., Lamanda, N., Blazy, J.-M., Valdés-Gómez, H. and Gary, C. 2015. Analysis of ecosystem services trade-offs to design agroecosystems with perennial crops. *Agron. Sustain. Dev.*, 35, 1373–90.
Rice, R. and Greenberg, R. 2000. Cacao cultivation and the conservation of biological diversity. *Ambio*, 29(3): 167–73.
Rolim, G. S. and Chiarello, A. G. 2004. Slow death of Atlantic forest trees in cocoa agroforestry in southeastern Brazil. *Biodivers. Conserv.*, 13: 2679–94.
Rousseau, G. X., Deheuvels, O., Rodriguez Arias, I. and Somarriba, E. 2012. Indicating soil quality in cacao-based agroforestry systems and old-growth forests: The potential of soil macrofauna assemblage. *Ecol. Indic.*, 23: 535–43. doi:10.1016/j.ecolind.2012.05.008
Ruf, F. O.. 2011. The myth of the complex cocoa agroforest: The case of Ghana. *Human Ecol.* 39(3): 373–88.
Ryan, D., Bright, G. and Somarriba, E. 2009. Damage and yield change in cocoa crops due to the harvesting of timber shade trees in Talamanca, Costa Rica. *Agroforest. Syst.*, 77(2): 97–106.
Saj, S., Jagoret, P. and Todem Ngogue, H. 2013. Carbon storage and density dynamics of associated trees in three contrasting *Theobroma cacao* agroforests of Central Cameroon. *Agroforest. Syst.*, 87: 1309–20. DOI 10.1007/s10457–013-9639-4.
Saj, S., Durot, C., Mvondo Sajouma, K., Tayo Gamo, T. and Avana-Tientcheu, M. L., 2017. Contribution of associated trees to long-term species conservation, carbon storage and sustainability: A functional analysis of tree communities in cacao plantations of Central Cameroon. *Int. J. Agr. Sustain.*, 15(3): 282–302. http://dx.doi.org/10.1080/14735903.2017.1311764.
Salgado-Mora, M. G., Ibarra-Nunez, G., Macias-Samano, J. E., Lopez-Baez, O. 2007. [Diversity of shade trees in cacao plantations of Soconusco, Chiapas, Mexico]. *Interciencia* 32, 763–8.
Sambuichi, R. H. R. 2002. Fitossociologia e diversidade de especies arbóreas em cabruca (mata atlántica raleada sobre plantacao de cacau) na regiao sul da Bahia, Brasil. *Acta Bot. Bras.*, 16(1): 89–101.
Sambuichi, R. and Haridasan, M. 2007. Recovery of species richness and conservation of native Atlantic forest trees in the cocoa plantations of southern Bahia in Brazil. *Biodivers. Conserv.*, 36: 3681–701.
Sambuichi R. H. R., Vidal, D. B., Piasentin, F. B., Jardim, J. G., Viana, T. G., Menezes, A. A., Mello, D. L. N., Ahnert, D. and Baligar, V. C. 2012. Cabruca agroforests in southern Bahia, Brazil: tree component, management practices and tree species conservation. *Biodivers. Conserv.*, 21(4): 1055–77.
Schneider, M., Andres, C., Trujillo, G., Alcon, F., Amurrio, P., Perez, E., Weibel, F. and Milz, J. 2017. Cocoa and total system yields of organic and conventional agroforestry vs. monoculture systems in a long-term field trial in Bolivia. *Exp. Agr.*, 53(3): 351-74. doi:10.1017/S0014479716000417.
Schroth, G. and da Mota, d. S. S. 2014. Agroforestry: Complex multistrata agriculture. *Ency. Agr Food Syst.*, 1: 195–207. doi:10.1016/b978-0-444-52512-3.00030-9.
Schroth, G., Harvey, C. A. and Vincent, G. 2014. Complex agroforests: Their structure, diversity, and potential role in landscape conservation. In: Schroth, G., da Fonseca, G. A. B., Harvey, C. A., Gascon, C., Vasconcelos, H. L. and Izac, A. M. N. (Eds), *Agroforestry and Biodiversity Conservation in Tropical Landscapes*. Island Press, Washington, USA, pp. 227–60.
Schroth, G., Jeusset, A., da Silva Gomez, A., Florence, C. T., Pinto Coelho, N. A., Faria, D. and Laderach, P. 2016. Climate friendliness of cocoa agroforests is compatible with productivity increase. *Mitig. Adapt. Glob. Change*, 21: 67–80.
Schwendenmann, L., Veldkamp, E., Moser, G., Hölscher, D., Köhler, M., Clough, Y., Anas, I., Djajakirana, G., Erasmi, S., Hertel, D., Leitner, D., Leuschner, C., Michalzik, B., Propastin, P., Tjoa, A., Tscharntke, T. and van Straaten, O. 2010. Effects of an experimental drought on the functioning of a cacao agroforestry system, Sulawesi, Indonesia. *Glob. Change Biol.*, 16(5): 1515–30. doi:10.1111/j.1365-2486.2009.02034.x
Siebert, S. F. 2002. From shade- to sun-grown perennial crops in Sulawesi, Indonesia: Implications for biodiversity conservation and soil fertility. *Biodivers. Conserv.*, 11, 1889–902.

Silva, C., Orozco, L., Rayment, M. and Somarriba, E. 2013. Conocimiento local sobre los atributos deseables de los árboles y el manejo del dosel de sombra en los cacaotales de Waslala, Nicaragua. *Agroforest. Ame.*, 49:51–60.
Smiley, G. L. and Kroschel, J. 2008. Temporal change in carbon stocks of cocoa–gliricidia agroforests in Central Sulawesi, Indonesia. *Agroforest. Syst.*, 73(3): 219–31.
Smiley, G. L. and Kroschel, J. 2009. Yield development and nutrient dynamics in cocoa-gliricidia agroforests of Central Sulawesi, Indonesia. *Agroforest. Syst.*, 78(2): 97–114. doi:10.1007/s10457-009-9259-1
Smith, M. L., Anderson, J. and Fladeland, M. 2008. Chapter 14: Forest canopy structural properties. In: Hoover, C. M. (Ed.), *Field Measurements for Forest Carbon Monitoring*. Springer, New York, USA, pp. 179–96.
Somarriba, E. 2002. Estimación visual de la sombra en cacaotales y cafetales. *Agroforest. Ame.*, 9(35–36): 86–94.
Somarriba, E. 2007. Cocoa and shade trees: Production, diversification and environmental services. *Glob. Res. Cocoa*, 11: 1–4.
Somarriba, E. and Beer, J. 2011. Productivity of *Theobroma cacao* agroforestry systems with timber or legume service shade trees. *Agroforest. Syst.*, 81(2), 109-21. doi:10.1007/s10457-010-9364-1
Somarriba, E. and Lachenaud, Ph. 2013. Successional cocoa agroforests of the Amazon – Orinoco – Guiana shield. *For. Trees Livel.*, 22(1): 51–9.
Somarriba, E., Valdivieso, R., Vásquez, W. and Galloway, G. 2001. Survival, growth, timber productivity and site index of *Cordia alliodora* in forestry and agroforestry systems. *Agroforest. Syst.*, 51: 111–18.
Somarriba, E., Quesada, F., Orozco, L., Cerda, R., Villalobos, M., Orozco, S., Astorga, C., Deheuvels, O., Say, E. and Villegas, R. 2011. The shade canopy of cocoa. Serie Técnica, Materiales de Extensión 5, CATIE, Turrialba, Costa Rica, 45p.
Somarriba, E., Beer, J., Alegre Orihuela, J., Andrade, H., Cerda, R., DeClerck, F., Detlefsen, G., Escalante, M., Giraldo, L. A., Ibrahim, M., Krishnamurthy, I., Mena, V. E., Mora-Delgado, J., Orozco, L., Scheelje, M. and Campos, J. J. 2012. Mainstreaming agroforestry in Latin America. In: Nair, P. K. R. and Garrity, D. P. (Eds), *Agroforestry: The Way Forward*. Advances in Agroforestry, 9. Springer, USA, pp. 429–53.
Somarriba, E., Cerda, R., Orozco, L., Deheuvels, O., Cifuentes, M., Dávila, H., Espin, T., Mavisoy, H., Ávila, G., Alvarado, E., Poveda, V., Astorga, C. and Say, E. 2013. Carbon stocks and cocoa yields in agroforestry systems of Central America. *Agr. Ecosyst. Environ.*, 173: 46–57.
Somarriba, E., Suárez-Islas, A., Calero-Borge, W., Villota, A., Castillo, C., Vílchez, S., Deheuvels, O. and Cerda, R. 2014. Cocoa – timber agroforestry systems: *Theobroma cacao - Cordia alliodora* in Central America. *Agroforest. Syst.*, 88: 1001–19. DOI 10.1007/s10457-014-9692-7.
Sonwa, D. J., Nkongmeneck, B. A., Weise, S. F., Tchatat, M., Adesina, A. A. and Janssens, M. J. J. 2007. Diversity of plants in cocoa agroforests in the humid forest zone of Cameroon. *Biodivers. Conserv.*, 16:2385–400.
Sonwa, D. J., Weise, S. F., Schroth, G., Janssens, M. J. J. and Howard-Yana, S. 2014. Plant diversity management in cocoa agroforestry systems in West and Central Africa – Effects of markets and household needs. *Agroforest. Syst.*, 88(6): 1021–34. doi:10.1007/s10457-014-9714-5
Steffan-Dewenter, I., Kessler, M., Barkmann, J., Bos, M., Buchori, D., Erasmi, S., Faust, H., Gerold, G., Glenk, K., Gradstein, S. R., Guhardja, E., Harteveld, M., Hertel, D., Höhn, P., Kappas, M., Köhler, S., Leuschner, C., Maertens, M., Marggraf, R., Migge-Kleian, S., Mogea, J., Pitopang, I. R., Schaefer, M., Schwarze, S., Sporn, S. G., Steingrebe, A., Tjitrosoedirdjo, S. S., Tjitrosoemito, S., Twele, A., Weber, R., Woltmann, L., Zeller, M. N. and Tscharntke, T. 2007. Tradeoffs between income, biodiversity, and ecosystem functioning during tropical rainforest conversion and agroforestry intensification. *PNAS*, 104(12): 4973–8.
Stenchly, K., Clough, Y. and Tscharntke, T. 2012. Spider species richness in cocoa agroforestry systems, comparing vertical strata, local management and distance to forest. *Agr. Ecosyst. Environ.*, 149, 189–94. doi:10.1016/j.agee.2011.03.021

Toledo-Hernández, M., Wanger, T. C. and Tscharntke T. 2017. Neglected pollinators: Can enhanced pollination services improve cocoa yields? A review. *Agr. Ecosyst. Environ.*, 247, 137–48.

Tondoh, J. E., Kouamé, F. N.g., Martinez Guéi, A., Sey, B., Wowo Koné, A. and Gnessougou, N. 2015. Ecological changes induced by full-sun cocoa farming in Côte d'Ivoire. *Glob. Ecol. Conserv.*, 3: 575–95.

Torquebiau, E. 2000. A renewed perspective on agroforestry concepts and classification. *Life Sci.*, 323: 1009–17.

Tscharntke, T., Clough, Y., Bhagwat, S. A., Buchori, D., Faust, H., Hertel, D., Ischer, D. H., Juhrbandt, J., Kessler, M., Perfecto, I., Scherber, C., Schroth, G., Veldkamp, E. and Wanger, T. C. 2011. Multifunctional shade-tree management in tropical agroforestry landscapes. *J. Appl. Ecol.*, 48(3): 619–29.

Utomo, B., Prawoto, A. A., Bonnet, S., Bangviwat, A. and Gheewala, S. H. 2016. Environmental performance of cocoa production from monoculture and agroforestry systems in Indonesia. *J. Clean. Prod.*, 134, 583–91. doi:10.1016/j.jclepro.2015.08.102

Vaast, P. and Somarriba, E. 2014. Trade-offs between crop intensification and ecosystem services: The role of agroforestry in cocoa cultivation. *Agroforest. Syst.*, 88(6): 947–56. doi:10.1007/s10457-014-9762-x

Van Noordwijk, M., Coe, R. and Sinclair, F. L., 2016. Central hypotheses for the third agroforestry paradigm within common definition. Working Paper No 233. World Agroforestry Center (ICRAF) Southeast Asia Regional Program, Bogor, Indonesia.

Vebrova, H., Lojka, B., Husband, T. P., Zans, M. E. C., Van Damme, P., Rollo, A. and Kalousova, M. 2013. Tree diversity in cacao agroforests in San Alejandro, Peruvian Amazon. *Agroforest. Syst.*, 88: 1101–15.

Vernon, A. J. and Morris, M. G. 1964. Variability in West African peasant cocoa farms: Time studies on Agodi cocoa fields. *J. Hortic. Sci.*, 39: 241–53.

Wade, A. S. I., Asase, A., Hadley, P., Mason, J., Ofori-Frimpong, K., Preece, D., Spring, N. and Norris, K. 2010. Management strategies for maximizing carbon storage and tree species diversity in cocoa-growing landscapes. *Agr. Ecosyst. Environ.*, 138(3–4): 324–34. doi:10.1016/j.agee.2010.06.007

Waldron, A., Justicia, R., Smith, L. and Sanchez, M., 2012. Conservation through chocolate: A win-win for biodiversity and farmers in Ecuador's lowland tropics. *Conserv. Lett.*, 1: 1–9.

Wu, J., Liu, W., Chen, C. and Bennett, J. 2016. Can intercropping with the world's three major beverage plants help improve the water use of rubber trees? *J. Appl. Ecol.*, 53(6): 1787–99. doi:10.1111/1365-2664.12730

Zapfack, L., Engwald, S., Sonke, B., Achoundong, G. and Birang, A. M. 2002. The impact of land conversion on plant biodiversity in the forest zone of Cameroon. *Biodivers. Conserv.* 2: 2047–61.

Zhang, W., Ricketts, T. H., Kremen, C., Carney, K. and Swinton, S. M. 2007. Ecosystem services and dis-services to agriculture. *Ecol. Econ.*, 64: 253–60.

Zomer, R. J., Trabucco, A., Coe, R. and Place, F. 2009. Trees on farm: Analysis of global extent and geographical patterns of agroforestry. Working Paper No. 89. ICRAF, World Agroforestry Centre, Nairobi, Kenya, 72p.

Zuidema, P. A., Leffelaar, P. A., Gerritsma, W., Mommer, L. and Anten, N. P. R. 2005. A physiological production model for cocoa (*Theobroma cacao*): Model presentation, validation and application. *Agr. Syst.*, 84(2): 195–225.

Chapter 21

Organic cocoa cultivation

Amanda Berlan, De Montfort University, UK

1 Introduction

In the last two decades, there has been an increase in demand for organically produced goods and more support for organic agriculture, which the Codex Alimentarius[1] defines as:

> a holistic production management system which promotes and enhances agro-ecosystem health, including biodiversity, biological cycles, and soil biological activity. It emphasizes the use of management practices in preference to the use of off-farm inputs, taking into account that regional conditions require locally adapted systems. This is accomplished by using, where possible, agronomic, biological, and mechanical methods, as opposed to using synthetic materials, to fulfill any specific function within the system (FAO/WHO Codex Alimentarius Commission 1999).[2]

Research into organic production has shown that it is able to outperform conventional farming on many sustainability metrics, such as soil fertility, water infiltration and retention capacity, nitrogen uptake by crops, energy use and efficiency and protecting biodiversity (Ponisio et al. 2015).

1. The Codex Alimentarius is a widely implemented set of global standards and guidelines established by the FAO/WHO Food Standards Programme to protect consumer health and promote fair practices in food trade.
2. http://www.fao.org/docrep/meeting/X0075e.htm

http://dx.doi.org/10.19103/AS.2017.0021.30

The global market for organic food and drink products has enjoyed a steady growth, with retail sales increasing from US$15.2 billion in 1999 to US$80 billion in 2014 (Sahota 2016). North America and Europe are the biggest markets, making up over 90% of organic food and drink sales (Sahota 2016). As a crop, cocoa can be grown successfully using organic methods, and demand for organic chocolate has risen in line with the overall growth in the organic market. Byers et al. state that 'global sales of organic chocolate alone (not including other cocoa products) [reached] US$304 million in 2005, a 75% rise over its level of 2002' (2008: 51). ICCO (2018), citing *Euromonitor International*, reports that global organic chocolate sales grew from an estimated US$171 million in 2002 to US$304 million in 2005.[3] Mintel/Food Navigator.com states that the demand for organic chocolate-based confectionery launches in Europe increased from 7% in 2010 to 11% in 2015.[4] These figures show that there is a growing market for organic chocolate, and therefore organically grown cocoa.

2 Production trends in organic cocoa

This section reviews what is known about organic production trends. However, there is a lack of data and conflicting evidence on this subject which the International Cocoa Organisation (ICCO) summarises as follows:

> It is very difficult to find reliable statistical data for organic markets. The market is made up of a small number of traders who are reluctant to divulge price and volume information. Data collection by NGOs and other organizations is less than optimal and official statistics do not distinguish conventional from organic certified cocoa, except in the Dominican Republic (whose data is incomplete). The competent national authorities in Europe (e.g. DEFRA in the United Kingdom, BDE in Germany and DPEI in France) do not collect data on imported volumes of organic cocoa but only data on expected volumes corresponding to the import permits issued (ICCO 2006).

The difficulty of finding reliable trade data on organic cocoa is echoed in other sources (Pay 2009; Byers et al. 2008). Byers et al. state: 'The data are incomplete and very fragmented. Worse, the reliability of the few available estimates may be questioned' (2008: 51). The absence of official statistics on cocoa categorised as 'certified' (which includes organic certification, Fairtrade, etc.) is due to a combination of factors such as the small volumes produced/certified/sold, the many different types of cocoa products (such as beans, liquor, powder, butter, paste, chocolate) and the fact that stocks can be held, thus creating a disparity between volumes produced and traded.

According to the ICCO (2018), the volume of organically certified cocoa produced is estimated to be around 15 500 MT, which is in the region of 0.5% of total cocoa production. However, this figure does not seem to have been reviewed recently and has been questioned for some time (Berlan and Bergés 2013). A variety of countries produce cocoa which is certified as organic. According to ICCO (2018), these include leading producers such as the Dominican Republic, Peru and Mexico, as well as Madagascar, Tanzania, Uganda, Belize, Bolivia, Brazil, Costa Rica, El Salvador, Nicaragua, Panama,

3. https://www.icco.org/about-cocoa/chocolate-industry.html
4. https://www.foodnavigator.com/Article/2016/02/17/Premium-priced-poverty-Organic-cocoa-shortages-loom-as-farmers-reluctant-to-meet-demand#

Venezuela, Fiji, India, Sri Lanka and Vanuatu. However, this list does not include relatively recent entrants into the organic cocoa market, such as Ghana which is said to have over 10 000 ha under organic cocoa cultivation made up of more than 1000 organic certified farms (Djokoto et al. 2016).

While the Dominican Republic has historically been the leading exporter of organically certified cocoa by a significant margin, 'estimates of its production and exports differ widely across sources' (Byers et al. 2008: 53), with estimates ranging between 5000 and 14 350 MT annually (Byers et al. 2008: 53). More broadly, because Peru has significantly expanded its output, and Dominican productivity has been lower in 2016/2017 due to its adverse climate, it is not clear whether the Dominican Republic still remains ahead of Peru and other countries in terms of organic cocoa production.

The lack of current and reliable statistics notwithstanding, there is data on the land area used for organic cocoa production which provides some insights into production trends (Lernoud and Willer 2016). The data supports the view that organic cocoa represents only a small part of global production. Figures indicate that only 2.5% of the total global area for cocoa production is used for organic cultivation which represents 250 000 ha of the 10 million ha of land used globally for cocoa growing (Lernoud and Willer 2016). By contrast, the land used for organic coffee production represents almost 8% of the total production area for coffee. The largest organic cocoa areas are the Dominican Republic (120 315 ha), Peru (over 25 000 ha) and Mexico (19 000 ha) (Lernoud and Willer 2016). There is a significant geographical concentration as regards the area of organic cocoa production as over 85% of the world's organic cocoa area is located in Latin America (Lernoud and Willer 2016). The organic cocoa area has increased almost fivefold since 2004 and has therefore increased faster than the organic land area for most other crops (Lernoud and Willer 2016). However, some of this increase can be attributed to the improved availability of data (Lernoud and Willer 2016). Finally, Lernoud and Willer (2016) report that at least 3% of the organic cocoa land area was under conversion in 2014, suggesting that an increased amount of organic cocoa will be produced in the future.

A key issue is the important distinction between cocoa which is grown organically and cocoa which is certified as organic. Due to the lack of inputs such as synthetic pesticides and fertilisers (which are not used due to costs, lack of training and/or availability and/or fears about their impact on human health), cocoa is often grown organically though not necessarily certified as such. For example, a 2001 STCP survey revealed that over 50% of cocoa producers in Ghana used no agrochemicals and were *de facto* organic producers (Victor et al. 2010). Similarly, Ayenor et al. (2004) reported that many cocoa farms in Ghana had not been sprayed for over 20 years and were 'organic by default' (p. 263). They argued that: 'the potential for producing organic cocoa exists but there is a need to organise and build the capacity of interested farmers and provide them with the necessary research and extension support to increase yield for sustainable production' (Ayenor et al. 2004). As a result of not using chemical inputs, or because of insufficient demand for certified cocoa, this *de facto* organic cocoa can end up being sold on the conventional market.

3 Organic certification

In order to be certified, organic producers must comply with the relevant national and international regulations (such as European Council Regulation No 834/2007 on organic production and labelling of organic products)[5] and the requirements of certifying bodies in importing countries (such as BCS Öko-Garantie in Germany or USDA Organic in the USA).

There are a number of challenges to certification, including transparency. While one might expect that certifiers would make their fees easily available to producers and other relevant stakeholders, an European Union-commissioned study found that access to such information was highly variable (Jespersen 2011). The percentage of certification bodies providing public price lists varied from 67% in the Czech Republic, 50% in Switzerland, 44% in Italy and the UK, 20% in Turkey and 14% in Germany (Jespersen 2011). Organic operators therefore have to contact certifiers to obtain quotes for inspection and certification fees. However, this process is complicated by the fact that fees are calculated in a variety of ways by different control bodies, making it difficult for external stakeholders to work out and compare cost-effectiveness (Jespersen 2011). Some certifiers have separate certification and inspection fees, but others do not. Some of them have fixed fees or a combination of fixed and variable fees, depending on factors such as the time spent on control, the area, the type of crops and animals or the organisation's annual turnover. There are also administrative costs to be factored in such as travel costs and other types of costs which may or may not be included in the fee or hourly rate (Jespersen 2011).

A second factor is the cost of certification. According to the Organic Standard (TOS 2001, cited in ICCO 2006), the average certification fees at farm level represented 3% of farm turnover. However, ICCO (2006) points out that:

> This may underestimate costs for cocoa farmers, as they are subject to additional costs due to the international aspect and the usually small size of their business. In addition to certification fees, the farmer organizations have to bear additional administrative costs, additional labour costs (as organic production is more labour-intensive than conventional production) and opportunity costs associated to the loss in yields after discarding synthetic inputs and converting their operations from conventional to organic production. Before restoration to a full organic system, pest suppression and fertility problems are common. The degree of yield loss varies and depends on the inherent biological attributes of the farm, farmer expertise and the extent to which synthetic inputs were used under the previous management regime (p. 5).

This suggests that the real costs may be a much higher percentage of many farms' turnover. From an organisational perspective, costs are also high. The USDA reports that in 2010, the average cost for a desk audit to verify the documentation provided for certification was $4428.[6] The average costs for an onsite assessment to become a foreign Accredited Certifying Agent (ACA) for USDA Organic was $24 082.[7] These figures are based on the National Organic Programme charging $108/h to prepare, conduct and write the desk audit, and carry out onsite assessments, as well as all costs for the NOP staff (such as airfare, hotel, meals, vehicle rental, fuel, parking, tolls, any incidentals based on location, etc.) while travelling to the site.

Once approved, farmer organisations/exporters are accredited for a period of 5 years though a review is conducted after two and a half years. Full onsite assessments (subject to the above costs) after 5 years are required for the full renewal of the ACA.[8] As regards

5. These can be found at http://eur-lex.europa.eu/LexUriServ/LexUriServ.do?uri=OJ:L:2007:189:0001:0023:EN:PDF
6. https://www.ams.usda.gov/services/organic-certification/faq-becoming-certifying-agent
7. Ibid.
8. Ibid.

renewal fees, these are equally variable and at times complex. For example, the UK Soil Association has an annual renewal fee following the initial certification which is a variable proportion of total organic certified sales. For organic certified sales between £236 001 and £2 175 000, the certification fee is 0.003 x organic certified sales, whereas the fee is 0.0015 multiplied by organic certified sales between £2 175 001 and £4 050 000.[9]

A third challenge is the need for multiple certifications. In the case of cocoa, the costs of individual certification schemes represent only one part of the overall costs that organisations will incur because producer groups/exporters are extremely unlikely to be financially sustainable selling certified cocoa to only one market. In order to secure a large enough market for their organic beans, most farmer organisations/exporters have multiple certifiers (Berlan and Bergés 2013), all of which have costs and associated requirements.

As an example, one of the producer organisations which provided data for the present chapter reported having three organic certifiers for their cocoa and having paid almost US$50 000 in audits/certification renewal costs in the past year alone. This did not include organic transaction certificates which cost US$50 and are required for every shipment of organic cocoa. If one considers the costs of the initial certification, all subsequent audits/renewals and the cost of organic transaction certificates, figures are substantially higher, even without including additional expenses relating to the actual cultivation of organic cocoa. It is also worth noting that the organisation which provided the information had been certified by two out of their three organic certifiers for over 20 years, suggesting very high overall cumulative costs for certification over time.

Figures are higher still if one considers the annual fees connected with Fairtrade certification. While Fairtrade certification is obviously distinct from organic certification, the two often go hand in hand (thereby incurring two sets of certification costs) in order to meet the requirements of ethically minded consumers in developed markets who seek chocolate which is produced both in a socially and environmentally sustainable way. In 2017, the cost of Fairtrade certification was US$5000, and this did not include the costs of the actual audits/inspection processes or the costs of compliance at field level (such as hiring technicians to ensure standards are being met). One of the informants who provided information for this chapter stated that a significant part of the premium earned from certification was spent on ensuring compliance with the standards rather than on social development. Therefore, it is clear that certification places very significant financial burdens on producers in developing countries and this has to be carefully balanced against the returns of such investments.

4 Pricing issues in organic cocoa production

In addition to opening up particular trading opportunities, the most significant benefit for exporters or cooperatives of being organically certified is the higher price of organic over conventional cocoa. However, as in many other issues relating to organic cocoa, the subject of price is not straightforward.

For certain types of cocoa (both organic and conventional), fixed premiums are paid. As an example, the premium for Fairtrade and organically certified cocoa is US$200/MT

9. https://www.soilassociation.org/media/.../fees-for-uk-food-businesses-2017-18.docx

above the minimum fixed price of US$2300/MT.[10] For comparison, the Fairtrade minimum price for conventional beans is US$2000/MT with a premium of US$200/MT.[11] However, the premiums paid for non-Fairtrade organically certified cocoa are significantly more variable. Pay (2009) reports:

> Because of the volatility of prices, as well as the difficulties of estimating premiums along the supply chain, price premiums for organic cocoa fluctuate considerably according to the source of information used. ICCO (2006) indicates a premium of US$100–300 per tonne, while Liu (2008) indicates premiums of up to US$1,600 in 2006. Other sources indicate an organic premium of between 10 and 40 percent, or 10 and 50 percent, over non-organic cocoa (p. 7).

Berlan and Bergés (2013) examined the different premiums paid for organic cocoa from the Dominican Republic and found that the premiums paid between 2004 and 2011 ranged between US$195/MT and US$1600/MT. For comparison, organic cocoa beans from Peru were reported to have fetched an average organic premium of US$3331 above the conventional price of US$2354/MT in 2006, making the total price for organic beans US$5684/MT (Byers et al. 2008). This is likely due to premium market chocolatiers seeking out organically produced beans with very specific flavour profiles, and therefore being willing to pay much higher premiums. However, at such high cost, it is unlikely (though theoretically possible) that large volumes were being bought.

These figures illustrate the immense variability in prices received for organic cocoa and how these varied depending on buyers, country of origin and from year to year. In doing so, they also illustrate the variable financial sustainability of organically certified cocoa cultivation and why certain statements relating to the financial returns of organic cocoa cultivation must be treated with caution. As an example, the German association Naturland claims that:

> Organically produced cocoa beans also command a higher price in the international marketplace than conventional ones (ICCO, 2006). Therefore, the reduced input costs and the improved prize [*sic.*] per ton of organic beans leads to a higher farm income, overall (Naturland 2014: 10).

Such statements only provide a very partial picture of the financial returns of organic cocoa production. It does not show their variability or reflect the fact that yields from organic production can be lower than those from conventional cocoa production or the fact that even when premiums are high, organic production incurs costs such as fees to certification bodies. Therefore, they require some contextualisation in order not to raise unrealistic expectations, especially if being used to promote organic cultivation to producers in developing countries.

It is also worth remembering that the financial sustainability of organically certified cocoa is to a large extent determined by the willingness of the consumer to pay high premiums for organic chocolate and depends on these high margins actually being passed on to the producers. As regards the latter, research on the premium craft chocolate market has shown that this is not always the case (Martin 2017).

10. https://www.fairtrade.net/standards/price-and-premium-info.html
11. Ibid.

5 Organic yields and cultivation methods

There is much evidence indicating that organic agriculture typically generates lower yields than conventional cultivation (Seufert et al. 2012), with some studies indicating a yield gap of 20% between conventional and organic production (De Ponti et al. 2012). However, other studies highlight the potential of organic agriculture to significantly improve yields in comparison with conventional cultivation: 'appropriate investment in agroecological research to improve organic management systems could greatly reduce or eliminate the yield gap for some crops or regions' (Ponisio et al. 2015: 1).

Conventional/organic yields have been compared in cocoa production. Onumah et al. (2013) found conventional production to be more technically efficient whilst highlighting the potential for increasing the efficiency of organic production through larger-scale cultivation. Some authors have argued that organic farms can achieve higher yields and overall annual family income, though this may be due to organic farmers participating in more training on cocoa cultivation, rather than to organic cultivation *per se* (Jacobi et al. 2015). This evidence does, however, suggest that organic cocoa production is viable if right amounts of extension support, tree seedlings, capacity building and other measures are provided. A brief comparison between conventional and organic cultivation is provided in Table 1.

There is also a growing body of evidence regarding 'frugal' organic innovations in cocoa growing which are effective and gaining in popularity such as the use of chicken manure

Table 1 Conventional and organic cocoa cultivation practices

Practice	Conventional practices	Organic practices
Planting	Spacing: 3 × 3 m. Seedlings from clean nurseries or planting at stake	Spacing: 3 × 3 m. Seedlings from only organic nurseries
Weed control	Manual weed control, use of cover crops and recommended chemicals	Weed regularly manually. Use cover crops. No chemical weed control
Pests and diseases control	Remove mistletoes, dead branches and black pods	Remove mistletoes, dead branches and black pods
	Collect and burn away from the farm diseased pods once a week to prevent spread of black pod disease. When necessary to bury on the field, bury 60 cm deep. Remove pods with even small sign of black pod attack. Apply recommended chemicals. Dosage 85–500 mL/ha depending on pesticide	Collect and burn away from the farm diseased pods once a week to prevent spread of black pod disease. When necessary to bury on the field, bury 60 cm deep. Remove pods with even small sign of black pod attack Spray with neem tree extract when sanitation practices are inadequate Apply 40 L/acre = 3 mistblowers/acre Wear protective clothing
	Wear protective clothing	Other permitted products can be used
Fertility management	Apply NPK fertilisers such as Asaase Wura and Cocofeed. Others are triple super phosphate and ammonium sulphate	Improve fertility with use of cover crops, leguminous plants, compost making, farm yard manure and planting shade trees. Other permitted products can be used

Source: Djokoto et al. (2016).

in Côte d'Ivoire (Ruf et al. n.d.) or goat manure in Indonesia (Guest 2017; Gunawan and Budisatria 2016). Using such methods is distinct from full organic production. However, the cheapness and efficacy of organic cocoa cultivation methods such as these are worthy of note as they could help to bolster production of conventional cocoa while minimising costs and environmental impact.

More widely, organic cocoa cultivation can make a significant contribution to environmental and conservation goals, as it is a key requirement of organic agriculture that crops are produced in keeping with ecological systems and cycles (ICCO 2006). Daniels (2006) states 'Cocoa has traditionally been a destroyer of tropical forests' though she adds the following statement from CATIE (2000) which argues that:

> Organic cacao provides a significant alternative, at the landscape-scale, as an environmentally sustainable, corridor compatible, land use with the potential to meet economic needs while preserving forest remnants (cited in Daniels 2006, p. 21).

Though not a requirement for organic certification, agroforestry has been seen as a key component for sustainable organic cocoa production (Nunoo et al. 2014) and a number of organic cocoa farms promote growing cocoa in agroforestry conditions. There is a significant body of literature on the benefits of agroforestry systems in cocoa cultivation (Ruf 2011; Bisseleua et al. 2009; Utomo et al. 2016; Vanhove et al. 2016; Vaast and Somarriba 2014), for example in mitigating the severity of cocoa swollen shoot virus disease (Andres et al. 2018).

Agroforestry-based organic cultivation is often contrasted with full-sun-grown cocoa farming which has become common in key origin countries such as Ghana (Nunoo et al. 2014), Indonesia (Ruf 2011) and Côte d'Ivoire, where it now dominates production (Tondoh et al. 2015). Cocoa has traditionally been grown under shade conditions, with shade being regarded as a general requirement in the successful establishment of cocoa (Wood and Lass 1985). Shade trees have been shown to have a variety of benefits such as helping to maintain soil fertility on cocoa farms (Hartemink 2005), boosting carbon sequestration and storage (Abou Rajab et al. 2016) and having the potential to mitigate climate change (Somarriba et al. 2013). The reduction or elimination of shade has led to concerns about loss of biodiversity, increased deforestation, high input demands, a shorter productive life for the tree, lower yields and food security over time (Belsky and Siebert 2003; Bisseleua et al. 2009; Nunoo et al. 2014). Full-sun cocoa is also associated with problems such as increased outbreaks of pests and diseases (Andres et al. 2016; Daniels 2006; Vanhove et al. 2016).

However, some scholars argue that the benefits of cocoa agroforestry systems for soil fertility and carbon sequestration are more limited than is believed and may not be sufficient to outweigh its costs (Blaser et al. 2017). It is also important to note that more intensive conventional methods of production may have environmental benefits such as reducing deforestation (Gockowski and Sonwa 2010), particularly in countries such as Indonesia where 'developing new agroforests now often means destroying protected forests' (Kusters et al. 2008, cited in Ruf 2011, p. 373). Ruf argues in favour of a polyculture system based on a rotation of tree crops due to his belief that:

> Maintaining biodiversity in cocoa plantations no longer involves reprieving or reproducing extensive systems developed at a time when forests were abundant. Complex cocoa agroforests are an anachronism, and are now in the process of becoming a myth. (2011: 386)

This debate about the role of agroforestry will continue.

6 Pest and disease management

As shown in Table 1, the cultivation of organic cocoa, just like the cultivation of conventional cocoa, requires adequate methods for controlling pests and diseases as these are well known to cause extensive losses and undermine the sustainability of both organic and conventional cocoa production (Ayenor et al. 2007; Berlan and Bergés 2013). In organic farming, chemical inputs are not permitted, but there is a range of other options. Naturland (2014) recommends using crop diversity, biological pest control agents (such as herbal extracts and beneficial fungi and bacteria species) and pruning to control pests and diseases. Ayenor et al. (2007) investigated organic capsid control methods including the use of sex pheromone traps, crude aqueous neem (*Azadirachta indica*) seed extracts and the use of ants (*Oecophylla longinoda*) as biological control agents against capsids. They concluded: 'All the three methods were effective and compatible; hence, they can be used in an integrated pest management strategy for cocoa, including organic production in Ghana' (p. 85).

The effectiveness of some of these methods, such as the use of ants, is supported by other studies. For example, Wijngaarden et al. (2007) reported that the black strain of *O. longinoda* was a particularly aggressive biological control agent against capsids and that the number of cocoa pods damaged was about 50% lower for trees with this type of ant compared to trees without ants. CABI (2008) also found that black ants in particular were effective, provided they were present in high numbers. Wijngaarden et al. (2007) recommended introducing ants in plantations where they were absent and reinforcing them in plantations where they were low in number. Such an intervention would represent a cheap and environmentally friendly method of achieving some degree of pest control not only on organic farms but also on conventional ones. However, this would only prove to be practically possible on non-sprayed conventional farms because synthetic pesticides wipe out ant colonies.

While the research undertaken on the effectiveness of neem *(A. indica)* to control the mirids (*Sahlbergella singularis* and *Distantiella theobroma)* which attack cocoa pods and shoots is not extensive, it is encouraging in terms of its results (Asogwa et al. 2010; Dormon et al. 2007). As an example, N'Guessan et al. (2006, cited in Asogwa et al 2010) recorded a cocoa mirid mortality rate of 88.2%, 96.6% and 98.5% with neem seed aqueous crude extracts at 10%, 20% and 30%, respectively, 96 h after treatment. While in some cases the mirid mortality rate is still lower than that achieved by conventional insecticides, a straightforward comparison may not be appropriate 'since the effectiveness of neem-based compounds is known to rely more on their ability to repel and inhibit feeding, growth and reproduction, rather than on direct insect kill' (Asogwa et al. 2010: 4). While neem benefits from recognition among some farmers, one of the drawbacks of using it is that it is not easily available in certain geographical locations and communities lack the infrastructure for producing and distributing it (Dormon et al. 2007; CABI 2008). Therefore, its adoption on a wide scale would require not only some degree of farmer education but also a significant amount of practical support.

7 Conclusions and future trends

This chapter has provided only a snapshot of organic cocoa and how it can be produced sustainably. However, it is clear that important lessons can still be learnt from the material presented, and that in some cases, these extend beyond organic cocoa. If we return to the example of capsids, one major advantage of using ants to control them is that cocoa farmers already had faith in this method (Wijngaarden et al. 2007; Ayenor et al. 2004). While this may not seem like a compelling reason, it is of relevance to the present discussion on the sustainable organic cultivation of cocoa (and cocoa growing more broadly) in a number of ways.

First, it is important to place the notion of farmer acceptance in the wider context of cocoa production. As a crop, cocoa has unfortunately become associated (both in the present time and historically) with ineffective top-down approaches, low uptakes of agricultural extension advice/new technologies and ongoing low yields relative to their potential (Clarence-Smith 2000; Berlan 2017; Ayenor et al. 2004; Guest 2017). Therefore, any solutions which already have some degree of acceptance among farmers, and which are inexpensive and practically achievable, warrant further consideration (Ayenor et al. 2007).

Secondly, the example of ants illustrates the importance of developing a wider integrated pest management strategy, which also includes other components such as the removal of black pods, weeding and pruning (Wijngaarden et al. 2007). This represents a significant challenge because it requires farmers to be fully committed and motivated, physically able, supported through good infrastructure and adequately rewarded through good prices and appropriate incentives, to name a few key factors. On the issue of price alone, irrespective of the high premiums which can make organic production/certification attractive, the market can still be volatile and unpredictable – and therefore risky and potentially unsustainable. For this reason, a sustainable organic sector needs to go beyond good cultivation practices and tackle broader market conditions and structures as these will otherwise continue to undermine its growth. This would also help to make cocoa more appealing to the younger generations on which the future of cocoa production depends. The fact that there is a chronic shortage of data on the organic cocoa sector also hinders its sustainability by making it difficult to discern particular trends, address challenges and exploit key opportunities.

Beyond this, it is important to recognise that in spite of the urgent and widely acknowledged need for agricultural production systems to be made more sustainable, this is unlikely to happen unless there is a financial case for doing so, and therein lies a more difficult challenge when it comes to organic cocoa. As has been shown, organic cocoa cultivation offers some distinct environmental benefits, but the organic cocoa market offers risk and opportunity in equal measures. Farmer cooperatives/private exporters *can* secure high premiums for organically produced cocoa, but these are not guaranteed. By contrast, securing the financial returns from organic certification involves an established requirement to grow the crop in a way that precludes any conventional 'shortcuts' such as synthetic pesticides and fertilisers and a guarantee of significant financial outlays through often complex and at times opaque foreign certification schemes. Aside from the power dynamics inherent to such trading relationships, it is necessary to recognise that for any business, successfully operating within such an environment – let alone thrive and succeed in it – is going to prove exceptionally challenging. Such a market requires a level of business acumen, resources and an enabling environment which many of the producer cooperatives located in Third-World countries (and dealing with weak infrastructure, the needs of poor communities and very limited resources) simply do not have. Therefore, there is also a fundamental need for more business capacity building if organic cocoa is to have a truly sustainable future. Such considerations may seem far removed from discussions of disease management, yields or shade cover. However, this case study has shown that the challenge of feeding a growing population in an increasingly stretched natural environmental is going to require wide-ranging and innovative thinking encompassing many different areas and disciplines.

8 Where to look for further information

The various stakeholders involved in the organic cocoa sector are represented by IFOAM Organics International which has acted as the global organic umbrella organization since 1972. IFOAM Organics International brings together over 1000 members from 130 countries, organises events, implements projects and provides training and guidance on organic standards and regulations. Readers interested in this topic are encouraged to consult their website (https://ifoam.bio/) and to look at the publications produced by the Research Institute of Organic Agriculture (FiBL) (https://www.fibl.org/en/homepage.html) and the Centre for Agriculture and Biosciences International (CABI) (https://www.cabi.org/). Both of these organizations are international, independent and not-for-profit research institutions. CABI does not specialise in organic production, but many of their projects and training programmes are aligned with organic cultivation methods. FiBL focuses exclusively on organic agriculture. There is no single authoritative guide on the subject of organic cocoa but the annual report produced by FiBL and IFOAM Organics International *The World of Organic Agriculture* provides the most extensive source of information on trends in organic cocoa production and other crops.

There are also a number of older publications which provide relevant data on aspects of the cocoa market, such as the report produced by the International Centre for Research in Organic Food Systems (ICROFS) on organic certification in selected European countries (Jespersen 2011). Readers wishing to understand the basic principles of legislation and certification as they relate to organic cocoa can consult the 2006 ICCO report *A study on the market for organic cocoa* (ICCO 2006).

9 References

Abou Rajab, Y., Leuschner, C., Barus, H., Tjoa, A. and Hertel, D. (2016), 'Cacao cultivation under diverse shade tree cover allows high carbon storage and sequestration without yield losses', *PLoS ONE*, vol. 11, no. 2, p. e0149949. doi:10.1371/journal.pone.0149949.

Andres, C., Comoé, H., Beerli, A., Schneider, M., Rist, S. and Jacobi, J. (2016), 'Cocoa in monoculture and dynamic agroforestry', in Lichtfouse, E. (Ed.), *Sustainable Agriculture Reviews*, Switzerland: Springer International Publishing, chapter 3, pp. 121–53.

Andres, C., Blaser, W. J., Dzahini-Obiatey, H. K., Ameyaw, G. A., Domfeh, O. K., Awiagah, M. A., Gattinger, A., Schneider, M., Offei, S. K. and Six, J. (2018), 'Agroforestry systems can mitigate the severity of cocoa swollen shoot virus disease', *Agriculture, Ecosystems & Environment*, vol. 252, pp. 83–92.

Asogwa, E. U., Ndubuaku, T. C. N., Ugwu, J. A. and Awe, O. O. (2010), 'Prospects of botanical pesticides from neem, *Azadirachta indica* for routine protection of cocoa farms against the brown cocoa mirid – *Sahlbergella singularis* in Nigeria', *Journal of Medicinal Plants Research*, vol. 4, no. 1, pp. 1–6.

Ayenor, G. K., Röling, N. G., Padi, B., Van Huis, A., Obeng-Ofori, D. and Atengdem, P. B. (2004), 'Converging farmers' and scientists' perspectives on researchable constraints on organic cocoa production in Ghana: Results of a diagnostic study', *NJAS-Wageningen Journal of Life Sciences*, vol. 52, pp. 261–84.

Ayenor, G. K., Van Huis, A., Obeng-Ofori, D., Padi, B. and Röling, N. G. (2007), 'Facilitating the use of alternative capsid control methods towards sustainable production of organic cocoa in Ghana', *International Journal of Tropical Insect Science*, vol. 27, no. 2, pp. 85–94.

Belsky, J. and Siebert, S. (2003), 'Cultivating cacao: Implications of sun-grown cocoa on local food security and environmental sustainability', *Agriculture and Human Values*, vol. 20, pp. 277–85.

Berlan, A. (2017), 'What works? A review of the contribution academic knowledge can make to building a stronger and more resilient cocoa economy', presented at the *International Symposium on Cocoa Research*, Lima, Peru, 13–17 November 2017.

Berlan, A. and Bergés, A. (2013), Cocoa production in the Dominican Republic: Sustainability, challenges and opportunities. Report of findings to Green and Black's.

Bisseleua, D. H. B., Missoup, A. D. and Vidal, S. (2009), 'Biodiversity conservation, ecosystem functioning and economic incentives under cocoa agroforestry intensification', *Conservation Biology*, vol. 23, no. 5, pp. 1176–84.

Blaser, W. J., Oppong, J., Yeboah, E. and Six, J. (2017), 'Shade trees have limited benefits for soil fertility in cocoa agroforests', *Agriculture, Ecosystems & Environment*, vol. 243, pp. 83–91.

Byers, A., Giovannucci, D. and Liu, P. (2008), *Value-Adding Standards in the North American Food Market: Trade Opportunities in Certified Products for Developing Countries* (Liu, P., ed.), Rome: FAO.

CABI (2008), GRO-Cocoa global research on cocoa – Working with and for farmers, issue 13 June 2008, pp. 1–8. Available at: https://www.cabi.org/Uploads/CABI/projects/gro-cocoa-issue-13-jun-2008.pdf

Clarence-Smith, W. G. (2000), *Cocoa and Chocolate, 1765–1914*, London: Routledge.

Daniels, S. (2006), *Developing Best Practice Guidelines For Sustainable Models of Cocoa Production to Maximize Their Impacts on Biodiversity Protection*, Vietnam: World Wildlife Fund Vietnam, pp. 1–58.

De Ponti, T., Rijk, B. and van Ittersum, M. K. (2012), 'The crop yield gap between organic and conventional agriculture', *Agricultural Systems*, vol. 108, pp. 1–9.

Djokoto, J. G., Owusu, V. and Awunyo-Vitor, D. (2016), 'Adoption of organic agriculture: Evidence from cocoa farming in Ghana', *Cogent Food & Agriculture*, vol. 2, pp. 1–15, Article ID: 1242181.

Dormon, E. N. A., van Huis, A. and Leeuwis, C. (2007), 'Effectiveness and profitability of integrated pest management for improving yield on smallholder cocoa farms in Ghana', *International Journal of Tropical Insect Science*, vol. 27, no. 1, pp. 27–39.

FAO/WHO Codex Alimentarius Commission (1999). http://www.fao.org/docrep/meeting/X0075e.htm. (accessed 22 June 2018).

Gockowski, J. and Sonwa, D. (2010), 'Cocoa intensification scenarios and their predicted impact on CO_2 emissions, biodiversity conservation, and rural livelihoods in the Guinea rain forest of West Africa', *Environmental Management*, vol. 48, no. 2, pp. 307–21.

Guest, D. (2017), 'Labour as a constraint to technology adoption: Understanding the role of farmer health', presented at the *International Symposium on Cocoa Research*, Lima, Peru, 13–17 November 2017.

Gunawan and I. G. S. Budisatria (2016), 'Technology innovation in cocoa-goats integration system for increasing of productivity and farmers income in Kulon Progo Regency, Yogyakarta Special Region Province, Indonesia', *Asian Journal of Animal Sciences*, vol. 10, pp. 273–9. Available at http://scialert.net/fulltext/?doi=ajas.2016.273.279&org=12

Hartemink, A. E. (2005), 'Nutrient stocks, nutrient cycling, and soil changes in cocoa ecosystems: A review', *Advances in Agronomy*, vol. 86, pp. 227–53. doi:10.1016/s0065-2113(05)86005-5

ICCO (2006), A study on the market for organic cocoa, *Executive Committee 130th Meeting*, London, 12–15 September 2006. Available at: https://www.icco.org/sites/www.roundtablecocoa.org/documents/EX-130-10%20-%20ICCO%20-%20Study%20on%20Market%20for%20Organic%20Cocoa.pdf

ICCO (2018). https://www.icco.org/about-cocoa/chocolate-industry.html. (accessed 22 June 2018).

Jacobi, J., Schneider, M., Pillco Mariscal, M., Huber, S., Weidmann, S., Bottazzi, P. and Rist, S. (2015), 'Farm resilience in organic and non-organic cocoa farming systems in Alto Beni, Bolivia', *Agroecology and Sustainable Food Systems*, vol. 39, pp. 798–832.
Jespersen, L. M. (2011), Organic certification in selected European countries: Control fees and size of the sector, Tjele, Denmark: International Centre for Research in Organic Food Systems (ICROFS).
Lernoud, J. and Willer, H. (2016), 'Organic agriculture worldwide: Current statistics on organic agriculture worldwide: Area, producers, markets, and selected crops', *The World of Organic Agriculture. Statistics and Emerging Trends 2016*, Frick: Research Institute of Organic Agriculture (FiBL) and Bonn: IFOAM – Organics International, pp. 33–117.
Martin, C. D. (2017), 'Sizing the craft chocolate market', *Fine Cacao and Chocolate Institute* (blog), 31 August 2017, https://chocolateinstitute.org/blog/sizing-the-craft-chocolate-market/
Naturland (2014), *How to Grow Organic Cocoa: An Illustrated Handbook on Organic Principles of Cocoa Production*, Gräfelfing, Germany: Naturland – Association for Organic Agriculture.
Nunoo, I., Owusu, V. and Darko, B. O. (2014), 'Cocoa agroforestry a bridge for sustainable organic cocoa production', in Rahmann, G. and Aksoy, U. (Eds), *Proceedings of the 4th ISOFAR Scientific Conference. 'Building Organic Bridges', at the Organic World Congress 2014*, 13–15 October, Istanbul, Turkey.
Onumah, J. A., Onumah, E. E., Al-Hassan, R. M. and Brümmer, B. (2013), 'Meta-frontier analysis of organic and conventional cocoa production in Ghana', *Agricultural Economics*, vol. 59, no. 6, pp. 271–80. Available at: http://www.agriculturejournals.cz/publicFiles/94078.pdf
Pay, E. (2009), The market for organic and fair-trade cocoa. Study prepared in the framework of FAO project GCP/RAF/404/GER 'Increasing incomes and food security of small farmers in West and Central Africa through exports of organic and fairtrade tropical products', Rome: FAO.
Ponisio, L. C., M'Gonigle, L. K., Mace, K. C., Palomino, J., de Valpine, P. and Kremen, C. (2015), 'Diversification practices reduce organic to conventional yield gap', *Proceedings B, Royal Society of Publishing*, vol. 282, no. 1799, p. 20141396.
Ruf, F. O. (2011), 'The myth of complex cocoa agroforests: The case of Ghana', *Human Ecology*, vol. 39, no. 3, pp. 373–88.
Ruf, F., Kla, A. G., Dja, K. and Kiendré, J. (n.d.), *Chicken Manure in the Cocoa Plantations of Côte d'Ivoire*, Montpellier, France: UMR Innovation/SupAgro/Inra/Cirad. Available at: https://www.researchgate.net/profile/Francois_Ruf2/publication/285055054_Chicken_manure_in_the_cocoa_plantations_of_Cote_d'Ivoire_An_agro-ecological_and_social_revolution_a_'frugal'_village_innovation/links/565b313908aeafc2aac60e20/Chicken-manure-in-the-cocoa-plantations-of-Cote-dIvoire-An-agro-ecological-and-social-revolution-a-frugal-village-innovation.pdf
Sahota, A. (2016), 'The Global market for organic food and drink', in *The World of Organic Agriculture. Statistics and Emerging Trends 2016*, Frick: Research Institute of Organic Agriculture (FiBL) and Bonn: IFOAM – Organics International, pp. 133–9.
Seufert, V., Ramankutty, N. and Foley, J. A. (2012), 'Comparing the yields of organic and conventional agriculture', *Nature*, vol. 485, pp. 229–32.
Somarriba, E., Cerda, R., Orozco, L., Cifuentes, M., Dávila, H., Espin, T., Mavisoy, H., Ávila, G., Alvarado, E., Poveda, V., Astorga, C., Say, E. and Deheuvels, O. (2013), 'Carbon stocks and cocoa yields in agroforestry systems of Central America', *Agriculture, Ecosystems & Environment*, vol. 173, pp. 46–57.
Tondoh, J. E., N'Guessan Kouamé, F., Guéi, A. M., Sey, B., Koné, A. W. and Gnessougou, N. (2015), 'Ecological changes induced by full-sun cocoa farming in Côte d'Ivoire', *Global Ecology and Conservation*, vol. 3, pp. 575–95.
Utomo, B., Prawoto, A. A., Bonnet, S., Bangviwat, A. and Gheewala, S. H. (2016), 'Environmental performance of cocoa production from monoculture and agroforestry systems in Indonesia', *Journal of Cleaner Production*, vol. 134, Part B, pp. 583–91.
Vaast, P. and Somarriba, E. (2014), 'Trade-offs between crop intensification and ecosystem services: The role of agroforestry in cocoa cultivation', *Agroforestry Systems*, vol. 88, no. 6, pp. 947–56. doi:10.1007/s10457-014-9762-x

Vanhove, W., Vanhoudt, N. and Van Damme, P. (2016), 'Effect of shade tree planting and soil management on rehabilitation success of a 22-year-old degraded cocoa (*Theobroma cacao* L.) plantation', *Agriculture, Ecosystems & Environment*, vol. 219, pp. 14–25.

Victor, A.-S., Gockowski, J., Agyeman, N. F. and Dziwornu, A. K. (2010), 'Economic cost-benefit analysis of certified sustainable cocoa production in Ghana', paper submitted to the *3rd Conference of African Association of Agricultural Economists, Africa and the Global Food and Financial Crisis*, 19–23 September 2010, Cape Town.

Wijngaarden, P. M., van Kessel, M. and van Huis, A. (2007), 'Oecophylla longinoda (Hymenoptera: Formicidae) as a biological control agent for cocoa capsids (Hemiptera: Miridae)', in *Proceedings of the Netherlands Entomological Society Meeting*, vol. 18, pp. 21–30.

Wood, G. A. R. and Lass, R. A. (1985), *Cocoa*, 4th edition, Oxford and New York: Blackwell Science.

Chapter 22

Cocoa sustainability initiatives: the impacts of cocoa sustainability initiatives in West Africa

Verina Ingram, Yuca Waarts and Fedes van Rijn, Wageningen University, The Netherlands

1 Introduction

The cocoa sector faces a number of deeply embedded, interrelated challenges, including old trees with low productivity; low farmer and worker incomes; pest and disease attacks; political instability in many of the main countries of origin; persistent poor labour conditions and negative environmental impacts, such as deforestation, soil degradation and pollution; tight relationship between supply in the face of growing demand and long-term cyclical recession and expansion booms affecting global market and farm gate prices (Ruf and Siswoputranto 1995; Nkamleu et al. 2010; Matissek et al. 2012; Fountain and Hutz-Adams 2015). Organising smallholders into groups and undertaking voluntary initiatives, such as certification, often accompanied by training and improved access to agricultural inputs, have been proposed to address some of these multiple, long-running problems. These sustainability interventions have been implemented largely by international traders in the cocoa value chain, government agencies, certification organisations and non-governmental organisations (NGOs). But do they work? What has been the impact of the multitude of sustainability-based interventions in the last decade on different types of farmers, and how to untangle the impacts of different interventions?

http://dx.doi.org/10.19103/AS.2017.0021.31

2 Historical context for the rise of voluntary sustainability standards

The majority of cocoa is produced on small farms with average size varying from 2.8 ha (Alonghi 2011) to 3.7 ha (KPMG 2012). An estimated 20% of such farmers are members of associations or cooperatives (ICCO 2014). Farms generally have low and decreasing productivity rates of 300–500 kg/ha (Ruf 2007; Mejia 2011; KPMG 2012). Decades of work on breeding and developing high-yielding, disease- and pest-resistant and climate-proof varieties; replacing old trees; and providing more appropriate farm inputs (fertilisers and crop protection products) has resulted in only marginal yield improvements, particularly in the main West African producing countries (Wessel and Quist-Wessel 2015). Farm to market infrastructure is generally lacking, as is finance for inputs and support for higher yielding, disease-resistant varieties for replanting and regenerating cocoa trees.

The cocoa value chain continues to be characterised by generally small-scale farmers in tropical countries, dominated by West Africa (73% of total production estimated in 2015), followed by southern America (16%) and Asia and Oceania (10.2%) (ICCO 2016), selling their produce to individual traders or cooperatives, who sell them to traders or major exporters. The products sold by traders to major confectionary and production companies such Mars, Mondelēz and Nestlé are further processed into food and cosmetic products (Abbott et al. 2005). Profits from farm gate prices for farmers and farmer groups are generally low compared to those from value-adding process and profits made on semi-processed and final products (FAO 2007; Kessler et al. 2012; Quarmine et al. 2014; Squicciarini and Swinnen 2016).

Although the context of cocoa farming has changed little, virtually every dimension of international cocoa value chains has changed in the last two decades (Wilcox and Abbott 2004). In major producing countries such as Ivory Coast, Ghana, Indonesia and Cameroon until 1990, activities related to exports, market power and price setting were shared between exporters and the government (Ton et al. 2008). Credit, pricing, export licensing and consumer behaviour are now largely determined by the private sector. Although there has been an overall 10% increase in production from 2003 to 2013 (ICCO 2013), sufficient quality supply has increasingly become an issue due to increasing demand (FAO 2007; Eberhard Krain 2011). Supply variability is due to large annual fluctuations, caused by multiple factors, particularly the weather. Quality has generally been increasing, due to training, increasing use of drying equipment and market and regulatory standards. In the past the governance of production and quality aspects, input credit and supply, extension services and market infrastructure has often been state-controlled in the cocoa-producing countries. Public bodies in producer countries used globally recognised quality standards and territorial (ecological and technically based) reputations to enhance their position in the global market. However, governments of cocoa-producing countries have gradually lost their ability to manage the international cocoa market and shape their own domestic markets. Since the late 1990s market-based corporate governance and price negotiation system has developed in many production countries with the breakdown of national institutions, failed cocoa harvests and pressure from the World Bank and International Monetary Fund for economic structural adjustment. Foreign companies used the room to increase investments and their integration and position in the chain. Exporters (including major traders such as Cargill, Barry Callebaut, Olam and Armajaro) were then free to buy and sell based on London market price. The fully liberalised system has left farmers exposed to the international cocoa price set in London. This resulted in reforms to the sector in Ghana and Ivory Coast in 1999 and 2012, respectively, including privatising buying and setting minimum export prices (Vellema et al. 2006; Vellema and Laven 2012). Fluctuations in international prices are adjusted mainly in the processing, manufacturing and distribution stages of the chain, with increases in cocoa prices more fully and rapidly transmitted to consumers than decreases: a manifestation of the market power of traders and chocolate manufacturers (Jean-François and Bonjean 2014).

In the global cocoa chain an increasing trend towards concentration can be observed in the last decade. Many of the cocoa and chocolate companies have merged and been taken over by competitors at both vertical and horizontal levels. Horizontal integration is a strategy where a company creates or acquires production units which are similar – either complementary or competitive. Vertical integration is where activities up and down a value chain are owned by that company. Often each organisation in a chain produces or engages in different product or market-specific activities in a chain. The strongest concentration has occurred in processing of beans to ingredients for chocolate (butter, powder and liquor). Two processors, Barry Callebaut and Cargill (after Cargill's merger with ADM), now account for 70–80% of cocoa processing worldwide. Other companies dominating the processing sector include Blommer, Ecom, Cargill, Olam, Barry Callebaut (who purchased Petra Foods cocoa ingredients in 2015) and Armajaro (whose cocoa business was taken over by Ecom in 2014). Globally, between 60% and 80% of cocoa revenue is in the hands of seven traders and grinders (Barry Callebaut, Cargill, Olam, Ecom, Touton, Blommer and Continaf). The six biggest chocolate manufacturers transform 40% of worldwide chocolate products (Fountain and Hutz-Adams 2015). This market concentration by buyers, processors and manufacturers weakens the positions of largely unorganised farmers (Fountain and Hutz-Adams 2015). Demand for chocolate and cocoa has also changed and despite large fluctuations in demand and supply (Ruf and Siswoputranto 1995), an upward trend is apparent with an annual growth rate of 3.3 from 2002 to 2011 (ICCO 2012b). There is strong growth in Russia, Brazil, China, India and also in countries of origin (ICCO 2012b; Naprta 2015; Squicciarini and Swinnen 2016).

These institutional, economic and market changes and restructuring in the chain have created space for innovation, particularly in creating new relationships between stakeholders in the value chain and introducing voluntary sustainability initiatives (as opposed to the wealth of tax, health, cultivation, manufacturing, health and food safety and composition regulations in both origin and consumer countries) by different types of organisations in the value chain.

3 Sustainability initiatives

Sustainability, defined broadly in the terms embraced by the United Nations' sustainable development goals, includes ending poverty and hunger; improving health and education and access to water and sanitation; reducing inequalities; decent work and economic growth combating climate change; responsible consumption and production; and protecting and restoring marine and terrestrial ecosystems and reversing degradation and biodiversity loss. Four types of sustainability initiatives have been found in the cocoa value chain (Ingram et al. 2016a):

1 Platforms, networks and associations refers to groups of private, public, research and/or civil society (CSO) or NGOs collaborating on a common goal of sustainability with a declared policy or programme and plan of action.
2 Voluntary sustainability standards (VSSs) refers to sustainability standards agreed for business-to-business transactions as well as those used for business-to-consumer transactions such as UTZ, Rainforest Alliance and FairTrade; management system standards as well as product standards such as ISO 14001 – whether owned by government, NGOs or business concerns, so long as they are voluntary in nature; environmental product declarations, such as the Dutch Albert Heijn supermarket's 'healthy choice' label, as well as independently certified labels such as Organic, and labels developed by producers or retailers as well as those with independent governance (Salmon 2002).
3 Individual corporate initiatives are a form of corporate social responsibility and self-regulation, whereby a business monitors and ensures its active compliance with the spirit of the law, ethical standards and national or international norms. A firm may engage in actions that appear to further some social good, beyond the interests of the firm and that which is required by law (McWilliams and Siegel 2001).
4 NGO and CSO campaigns have aimed to raise awareness, such as the TCC, which pushed many companies to work together and created for animo changes on a sector-wide level.

As Fig. 1 shows, the majority of sustainability initiatives in the cocoa value chain started in the mid-2000s and have significantly upscaled, mainly by replication of similar projects and initiatives by different companies, NGOs and service providers, and outscaling into more cocoa-producing communities and countries since 2008. Precise figures of the numbers of farmers and cooperatives participating and volumes produced are not available, as farmers participate in more than one trader's initiative and in several voluntary certification schemes at one time. Corporate data are unfortunately also not comparable as definitions of throughput volumes and, in particular, farmers and producer organisations participating in initiatives differ widely, and data are not available and comparable either for given years or for specific geographic locations.

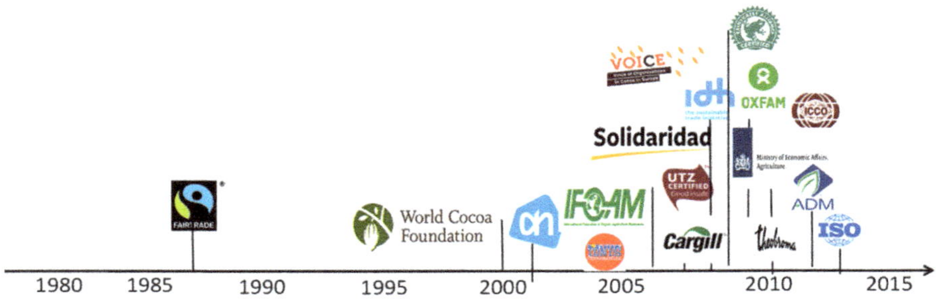

Figure 1 Cocoa sustainability initiatives timeline.

3.1 Platforms, networks and associations

The last two decades have seen a rise in the number and activities of platforms, networks and associations engaged in cocoa and in particular with an explicit focus on sustainability. Platforms have been established to bring together organisations from different segments of the value chain in the hope of encouraging collective discussions and addressing specific sustainability issues. Examples of such platforms include the World Cocoa Foundation (WCF) and its Cocoa Action, and the Sustainable Trade Initiative (IDH). Networks include International Cocoa Initiative (ICI); Women in Cocoa & Chocolate Network (WINCC), the international cocoa researchers' network – a linked group; and CacaoNet, a global network for cacao genetic resources, and the Voice Network of NGOs and trade unions concerned with sustainability of cocoa. Networks have tended to bring together stakeholders from different backgrounds, disciplines and sectors in the value chain. International trade associations and inter-governmental organisations increasingly act and speak on behalf of their members and have increasingly consolidated to have national, regional and global voice, representing all segments of the chain, for example, the International Cocoa Organisation (ICCO), representing governments which in the mid-2000 highlighted sustainability issues in the sector with a number of studies, discussion documents, a global cocoa agenda roadmap and declarations made during international conferences. In the private sector, the Association for the Bakery and Confectionary Industry [Vereniging voor de Bakkerijen Zoetwarenindustrie (VBZ)] has made corporate social responsibility a priority for manufacturers of cookies, candy and chocolate, focusing on the raw material supply chain, innovation and employees and has been supporting its members to adopt sustainable practices. The nested scale of associations is illustrated by the VBZ, a member of the Association of Chocolate, Biscuit and Confectionery Industries of Europe (CAOBISCO), and European Cocoa Association (ECA), a trade association grouping the major cocoa bean traders and processers, warehousing and related logistical activities in Europe. Cocoa farmer organisations have also seen some of the most rapid revitalisation in the last decade, particularly due to the role that farmer groups play in certification and feelings of marginalisation in the face of industry consolidation, and many new cooperatives and associations have been created or re-stimulated with the support of traders, leading to (revitalised) national associations such as the Federation Ivoirienne des Producteurs de Café Cacao, Ivory Coast and Cocoa Organic Farmers Association Ghana, and the umbrella World Cocoa Farmers Organisation (WCFO) and World Cocoa Producers Organization, representing cocoa farmers globally. The re-emergence of local, national and international cocoa farmer organisations and their consolidation, such as the merger of the International Cocoa Farmers Organisation (ICCFO) with the WCFO, hopes to counterbalance the consolidation of power and voice among traders and manufacturers. All these types of organisations have engaged in European and International forums related to their member's contributions to a sustainable cocoa economy. This has been manifested practically – such as starting their own programmes, studies and supporting members, as well as engaging and lobbying in (inter)national forums such as the World Cocoa Conferences. The support of both international trade associations and inter-governmental organisations has been instrumental in the upscaling of voluntary certification schemes and standards. Vertical upscaling has occurred as projects have become institutionalised through corporate and sector-wide programmes and standard practices for both the Dutch importers, processors, manufactures and retailing companies and their farmer suppliers.

3.2 Voluntary sustainability standards

Among VSS, four standards (UTZ Certified, Rainforest Alliance, Fairtrade and Organic) dominate and have increasingly been adopted by cocoa trading and chocolate manufacturing and retail companies as a way to achieve more sustainable practices in their operations and their supply chains. Combined, they certified a minimum of 1.2 million ha and a maximum of 2.7 million ha in 2013 (average 2 million ha), mainly in the largest cocoa-producing countries in West Africa, Brazil, Peru, Honduras, Dominican Republic, Indonesia and Vietnam, (Potts et al. 2014; International Trade Centre (ITC) 2015). A new standard, the Sustainable Cocoa Standard, is currently being developed concurrently by national committees, the European Committee for Standardisation (CEN) and International Standards Organisation (ISO). Outscaling has occurred with an increase in the number of certification schemes for cocoa over time, since Fairtrade began certifying cocoa in 1987. UTZ-certified cocoa grew sevenfold in the period 2010–14 and accounted for 15% of the global cocoa area and thus the largest VSS-certified cocoa area, with Rainforest Alliance accounting for the largest area under cultivation (2008–13) (International Trade Centre (ITC) 2015). Three certification schemes (UTZ, Fairtrade and Rainforest Alliance) dominate the voluntary standards adhered to in terms of volumes produced and numbers of farmers covered, with a fourth standard (ISO-CEN) currently under development. The organisations who initiated VSS, such as UTZ, were smaller companies, and NGOs, such as Rainforest Alliance, translated knowledge gained from cultivating timber, coffee and banana to cocoa, and Fairtrade/Max Havelaar, which also grew from working with coffee farmers. The Dutch government has participated in the member NEN meetings. The current standards in operation have been set without any direct role of governments, founded largely by NGOs, with some business interests. The four main schemes initially had different focuses on environmental and social issues: Fairtrade on ethical issues, Rainforest Alliance on deforestation, IFOAM Organic on pesticide free farming and UTZ on mainstreaming sustainability in farming practices. All of the four main standards have a strong focus on farmer practices, with chain of custody approaches, traceability and transparency also playing an important role, reflecting concerns by consumers and NGOs about how the chain is governed. A comparison in 2014 of the main certification schemes' global standards shows more harmonisation, with 47% of the 15 main environmental, social and economic issues and 27% of the 346 criteria commonly shared (ICCO 2014). The revisions to UTZ and Rainforest Alliance standards in 2014, and the standards drafted by ISO/CEN, show that harmonisation towards an even more common framework has further increased, with the ISO standard promoting itself as 'one standard to bind them' (Neiburg 2016). Most of the major traders/exporters and processors have now adopted existing certification schemes, indicative of replication scaling among stakeholders in the chain. Small-scale, speciality, fine-flavoured cocoa and chocolate producers have been less involved in certification schemes, partly due to the costs involved, as well as different marketing and consumer engagement strategies (Abdulmalik et al. 2016). Complaints by industry, farmers and support organisations about the complexity and proliferation of certification standards have also forced different platform organisations in the chain [such as ISEAL Alliance (ISEAL Alliance 2010) and ICCO] to understand differences and complementarities between standards, and harmonise requirements such as auditing and data provision to reduce the cost and time of certification (KPMG 2012; Resolve Inc. 2012; ICCO 2014). Vertical upscaling has occurred as certification has become mainstreamed and institutionalised through standard practices for the cocoa chain from Dutch importers, processors, manufactures and retailing companies to their farmer suppliers, as shown in Fig. 2.

3.3 Corporate sustainability initiatives

The most varied type of sustainability initiatives are individual corporate initiatives, programmes and projects. Companies demonstrate their commitment to addressing sustainability and scaling up their initiatives in different ways, often combining corporate programmes with certification, using one or multiple third-party voluntary certification schemes to demonstrate their own practices and those of their suppliers. Corporate-based sustainability initiatives including own brands and programmes appear to be the result of the short chain between trader, manufactures and consumers. Many smaller companies are not involved in platforms or certification standards. Instead they voice their opinion through professional trade associations or their own branding and approach to addressing the sustainability side of business. Large traders and processors such as Cargill and ADM have developed their own certification standards to increase the possibilities for compliance at a lower cost while protecting their own corporate interests. Large manufacturing and retail companies such as Unilever and AHold have also set their own sourcing guidelines which place stringent requirements on their suppliers.

Some of the earliest corporate programmes and projects focused on setting up certification schemes, such as PUK and UTZ, often in partnership with CSOs and via consumer country government agencies (such as the Swiss, Dutch and German), and partnerships with CSOs such as Cargill (with Solidaridad), Continaf, Ferrero and Petra Foods (with Oxfam). Companies have also partnered with international NGOs (for example, Care International and VSO) and local NGOs in countries of origin, as well as with national governments (such as Cocobod and extension services in Ghana, the Conseil du Café Cacao in Ivory Coast). Some initiatives by the large trading and manufacturing companies have been supported by governments such as the Dutch. Most corporate initiatives first focused on 'sustainable cocoa farming', a wide-ranging term which intertwines disseminating techniques (through classroom and field training, demonstration plots and pilots) to raise productivity via good agricultural practices (GAPs), quality and some small-scale public-relations-oriented community development activities. In the last few years more initiatives have focused on professionalising farmer organisations and farmers'

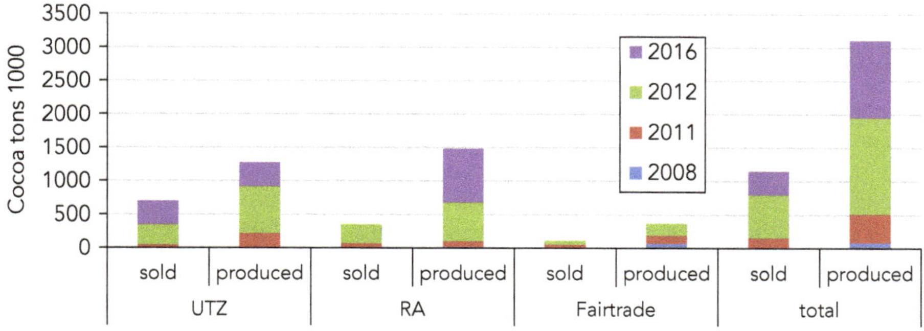

Figure 2 Scaling up of certified cocoa production and sales globally, 2008–16. Source: Fountain and Hutz-Adams (2015).

access to and supply of inputs including credit. The initiatives have various characteristics in common as discussed below:

Most corporate initiatives have been implemented or scaled up horizontally and vertically significantly, doubling or tripling the number of farmers included in programmes and the financial costs of providing services in the period from 2007 to 2010 to date. Although a few initiatives existed prior to 2007, they were much smaller and seen mainly as *ad hoc* activities, than sustainability programmes, as seen today. Data about the numbers of farmers, the organisations, activities and costs of these programmes are difficult to interpret, and financial data on activities tend to be reported in global terms in corporate reports and are not broken down per farmer or producer organisation, or compared over time. For example, Mondelēz Cocoa Life programme started in 2012, building on support to farmers commencing in 2008 in Ghana and growing from 38 000 farmers in 2014 in six countries, and aims to reach 200 000 farmers in the six countries by 2022. Cargill's Cocoa Promise programme which started in 2012 developed from activities in Ghana and Ivory Coast, dating back to 2003. A total of 60 000 farmers were supported in the Ivory Coast in 2008, which grew to 115 000 farmers in the two countries in 2014.

The majority of the initiatives have used and scaled up activities by supporting certification schemes at farmer group level since 2007; often this has meant also supporting farmers to organise into groups, and providing support to develop the management, business and organisational skills of farmer groups. Many corporate initiatives have recently slowed down on scaling up and turned to scaling out and horizontal scaling, increasing the types of services and support provided to farmers and producer organisations. Most corporate initiatives have involved significant scaling up (extending and replicating) of the support provided by Dutch-based traders, processers and buyers to small-scale cocoa farmers, almost exclusively via their producer organisations (cooperatives and unions). In the last four years, certification has increasingly been accompanied by scaling up access for farmer organisations to (directly or via providers) inputs such as approved fertilisers, pesticides and credit. Corporate sustainability initiatives have out-scaled in the last two years to include not just GAP but also management and organisational aspects of producer organisations. Companies have been increasingly vertically and horizontally scaling up and replicating their sustainability initiatives, some with support of public sector – the Dutch IDH, German GIZ, Swiss Contact and internationally via ICCO and through making more direct contact at farmer level with governments of producer countries – for example, contacting government agencies in Ghana and Ivory coast to provide training to farmers.

3.4 NGO and CSO initiatives

Strong NGO and CSO lobby and advocacy have played an important role in highlighting the main sustainability issues. They have accomplished this through targeted public campaigns directed at both consumers and companies. Examples include the agreement signed in 1995 by the US State Department and ILO, and news items by the BBC and CNN, and subsequent campaigns regarding slave and child labour in the cocoa sector, dating from the mid-1990s and Oxfam's 'chocolate letter' campaign in 2009. The creation, intensification and/or upscaling of sustainability initiatives shows a degree of correlation with consumer awareness campaigns. CSOs are of the opinion that industry crises caused by campaigns and negative press coverage (such as child labour) generate consumer awareness, which results in pressure to transition to more sustainable supply strategies and government involvement. The approach of the vocal social and environmental movement concerning cocoa and chocolate – exemplified by NGOs such as Oxfam Novib, Solidaridad, the Tropical Commodity Coalition (TCC), Voice Network and trade unions (such as the Dutch FNV) – has changed. In contrast to using traditional opposition methods and campaigning to highlight critical issues such as labour conditions, environmental awareness and living wages, they now tend to take a more partnering approach to exert more influence within the cocoa chain and to mainstream sustainable cocoa initiatives. As an example, the TCC encouraged companies to work together and enabled changes on a sector-wide level in the Netherlands, such as the 2010 Letter of Intent for sustainable cocoa (Ministerie van Landbouw Natuur en Vodelselkwaliteit 2010; Logatcheva and Ingram 2014). The commitment is comprised of convening a working group of companies, trade unions, NGOs and governmental bodies who then signed a declaration of intent stating that by 2025 all chocolate sold in the Netherlands will be sustainable. The commitment was instrumental in bringing together multiple stakeholders and stimulating them to both adopt certification and examine other ways of making cocoa more sustainable. The 'consensus based' approach, particularly in the form of a voluntary agreement with a facilitating role of the government, often leading to public–private and CSO partnerships has been a common approach in environmental policy-making during the last two decades to improve the sustainability of international commodity value chains (Vermeulen and Seuring 2009).

3.5 Interlinking initiatives

There is a high degree of interlinkage between the three types of sustainability initiatives. Platforms, networks and associations have been intensely used to both launch initiatives and support the implementation and replication for the upscaling of corporate and voluntary certification, for example, certification standards partnering with traders and international NGOs, service providers and consultants such as Cargill, Solidaridad, UTZ and IDH. Traders have collaborated with certifiers, for example, Cargill with UTZ and Rainforest Alliance, and private sector companies have partnered with research organisations, such as the Sustainability Consortium (TSC) and the community of practice learning groups financed by IDH.

4 Case study: the impacts of cocoa sustainability initiatives in West Africa

This section introduces a case drawing on studies on the impact of voluntary certification and associated sustainability initiatives on cocoa farmers and their cooperatives in Ghana from 2011 to 2014 (Waarts et al. 2013, 2015; Ingram et al. 2016b) and in Ivory Coast from 2013 to 2016 (Ingram et al. 2013, 2014b,c,d, 2016b).

4.1 Approach

The impact evaluation was based on quantitative and qualitative interviews, accompanied by the development of impact logics for each country with a certifying organisation (UTZ Certified), cocoa traders and a financing organisation (IDH, the Initiative for Sustainable Trade), followed by two rounds of interviews (in 2012 and 2015 in Ghana, and 2013 and

2017 in Ivory Coast) with 725 in 97 cooperatives in Ivory Coast in 2013, and 385 farmers in six cooperatives in Ghana in 2012 and 352 farmers in six cooperatives in 2015 who were participating in UTZ Certified certification for different periods of time and situated in the three main agro-ecological zones in both countries. In Ghana 120 farm labourers and 41 farmers were also interviewed in 2015 specifically about their working conditions. Certified farmers were compared to non-certified farmers (55 in Ivory Coast and 127 in Ghana in 2012 and 117 in 2015) who had not participated in the certification programme, were cooperative members, and situated in the same three agro-ecological zones at least 10 km from certified cooperatives. The size of 99 farms was measured in Ivory Coast. In-depth interviews were also conducted with six cooperative managers (six in Ivory Coast), 12 traders (seven in Ivory Coast and five in Ghana) and four support organisations (two in Ivory Coast and two in Ghana, including COCOBOD, the Ghanaian Cocoa Board). A total of 18 focus group meetings were held with villages in the communities where farmers and their cooperatives are located (ten in Ivory Coast in 2013 and eight in Ghana in 2015), and in Ivory Coast, with eight school teachers to obtain more qualitative information on interventions and impacts.

In both countries, fifteen environmental, economic and social indicators were used to measure the outcomes of the plans implemented between 2008 and 2013. These correspond to UTZ-certified 'Better farming, better future' theory of change. Statistical and qualitative analyses of the indicators at farm household level were then conducted. Using the results of the interviews, farmer's perceptions of changes in the indicators were analysed. Comparisons of the indicators were made to see whether any significant differences could be found by comparing: farmers participating in the programme for different periods of time (ranging from zero to five years); farmers located in different agro-ecological zones; farmers participating in the UTZ programme and farmers not participating (the comparison group); and farmers participating in the UTZ programme who are certified and those not-yet certified. The results were benchmarked to grey and literature data about the indicators were peer reviewed to provide context and comparison with other cocoa farmers in each country. Preliminary results were presented and verified at meetings in Ivory Coast with cooperative managers and representatives of traders, NGOs in Abidjan and Amsterdam in October 2013, and in Ghana in June 2012 and 2015. During these meetings an assessment of external influences which could affect farmers' performance on the indicators, such as the Ivorian government's reform of the sector and the weather effects on cocoa production, was also made.

Quantitative data were analysed in two ways. First, farmers in the baseline studies were grouped and an agglomerative hierarchical cluster analysis was performed to classify clustering into different groups based on the indicators using Euclidean distance, Ward's linkage method of aggregation and analysis of variance. Second, a difference-in-difference analysis was used to compare changes in indicators over time (2011 and 2014 in Ghana) for UTZ-certified farmers compared with non-certified farmers. A t-test was used to verify the statistical significance of the difference between years and the difference-in-difference between UTZ-certified and uncertified farmer groups. For each indicator, statistical analyses, including t-tests and regression analyses, were conducted to examine the impact, to test whether the impact might be attributed to different factors and to gain insight into the robustness of the results. Propensity score matching tests (kernel matching, nearest neighbour and stratification methods) were used to assess the influence of specific interventions: whether farmers participated in trainings, total number of trainings they participated in, and whether or not farmers are lead farmers in the programme.

4.2 Results: impacts of cocoa sustainability initiatives in West Africa

A cluster analysis of the farmers participating in the baseline studies in Ghana and Ivory Coast (van Rijn et al. 2015) identified five main 'types' of farmers, with clear profiles in terms of their age, farm size, income farm ownership, costs, location in different agro-ecological zones, length of time participating in certification schemes and knowledge and adoption of GAPs. These clusters were based on statistically significant meaningful differences in cocoa production: in Ghana yields ranged from 0.83 to 4.47 bags/acre with an overall average of 2.03 bags/acre. For 'professional farmers' with average yields of 4 bags/acre, typified having larger than average family size, putting in a higher number of labour hours into their farm, good knowledge and implementation of GAPs, invests a lot in their farm and with a high net income. The higher than average (2.6 bags/acre) yielding 'sustainable professionals' had high levels of good knowledge and implementation of GAPs, likely to own the land they farm, above average education levels, with a heavy reliance on cocoa for their household income and above average net income from cocoa. In the Ivory Coast, yields ranged from 350 to 1783 kg/ha, with an overall average of 499 kg/ha. The cluster of farmers with above average yields could be seen as 'Professional investors': typically men In their late forties, efficient workers, with top yields of on average 1077 kg/ha and high income, excellent income/cost ratio due to high turnover and low costs, large farms, mostly found in the excellent agro-ecological zone, who own most of the land farmed, had the longest participation in project activities, and focussed more on production and less on GAPs. These differences in farmer profiles could be used to aid traders, NGOs and extension agents in differentiating the services and support offered to farmers, helping them target different types of with services (such as training, credit, technology and certification support) which may be more likely to benefit them.

4.2.1 Ghana

In Ghana, the impact assessment (Waarts et al. 2013, 2015, Ingram et al. 2016b) comparing the differences between certified and non-certified farmers performances in 2011 and 2014 showed that predominantly older men participated in interventions, attributed to cultural norms regarding farm ownership and cooperative membership, and as farm managers are on average 46 years old. All farmers improved their knowledge and implementation of GAPs by about 30%, but productivity levels remain low for all farmers, and the quality of cocoa has deteriorated since 2011. No differences were found between certified and uncertified farmers with regard to changes in knowledge and the implementation of GAPs, nor were there differences in terms of changes in productivity and quality over time. Cocoa farming continued to generate low incomes, although profitability increased per hectare for all farmers but decreased per day spent by family labour. Price increases in 2014 offset the rising costs of production (especially for hired labour) and increased cocoa profitability per hectare for all farmers. However, cocoa farming became less attractive in terms of profit per day spent by family labour, as it decreased over time, compared to wage levels for hired labour. No differences were found between groups. Income from cocoa and total household income increased for all farmers but remains low. Both the income from cocoa and the total household income have increased over time, and we see similar increases for both certified and uncertified farmers. Per person, income from cocoa is slightly lower than the USD1.25 poverty line in 2014, while the total household income is slightly higher than this poverty line per household member (USD1.40). The minor difference between income from cocoa and total household income indicates cocoa farmers' continued high dependency on income from cocoa.

A positive impact was found regarding working conditions, awareness of child labour issues and health. Certified farmers are more aware of what children should not do on the farm, as well as the benefits of going to school, than uncertified farmers. Children did spend time on cocoa farm activities, but farmers stressed that their work on the farm is important, so that they can learn about cocoa production. Almost all children (97% of children aged 6–11 and 93% of children aged 12–17) go to school (in 2014). We see no difference over time between certified and uncertified farmers regarding school enrolment for both age groups. As we do not know whether farm activities were done by children instead of going to school, we cannot determine whether these activities should be considered 'child labour'. Certified farmers mentioned that certification has had a positive effect on their health, changes not indicated by uncertified farmers. A few (8%) certified farmers did not comply with the UTZ Code of Conduct regarding child labour and children under the age of 18 performed hazardous tasks at 19 certified farms in 2014, albeit these activities were minimal: children on average spent less than one day per year performing hazardous activities. None of the children under the age of 18 were involved in pest or disease control in either 2011 or 2014. No difference over time was found between certified and uncertified farmers regarding the time children under the age of 18 spent on doing hazardous tasks. Nine certified farmers (4%) had children under the age of 14 assist them in hazardous activities in 2014, compared to seven uncertified farmers (6%). The total yearly time spent is very low for both groups, 0.15 days versus 0.60 days per year on average, respectively.

Reasons why certification has not left a mark yet in Ghana could be that training uncertified farmers allowed them to develop in a similar way as certified farmers. The impact of input services provided by traders also probably has not taken effect yet, as not all farmers use these services, and it takes time for fertiliser use and planting seedlings to affect productivity. Unfavourable climatic conditions were indicated to negatively influence productivity in 2014, as this affected both uncertified and certified farmers and the changes in the productivity cannot be attributed to certification. For the whole sample, enhanced knowledge on GAPs is associated with better GAP implementation. Farmers who improved GAP implementation were more likely to have improved productivity than farmers who had not. There was a positive association between improved knowledge levels and improved profit per kg of cocoa. Farmers who improved their GAPs were more likely to have lower profit per kg. Improvements in efficiency were associated with lower profitability per hectare, and increased productivity per hectare was associated with higher profitability, due to the 2014 cocoa price increase.

4.2.2 Ivory Coast

Voluntary certification schemes have been implemented in the Ivory Coast through partnerships between UTZ certification and eight traders, seven also have partnerships with Rainforest Alliance and two with FairTrade. The traders have working relationships and partnerships with cooperatives from which they purchase beans. All of the traders have their own corporate social responsibility programmes. Of the farmers participating in the UTZ programme, 21% were also Rainforest Alliance-certified and 2% were both UTZ and FairTrade-certified. This reflects the general trend: as of June 2012, 51% of 86 UTZ-certified cooperatives had multiple certifications. Both prior to and during the UTZ certification programme there have been and continue to be over 40 sustainability interventions (Hatløy et al. 2012; Ingram et al. 2014b,c,d, 2016b); see Table 1. Many interventions occur

Table 1 Overview of sustainability initiatives in the cocoa sector in Ivory Coast, 2008–16

Main implementing organisation(s)	Sustainability initiative
Network, association and platform initiatives	
World Cocoa Foundation (WCF)	Livelihood programme Cocoa link WCF empowering cocoa households with opportunities and education solutions (ECHOES) WCF African Cocoa Initiative (WCF/ACI) is a public-private partnership, bringing together WCF, cocoa industry members, the Sustainable Trade Initiative (IDH) and U.S. Agency for International Development through its Global Development Alliance Certification Capacity Enhancement (CCE) project African Cocoa Initiative (ACI) African Cocoa Initiative (ACI)
International Cocoa Organisation (ICCO)	Capacity Building Programme on Pesticides Residues and other Harmful Substances in Cocoa in Africa Cocoa productivity and quality improvement: a participatory approach Analysis of the value chain in cocoa-producing countries Cocoa germplasm utilisation and conservation: a global approach Improvement of cocoa marketing and trade in liberalizing cocoa-producing countries supply chain management for total quality cocoa: pilot phase Pilot project on price risk management for cocoa farmers Preventing and managing the spread of cocoa pests and pathogens: lessons from the witches' broom disease Cocoa of Excellence: promoting diverse high-quality cocoa origins SPS capacity building in Africa to mitigate the harmful effects of pesticide residues in cocoa and to maintain market access
National Confectioners Association, CAOBISCO, ECA & farmers	Regional trade associations and their memberships
UNDP Green Commodities Facility	Green Commodities Facility, Cote D´lvoire Sustainable Cocoa Initiative NORAD, World Cocoa Foundation (WCF), International Cocoa Initiative (ICI), Echoes – Youth Education and Livelihoods Programme, UNDP and the Associations of Chocolate Manufacturers from Denmark, Finland, Norway and Sweden
ILO	International Cocoa Initiative
Voluntary standards schemes	
UTZ (+ Solidaridad)	UTZ certification
Rainforest Alliance (+ GIZ)	Rainforest Alliance certification
Fairtrade (+ Agro Eco Louis Bolk Institute & Rabobank, the Dutch structure Control Union for organic certification and FAIR TRADE)	FairTrade certification

(*Continued*)

Table 1 (*Continued*)

Main implementing organisation(s)	Sustainability initiative
Organic (+ Agro Eco Louis Bolk Institute)	Organic certification
Corporate sustainability initiatives	
Cargill, ADM, Barry Callebaut, Armajaro-CI, Outspan, Ecom, CEMOI & farmers	Corporate programmes with consultants, cabinets, ANADER
Olam International and Blommer Chocolate & farmers	Alliance between cocoa farmers in Ivory coast, Olam International and Blommer Chocolate
Mondelez (Cadbury), Conseil du Café Cocoa (CCC), CARE farmers	Cocoa Life programme to help farmers increase sustainable cocoa production and create thriving communities
Nestlé & farmers	Cocoa Plan, action plan responsible sourcing CNRA under the initiative of creating added value
Kraft Foods and Hans Neumann Stiftung & farmers	Sustainability alliance with Rainforest Alliance Market-oriented promotion of certified sustainable cocoa
Mars & farmers	Sustainable Cocoa Initiative [cocoa development centers (CDC) and cocoa village clinics (CVC)]: rehabilitation of old and ageing farms with good planting material, soil fertility management, solid agricultural practices including pest and disease control IMPACT project with government of CdI, ICI, AIECA, AFRICARE, SOCODEVi, STCP, Rainforest Alliance, IFESH, INADES, BFCD
ADM, Barry Callebaut, Cargill, Ferrero, The Hershey Company, Kraft Foods, Mars Incorporated and Nestlé & farmers	Framework of action: Harkin-Engel protocol (responsible cocoa) and industry partnership and public certification: development of a public certification process.
Partnerships	
Sustainable trade initiative (IDH) & private sector partners (Cargill, Ecom, Barry Callebaut CEMOI, Mars) UTZ, Rainforest Alliance	Cocoa improvement programme, Cocoa Productivity and Quality Programme (CPQP), soil fertility programme
Signatories include governments and representatives of the cocoa industry and witnesses include social activists, NGOs and labour unions	International Cocoa Initiative (ICI) to eliminate the worst forms of child labour and forced labour and the Harkin Engel protocol
Mars incorporated, Hershey company, Kraft foods and Armajaro trading	Agriculture Development Programme (CAADP)

Main implementing organisation(s)	Sustainability initiative
Institut Européen de Coopération et Développement IECD/ Cargill/MAH, Dutch Ministry of Agriculture, ANADER IECD, PEFACI; Ministry of Agriculture, Department of Animal Production, Ministry of Education, Plate forme des Ecoles Familiales Agricoles de Côte d'Ivoire (PEFACI)	Projet Ecoles Familiales Agricoles (EFA)
GTZ/GIZ, USAID, ANADER, STCP, Kraft, Armajaro	Market-oriented promotion of certified sustainable cocoa production Côte d'Ivoire (2005–09)
International Institute for Tropical Agriculture (IITA) + USAID, Primature, MINAGRI, CGFCC, FIRCA,GEPEX, ANADER, CNRA ONG, BFCG, INADES, SOCODEVI, Rainforest Alliance, BFCD; GTZ, Technoserve	Sustainable Tree Crop Program (STCP)
CIRAD & CEMOI	Creation of the cocoa centre of fermentation and sun drying
ICRAF (World Agroforestry Center)	Vision for change farmer training programme
Payson Center for International Development and Technology, Tulane University	Annual Survey of Child Labor in the cocoa-growing areas of Ivory coast and Ghana
FAFO	Research program on Trafficking and Child Labour. Child labour and cocoa production in West Africa
	Côte d'Ivoire Sustainable Cocoa Initiative (CISCI)
Government initiatives	
GIZ	Programme de Développement Economique en Milieu Rural (PRODEMIR)
USAID	Towards child labour free cocoa growing communities in Ivory Coast and Ghana through an integrated area based approach
Ivory Coast Exportation Professional Association (APEXCI), Cocoa & Coffee Interprofessional Board (CICC), Raw Materials Interministerial Board (CIMP), CAISTAB	Implements national development plan and regulates all activities of coffee-cocoa sectors
Ministry of Agriculture	Fonds Interprofessionnel pour la Recherche et le Conseil Agricole (FIRCA)

(*Continued*)

Table 1 (*Continued*)

Main implementing organisation(s)	Sustainability initiative
Cocoa and Coffee Management Council/Conseil du Café Cocoa (CCC)	Ivory Coast quality cocoa control programme National programme of fight against disease of the cocoa swollen shoot
Centre National de Recherche Agronomique (CNRA)	National Agricultural Centre conducting agronomical research
SOCODEVI, ANADER; NGOs; cooperatives	Mutual and Cooperative Partnership Programme (PPCM)
National Agency for Rural Development (ANADER)	Extension services, promotion of farmer's skills and entrepreneurship by designing and implementing appropriate tools and conducting agricultural extension services. Fight against disease swollen shoot (pilot project) Project certified sustainable cocoa production
Information Programme on the Cocoa and Coffee Markets (PRIMAC)	Programme for the intensification of local processing 50% of the overall cocoa production in the year 2005, etc.
Ministry of Agriculture	Master plan for agricultural development 1992–2015 (PDDA)
Comité de Gestion de la Filière *Café Cocoa* (CGFCC)	Cocoa-related institutions
Ministry of Agriculture	Member of COPAL (alliance of cocoa-producing countries), COPAL activities
NGO initiatives	
Oxfam	Behind the brands – cocoa case studies
World Vision	Anti-child labour campaigns
Solidaridad	Cocoa improvement programme

both nationally and on a very local scale, implemented by multiple partners. There is remarkable similarity between these interventions and those implemented as part of UTZ and Rainforest certification, which makes it difficult to attribute changes specifically to certification. Specific activities implemented by traders participating in the UTZ certification programme included cooperative capacity building, farmer training, farmer development, financial support, community development and processing. Strong differences were found between traders' approaches to implementing certification as a standalone activity or as part of a package of activities.

In the Ivory Coast, the baseline studies (Ingram et al. 2014b,c,d, 2016b) comparing certified with non-certified farmers indicated a similar situation to Ghana of mainly older men involved in cooperatives and certification. The rapid upscaling of certification and related activities since 2008 was found, with most farmers receiving training and certification and a proportion benefitting from access to crop protection products, fertilisers and seedlings, and community and social programmes. Perceptions were that the quality of training has been variable and lacks minimum standards. Generally, higher knowledge levels and improved implementation of record keeping were positively related with increases in farmer productivity. Overall, higher knowledge levels were positively related with improved implementation of GAPs. For specific GAPs (waste management, soil management, water and biodiversity protection), no positive correlation was found. No relationship was found between the implementation of GAPs and post-harvest practices and bean quality. Yields and profitability at farmer and cooperative levels are low. By revitalising the sector, certification appears to contribute to making cocoa farming more attractive. Other outcomes of certification were that it provided a means to rapidly upscale sustainable (certified) cocoa production and allowed farmers to access certified markets where they can benefit from premium prices. Certification promoted professional producer associations which farmers perceive as providing a range of benefits, although farmers and cooperatives indicated some of these could be improved. Farmers indicated that the certification and associated interventions led to increased productivity and income. Eighty two per cent of farmers indicated an improvement in their living conditions since participating in certification and 92% of farmers indicated positive changes after certification. Farmers believe that higher incomes from increased participation led to a larger proportion of the cocoa income given to their spouses, being used to meet basic family needs and for children's schooling. Farmers' incomes appeared to increase with certification, but farmers had concerns about the long-term viability of cocoa farming and possible discontinuation of the premium for certified cocoa. About 50% of farmers said income increased since certification, with net household income from cocoa for certified farmers on average 1 535 000 CFA (2343€) and for non-certified farmers 1 318 000 CFA (2013€) in 2012. The longer farmers participated in the UTZ programme, the higher net income they earned. Programme participants and UTZ-certified farmers do not earn a significantly higher net income than non-certified and control group farmers. Average production costs for UTZ programme participants were 152 CFA/kg, compared to 129 CFA/kg for farmers not participating. The extent of participation in the programme did not influence production costs. Average production of UTZ programme participants was 453 kg/ha, for the comparison farmers 329 kg/ha, with certified farmers having higher yields (467 kg/ha) than non-certified farmers (315 kg/ha). Cocoa farming formed on average 79% of total gross household income and was the only or the main source of cash income for most farmers.

4.3 Conclusion

The case study shows that sustainability interventions in the cocoa sector, such as voluntary sustainability standards, predominantly UTZ certification and Rainforest Alliance certification accompanied by supporting corporate initiatives, have been extensively scaled up by international traders in Ivory Coast and Ghana. Voluntary standards were later combined with initiatives financed and implemented by government agencies, traders' corporate initiatives, by CSOs and NGOs and network, platform and associations. The outcomes for farmers' incomes, cocoa productivity, farmer's and farm worker's living and working conditions, agricultural practices and environmental effects have been mixed and generally modest, compared over time and compared to outcomes experienced by non-certified farmers in similar agro-ecological regions. The five main farmer clusters found indicate that as most initiatives have been aimed either at all farmers in a producer group, or 'self-selecting' farmers who respond to participate to an initiative, a possibly more effective option could be to invite only targeted farmers with specific profiles to participate in sustainability initiatives.

5 Future trends

Trends matter

Identifying trends can contribute to solving some of the challenges confronting the cocoa sector. Confronted with global forces shaping the cocoa value chain, many stakeholders assume that their ability to shape the future of this sector is minimal. Whilst changing a demographic trend or widespread shift in consumer consciousness can be difficult, reacting to such forces or, better, anticipating them can be advantageous. Ignoring trends can be perilous.

5.1 Trends in sustainability initiatives

Sustainability initiatives in the cocoa value chain have been dramatically upscaled through replication, extension and horizontal and vertical scaling. This has resulted in an increased number of private sector, as well as civil society, organisations adopting and supporting sustainability initiatives. The coverage of these initiatives has also been significantly scaled up, in terms of the market share of 'sustainably produced' products, production volumes, the numbers of farmers and their organisations producing cocoa in a sustainable way, and the area of cocoa produced using sustainable production methods. The location and extent to which market and technical information is disseminated along the chain has been a key element in determining the balance of power between farmers and processors, with power mostly concentrated among the large processors and traders.

The VSS emerged in the shade-grown production systems of Latin America (notably Organic production), Fairtrade, Rainforest Alliance and UTZ cocoa and rapidly expanded into mainstream markets in full-sun production systems in West Africa and Indonesia. This development shows promise for the continued growth and relevance of VSS. It also indicates a convergence in the two challenges of systemic poverty and low cocoa productivity. Poverty fuels, among other things, unacceptable labour practices. Supply is limited by old tree age and exhausted soils, threatening overall global supply. As cocoa purchase is dominated by the global North with a history of promoting VSS (notably the Netherlands – but expanding to other among consumer countries) this points towards continued growth prospects for standard-compliant cocoa (Potts et al. 2014). With cocoa supplies tight and overall prices relatively high, the current market situation provides an opportunity for increased investment in the sector towards more sustainable practices, and high prices allow costs for transitioning to sustainable practices to be absorbed and not passed onto consumers (Potts et al. 2014). However, to date the largest share of costs of the transition to sustainable cocoa production has been borne by farmers, with a proportion of costs carried by traders (Victor et al. 2010; Kessler et al. 2012; Beukering et al. 2013; KPMG Sustainability 2013; Waarts and Kuit 2014). This is despite the largest share of value adding remaining within the later stages in the chain in manufacturing (32%), retail and taxes (44%), in contrast to 6.6% of value with farmers (Fountain and Hutz-Adams 2015). There remains also a significant gap between existing production levels of standard-compliant cocoa and actual sales, and the implementation of commitments to purchase sustainable cocoa (Fountain and Hutz-Adams 2015).

Sustainability initiatives have resulted in subtle changes in power, with an increase in the provision of technical information disseminating GAPs through strengthened farmer organisations. As more large-scale professional farmers emerge, particularly in Latin America, different models have arisen, with a focus on sustainable high-quality, speciality niche products and bean-to-bar brands (Naprta 2015; Abdulmalik et al. 2016).

The majority of sustainability initiatives are directed at countries supplying the consumer markets in Europe, North America and Australia. This is the result of the importance placed upon sustainability issues in Europe, as a consequence of increasing consumer awareness. In emerging wholesale and consumer markets (e.g. Brazil, China, India and other emerging economies) sustainability issues often do not have the same level of importance, which decreases the motivation to upscale sustainability initiatives. One clear trend has been the harmonisation between standards and a move to 'certification plus', looking beyond certification as the silver bullet to solve all of the challenges in cocoa production (Barrientos 2011) to certification plus and a broad range of service delivery by traders and partners, including origin and consumer governments. Governments, particularly in consumer countries, have been forced by consumer pressure to contribute to a more sustainable cocoa value chain, mainly through facilitating and mandating policy instruments, rather than regulatory mechanisms (Ingram et al. 2016a). There are also debates about how and if certification and services marginalise some types of farmers, especially the poorest and should be more inclusive – not just the largely male farm owners, but also women, youths and workers.

Some standards in some situations have created welfare gains (Potts et al. 2014) but also distributed rent which induces lobbying by interest groups to set the standards at their preferred level. The number of sustainability initiative impact studies has increased and the ways and metrics in which impacts are being measured are being harmonised to be comparable (such as the World Cocoa Foundation's CocoaMAP cocoa measurement and progress framework based on key performance indicators, work springing from ICCO's Direct Dialogue Workshop on Certification in 2014 (ICCO 2014), and work by ISEAL Alliance on common core indicators and the Committee on Sustainability Assessment) (Blonk et al. 2010; COSA 2013; Maytak 2014; World Cocoa Foundation 2016). However, only a limited number of studies from around the world are available that can be used to reach conclusions on the impact of certification: with reported, observed and measured impacts not being solely attributed to certification, as contract arrangements between traders and farmers also play a role in achieving social, economic and environmental impacts (Kuit and Waarts 2014). Because the counterfactual ('what would have happened with the sustainability initiative') is seldom rigorously evaluated, most existing research cannot say much about possible causation. A number of studies of sustainability initiatives, largely certification in the main producer countries, provide contradictory evidence (COSA 2013). These factors combine to make it difficult to distinguish socially desirable standards from those resulting from political rent-seeking (Swinnen 2016). Evidence on the impacts of the three other types of initiatives: partnerships, corporate and NGO, is largely either lacking or *ad hoc*, and often not scientifically robust. Until (Swinnen 2016) comparable research on the impacts of sustainability initiatives can be measured, most of the fact-based conversations in the sector will revolve around estimates of cocoa household incomes, making it hard to accurately indicate the gap between current and desired income sufficient for a decent standard of living (Fountain and Hutz-Adams 2015). These discussions have led to more individual farmer-specific targeted interventions that continue to revolve around voluntary sustainability standards and certification but also look at livelihood diversification. The extent to which this is counterproductive for trader interests in securing supply is however questionable. With the realisation that certification cannot solve some of the most controversial problems such as farm gate pricing, child and slave labour, policies and regulatory mechanisms are being explored. There are also increasing attempts to link sustainability initiatives, especially farm-based certification, to other larger spatial-scale-oriented sustainability schemes such as landscape approaches, zero deforestation initiatives and reducing emissions from deforestation and degradation (REDD+).

Standards have played an important role in food trade for long. Their rapid growth has triggered vigorous debates on their impacts on international trade and development, with an argument that standards form non-tariff barriers to trade, resulting in differing adoption globally and reluctance of governments to make adherence to standards a legal requirement, instead of relying on facilitating them as voluntary, corporate social responsibility measure (ITC 2012). A result is that sustainability practices in the field can differ widely [Ruben 2008; Resolve Inc. 2012; International Trade Centre (ITC) 2010].

5.2 Research trends

A survey of cocoa research (Ingram et al. 2014a) highlights that there are research gaps in genetics and breeding, soil fertility, productivity and climate change. Areas for collaboration noted by respondents include sustainability indicators, standards and guidelines, data sharing and publishing data. The respondents indicated need to assess why over 50 years research on cocoa has not been able to solve the challenges and create more impact to date; the need to prioritise research themes, make data more widely available and translate data into (more) user-friendly formats; provide and maintain a map/inventory of which research institutions and researchers are doing what, where and with whom, to link up cocoa research with other related and applicable research fields and to collaborate on contributing to key performance indicators (KPIs) developed by national platforms, making appropriate recommendations when required. The survey revealed the priority research lines (and their links to the themes in the 2012 Abidjan Declaration (ICCO 2012a), shown in Table 2.

The challenges in the sector combined with growth in demand and new markets have encouraged researchers to look again at persistent problems such as tree genetics, improved soil fertility, pests and diseases and productivity. They have also opened up new avenues for more multi-disciplinary collaborations and research on cocoa and chocolate economics and markets – particularly for niche, high-quality and speciality chocolate products; climate-smart cocoa; living income for cocoa farmers and workers; health aspects – such as the growing preference for as dark chocolate; credibility of sustainability claims; transparency on traceability in the chain – linked to both food and safety; sustainable production conformity with certification standards and a particular focus on labour rights (child and slave labour).

6 Conclusion

Sustainability initiatives, particularly voluntarily introduced by companies, NGOs, CSOs and groups, associations and partners of one or several types of organisations, have developed in response to the multiple problems and challenges involved in producing cocoa in an economically, socially and environmentally sustainable way. Such initiatives have sometimes been mandated, supported or facilitated by governments, seeking to avoid regulations, particularly given global trade nature of the cocoa chain and fear of introducing tariffs or non-tariff barriers to internationally traded cocoa and chocolate products. Despite a growing number of studies on the impact of different initiatives, particularly scientifically robust impact evaluations of VSS, until comparable research on impacts of sustainability initiatives is produced, evidence of how effective and efficient sustainability initiatives actually are in solving the entwined social, economic and

Table 2 Priority research themes in cocoa

Priority	Research lines	Key themes
1	*Vulnerability and productivity* Production sustainability	• Production systems, performance and risk management • Production system diagnostics and gaining productivity increases • Breeding, productivity, resistance and quality • Pest and disease spread, control and monitoring, management, and early warning systems • Fertiliser and pesticide usage and effects • Land tenure and impact of technology transfer • Increasing productivity (optimising pruning) • Exchange of (new) germplasm by grower governments and improving supply of vegetative material/seedlings
2	*Diversity and sustainability of production* Production sustainability	• Sustainable intensification (including labour requirements) and optimising subsequent effect for example, upscaling, finance • Assimilating cocoa into total food crop systems • Genetics, national and local genetic variations • Sensory and organoleptic attributes and regional specificities • Sustainability of production (certification standards and costs) and costs (true price for farmers, industry and governments) • Diversification (crop/income) combined with intensification, impacts (yields, incomes) and costs • Effect of reducing child labour and optimising children's education • Strategies to increase youth farming/involvement • Social, economic and ecological diversity and resilience farmers, production systems, environment, organisations and chains
3	*Market transformation and transparency* Sector strategic management	• Market/chain interactions and economics • Chain of custody and certification • Rural transformation • Increasing gender equity in cocoa farming and inclusiveness of benefits • Transparency of profits and prices along the chain • Governance structures to increase transparency
4	*Quality and new business models* Consumption & chain sustainability	• Quality prognosis and improvement • Risk management (farmer and industry, farm economics, financing) • Production, processing and chain models, their efficiency and impacts • Models to effectively and efficiently upscale access to farm inputs, credit and other services • Models to develop the fine flavour market further • Perceptions (farmers, industry, researchers) of intervention logic of new models • Social-economic aspects of cocoa production[1] • Understanding consumer demands • Value adding at producer country level • Increasing/improving use of cocoa by-products • Enhancing the use of 'big data' and sharing data – particularly for monitoring and evaluation and impact assessment • Why do differences in implementation occur in practice? • Impact of research

(*Continued*)

Table 2 (*Continued*)

Priority	Research lines	Key themes
5	*Service delivery and concerted action* Industry chain sustainability & Sector strategic management	• Effectiveness of service delivery to smallholders – including extension services and ability/capacity to deliver services • Models of and impact of concerted and collective action • In-field practical training on the application of (new/improved) technologies • Governance models impact on service delivery and concerted action on production and quality • Scale and connectivity

[1] Participants indicated that socioeconomic aspects should be a higher priority than indicated in the survey results.

environmental problems besetting the cocoa sector is difficult to assess. This knowledge gap calls for more concerted, multi-disciplinary collaboration among researchers and the industry to provide a much needed reality check of what is working, where, when, how and why in terms of sustainability initiatives. Better information on the impacts of sustainability initiatives can help drive uptake, with VSS being one avenue to provide a standardised, credible, verifiable roadmap for sustainability which can result in measurable progress. This has two implications – particularly for scientists – of the need for much more intensive collaboration with public, private and civil sector to gather evidence about on-the-ground impacts, high-quality outcomes and measureable progress. The second implication is for policymakers – again supported by the research community – relating to the need to experiment with a wider range of instruments to increase cocoa value chain sustainability alongside voluntary initiatives. For example, consumer awareness campaigns help stimulate demand for sustainable products. Fiscal and monetary instruments – such as lowering trade barriers and providing tax advantages for sourcing and supplying sustainable certified products – are little used. They could, however, stimulate further upscaling and ensure all stakeholders pay the costs of sustainability initiatives, and provide more certainty about the level of return on investments in sustainability schemes. Currently no legislative instruments favour imports of sustainable cocoa, due to the complexities of world trade agreements; however they could create a more level playing field for importers and exporters. A wider range of instruments could contribute to upscaling, particularly when applied more at international and pan-regional level.

7 Where to look for further information

The ISEAL Alliance is an umbrella organisation open to all multi-stakeholder sustainability standards and accreditation bodies that demonstrate their ability to meet the ISEAL Codes of Good Practice and accompanying requirements, and commit to learning and improving. Many of the largest cocoa voluntary sustainability standards are members of ISEAL. ISEAL has developed codes of good practice for standards on credibility principles, standard-setting, impacts, assurance, independent evaluators and in claims and labelling.

The ITC Standards Map provides searchable overviews of 234 standards, codes of conduct and audit protocols worldwide addressing sustainability in supply chains and currently has data on 68 standards concerned with cocoa.

The Sustainability Impacts Learning Platform is a collaboration between the Food Lab, ISEAL and WWF and is intended to facilitate increased knowledge-sharing, collective learning and collaboration towards increased sustainability in global supply chains. Researchers and practitioners active in the field, including cocoa, are encouraged to contribute their studies to an interactive, searchable global map and use it to learn from and connect with others doing similar work.

The International Cocoa Organization ICCO is a global organisation composed of cocoa-producing and cocoa-consuming member countries. Recently relocated to Abidjan, Ivory Coast, ICCO was established in 1973 to put into effect the first International Cocoa Agreement negotiated in Geneva at a United Nations International Cocoa Conference. There have since been seven agreements, the 7th International Cocoa Agreement came into force provisionally in 2012. Their website provides cocoa production and price statistics, latest global news, basic information on growing and processing cocoa, pests and diseases, health and nutrition, cocoa trading and a searchable database on projects supported by ICCO.

Cocoa Connect is a digital platform to share and learn about sustainable cocoa production. Cocoa Connect is a knowledge broker and provides space to learn about the socio-economic domains of sustainable cocoa: farmers and production, chains and relations and markets and policies and acquire more technical knowledge on crops and soil. The site contains documents in an online repository, news items and matchmaking between organisations and individuals and host working groups on gender and fine flavour.

The LinkedIn International Cocoa Researchers Group provides a space for researchers from all disciplines conducting research on cocoa to link up to ensure the visibility of scientific evidence about cocoa, and provides a channel for its application to contribute to solve challenges in the cocoa sector. The group aims to bring together researchers at international fora and dedicated research conferences as well as regional and thematic exchanges.

The World Cocoa Foundation is an international membership organisation that promotes sustainability in the cocoa sector by providing cocoa farmers with the support they need to grow more quality cocoa and strengthen their communities. Representing more than 80% of the global cocoa market, WCF's programmes work with farmers in Africa, Southeast Asia and the Americas. The knowledge centre provides a searchable range of media resources, including research documents and updates, resource manuals, reports, videos, meeting minutes and presentations.

8 References

Abbott, P., M. Wilcox and W. A. Muir (2005). *Corporate Social Responsibility in International Cocoa Trade*. West Lafayette: Purdue University.

Abdulmalik, M., V. Ingram, K. Logchageva and A. Laven (2016). The 'middle' chocolate market in the Netherlands - A market differentiation study. In: *Chocoa 2016 Conference on Sustainable Cocoa: Preparing for the Next Generation*. Amsterdam, The Netherlands, February 2016.

Alonghi, T. (2011). Côte d'Cocoa the political and social effects of Côte d'Ivoire's cocoa sector. *Jackson School Focus* 2(2): 62.

Barrientos, S. (2011). Beyond fair trade: Why are mainstream chocolate companies pursuing social and economic sustainability in cocoa sourcing. Paper to ILO/IFC better work conference. Oxford: Oxford University Press.

Beukering, P. v., M. v. Drunen and O. Kuik (2013). *Valuing Economic Costs and Benefits of the Supply Chain of Cocoa, Soy and Palm Oil*. The Netherlands: Environmental Assessment Agency (PBL), Institute for Environmental Studies, p. 104.

Blonk, H., J. Scholten and R. Broekema (2010). *Measuring the Sustainability Performance of Agro-food Chain Initiatives: A Method for Estimating the Potential Sustainability Performance of the Initiatives in Which Transforum Participated*. Gouda, The Netherlands: Blonk Consultants.

COSA (2013). *The COSA Measuring Sustainability Report Coffee and Cocoa in 12 Countries*. Philadelphia, PA: The Committee on Sustainability Assessment, p. 100.

Eberhard Krain, E. M., E. Konan and E. Servat (2011). Trade and pro-poor growth: Introducing Rainforest Alliance certification to cocoa production in Côte d'Ivoire. Project, AfT category 2 (CRS code 25010). Eschborn, Germany: Deutsche Gesellschaft für Internationale Zusammenarbeit (GIZ) GmbH.

FAO (2007). Governance, coordination and distribution along commodity value chains. FAO Commodities and Trade Proceedings. Rome: Trade and Markets Division, Food and Agriculture Organization of the United Nations, p. 297.

Fountain, A. C. and F. Hutz-Adams (2015). *Cocoa Barometer*. The Netherlands: VOICE Network, FNV Mondiaal, Südwind, HIVOS, Solidaridad, p. 48.

Hatløy, A., T. A. Kebede, P. J. Adeba and C. Elvis (2012). Towards Côte d'Ivoire Sustainable Cocoa Initiative (CISCI). Baseline study report. Technical report. Norway: Fafo Institute for Applied International Studies.

ICCO (2012a). Abidjan cocoa declaration. In: *Abidjan World Cocoa Conference*, Abidjan, Côte d'Ivoire, p. 2.

ICCO (2012b). The world cocoa economy: Past and present. In: *EX/146/7 Executive Committee 146th Meeting*, 18–21 September 2012. London: ICCO, p. 43.

ICCO (2013). World cocoa bean production, grindings and stocks. Cocoa year 2012/13. 02 December 2013. ICCO Quarterly Bulletin of Cocoa Statistics, Vol. No. 4, XXXIX(4).

ICCO (2014). Outcome of the second international cocoa certification workshop. Working discussion document sustainable cocoa and certification. Zurich, Switzerland: Radisson Blu.

ICCO (2016). World cocoa bean production, grindings and stocks. ICCO Quarterly Bulletin of Cocoa Statistics XLII (3, Cocoa year 2015/16).

Ingram, V., Y. Waarts, L. Ge and G. Ton (2013). Baseline & impact assessments & lessons learnt: UTZ certified Ghana and Ivory Coast. In: *ICCO International Workshop on Cocoa Certification*. Yaoundé, Cameroon, p. 22.

Ingram, V., D. Jansen and L. Judge (2014a). Developing strategic research agendas for cocoa and coffee. In: *World Cocoa Conference*. Amsterdam: ICCO, p. 45.

Ingram, V., Y. Waarts, L. Ge, S. van Vugt, W. L. L. Puister-Jansen, F. Ruf and R. Tanoh (2014b). *Towards Sustainable Cocoa: Assessment of Cargill and Solidaridad Farmer Support Activities in Ivory Coast 2008–2012*. Den Haag, The Netherlands: LEI Wageningen UR, p. 24.

Ingram, V., Y. Waarts, L. Ge, S. van Vugt, L. Wegner, L. Puister-Jansen, F. Ruf and R. Tanoh (2014c). *Impact of UTZ Certification of Cocoa in Ivory Coast Assessment Framework and Baseline*. Den Haag, The Netherlands: LEI Wageningen UR, p. 16.

Ingram, V., Y. Waarts, L. Ge, S. van Vugt, L. Wegner, L. Puister-Jansen, F. Ruf and R. Tanoh (2014d). *Impact of UTZ Certification of Cocoa in Ivory Coast Assessment Framework and Baseline*. Den Haag, The Netherlands: LEI Wageningen UR, p. 174.

Ingram, V., M. Luskova, L. Judge, J. van den Berg and S. v. Berkum (2016a). Upscaling sustainability initiatives in international commodity chains. Examples in the cocoa, coffee and soy chains. Wettelijke Onderzoekstaken Natuur & Milieu, Technical Report 6. Wageningen, The Netherlands: LEI Wageningen UR, p. 93.

Ingram, V., F. van Rijn, Y. Waarts and M. Selten (2016b). Cocoa farmer models and practices in Ghana and Ivory Coast. In: *World Cocoa Conference 2016*. Bavaro, Dominican Republic.

International Trade Centre (ITC) (2010). Market access, transparency and fairness in global trade: Export impact for good 2010. Geneva: ITC, p. 144.

International Trade Centre (ITC) (2012). The interplay of public and private standards literature review series on the Impacts of private standards – Part III. Technical Paper. Geneva: ITC. MAR-11-215.E, p. 55.
International Trade Centre (ITC) (2015). The state of sustainable markets: Statistics and emerging trends 2015. Geneva: ITC, xviii, p. 148.
ISEAL Alliance (2010). The ISEAL 100: A survey of thought leader views on sustainability standards 2010. London: ISEAL Alliance.
Jean-François, B. and C. A. Bonjean (2014). Chocolate price fluctuations may cause depression: An analysis of price pass-through in the cocoa chain. Working Paper 20. CEDRI. EconPapers.
Kessler, J. J., J. Brons, L. Braam, M. v. Kuijk and P. Pelders (2012). Social & economic effects of value chains of tropical agro-commodities & sustainability initiatives. Final report. Amsterdam, The Netherlands: Aidenvironment - Planbureau voor de Leefomgeving (PBL), p. 192.
KPMG (2012). Cocoa Certification: A study on the costs, advantages and disadvantages of cocoa certification. The Netherlands: The International Cocoa Organization (ICCO), p. 99.
KPMG Sustainability (2013). Improving smallholder livelihoods: Effectiveness of certification in coffee, cocoa and cotton. SUSTAINEO. 126-2013\BH\JT\KS\bk, p. 65.
Kuit, M. and Y. Waarts (2014). *Small Scale Farmers and Standards: Is There a Business Case? Costs and Benefits of Standard Certification Verification Systems for Small Scale producers of Cocoa, Coffee, Cotton, Fruit and Vegetable Sectors*. Wageningen, The Netherlands: CTA/LEI.
Logatcheva, K. and V. J. Ingram (2014). *Monitoring the Dutch Cocoa Letter of Intent*. Den Haag, The Netherlands: Wageningen UR.
Matissek, R., J. Reinecke, O. Von Hagen and S. Manning (2012). Sustainability in the cocoa sector-review, challenges and approaches. *Moderne Ernaehrung Heute, Official Journal of the Food Chemistry Institute of the Association of the German Confectionery Industry*, 1–27.
Maytak, L. (2014). *Discussion and Comparison of Farm Level Data Collection Methods for Five Cocoa Studies in Cote d'Ivoire*. Washington: Prepared for Cocoa Collaborative Learning Group facilitated by the Sustainable Food Lab, p. 26.
McWilliams, A. and D. Siegel (2001). Corporate social responsibility: A theory of the firm perspective. *Academy of Management Review* 26(1): 117–27.
Mejia, R. (2011). *Increasing Productivity of Cocoa in Côte d'Ivoire*. Ghana, Indonesia and Colombia: Purdue University.
Ministerie van Landbouw Natuur en Vodelselkwaliteit (2010). Intentieverklaring - 4 maart 2010 'Duurzame cacaoconsumptie -en productie'. Den Haag, The Netherlands: Rijksoverhied. Agroketens en Visserij, p. 5.
Naprta, M. (2015). Discussing chocolate market trends. *Agro Food Industry Hi-Tech* 26(1): 24–7.
Neiburg, O. (2016). One standard to bind them: Sustainable cocoa framework to go live in 2016. Confectionery News (21 October 2014).
Nkamleu, G. B., J. Nyemeck and J. Gockowski (2010). Technology gap and efficiency in cocoa production in West and Central Africa: Implications for cocoa sector development. Tunis, Tunisia: African development bank.
Potts, J., M. Lynch, A. Wilkings, G. Huppe, M. Cunningham and V. Voora (2014). *The State of Sustainability Initiatives Review 2014: Standards and the Green Economy*. Winnipeg and London: International Institute for Sustainable Development and International Institute for Environment and Development.
Quarmine, W., R. Haagsma, A. v. Huis, O. Sakyi-Dawson, D. Obeng-Ofori and F. Asante (2014). Did the price-related reforms in Ghana's cocoa sector favour farmers? *International Journal of Agricultural Sustainability* 12(3): 248–62.
Resolve Inc. (2012). *Toward Sustainability. The Roles and Limitations of Certification*. Washington DC: Steering Committee of the State-of-Knowledge Assessment of Standards and Certification.
Ruben, R. (2008). *The Impact of Fair Trade*. Wageningen, The Netherlands: Wageningen Academic Publishers.
Ruf, F. (2007). *The Cocoa Sector: Expansion, or Green and Double Green Revolutions*. London: ODI, p. 3.

Ruf, F. and P. Siswoputranto (1995). *Cocoa Cycles: The Economics of Cocoa Supply*. Cambridge: Woodhead Publishing.

Salmon, G. (2002). *Round Table on Sustainable Development: Voluntary Sustainability Standards and Labels (VSSLs): The Case for Fostering Them*. Paris: Organisation for Economic Co-operation and Development, p. 14.

Squicciarini, M. P. and J. Swinnen (2016). *The Economics of Chocolate*. Oxford: Oxford University Press.

Swinnen, J. (2016). Economics and politics of food standards, trade, and development. *Agricultural Economics* 47(S1): 7–19.

Ton, G., J. L. F. Hagelaar, A. Laven and S. Vellema (2008). *Chain Governance, Sector Policies and Economic Sustainability in Cocoa: A Comparative Analysis of Ghana, Côte d'Ivoire, and Ecuador*. Wageningen, The Netherlands: Wageningen International, p. 40.

van Rijn, F., M. Kuit, Y. Waarts and V. Ingram (2015). *Cluster Analysis Among UTZ Certified Cocoa Farmers in Ghana and Ivory Coast*. Den Haag, The Netherlands: LEI Wageniingen UR.

Vellema, S. and A. Laven (2012). Chain governance, sector policies and economic sustainability in cocoa: How to bring the state back in? In: *Proceedings of the First Conference on Economics and Politics of Chocolate*, Leuven Belgium, 16–19 September 2012, KU Leuven.

Vellema, S., L. Admiraal and O. v. d. Valk (2006). *Quality Control in Cross-border Agro-based Supply Chains: Modes of Regulation in Coffee, Cocoa, Bananas, Palm oil, Timber and Aquaculture*. The Hague, The Netherlands: Agricultural Economics Research Institute (LEI).

Vermeulen, W. J. and S. Seuring (2009). Sustainability through the market – the impacts of sustainable supply chain management: Introduction. *Sustainable Development* 17(5): 269–73.

Victor, A.-S., J. Gockowski, N. F. Agyeman and A. K. Dziwornu (2010). Economic cost-benefit analysis of certified sustainable cocoa production in Ghana. In: *Proceedings of the 3rd African Association of Agricultural Economists (AAAE) and 48th Agricultural Economists Association of South Africa (AEASA) Conference*, Cape Town, South Africa, 19–23 September 2010.

Waarts, Y. and M. Kuit (2014). *Costs and Benefits of Certification for Small-scale Producers in Cocoa, Coffee, Cotton, Fruit and Vegetables*. Wageningen, The Netherlands: LEI/CTA, p. 50.

Waarts, Y., L. Ge, G. Ton and J. v. d. Mheen (2013). A touch of cocoa, Baseline study of six UTZ-Solidaridad cocoa projects in Ghana. LEI report 2013-048. Wageningen, The Netherlands: LEI Wageningen UR.

Waarts, Y., V. Ingram, V. Linderhof, L. Puister-Jansen, F. v. Rijn and R. Aryeetey (2015). Impact of UTZ certification on cocoa producers in Ghana 2011 to 2014. Den Haag, The Netherlands: LEI Wageningen UR, p. 50.

Wessel, M. and P. M. F. Quist-Wessel (2015). Review: Cocoa production in West Africa, a review and analysis of recent developments. *NJAS - Wageningen Journal of Life Sciences* 74–75, 1–7.

Wilcox, M. D. and P. C. Abbott (2004). Market power and structural adjustment: The case of West African cocoa market liberalization. In: *American Agricultural Economics Association Annual Meeting*, Denver, Colorado, 1–4 August.

World Cocoa Foundation (2016). Measuring what matters: Under the CocoaAction strategy, WCF is leading a industry commitment to measure progress towards sustainability in the cocoa value chain. Retrieved 13 December 2016, from http://www.worldcocoafoundation.org/measuring-success/

Chapter 23

Supporting smallholders in achieving more sustainable cocoa cultivation: the case of West Africa

Paul Macek, World Cocoa Foundation[1], USA; Upoma Husain and Krystal Werner, Georgetown University, USA

1 Introduction

The global cocoa and chocolate value chain supports the lives of an estimated 40–50 million people. The entire cocoa industry is heavily reliant on smallholder production models that include approximately 4–6 million smallholder farmers concentrated in a few countries in West Africa, and, to a lesser extent, in Southeast Asia and Latin America. Though the market for cocoa continues to grow, supply is projected to plateau or even decline in the principal cocoa-producing countries in West Africa. There has been little innovation in cocoa production since cocoa was first introduced during the colonial era and productivity remains low. The growth in West African production has depended on the expansion of cocoa farms into uncultivated land to boost yields rather than shifts in cultivation methods (Kolavalli et al. 2016). In late 2016, cocoa prices which had grown steadily from 2005 to August 2016 began a sharp decline and eventually decreased by over one-third, putting additional pressure on producer governments and smallholder cocoa farmers in West Africa. Governments of cocoa-producing countries and the leading private companies in

1. The views expressed in this chapter are those of the authors and do not necessarily reflect the views of the World Cocoa Foundation, or its member companies.

http://dx.doi.org/10.19103/AS.2017.0021.32

cocoa trading and chocolate manufacturing have recognized the urgent need to address the threats involved in cocoa production and coordinate their efforts to develop solutions to the ongoing and emerging challenges to sustainability of the cocoa sector (PARDI 2012).

Poverty and inequality are endemic to many cocoa-growing communities and perpetuated by various factors, including limited access to basic education, disempowerment of women and child labour. Many cocoa farms have become unproductive, due to ageing trees, poor agricultural practices, a heavy burden of disease and related social and economic barriers faced by smallholder cocoa farmers. Farmers often lack access to improved planting material and other agricultural inputs as well as the training and tools needed to revitalize ageing farms with declining soil fertility. As a result, yields and profitability of the cocoa farms are low, lessening the appeal of cocoa farming particularly for the younger generation.

The declining productivity of farms, combined with increasing scarcity of suitable land for new plantings, has spurred deforestation. The growth of cocoa production has depended on geographic expansion of areas under cultivation, rather than through intensification, which has contributed to widespread deforestation in Côte d'Ivoire and Ghana. New plantings have come at the expense of existing rainforest. Changes in climate, such as erratic rainfall, drought and rising average temperatures, have increasingly visible impacts across large tracts of cocoa farming regions, affecting yields, incidence of pests and diseases as well as the number of harvested cocoa pods per tree and the weight and quality of cocoa beans (Abdulai et al. 2018). Private cocoa companies have had a progressively central role in addressing the complex set of environmental and socio-economic challenges of cocoa farming, employing internal standards and voluntary certification schemes to promote sustainability in the supply chain. While there is some evidence that certification has had an overall positive effect on individual farmer incomes, especially through the premiums paid to farmers who manage more than 1 ha of cocoa and are already members of an existing cooperative, there is little data demonstrating that these efforts have had a significant impact on root causes of major problems, such as gender inequality, child labour, weak and non-existent farmers organizations or deforestation (KPMG 2012; World Bank 2017).

West Africa remains the world's most important source of cocoa beans, accounting for nearly 70% of global production, but threats to the sector's sustainability are even more pronounced in the region. The estimated 2 million smallholder cocoa farmers that depend on cocoa as their primary source of income are price takers in the marketplace; they have little to no market power and are particularly vulnerable to fluctuations in the market and directly affected by prices and exchange rates originating from the global futures market (Kolavalli et al. 2016). They remain at low levels of social and economic development with little opportunity to significantly better their situation without changes to the current production model and the policy environment.

Fundamental shifts to this production model require the collaboration of the public and private sectors. Recognizing this, governments, research institutions, leading cocoa companies and other actors from across the cocoa industry have initiated collaborative efforts, often in the form of public–private partnerships and sector-wide approaches. One early example of this approach was the Sustainable Tree Crops Programme (STCP) formed in 2001. STCP gave diverse stakeholders a common vision of improving the economic, social and environmental conditions for cocoa smallholders and made diverse expertise, financial resources and delivery systems available for this shared cause. STCP successfully piloted and expanded Farmer Field Schools (FFS) methodology, an adaptive, participatory model designed to build capacity of farmers through the reliance on local 'lead' farmers who would train fellow farmers (IITA 2005). While STCP's efforts with FFS demonstrated success in improving smallholder conditions on a small scale and has reached more farmers than expected (IITA 2005), scaling the model has shown mixed results due to limited host government buy-in (Muilerman et al. 2018). This example shows the possibility of significant shifts through collaboration and strong private sector support, while also demonstrating the importance of real public–private sector dialogue and collaboration in scaling solutions, sufficiently adapting them to different contexts, institutional limitations and social-political climates (Muilerman et al. 2018).

This chapter looks at changes in global cocoa production and the resulting challenges to sustainability across three interconnected and complex sets of problems – economic, social and environmental. The chapter then reviews the range of initiatives that have been developed to support smallholders, focusing on new, innovative approaches being championed as the private sector takes an increasingly prominent role in cocoa sustainability in West Africa.

2 Changes in cocoa production

The structure and geographic concentration of cocoa production has changed significantly over the last century. In 1900, the worldwide production of cocoa was 115 000 MT and the top three producers of cocoa were Ecuador (23 000 MT), Brazil (18 000 MT) and Nigeria (17 000 MT) (Wood and Lass 2008). By 2014–15, worldwide production had increased to over 4.25 million MT, and the three largest producers were Côte d'Ivoire (1.796 million MT), Ghana (0.74 million MT) and Indonesia (0.33 million MT) (ICCO 2013, 2015, 2017). Côte d'Ivoire and Ghana together accounted for more than 60 per cent of global production. The twentieth century thus saw the structure and organization of global cocoa production fundamentally shift from a large plantation model in Latin America and the Caribbean region to one dominated by poor, smallholder farming in West and Central Africa (Wood and Lass 2008).

During the last two decades, the growth in cocoa production in West Africa accelerated even further because of the expansion of cocoa production into available forests and government policies favourable to cocoa farming. These policies, combined with rising farm gate prices, saw cocoa production from West Africa increase significantly between 1970 and 1990 and again between 2000 and 2016. Still, real cocoa prices today remain well below their historical peaks in the late 1970s and early 1980s (LMC 2018) and at almost half of the value of cocoa sold in the 1950s, demonstrating a continuous erosion in real prices. By the end of the twentieth century, global demand for cocoa and cocoa products relied heavily on African production, where farm productivity stagnated across an ever-widening landscape of poor, smallholder farming concentrated in the Côte d'Ivoire and Ghana.

3 Agronomic and environmental challenges

While global cocoa production was growing in response to rising demand and trade liberalization, the footprint of cocoa production was spreading geographically in the top producing countries of Côte d'Ivoire and Ghana. Traditional shaded production systems with a higher number of shade trees gave way to full-sun production systems reliant on new Amazonian hybrid seed varieties that approximated monoculture (Ruf 2011; Wessel and Quist-Wessel 2015). The move away from shaded cocoa to full-sun production systems gave a short-term boost in the cocoa harvest but left smallholder cocoa farms in West Africa more vulnerable to pests and diseases and heavily reliant on continuous weeding, pruning and chemical inputs. Today, estimates of productivity remain extremely low at 400 kg/ha in Ghana and between 500 and 600 kg/ha in Côte d'Ivoire (Wessel and Quist-Wessel 2015).

The cocoa sector in West Africa today is characterized by low-input, extensive cultivation methods (Wessel and Quist-Wessel 2015). Although production has increased across the region during the past 15 years, this reflected an expansion in the area under cultivation, rather than any significant improvement in farmers' yields per hectare. Farmers have sought to increase their output by expanding into fertile forestland, instead of shifting to high-input cultivation or significantly investing in replacing old cocoa trees to enhance production on fallow land and established cocoa farms. In some cases, income generated by cocoa has been put towards the purchase of new land for cultivation (Asare et al. 2014). This migratory pattern of expansion has been a major driver of deforestation throughout the region, clearing significant tracts of the Guinean Rainforest, which by 2000 was reduced to 18% of its original area (Gockowski and Sonwa 2010). Between 2000 and 2015, UNDP estimates Côte d'Ivoire lost its forest at a rate of 2.69% annually. By 2015, only 11% or 3.4 million ha of forest remained in Côte d'Ivoire (Mondelēz, MINSEED, REDD+, United Nations Development Program 2017). Ghana experienced a similar expansion of farming with an average loss of 2% of forests over the last several decades. In both countries, cocoa remained an important driver in recent deforestation (Fountain and Huetz-Adams 2018).

Full-sun cultivation methods require greater investments in agricultural inputs, such as fertilizers and pesticides, as well as more labour-intensive activities, including weeding (Obiri et al. 2007). However, limited knowledge and uptake of good agricultural practices including continued low and inconsistent use of fertilizer and failure to appropriately prune has ensured stagnant or declining productivity – a reality that has put even greater pressure on the remaining frontier forest regions (Gockowski et al. 2013). With little available forest to expand into, farmers in both countries now face the challenge of increasing the output of an ageing tree stock. This effort has even greater labour and input requirements, and often requires full replacement, driving up overall production costs (Ruf 2011). With little forest remaining, West African countries are left with little choice than to incentivize intensification of cocoa production using fertilizers and improved varieties in an effort to increase cocoa income (Gockowski and Sonwa 2010).

One major threat to the region's cocoa production is the spread of the cocoa swollen shoot virus disease, caused by the cocoa swollen shoot virus (CSSV), genus Badnavirus, which is primarily transmitted through several species of mealybugs. Ameyaw et al. (2014) explain that in its most virulent form, the virus can substantially reduce yield by about 70% and cause death of the cocoa tree within 2–3 years. Ghana's Cocoa Board (Cocobod) reports that national CSSV prevalence rate is approximately 16.5% across the major producing areas, with significant regional variation (Cocobod Health and Extension Division 2018). Estimates for the prevalence of CSSV in Côte d'Ivoire remain scant but the Conseil du Café-Cacao, the national regulator, continues to prioritize its containment and a strategy for cutting out infected areas. Recently, both countries have initiated policies to coordinate their efforts through a new partnership, signalled by the signing of a joint declaration for cooperation and several high-level meetings (Republic of Côte d'Ivoire and Republic of Ghana 2018).

To address threats to the sector's environmental sustainability, private sector cocoa companies, with support from an international community eager to reduce carbon emissions have become involved in new initiatives to reduce deforestation and revitalize the natural environment in cocoa regions (Roth et al. 2017). For the first time, public and private sector actors are coming together to form unique, sector-wide partnerships to address shared environmental concerns affecting cocoa production. The formation of the Consumer Goods Forum, a network of over 400 companies, has made commitments to achieve deforestation-free supply chains, establishing the foundation for the Tropical Forest Alliance, a public–private partnership focused on preventing global deforestation associated with the supply chains of commodities such as cocoa and palm oil. In conjunction with the Prince of Wales's International Sustainability Unit, the Dutch Sustainable Trade Initiative and the World Cocoa Foundation, a voluntary grouping of over two dozen companies came together in 2017 to pledge to end deforestation in the cocoa supply chain in Ghana and Côte d'Ivoire under the Cocoa and Forests Initiative (World Cocoa Foundation 2017).

4 Social and economic challenges

The history of forced labour in cocoa cultivation in West Africa goes back to the 1800s. Even after Portugal abolished slavery in its colonies in 1876, plantation owners and government officials found legal loopholes which enabled them to continue employing slave labour throughout the nineteenth century until William Cadbury, director of the chocolate manufacturer Cadbury Brothers Limited, boycotted Portuguese cocoa and persuaded two other chocolate firms to do the same (Clarence-Smith and Gervase 1993; Higgs 2012). Problems with labour exploitation in the cocoa industry persist to this day. In 2000, the industry was shocked by a British Broadcasting Corporation documentary uncovering the plight of children working on West African cocoa farms. This triggered a wave of initiatives by governments, cocoa manufacturers and non-governmental organizations, in an effort to address the problem.

One such initiative was the Harkin–Engel Protocol, in which a group of leading cocoa and chocolate companies committed to ending the worst forms of child labour. Since the signing of the Harkin–Engel Protocol by the major chocolate manufacturers and others in September 2001, much attention has been given to eliminating child labour from the cocoa supply chain. The International Cocoa Initiative (ICI) was founded in 2002 within the framework of the Harkin–Engel Protocol, working with the cocoa industry, civil society, farmers' organizations, communities and national governments in cocoa-producing countries to combat child labour in the supply chain. These efforts include increased education of cocoa farmers on the dangers of child labour and the incorporation of improved monitoring systems into company activities and by the producer governments. However, progress has been slow. In 2015, a study by the Payson Center at Tulane University found an increase in child labour commensurate with the expansion of production. The study estimated a 21% increase in child labour in 2013–14 compared to previous survey completed 5 years earlier in 2008–09 (Payson Center 2015). According to analysis by ICI, an estimated four-fifths of this increase was due to the net increase in households active in cocoa production, suggesting that the flood of producers into the cocoa sector in recent years was an important driver in the observed increases (International Cocoa Initiative 2015).

Despite the magnitude of these increases, the statistics masked some important signs of progress as an excellent overview of the report by the ICI points out (ICI 2015). The Payson Center highlighted the complexity of this issue in the 2015 report. They wrote that 'while the absolute numbers of these children increased, labour force participation in the cocoa sector of children within cocoa producing households fell in both countries, and children in these households were less likely to be exposed to child labour and most types of hazardous work' (Payson Center 2015, p. 77). Similarly, the report also noted an overall decrease in the prevalence of hazardous cocoa work across the two countries. For example, in Ghana, hazardous work decreased from 61.2% to 54.6% and in Côte d'Ivoire, hazardous work dropped from 41.6% to 35.7% between the two surveys. The percentage of children exposed to multiple hazardous tasks also decreased slightly (by an average of 5.2% across Ghana and Côte d'Ivoire) between the 2008–09 survey and 2013–14 surveys. One of the important responses to address child labour on cocoa farmers has been to support and promote access to primary education. The Payson Center report reassuringly showed that there were important increases in child enrolment in primary education in both Côte d'Ivoire and Ghana. Using the Payson Center survey data, ICI estimated that a total of 651 757 additional children in cocoa-growing areas were attending school in both countries (Payson Center 2015; ICI 2015).

The phenomenon of child labour is closely linked to the realities of poverty among cocoa-producing households (Bureau of Labor Affairs and US Department of Labor 2015; Fountain and Hütz-Adams 2018). In sub-Saharan Africa, vulnerable families continue to rely on child labour to cope with poverty and economic shocks in the absence of social safety nets. As has been noted, with little or no market power, low levels of social and economic development as well as dependence on cocoa for income, an estimated 2 million smallholder cocoa farmers are particularly vulnerable to price fluctuations in the marketplace (Oomes et al. 2016). The recent decline in real prices has also brought into sharp relief a growing discussion on the concept of a living income for cocoa farmers (Fountain and Hütz-Adams 2015, 2018).

Smallholders have expressed growing concerns about market volatility, low and declining real prices and the erosion of producer share in the value chain (ul-Haque 2004). As a result, most cocoa farmers live in poverty on an income of less than $2 per day, while many cocoa farming families in Ghana and Côte d'Ivoire survive on real incomes of about $0.40 per dependent per day (International Labour Rights Forum 2014). The average cocoa farmer in Ghana earns $0.84 per day, while the typical small farmer in Côte d'Ivoire earns just $0.50 per day (Fountain and Hütz-Adams 2015). A recent survey by Fairtrade International estimated that a living income for a cocoa farmer in Côte d'Ivoire was $2.58 per day in 2018, although the average cocoa farmer only earned one-third of that amount (Rusman et al. 2018).

Both Ghana and Côte d'Ivoire have government-run marketing boards for cocoa that set a minimum farm gate price for producers to stabilize the sector and garner political support. Ghana's Cocobod guarantees farmers approximately 70% of the net free on board (FOB) price while the Conseil du Café-Cacao in Côte d'Ivoire, which emerged in 2011 as part of broader reforms required as a precondition for International Monetary Fund-backed debt relief, also guarantees a minimum farm gate price to producers and seeks to achieve at least 60% of the net FOB price (Asante-Poku and Angelucci 2013). The FOB price reflects the world market price which is set by the international commodity exchanges in London and New York. As the Cocoa Barometer 2018 notes, a complex web of variables determine the global market prices, including 'the relationship between demand, stocks, and current and future supply. Traders pay slightly different prices for cocoa from different countries due to requirements concerning quality and delivery date (Fountain and Hütz-Adams 2018, p. 7)'.

Other challenges faced by smallholder cocoa farmers in West Africa are the lack of farmer organization and poor bargaining power. Many cocoa farmers in Ghana and Côte d'Ivoire operate outside of cooperatives. It is estimated that between 80% and 85% in Ghana and about 50% in Côte d'Ivoire operate through formal producer groups (US Department of Labor 2013; GEFAK 2015). This lack of farmer organization in a relatively informal sector involving many intermediaries makes it difficult to deliver valuable farm services let alone control and monitor the conditions at the base of production. As a result, cocoa farmers are often price takers in the market and receive few, if any, services that would allow them to access better training and inputs. These factors further constrain the ability to achieve higher levels of livelihood diversification or more professionalization – two important preconditions for poverty alleviation.

Furthermore, female farmers, despite playing a significant role in cocoa production, largely miss out on the benefits from most services. According to one survey, female farmers in Ghana had 25–30% less income and productivity than those of their male counterparts due to lower access to training, farm inputs and financial services (Hiscox and Goldstein 2014). Women tend to have much lower participation in cocoa cooperatives; their contribution in the cocoa supply chain often goes unrecognized, and they receive lower compensation for their labour (Fair labor Association 2012; Cocoa Life 2014; Marston 2016). In Côte d'Ivoire, for example, women earned only 21% of the income generated from cocoa, despite making up approximately 68% of the labour force (African Development Bank 2015).

5 Initiatives to support smallholders

Cocoa consumers in high-income countries have shown a growing concern for the safety, quality as well as socio-economic and environmental impacts of cocoa production on farmers and communities (Potts et al. 2014; Chocolate Manufacturers Association 2001). Consequently, there has also been a shift to expand the way in which businesses account for these impacts by measuring their sustainability efforts through measuring their impact on people, planet and profits. This means measuring not just profit and loss (traditional business accounting) but also social and environmental impacts (Elkington 1999).

5.1 Voluntary sustainability standards

Against this backdrop, several new movements, such as voluntary sustainability standards (VSS), company-owned and branded sustainability programmes and civil society campaigns have emerged to advocate for specific solutions. These consumer-facing initiatives have sought to secure a more sustainable production base for cocoa while addressing some of the socio-economic, environmental and agricultural challenges in the cocoa sector. Cocoa consumers in high-income countries are more and more interested in the socio-economic and environmental impacts of cocoa production on farmers and communities (Potts et al. 2014). Some approaches have looked at specific issues, such as child labour within the Harkin–Engel Protocol or the concerns over better farmer training, promising to make smallholder production a more profitable venture as is the case in Rainforest, UTZ and Fairtrade certification for cocoa (Potts et al. 2010, 2014).

These concerns have resulted in the growth of VSS which aim to improve farmer income levels, farm productivity, terms of trade for producers, and to alleviate poverty (Mitiku et al. 2017). Starting in the early 2000s, VSS covered only 3% of global production in 2003 but, by 2012, they covered an estimated 22% of production (Potts et al. 2014). Despite the rapid growth in VSS and the volume of cocoa they cover, a portion of certified cocoa continues to be sold as ordinary cocoa – meaning that the total production of certified cocoa often exceeds the market demand. Despite the challenges that VSS face, they continue to represent an early and important private sector response to sustainability challenges in the sector.

5.2 Public–private partnerships and company sustainability programmes

Many of the recent efforts designed to address cocoa sustainability have been conceived as public–private partnerships that bring together producer governments, industry and civil society to identify collective action and shared solutions. One of the earliest examples of this was the STCP started in 2001, which focused on environmentally and socially responsible tree crop cultivation in West and Central Africa within a wider context of innovation and collaboration. Its aim was to increase farmers' income and combat child labour. The STCP was funded by private cocoa and chocolate companies, formed in collaboration with USAID and the World Cocoa Foundation, and managed by the International Institute of Tropical Agriculture. It focused on adapting the United Nations Food and Agriculture Organization's FFS methodology for training cocoa farmers in good agricultural practices. FFS training in Integrated Crop and Pest Management for farmers in Côte d'Ivoire, Ghana, Nigeria and Cameroon showed on average 15–40% higher yields, and a reduction in pesticide use of 10–20% (Gockowski et al. 2013).

In 2014, the Ivorian Ministry of Agriculture and Rural Development launched the Programme Quantité-Qualité-Croissance (2QC) to address high tree mortality and declining yields by improving and intensifying coffee and cocoa production systems. The project aimed to replant 800 000 ha (including 150 000 ha affected by swollen shoot disease) of cocoa area with improved planting material and rehabilitate 1 000 000 ha of cocoa with proper management and input by 2023 (Ministère de l'Agriculture République de Côte d'Ivoire 2014). In early 2018, The Ghana Cocobod also launched financial schemes to raise up to $250 million for a large-scale cocoa replanting programme (Oxford Business Group 2018).

Private actors are also filling in gaps and providing services, such as access to appropriate financial services and the distribution of hybrid planting material, which the government in some cases has limited capacity to provide in sufficient quantities. The Cocoa Abrabopa Association in Ghana, for example, has introduced a programme to increase the use of fertilizers, insecticides and fungicides by cocoa farmers. To address the farmers' lack of access to the credit required to improve yields and farm quality, agrochemicals have been provided to cocoa farmers on a seasonal credit basis through creative risk sharing mechanisms among the farmers, the association and the agricultural input provider. The loan recovery rate has been around 90%, and farmers have reported a significant increase in yield, as well as in income (Opoko et al. 2009).

As an industry effort to address the issue of old and diseased trees, the Cocoa Rehabilitation and Intensification Programme co-funded by the Netherlands Embassy in Ghana and the private sector, has developed Rural Service Centres (RSCs) which offer agricultural inputs as well as agronomic training and support on an ongoing basis to farmers with limited access to resources. Though the effectiveness of the RSCs is still unclear, this type of partnership demonstrates the cross-sector willingness to find new and creative solutions to the challenges of rural service delivery. Other public–private partnerships have rapidly emerged, such as the Cocoa Livelihoods Programme and the African Cocoa Initiative – two projects of the World Cocoa Foundation and its member companies that bring together the cocoa and chocolate industry around common challenges of poverty alleviation and food security with the support of public and private development partners. In 2015, the German Initiative on Sustainable Cocoa, a multi-stakeholder initiative started by the German Federal Ministry for Economic Cooperation and Development and the cocoa and chocolate industry, launched a new programme – *Pro-Planteur* – in partnership with the Ivorian government to address environmental, social and economic challenges facing 20 000 cocoa farmers (GEFAK 2015).

Likewise, in 2015, the nine leading cocoa and chocolate companies agreed to align their sustainability efforts under a common set of objectives to realize greater impact at a faster pace. This unprecedented act of cooperation launched under the auspices of the World Cocoa Foundation sought to prioritize key interventions in cocoa productivity and community development to both modernize and professionalize the cocoa sector, while also providing impetus to increase productivity and better social development outcomes for 300 000 cocoa farmers living in an estimated 1200 cocoa-growing communities of Côte d'Ivoire and Ghana. This type of effort, to realize collective impact, challenges businesses to deliver social benefits and economic value to society (Kramer et al. 2016). Building on their earlier successful collaboration under CocoaAction, the major cocoa and chocolate companies came together again in 2017 to make a public commitment to collaboratively end deforestation and forest degradation in the cocoa supply chain. This landmark agreement has grown to include 35 companies as of March 2018 including a number of consumer goods and retail companies (Kroeger et al. 2017; World Cocoa Foundation 2018).

These collective efforts are increasingly delivered through the equally noteworthy and impressive growth of individual company-initiated cocoa and chocolate sustainability programmes (see Table 1). It is not uncommon today for cocoa or chocolate companies to have their own branded sustainability commitments that reflect their corporate thinking and vision of sustainability while also encompassing the totality of their joint efforts with the rest of industry. The realization of these commitments may be done through their contracted suppliers, their own staff or their partner organizations, which may include contracted non-profit organizations and other specialized stainability service providers. The cocoa trading and processing companies, such as Barry Callebaut, Blommer Chocolate, Cargill, Olam and Touton, for example, will have the added dimension of not only implementing sustainability programmes to meet their own corporate commitments but also those of branded chocolate manufacturers to which they supply cocoa and chocolate products. By working through the 'supply chain' network of relationships, cocoa companies have made these investments a critical part of their business model and their corporate values.

5.3 Efforts to enhance women's empowerment and eliminate child labour

Several recent private sector initiatives to address the barriers faced by women in cocoa supply chain are noteworthy. In 2014, the Walmart Foundation and World Cocoa

Table 1 Company sustainability programmes

Company	Programme on sustainability	Year launched	Sources (Accessed 15 April 2018)
Barry Callebaut	Forever Chocolate	2016	https://www.barry-callebaut.com/forever-chocolate
	Cocoa Horizon Foundation	2014	https://www.cocoahorizons.org/
Blommer Chocolate	GrowCocoa (with Olam)	2012	http://www.blommer.com/sustainable-origins-growcocoa.php
	Sustainable Origins	2012	http://www.blommer.com/sustainable-origins.php
Cargill	Cocoa Promise	2012	https://www.cargill.com/sustainability/cargill-cocoa-promise
Cemoi	Transparence Cacao	2015	https://group.cemoi.com/our-commitments/transparence-cacao/
ECOM Agrotrade	Ecom Sustainability Partnership	2013	http://www.blommer.com/sustainable_origins_programs.html
	Ecom Sustainability Partnership (with Blommer)	2012	http://www.blommer.com/sustainable_origins_programs.html
Ferrero	Individual pledges to end deforestation and slavery; 100% traceability	Various	https://www.ferrerocsr.com/?lang=en
Guittard Chocolate	Cultivate Better	2018	https://www.guittard.com/cultivate-better/honorable-sourcing
Hershey Company	Cocoa for Good	2018	https://www.thehersheycompany.com/en_us/responsibility/good-business/creating-goodness/cocoa-for-good.html
Mars Inc.	Vision for Change	2010	http://www.mars.com/global/about-us/policies-and-practices/cocoa-policy
	Sustainable in a Generation	2017	http://www.mars.com/global/sustainable-in-a-generation
Mondelēz	Cocoa Life	2012	https://www.cocoalife.org/
Nestlé	Cocoa Plan	2009	http://www.nestlecocoaplan.com/
Olam	GrowCocoa (with Blommer)	2012	http://olamgroup.com/products-services/confectionery-beverage-ingredients/cocoa/sustainability/growcocoa/
	Olam Livelihood Charter	2010	http://49tmko49h46b4e0czy3rlqaye1b.wpengine.netdna-cdn.com/wp-content/uploads/2017/09/Olam-Livelihood-Charter-report-2016-2.pdf
Touton	Sustainable Sourcing	2015	https://touton.fr/news/touton-s-first-sustainable-sourcing-report

Foundation collaborated to promote gender equality within the value chain by closing the gender-based yield gap in four cocoa-growing countries. By the end of 2017, 7163 female farmers had experienced income increases (World Cocoa Foundation 2017). The System Innovation for Women's Economic Empowerment programme is a multi-stakeholder initiative that takes a 'social lab' approach to conduct action research to understand the barriers to women's economic empowerment and implement innovative pilot projects to address them (Oxfam 2017). In Nigeria, the Farmer's Development Union has been adapting the Gender Action Learning System, which uses picture-based diagramming tools and exercises to raise awareness about gender to build understanding of gender issues between women and men in cocoa communities (Marston 2016).

In 2012, the ICI together with Nestlé started piloting what was to become a child labour monitoring and remediation system for cocoa farmers in Côte d'Ivoire. By mid-2017, the initiative reached more than 48 000 cocoa farmers in 75 cooperatives and was monitoring over 40 000 children. The approach relies on community-level monitoring to identify cases of child labour and provides remediation measures appropriate for the specific case, such as access to school or income-generating activities for the family. The programme's scale-up continues and is providing valuable learning for the cocoa industry on how to approach this complicated issue (Nestlé Cocoa Plan and International Cocoa Initiative 2017).

6 Conclusion

This chapter has sought to provide an overview of the key considerations and the long-standing and emerging challenges that the cocoa and chocolate industry faces in achieving a more sustainable cocoa and chocolate supply chain. Clearly some of the intractable problems facing the sector have only worsened over time: child labour, stagnating productivity and persistent poverty among smallholders. These established problems have combined with more recent issue, such as the sharp decline in the world market price for cocoa, the impact of climate change on cocoa production and the acceleration of cocoa-driven deforestation in the principal region of global cocoa production – West Africa – to amplify the complexity and magnitude of the problems the sector faces today.

This chapter has highlighted three important trends that demonstrate the potential for the sector to swiftly and comprehensively address some of the challenges. First, the growth in voluntary certification schemes and the introduction of international standards for ethically and responsibly sourced cocoa has helped to engage consumers and provided important benchmarks for measuring progress. Second, the last two decades have witnessed a deepening effort to build public–private partnerships and coalitions to address the sector's multi-faceted and complex challenges. Lastly, the near universal growth in company owned and led stainability programs combined with a growing number of joint company efforts aimed at measurable impact positions the cocoa sector as leading industry in responsible sourcing. These promising trends also point to areas where additional research and attention are needed, as discussed below.

6.1 Voluntary certification and standards

Voluntary certification and standards are being clarified and redefined. Several important developments have occurred that suggest an evolution in the way in which voluntary certification schemes and standards operate. The international community through the European Committee for Standardization and the International Standards Organization (ISO) and a group of national ISO member committees in Côte d'Ivoire (CODINORM), Ghana and the Netherlands has been developing a new cocoa standard that seeks to establish a robust and sustainability-focused international cocoa standard. The new ISO standard is scheduled to come into effect in 2019.

Another developing trend in this space has been the internalization of sourcing standards by leading chocolate companies, linking them directly with their own sustainability programmes and capacity-building. Companies such as Nestlé have removed the consumer-facing UTZ certification logo from products in the United Kingdom. Another major global manufacturer, Mondelēz, announced a new partnership with Fairtrade, under which Fairtrade will act as an implementing partner organization to Mondelēz's Cocoa Life sustainability programme. And in perhaps the most dramatic recent development, the two main VSSs in cocoa, namely UTZ and the Rainforest Alliance, merged into a single organization in 2018 (Sustainable Brands 2017). At the same time in 2017, several VSS and other partners announced a new Accountability Framework designed to improve the credibility of claims made by their members, by defining best practices for 'traceability, management of non-compliances, monitoring, verification, reporting, and other key implementation functions...' (Accountability Framework 2017). While it is too early to gauge the specific impact that these developments will have on overall cocoa farmer sustainability, it ushers in a new era for alignment of sustainability standards and will play an important role in the promotion of specific farm-level management practices and farmer professionalization.

6.2 The growing importance of living income

From the outset, VSSs aimed at creating a premium product that would attract higher value on the market and pay a premium price to producers. For many cocoa farmers, certification is inextricably linked with obtaining a premium payment, above the farm gate price for ordinary cocoa. However, experience has shown that certification premiums alone do not equate with a living income. Progress in achieving a living income in the cocoa sector will require continued effort to support more diversified livelihoods, including the development of non-cocoa income streams at the farm level. There is potential to develop agroforestry systems that combine cocoa, fruit and other tree plantings and inter-cropping of cocoa and food crops for greater and more diversified farmer income. A small but growing number of cocoa and chocolate companies, such as Divine Chocolate, Uncommon Cocoa and Tony's Chocolony, are also developing novel approaches for determining cocoa prices and premiums that warrant further consideration in an effort to increase incomes for cocoa farmers.

6.3 The question of price and supply and demand

The last two years have exposed serious risks to sustainability from price fluctuations, market volatility and the supply and demand for cocoa. Mitigating these risks will require good management of the long-term supply of cocoa to overcome the current oversupply. Initiatives designed to support farmer and farmer-group professionalization through the development of strong cooperatives and private sector farm services, coupled with sensible policy reforms, will help to create the capacity and enabling environment for the emergence of a modern cocoa sector. Professional farmers capable of producing more cocoa on less land will reap the benefits of increased productivity and income diversification. At a national level, increases in cocoa production must be complemented by better land-use planning and enforcement of national environmental regulations to ensure that cocoa is grown only in environmentally sustainable areas. Finally, the cocoa and chocolate industry needs to continue to invest in market development, in both established markets as well as in emerging economies like India, China and Brazil, where expanding middle classes have fuelled the growth in global chocolate consumption in recent years.

Faced with the challenges of structural oversupply and persistent poverty, public and private actors have a unique opportunity to deepen their collaboration and renew their joint efforts. Recent work to tackle deforestation and climate change provides an excellent blueprint for the kind of bold actions that are needed to overcome the sustainability challenges of the sector. While some of these joint initiatives are still relatively new and may lack the dedicated staff and resources to function effectively, they nevertheless embody the renewed sense of urgency and purpose that characterizes current collaboration between the public and private sectors (Greenberg and Root 2016).

As the challenges to cocoa sustainability have mounted, the national marketing boards of Côte d'Ivoire and Ghana have begun to consider a backlog of necessary policy and regulatory reforms designed to meet fiscal and budgetary constraints and better serve their countries and their smallholder farmers. The new government in Ghana recently announced its intention to reform the input supply process, potentially scrapping or at least reducing the long-standing national subsidy programme for inputs. The government of Ghana has also begun to highlight other obstacles to improved productivity: the spread of CSSV; the impact of customary property rights and land tenure on cocoa farmers; and the challenges faced by youth trying to enter the sector. Similarly, in Côte d'Ivoire, the government announced a new national forestry code designed to modernize and update the national classification system for the forests – a process which will directly affect cocoa production, especially when sourced out of protected forests and reserves.

7 Future trends

Future research in this field will need to grapple with several new and complex issues – pricing and premiums, national policies and deforestation – that are at the core of the global sustainability discussion in cocoa and other soft commodities. Research into voluntary price setting models that allow producers to at least meet the costs of production while also allowing companies to voluntarily incentivize producers to provide a premium product are desperately needed. A critical review of the advantages and disadvantages of various pricing schemes would help to define the benefits and challenges of different approaches, providing a greater comparison of these alternative systems against the prevailing pricing methods. There is also an emerging focus on understanding and promoting the concept of a 'living income' within the cocoa value chain which draws attention to the importance of on-farm and off-farm income and the need for crop and livelihood diversification to overcome poverty among smallholder farmers.

We can also anticipate a strong focus in future research on the importance of policy reform with the aim of maximizing the return to the smallholder farmer. This 'farmer centric' research would also highlight the political economy of the cocoa sector and the share that goes to producers versus the share enjoyed by other actors in the supply chain. By drawing attention to areas of underinvestment in the sector, future research can demonstrate how certain policies can help achieve multiple aims such as alleviating poverty, empowering women or combatting deforestation. Policy reform, especially in a context of retreating prices, will need to demonstrate the ability to support capital and infrastructure investments while simultaneously reducing the costs of inputs through the use of private sector input provision and the reduction of state-sponsored agricultural input subsidies.

In the past few years, climate change and the related impact of cocoa production on the natural environment have received a lot of attention. This trend is expected to continue as the importance of maintaining natural forests and expanding forest coverage grows in West Africa. Suggested areas of research include the role of agroforestry and mixed production models in delivering optimal agricultural productivity, income diversification and valuable ecosystem services. Another area of great concern will be how to scale up and utilize more systematically climate sensitive production practices that optimize shade cover and assist in moisture retention while also protecting cocoa farms from drought, erratic weather and pests and diseases in an integrated and sustainable fashion – all practices that are often referred to as climate-smart practices.

8 Where to look for further information

For additional information on emerging issues in the cocoa sector, the World Cocoa Foundation (http://www.worldcocoa.org) has a comprehensive website and excellent collection of reports and other relevant data on the sector. For official statistics and the formal policy deliberations between producing and consuming countries, readers can visit the website of The International Cocoa Organization (ICCO) which is the UN-affiliated international organization created to implement the series of International Cocoa Agreements, the first of which came into force in 1972 (https://www.icco.org/). Commodity trade statistics on cocoa can also be found in the UN Comtrade Database (https://comtrade.un.org/data/) and the USDA Foreign Agricultural Service's Global Agricultural Trade System (GATS) database (https://apps.fas.usda.gov/Gats/default.aspx).

Further information and resources on child labour and other sustainability issues can also be found at the International Cocoa Initiative (https://cocoainitiative.org/), and the Sustainable Trade Initiative (IDH) websites (https://www.idhsustainabletrade.com/sectors/cocoa/). The Sustainability Standards Database at the International Trade Centre offers information on standards for environmental protection, worker and labour rights, economic development, quality and food safety, as well as business ethics (http://www.intracen.org/itc/market-info-tools/voluntary-standards/standardsmap/). Additional sustainability standards relevant to cocoa agribusiness and investments can be accessed from International Finance Corporation's (IFC) sustainability framework (http://www.ifc.org/sustainabilityframework) and UNEP's environmental, social and economic sustainability framework (https://www.unenvironment.org/resources/report/un-environments-environmental-social-and-economic-sustainability-framework). The three major certification schemes in cocoa industry are Fairtrade, Rainforest Alliance and UTZ Certified. They operate in similar ways and have a key objective to promote sustainable practices in the cocoa supply chain and improve the livelihoods of farmers in producing countries; but differ in their specific focus and requirements. More information can be found on their respective websites (https://www.fairtrade.net; https://utz.org; and https://www.rainforest-alliance.org). It should be noted that as of January 2018, UTZ Certified and Rainforest Alliance have merged into a single organization.

For more information regarding national policy and production context in West Africa, readers can look to the regulatory bodies governing the cocoa sector in these countries – the Ghana Cocoa Board (Cocobod) (https://www.cocobod.gh/) and the Conseil du Café-Cacao in Côte d'Ivoire (http://www.conseilcafecacao.ci/) – both websites provide valuable and updated information about the national production, prices and policies.

9 References

Abdulai, I., Jassogne, L., Graefe, S., Asare, R., Van Asten, P., Läderach, P. and Vaast, P. (2018), 'Characterization of cocoa production, income diversification and shade tree management along a climate gradient in Ghana', *Plos One* 13(4), e0195777.

Accountability Framework (2017), 'Communications and Claims Related to the Accountability Framework', *Accountability Framework* (online), Available at: https://accountability-framework.org/claims/ (accessed on 2 December 2017).

African Development Bank (2015), *Economic Empowerment of African Women through Equitable Participation in Agricultural Value Chains,* Abidjan, Côte d'Ivoire: African Development Bank.

Ameyaw, G. A., Dzahini-Obiatey, H. K. and Domfeh, O. (2014), 'Perspectives on cocoa swollen shoot virus disease (CSSVD) management in Ghana', *Crop Protection* 65, 64–70.

Asante-Poku, A. and Angelucci, F. (2013), 'Analysis of Incentives and Disincentives for Cocoa in Ghana', Technical Note Series, Monitoring African Food and Agricultural Policies project (MAFAP), Food and Agriculture Organization.

Asare, R., Afari-Sefa, V., Osei-Owusu, Y. and Pabi, O. (2014), 'Cocoa agroforestry for increasing forest connectivity in a fragmented landscape in Ghana', *Agroforestry Systems* 88, 1143–56.

Bureau of Labor Affairs and US Department of Labor (2015), *Findings on the Worst Forms of Child Labor*, Washington DC: USDOL.

Chocolate Manufacturers Association (2001), Protocol for the growing and processing of cocoa beans and their derivative products in a manner that complies with ILO Convention 182 concerning the prohibition and immediate action for the elimination of the worst forms of child labor.

Clarence-Smith and Gervase, W. (1993), 'Cocoa plantations and coerced labor in the Gulf of Guinea, 1870–1914', in *Breaking the Chains: Slavery, Bondage, and Emancipation in Modern Africa and Asia*, edited by M. A. Klein, Madison, WI: University of Wisconsin Press.

Cocoa Life (2014), 'Gender Assessment Report Care International in Côte d'Ivoire', Available at: http://www.cocoalife.org/progress/gender-assessment-report-care-international-in-cote-divoire (accessed 4 February 2018).

Cocobod Health and Extension Division (2018), 'Programmes and Activities Main Focus', A presentation given on 10 April 2018 at *Private Stakeholders Meeting*, Accra, Ghana.

Elkington, J. (1999), *Cannibals with Forks: The Triple Bottom Line of 21st Century Business*, Oxford, UK: Capstone Publishing Ltd.

Fair Labor Association (2012), 'Sustainable Management of Nestlé's Cocoa Supply Chain in the Ivory Coast: Focus on Labor Standards', Available at: http://www.fairlabor.org/sites/default/files/documents/reports/cocoa-report-final_0.pdf (accessed 12 December 2017).

Fountain, A. and Hütz-Adams, F. (2015), '2015 Cocoa Barometer', Available at: http://www.cocoabarometer.org/Cocoa_Barometer/Download_files/Cocoa%20Barometer%202015%20.pdf (accessed 12 December 2017).

Fountain, A. and Hütz-Adams, F. (2018), '2018 Cocoa Barometer', Available at: http://www.cocoabarometer.org/Cocoa_Barometer/Download_files/2018%20Cocoa%20Barometer%20180420.pdf (accessed 20 March 2018).

GEFAK mbH (2015), 'Study on the State of Farmer Cooperatives in the Cocoa Sector of Côte d'Ivoire', Draft final report, Marburg, Germany, Available at: https://www.kakaoforum.de/fileadmin/Redaktion/Studien/GISCO_COOP_Report_GEFAK_final.pdf (accessed 12 December 2017).

Gockowski, J. and Sonwa, D. (2010), 'Cocoa intensification scenarios and their predicted impact on CO_2 emissions, biodiversity conservation, and rural livelihoods in the Guinea rain forest of West Africa', *Environmental Management* 48(2), 307–21.

Gockowski, J., Afari-Sefa, V., Sarpong, D. B., Osei-Asare, Y. B. and Agyeman, N. F. (2013), 'Improving the productivity and income of Ghanaian cocoa farmers while maintaining environmental services: What role for certification?' *International Journal of Agricultural Sustainability* 11(4), 331–46.

Greenberg, D. and Root, C. (2016), *African Cocoa Initiative Final Performance Evaluation Report*, Washington DC: USAID.

Higgs, C. (2012), *Chocolate Islands: Cocoa, Slavery, and Colonial Africa*, Columbus, OH: Ohio University Press.

Hiscox, M. and Goldstein, R. (2014), *Gender Inequality in the Ghanaian Cocoa Sector*, Cambridge, MA: Harvard University.

International Cocoa Initiative (2015), ICI synthesis and review of the final Tulane report 2013/14.

International Cocoa Organization (ICCO) (2013), Quarterly Bulletin of Cocoa, Statistics, Volume XLI No.1, Cocoa Year 2011/12.

International Cocoa Organization (ICCO) (2015), Quarterly Bulletin of Cocoa, Statistics, Volume XLI No.2, Cocoa Year 2014/15.

International Cocoa Organization (ICCO) (2017), Quarterly Bulletin of Cocoa, Statistics, Volume XLIII No.1, Cocoa Year 2016/17.

International Institute of Tropical Agriculture (IITA) (2005), The Sustainable Tree Crop Program (STCP): External review synopsis, April 2005.

International Labor Rights Forum (2014), 'The Fairness Gap', Available at: http://www.laborrights.org/sites/default/files/publications/Fairness%20gap_low_res.pdf (accessed 18 February 2017).

Kolavalli, S., Vigneri, M. and Gockowski, J. (2016), *The Cocoa Coast: The Board Managed Cocoa Sector in Ghana*, Washington DC: International Food Policy Research Institute (IFPRI).

KPMG (2012), *Cocoa Certification A Study on the Costs, Advantages and Disadvantages of Cocoa Certification*, the Netherlands: The International Cocoa Organization (ICCO), p. 99.

Kramer, M. and Pfitzer, M. W. (2016), 'The ecosystem of shared value', *Harvard Business Review* 94(10), 80–9.

Kroeger, A., Bakhtary, H., Haupt, F. and Streck, C. (2017), *Eliminating Deforestation from the Cocoa Supply Chain*, Washington DC: World Bank, Available at: http://documents.worldbank.org/curated/en/876071495118818649/pdf/115144-REVISED-20170530-Cocoa-final-updated.pdf (accessed 12 February 2018).

LMC (2018), Cocoa: The global market Outlook for beans, butter, liquor and powder, 2018 Brochure.

Marston, A. (2016), *Women's Rights in the Cocoa Sector: Examples of Emerging Good Practice*, Oxford, UK: Oxfam.

Ministre de l'Agriculture et du Développement Rural, République de Côte d'Ivoire, (2014), 'Programme Quantité-Qualité-Croissance', Available at: http://www.conseilcafecacao.ci/index.php?option=com_k2&view=item&id=313:programme-quantite-qualite-croissance-2qc-2014-2023&Itemid=18 (accessed 26 March 2018).

Mitiku, F., de Mey, Y., Nyssen, J. and Maertens, M. (2017), 'Do private sustainability standards contribute to income growth and poverty alleviation? A comparison of different coffee certification schemes in Ethiopia', *Sustainability* 9(2), 246.

Mondelēz, MINSEED, REDD+, United Nations Development Program (2017), *Forest Friendly Cocoa in Côte d'Ivoire*, UNDP, New York, NY.

Muilerman, S., Wigboldus, S. and Leeuwis, C. (2018), *Scaling and Institutionalisation within Agricultural Innovation Systems: The Case of Cocoa Farmer Field Schools in Cameroon*, Wageningen, the Netherlands: Wageningen University.

Nestlé (2017), *'Nestlé Cocoa Plan: Tackling Child Labor 2017 Report'*, available at: https://www.nestlecocoaplanreport.com/sites/default/files/2017-10/NestleCocoaPlanReport2017_EN_0.pdf (accessed 28 April 2018).

Nestlé and the International Cocoa Initiative (2017), *Tackling Child Labour: 2017 Report*, Nestlé, Vevey, Switzerland.

Obiri, B. D., Bright, G. A., McDonald, M. A., Anglaaere, L. A. and Cobbina, J. (2007), 'Financial analysis of shaded cocoa in Ghana', *Agroforestry Systems* 71(2), 139–49.

Oomes, N., Tieben, B., Laven, A., Ammerlaan, T., Appelman, R., Biesenbeek, C. and Buunk, E. (2016), *Market Concentration and Price Formation in the Global Cocoa Value Chain*, Amsterdam, the Netherlands: SEO Amsterdam Economics.

Opoko, E., Dzene, R., Caria, S., Teal, F. and Zeitlin, A. (2009), 'Impacts of group-based microfinance in agriculture: Evidence from Ghana's Cocoa Abrabopa Association', *Annual Conference of Center for the Study of African Economies (CSAE)*, Oxford, England.

Oxfam (2017), *System Innovation for Women's Economic Empowerment – Internal Report for the Inception Research*, Oxford, UK: Oxfam.

Oxford Business Group (2018), 'Ghana to Launch Funding Scheme to Protect Cocoa Industry from Plant Disease', Available at: https://oxfordbusinessgroup.com/news/ghana-launch-funding-scheme-protect-cocoa-industry-plant-disease (accessed 26 March 2018).

Pacific Agribusiness Research & Development Initiative (PARDI) (2012), *Cocoa Chain Review*, Australia: PARDI.

Payson Center, Tulane University (2015), Final report: 2013/14 survey research on child labour in West African cocoa-growing areas, New Orleans, USA: Payson Center for International Development and Technology Transfer.

Potts, J., Van der Meer, J. and Daitchman, J. H. (2010), *The State of Sustainability Initiatives Review 2010. Sustainability and Transparency* (1st Edition), Winnipeg and London: International Institute for Sustainable Development (IISD) and International Institute for Environment and Development (IIED).

Potts, J., Lynch, M., Wilkings, A., Huppé, G., Cunningham, M. and Voora, V. (2014), *The State of Sustainability Initiatives Review. Standards and the Green Economy* (1st Edition), Winnipeg and London: International Institute for Sustainable Development (IISD) and International Institute for Environment and Development (IIED).

Republic of Côte d'Ivoire and Republic of Ghana (2018), 'Déclaration d'Abidjan: La Côte d'Ivoire et le Ghana ensemble pour faire face aux défis de l'économie cacaoyère', 26 March 2018, Abidjan, Côte d'Ivoire.

Roth, M., Antwi, Y. and O'Sullivan, R. (2017), *Land and Natural Resource Governance and Tenure for Enabling Sustainable Cocoa Cultivation in Ghana*, Washington DC: USAID Tenure and Global Climate Change Program.

Ruf, F. O. (2011), 'The myth of complex cocoa agroforests: The case of Ghana', *Human Ecology* 39(3), 373–88.

Rusman, A., de Adelhart Toorop, R., de Boer, J. and de Groot Ruiz, A. (2018), Cocoa farmer income. The household income of cocoa farmers in Côte d'Ivoire and strategies for improvement, True Price/Fairtrade, April 2018.

Sustainable Brands (2017), 'Rainforest Alliance, UTZ Merge to Create Single Agriculture Sustainability Standard', accessed on 11 November 2017, Available at: http://www.sustainablebrands.com/news_and_views/collaboration/sustainable_brands/rainforest_alliance_utz_merge_create_single_agricult (accessed on 30 April 2018).

ul-Haque, I. (2004). 'Commodities under Neo-Liberalism: the Case of Cocoa', G24 Discussion Papers 25, United Nations Conference on Trade and Development. Available at: http://www.unctad.org/en/Docs/gdsmdpbg2420041_en.pdf (accessed 20 April 2018).

US Department of Labor (2013), 'The Child Labor Cocoa Coordinating Group: Annual Report 2012', Available at: https://www.dol.gov/ilab/issues/child-labor/cocoa/2012-CLCCG-Report.pdf (accessed 20 April 2018).

Wessel, M. and Quist-Wessel, P. F. (2015), 'Cocoa production in West Africa, a review and analysis of recent developments', *NJAS - Wageningen Journal of Life Sciences* 74–75, 1–7.

Wood, G. and Lass, R. (2008), *Cocoa*, New York, NY: John Wiley & Sons.

World Cocoa Foundation (2017), 'Empowering Women Cocoa Farmers in West Africa: An industry experience in women's empowerment through public private partnerships', Available at: http://www.worldcocoafoundation.org/wp-content/uploads/files_mf/1512401062EmpoweringwomencocoafarmersinWestAfricaFINALVERSIONPUBLIC.pdf (accessed 23 September 2017).

World Cocoa Foundation (2018), 'Cocoa and Forest Initiative', Available at: http://www.worldcocoafoundation.org/cocoa-forests-initiative-statement-of-intent/ (accessed 15 April 2018).

Index

Lightning Source UK Ltd.
Milton Keynes UK
UKHW02n0104100818
326999UK00001B/3/P

9 781786 761682